MOLECULES ENGINEERED AGAINST ONCOGENIC PROTEINS AND CANCER

MOLECULES ENGINEERED AGAINST ONCOGENIC PROTEINS AND CANCER

E. J. Corey
Harvard University
Cambridge, USA

Yong-Jin Wu
Small Molecule Drug Discovery
Bristol Myers Squibb
New York, USA

WILEY

Copyright © 2024 by John Wiley & Sons, Inc. All rights reserved.

Published by John Wiley & Sons, Inc., Hoboken, New Jersey.
Published simultaneously in Canada.

No part of this publication may be reproduced, stored in a retrieval system, or transmitted in any form or by any means, electronic, mechanical, photocopying, recording, scanning, or otherwise, except as permitted under Section 107 or 108 of the 1976 United States Copyright Act, without either the prior written permission of the Publisher, or authorization through payment of the appropriate per-copy fee to the Copyright Clearance Center, Inc., 222 Rosewood Drive, Danvers, MA 01923, (978) 750-8400, fax (978) 750-4470, or on the web at www.copyright.com. Requests to the Publisher for permission should be addressed to the Permissions Department, John Wiley & Sons, Inc., 111 River Street, Hoboken, NJ 07030, (201) 748-6011, fax (201) 748-6008, or online at http://www.wiley.com/go/permission.

Trademarks: Wiley and the Wiley logo are trademarks or registered trademarks of John Wiley & Sons, Inc. and/or its affiliates in the United States and other countries and may not be used without written permission. All other trademarks are the property of their respective owners. John Wiley & Sons, Inc. is not associated with any product or vendor mentioned in this book.

Limit of Liability/Disclaimer of Warranty: While the publisher and author have used their best efforts in preparing this book, they make no representations or warranties with respect to the accuracy or completeness of the contents of this book and specifically disclaim any implied warranties of merchantability or fitness for a particular purpose. No warranty may be created or extended by sales representatives or written sales materials. The advice and strategies contained herein may not be suitable for your situation. You should consult with a professional where appropriate. Further, readers should be aware that websites listed in this work may have changed or disappeared between when this work was written and when it is read. Neither the publisher nor authors shall be liable for any loss of profit or any other commercial damages, including but not limited to special, incidental, consequential, or other damages.

For general information on our other products and services or for technical support, please contact our Customer Care Department within the United States at (800) 762-2974, outside the United States at (317) 572-3993 or fax (317) 572-4002.

Wiley also publishes its books in a variety of electronic formats. Some content that appears in print may not be available in electronic formats. For more information about Wiley products, visit our web site at www.wiley.com.

Library of Congress Cataloging-in-Publication Data applied for

Hardback ISBN: 9781394207084

E-pdf: 9781394207091

E-pub: 9781394207107

Cover Design: Wiley
Cover Image: Courtesy of E.J. Corey and Yong-Jin Wu

Contents

	Preface		VII
Chapter 1.	Introduction		1
1.1	Types of Protein Kinases		1
1.2	Protein Kinase Domains		1
1.3	ATP-Binding Site		2
1.4	Types of Kinase Inhibitors		3
1.5	Brief History of Small-molecule Kinase Inhibitors		5
1.6	Peak 12-Month Sales for Leading Kinase Inhibitors		7
1.7	Approved Kinase Inhibitors		7
Chapter 2.	BCR-ABL Inhibitors		18
2.1	Imatinib*	(1)	19
2.2	Nilotinib*	(2)	24
2.3	Dasatinib*	(3)	27
2.4	Bosutinib*	(4)	30
2.5	Ponatinib*	(5)	33
2.6	Olvermbatinib**	(6)	37
2.7	Asciminib*	(7)	38
Chapter 3.	BTK Inhibitors		43
3.1	Ibrutinib*	(8)	45
3.2	Acalabrutinib*	(9)	51
3.3	Zanubrutinib*	(10)	54
3.4	Tirabrutinib**	(11)	57
3.5	Orelabrutinib**	(12)	58
Chapter 4.	EGFR/HER Family Inhibitors		59
4.1	Gefitinib*	(13)	61
4.2	Erlotinib*	(14)	67
4.3	Icotinib**	(15)	72
4.4	Afatinib*	(16)	74
4.5	Dacomitinib*	(17)	77
4.6	Osimertinib*	(18)	80
4.7	Mobocertinib*	(19)	86
4.8	Lapatinib*	(20)	90
4.9	Tucatinib*	(21)	93
4.10	Neratinib*	(22)	95
Chapter 5.	VEGFR/Multikinase Inhibitors		97
5.1	Sorafenib*	(23)	99
5.2	Regorafenib*	(24)	104
5.3	Sunitinib*	(25)	106
5.4	Pazopanib*	(26)	112
5.5	Axitinib*	(27)	114
5.6	Nintedanib*	(28)	117
5.7	Apatinib**	(29)	121
5.8	Lenvatinib*	(30)	122
5.9	Tovozanib*	(31)	125
Chapter 6.	CDK4/6 Inhibitors		127
6.1	Palbociclib*	(32)	129
6.2	Ribociclib*	(33)	136
6.3	Abemaciclib*	(34)	139
6.4	Trilaciclib*	(35)	142
Chapter 7.	JAK Inhibitors		144
7.1	Tofacitinib*	(36)	147
7.2	Baricitinib*	(37)	151
7.3	Peficitinib**	(38)	153
7.4	Upadacitinib*	(39)	158
7.5	Delgocitinib**	(40)	161
7.6	Filgotinib**	(41)	163
7.7	Abrocitinib*	(42)	166
7.8	Ruxolitinib*	(43)	170
7.9	Fedratinib*	(44)	173
7.10	Pacritinib*	(45)	175
7.11	Ritlecitinib#	(46)	177
7.12	Brepocitinib#	(47)	181
7.13	Ropsacitinib#	(48)	184
Chapter 8.	Allosteric TYK2 Inhibitors		187
8.1	Deucravacitinib*	(49)	189
Chapter 9.	ALK/multikinase Inhibitors		195
9.1	Crizotinib*	(50)	197
9.2	Ceritinib*	(51)	202
9.3	Alectinib*	(52)	205
9.4	Brigatinib*	(53)	207
9.5	Lorlatinib*	(54)	210
Chapter 10.	BRAF/Multikinase Inhibitors		214
10.1	Vemurafenib*	(55)	216
10.2	Dabrafenib*	(56)	222
10.3	Encorafenib*	(57)	225
Chapter 11.	MEK Inhibitors		227
11.1	Trametinib*	(58)	228
11.2	Cobimetinib*	(59)	232
11.3	Binimetinib*	(60)	235
11.4	Selumetinib*	(61)	237
Chapter 12.	RET/Multikinase Inhibitors		240
12.1	Vandetanib*	(62)	242

	12.2	Cabozantinib*	(63)	245
	12.3	Selpercatinib*	(64)	247
	12.4	Pralsetinib*	(65)	251
Chapter 13.		FGFR Inhibitors		253
	13.1	Erdafitinib*	(66)	255
	13.2	Pemigatinib*	(67)	260
	13.3	Infigratinib*	(68)	263
	13.4	Futibatinib*	(69)	265
Chapter 14.		PI3K Inhibitors		267
	14.1	Alpelisib*	(70)	269
	14.2	Idelalisib*	(71)	273
	14.3	Duvelisib*	(72)	277
	14.4	Umbralisib*	(73)	279
	14.5	Copanlisib*	(74)	281
Chapter 15.		TRK/Multikinase Inhibitors		284
	15.1	Larotrectinib*	(75)	285
	15.2	Entrectinib*	(76)	288
	15.3	Repotrectinib#	(77)	291
Chapter 16.		MET Inhibitors		294
	16.1	Capmatinib*	(78)	295
	16.2	Tepotinib*	(79)	297
Chapter 17.		KIT/PDGFR/Multkinase Inhibitors		299
	17.1	Avapritinib*	(80)	301
	17.2	Ripretinib*	(81)	304
Chapter 18.		FLT3 Inhibitors		306
	18.1	Midostaurin*	(82)	308
	18.2	Gilteritinib*	(83)	313
Chapter 19.		mTOR Inhibitors		315
	19.1	Sirolimus* and Analogs	(84)	317
Chapter 20.		Other Kinase Inhibitors		322
	20.1	Netarsudil*	(85)	324
	20.2	Belumosudil*	(86)	326
	20.3	Fostamatinib*	(87)	328
	20.4	Pexidartinib*	(88)	331
Chapter 21.		KRAS Inhibitors		335
	21.1	Sotorasib*	(89)	337
	21.2	Adagrasib*	(90)	346
	21.3	JDQ443#	(91)	350
Chapter 22.		An Overview of the Discovery Process for Medically Useful Inhibitors of Oncogenic Protein Kinases		353
	22.1	High-quality Leads		353
	22.2	Integrating Substructures from Different High Quality Leads or Established Inhibitors		355
	22.3	Variation of Hinge-binding Nucleus		357
	22.4	Macrocyclization		359
	22.5	Fragment-based Approach		360
	22.6	Covalent Inhibitors		361
	22.7	Strategic Structural Modification of Prior Drugs		362
	22.8	Exploiting a Specific Kinase Pocket to Optimize Selectivity		364
	22.9	Solvent-exposed Appendages to Enhance Solubility and PK Properties		367
Chapter 23.		Targeted Molecular Anticancer Therapies – Successes and Challenges		368
	23.1	The Beginning		368
	23.2	Further Developments		368
	23.3	Biomarker-driven Drug Development		369
	23.4	Mitigation of Drug Resistance		370
	23.5	Miscellaneous Approaches		371
	23.6	Discovery Chemistry		373
Appendix 1.		First FDA Approvals by Year		374
Appendix 2.		Kinase/KRAS Inhibitors in Development		375
Appendix 3.		Visualization of Differentially Expressed Kinases in Cancer		378
Appendix 4.		M & A Transactions Driven by Oncology-focused Kinase and KRAS Inhibitors		379
Appendix 5.		Alphabetic List of Oncogenic Protein Inhibitors		380

*   Approved by FDA
**   Approved in other countries
\#   To be approved

Preface

This book documents a series of recent scientific advances that have transformed one of the most challenging areas of medicine – cancer treatment. Remarkably, in the last 30 years, an entirely new kind of molecular approach to cancer therapy has emerged in the form of almost 80 new FDA-approved synthetic compounds which directly target the mutated biomolecules that cause cancer. In addition, there are more than 250 new tumor-directed anti-cancer molecules now in clinical trials. For comparison, in 1950, there were only a total of 50 FDA-approved molecular drugs! These new compounds are now being used to treat a variety of cancers that were previously considered hopeless. They are totally different from the cytotoxic anticancer agents that were developed in the latter half of the 20th century, which were toxic to both normal and tumor cells.

Such astonishing progress was no accident. It was enabled by a broad range of fundamental scientific research in many areas, including discoveries showing that cancer growth and proliferation are driven by mutated genes (oncogenes) and the mutated proteins (oncoproteins) that they encode. Oncoproteins cause cancer because they evade the crucial biochemical controls that regulate cell growth and division. Oncogenes can result from viral infection, foreign carcinogens, radiation, or the random mutations of aging. Most (but not all) known oncoproteins fall into the kinase class of enzymes that regulate protein structure, activity, and function by attaching a phosphate group to the protein amino acid components, tyrosine, serine, or threonine. Kinases are also extraordinarily important to most other life processes, including brain and organ function, reproduction and immunity. The human genome encodes 518 protein kinases, the "kinome" (Fig. 1).

Among the other areas of science that were critical to the new wave of discovery behind these miracle anticancer molecules are the following: (1) genomic analysis of tumor tissues and cells; (2) biochemical advances in cell regulation and signaling pathways; (3) the production of biosynthetic oncoproteins; (4) technology for very high-throughput screening of synthetic molecules to evaluate kinase inhibition; (5) determination of the three-dimensional structure of the target oncoprotein, e.g., by X-ray crystallography; (6) structural design of possible molecular inhibitors using the logic of medicinal chemistry and the data from screening other candidate molecules, further guided by computational analysis of therapeutic molecule–protein interactions; (7) sophisticated methods of chemical synthesis that provide access to new molecular structures; (8) rapid evaluation of metabolism, pharmacokinetics, tissue distribution, and other pharmacological parameters for each active candidate; and (9) identification of robust and validated biomarkers to find the right patients for the targeted therapy.

Much remains to be discovered in the area of anticancer molecules engineered to inhibit oncoproteins. One can realistically hope that one day – possibly before the end of this century – the many forms of malignant proliferative disease will be curable or manageable, especially because of complimentary advances in immunotherapy using monoclonal antibodies, vaccines and engineered anticancer cells.

We are grateful to Dr. Joanne Bronson for continuous advice and encouragement, to Brian Venables, Matthew Patton, Drs. Richard Hartz, Xiaojun Han, and Zhaoming Xiong for proofreading, and Dr. Stephen Wrobleski for suggestions on Chapter 8. We also thank Dr. Erik Vik for generating the cover graphic of the imatinib–ABL complex as well as several co-crystal structures, Drs. Sirish Kaushik Lakkaraju and Yilin Meng for drawings of other co-crystal structures. Dr. Shunying Liu, a visiting scholar at Harvard (2018), assisted EJC at the start of this project. Thanks also to Dr. Heather Zhang for helpful discussions on the discovery of sunitinib.

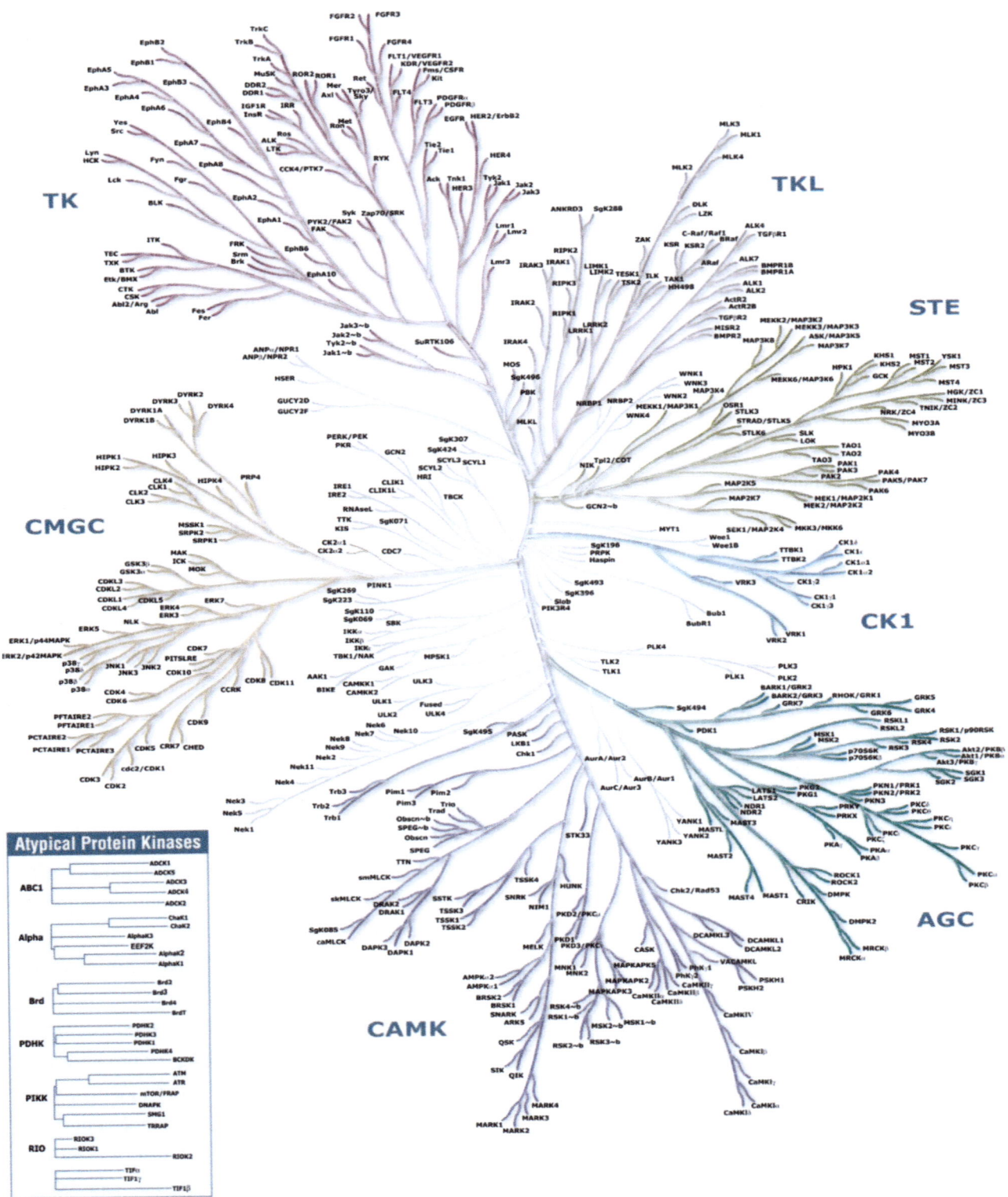

Figure 1. The human kinome map of 518 human kinases.
(Kinome illustration by Cell Signaling Technology).

Chapter 1. Introduction

Research over the past four decades has revealed that cancers are caused by mutations which favor cellular dysregulation, proliferation, and further mutation. Mutated protein kinases play a major role in the disease process. Normally, these kinases are central to cellular regulation by signal transduction, the transfer of a signal from the cell membrane to the nucleus, and also cell-cycle control and cell division. Protein kinases catalyze the transfer of the terminal phosphate group from adenosine triphosphate (ATP) to the hydroxyl group of a serine, threonine, or tyrosine residue of the kinase itself, another kinase or another protein substrate (Fig. 1). These kinases may adopt a number of conformations which may be active (i.e., on the catalytic pathway) or inactive.[1]

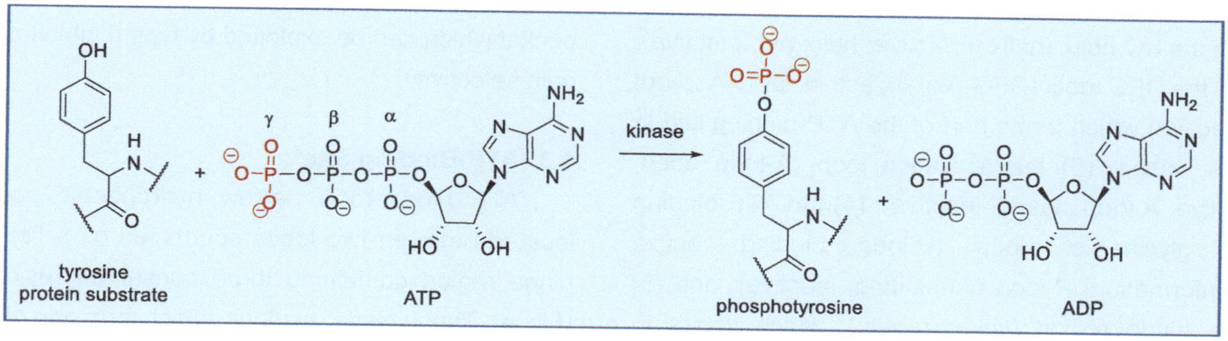

Figure 1. Protein kinase-catalyzed transfer of a γ-phosphoryl group from ATP onto target tyrosine side chain of a protein substrate.

1.1 Types of Protein Kinases[2]

The human kinome contains 518 known protein kinases: 478 typical and 40 atypical kinases. According to the sequence similarity of their kinase domains, typical kinases are divided into eight groups: TK (tyrosine kinase, 90), TKL (tyrosine kinase-like kinase, 43), STE (STE7-, STE11-, and STE20-related kinase, 47), CK1 (casein kinase 1, 12), AGC (protein kinase A/G/C-related kinase, 63), CAMK (Ca2+/calmodulin-dependent kinase, 74), CMGC (CDK/MAPK/GSK/CDK like-related, 61), and RGC (receptor guanylyl cyclase, 5). The 40 atypical protein kinases include PI3K (lipid kinase) and mTOR (serine/threonine protein kinase).

TKs are further classified as RTKs receptor tyrosine kinases (RTKs) and non-RTKs. RTKs are transmembrane proteins with an extracellular and intracellular kinase domain. They become activated upon dimerization, triggered by the binding of their ligands to the extracellular domain. RTKs have been extensively exploited as anticancer therapeutics, for example, EGFR/HER family proteins, VEGFR, TRK, ALK, ROS1, FLT3, PDGFRs, and FGFRs. Non-RTKs include SRC, ABL, and JAK.

The TKL family consists of serine/threonine kinases, with sequences similar to those of the TK group, including IRAK, BRAF, LIMK, and TGFβ. Three families of STE kinases have been found: STE20 (MAP4K), STE11 (MAP3K), and STE7 (MEK), while CDKs, MAPK, GSK, and CDK-like kinases constitute the CMGC group.

1.2 Protein Kinase Domains

The 518 human protein kinases share a number of common features (depicted in Fig. 2), including (1) separate N- and C-terminal lobes (domains) connected by an intervening peptide chain (linker or hinge region) which also provides a binding site for the universal phosphoryl donor ATP; (2) availability of an active conformation which can bind ATP and accelerate phosphorylation of the specific protein which is downstream on the signaling pathway; and (3) a conserved sequence of aspartate (D), phenylalanine (F), and glycine (G) in the ATP-

Molecules Engineered Against Oncogenic Proteins and Cancer, First Edition. E. J. Corey and Yong-Jin Wu.

binding region of the linker (DFG motif) located on the N-terminal side of the activation loop. Many kinases can also adopt other conformations which do not bind ATP or catalyze phosphorylation of the downstream protein.

The structure of BCR-ABL kinase is depicted in Figure 2 which shows the relative locations of the various subregions. It is typical of the large family of kinases. Conformational transitions between kinase states are orchestrated by the movement of five conserved structural motifs in the catalytic domain: (1) the αC helix (helix in, active; helix out, inactive); (2) the DFG motif (DFG-Asp-in, active; DFG-Asp-out, inactive) which forms part of the ATP-binding site in the kinase; (3) the activation loop (A-loop open, active; A-loop closed, inactive); (4) the ATP-binding of glycine-rich loop (P-loop binding, active conformation; P-loop nonbinding, inactive); and (5) the hinge region (linker region), which varies in location between the N-terminal domain (N-lobe) and the C-terminal domain (C-lobe).[3]

ATP-binding site (DFG-Asp-in, or DFG-in for simplicity) and coordinates two Mg^{2+} ions. Another hallmark is that the αC helix located on the N-terminal domain is rotated inward toward the active site (αC-in). A different and inactive conformation involves a translocation of the positions of the Asp and Phe residues, which moves the aspartate away from the ATP-binding site by ~5 Å to generate a catalytically incompetent state, the DFG-Asp-out (or DFG-out) state. The resulting conformation has a new allosteric pocket adjacent to the ATP-binding pocket which can be exploited by type II inhibitors to gain selectivity.

1.3 ATP-Binding Site[4-7]

ATP binds to a narrow hydrophobic pocket located between two lobes connected by a flexible hinge region containing three conserved residues (Fig. 4). These hinge residues, which form part of the ATP-binding pocket, engage several key H-bonds with the heteroaromatic adenine ring of ATP (Fig. 3). The hinge region is the primary target of most ATP-competitive kinase inhibitors possessing a hinge-binding motif. These inhibitors are designed to form some or all of the same key hydrogen-bonding acceptor and donor interactions with the hinge backbone atoms as adenine of ATP, thereby competitively displacing ATP from the binding site.

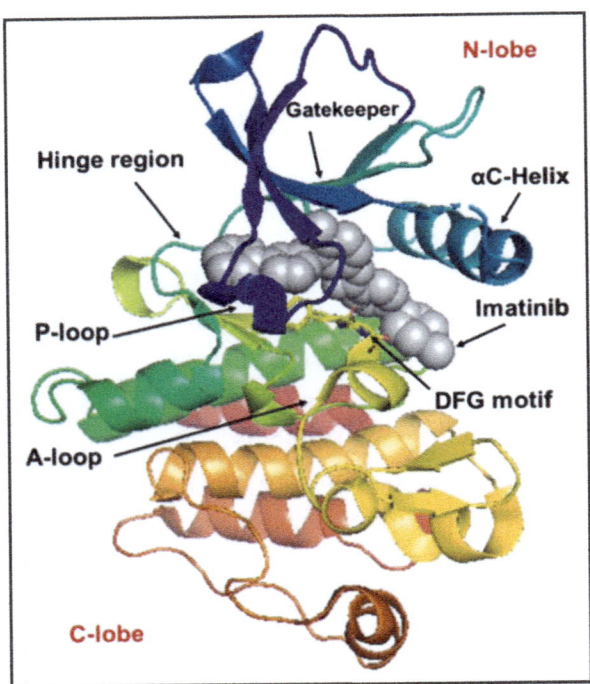

Figure 2. X-ray crystal structure of BCR-ABL with imatinib bound and with key structural motifs highlighted. Adapted from PDB: 2HYY.

One hallmark of an active conformation is that the aspartate of the DFG motif points toward the

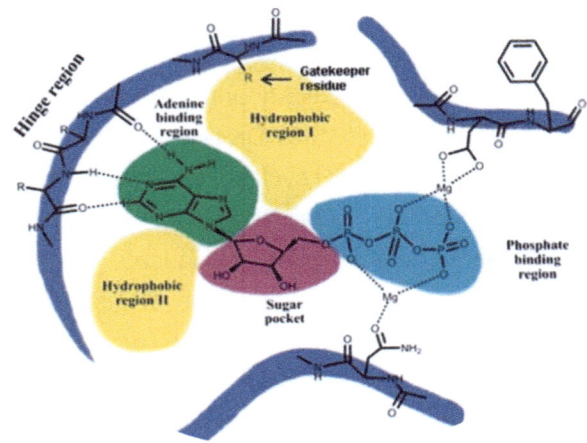

Figure 3. ATP-binding site of a typical protein kinase. Adapted from *ACS Chem. Biol.* **2013**, *8*, 1044–1052.

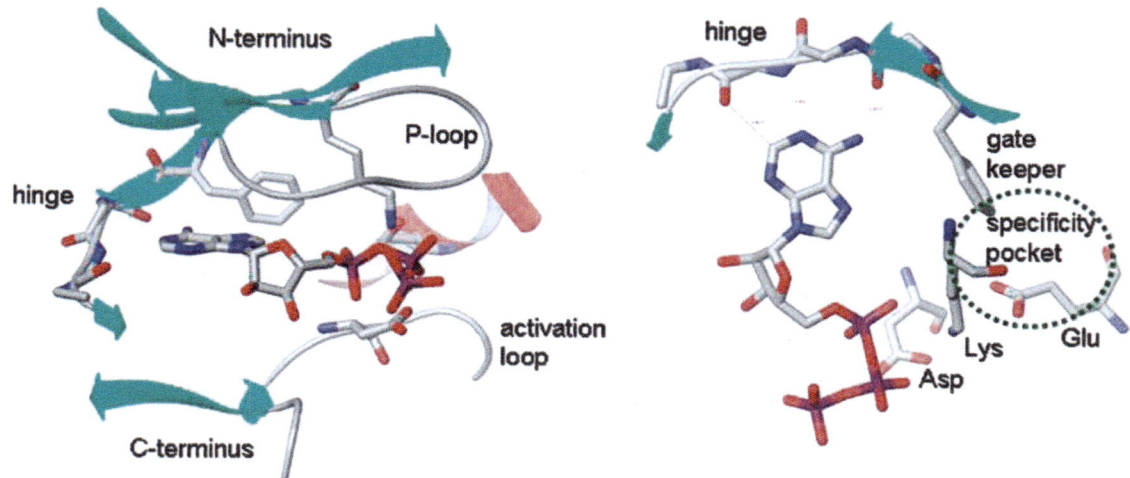

Figure 4. ATP-binding site. Left: ATP binds between the N- and C-termini. The P-loop forms the roof, and a C-terminal β-sheet covers the floor. Right: Gatekeeper and conserved lysine, glutamate (of αC-helix), and Asp of the DFG motif control access to specificity pocket. Adapted with permission from *J. Med. Chem.* **2008**, *51*, 5149–5171.

Determinants of the selectivity in kinases include the conformation of the DFG motif at the start of the A-loop and the conformation of the P-loop.[8] Both DFG-Asp-out and DFG-Asp-in conformations have been observed in binding with various selective kinases (Fig. 5). The P-loop can contribute to this selectivity by forming extensive contacts with the inhibitor.

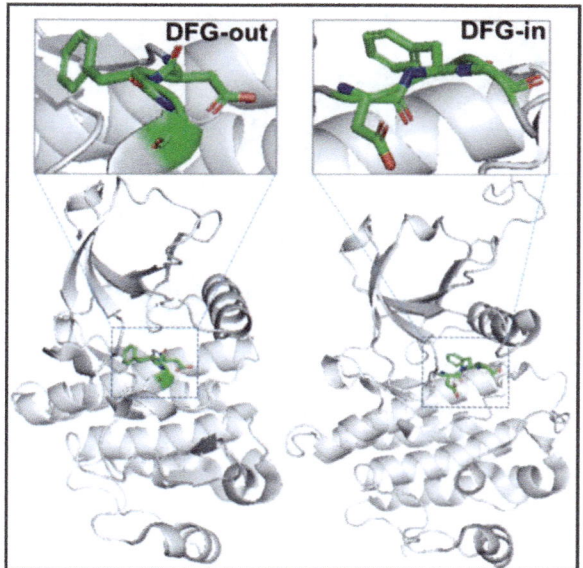

Figure 5. The DFG-Asp-out or DFG-Asp-in conformations in ABL (PDB ID: 2HYY and 2GQG).

Kinase domains contain a gatekeeper residue, located at the beginning of the hinge region. It is the single most important residue in the ATP-binding site, which partially or fully blocks a hydrophobic region deep in the ATP-binding pocket, depending on the size/volume of the gatekeeper's side chain. The gatekeeper residue also contributes to the selectivity for small-molecule kinase inhibitors.

1.4 Types of Kinase Inhibitors[9–16]

In general, there are two types of kinase inhibitors: covalent and non-covalent inhibitors. The latter can be divided into four major classes according to their binding mode.

Type I inhibitors (such as dasatinib, Figs. 6, 7, 10) bind to the DFG-in, αC-in active conformation of the kinase with the Asp of the DFG motif pointing into the ATP-binding pocket. Type 1 inhibitors take advantage of the gatekeeper residue for selectivity. The majority of FDA-approved kinase inhibitors are type I inhibitors, e.g., gefitinib, erlotinib and sunitinib.

Type II inhibitors (such as imatinib, Figs. 6, 7, 10) bind to the DFG-out, αC-out inactive conformation with the Phe of the DFG motif pointing into the ATP-binding pocket. In the DFG-out conformation, Phe and Asp have switched positions relative to the DFG-in conformation, resulting in the opening of a new hydrophobic allosteric pocket adjacent to the ATP site. The additional binding to this allosteric pocket enhances the selectivity of type

II inhibitors as compared with type I. The type II class includes nilotinib and sorafenib.

Type I1/2 is between type I and type II. The EGFR inhibitor lapatinib falls into this class which binds to a DFG-in, αC-out inactive conformation (Figs. 8, 9). In this case, the EGFR adopts a DFG-in conformation, typical of an active kinase; however, the αC helix is pushed out (αC-out) by the fluorobenzyl substituent of the inhibitor, resulting in the disruption of the ion pairing between the active site Lys and the Glu from the αC helix and an inactive conformation. As a result, the fluoro-benzyl substituent extends into an allosteric pocket opened by the outward, inactive position of the αC helix. Vemurafenib is another type I1/2 inhibitor that binds to a DFG-in, αC-out inactive conformation of BRAF.

Type III inhibitors bind to an allosteric pocket next to the ATP-binding pocket (Fig. 10), with no hydrogen bond interactions with the hinge residues. They induce conformational changes in the activation loop, forcing the αC helix to adopt an inactive conformation. The type III class includes MEK inhibitors such as trametinib and binimetinib.

Type IV inhibitors bind to any allosteric sites distant from the ATP-binding pocket (Fig. 10). They induce conformational changes to lock the protein into an inactive state. For example, asciminib is a type IV allosteric inhibitor of BCR-ABL that binds to a myristoyl site of the BCR-ABL protein.

Covalent kinase inhibitors (CKIs) usually target an active site cysteine in or around the ATP-binding site. Covalent inhibition is typically irreversible, but can be reversible, depending on the nature of the electrophilic warhead. CKIs have been developed for various protein kinases, including FGFR, VEGFR-2 and BTK. Covalent, irreversible KRAS(G12C) inhibitors have recently been brought to market.

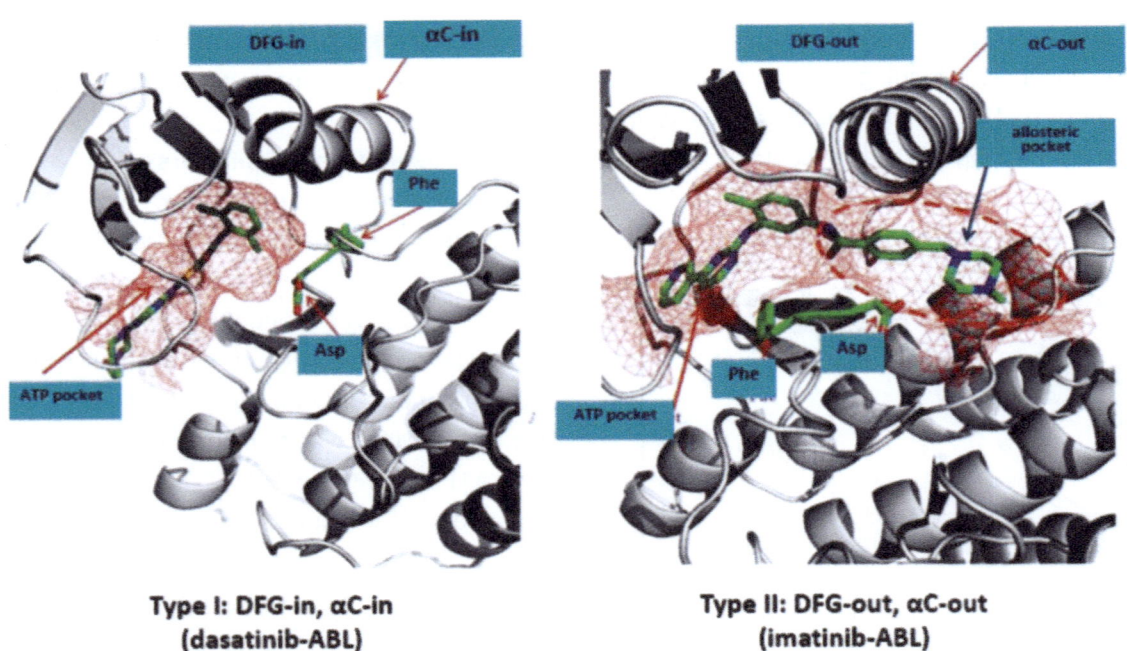

Figure 6. Type I vs. type II inhibitors. Adapted with permission from *J. Med. Chem.* **2015**, *58*, 466–479.

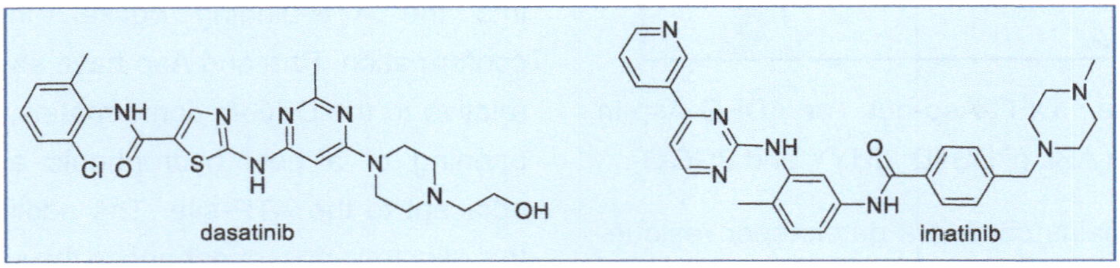

Figure 7. Structures of dasatinib and imatinib.

Introduction

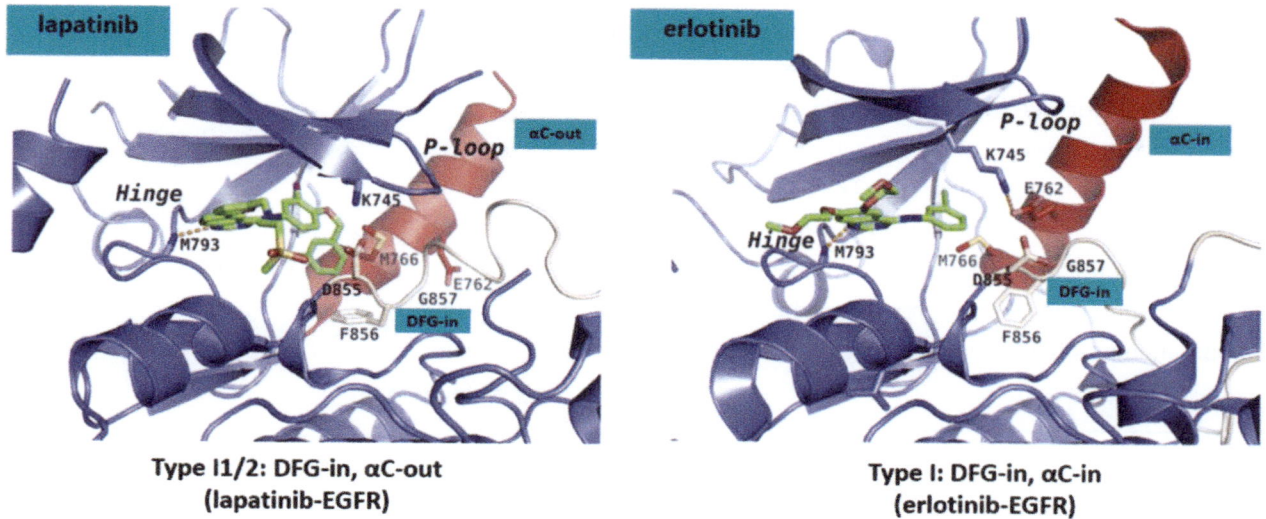

Figure 8. Type I1/2 vs. type I inhibitors. Adapted from *J. Biol. Chem.* **2011**, *286*, 18756–18765.

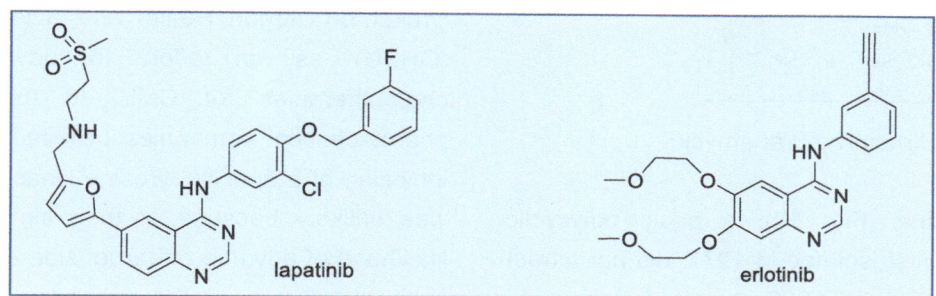

Figure 9. Structures of lapatinib and erlotinib.

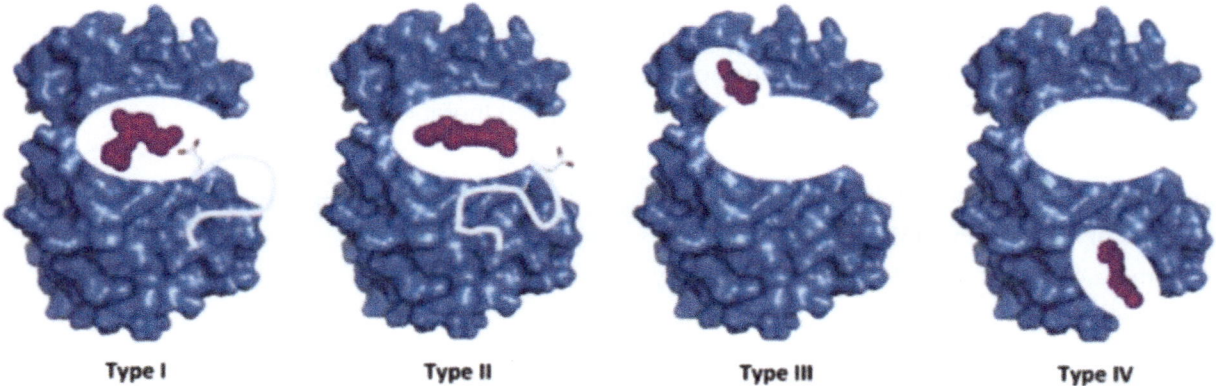

Figure 10. Cartoon illustration of 4 types of inhibitors. Type I: DFG-in active conformation; Type II: DFG-out inactive conformation; Type III: allosteric pocket close to ATP-binding site; Type IV: allosteric pocket remote from the ATP site. Adapted from *Trends Pharmacol. Sci.* **2015**, *36*, 422–439.

1.5 Brief History of Small-molecule Kinase Inhibitors

In 1981, a screening project with a variety of natural products for antitumor activity by the U.S. National Cancer Institute (NCI) identified rapamycin (Fig. 11) as a unique anticancer agent. Against the 60 tumor cell line panel, rapamycin potently inhibited the growth of a number of solid tumors, a significant finding at the time because it was the very first non-toxic cytostatic agent. Approximately 20 years after the discovery of its antitumor activity, it was determined that the target of rapamycin is the protein mTOR. Binding of rapamycin to mTOR blocks the PI3K/AKT/mTOR signal transduction pathway, and the complete structure of mTORC1 was elucidated in 2005. Rapamycin was approved by the FDA in 1999 as an immunosuppressant drug.

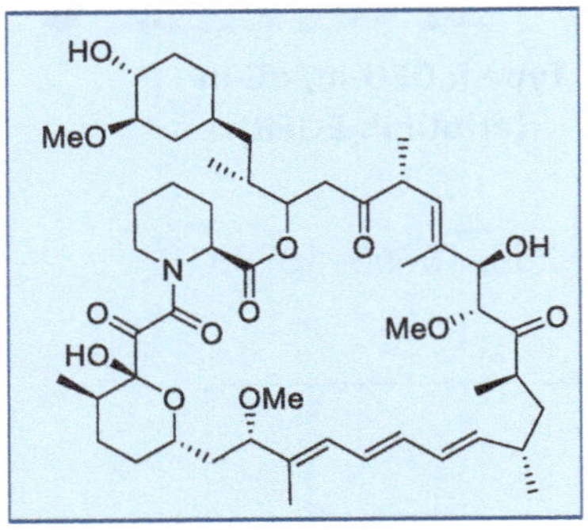

![Structure of rapamycin]

Figure 11. Structure of rapamycin.

Staurosporine (Fig. 12), a basic polycyclic natural product first isolated in 1977, did not attract significant attention until 1986 when it was shown to be the first broad spectrum protein kinase inhibitor. Staurosporine itself is too toxic to be used as a drug. Subsequent medicinal chemistry efforts led to the N-benzoyl derivative, later known as midostaurin, which displayed improved selectivity. Because of its broad kinase inhibition profile and its efficacy to shrink tumors in murine tumor xenograft models, midostaurin entered into clinical trials in 1991 – four years ahead of imatinib (in 1995) – for the treatment of solid tumors. However, its potent antitumor activity did not translate into clinical efficacy. Unfortunately, there was no biomarker at the time to monitor its target proteins and it was difficult to predict who might benefit from the drug. In 2001, midostaurin was identified as a potent FLT3 tyrosine kinase inhibitor by a collaborative effort between the Dana-Farber Cancer Institute and Novartis to discover inhibitors of mutant FLT3-positive AML in cellular assays. It was this finding that led to the repurposing of a twice failed drug for FLT3 mutant AML. Midostaurin was approved by the FDA in 2017 for newly diagnosed FLT3-mutated AML.

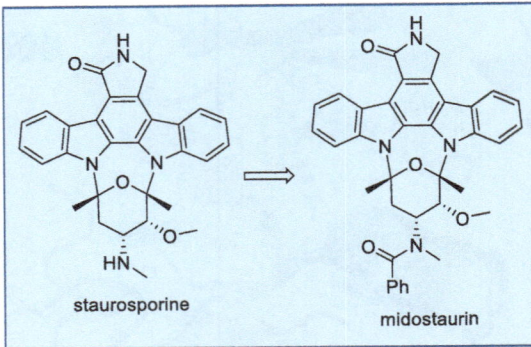

Figure 12. From staurosporine to midostaurin.

The modern era of protein kinase drug discovery was initiated in 1988 at Ciba-Geigy by biochemist Nick Lydon in collaboration with Brian Druker of Oregon Health and Science University (OHSU) as an effort to develop targeted chemotherapies for CML. At the time, most pharmaceutical companies believed that selective inhibition of a specific tyrosine kinase among >500 was unlikely because of their similarity and the likelihood of adverse off-target side effects. Despite such skepticism of the feasibility of developing selective kinase inhibitors, the joint effort was successful and ultimately led to the antileukemic BCR-ABL oncoprotein inhibitor imatinib (Fig. 13) as a preclinical candidate in 1995. The discovery of imatinib ushered in the era of kinase inhibitors as targeted therapies.

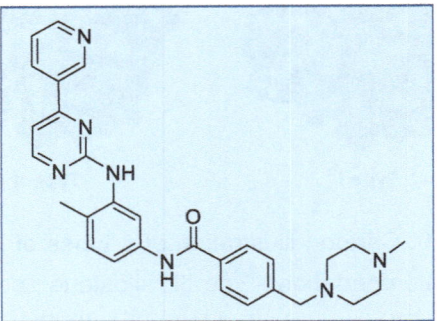

Figure 13. Structure of imatinib.

In 1991 – three years after Ciba-Geigy started the CML program, Sugen was founded with the mission of developing selective protein kinase inhibitors. Scientists at Sugen were seeking small-molecule VEGFR inhibitors to block the VEGF/VEGFR signaling pathway and subsequently attenuate blood vessel formation required for tumor growth, while Genentech was focusing on monoclonal antibodies that bind to VEGFs to prevent activation of their receptors VEGFRs. Sugen's early multikinase inhibitors that cross-inhibit VEGFRs were disappointing in clinical trials due to poor PK, but subsequent optimizations led to sunitinib (Fig. 14), which was approved in 2006 for gastrointestinal stromal tumor (GIST) and renal cell carcinoma (RCC), following only one month after Bayer's sorafenib (Fig. 14) which was initially identified as a RAF inhibitor and subsequently shown also to inhibit several VEGFRs.

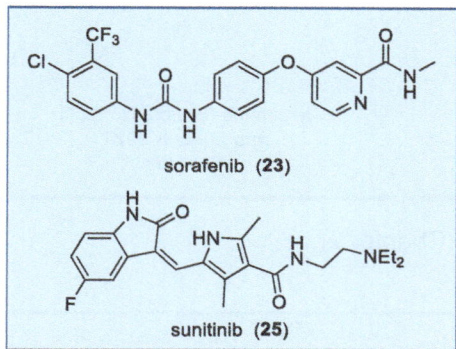

Figure 14. Structures of sorafenib and sunitinib.

1.6 Peak 12-Month Sales for Leading Kinase Inhibitors

1. Ibrutinib (BTK): $9.7 billion (2021)
2. Palbociclib (CDK4/6): $6.0 billion (2022)
3. Osimertinib (EGFR): $5.4 billion (2022)
4. Imatinib (BCR-ABL): $4.7 billion (2016)
5. Ruxolitinib (JAK1/2): $3.97 billion (2022)
6. Nintedanib (PDGFR/VEGFR/FGFR): 3.6 billion (2022)
7. Tofacitinib (pan-JAK): $2.5 billion (2020/21)
7. Abemaciclib (CDK4/6): $2.5 billion (2022)
9. Dasatinib (BCR-ABL): $2.1 billion (2019)
9. Nilotinib (BCR-ABL): $2.1 billion (2021)
9. Everolimus (mTOR): $2.1 billion (2021)
9. Acalabrutinib (BTK): $2.1 billion (2022)
13. Cabozantinib (RET): $1.9 billion (2022)
14. Upadacitinib (JAK1): $1.8 billion (2022)
15. Erlotinib (EGFR): $1.5 billion (2014)
15. Alectinib (ALK): $1.5 billion (2022)
17. Sunitinib (VEGFR): $1.2 billion (2014)
17. Ribociclib (CDK4/6): $1.2 billion (2022)
19. Sorafenib (VEGFR): $1 billion (2014)
20. Dabrafenib (BRAF)/trametinib (MEK): $0.9 billion each (2022)

1.7 Approved Kinase Inhibitors

As of January 01, 2023, there are 78 FDA-approved small-molecule kinase inhibitors (Figs. 15-34) and two KRAS(G12C) inhibitors (Fig. 35) (also see Appendix 1: First FDA Approvals by Year) for the treatment of a variety of diseases, especially cancer. All these inhibitors are oral medications except for netarsudil, which is available as an ophthalmic solution (eye drops), and temsirolimus, copanlisib, and trilaciclib that are administered intravenously.

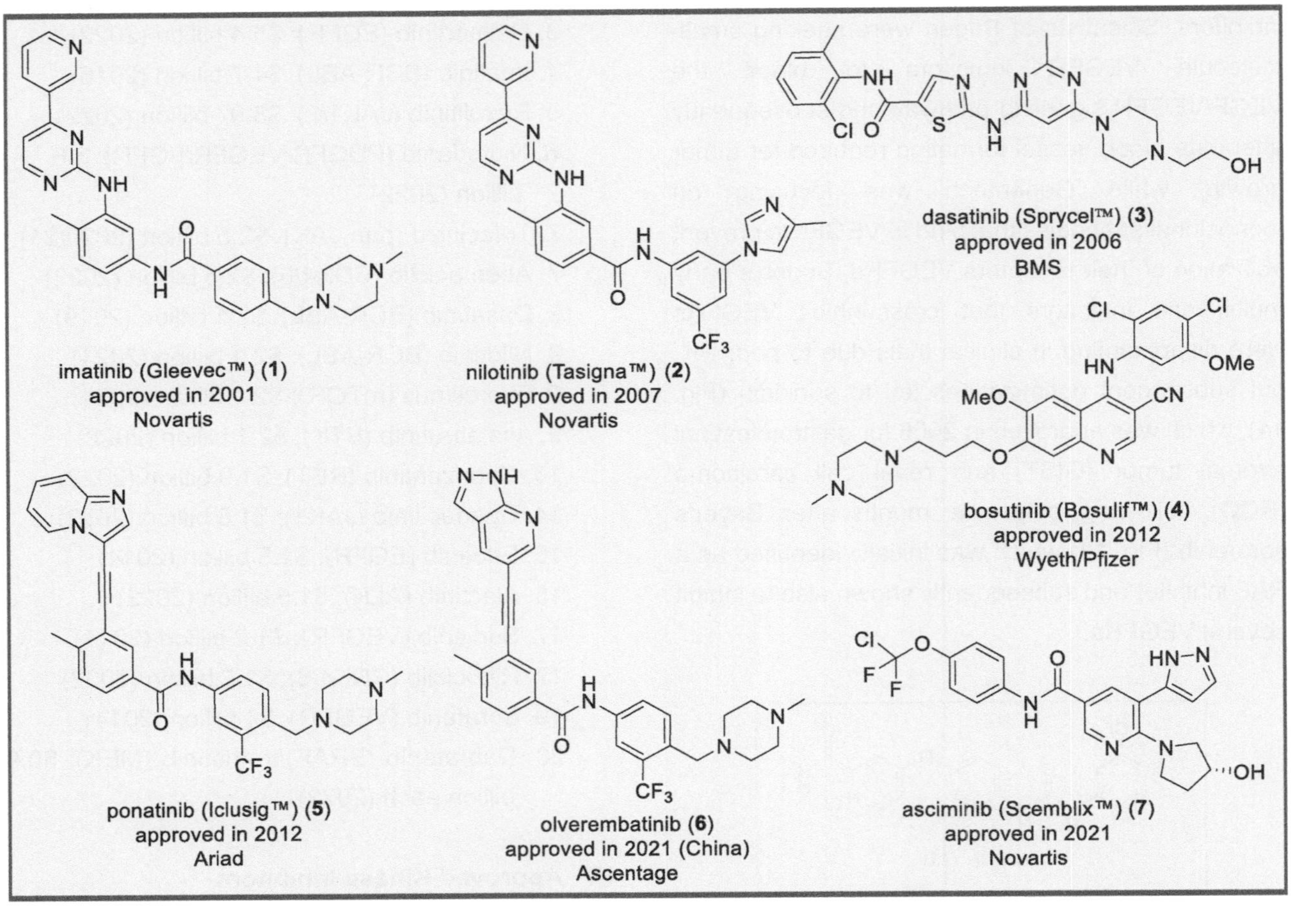

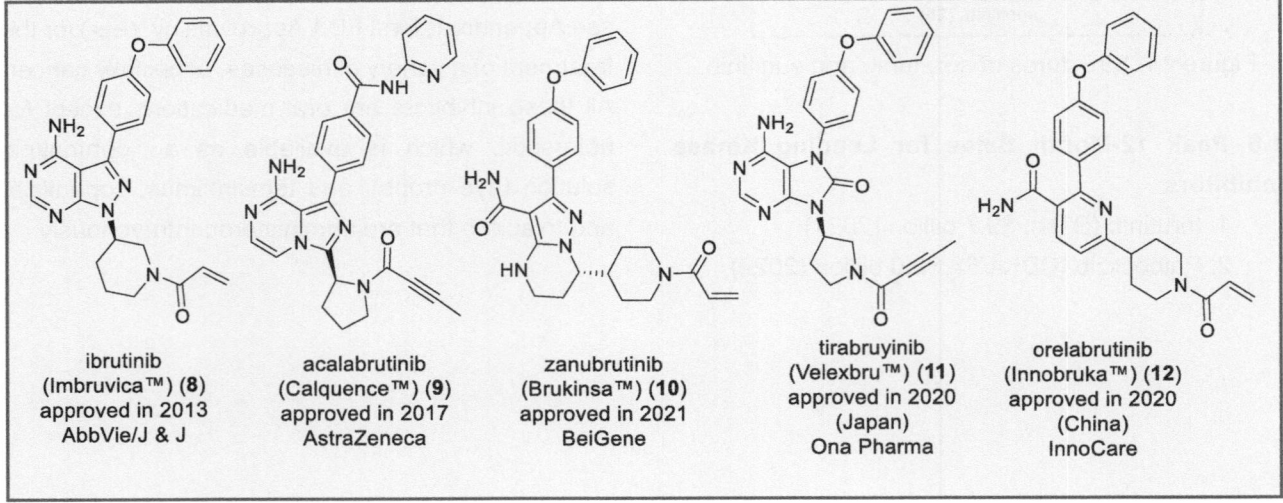

Figure 15: BCR-ABL inhibitors (Chapter 2).

Figure 16. BTK inhibitors (Chapter 3).

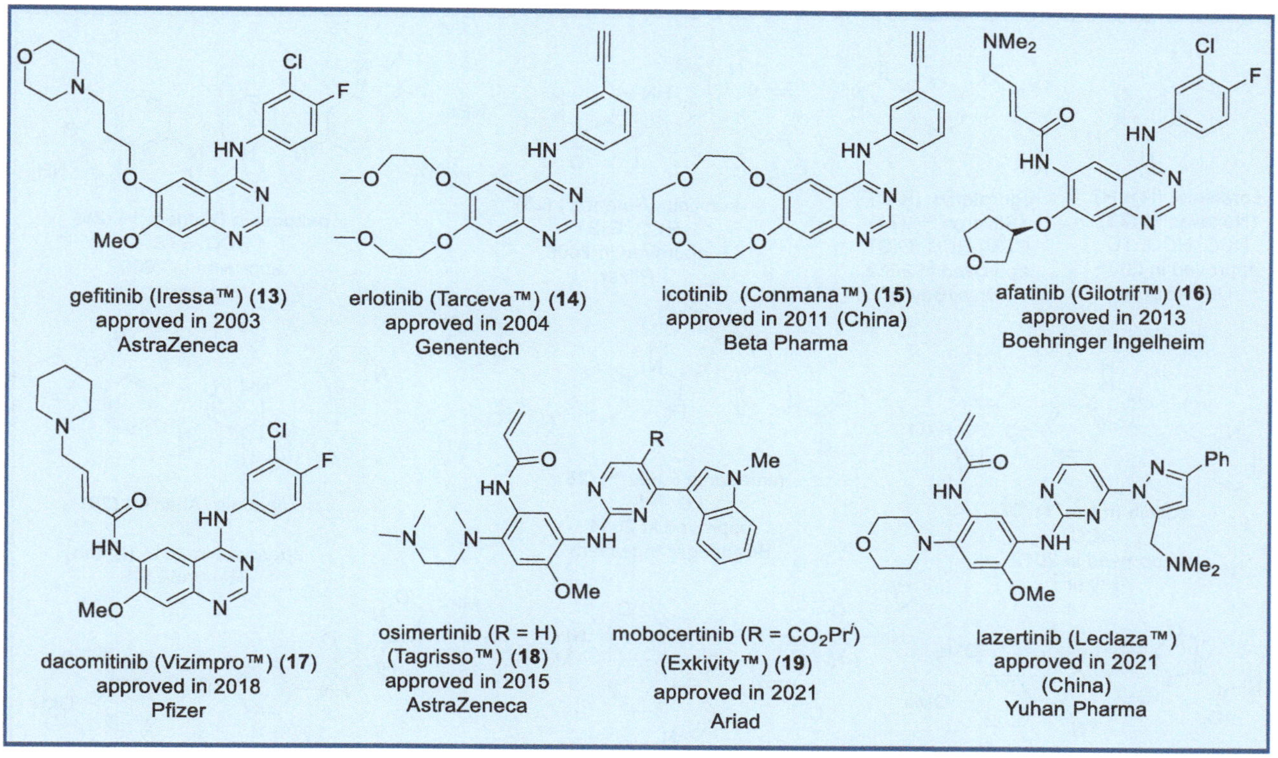

Introduction

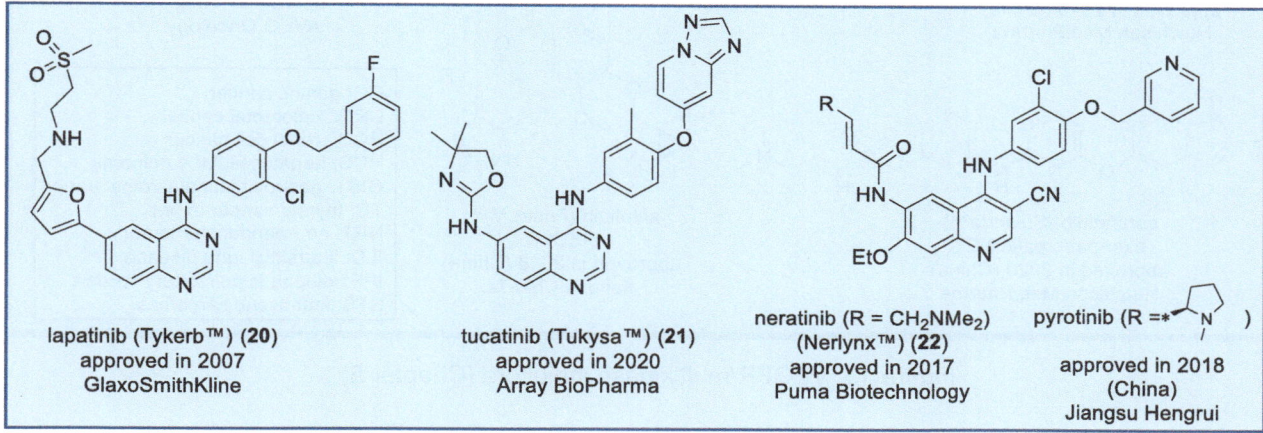

Figure 17. EGFR inhibitors for NSCLC (Chapter 4).

Figure 18. HER2/EGFR inhibitors for HER2+ breast cancer (Chapter 4).

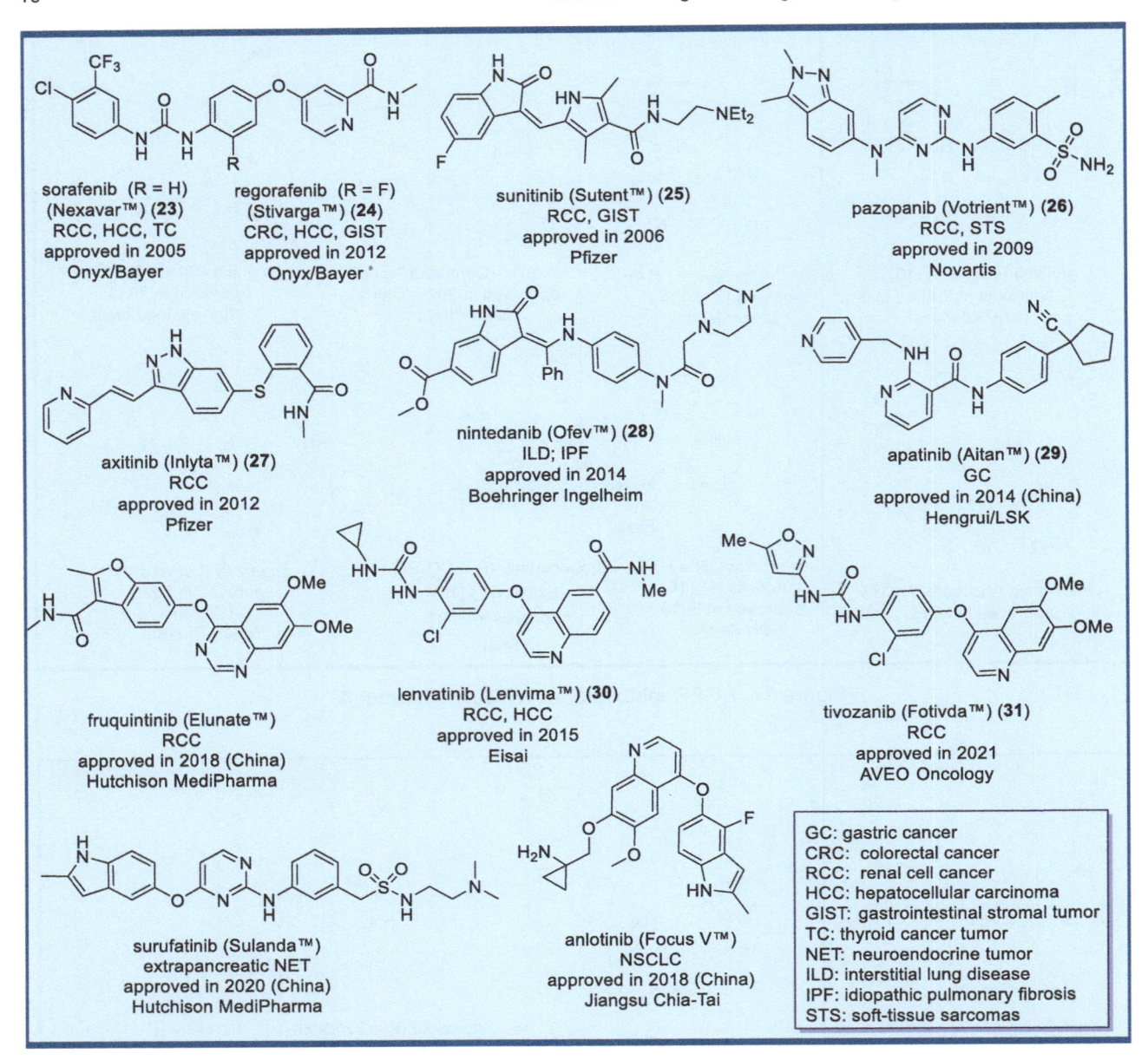

Figure 19. VEGFR/multikinase inhibitors (Chapter 5).

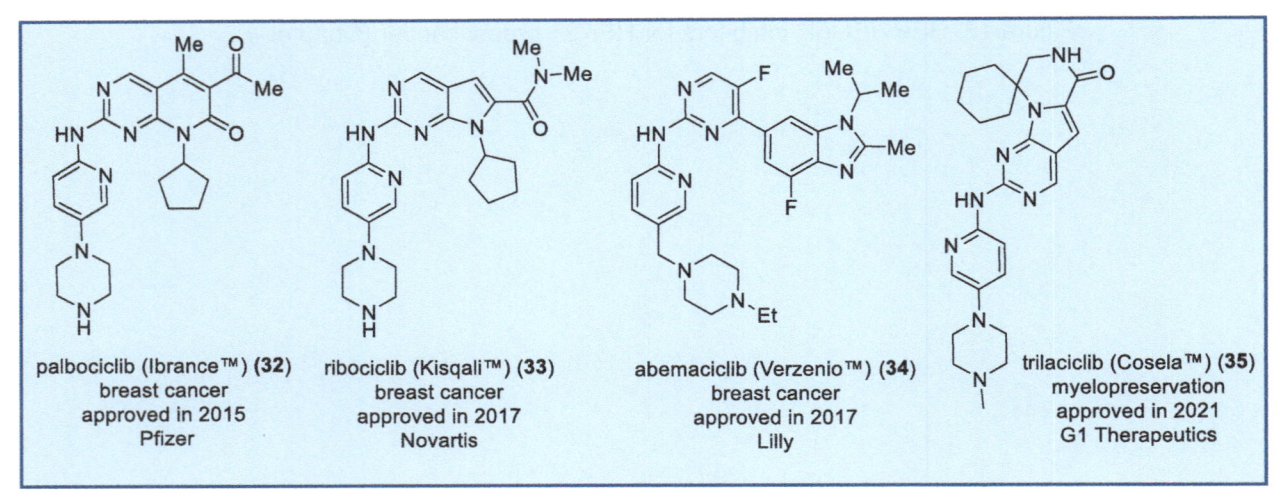

Figure 20. CDK4/6 inhibitors (Chapter 6).

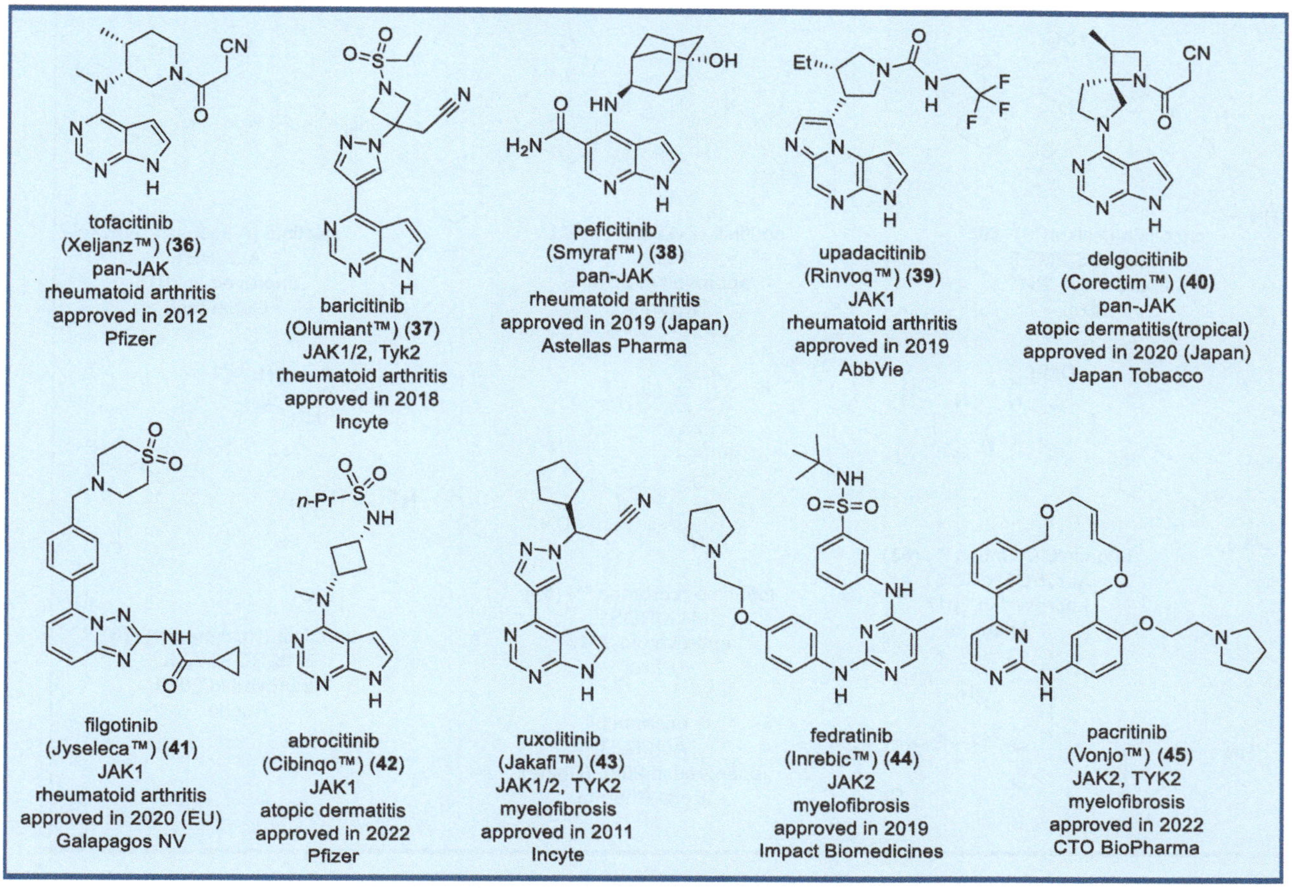

Introduction

Figure 21. JAK inhibitors (Chapter 7).

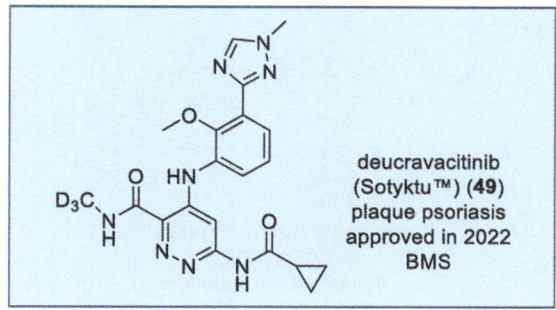

Figure 22. Allosteric TYK2 inhibitor (Chapter 8).

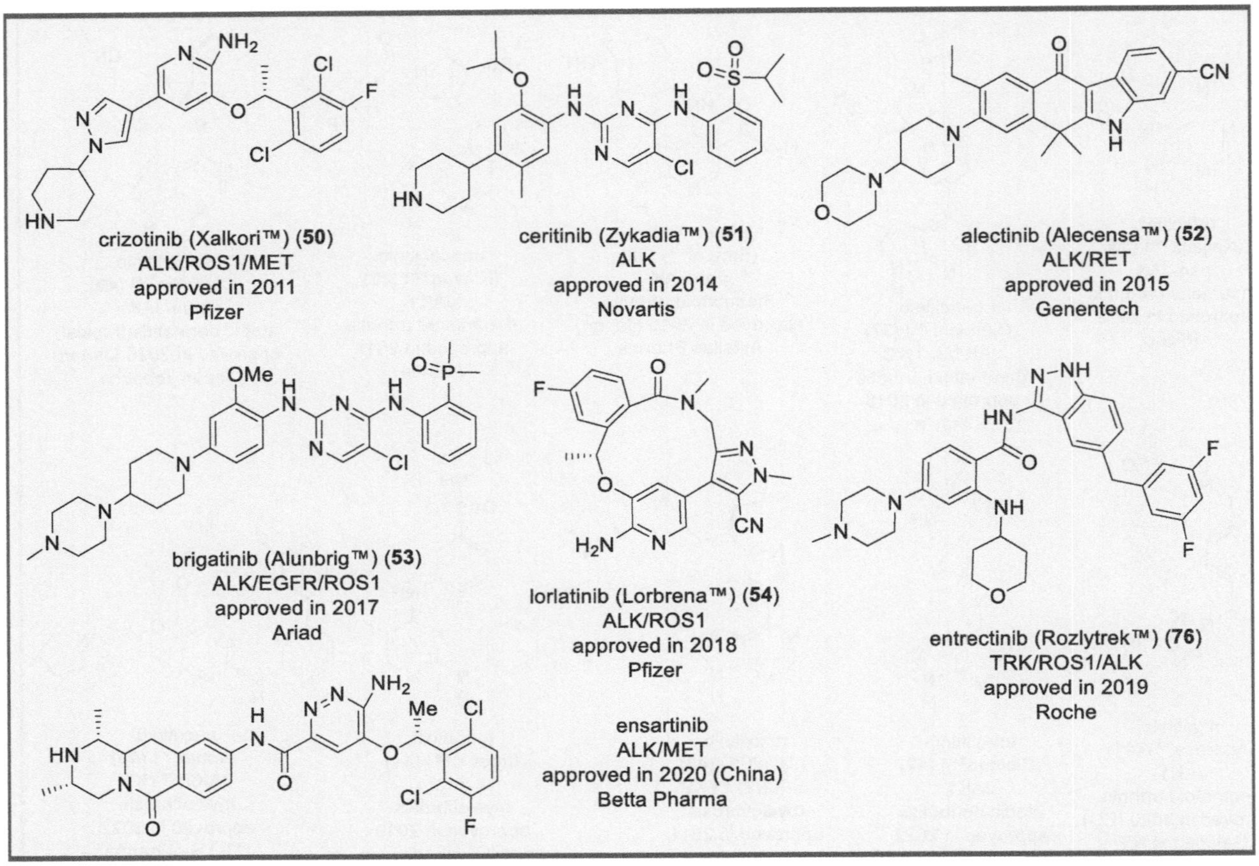

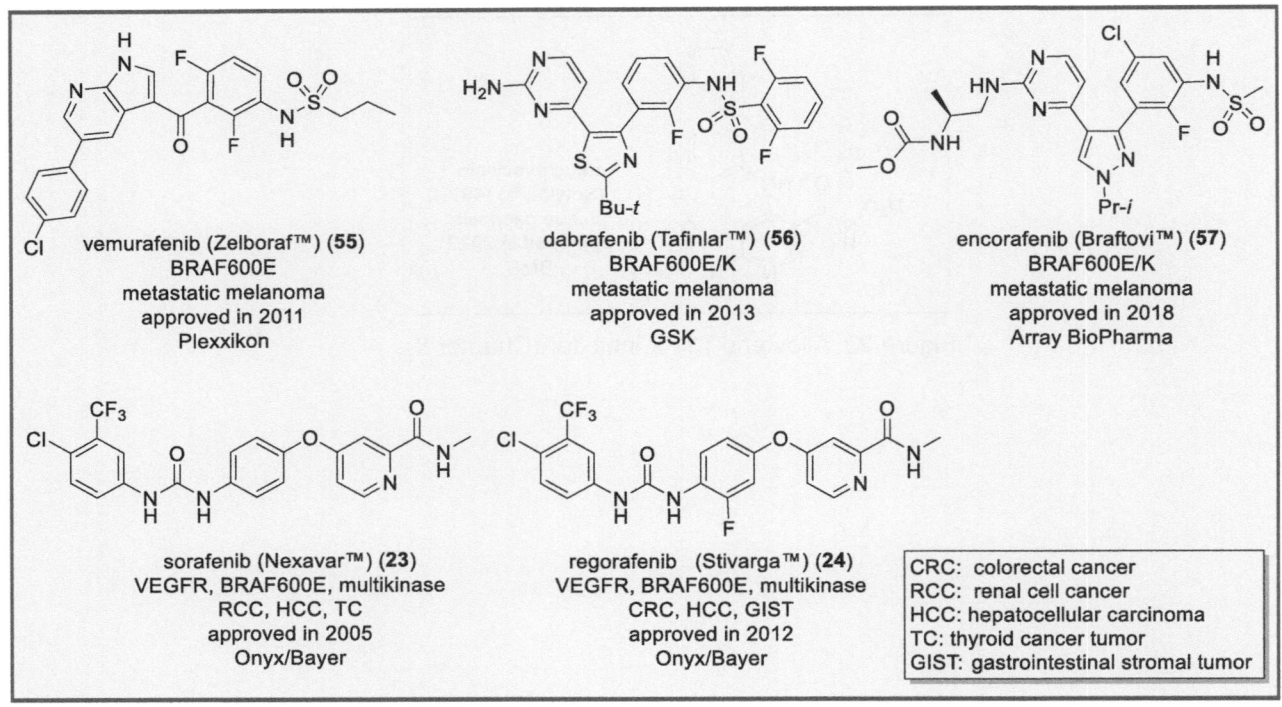

Figure 23. ALK/multikinase inhibitors for NSCLC (Chapter 9).

Figure 24. BRAF inhibitors for various cancers (Chapter 10).

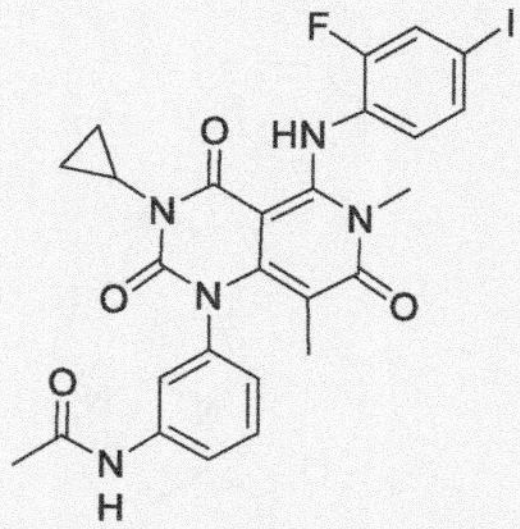

trametinib (Mekinist™) (**58**)
melanoma
approved in 2013
GSK

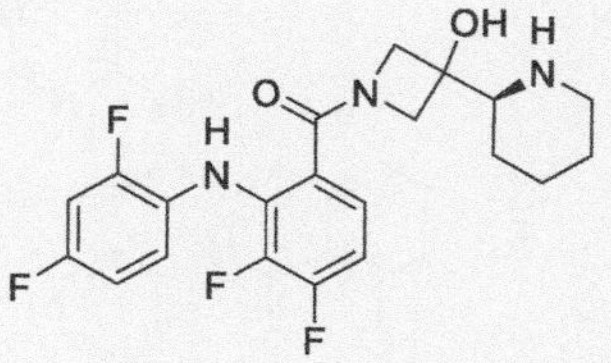

cobimetinib (Cotellic™) (**59**)
melanoma
approved in 2015
Roche

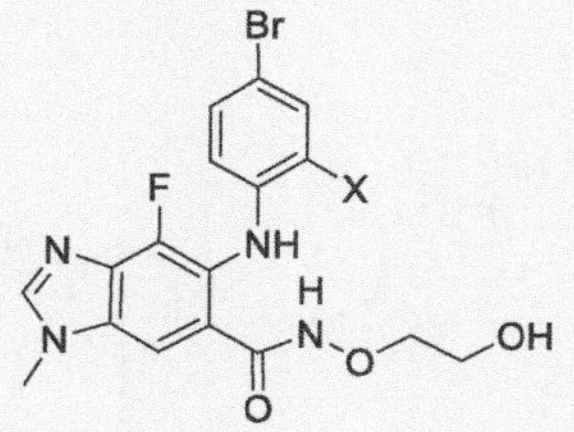

binimetinib (X = F)
(Mektovi™) (**60**)
melanoma
approved in 2018
Pfizer

selumetinib (X = Cl)
(Koselugo™) (**61**)
neurofibromatosis
approved in 2020
AstraZeneca

Figure 25. MEK inhibitors (Chapter 11).

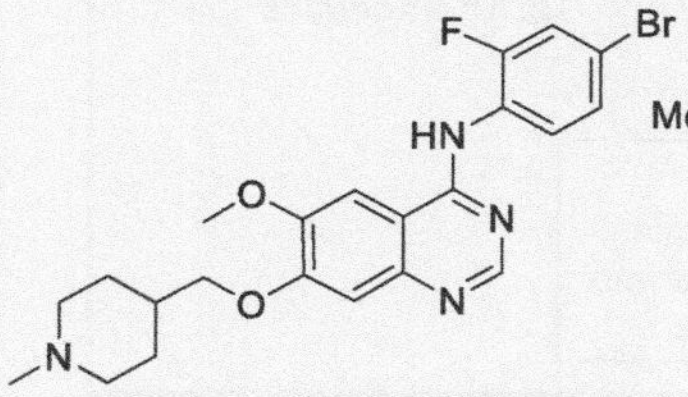

vandetanib (Caprelsa™) (**62**)
RET/VEGFR/EGFR
MTC
approved in 2011
AstraZeneca

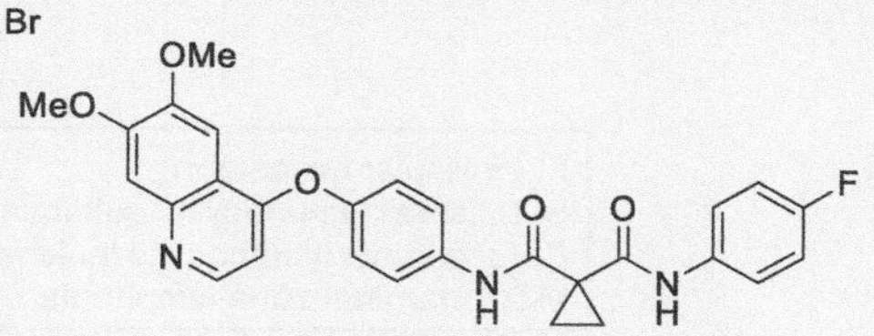

cabozantinib (Cometriq™) (**63**)
RET/VEGFR/MET
MTC, RCC, HCC
approved in 2012
BMS

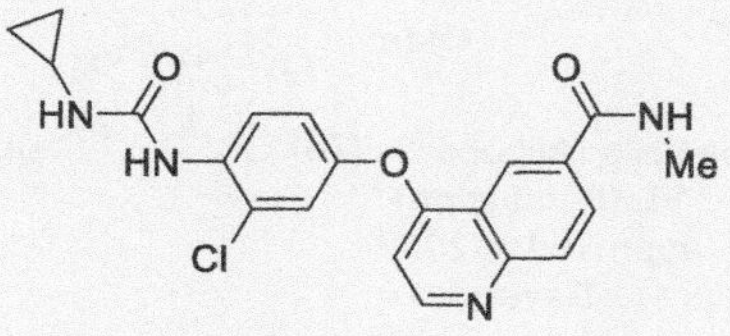

lenvatinib (Lenvima™) (**30**)
VEGFR/RET/EGFR
RCC, HCC, TC
approved in 2015
Eisai

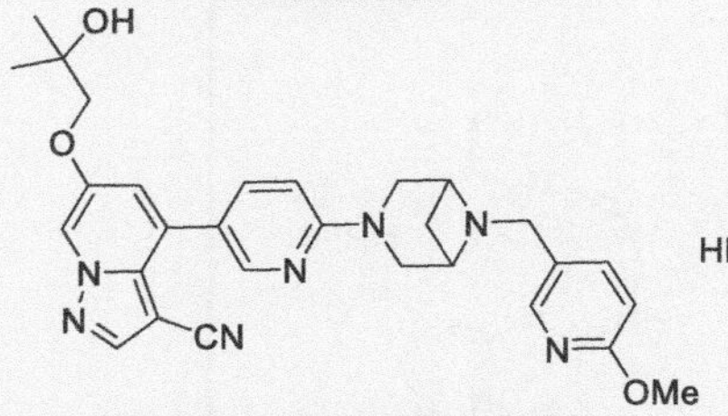

selpercatinib (Retevmo™) (**64**)
RET selective
NSCLC, MTC
approved in 2020
Lilly

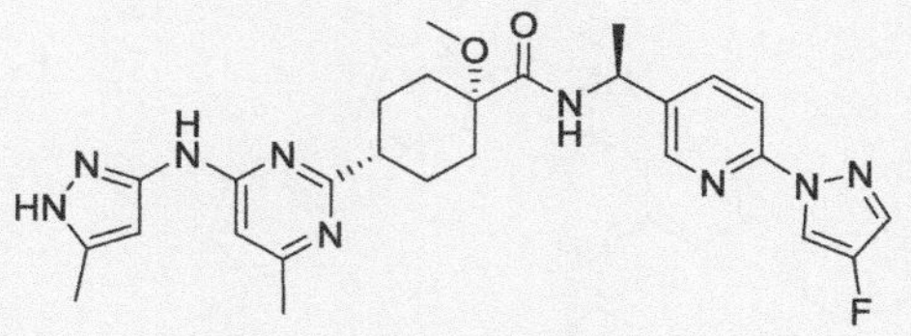

pralsetinib (Gavreto™) (**65**)
RET selective
NSCLC, MTC
approved in 2020
Blueprint Medicines

TC: thyroid cancer
MTC: medullary thyroid cancer
RCC: renal cell carcinoma
HCC: Hepatocellular carcinoma
NSCLC: Non-small cell lung cancer

Figure 26. RET/multikinase inhibitors (Chapter 12).

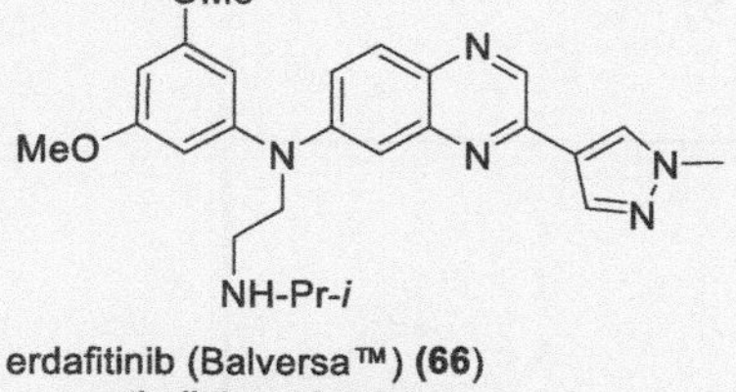

erdafitinib (Balversa™) (**66**)
urothelial carcinoma
approved in 2019
Janssen

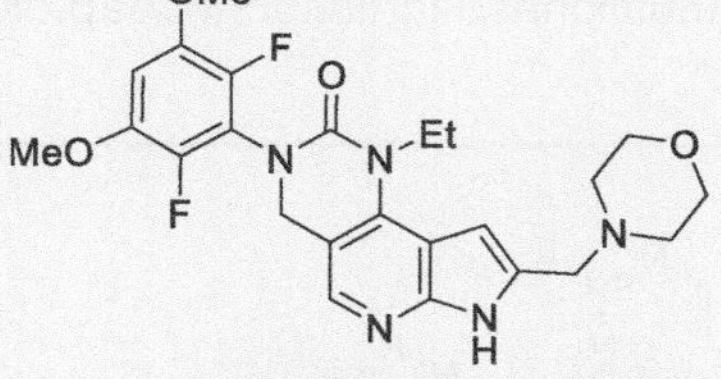

pemigatinib (Pemazyre™) (**67**)
cholangiocarcinoma
approved in 2020
Incyte

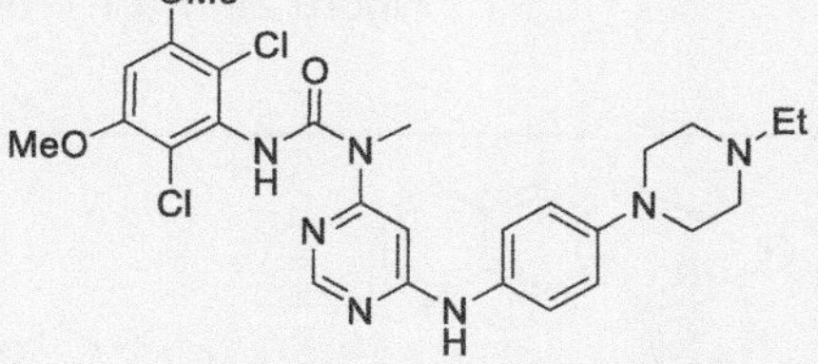

infigratinib (Truseltiq™) (**68**)
cholangiocarcinoma
approved in 2021
Novartis

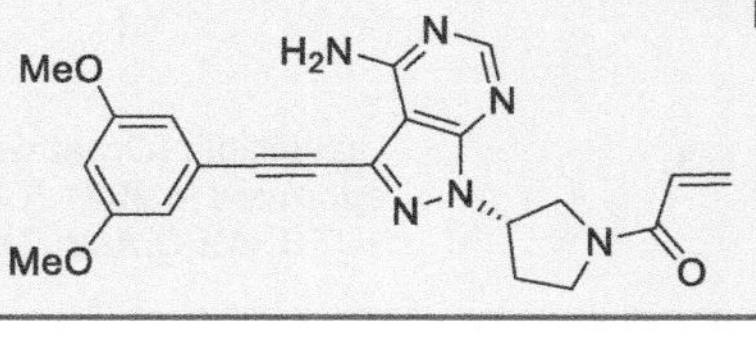

futibatinib (Lytgobi™) (**69**)
approved in 2022
cholangiocarcinoma
Taiho

Figure 27. Pan-FGFR inhibitors (Chapter 13).

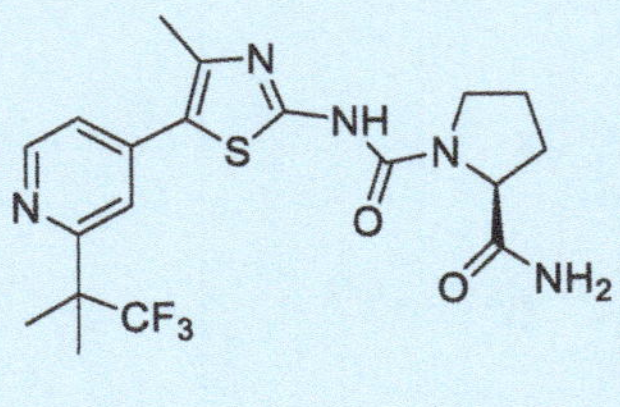

alpelisib (Piqray™) (**70**)
breast cancer
approved in 2019
Novartis

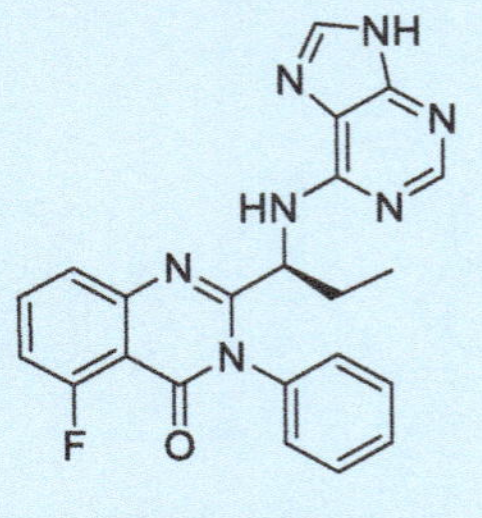

idelalisib (Zydelig™) (**71**)
CLL, SLL, FL
approved in 2014
Gilead

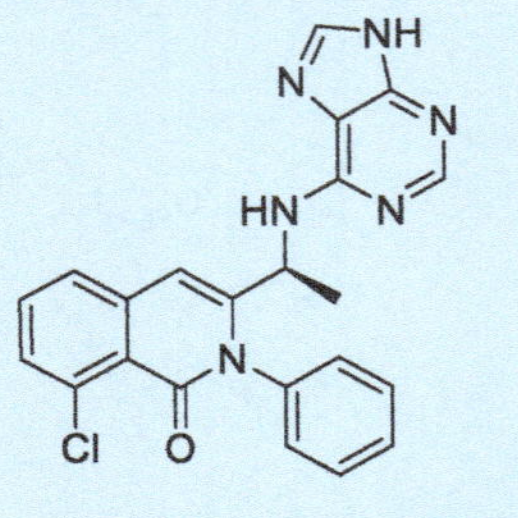

duvelisib (Copiktra™) (**72**)
CLL, SLL, FL
approved in 2018
Verastem Oncology

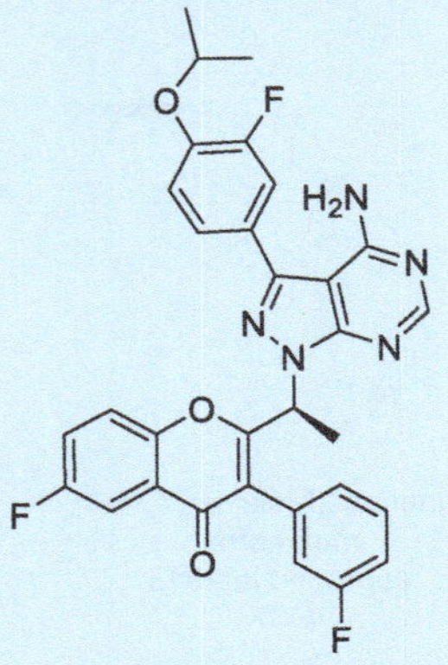

umbralisib (Ukonig™) (**73**)
MZL, FL
approved in 2021
TG Therapeutics

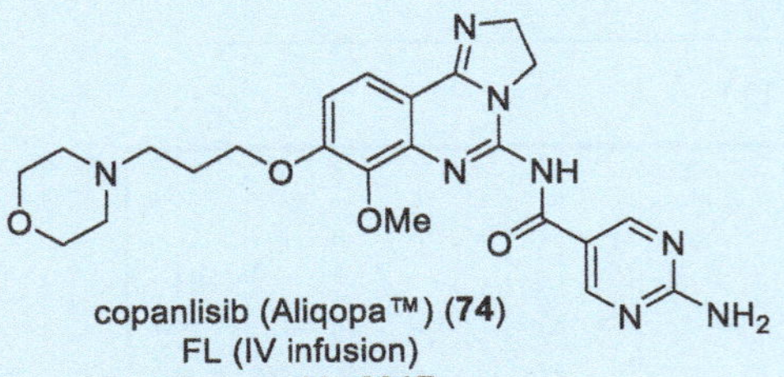

copanlisib (Aliqopa™) (**74**)
FL (IV infusion)
approved in 2017
Bayer

FL: follicular lymphoma
SLL: small lymphocytic leukemia
CLL: chronic lymphocytic leukemia
MZL: marginal zone lymphoma

Figure 28. PI3K inhibitors (Chapter 14).

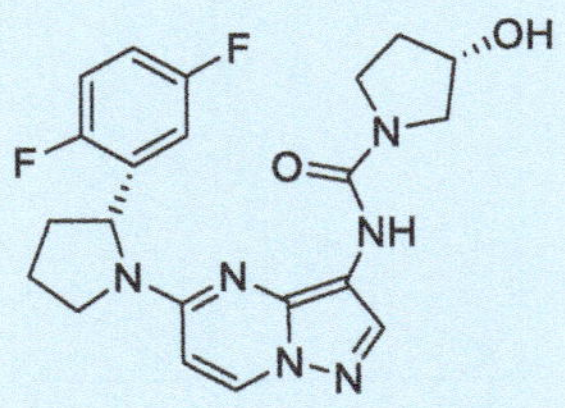

larotrectinib (Vitrakvi™) (**75**)
pan-TRK
NTRK+ solid tumors
approved in 2018
Loxo/Bayer

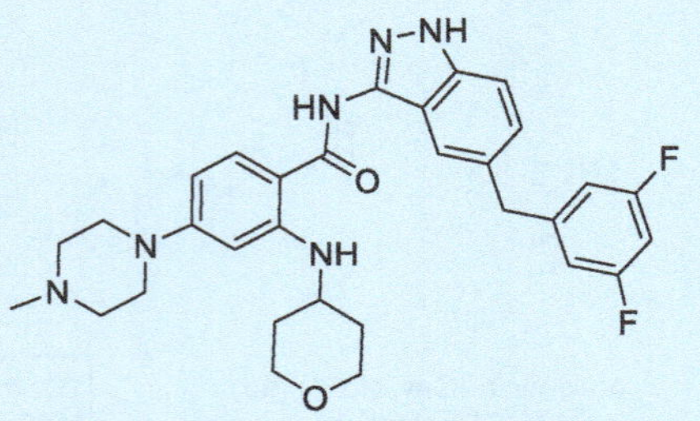

entrectinib (Rozlytrek™) (**76**)
pan-TRK/ROS1/ALK
NTRK+ solid tumors
ROS1+ NSCLC
approved in 2019
Roche

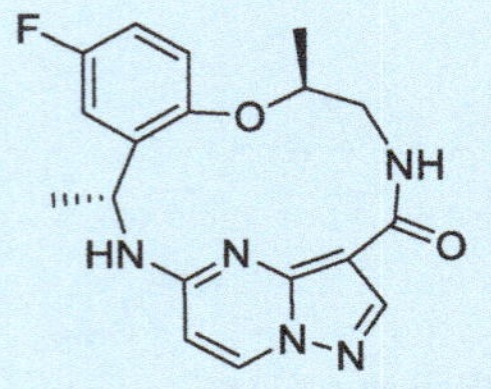

repotrectinib (**77**)
pan-TRK/ROS1/ALK
ROS1+ NSCLC
FDA approval
expected in June 2023
TP Therapeutics

Figure 29. Pan-TRK/multikinase inhibitors (Chapter 15).

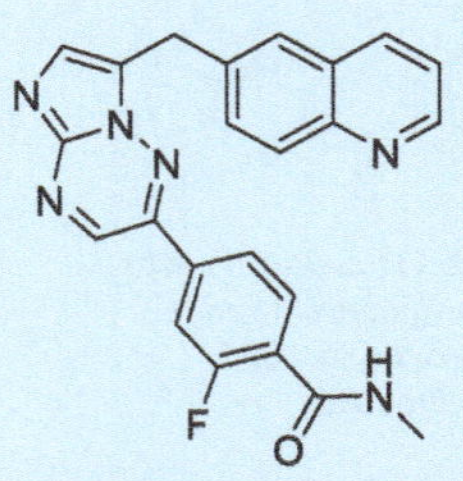

capmatinib (Tabrecta™) (**78**)
approved in 2020
Incyte

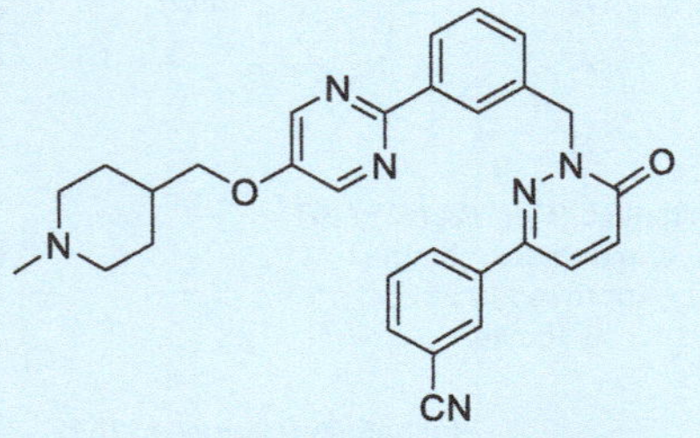

Tepotinib (Tepmetko™) (**79**)
approved in 2021
EMD Serono

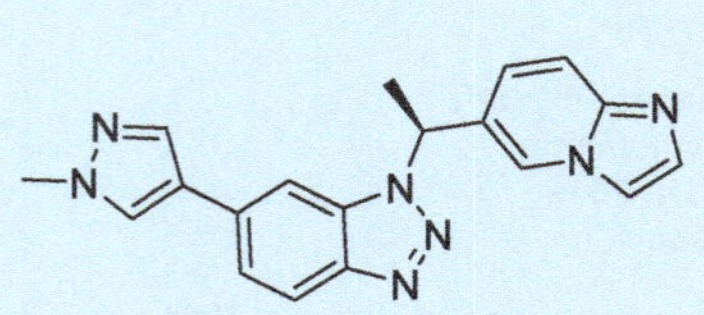

savolitinib (Orpathys™)
approved in 2021 (China)
HUTCHMED/AstraZeneca

Figure 30. MET inhibitors for MET-altered NSCLC (Chapter 16).

imatinib (Gleevec™) (**1**)
BCR-ABL/KIT/PDGFR
CML, GIST
approved in 2001
Novartis

sunitinib (Sutent™) (**25**)
VEGFR/KIT/PDGFR
RCC, GIST
approved in 2006
Pfizer

regorafenib (Stivarga™) (**24**)
VEGFR/KIT/PDGFR
CRC, HCC, GIST
approved in 2012
Onyx/Bayer

avapritinib (Ayvakit™) (**80**)
KIT/PDGFR
GIST; SM
approved in 2020
Blueprint Medicines

ripretinib (Qinlock™) (**81**)
KIT/PDGFR
GIST
approved in 2020
Deciphera

GIST: gastrointestinal stromal tumor
SM: systemic mastocytosis
CRC: colorectal cancer
HCC: hepatocellular carcinoma
CML: chronic myelogenous leukemia

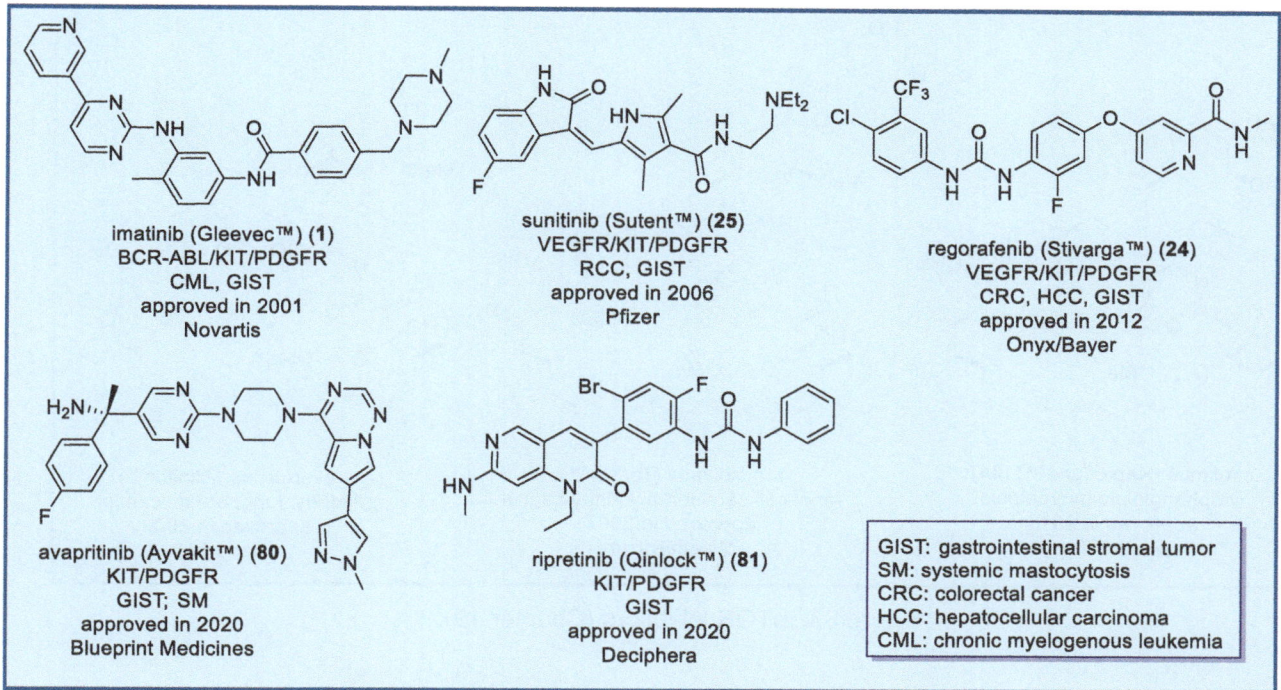

Figure 31. KIT/PDGFR/multikinase inhibitors (Chapter 17).

midostaurin (Rydapt™) (**82**)
FLT3/KIT
approved in 2017
Novartis

gilteritinib (Xospata™) (**83**)
FLT3/AXL
approved in 2018
Astellas

quizartinib
FLT3
approved in 2019 (Japan)
Daiichi Sankyo

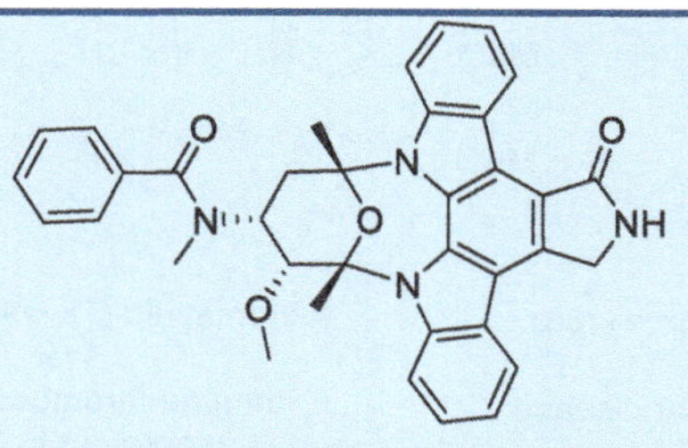

Figure 32. FLT3 inhibitors for FLT3+ AML (Chapter 18).

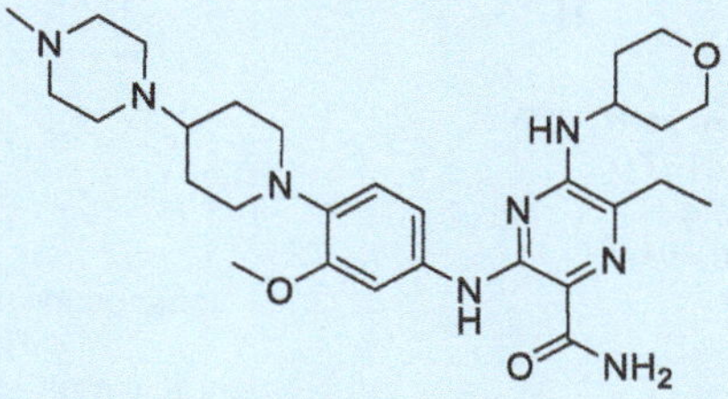

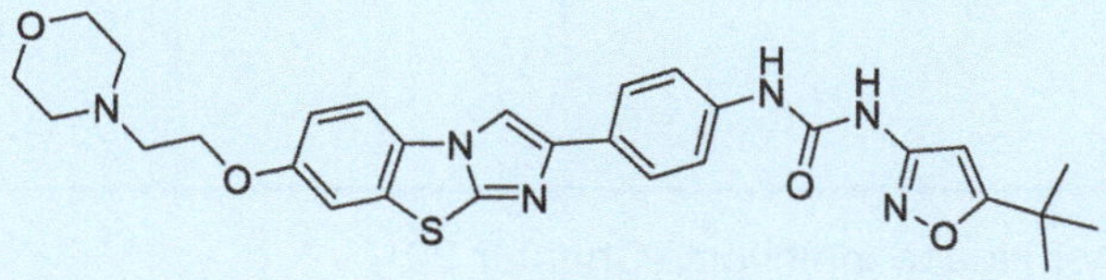

sirolimus (Rapamune™) (**84**)
lymphangioleiomyomatosis
approved in 2015
Wyeth/Pfizer

temsiolimus (Torisel™)
renal cell carcinoma, kidney cancer
approved in 2007
Wyeth/Pfizer

everolimus (Afinitor™)
kidney, lung, breast cancer
approved in 2009
Novartis

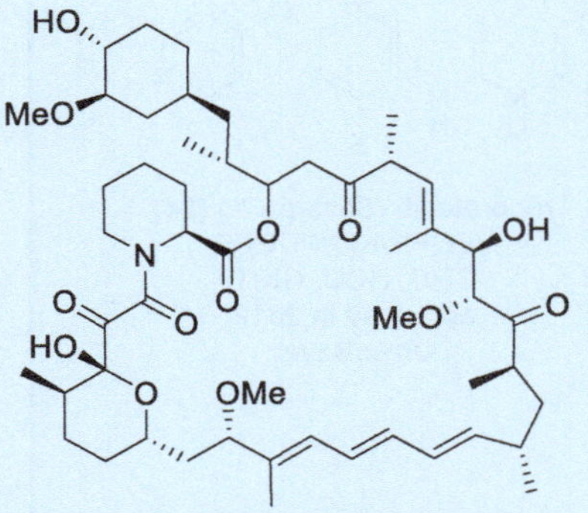

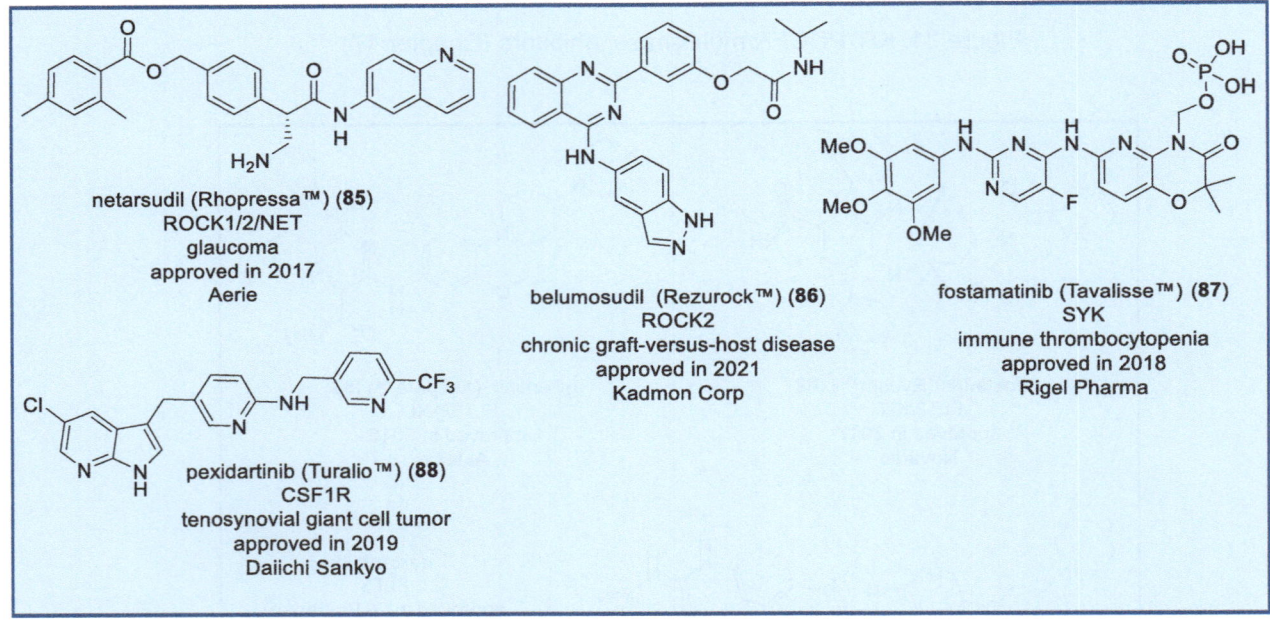

Figure 33. mTOR inhibitors (Chapter 19).

netarsudil (Rhopressa™) (**85**)
ROCK1/2/NET
glaucoma
approved in 2017
Aerie

belumosudil (Rezurock™) (**86**)
ROCK2
chronic graft-versus-host disease
approved in 2021
Kadmon Corp

fostamatinib (Tavalisse™) (**87**)
SYK
immune thrombocytopenia
approved in 2018
Rigel Pharma

pexidartinib (Turalio™) (**88**)
CSF1R
tenosynovial giant cell tumor
approved in 2019
Daiichi Sankyo

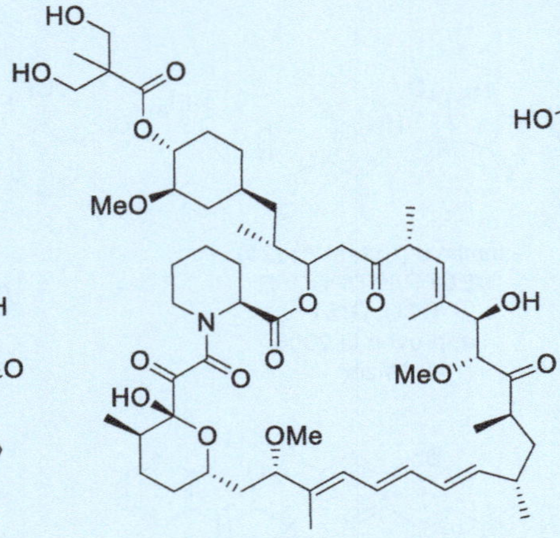

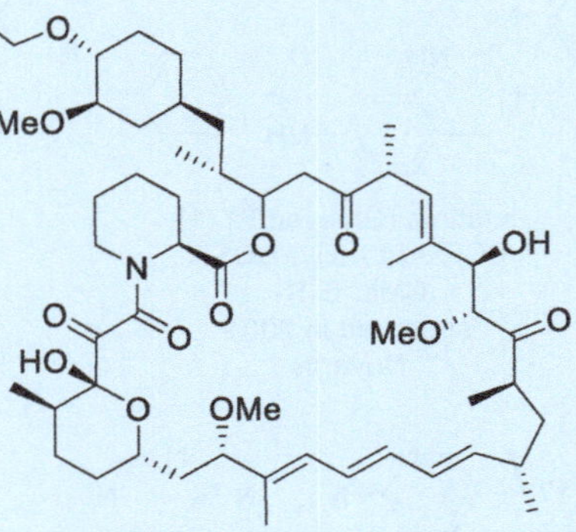

Figure 34. Other kinase inhibitors (Chapter 20).

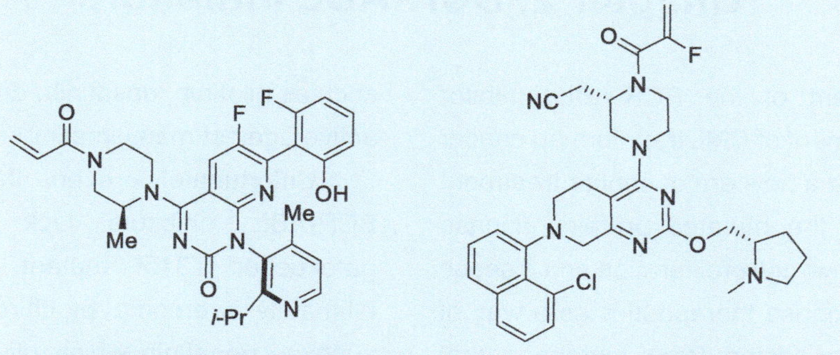

sotorasib (Lumakras™) (**89**)
G12C
approved in 2021
Amgen

adagrasib (Krazati™) (**90**)
G12C
approved in 2022
Mirati

Figure 35. KRASG12C inhibitors (Chapter 21).

1. L. A. Smyth, I. Collins. Measuring and interpreting the selectivity of protein kinase inhibitors. *J. Chem. Biol.* **2009**, *2*, 131–151.
2. T. O'Hare, R. Pollock, E. P. Stoffregen, J. A. Keats Wang B, Wu H, Hu C, Wang H, Liu J, Wang W, Liu Q. An overview of kinase downregulators and recent advances in discovery approaches. *Signal Transduct. Target Ther*. **2021**, *6*, 423.
3. S. R. Hubbard, J. H. Till. Protein tyrosine kinase structure and function. *Annu. Rev. Biochem.* **2000**, *69*, 373–398.
4. S. Kothiwale, C. M. Borza, E. W. Lowe, A. Pozzi, J. Meiler. Discoidin domain receptor 1 (DDR1) kinase as target for structure-based drug discovery. *Drug Discov.Today*. **2015**, 20, 255–261.
5. A. K. Ghose, T. Herbertz, D. A. Pippin, J. M. Salvino, J. P. Mallamo. Knowledge based prediction of ligand binding modes and rational inhibitor design for kinase drug discovery. *J. Med. Chem*. **2008**, *51*, 5149–5171.
6. R. Urich, G. Wishart, M. Kiczun, A. Richters, N. Tidten-Luksch, D. Rauh, B. Sherborne, P. G. Wyatt, R. Brenk. De novo design of protein kinase inhibitors by in silico identification of hinge region-binding fragments. *ACS Chem. Biol.* **2013**, *8*, 1044–1052.
7. P. Mukherjee, J. Bentzien, T. Bosanac, W. Mao, M. Burke, I. Muegge. Kinase crystal miner: A powerful approach to repurposing 3D hinge binding fragments and its application to finding novel bruton tyrosine kinase inhibitors. *J. Chem. Inf. Model.* **2017**, *57*, 2152–2160.
8. G. Scapin. Protein kinase inhibition: Different approaches to selective inhibitor design. *Curr. Drug Targets.* **2006**, *7*, 1443–1454.
9. I. Abdelbaky, H. Tayara, K. T. Chong. Prediction of kinase inhibitors binding modes with machine learning and reduced descriptor sets. *Sci. Rep.* **2021**, *11*, 706.
10. L. K. Gavrin, E. Saiah. Approaches to discover non-ATP site kinase inhibitors. *MedChemComm.* **2013**, *4*, 41–51.
11. P. Y. Lee, Y. Yeoh, T. Y. Low. A recent update on small-molecule kinase inhibitors for targeted cancer therapy and their therapeutic insights from mass spectrometry-based proteomic analysis. *The FEBS J.* **2022**, 10.1111/febs.16442. Advance online publication.
12. R. S. Vijayan, P. He, V. Modi, K. C. Duong-Ly, H. Ma, J. R. Peterson, J. R., Dunbrack, R. M. Levy. Conformational analysis of the DFG-out kinase motif and biochemical profiling of structurally validated type II inhibitors. *J. Med. Chem.* **2015**, *58*, 466–479.
13. D. Fabbro, S. W. Cowan-Jacob, H. Moebitz H. Ten things you should know about protein kinases: IUPHAR Review 14. *Br. J. Pharmacol.* **2015**, *172*, 2675-700.
14. D. E. Heppner, M. J. Eck. A structural perspective on targeting the RTK/Ras/MAP kinase pathway in cancer. *Protein Sci.* **2021**, *30*, 1535–1553.
15. K. Aertgeerts, R. Skene, J. Yano, B. C. Sang, H. Zou, G. Snell, A. Jennings, K. Iwamoto, N. Habuka, A. Hirokawa, T. Ishikawa, T., Tanaka, H. Miki, Y. Ohta, S. Sogabe. Structural analysis of the mechanism of inhibition and allosteric activation of the kinase domain of HER2 protein. *J. Biol. Chem.* **2011**, *286*, 18756–18765.
16. P. Wu, T. E. Nielsen, M. H. Clausen. FDA-approved small-molecule kinase inhibitors. *Trends Pharmacol. Sci.* **2015**, *36*, 422–439.

Chapter 2. BCR-ABL Inhibitors

The development of the BCR-ABL inhibitor imatinib for the treatment of CML transformed cancer research and opened a new era of cancer treatment based on targeting the mutated proteins (kinase oncoproteins) that drive cell proliferation and disease. The concept of antikinase therapeutics as a way of dealing with tumors arising from mutations that cause abnormal cell signaling and proliferation was validated – as was the use of protein structural analysis by X-ray diffraction to guide molecular design.

The clinical application of imatinib and other kinase inhibitors revealed that many patients eventually develop resistance, usually associated with mutations in the ABL kinase domain that either directly or indirectly affect the binding affinity of imatinib to ABL. Such resistance prompted development of second-generation ABL inhibitors such as nilotinib, dasatinib, and bosutinib, which are active against many imatinib-resistant mutants.

Unfortunately, even the second-generation BCR-ABL inhibitors lack potency against a gatekeeper T315I mutant. However, that was ultimately overcome by third-generation inhibitors such as ponatinib which utilizes a unique ethynyl linker to reduce steric clash with the bulky leucine side chain of the mutated gatekeeper residue I315. Because ponatinib can cause cardiovascular side effects, asciminib, a fourth-generation TKI for CML, has been developed. It locks the ABL protein into an inactive conformation by binding to a myristoyl site, thus overcoming drug resistance arising from ATP-binding site mutations.

This chapter describes the discovery of the following BCR-ABL inhibitors.

Molecules Engineered Against Oncogenic Proteins and Cancer, First Edition. E. J. Corey and Yong-Jin Wu.
© 2024 John Wiley & Sons Inc. Published 2024 by John Wiley & Sons Inc.

2.1 Imatinib (Gleevec™) (1)

Structural Formula	Space-filling Model	Brief Information
$C_{29}H_{31}N_7O$; MW = 494 logP = 3.1; tPSA = 85		Year of discovery: 1993 Year of introduction: 2001 Discovered by: Ciba-Geigy Developed by: Novartis Primary target: BCR-ABL Binding type: II (ABL) Class: non-receptor tyrosine kinase Treatment: CML, ALL, GIST Other name: STI571 Oral bioavailability = 98% Elimination half-life = 20 h Protein binding = 95%

● = C   ○ = H   ● = O   ● = N   ● = S   ● = F   ● = Cl   ● = Br   ● = I

Imatinib was first approved in 2001 as first-line treatment for Philadelphia chromosome-positive (Ph+) chronic myelogenous leukemia (CML) in adults and children and later was also cleared for the treatment of other types of cancers including acute lymphocytic leukemia (ALL), gastrointestinal stromal tumors (GISTs), systemic mastocytosis, and myelodysplastic syndrome.[1–3]

CML is characterized by a mutated chromosome (the "Philadelphia chromosome", see below) resulting from a translocation between chromosome 9 and 22 which leads to the BCR-ABL fusion gene. This gene produces an abnormal BCR-ABL tyrosine kinase protein, which causes CML cells to grow and reproduce uncontrollably. Imatinib was the first drug to specifically target this protein. It was touted as a "magic bullet" to target CML because it changed a once deadly blood cancer into a manageable chronic disease. Although it does not cure cancer, simply taking one pill a day by mouth for life enabled the patients to live nearly normal lives.

Because of its efficacy in treating CML as well as its favorable safety and tolerability profile, imatinib quickly became the drug of choice for treatment of CML and other types of cancer. Before its patent expired in the United States in 2016, its annual sales reached $4.7 billion (vs. $9.7 billion from AbbVie's ibrutinib in 2020, a BTK inhibitor for CLL, and $5.4 billion from Pfizer's palbociclib, a CDK4/6 inhibitor for breast cancer). Imatinib is on the World Health Organization's Model List of Essential Medicines. It has been estimated that the discovery and development of imatinib and its related drugs has created $143 billion in societal value. Imatinib not only revolutionized CML treatment but also opened the field of targeted molecular therapies for cancer.[4]

Background[4–6]

The basic cancer research that led to imatinib spanned over four decades, starting in the 1950s, when the role of genes in cancer was unknown. In 1960, Peter Nowell and David Hungerford at the University of Pennsylvania examined cancer cells from several CML patients and noticed a common feature in their genetic material – one of the 46 chromosomes was abnormally short. This tiny chromosome was later known as "the Philadelphia chromosome" named after the city where they discovered it. Further research showed that 95% of patients with CML carry the Philadelphia chromosome.

As the techniques for studying DNA improved in the 1970s, it was discovered that the Philadelphia chromosome forms when two chromosomes break, and the resulting pieces hybridize. More specifically, two genes BCR and ABL, which are normally separated, are fused together to form a new hybrid gene (fusion gene) called BCR-ABL. In the late 1980s, it was shown that the formation of the abnormal BCR-ABL proteins (encoded by the fused gene) in blood cells causes CML.

In 1988, Ciba-Geigy led by biochemist Nick Lydon started a collaboration with Brian Druker of Oregon Health and Science University to develop targeted chemotherapies for CML. At the time, it was commonly believed that tyrosine kinases were too similar and required for too many essential cellular functions to be selectively and safely inhibited in vivo. Despite such skepticism of the feasibility of developing selective kinase inhibitors, the joint effort was successful and ultimately led to STI571, later known as imatinib, as a preclinical candidate in 1995. In 1996, Ciba-Geigy merged with Sandoz to form Novartis, and the new management, who seriously doubted that imatinib would work against CML, reluctantly agreed to go forward with phase I trials. Gratifyingly, the results were miraculous: a positive response with every patient receiving the effective dose.

Unfortunately, at that time, Novartis ran out of sufficient supply of the drug for phase II trial, and in the meantime, there were too many patients waiting to participate in the trials. Thus, they petitioned directly to the CEO for greater access, and as a result, Novartis expedited the phase II studies. Based on the impressive phase II data, the FDA approved the drug in 2001 after only a 2.5-month review time, making this the fastest approval to market for any cancer treatment. Imatinib went to market initially for CML – a relatively small disease affecting 50,000 patients, but it was later approved for more than 10 indications and has been used by >200,000 patients worldwide.

BCR-ABL in CML[7-12]

The Abelson (ABL) protein (a proto-oncogene) is a non-receptor tyrosine kinase that normally shuttles between the nucleus and the cytoplasm and plays an important role in regulating cell growth and survival. In normal cells, ABL exists in an autoinhibited inactive conformation due to the interaction of its Myr-bound cap with an allosteric site at the base of the catalytic domain (for a detailed discussion on autoinhibition, see Section 2.7); however, in 90–95% of CML patients, the gene encoding ABL is fused with the breakpoint cluster region (BCR) gene, resulting in the formation of an oncoprotein (BCR-ABL) which is localized within the cytoplasm. In the process of gene fusion with BCR, all of the ABL domains (including the kinase domain) remain intact except the N-terminal cap region which is essential for autoinhibition. The cap region is replaced with the BCR-derived portion including an N-terminal coiled-coil oligomerization domain. This fusion essentially removes the myristate autoregulatory function of ABL (due to loss of the cap region), and the newly incorporated N-terminal coiled-coil domain initiates oligomerization of the fused protein and subsequent autophosphorylation, leading to a sustained activation of ABL kinase. This constitutively activates several cytokine signal transduction pathways that stimulate hyperproliferation of the CML cells and prevent apoptosis in hematopoietic cells (Fig. 1).

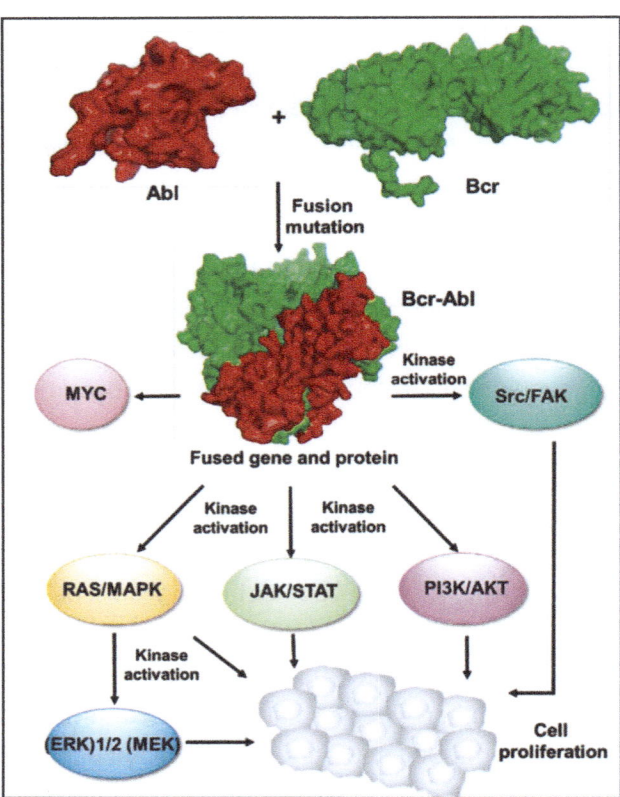

Figure 1. Mode of action of imatinib.

Imatinib binds close to the ATP-binding site of BCR-ABL and locks it in a closed inactive conformation which excludes ATP. This shuts down its sustained signaling, thus blocking leukemic cell growth.[13]

Discovery of Imatinib[14–20]

The discovery of imatinib, which predated the availability of structural information on BCR-ABL, started with extensive screening of the Ciba-Geigy compound libraries against protein kinase C (PKC).[14] This effort identified a phenylaminopyrimidine **1** as an attractive lead compound considering its "lead-like" properties and a high potential for structural diversity, allowing rapid analog synthesis for optimization of potency and selectivity (Fig. 2). Although compound **1** lacked in cellular activity, attachment of a 3'-pyridyl group at the 3'-position of the pyrimidine gave **2** which showed strong PKC inhibition in cellular assays. Incorporation of an N-acylamino group on the phenyl ring gave **3** which provided inhibitory activity against tyrosine kinases, such as the BCR-ABL kinase. At this point, SAR studies showed a striking observation: a simple substitution at position 6 of the diaminophenyl ring abolished PKC inhibitory activity. For example, a methyl substitution (as in **4**) removed PKC activity, while maintaining or even increasing activity against protein tyrosine kinases. However, the first series of selective inhibitors represented by **4** with simple amide side chains showed poor oral bioavailability and low aqueous solubility due to high lipophilicity. These deficiencies could be overcome by attaching a water-solubilizing N-methylpiperazine group (to give **5**). To avoid the mutagenic potential of the aniline moiety in **5**, a methylene spacer was inserted between the phenyl ring and the nitrogen atom which resulted in imatinib as a preclinical candidate.

Imatinib exhibits good unbound fraction in plasma (5%), excellent oral bioavailability (98%) and long terminal half-life (20 h), compatible with once-daily dosing of 400 mg for CML.

In vitro screening against a panel of protein kinases reveals that imatinib inhibits the autophosphorylation of essentially three kinases: BCR-ABL, KIT and the platelet-derived growth factor (PDGF) receptor. It is a weak inhibitor of about 30 other protein kinases that were tested.[14]

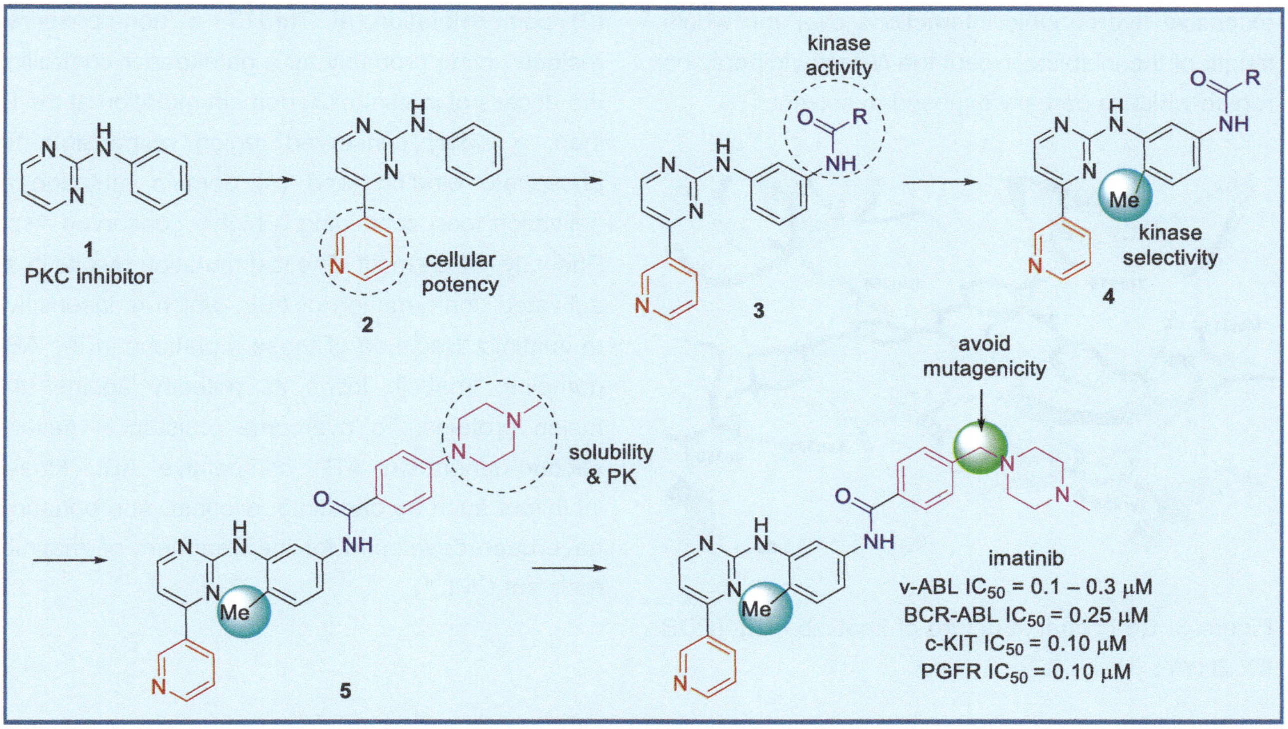

Figure 2. Structural optimization of imatinib.

The inhibition of autophosphorylation of BCR-ABL correlated well with its antiproliferative activity. Incubation with submicromolar concentrations of imatinib selectively induced apoptosis in BCR-ABL-positive cell lines, and induced cell killing in primary leukemia cells from patients with Ph+ CML and ALL.[14] The submicromolar plasma concentrations can be readily achieved and sustained in humans by once-daily administration of imatinib at 400 mg.[20]

Binding Mode[21,22]

Imatinib binds with high specificity to an inactive form of the ABL kinase, resulting in high kinase selectivity (Fig. 3). Interestingly, the *N*-methylpiperazine group, which was installed to increase solubility, also interacts strongly with ABL via hydrogen bonds to the backbone carbonyl group of Ile360 and His361. The co-crystal structure of imatinib with ABL shows additional four hydrogen bonds between (1) the pyridine ring N atom and the backbone NH of Met318 in the hinge region; (2) the anilino NH and the side chain of the gatekeeper residue Thr315; (3) the amide NH and the side chain of Glu286 from helix C; and (4) the amide carbonyl and the backbone NH of Asp381 (Fig. 4). The hydrogen bonds are also complemented by extensive hydrophobic interactions over the whole length of the inhibitor except the *N*-methylpiperazine region which is partially exposed to solvent.

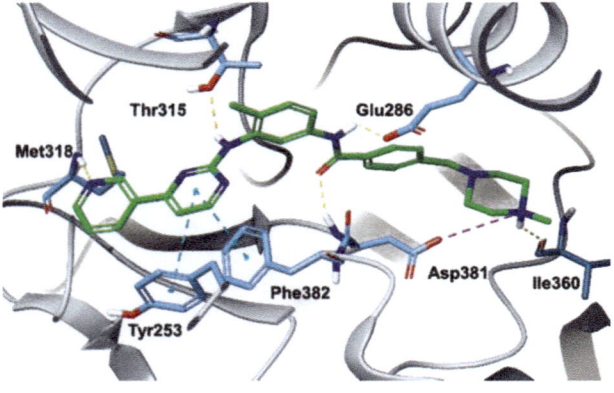

Figure 3. Co-crystal structure of imatinib–ABL (PDB ID: 2HYY).

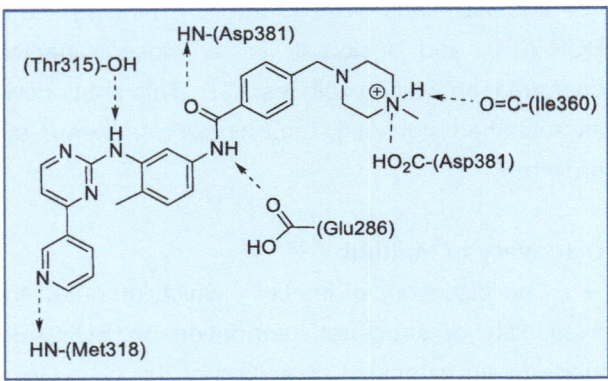

Figure 4. Summary of ABL–imatinib interactions based on X-ray co-crystal structural data.

Imatinib Resistance[23,24]

Despite its remarkable success in the treatment of CML, about 10% of patients who initially respond to imatinib subsequently develop resistance. The resistance can arise from (1) point mutation (missense mutation) within the ABL kinase domain; (2) massive overexpression of BCR-ABL; and (3) overexpression of the P-glycoproteins which remove the drug from cell. So far, point mutations within the ABL kinase domain that interfere with imatinib binding account for a majority of imatinib resistance.

The ABL kinase mutations are clustered in four regions: (1) point mutations at Met351 and Glu355; (2) point mutation at Thr315, a non-conserved residue acting probably as a gatekeeper controlling the access of imatinib; (3) domain mutation at the P-loop, a highly conserved region responsible for phosphate binding; and (4) domain mutation at activation loop containing a highly conserved Asp–Phe–Gly (DFG) motif. The last mutation results in an activated conformation of ABL, which is insensitive to imatinib. Because of these mutations in the ABL domains, imatinib loses its potency against the fusion proteins. To overcome resistance, several second-generation ATP competitive ABL kinase inhibitors such as dasatinib, nilotinib, and ponatinib have been developed for the treatment of imatinib-resistant CML.[4]

1. M. D. Moen, K. McKeage, G. L. Plosker, M. A. Siddiqui. Imatinib: a review of its use in chronic myeloid leukaemia. *Drugs*. **2007**, *67*, 299–320.
2. B. J. Druker, M. Talpaz, D. J. Resta, B. Peng, E. Buchdunger, J. M. Ford, N. B. Lydon,, H. Kantarjian, R. Capdeville, S. Ohno-Jones, C. L. Sawyers. Efficacy and safety of a specific inhibitor of the BCR-ABL tyrosine kinase in chronic myeloid leukemia. *N. Engl. J. Med.* **2001**, *344*, 1031–1037.
3. D. M. Ross, T. P. Hughes. Cancer treatment with kinase inhibitors: what have we learnt from imatinib? *Br. J. Cancer* **2004**, *90*, 12–19.
4. F. Rossari, F. Minutolo, E. Orciuolo. Past, present, and future of Bcr-Abl inhibitors: From chemical development to clinical efficacy. *J. Hematol. Oncol.* **2018**, *11*, 84.
5. C. Dreifus. Researcher behind the drug Gleevec. The New York Times. November 02, 2009.
6. How Gleevec Transformed Leukemia Treatment – NCI website.
7. N. Dölker, M. W. Górna, L. Sutto, A. S. Torralba, G. Superti-Furga, F. L. Gervasio. The SH2 domain regulates c-Abl kinase activation by a cyclin-like mechanism and remodulation of the hinge motion. *PLoS Comput. Biol.* **2014**, *10*, e1003863.
8. P. W. Manley, L. Barys, S. W. Cowan-Jacob. The specificity of asciminib, a potential treatment for chronic myeloid leukemia, as a myristate-pocket binding ABL inhibitor and analysis of its interactions with mutant forms of BCR-ABL1 kinase. *Leuk. Res.* **2020**, *98*, 106458.
9. Y. Choi, M. A. Seeliger, S. B. Panjarian, H. Kim, X. Deng, T. Sim, B. Couch, A. J. Koleske, T. E. Smithgall, N. S. Gray. N-myristoylated c-Abl tyrosine kinase localizes to the endoplasmic reticulum upon binding to an allosteric inhibitor. *J. Biol. Chem.* **2009**, *284*, 29005–29014.
10. A. A. Wylie, J. Schoepfer, W. Jahnke, S. W. Cowan-Jacob, A. Loo, P. Furet, A. L. Marzinzik, X. Pelle, J. Donovan, W. Zhu, S. Buonamici, A. Q. Hassan, F. Lombardo, V. Iyer, M. Palmer, G. Berellini, S. Dodd, S. Thohan, H. Bitter, S. Branford, W. R. Sellers. The allosteric inhibitor ABL001 enables dual targeting of BCR-ABL1. *Nature*, **2017**, *543*, 733–737.
11. F. Carofiglio, D. Trisciuzzi, N. Gambacorta, F. Leonetti, A. Stefanachi, O. Nicolotti. Bcr-Abl Allosteric Inhibitors: Where We Are and Where We Are Going to. *Molecules*. **2020**, 25, 4210.
12. J. Y. Wang. The capable ABL: What is its biological function? *Mol. Cell. Bio.* **2014**, *34*, 1188–1197.
13. T. Schindler, W. Bornmann, P. Pellicena, W. T. Miller, B. Clarkson, J. Kuriyan. Structural mechanism for STI571 inhibition of Abelson tyrosine kinase. *Science*. **2000**, *289*, 1938–1942.
14. R. Capdeville, E. Buchdunger, J. Zimmermann, A. Matter. Glivec (STI571, imatinib), a rationally developed targeted anticancer drug. *Nat. Rev. Drug Disc.* **2002**, *1*, 493–502.
15. J. Zimmermann, E. Buchdunger, H. Mett, T. Meyer, N. B. Lydon. Potent and selective inhibitors of the ABL-kinase: Phenylaminopyrimidine (PAP) derivatives. *Bioorg. Med. Chem. Lett.* **1997**, *7*, 187–192.
16. J. Zimmermann, E. Buchdunger, H. Mett, T. Meyer, N. B. Lydon, P. Traxler. Phenylamino-pyrimidine (PAP) derivatives: A new class of potent and highly selective PDGF-receptor autophosphorylation inhibitors. *Bioorg. Med. Chem. Lett.* **1996**, *6*, 1221–1226.
17. H. Gündüz, Y. Özlü, S. Yalçin. Process for the preparation of imatinib base. US Patent US 8252926, **2010**.
18. J. Zimmermann. Pyrimidine derivatives and processes for the preparation thereof, US Patent 5521184, **1996.**
19. E. Buchdunger, J. Zimmermann, H. Mett, T. Meyer, M. Muller, B. J. Druker, and N. B. Lydon. Inhibition of the Abl protein-tyrosine kinase in vitro and in vivo by a 2-phenylaminopyrimidine derivative. *Cancer Res.* **1996**, *56*, 100-104.
20. R. Roskoski. STI-571: An anticancer protein-tyrosine kinase inhibitor. *Biochem. Biophys. Res. Commun.* **2003**, *309*, 709–717.
21. S. W. Cowan-Jacob, G. Fendrich, A. Floersheimer, P. Furet, J. Liebetanz, G. Rummel, P. Rheinberger, M. Centeleghe, D. Fabbro, P. W. Manley. Structural biology contributions to the discovery of drugs to treat chronic myelogenous leukaemia. *Acta Cryst. Sect. D*, **2007**, *63*, 80–93.
22. S. Agnello, M. Brand, M. F. Chellat, S. Gazzola, R. Riedl, R. A structural view on medicinal chemistry strategies against drug resistance. *Angew. Chem. Int. Ed. Engl.* **2019**, *58*, 3300–3345.
23. C. B. Gambacorti-Passerini, R.H. Gunby, R. Piazza, A. Galietta, R. Rostagno, L .Scapozza. Molecular mechanisms of resistance to imatinib in Philadelphia-chromosome-positive leukaemias. *Lancet Oncol.* **2003**, *4*, 75–85.
24. A. Hochhaus, S. Kreil, A. S. Corbin, P. La Rosée, M. C. Müller, T. Lahaye, B. Hanfstein, C. Schoch, N. C. Cross, U. Berger, H. Gschaidmeier, B. J. Druker, R. Hehlmann. Molecular and chromosomal mechanisms of resistance to imatinib (STI571) therapy. *Leukemia* **2002**, 16, 2190–2196.

2.2 Nilotinib (Tasigna™) (2)

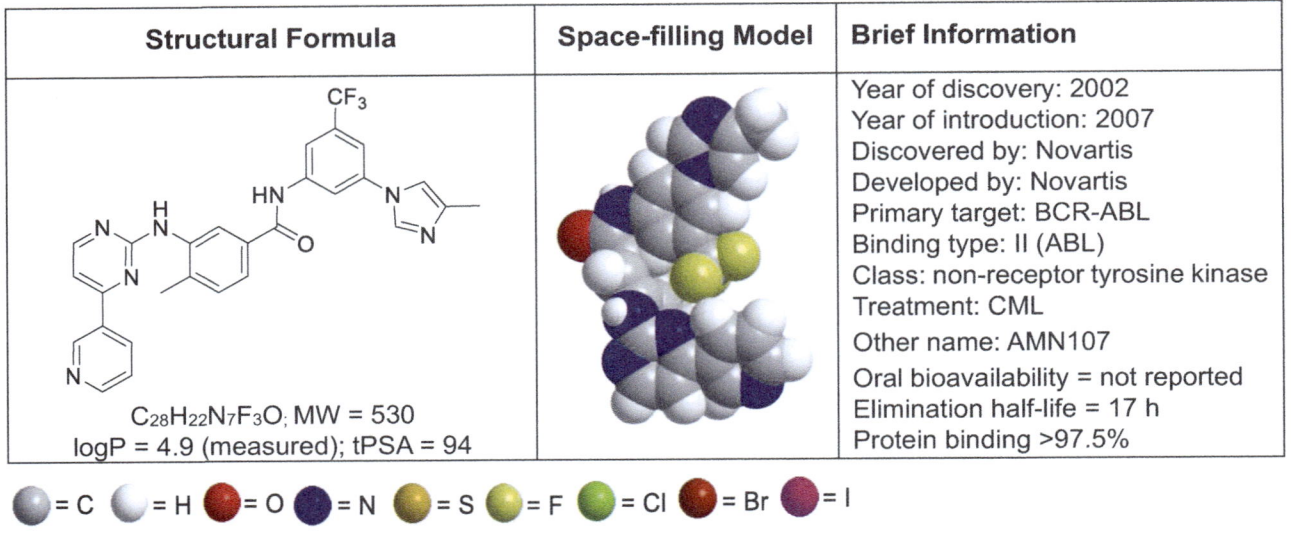

Nilotinib, a second-generation BCR-ABL tyrosine kinase inhibitor (TKI), was first approved in 2007 for the treatment of chronic phase and accelerated phase Philadelphia chromosome positive (Ph+) chronic myelogenous leukemia (CML) in adult patients resistant to or intolerant to prior treatment including imatinib, and in 2010 as a first-line treatment for CML in chronic phase. In 2018, its indications were expanded to the first- and second-line pediatric patients one year of age or older with Ph+ CML in the chronic phase.[1–3]

The introduction of the BCR-ABL inhibitor imatinib by Novartis in 2001 revolutionized the treatment of patients with CML; nonetheless, its use is frequently limited by the emergence of intolerance or resistance. In addition, it is not curative, so treatment needs to continue for the rest of a patient's life. These prompted the development of nilotinib again by Novartis in 2007 as a highly potent and selective inhibitor of the wild-type and mutated forms of BCR-ABL. It maintains activity against the majority of imatinib-resistant mutants and has fewer off-target adverse effects. Most importantly, because of its broad coverage of resistant mutants, some CML patients can achieve sustained response, and therefore can safely stop taking nilotinib. This has made nilotinib one of the most successful second-generation BCR-ABL TKIs with annual sales reaching $2.06 billion in 2021.

Discovery of Nilotinib[4–8]

The second generation of BCR-ABL inhibitors from Novartis, represented by **1** (Fig. 1), were developed starting with the imatinib structure by replacing the N-methylpiperazine group while retaining the rest of the molecule (i.e., A, B, C, D rings, Fig. 1) and reversing the amide functionality, a classic strategy in drug development. The reversed amide group was thought suitable for hydrogen-bonding to the side chain carboxylate of Glu286 and simultaneously with Asp381 (see co-crystal structures in Fig. 1 and the binding mode section). Subtle conformational changes in the protein induced by the reverse amide inhibitor could enable the D ring substituents to bind at sites different from that occupied by the N-methylpiperazine ring of imatinib, possibly resulting in enhanced binding. Based on this design concept, numerous analogs of formula **1** with various mono- and bis-phenyl substituents were prepared, leading to nilotinib in which the 4-methylimidazole is attached to the D ring to reduce lipophilicity and enhance aqueous solubility. Although the imidazole subunit of nilotinib can be potentially bound to CYP3A4 via coordination to the heme iron to cause CYP3A4 inhibition and drug–drug interactions, in this case, that turned out not to be problematic.[9]

Due to the absence of the basic *N*-methylpiperazine group, nilotinib is significantly less basic and more lipophilic (logP = 4.9) than imatinib (logP = 3.1). Consequently, imatinib exhibits good aqueous solubility, whereas nilotinib as its hydrochloride salt is only moderately soluble (0.29 mg/mL). This led to solubility-limited oral absorption in patients, and therefore, it is administered orally twice daily at 400 mg (vs. once daily at 400 mg for imatinib), which provides steady-state plasma trough levels of 1.95 µM, well above the EC_{50} concentrations required to inhibit the proliferation of a majority of imatinib-resistant mutants in vitro.[8]

Nilotinib is 20–50-fold more potent than imatinib against BCR-ABL in various cellular assays and proved to be effective in blocking proliferation of BCR-ABL dependent cell lines derived from CML patients. In addition, nilotinib demonstrated 10–20-fold improved potency over imatinib in inhibiting the autophosphorylation of BCR-ABL on Tyr177, which is involved in CML pathogenesis through regulation of several pathways, including the activation of phosphatidylinositol-3 kinase (PI3K) and RAS/ERK. Unfortunately, as with imatinib, nilotinib lacks activity against T315I mutants because it binds near the T315 residue. The mutation of threonine to isoleucine removes the critical hinge-binding of the threonine side chain hydroxyl with the pyrimidinyl NH, and furthermore, the greater steric bulk of the isoleucine side chain also prevents binding.

The T315I mutation has been overcome by the development of the third and fourth generation BCR-ABL inhibitors, such as ponatinib and asciminib, which have no specific binding interactions in the T315I region (see Sections 2.5 and 2.7).

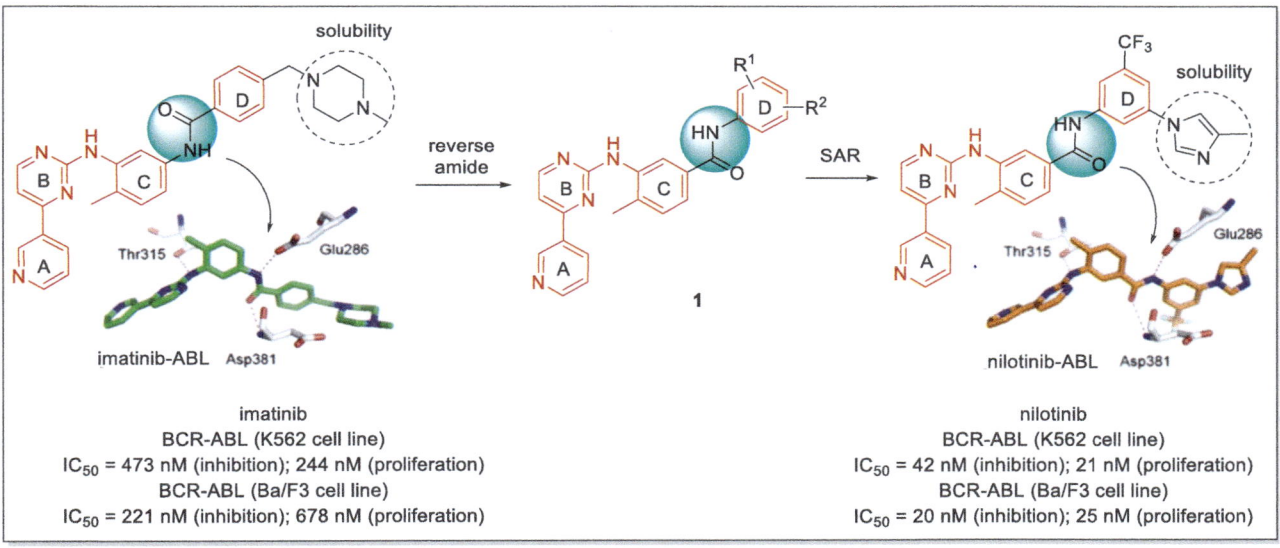

Figure 1. From imatinib to nilotinib. Co-crystal structures adapted with permission from *Bioorg. Med. Chem.* **2010**, *18*, 6977–6986.

Binding Mode[6,7]

Both imatinib and nilotinib bind the inactive conformation of ABL, but the latter allows a better topographical fit (Fig. 2). As compared to six H-bonds with imatinib (Figs. 2, 3), nilotinib makes only four H-bond interactions with the ABL kinase domain, involving the pyridyl N and the backbone amide NH of Met318; the anilino NH and the side chain hydroxyl of Thr315; the amido NH and side chain carboxylate of Glu286; and the amide carbonyl with the backbone NH of the Asp381. The reverse amide group of nilotinib hydrogen bonds to the side chain carboxylate of Glu286 and the backbone NH of Asp381 similarly to the amide in imatinib as the position of the Glu286 side chain of ABL kinase shifts and twists slightly to interact with the reverse amide.

These changes lead to the 4-methylimidazole and the CF₃ group as being highly effective alternative back pocket binders for the *N*-methylpiperazine group of imatinib.

Superposition of nilotinib (magenta) bound to ABL M351T (orange) and imatinib (green) bound to ABL (yellow) (Fig. 2) indicates that the hinge-binding segments prior to the amide carbonyl overlap well, whereas there are major differences in the back pocket interactions. While the basic *N*-methylpiperazine of imatinib is partially exposed to solvent, forming two H-bonds to the backbone carbonyl group of Ile360 and His361, the methyl-imidazole group and the trifluoromethyl group of nilotinib pack into a unique hydrophobic pocket, making substantial interactions with the ABL kinase domain of the BCR-ABL protein.

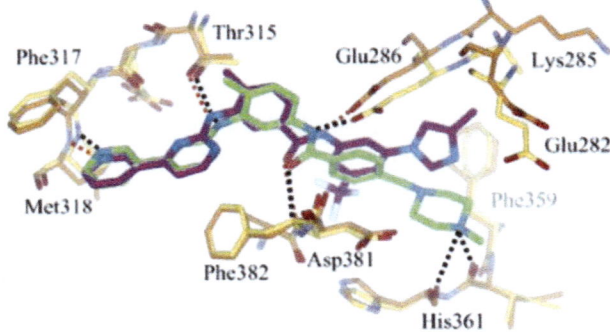

Figure 2. Superposition of nilotinib (magenta) bound to ABL M351T (orange) and imatinib (green) bound to ABL (yellow). Adapted with permission from *Cancer Cell*. **2005**, *7*, 129–141.

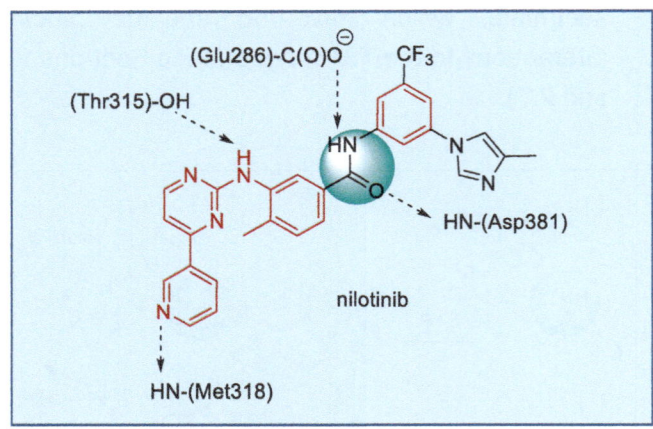
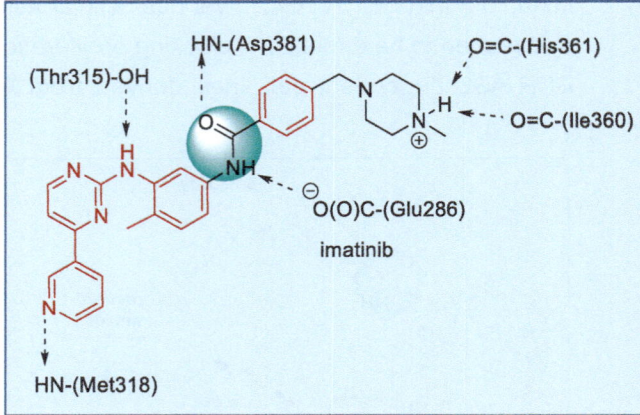

Figure 3. Summary of nilotinib–ABL M351T interactions (left) vs. imatinib-ABL interactions (right), based on co-crystal structural data.

1. E. Jabbour, J. Cortes, H. Kantarjian. Nilotinib for the treatment of chronic myeloid leukemia: An evidence-based review. *Core Evid*. **2009**, *4*, 207–213.
2. M. Breccia, G. Alimena. Nilotinib: A second-generation tyrosine kinase inhibitor for chronic myeloid leukemia. *Leukemia Res*. **2010**, *34*, 129–134.
3. J. Y. Blay, M. von Mehren. Nilotinib: A. novel. selective tyrosine kinase inhibitor. *Semin. Oncol*. **2011**, *38*, suppl. 1, S3–S9.
4. W. Breitenstein, P. Furet, S. Jacob, P. W. Manley. Inhibitors of tyrosine kinases. US patent 7169791, **2007**.
5. P. Manley, S. W. Cowan-Jacob, J. Mestan. Advances in the structural biology, design and clinical development of Bcr-Abl kinase inhibitors for the treatment of chronic myeloid leukaemia. *Biochim. Biophys. Acta*. **2005**, *1754*, 3–13.
6. P. Manley, N. Stiefl, S. W. Cowan-Jacob, S. Kaufman. J. Mestan, M. Wartmann, M. Wiesmann, R. Woodman, N. Gallagher. Structural resemblances and comparisons of the relative pharmacological properties of imatinib and nilotinib. *Bioorg. Med. Chem*. **2010**, *18*, 6977–6986.
7. E. Weisberg, P. W. Manley, W. Breitenstein, J. Brueggen, S. W. Cowan-Jacob, A. Ray, B. Huntly, D. Fabbro, G. Fendrich, E. H. -Meyers, A. L. Kung, J. Mestan, G. Q. Daley, L. Callahan, L. Catley, C. Cavazza, M. Azam, D. Neuberg, R. D. Wright, D. G. Gilliland, J. D. Griffin. Characterization of AMN107, a selective inhibitor of native and mutant Bcr-Abl. *Cancer Cell*. **2005**, *7*, 129–141.
8. P. W. Manley, N. J. Stiefl. Progress in the discovery of BCR-ABL kinase inhibitors for the treatment of leukemia. *Top. Med. Chem*. **2018**, *28*, 1–38.
9. A. Haouala, N. Widmer, M. A. Duchosal, M. Montemurro, T. Buclin, L. A. Decosterd. Drug interactions with the tyrosine kinase inhibitors imatinib, dasatinib, and nilotinib. *Blood*. **2011**, *117*, e75–e87.

2.3 Dasatinib (Sprycel™) (3)

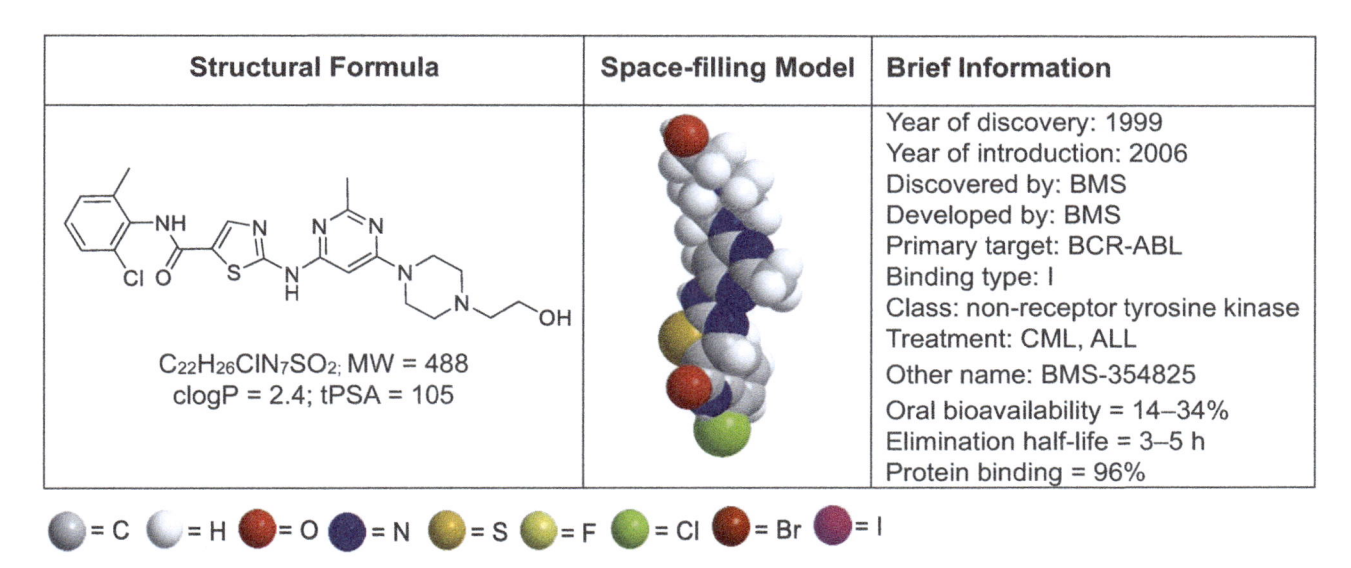

Structural Formula	Space-filling Model	Brief Information
$C_{22}H_{26}ClN_7SO_2$; MW = 488 clogP = 2.4; tPSA = 105		Year of discovery: 1999 Year of introduction: 2006 Discovered by: BMS Developed by: BMS Primary target: BCR-ABL Binding type: I Class: non-receptor tyrosine kinase Treatment: CML, ALL Other name: BMS-354825 Oral bioavailability = 14–34% Elimination half-life = 3–5 h Protein binding = 96%

● = C   ○ = H   ● = O   ● = N   ● = S   ● = F   ● = Cl   ● = Br   ● = I

Dasatinib, a second-generation BCR-ABL tyrosine kinase inhibitor (TKI), was first approved in 2006 for the treatment of chronic phase and accelerated phase Philadelphia chromosome positive (Ph+) chronic myelogenous leukemia (CML) in adult patients resistant to or intolerant to prior treatment including imatinib; in 2010 as a first-line treatment for CML in chronic phase; and in 2017 for children with Ph+ CML in the chronic phase. Its indications were expanded in 2019 to children with newly diagnosed Ph+ acute lymphoblastic leukemia (ALL) in combination with chemotherapy.[1–4]

The development of resistance in most patients treated with imatinib predominantly results from point mutations of the BCR-ABL gene and the activation of alternate signaling pathways such as SRC family kinases. Dasatinib overcomes most of these resistance mechanisms by targeting the active DFG-in conformation of ABL (while imatinib binds to the DFG-out inactive conformation) as well as all SRC isoforms (members of the SRC family of non-receptor TK's may also be oncogenic).[5] It is active against all imatinib-resistant mutants except for the T315I mutant, offering an option for the treatment of CML or Ph+ ALL patients resistant to prior therapy. Thanks to its favorable safety profile and efficacy against most BCR-ABL mutants, dasatinib, along with nilotinib, has improved upon imatinib's remarkable success in treating CML, achieving annual sales of $2.1 billion three years in a row from 2019 (vs. $2.06 billion in 2021 for Novartis' nilotinib).

Discovery of Dasatinib[6–8]

The discovery of imatinib by Novartis as a BCR-ABL inhibitor for the treatment of CML prompted other pharmaceutical companies to evaluate their own kinase inhibitors against ABL. This approach led to the development of dasatinib, which was originally identified as an inhibitor of lymphocyte-specific kinase (LCK) from an immunology drug discovery program at Bristol Myers Squibb (BMS).

Because LCK, a member of the SRC-family of cytosolic kinases, plays an important role in T-cell signaling, it was thought that LCK inhibitors may serve as immunosuppressive agents to treat autoimmune and anti-inflammatory diseases.

High-throughput screening identified aminothiazole **1** (Fig. 1) as a weak LCK inhibitor in biochemical and cell-based assays (hLCK IC_{50} = 5 µM; T-cell proliferation IC_{50} >10 µM), and subsequent SAR studies resulted in the des-methyl thiazole cyclopropylamide **2** with hLCK IC_{50} of 35 nM.[7] Surprisingly, the methyl thiazole analog **3**, derived from the original lead compound **1**, lost potency, a striking SAR finding that helped the program move forward. Because molecular docking of **2** with LCK showed no direct H-bond interaction with the cyclopropyl amide carbonyl, the cyclopropyl carboxamide was then replaced by various constrained "amide mimetics", such as heterocycles, to further improve potency. This effort gave rise to the 2″,6″-dimethyl-4″-pyrimidinyl-substituted analog **4** as a potent LCK inhibitor (hLCK IC_{50} = 1 nM; T-cell IC_{50} = 80 nM). Finally, introduction of a 4-hydroxyethylpiperazinyl side chain at the 2″-position of the pyrimidine gave dasatinib with ~30-fold improved potency on T-cells (hLCK IC_{50} = 0.4 nM; T-cell IC_{50} = 3 nM). The significant increase in cellular potency can be partially attributed to the improvement of the physicochemical properties and/or cell permeability due to incorporation of a polar side chain.

Dasatinib was later found to inhibit BCR-ABL with 325- and 16-fold enhanced potency over imatinib and nilotinib, respectively, in biochemical assays.[8,9] It is highly active against all imatinib-resistant mutants with the exception of the T315I mutant because of the close contact with this residue (see binding mode). In addition, dasatinib also demonstrated significant activity against KIT and PDGFRβ.[8]

The T315 mutations have been addressed by the development of the third- and fourth-generation BCR-ABL inhibitors, such as ponatinib and asciminib, which have no specific binding interactions in the T315 region (see Sections 2.5 and 2.7).

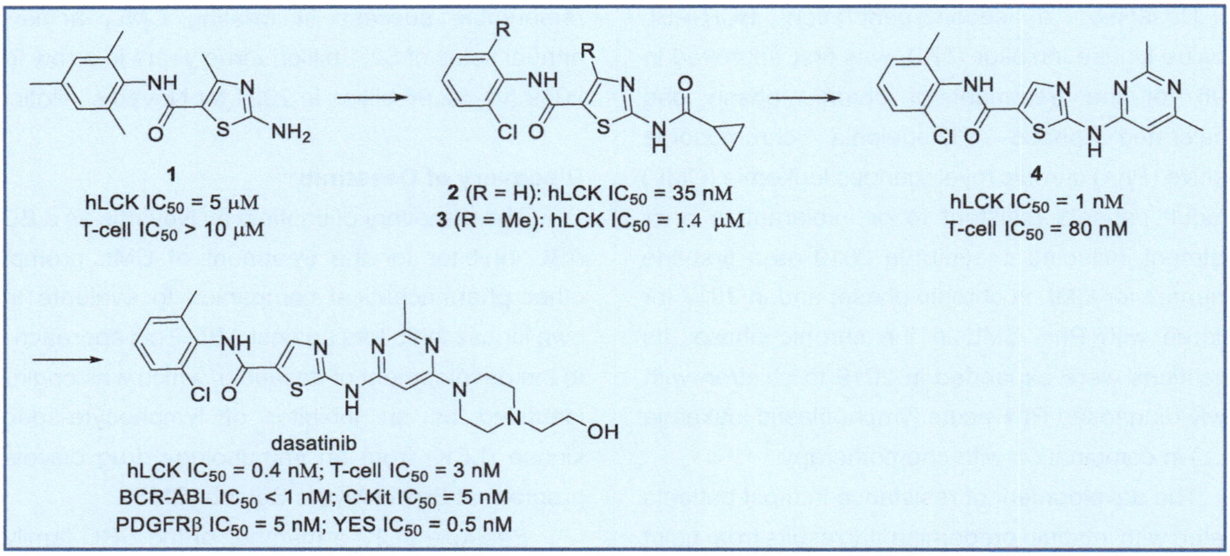

Figure 1. Structural optimization of dasatinib.

Binding Mode[5,8]

In the co-crystal structures of the two molecules of dasatinib (green and blue) in the asymmetric unit cell with ABL, it binds to an active DFG-in conformation of ABL (Fig. 2). The aminothiazole core, which is orthogonal to the 2-chloro-6-methylphenyl moiety, occupies the site that normally binds the adenine group of ATP. The pyrimidine NH and the thiazole nitrogen interact with the backbone amide carbonyl and NH of the hinge region Met318, and the carboxamide NH forms a H-bond with the side chain hydroxyl of Thr315. The terminal hydroxyethyl group interacts differently with each of the two molecules in the asymmetric unit cell: (1) via a hydrogen bond between the hydroxyethyl group and the backbone carbonyl of Tyr320 (Figs. 2, 3); and (2) with the hydroxyethyl group pointing in the opposite direction and approaching the solvent front.

By binding strongly to the active conformation of ABL within the ATP-binding site (vs. the inactive conformation for imatinib), dasatinib efficiently competes with ATP. Due to a different binding mode from that of imatinib, it maintains cellular activity against most imatinib-resistant BCR-ABL mutations (but not the T315I mutation due to the loss of a hydrogen bond with T315 and increased steric clash with the bulky isobutyl chain of 315I) (Figs. 2, 3). Unfortunately, binding to the more highly conserved active conformation results in reduced selectivity relative to imatinib, and as a result, dasatinib inhibits many additional protein kinases at physiologically relevant concentrations.

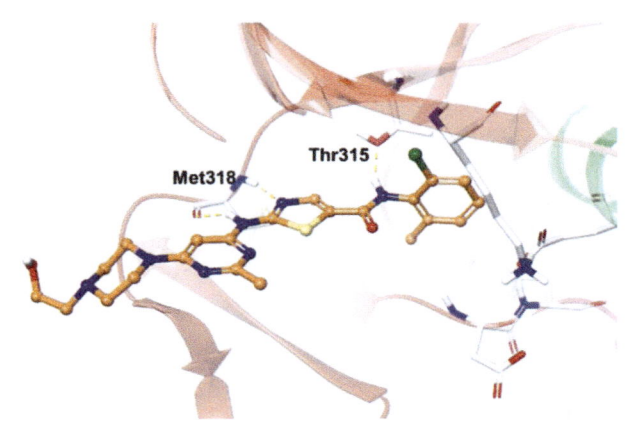

Figure 2. Co-crystal structure of dasatinib–ABL (PDB ID: 2GQG).

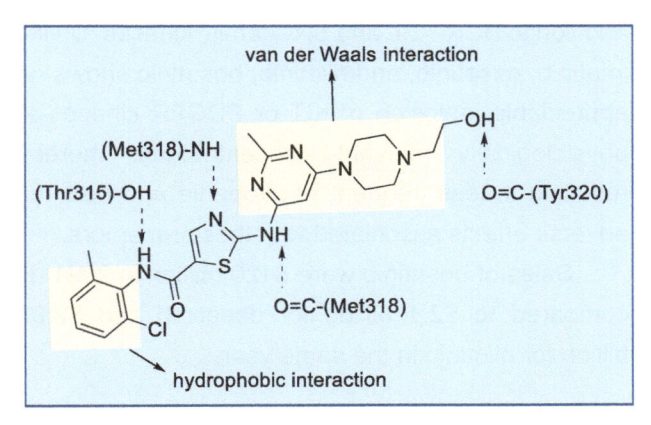

Figure 3. Summary of dasatinib–ABL interactions based on X-ray co-crystal structural data.

1. L. Han, J. J. Schuringa, A. Mulder, E. Vellenga. Dasatinib impairs long-term expansion of leukemic progenitors in a subset of acute myeloid leukemia cases. *Ann. Hematol.* **2010**, *89*, 861–871E.
2. D. G. Aguilera, A. M. Tsimberidou. Dasatinib in chronic myeloid leukemia: a review. *Ther. Clin. Risk. Manag.* **2009**, *5*, 281–289.
3. G. Wei, S. Rafiyath, D. Liu. First-line treatment for chronic myeloid leukemia: dasatinib, nilotinib, or imatinib. *J. Hematol. Oncol.* **2010**, *3*, 47.
4. A. Olivieri, L.Manzione. Dasatinib: a new step in molecular target therapy. *Ann. Oncol.* **2007**, *18*, 42–46.
5. J. S. Tokarski, J. Newitt, C. Y. J. Chang, J. D. Cheng, M. Wittekind, S. E. Kiefer, K. Kish, F. Y. F. Lee, R. Borzilerri, L. J. Lombardo, D. Xie, Y. Q. Zhang, H. E. Klei. The structure of dasatinib (BMS-354825) bound to activated Abl kinase domain elucidates its inhibitory activity against imatinib-resistant Abl mutants. *Cancer Res.* **2006**, *66*, 5790–5797.
6. J. Das, R. Padmanabha, P. Chen, D. J. Norris, A. M. P. Doweyko, J. C. Barrish, J. Wityak. Cyclic protein tyrosine kinase inhibitors. US patent 6596746, **2003**.
7. J. Das, P. Chen, D. Norris, R. Padmanabha, J. Lin, R. V. Moquin, Z. Q. Shen, L. S. Cook, A. M. Doweyko, S. Pitt, S. H. Pang, D. R. Shen, Q. Fang, H. F. de Fex, K. W. McIntyre, D. J. Shuster, K. M. Gillooly, K. Behnia, G. L. Schieven, J. Wityak, J. C. Barrish. 2-Aminothiazole as a novel kinase inhibitor Template. Structure–Activity Relationship Studies toward the Discovery of N-(2-Chloro-6-methylphenyl)-2-[[6-[4-(2-hydroxyethyl)-1-piperazinyl)]-2-methyl-4-pyrimidinyl]amino)]-1,3-thiazole-5-carboxamide (Dasatinib, BMS-354825) as a Potent pan-Src Kinase Inhibitor. *J. Med. Chem.* **2006**, *49*, 6819–6832.
8. L. J. Lombardo, F. Y. Lee, P. Chen, D. Norris, J. C. Barrish, K. Behnia, S. Castaneda, L. A. Cornelius, J. Das, A. M. Doweyko, C. Fairchild, J. T. Hunt, I. Inigo, K. Johnston, A. Kamath, D. Kan, H. Klei, P. Marathe, S. Pang, R. Peterson, R. M. Borzilleri. Discovery of N-(2-chloro-6-methylphenyl)-2-(6-(4-(2-hydroxyethyl)-piperazin-1-yl)-2-methylpyrimidin-4-ylamino)thiazole-5-carboxamide (BMS-354825), a dual Src/Abl kinase inhibitor with potent antitumor activity in preclinical assays. *J. Med. Chem.* **2004**, *47*, 6658–6661.
9. P. W. Manley, S. W. Cowanjacob, J. Mestan. Advances in the structural biology, design and clinical development of Bcr-Abl kinase inhibitors for the treatment of chronic myeloid leukaemia. *Biochim. Biophys. Acta.* **2005**, *1754*, 3–13.

2.4 Bosutinib (Bosulif™) (4)

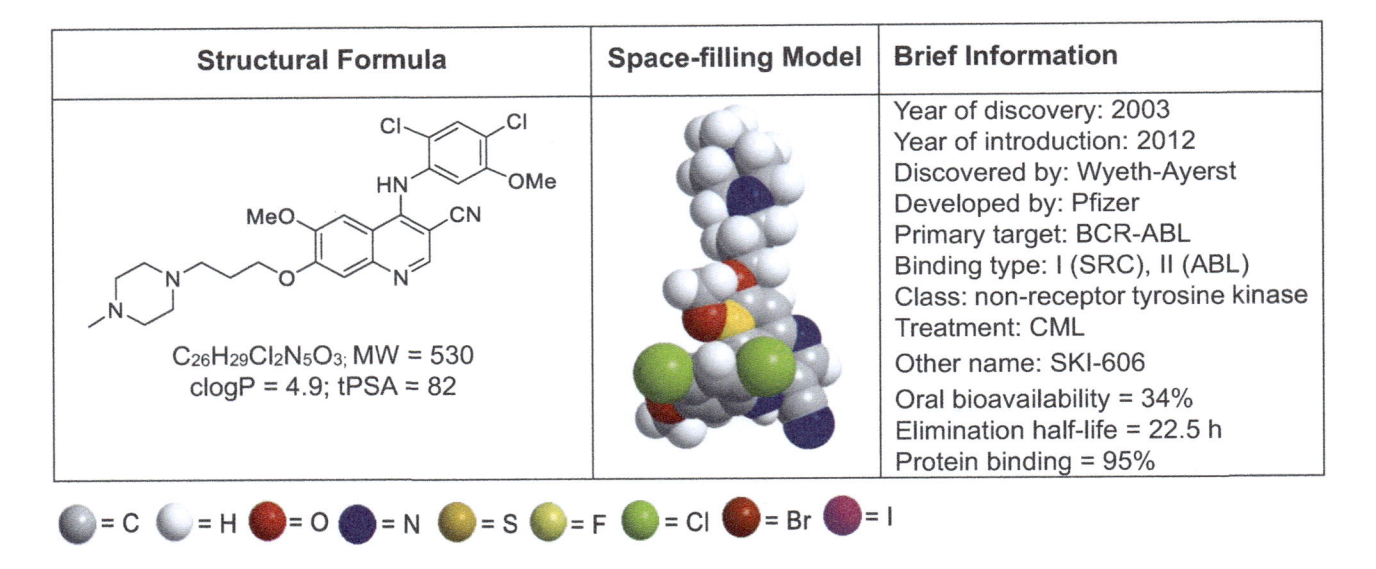

Structural Formula	Space-filling Model	Brief Information
$C_{26}H_{29}Cl_2N_5O_3$; MW = 530 clogP = 4.9; tPSA = 82		Year of discovery: 2003 Year of introduction: 2012 Discovered by: Wyeth-Ayerst Developed by: Pfizer Primary target: BCR-ABL Binding type: I (SRC), II (ABL) Class: non-receptor tyrosine kinase Treatment: CML Other name: SKI-606 Oral bioavailability = 34% Elimination half-life = 22.5 h Protein binding = 95%

● = C   ● = H   ● = O   ● = N   ● = S   ● = F   ● = Cl   ● = Br   ● = I

Bosutinib, a second-generation BCR-ABL tyrosine kinase inhibitor (TKI), was first approved in 2012 for patients with chronic, accelerated or blast phase Philadelphia chromosome positive (Ph+) chronic myelogenous leukemia (CML) who are resistant to or who cannot tolerate other therapies, including imatinib. Bosutinib was later approved in 2017 as a first-line treatment for CML in chronic phase.[1–3]

The discovery that the BCR-ABL kinase inhibitor imatinib is a highly effective treatment for CML transformed a once deadly blood cancer into a manageable chronic disease. However, some patients develop imatinib resistance, frequently caused by point mutations in the kinase domain of the BCR-ABL enzyme that reduce sensitivity towards imatinib. Bosutinib is the third- and last second-generation ABL kinase inhibitor launched in 2012 for the treatment of imatinib-resistant CML, following BMS' dasatinib in 2006 and Novartis' nilotinib in 2007.

Bosutinib was built upon a distinct anilinoquinoline scaffold, which is similar to the anilinoquinazoline commonly found in EGFR inhibitors (e.g., gefitinib) but different from the aminopyridine in imatinib (Fig. 1). Thanks to its unique structure, bosutinib overcomes imatinib-resistant BCR-ABL mutants (except the gatekeeper T315I mutant) by binding both the DFG-in active and the DFG-out inactive conformation (while imatinib only targets an inactive conformation). However, as a result of its EGFR hinge-binding element, bosutinib also inhibits EGFR (including L858R and L861Q), in addition to BCR-ABL and SRC family kinases. Unlike imatinib, dasatinib, and nilotinib, bosutinib shows no appreciable inhibition of KIT or PDGFR kinases at physiologically relevant concentrations, thereby resulting in less frequent neutropenia and oedema adverse effects associated with these receptors.[1–3]

Sales of bosutinib were $120 million in 2021 as compared to $2.1 billion for dasatinib and $2.06 billion for nilotinib in the same year.

Discovery of Bosutinib[4–9]

As with dasatinib, bosutinib was initially identified as a potent inhibitor of the SRC-family of kinases. Only later after the discovery of imatinib for the treatment of CML was it found to be an inhibitor also of BCR-ABL.

The proto-oncogene c-SRC (SRC) is the first reported oncogene which encodes the first non-receptor tyrosine kinases, the SRC family kinases. Because these kinases play an important role in tumor progression, invasion, and metastasis in human cancer, their inhibition could be important for the treatment of various diseases, including cancer.

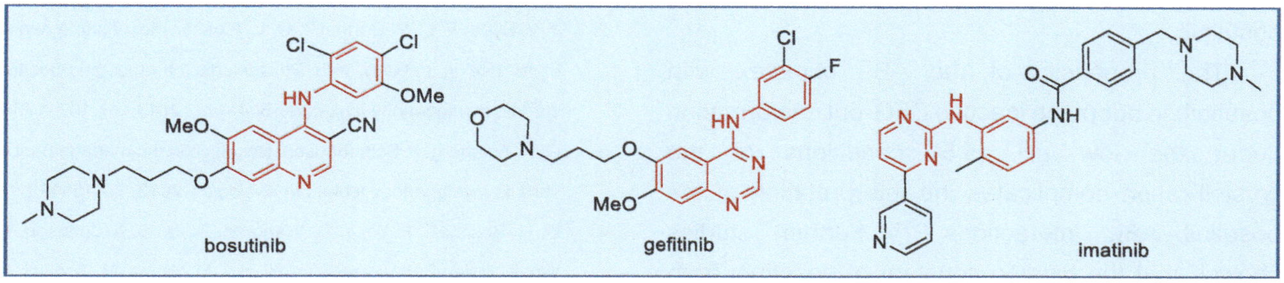

Figure 1. Structures of bosutinib, gefitinib and imatinib.

High-throughput screening of the Wyeth-Ayerst compound library identified cyano-quinoline **1** as a potent SRC inhibitor with an IC$_{50}$ of 30 nM (Fig. 2).[4] Simply adding a methoxy to the aniline increased potency by ~8-fold (**2**, IC$_{50}$ = 4.3 nM), and the activity was further improved to 0.8 nM by replacing 7-methoxy of the anilinoquinoline core with 3-morpholinopropoxy group (to give **3**). However, intraperitoneal injection of **3** to mice at 50 mpk showed only marginal plasma exposure (0.10 μg/mL at both 1 h and 4 h) likely due to high lipophilicity (clogP = 5.9). The lipophilicity was reduced by replacement of the morpholine moiety with a more water-solubilizing methylpiperazine to give bosutinib (clogP = 4.9) with substantially enhanced plasma exposure (3.0 μg/mL at 1 h and 0.9 μg/mL 4 h).

Bosutinib is a potent dual SRC/ABL inhibitor with IC$_{50}$ of 1.2 and 1.0 nM in cell-free assays, respectively. It potently inhibits SRC-dependent cell proliferation with an IC$_{50}$ of 100 nM and is active against many BCR-ABL mutations associated with imatinib resistance (with the exception of T315I).

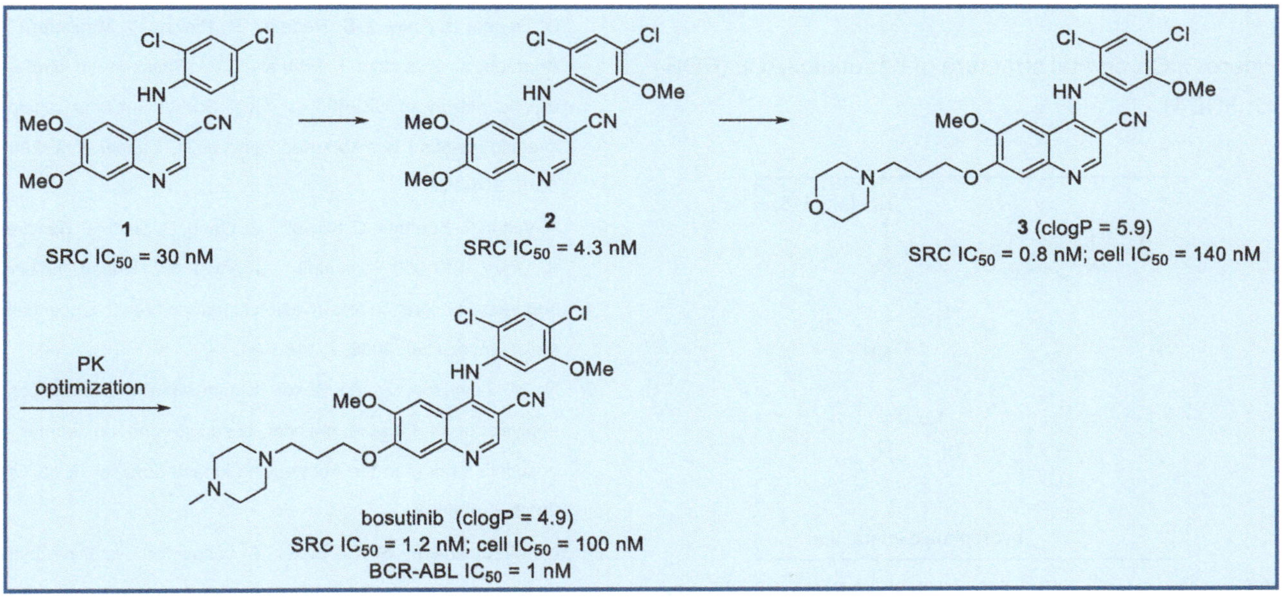

Figure 2. Structural optimization of bosutinib.

Binding Mode[10,11]

In the bosutinib-ABL co-crystal structure (Fig. 3), the inhibitor binds to a "DFG-out" inactive conformation that restricts important interactions with ATP. The quinoline moiety occupies the ATP-binding pocket, forming a hydrogen bond between the quinolone nitrogen and the backbone amide NH of Met318 in the hinge region (Fig. 4). The side chain of T315 is involved in extensive van der Waals contacts with both the cyano group and the 5-methoxy group of the aniline ring, while the side chain of V299 is also in van der Waals contact of the nitrile group. This binding model rationalizes the resistance of the T315I and V299L mutants to bosutinib.

The propensity of the ABL complex with bosutinib to adopt the inactive DFG-out conformation under the low pH (5.5) conditions of the crystallization complicates the interpretation of the bosutinib–ABL interactions.[10,11] Further studies showed that the binding constant of bosutinib from the phosphorylated ABL is the same as that from the unphosphorylated ABL, indicating that bosutinib, unlike imatinib, can bind to both the DFG-in and DFG-out conformations of ABL (Fig. 3).[11] This unique dual binding mode may explain its activity against imatinib-resistant BCR-ABL mutants.

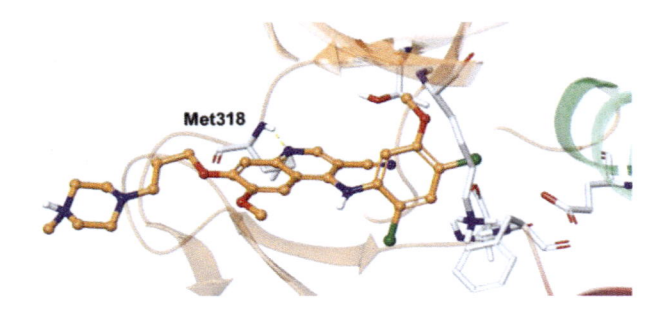

Figure 3. Co-crystal structure of bosutinib–ABL (PDB ID: 3UE4).

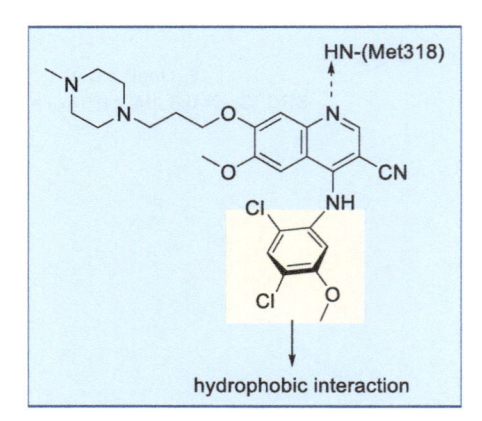

Figure 4. Summary of bosutinib–ABL interactions based on co-crystal structural data.

1. V. A. Keller, T. H. Brummendorf. Novel aspects of therapy with the dual Src and Abl kinase inhibitor bosutinib in chronic myeloid leukemia. *Expert Rev. Anticancer Ther.* **2012**, *12*, 1121–1127.
2. Y. Y. Syed, P. L. McCormack, G. L. Plosker. Bosutinib: A review of its use in patients with Philadelphia chromosome-positive chronic myelogenous leukemia. *BioDrugs.* **2014**, *28*, 107–120.
3. G. K. Amsberg, P. Schafhausen. Bosutinib in the management of chronic myelogenous leukemia. *Biologics.* **2013**, *7*, 115–122.
4. D. H. Boschelli, F. Ye, Y. D. Wang, M. Dutia, S. L. Johnson, B. Wu, K. Miller, D. W. Powell, D. Yaczko, M. Young, M. Tischler, K. Arndt, C. Discafani, C. Etienne, J. Gibbons, J. Grod, J. Lucas, J. M. Weber, F. Boschelli. Optimization of 4-phenylamino-3-quinolinecarbonitriles as potent inhibitors of Src kinase activity. *J. Med. Chem.* **2001**, *44*, 3965–3977.
5. P. W. Manley, S. W. Cowan-Jacob, J. Mestan. Advances in the structural biology, design and clinical development of Bcr-Abl kinase inhibitors for the treatment of chronic myeloid leukemia. *Biochimica et biophysica acta.* **2005**, *1754*, 3–13.
6. A. Wissner, B. D. Johnson, M. F. Reich, M. B. Floyd, Jr., D. B. Kitchen, H. R. Tsou. Substituted 3-cyano quinolines, US Patent US 6002008.
7. B. Raj, Ramanjaneyulu, J. M. Reddy, Chaturvedi, V. K. Shrawat,. Process for preparation of bosutinib. *Patent WO 2015198249*.
8. M. Puttini, A. M. Coluccia, F. Boschelli, L. Cleris, E. Marchesi, A. D. Deana, S. Ahmed, S. Redaelli, R. Piazza, V. Magistroni, F. Andreoni, L. Scapozza, F. Formelli, C. G. Passerini. In vitro and in vivo activity of SKI-606, a novel Src-Abl inhibitor, against imatinib-resistant Bcr-Abl+ neoplastic cells. *Cancer Res.* **2006**, *66*, 11314–22.
9. A. Vultur, R. Buettner, C. Kowolik, W. Liang, D. Smith, F. Boschelli, R. Jove. SKI-606 (bosutinib), a novel Src kinase inhibitor, suppresses migration and invasion of human breast cancer cells. *Mol. Cancer Ther.* **2008**, *7*, 1185–94.
10. N. M. Levinson, S. G. Boxer. Structural and spectroscopic analysis of the kinase inhibitor bosutinib and an isomer of bosutinib binding to the Abl tyrosine kinase domain. *PLoS One* **2012**, *7*, e29828.
11. N. M. Levinson, S. G. Boxer. A conserved water-mediated hydrogen bond network defines bosutinib's kinase selectivity. *Nature. Chem. Biol.* **2014**, *10*, 127–132.

2.5 Ponatinib (Iclusig™) (5)

Structural Formula	Space-filling Model	Brief Information
$C_{29}H_{27}F_3N_6O$; MW = 533 clogP = 5.8; tPSA = 64		Year of discovery: 2005 Year of introduction: 2012 Discovered by: Ariad Developed by: Ariad Primary target: BCR-ABL Binding type: II (ABL) Class: non-receptor tyrosine kinase Treatment: CML Other name: AP-24534 Oral bioavailability = not reported Elimination half-life = 24 h Protein binding = 99%

● = C ◯ = H ● = O ● = N ● = S ● = F ● = Cl ● = Br ● = I

Ponatinib, a third-generation BCR-ABL tyrosine kinase inhibitor (TKI), was first approved in December 2012 as a third-line treatment for CML patients. However, three months later, its approval was suspended, primarily due to safety concerns arising from the increased incidence of serious vascular occlusive events. Because there was no alternative treatment option at the time for CML patients with the T315I mutation, the suspension was lifted after a couple of months, and ponatinib was reintroduced for adult patients with T315I-postive CML (including accelerated, chronic, or blast phases) or those with T315I-positive Ph+ ALL with multiple safety measures regarding cardiovascular risk.[1-3]

The precise mechanism of ponatinib-induced adverse vascular effects remains unclear. However, its broad spectrum of kinase inhibition is thought to be the cause. It was shown that ponatinib-induced vascular toxicity was primarily mediated via the Notch-1 signaling pathway, and selective blockade of Notch-1 prevented ponatinib-induced vascular toxicity. Thus, the ponatinib-mediated cardiovascular side effects are believed to stem from the off-target impacts on platelet activation and endothelial cellular function.[4]

Ponatinib was developed to overcome the gatekeeper T315I mutation, which was insensitive to the second-generation BCR-ABL inhibitors such as dasatinib and nilotinib. It targets a DFG-out inactive conformation of the mutant kinase using a unique ethynyl linkage to evade a steric clash with the bulky isoleucine side chain of T315I mutant.[5,6] It is a pan-BCR-ABL inhibitor, active against the native enzyme and all clinically relevant mutants including T315I with IC_{50} values of 0.37–2.0 nM.[1-3]

Ponatinib inhibits multiple tyrosine kinases including FLT3, FGFR, and VEGFR family kinases in the single-digit nanomolar range. As with imatinib, nilotinib, and dasatinib, it also inhibits KIT and PDGFRα/β. It is likely that the relatively broad kinase specificity profile of ponatinib depends critically on the linear ethynyl linkage, which permits binding to hydrophobic residues of various kinases at the gatekeeper position.[1-3]

Although ponatinib is effective against T315I mutants, the dosage is limited by adverse events, so it is not ideal. More recently, asciminib, a fourth generation type IV TKI for CML, has been developed. It binds to a myristoyl site of the BCR-ABL protein and locks the protein into an inactive conformation through a mechanism distinct from those of all other orthosteric TKIs such as imatinib, thus overcoming drug resistance arising from ATP-binding site mutations (see Section 2.7).

Discovery of Ponatinib[7–10]

Ponatinib originated from AP23464, a purine-based dual SRC/ABL inhibitor that strongly inhibits SRC, wild-type BCR-ABL as well as BCR-ABL mutants (except T315I) with subnanomolar potencies (Fig. 1).[9] Many structural modifications were required to reach ponatinib.

To target the DFG-out conformation of the T315I mutant kinase,[6] the hydroxyphenethyl part of AP23464 was replaced by a DFG-out targeting diarylcarboxamide fragment taken from nilotinib, and the connecter was changed from CH_2CH_2 to CH=CH to avoid contact with the isoleucine side chain of the mutant T315I. In addition, the C2 cyclopentyl group of the purine core was removed because it is not accommodated in a DFG-out targeted conformation due to clashes with residues located in the glycine-rich P-loop. Furthermore, the non-pharmacophore phenyl-dimethylphosphine oxide group was replaced by a much smaller cyclopropyl to reduce lipophilicity and improve PK properties. Taken together, these structural modifications led to AP24163 which demonstrated modest potency against T315I in both biochemical and cell-based assays (IC_{50} = 478 and 422 nM, respectively). By contrast, nilotinib was completely inactive against this mutant due to a steric clash between the amino group on the pyrimidine ring and the larger gatekeeper isoleucine.

To further reduce steric repulsion with the bulky isoleucine side chain of the T315I mutant, the vinyl linker was replaced by C≡C, resulting in significant improvement in potency against T315I. Because these alkynyl analogs showed poor PK properties likely due to the purine N–C≡C cleavage to release the free purine segment, the N–C≡C subunit was replaced by C–C≡C to give the imidazopyridine analog **1** with much improved oral bioavailability in rats (42%). Despite its good potency in the biochemical assay (IC_{50} = 102 nM), its cellular activity remained modest (IC_{50} = 471 nM) possibly due to limited cellular membrane permeability.

Next, the imidazole of the terminal phenyl was eliminated, and the *N*-methylpiperazine group (copied from imatinib) was installed to increase solubility and enhance potency via hydrogen bonds to the backbone carbonyl group of Ile360 and His361, (as observed in imatinib). These modifications gave rise to **2** which showed an improvement in biochemical (IC_{50} = 56 nM vs. 102 nM for **1**) and cellular assays (IC_{50} = 26 nM vs. 471 nM for **1**). This compound exhibited favorable oral PK properties in both rats (F = 29%) and mice and was subsequently identified as the first orally active BCR-ABL T315I inhibitor (within this series) in a pilot efficacy study.

Because **2** is highly lipophilic (clogP = 6.7), carbon was replaced by nitrogen at various positions of the imidazopyridine core to reduce the lipophilicity. This modification led to ponatinib with imidazopyridazine as the head group, which reduced clogP by one unit and also increased the enzyme and cellular potency against the T315I mutant relative to **2**.

Ponatinib was shown to be active against all tested BCR-ABL mutants (including T315I), which were insensitive to either dasatinib or nilotinib. An accelerated mutagenesis screen for resistance showed that it completely suppressed the outgrowth of all mutants at a concentration of 40 nM.[10]

Binding Mode[9–11]

In the co-crystal structure of ponatinib in complex with the murine ABLT 315I kinase domain, the inhibitor binds in the DFG-out inactive conformation (Fig. 2). The imidazopyridazine core occupies the adenine pocket, the methylphenyl group binds to the hydrophobic pocket behind the gatekeeper residue, the trifluoromethylphenyl group fits the pocket induced by the DFG-out conformation of the protein, and the ethynyl linkage makes favorable van der Waals interactions with the I315 mutated residue.

The N1 nitrogen of the imidazopyridazine forms a hydrogen bond with the NH group of the ABL hinge Met318, and the amide NH group of the amide linker hydrogen bonds to the carboxylate side chain of Asp286, while the amide carbonyl oxygen of the linker hydrogen bonds to the amide NH of the DFG-Asp381. The N4 nitrogen of the methylpiperazine makes additional polar contacts with the carbonyl groups of Ile360 and His361 of the catalytic loop.

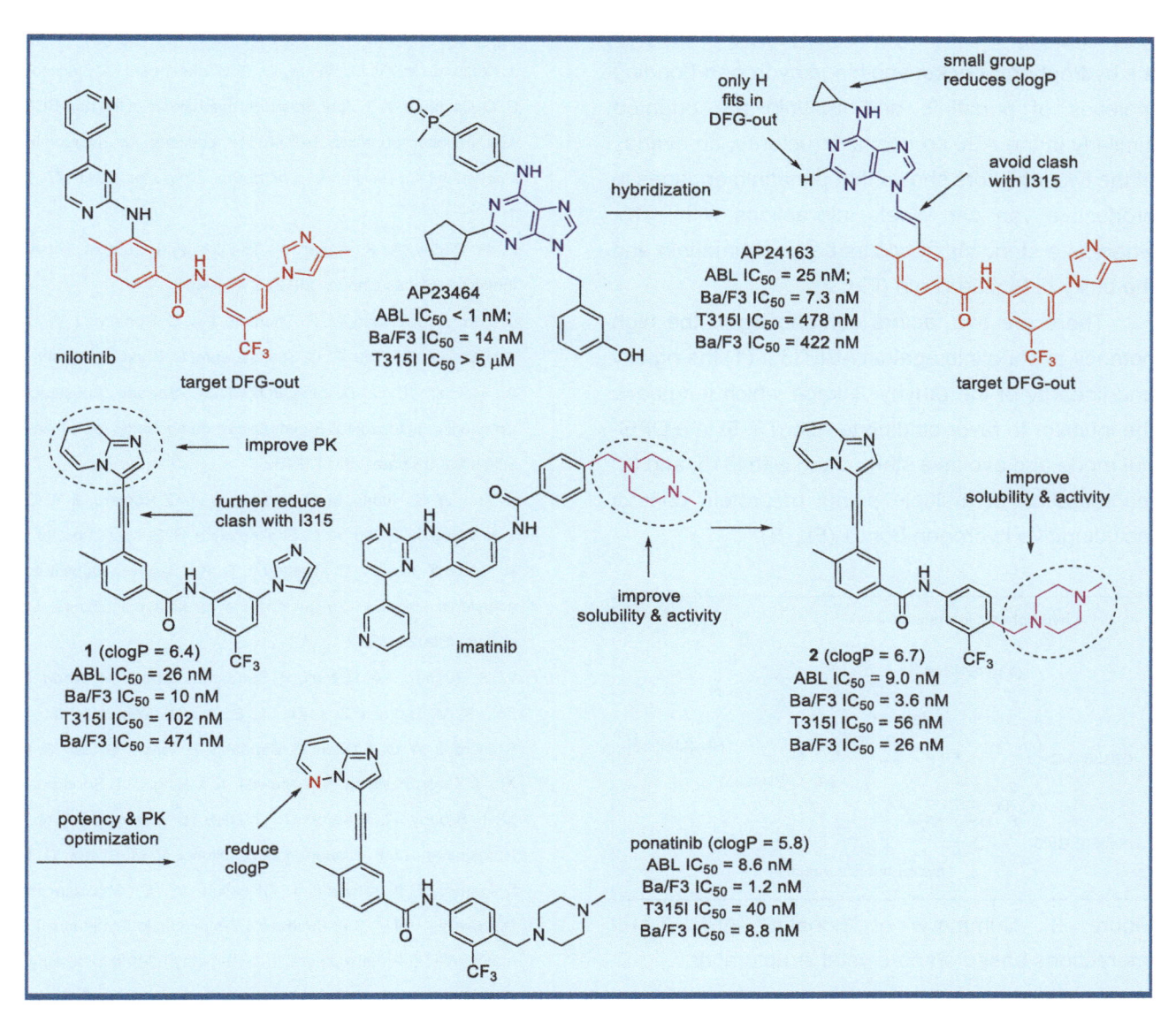

Figure 1. Structural changes leading from AP23464 to ponatinib.

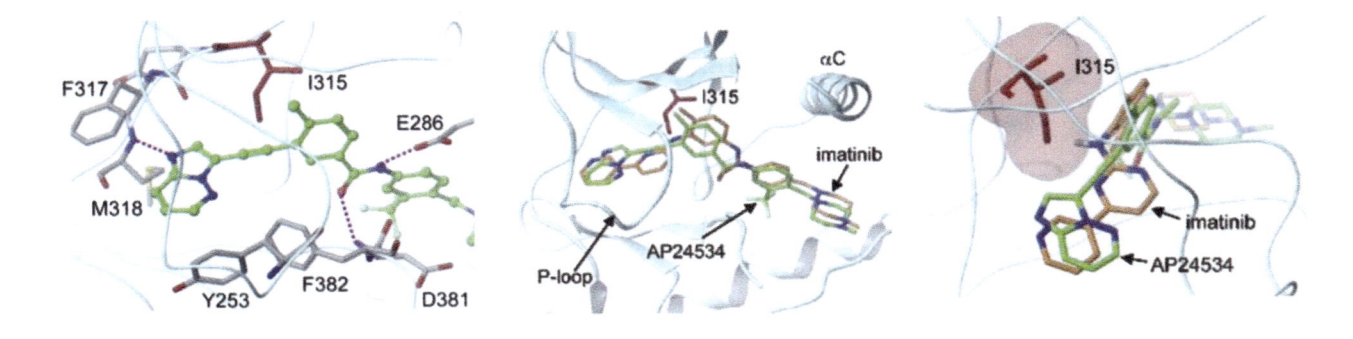

Figure 2. Left: Co-crystal structure of ponatinib (AP24534) in complex with the ABL T315I mutant kinase. Middle and Right: Superposition of imatinib and ponatinib (AP24534) highlighting the effect of the Thr to Ile mutation. The superposition was based on the Cα positions of ABL residues 312–321 in the T315I mutant and in native ABL kinase complexed with imatinib. Adapted with permission from *Cancer Cell.* **2009**, *16*, 401–412.

Although the methylphenyl groups occupying the hydrophobic pocket and hinge hydrogen-bonding moieties of ponatinib and imatinib are oriented similarly in the ABL co-crystal structures, an overlay of the two inhibitors shows that ponatinib engages in productive van der Waals interactions with I315, whereas a steric clash occurs between imatinib and the bulky I315 side chain (Fig. 2).

There are two factors contributing to the high potency of ponatinib against ABLTI35: (1) the rigidity and linearity of the ethynyl linkage which lengthens the inhibitor to favor binding to ABL T315I in a DFG-out mode and avoids a steric clash with I315; and (2) incorporation of multiple points of protein contact including five hydrogen bonds (Fig. 3).

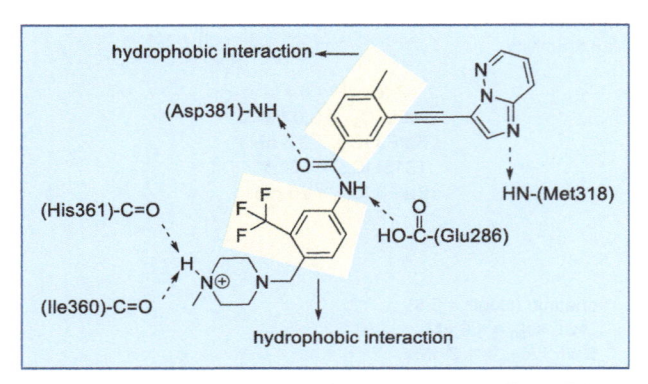

Figure 3. Summary of ponatinib–ABL T315I interactions based on co-crystal structural data.

1. J. E. Cortes, H. Kantarjian, N. P. Shah, D. Bixby, M. J. Mauro, I. Flinn, T. O. Hare, S. Hu, N. I. Narasimhan, V. M. Rivera, T. Clackson, C. D. Turner, F. G. Haluska, B. J. Druker, M. W. Deininger, M. Talpaz. Ponatinib in refractory philadelphia chromosome–positive leukemias. *N. Engl. J. Med.* **2012**, *367*, 2075–2088.

2. S. M. Hoy. Ponatinib: A review of its use in adults with chronic myeloid leukaemia or Philadelphia chromosome-positive acute lymphoblastic leukaemia. *Drugs.* **2014**, *74*, 793–806.

3. T. O'Hare, R. Pollock, E. P. Stoffregen, J. A. Keats O. M. Abdullah, E. M. Moseson, V. M. Rivera, H. Tang, C. A. Metcalf, R. S. Bohacek. Inhibition of wild-type and mutant Bcr-Abl by AP23464, a potent ATP-based oncogenic protein kinase inhibitor: implications for CML. *Blood.* **2004**, *104*, 2532–2539.

4. A. P. Singh P. Umbarkar, S. Tousif, H. Lal. Cardiotoxicity of the BCR-ABL1 tyrosine kinase inhibitors: Emphasis on ponatinib. *Int. J. Cardiol.* **2020**, *316*, 214–221.

5. T. J. Zhou, L. Commodore, W. S. Huang, Y. H. Wang, M. Thomas, J. Keats, Q. Xu, V. M. Rivera, W. C. Shakespeare, T. Clackson, D. C. Dalgarno, X. T. Zhu. Structural mechanism of the pan-BCR-ABL inhibitor ponatinib (AP24534): Lessons for overcoming kinase inhibitor resistance. *Chem. Biol. Drug Des.* **2011**, *77*, 1–11.

6. E. P. Reddy, A. K. Aggarwal. The Ins and Outs of Bcr-Abl Inhibition. *Genes Cancer.* **2012**, *3*, 447–454.

7. D. Zou, W. S. Huang, M. R. Thomas, J. A. C. Romero, J. W. Qi, Y. H. Wang, X. T. Zhu, W. C. Shakespeare, R. Sundaramoorthi, C. A. Metcalf III, C. D. Dalgarno, T. K. Sawyer. Substituted acetylenic imidazo[1,2-B]pyridazine compounds as kinase inhibitors. US patent 8114874B2.

8. D. Zou, W. S. Huang, M. R. Thomas, J. A. C. Romero, J. W. Qi, Y. H. Wang, X. T. Zhu, W. C. Shakespeare, R. Sundaramoorthi, C. A. Metcalf III, C. D. Dalgarno, T. K. Sawyer. Substituted acetylenic imidazo[1,2-A]pyridazines as kinase inhibitors. US patent US9029533B2.

9. W. S. Huang, C. A. Metcalf, R. Sundaramoorthi, Y. H. Wang, D. Zou, R. M. Thomas, X. T. Zhu, L. S. Cai, D. Wen, S. H. Liu, J. Romero, J. W. Qi, I. Chen, G. Banda, S. P. Lentini, S. Das, Q. H. Xu, J. Keats, F. Wang, S. Wardwell, Y. Y. Ning, J. T. Snodgrass, M. I. Broudy, K. Russian, T. J. Zhou, L. Commodore, N. I. Narasimhan, Q. K. Mohemmad, J. Iuliucci, V. M. Rivera, D. C. Dalgarno, T. K. Sawyer, T. Clackson, W. C. Shakespeare. Discovery of 3-[2-(Imidazo[1,2-b]pyridazin-3-yl)ethynyl]-4-methyl-N-{4-[(4-methylpiperazin-1-yl)-methyl]-3-(trifluoromethyl)phenyl}benzamide (AP24534), a potent, orally active pan-inhibitor of breakpoint cluster region-abelson (BCR-ABL) kinase including the T315I gatekeeper mutant. *J. Med. Chem.* **2010**, *53*, 4701–4719.

10. T. O. Hare, W. C. Shakespeare, X. T. Zhu, C. A. Eide, V. M. Rivera, F. Wang, L. T. Adrian, T. J. Zhou, W. S. Huang, Q. H. Xu, C. A. Metcalf III, J. W. Tyner, M. M. Loriaux, A. S. Corbin, S. Wardwell, Y. Y. Ning, J. A. Keats, Y. H. Wang, R. Sundaramoorthi, M. Thomas, D. Zhou, J. Snodgrass, L. Commodore, T. K. Sawyer, D. C. Dalgarno, M. W. N. Deininger, B. J. Druker, T. Clackson. AP24534, a pan-BCR-ABL inhibitor for chronic myeloid leukemia, potently inhibits the T315I mutant and overcomes mutation-based resistance. *Cancer Cell.* **2009**, *16*, 401–412.

11. R. Roskoski, Jr. Classification of small molecule protein kinase inhibitors based upon the structures of their drug-enzyme complexes. *Pharmacol. Res.* **2016**, *103*, 26–48.

2.6 Olverembatinib (6)

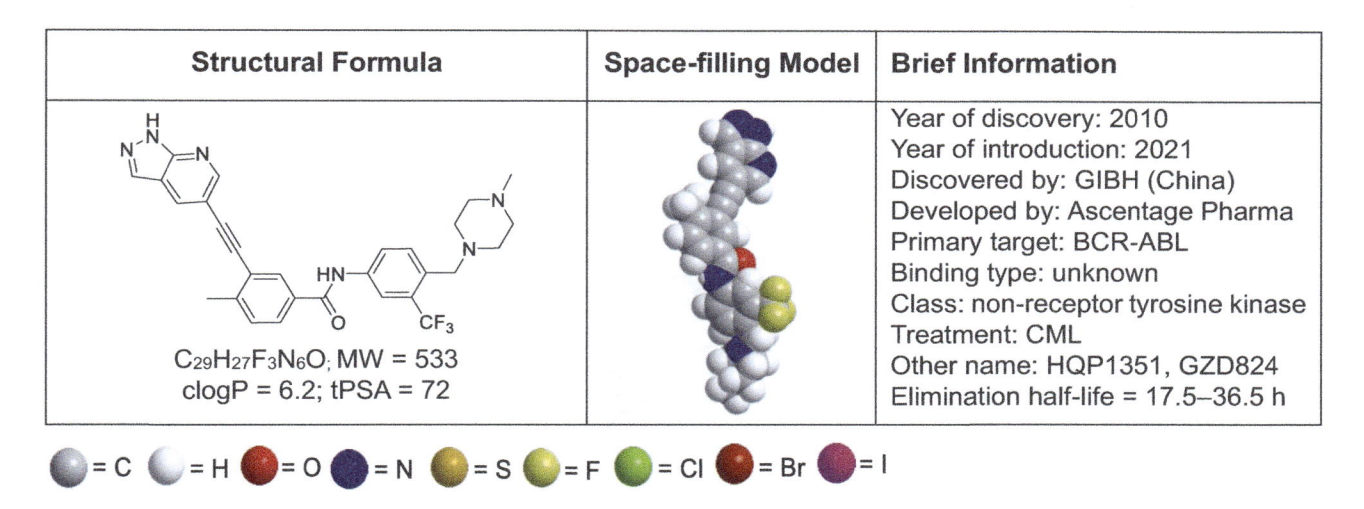

Structural Formula	Space-filling Model	Brief Information
$C_{29}H_{27}F_3N_6O$; MW = 533 clogP = 6.2; tPSA = 72		Year of discovery: 2010 Year of introduction: 2021 Discovered by: GIBH (China) Developed by: Ascentage Pharma Primary target: BCR-ABL Binding type: unknown Class: non-receptor tyrosine kinase Treatment: CML Other name: HQP1351, GZD824 Elimination half-life = 17.5–36.5 h

○ = C   ○ = H   ● = O   ● = N   ● = S   ● = F   ● = Cl   ● = Br   ● = I

Olverembatinib, another third-generation BCR-ABL tyrosine kinase inhibitor (TKI), was approved in 2021 in China for patients with T315I-postive chronic phase chronic myeloid leukemia (CML-CP) or accelerated-phase CML (CML-AP).[1]

Olverembatinib is identical to ponatinib except the hingebinding head group: 1H-pyrazolo[3,4-b]pyridine vs. imidazo[1,2-b]pyridazine (Fig. 1).[2,3] The 1H-pyrazolo[3,4-b]pyridine core of olverembatinib could form a hydrogen bond donor−acceptor network with the hinge region of BCR-ABL (vs. only H-bond donor for imidazo[1,2-b]pyridazine moiety of ponatinib) to achieve stronger binding than ponatinib (Fig. 1). Consistent with this hypothesis, olverembatinib showed 2- to 3-fold more potent inhibition of ABLWT and a panel of mutants than ponatinib. For example, the IC$_{50}$ values of ponatinib were 1.39, 1.08, and 1.43 nM against ABLT315I, ABLE255K, and ABLG250E, respectively, whereas the respective data for olverembatinib were 0.68, 0.27, and 0.71 nM. It also potently inhibited the proliferation of Ba/F3 cells that stably expressed the most refractory BCR-ABL T315I mutant, with an IC$_{50}$ value of 7.1 nM. Furthermore, it displayed a similar potency of inhibition against Ba/F3 cells expressing 14 other imatinib-resistant BCR-ABL mutants.[2]

Like ponatinib, olverembatinib also inhibits multikinases, including B-RAF, DDR1, FGFR, Flt3, KIT, PDGFRα/β, RET, and SRC.[2]

The recommended oral dosage of olverembatinib is 40 mg once every 2 days, while ponatinib is administered 45 mg orally once daily.

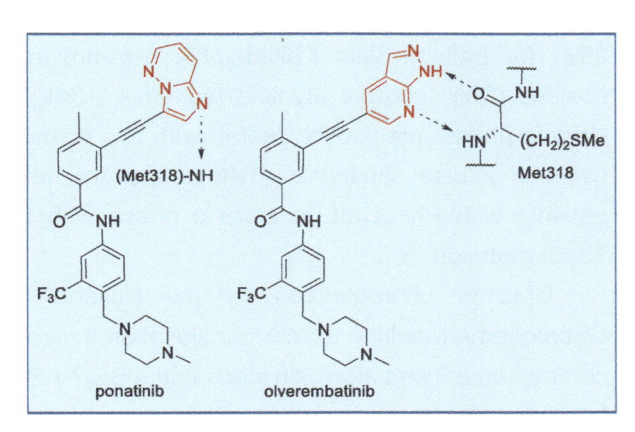

Figure 1. Ponatinib vs. olverembatinib.

1. S. Dhillon. Olverembatinib: First approval. *Drugs*, **2022**, *82*, 469–475.
2. X. Ren, X. Pan, Z. Zhang, D. Wang, X. Lu, Y. Li, D. Wen, H. Long, J. Luo, Y. Feng, X. Zhuang, F. Zhang, J. Liu, F. Leng, X. Fang, Y. Bai, M. She, Z. Tu, J. Pan, K. Ding. Identification of GZD824 as an orally bioavailable inhibitor that targets phosphorylated and nonphosphorylated breakpoint cluster region-Abelson (Bcr-Abl) kinase and overcomes clinically acquired mutation-induced resistance against imatinib. *J. Med. Chem.* **2013**, *56*, 879–894.
3. D. Wang, D. Pei, Z. Zhang, M. Shen, K. Luo, Y. Feng. Heterocyclic alkynyl benzene compounds and medical compositions and uses thereof. WO2012000304A1.

2.7 Asciminib (Scemblix™) (7)

Structural Formula	Space-filling Model	Brief Information
$C_{20}H_{18}N_5O_3ClF_2$; MW = 449 clogP = 3.3; tPSA = 99		Year of discovery: 2013 Year of introduction: 2021 Discovered by: Novartis Developed by: Novartis Primary target: BCR-ABL1 Binding type: IV Class: non-receptor tyrosine kinase Treatment: CML Other name: ABL001 Oral bioavailability = not reported Elimination half-life = 14.2 h Protein binding = 97%

● = C   ○ = H   ● = O   ● = N   ● = S   ● = F   ● = Cl   ● = Br   ● = I

Asciminib is the first-in-class Specifically Targeting the ABL1 Myristoyl Pocket (STAMP) inhibitor, which was granted accelerated approval in 2021 for patients with Philadelphia chromosome-positive (Ph+) chronic myeloid leukemia (CML) in chronic phase, previously treated with two or more tyrosine kinase inhibitors (TKIs), and for adult patients with Ph+ CML in chronic phase with the T315I mutation.[1,2]

CML is characterized by the Philadelphia chromosome resulting from a translocation between chromosome 9 and 22 which leads to the BCR-ABL1 fusion gene. This gene produces an abnormal BCR-ABL1 (also known as BCR-ABL) tyrosine kinase protein, which causes CML cells to grow and reproduce out of control. Imatinib was the first drug to specifically target this protein. It binds to the inactive conformation (DFG-out) of the kinase domain, thereby reducing the availability of the catalytically active conformation (DFG-in) required for ATP-binding. Imatinib was touted as a "magic bullet" to target CML because it changed a once deadly blood cancer into a manageable chronic disease and enabled the patients to live a nearly normal lifespan. Subsequently, more potent second-generation tyrosine kinase inhibitors (TKIs), dasatinib, nilotinib, and bosutinib, have been introduced to the market to treat CML. Like imatinib, these TKIs target the catalytic ATP-binding site of BCR-ABL1. Unfortunately, over the years, some patients have developed drug resistance primarily due to point mutations in the ATP-binding region of the BCR-ABL1 protein. The gatekeeper mutation T315I harboring a substitution of the threonine to isoleucine is the most common mechanism that prevents inhibitors from entrance to ATP-binding domain. Although this mutant is only susceptible to the third-generation TKI, ponatinib, the dosage is limited by adverse events. Against the backdrop of increasing drug resistance to orthosteric TKIs, asciminib has been developed as a backup therapy with a unique mechanism of action.

Asciminib binds to a myristoyl site of the BCR-ABL1 protein and locks the protein into an inactive conformation through a mechanism distinct from those of all other orthosteric TKIs such as imatinib, thus overcoming drug resistance arising from ATP-binding site mutations. Asciminib mimics the function of the myristoylated N-terminus of ABL1 and restores the natural autoinhibition of the ABL1b protein.[1]

Myristate Binding in ABL1b Autoinhibition[3–7]

The ABL protein 1 (ABL1) is a non-receptor tyrosine kinase that plays an important role in regulating cell growth and survival. There are two human ABL1 isoforms, ABL1a and 1b, which result from alternative splicing events. Both isoforms contain a N-terminal cap region, but the cap region of ABL1a is 19 amino acids shorter than that of ABL1b. In addition, ABL1b bears a myristate site (Myr-NH) at the extreme end of the amino-terminal segment. The cap region is followed by an SH3 domain, an SH2 domain, all of which act together to maintain the inactive kinase conformation (Figs. 1 and 2).

In the normal cells, ABL1 exists in an autoinhibited conformation due to the interaction of its Myr-bound cap with an allosteric site at the base of the catalytic domain as revealed by the crystal structure of the N-terminal portion of ABL1b. Binding of the N-terminal myristic acid group (which is covalently linked to the N-terminus of ABL1b) to a deep hydrophobic pocket in the C-terminal lobe of the kinase domain, also known as the "myristate binding site/pocket," induces a bend in the C-lobe α helix I, allowing the SH3 and SH2 domains along with the SH2-kinase linker to clamp around the kinase domain. This clamp reduces the conformational flexibility of the kinase domain and prevents the kinase from shifting to an active conformation, a process which requires the disassociation of the myristoyl group from the C-lobe of the kinase domain (Fig. 2). The critical contribution of the myristoylated N-cap to the autoinhibited state is confirmed by a mutation of the myristoylation signal sequence, which gives rise to a highly active kinase.

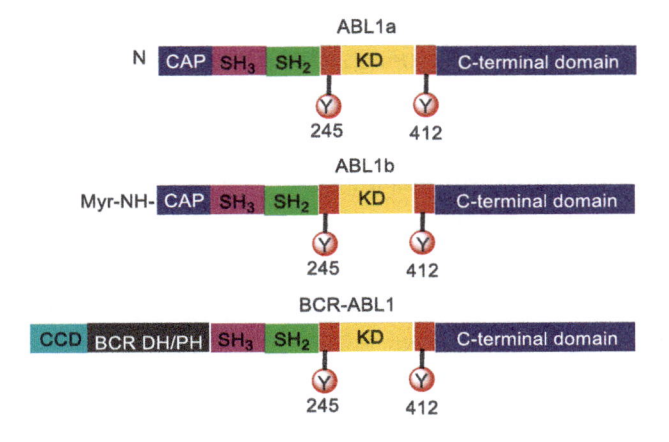

Figure 1. Schematic representation of ABL1a, ABL1b, and BCR-ABL1. CCD: coil-coiled domain.

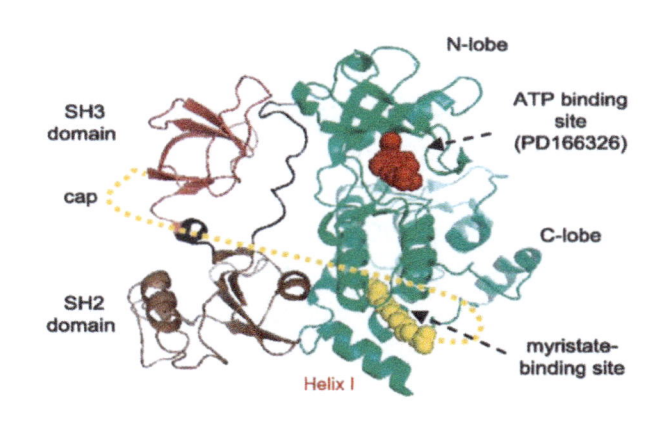

Figure 2. Ribbon representation of the ABL1 kinase N-terminal half residues, including the SH3, SH2, and kinase domains. The N-terminal cap is indicated by dotted lines. The helix I is bent in the autoinhibited inactive state. Adapted from *J. Biol. Chem.* **2009**, *284*, 29005–29014.

In the assembled autoinhibited form, the intramolecular myristate binding -is essential for the maintenance of the ABL1 inactive conformation. The structural basis of this natural autoinhibition is distinctly different from the typical kinase inactive DFG-out conformation which BCR-ABL catalytic site inhibitors (e.g., imatinib) bind. Most notable are the two different helix I conformations, which are bent in the assembled inactive state of ABL1b but linear and partially disordered in the absence of autoinhibition as shown in the co-crystal of imatinib in complex with the ABL1 SH1 domain (Fig. 3). The autoregulation of the kinase activity via myristate binding has inspired the search for small-molecule inhibitors that block ABL1 kinase by interacting with the myristoyl site (Fig. 4). Such allosteric inhibitors can provide several advantages, such as avoiding drug resistance due to ATP site mutations, greater kinase selectivity, and fewer off-target effects.

Loss of Autoinhibition in BCR-ABL1[3–7]

ABL1 is normally maintained in its inactive state in the cell; however, in 90–95% CML patients, the gene encoding ABL1 is fused with the breakpoint cluster region (BCR) gene, resulting in the formation of the aberrant fusion protein BCR-ABL1. In the process of gene fusion with BCR, all of the ABL1 domains–SH3, SH2, and kinase domains–remain intact except for the N-terminal cap region of ABL1b that contains the myristoylation site, which is replaced with the BCR-derived portion that includes an N-terminal coiled-coil (CC) oligomerization domain as well as DBL and pleckstrin homology domains (DH/PH) (Fig. 1). This fusion essentially eradicates the myristate autoregulatory function of ABL1b due to loss of the cap region, and the newly incorporated N-terminal coiled-coil domain initiates oligomerization of the fused protein, leading to a constitutive activation of ABL kinase, and consequently hyper-proliferation of the CML cells.

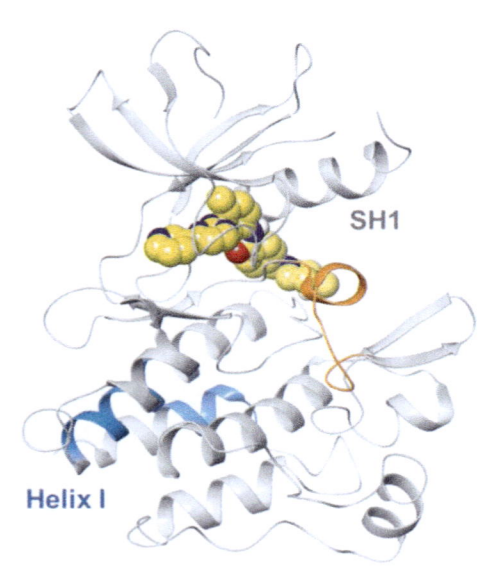

Figure 3. Structure of the ABL1 SH1 domain in complex with imatinib. The helix I is linear and partially disordered in the absence of autoinhibition (PDB ID: 2HYY).

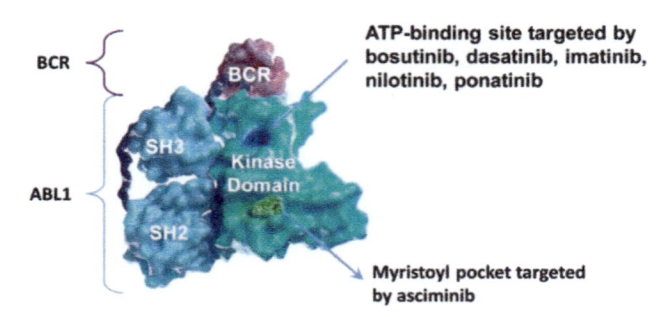

Figure 4. Different binding sites for orthosteric and allosteric inhibitors. Adapted with permission from *Leuk. Res.* **2020**, *98*, 106458.

Biomimetic Autoinhibition of BCR-ABL1: Discovery of Asciminib[8,9]

To seek ABL1 myristate pocket binder lead compounds, the ATP-site was blocked with imatinib, and the ABL1–imatinib complex was subjected to fragment screening using NMR spectroscopy. This effort identified ~30 hits from ~500 fragments, and representative fragments were crystallized as ternary fragment–imatinb–ABL1 complexes. The ternary structures showed the desired binding at the myristate pocket, but no bending of the helix I, a conformational change required for the inactive state. As a result, these fragment hits exhibited no functional activity in the cellular assays. Subsequently, similarity and pharmacophore searches were conducted on multiple ternary crystal structures, and the resulting hits were docked onto the bent helix I conformation of the myristate pocket. This process led to compound **1** with modest binding affinity (K_D = 6 µM) (Fig. 5); however, no functional activity was observed again due to lack of helix I bending. The bending conformation was finally achieved by removing the *meta*-chlorine and replacing the *para*-chlorine with OCF_3 to give **2**, which elicited modest cellular BCR-ABL1 (wild-type) inhibitory activity (GI_{50} = 8 µM). Removal of both the hydroxy group and the chlorine from the benzamide core as well as replacement of the *N*-methylpiperazine by a morpholine ring gave compound **3** with improved cellular potency (GI_{50} = 0.25 µM for BCR-ABL1 wt; 2.93 µM for T315I mutant). Substitution of the 3-morpholinomethyl with a weakly basic unsubstituted pyrimidine further increased potency, and conversion of the benzamide core to nicotinamide and attachment of a 3-hydroxypyrrolidine subunit onto the pyridine boosted aqueous solubility. These modifications led to nicotinamide **4** as a potent BCR-ABL1 inhibitor in the biochemical (IC_{50} = 2.3 nM) and cellular proliferation assays with wild-type BCR-ABL1 and T315I mutant (GI_{50} = 1.7 and 7.3 nM, respectively). Although compound **4** exhibited favorable BCR-ABL1 inhibitory activity and pharmacokinetic properties, the potential for hERG inhibition was problematic. Replacement of the pyrimidine with the pyrazole reduced the hERG liability, without compromising the kinase activity. The ABL1 activity was further increased by substituting CF_3 with a slightly larger CF_2Cl which fits snugly in the deepest part of the myristate pocket, and this led to asciminib with IC_{50} of 0.5 nM in the ABL1 biochemical assay. Although aryl chloromethyl ethers are generally reactive electrophiles, that activity is nullified by the two fluorines in $ArOCF_2Cl$. The potent enzyme activity was shown to translate into cellular potency as shown by its GI_{50} values of 1.0 and 25 nM for BCR-ABL1 wt and T315I mutant, respectively. Asciminib also maintained low nanomolar IC_{50} values against all other catalytic-site mutations.[4,6] In addition, it showed high selectivity against more than 60 kinases because it targets a unique myristoyl-binding site of ABL1, which is present in a limited number of kinases.[4,6]

Figure 5. Structural optimization of asciminib.

Binding Mode

X-ray crystallographic analysis of a ternary complex between asciminib, nilotinib and the ABL1[46-534] protein possessing T315I and Asp382Asn substitutions show that asciminib and nilotinib co-bind a single molecule of ABL1, occupying myristate binding pocket and ATP-binding site, respectively (Fig. 6).[4]

As shown in the ABL1 co-crystal structures bound to either myristate or asciminib (data not shown),[6] each binds similarly to ABL1 in a deep-pocket (myristate pocket) on the C-lobe of kinase domain to induce a conformational change involving the bending of helix I.

In the co-crystal structure of asciminib liganded to the ABL1–imatinib complex (Fig. 7), the inhibitor binds to the anticipated myristate pocket, forming one hydrogen bond between pyrazole free NH and the backbone carbonyl group of Glu481.

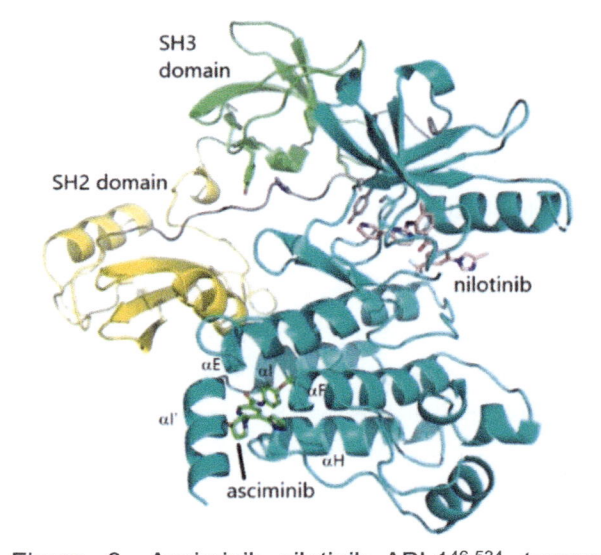

Figure 6. Asciminib–nilotinib–ABL1[46-534] ternary complex. asciminib (green) bound to the myristoyl pocket and nilotinib (magenta) bound to the ATP pocket. SH3 domain: green, SH2 domain: yellow, kinase or SH1 domain: cyan. Adapted with permission from Leuk. Res. **2020**, 98, 106458.

The chlorine atom of the OCF$_2$Cl group makes van der Waals and hydrophobic contacts with several side chain residues in the deep pocket. The inhibitor also interacts indirectly with the kinase via multiple water molecules (Figs. 7, 8).[9]

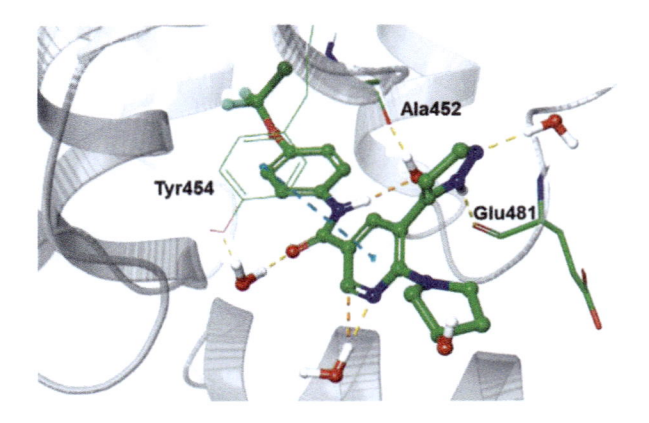

Figure 7. Crystal structure of ABL1–imatinib complex bound to asciminib (PDB ID: 5MO4).

Figure 8. Summary of asciminib–ALB1 interactions based on crystal structure of ABL1–imatinib complex bound to asciminib.

1. J. E. Cortes, T. P. Hughes, M. J. Mauro, A. Hochhaus, D. Rea, Y. T. Goh, J. Janssen, J. J. Steegmann, M. C. Heinrich, M. Talpaz, G. Etienne, M. Breccia, M. W. Deininger, P. D le Coutre, F. Lang, P. Aimone, F. Polydoros, S. Cacciatore, L. Stenson, D. Kim. Asciminib, a first-in-class STAMP inhibitor, provides durable molecular response in patients (pts) with chronic myeloid leukemia (CML) harboring the T315I mutation: Primary efficacy and safety results from a phase 1 trial. *Blood*, **2020**, *136*, Suppl. 1, 47-50,

2. E. D. Deeks. Asciminib: First approval. *Drugs*. **2022**, *82*, 219–226.

3. N. Dölker, M. W. Górna, L. Sutto, A. S. Torralba, G. Superti-Furga, F. L. Gervasio. The SH2 domain regulates c-Abl kinase activation by a cyclin-like mechanism and remodulation of the hinge motion. *PLoS Comput. Biol.* **2014**, *10*, e1003863.

4. P. W. Manley, L. Barys, S. W. Cowan-Jacob. The specificity of asciminib, a potential treatment for chronic myeloid leukemia, as a myristate-pocket binding ABL inhibitor and analysis of its interactions with mutant forms of BCR-ABL1 kinase. *Leuk. Res.* **2020**, *98*, 106458.

5. Y. Choi, M. A. Seeliger, S. B. Panjarian, H. Kim, X. Deng, T. Sim, B. Couch, A. J. Koleske, T. E. Smithgall, N. S. Gray. N-myristoylated c-Abl tyrosine kinase localizes to the endoplasmic reticulum upon binding to an allosteric inhibitor. *J. Biol. Chem.* **2009**, *284*, 29005–29014.

6. A. A. Wylie, J. Schoepfer, W. Jahnke, S. W. Cowan-Jacob, A. Loo, P. Furet, A. L. Marzinzik, X. Pelle, J. Donovan, W. Zhu, S. Buonamici, A. Q. Hassan, F. Lombardo, V. Iyer, M. Palmer, G. Berellini, S. Dodd, S. Thohan, H. Bitter, S. Branford, W. R. Sellers. The allosteric inhibitor ABL001 enables dual targeting of BCR-ABL1. *Nature*. **2017**, *543*, 733–737.

7. F. Carofiglio, D. Trisciuzzi, N. Gambacorta, F. Leonetti, A. Stefanachi, O. Nicolotti. Bcr-Abl allosteric inhibitors: Where we are and where we are going to. *Molecules*. **2020**, *25*, 4210.

8. W. Jahnke, R. M. Grotzfeld, X. Pellé, A. Strauss, G. Fendrich, S. W. Cowan-Jacob, S. Cotesta, D. Fabbro, P. Furet, J. Mestan, A. L. Marzinzik. Binding or bending: Distinction of allosteric Abl kinase agonists from antagonists by an NMR-based conformational assay. *J. Am. Chem. Soc.* **2010**, *132*, 7043–7048.

9. J. Schoepfer, W. Jahnke, G. Berellini, S. Buonamici, S. Cotesta, S. W. Cowan-Jacob, S. Dodd, P. Drueckes, D. Fabbro, T. Gabriel, J. M. Groell, R. M. Grotzfeld, A. Q. Hassan, C. Henry, V. Iyer, D. Jones, F. Lombardo, A. Loo, P. W. Manley, X. Pellé, P. Furet. Discovery of asciminib (ABL001), an allosteric inhibitor of the tyrosine kinase activity of BCR-ABL1. *J. Med. Chem.* **2018**, *61*, 8120–8135.

Chapter 3. BTK Inhibitors

Ibrutinib, originally regarded as just a research tool for screening purposes, was later shown clinically to be effective in the treatment of chronic lymphocytic leukemia (CLL), eventually having an impact comparable to imatinib on myelogenous leukemia (CML). It became the first irreversible inhibitor of Bruton tyrosine kinase (BTK) through covalently binding to the Cys481 residue in the BTK kinase domain. It has become the most successful kinase inhibitor (covalent or non-covalent) with sales totaling more than $26 billion as of the end of 2021. The success of ibrutinib demonstrated that the reactivity and selectivity of covalent inhibitors can be engineered to generate safe and effective therapeutic antikinase molecules (Fig. 1).

The first-in-class BTK inhibitor, ibrutinib, has been associated with several adverse effects, including rash and diarrhea, that have been attributed to the drug's potent off-target inhibition of the epidermal growth factor receptors (EGFR), as well as bleeding and atrial fibrillation. As a result, there has been much further research to develop BTK inhibitors with reduced inhibition of off-target kinases, such as those of the EGFR and TEC families, and also with improved pharmacokinetic properties. Those efforts have led to the development of improved and safer second-generation inhibitors, including, acalabrutinib zanubrutinib, and tirabrutinib (Fig. 1).[1,2]

BTK is also implicated in the underlying pathophysiology of autoimmune diseases such as rheumatoid arthritis as well as multiple sclerosis (which requires brain penetrance). Figure 2 shows several covalent BTK inhibitors currently in clinical trials for the treatment of these diseases.[1,2] Of these rilzabrutinib is a reversible covalent inhibitor, while the rest are irreversible. Reversible covalent inhibition is thought to be advantageous for kinase selectivity and safety.

To overcome the acquired resistance associated with the first- and second-generation BTK inhibitors as well as their potential liabilities due to selectivity, third-generation non-covalent reversible BTK inhibitors have been developed. Among these is pirtobrutinib (Fig. 3), which received accelerated approval from the FDA on January 27, 2023 for relapsed or refractory mantle cell lymphoma (MCL) after at least two lines of systemic therapy, including a BTK inhibitor. Figure 4 shows the other two non-covalent BTK inhibitors under development.

This chapter describes the discoveries of covalent BTK inhibitors **8–12** (Fig. 1). Pirtobrutinib is not included as it was approved right after completion of this book.

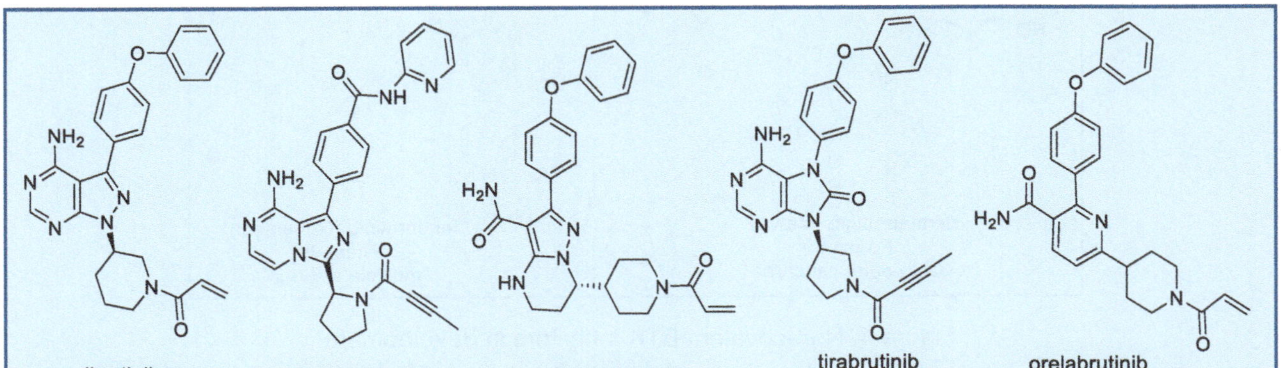

Figure 1. Approved covalent BTK inhibitors.

Molecules Engineered Against Oncogenic Proteins and Cancer, First Edition. E. J. Corey and Yong-Jin Wu.
© 2024 John Wiley & Sons Inc. Published 2024 by John Wiley & Sons Inc.

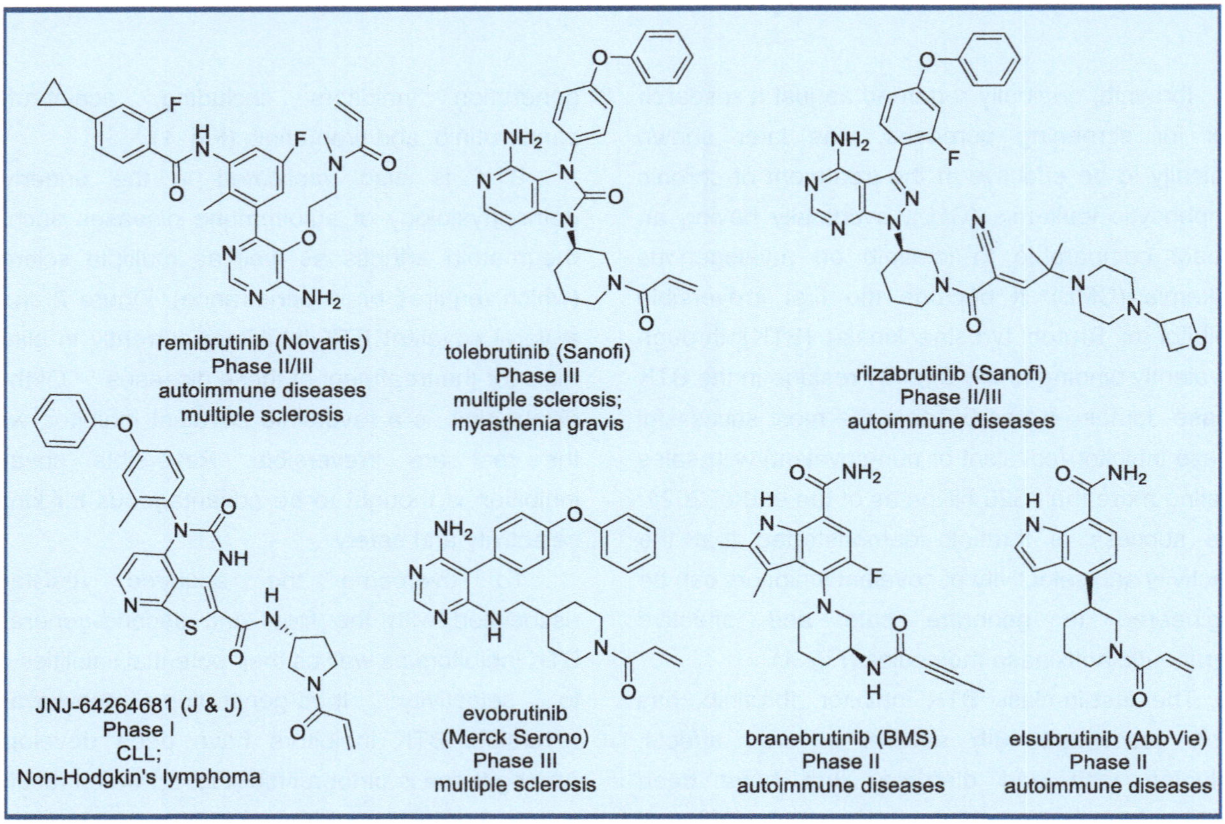

Figure 2. Covalent BTK inhibitors in development.

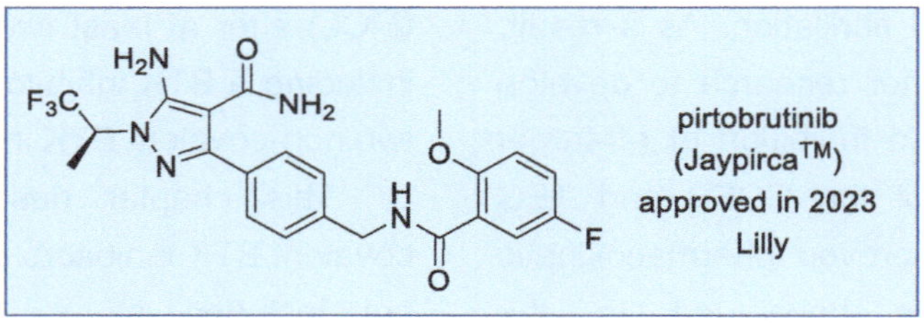

Figure 3. Approved non-covalent BTK inhibitor.

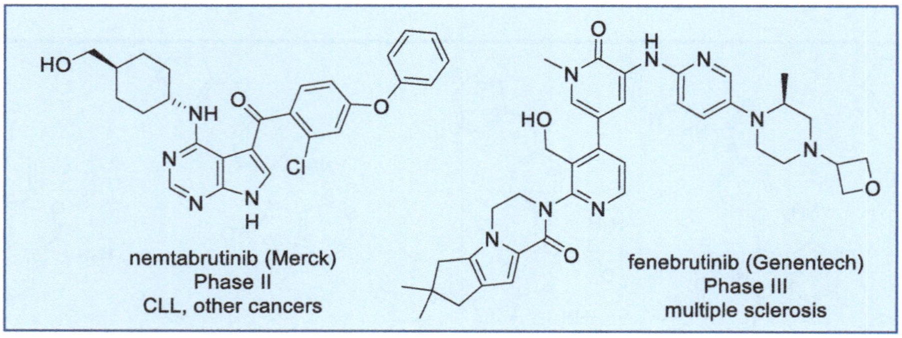

Figure 4. Non-covalent BTK inhibitors in development.

1. B. Tasso, A. Spallarossa, E. Russo, C. Brullo. The Development of BTK Inhibitors: A five-year update. *Molecules.* **2021**, *26*, 7411.
2. F. Ran, Y. Liu, C. Wang, Z. Xu, Y. Zhang, Y. Liu, G. Zhao, Y. Ling. Review of the development of BTK inhibitors in overcoming the clinical limitations of ibrutinib. *Euro. J. Med. Chem.* **2022**, *229*, 114009.

3.1 Ibrutinib (Imbruvica™) (8)

Structural Formula	Space-filling Model	Brief Information
$C_{25}H_{24}N_6O_2$; MW = 440 clogP = 4.0; tPSA = 96		Year of discovery: 2007 Year of introduction: 2013 Discovered by: Celera Genomics Developed by: Pharmacyclics/AbbVie Primary target: BTK Binding type: covalent Class: non-receptor tyrosine kinase Treatment: CLL, MCL Other name: PCI-32765 Oral bioavailability = 4% (fasted), 8% (fed) Elimination half-life = 3.2 h Protein binding = 97.3%

● = C ○ = H ● = O ● = N ● = S ● = F ● = Cl ● = Br ● = I

Ibrutinib, a first-in-class covalent inhibitor of Bruton's tyrosine kinase (BTK), has been approved in many countries to treat relapsed/refractory chronic lymphocytic leukemia (CLL) and more recently, as a frontline therapy for CLL. It has revolutionized the treatment of CLL and has been touted as "the imatinib of CLL." Ibrutinib has also been approved to treat Waldenström's macroglobulinemia, mantle cell lymphoma (MCL), marginal zone lymphoma, and chronic graft vs. host disease. It is on the World Health Organization's Model List of Essential Medicines and represents one of the most successful kinase inhibitors with approximately $9.8 billion in sales for 2021. The success of ibrutinib provides an instructive example of the importance of patience, persistence, thoughtfulness, optimism and serendipity in the quest for breakthrough medicines.

Background[1-3]

In 1952, Dr. Ogden Bruton published a ground-breaking research paper describing a boy unable to develop immunities to common childhood diseases and infections due to primary immunologic deficiency (PID),[4] later known as X-linked agammaglobulinemia (XLA). XLA is attributed to the lack of serum immunoglobulins (Ig) and circulating mature B cells. In 1993-more than four decades later-the molecular basis for XLA was deciphered: a mutation on the X chromosome (Xq21.3-q22) of a single gene which produces BTK.[5,6] Indeed, BTK mutation in mice led to X-linked immunodeficiency (XID), similar to the human phenotype XLA.[6] In people with XLA, the process of white blood cell formation does not lead to the type of mature B cells which are essential for the immune system to produce antibodies (known as immunoglobulins) which defend the body against infections. BTK is primarily responsible for mediating B cell development and maturation through a signaling pathway on the B cell receptor (BCR).

BTK, a 659 amino acid member of a subfamily of SRC-related cytoplasmic tyrosine kinase, possesses five different protein interaction domains: a pleckstrin homology (PH) domain, a proline-rich TEC homology (TH) domain, SRC homology (SH) domains SH3 and SH2, and a catalytic domain (Fig. 1). The PH domain binds to phosphatidylinositol lipids, such as PIP3, and recruits proteins to the cell membrane. Of special note are two critical tyrosine phosphorylation sites: TYR223 in the SH3 domain and TYR551 in the kinase domain. In the BCR signaling pathway, phosphorylation at TYR551 by SYK or LYN increases the catalytic activity of BTK and induces autophosphorylation at TYR223. Phosphorylation at both sites fully activates BTK.

BTK inhibitors, either covalent or non-covalent, bind to the BTK kinase domain and block the catalytic activity of BTK, thus preventing autophosphorylation at TYR223. Several covalent BTK inhibitors, including ibrutinib, acalabrutinib, zanubrutinib, and orelabrutinib, have been developed to target BTK in B cell malignancies. These covalent inhibitors have proved to be highly effective for the treatment of B cell malignancies, including CLL and MCL.

BTK in BCR/TLR and Chemokine Receptor Signaling[1-3]

BCR is always non-covalently associated with a disulfide-linked heterodimer, CD79A/B. When specific antigens bind to BCR, the SRC family protein tyrosine kinase LYN phosphorylates the immunoreceptor tyrosine-based activation motifs (ITAM) on the cytoplasmic tails of CD79A/B and generates docking sites for SYK, spleen tyrosine kinase. Concurrently, LYN also phosphorylates tyrosine residues in the cytoplasmic tail of the BCR co-receptor CD19, thereby activating PI3K, which is recruited to the BCR complex at plasma membrane by the cytoplasmic B cell adapter (BCAP). Then PI3K catalyzes the phosphorylation of the plasma membrane lipid phosphatidylinositol 4,5-bisphophate (PIP2) to generate the second messenger phosphatidylinositol (3,4,5)-trisphosphate (PIP3). PIP3 binds to the PH domain of BTK which causes attachment of BTK to the cell membrane where it is then fully activated via transphosphorylation at Tyr551 in the catalytic kinase domain by LYN or SYK. That is followed by autophosphorylation at Tyr223 within the SH3 domain.

The PIP3-BTK PH binding to attract BTK to membrane is critical because BTK is predominantly cytoplasmic and only transiently recruited into the membrane. The activated SYK also phosphorylates a B-cell linker protein (BLNK), which serves as a critical scaffold protein to bring SYK, BTK and PLCγ2 together to form micro-signalosomes, which consists of PI3K, BTK, BLNK and PLCγ2. PLCγ2 undergoes phosphorylation at TYR753, and TYR759 primarily through BTK. Upon activation, PLCγ2 hydrolyzes membrane PIP2 to release two potent secondary messengers, inositol triphosphate (IP3) and diacylglycerol (DAG).

IP3 diffuses into the cytosol, whereas DAG, a hydrophobic lipid, remains within the plasma membrane. IP3 binds to its receptor, a calcium channel located in the endoplasmic reticulum, resulting in the release of calcium into the cytosol and activation of the transcription factor NFAT (Nuclear Factor of Activated T-cells). NFAT is involved in many aspects of cancer and autoimmune disease.

DAG mediates activation of protein kinase Cβ (PKCβ), which induces activation of oncogenic transcription factor MYC and the transcription nuclear factor kappa B (NF-κB). (Fig. 2). NF-κB is important in inflammatory and immune responses and plays a critical role in the regulation of expression of many other genes related to cell survival, proliferation, and differentiation. MYC activation is associated with cancer initiation and maintenance.

The catalytic domain of BTK can also be activated via the antigen-independent toll-like receptor (TLR) and chemokine receptor signaling pathways. The activated BTK phosphorylates PLCγ2, inducing the downstream transcription of NF-κB, NFAT, and MYC as shown in the BCR signaling pathway (Fig. 2).

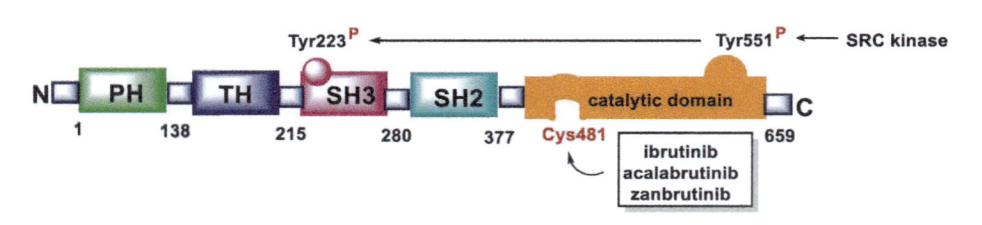

Figure 1. Structure of BTK.

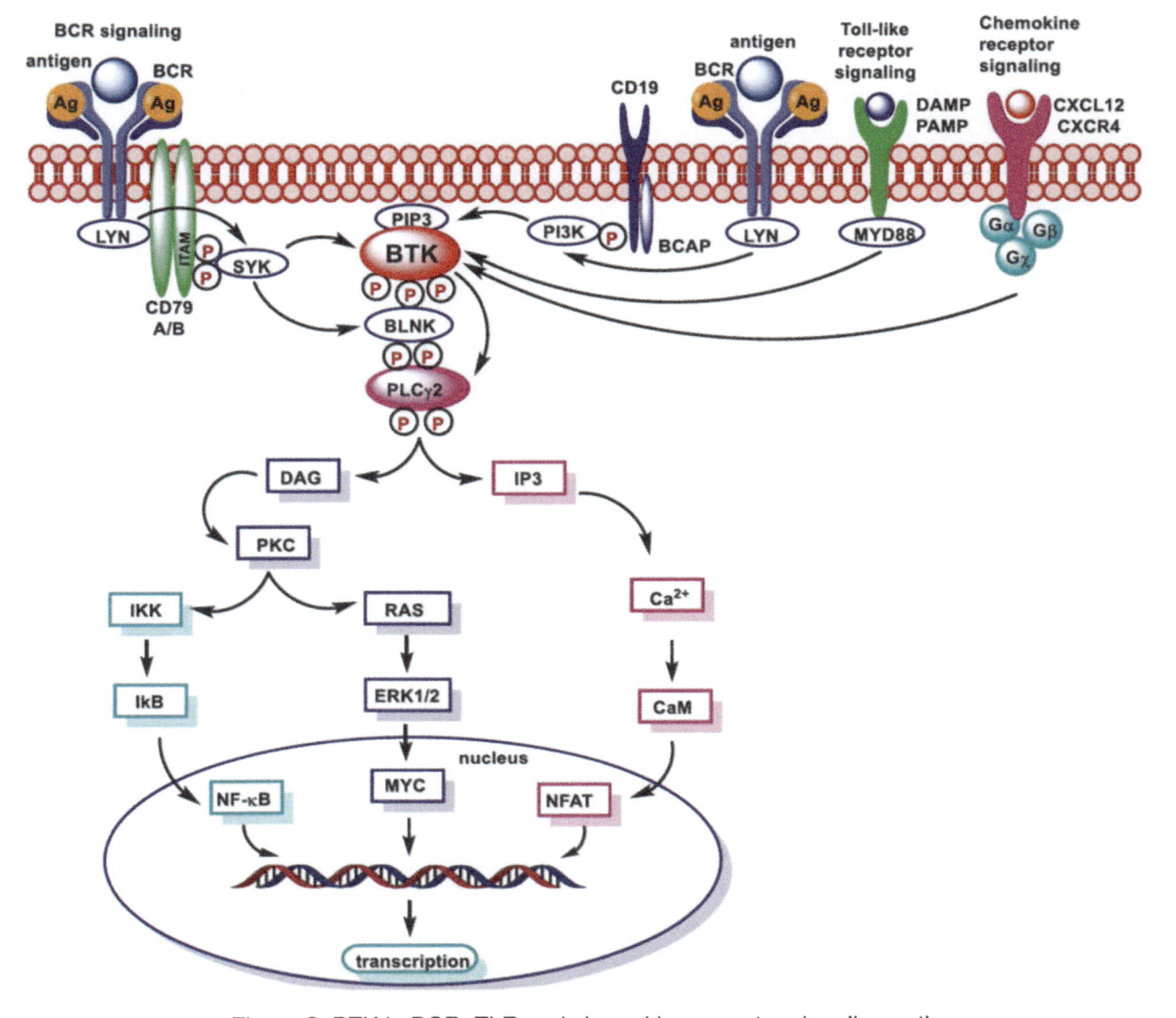

Figure 2. BTK in BCR, TLR and chemokine receptor signaling pathways.

Discovery of Ibrutinib[7,8]

In 1998, Celera genomics was founded with the mission of applying the human genome sequence (accomplished in 2000) and 3D protein structure to drug discovery. Subsequently, Celera initiated several drug discovery programs including inhibitors of BTK for the treatment of rheumatoid arthritis. As part of their efforts to screen noncovalent BTK inhibitors, Celera scientists desired to have a tool compound, a molecule that would bind to BTK covalently and serve as a fluorescent probe. This tool compound would facilitate selection of other compounds that could bind BTK tightly, but not covalently, because at the time covalent kinase inhibitors were deemed unacceptable due to potential severe off-target liabilities (even though some well-known drugs such as aspirin, penicillins, clopidogrel (Plavix) and esomeprazole (Nexium) are chemically reactive and form a covalent bond with their respective biological targets). Thus, a series of tool compounds was designed to target Cys481 in the ATP-binding domain of BTK through an irreversible covalent addition of SH to Michael acceptor warheads.[9] The lead compound **1**, an aminopyrazolopyrimidine-based diphenyl ether, showed potent, reversible inhibition against BTK and the SRC-family kinases. To transform **1** into covalent inhibitors, the cyclopentyl was replaced by 3-piperidine to give **2**, and the free amine was substituted with a variety of acrylamide warheads, which served as Michael acceptors to the thiol of Cys481 in the BTK catalytic domain. This work led to ibrutinib (as a racemate) with 10-fold improvement in BTK inhibition over the lead compound **1**.

Despite the limited selectivity between the two compounds in biochemical enzymology assays using purified kinases, ibrutinib (racemate) showed superior selectivity profile in cellular assays. In anti-IgM-stimulated Ramos cells, a human B cell line, ibrutinib (racemate) significantly reduced the phosphorylation of BTK's substrate PLCγ1 with an IC_{50} of 14 nM, while the LYN- and SYK-dependent phosphorylation at BTK TYR551 was inhibited to a much lesser extent with IC_{50} > 7.5 µM, demonstrating >500-fold BTK selectivity over LYN or SYK in cellular systems. In contrast, the reversible inhibitor **1** exhibited only 4-fold selectivity.

The enantiomerically pure ibrutinib was further tested in a screening panel of kinases that contain an active cysteine residue homologous to Cys481 in BTK, including epidermal growth factor receptor (EGFR) (IC_{50} = 12 nM), human epidermal growth factor receptor 2 (HER2) (IC_{50} = 22 nM), HER4 (IC_{50} = 0.6 nM), interleukin 2-inducible T-cell kinase (ITK) (IC_{50} = 12 nM), BMX (IC_{50} = 1 nM), Janus kinase 3 (JAK3) (IC_{50} = 22 nM), TEC (IC_{50} = 1 nM), and B lymphocyte kinase (BLK) (IC_{50} = 1 nM).[10,11] The impact of inhibiting these alternative kinases on the efficacy or toxicity of ibrutinib remains to be determined. Ibrutinib is also a reversible inhibitor of other kinases without a Cys481 homolog. The reversible kinase inhibition is likely to have limited impact in vivo due to short effective half-life of ibrutinib in humans (2–3 h). Thus, the fast irreversible binding to BTK together with rapid in vivo clearance results in improved BTK selectivity in vivo over the reversibly inhibited off-target kinases. Ibrutinib also showed dose-dependent oral efficacy in a mouse arthritis model. Overall, this tool compound displayed a more favorable profile than noncovalent reversible inhibitors.

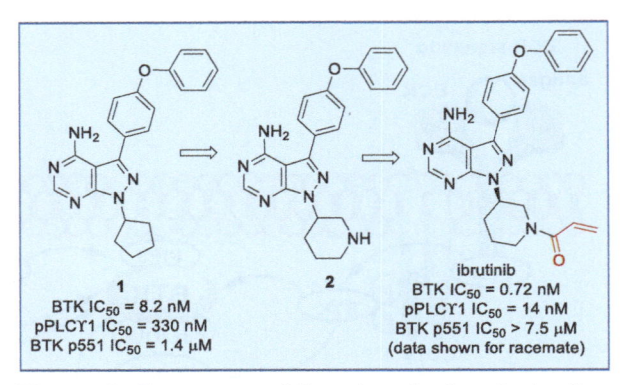

Figure 3. Summary of the chemical optimization of ibrutinib.

Despite progress with BTK and several other drug discovery programs, Celera failed to turn massive genomic data quickly into real drug molecules. In 2006, optimism evaporated, and Celera elected to concentrate on diagnostic testing, selling its early-stage pipeline including BTK inhibitors for only $6 million to Pharmacyclics, a struggling small company that was desperately trying to recover from the catastrophic failure of its sole clinical candidate, Xcytrin.

Pharmacyclics conducted studies of Celera's BTK inhibitors for both autoimmune diseases and B-cell cancers as well. At the time it was a challenge to find an animal model in which the tumor growth relied solely upon B-cell receptor activation, unlike most traditional B-cell cancer model systems. The only viable, but less traditional approach was spontaneously occurring lymphoma model in dogs, and even in this model, ibrutinib elicited only a partial response. Considering the covalent nature of inhibitor and non-robust in vivo activity in dogs, most pharmaceutical companies would have shelved this compound, but Pharmacyclics took a chance and quickly initiated a clinical study.

The results were extraordinary: the drug shrank certain tumors, especially some B-cell cancers, and yet caused no serious decrease in the production of normal healthy cells. On the basis of the impressive clinical outcome, ibrutinib was approved for treatment of CLL in 2013 and later was also cleared for the treatment of other types of cancers. While ibrutinib does not cure cancer, simply taking one pill a day by mouth for life can manage the disease for two to three years, or occasionally longer, thus extending the lives of terminally ill patients. In common with several other cancer drugs, ibrutinib is expensive: $90 per single pill, which comes to $98,550 to $131,400 per year per patient. Because it is not curative, it requires indefinite therapy.

In 2015, Pharmacyclics was acquired by AbbVie for $21 billion, representing one of the top 20 pharma deals ever (see Appendix 3). Since then, ibrutinib sales have totaled more than $26 billion as of the end of 2021, according to AbbVie's quarterly filings.

Ibrutinib became the second FDA approved covalent kinase inhibitor (CKI), just a few months after afatinib, a dual HER2/EGFR inhibitor. Ibrutinib has transformed the treatment of CLL and has validated BTK as a therapeutic target in this disease. Nevertheless, it has limitations, due to off-target toxicities, leading eventually to the emergence of second-generation BTK inhibitors with reduced adverse effects.

Binding Mode[3]

As shown in Figures 4 and 5, the 4-amino NH and N5 of the pyrazolopyrimidine ring in ibrutinib hydrogen bond with main chain atoms of Glu475 and Met477 in the hinge region, respectively. The pyrazolopyrimidine core is surrounded by side chain residues of Val416 in the glycine-rich P-loop, Tyr476 in the hinge, as well as residues Ala428 and Leu528. The terminal phenyl group projects into a hydrophobic pocket and engages in π–π stacking with the DFG residue Phe540 at the bottom. The diphenyl ether subunit is surrounded by αC-helix (Met449), DFG motif (Asp539), activation loop (Leu542) and residues Lys430, Ser538, Ile472, and Thr474. The partially solvent-exposed N-acryloyl piperidine moiety projects toward an opening of the active site that is situated between Leu408 and the Gly480-Cys481 moiety. The electrophilic acrylamide covalently links with the thiol of Cys481 at the edge of the pocket (Figs. 5, 6).

The acrylamide carbonyl oxygen hydrogen bonds with the backbone amide NH of Cys481 and also interacts indirectly with the N7 of pyrazolopyrimidine via a water molecule. The ibrutinib-bound BTK kinase domain adopts an inactive conformation: the activation loop collapses into the active site with the αC helix exposed.

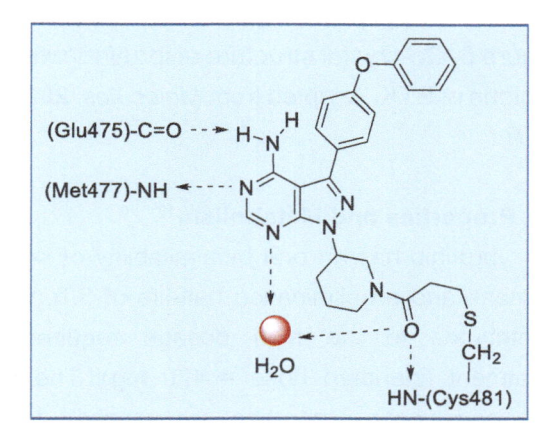

Figure 4. Summary of ibrutinib–BTK interactions based on co-crystal structure.

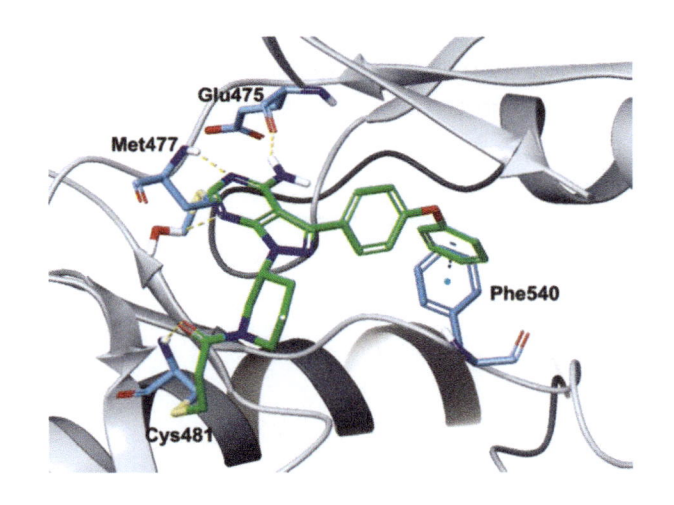

Figure 5. Co-crystal structure of ibrutinib with BTK (PDB ID: 5P9J).

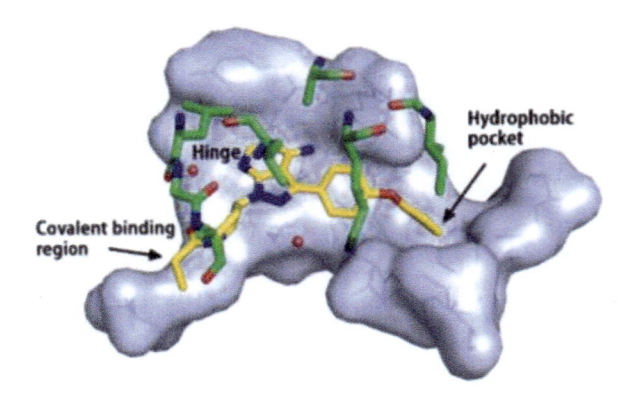

Figure 6. Co-crystal structure of ibrutinib with key regions of BTK. Adapted from *Molecules.* **2021**, *26*, 4907.

PK Properties and Metabolism[12]

Ibrutinib has an oral bioavailability of <4% in humans and an elimination half-life of 3 h, which contributes to the high dosage required for treatment (standard dose = 420 mg). The poor pharmacokinetic properties are ascribed to the rapid and extensive oxidative metabolism mediated by cytochrome P450 3A4. The metabolic degradation occurs via three major pathways: (1) hydroxylation of the phenyl to give **3**; (2) epoxidation of the ethylene on the acryloyl moiety followed by hydrolysis to afford dihydrodiol **4**; and (3) cleavage of the piperidine ring to furnish acid **5** (Fig. 7).

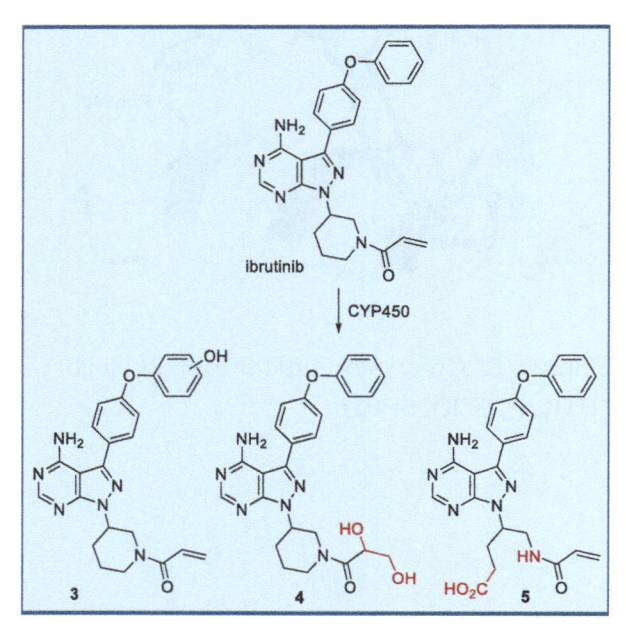

Figure 7. Metabolism of ibrutinib in humans.

1. R. W. Hendriks, S. Yuvaraj, J. Kil. Targeting Bruton's tyrosine kinase in B cell malignancies. *Nat Rev. Cancer.* **2014**, *14*, 219–232.
2. J. Liu, C. Chen, D. Wang, J. Zhang, T. Zhang. Emerging small-molecule inhibitors of the Bruton's tyrosine kinase (BTK): Current development. *Euro. J. Med. Chem.* **2021**, *217*, 113329.
3. D. Zhang, H. Gong, F. Meng. Recent advances in BTK inhibitors for the treatment of inflammatory and autoimmune diseases. *Molecules.* **2021**, *26*, 4907.
4. O. C. Bruton. Agammaglobulinemia. *Pediatrics.* **1952**, *9*, 722–728.
5. D. J. Rawlings, D. C. Saffran, S. Tsukada, D. A. Largaespada, J. C. Grimaldi, L. Cohen, R. N. Mohr, J. F. Bazan, M. Howard, N. G. Copeland. Mutation of unique region of Bruton's tyrosine kinase in immunodeficient XID mice. *Science.* **1993**, *261*, 358–361.
6. J. D. Thomas, P. Sideras, C. I. Smith, I. Vorechovský, V. Chapman, W. E. Paul. Colocalization of X-linked agammaglobulinemia and X-linked immunodeficiency genes. *Science.* **1993**, *261*, 355–358.
7. D. Shaywitz. The wild story behind a promising experimental cancer drug. *Forbes.* April 5, 2013.
8. J. C. Owens, L. M. Krieger. Pharmacyclics' 'miracle cure': A cancer drug. *The Mercury News.* December 26, 2015.
9. Z. Pan, H. Scheerens, S. J. Li, B. E. Schultz, P. A. Sprengeler, L. C. Burrill, R. V. Mendonca, M. D. Sweeney, K. C. Scott, P. G. Grothaus, D. A. Jeffery, J. M. Spoerke, L. A. Honigberg, P. R. Young, S. A. Dalrymple, J. T. Palmer. Discovery of selective irreversible inhibitors for Bruton's tyrosine kinase. *ChemMedChem.* **2007**, *2*, 58–61.
10. R. H. Advani, J. J. Buggy, J. P. Sharman, S. M. Smith, T. E. Boyd, B. Grant, K. S. Kolibaba, R. R. Furman, S. Rodriguez, B. Y. Chang, J. Sukbuntherng, R. Izumi, A. Hamdy, E. Hedrick, N. H. Fowler. Bruton tyrosine kinase inhibitor ibrutinib (PCI-32765) has significant activity in patients with relapsed/refractory B-cell malignancies. *J. Clin. Oncol.* **2013**, *31*, 88–94.
11. J. A. Burger, J. J. Buggy. Bruton tyrosine kinase inhibitor ibrutinib (PCI-32765). *Leuk. Lymphoma* **2013**, *54*, 2385–2391.
12. E. Scheers, L. Leclercq, J. de Jong, N. Bode, M. Bockx, A. Laenen, F. Cuyckens, D. Skee, J. Murphy, J. Sukbuntherng, G. Mannens. Absorption, metabolism, and excretion of oral ^{14}C radiolabeled ibrutinib: An open-label, phase I, single-dose study in healthy men. *Drug Metab. Dispos.* **2015**, *43*, 289–297.

3.2 Acalabrutinib (Calquence™) (9)

Structural Formula	Space-filling Model	Brief Information
$C_{26}H_{23}N_7O_2$; MW = 465 clogP = 1.6; tPSA = 116		Year of discovery: 2011 Year of introduction: 2017 Discovered by: Acerta Pharma Developed by: AstraZeneca Primary target: BTK Binding type: covalent Class: non-receptor tyrosine kinase Treatment: CLL, SLL, MCL Other name: ACP-196 Oral bioavailability = 25% Elimination half-life = 0.9 h Protein binding = 97.5%

● = C   ○ = H   ● = O   ● = N   ● = S   ● = F   ● = Cl   ● = Br   ● = I

The first-generation BTK inhibitor, ibrutinib, has transformed the treatment of chronic lyphocytic leukemia (CLL). Unfortunately, it has been associated with numerous adverse effects including rash and diarrhea, that have been attributed to the drug's potent off-target inhibition of the epidermal growth factor receptors (EGFR), as well as bleeding and atrial fibrillation. In addition, its high clearance and poor oral bioavailability contribute to high doses of ibrutinib required for clinical efficacy (standard dose = 420 mg, QD), which, in combination with the side effect profile, results in approximately 25% discontinuation rate.[1] As a result, efforts continued toward the development of BTK inhibitors with reduced inhibition of off-target kinases such as those of the EGFR and TEC families and also with improved pharmacokinetic properties. Those efforts led to the discovery of improved and safer second-generation inhibitors including, acalabrutinib zanubrutinib, and tirabrutinib.[2]

Discovery of Acalabrutinib[3-7]

In 2009, Merck acquired Schering-Plough (S-P) for $41 billion in a move to off-set impending patent expirations with approved S-P assets such as ezetimibe, but with no idea that it was also gaining access to a colossal future goldmine, the blockbuster PD1 (cell cycle) inhibitor pembrolizumab (Keytruda™). This inhibitor which had been discovered by Organon BioSciences (the human and animal health business of Dutch chemical giant Akzo-Nobel), had been acquired by S-P for $14 billion in 2007. Merck sidelined the PD1 asset, and it was not activated until very promising results were demonstrated with Bristol-Myers Squibb's rival cell cycle inhibitor, nivolumab (Opdivo™).

The S-P acquisition also gave Merck ownership of a covalent BTK inhibitor, acalabrutinib, via Organon. That BTK inhibitor, which featured a unique 2-butynamide warhead, was then out-licensed to two former Organon scientists who started a new company, Acerta, in 2013.

Acerta quickly profiled acalabrutinib and found it to have excellent selectivity against all kinases with cysteines in the ATP-binding site corresponding to Cys481 in BTK with the exception of BMX, HER4 and TEC (Table 1); however, it showed weaker BTK inhibition than ibrutinib (IC_{50} = 2.5 vs. 0.47 nM).[8] The improved selectivity can be attributed to three structural differences between the two inhibitors (Fig. 1): (1) Michael acceptor warhead (butynamide vs. acrylamide); (2) linker (pyrrolidine vs. piperdine); and (3) back pocket binder (pyridylamide vs. phenoxy ether). The pyrrolidine was linked to butynamide, directing the warhead in the proximity of the active Cys481 thiol for covalent binding. The pyridylamide adds two extra hydrogen bonds to the backbone residues, also contributing to its improved selectivity. These subtle modifications also account for improved pharmacokinetic properties.

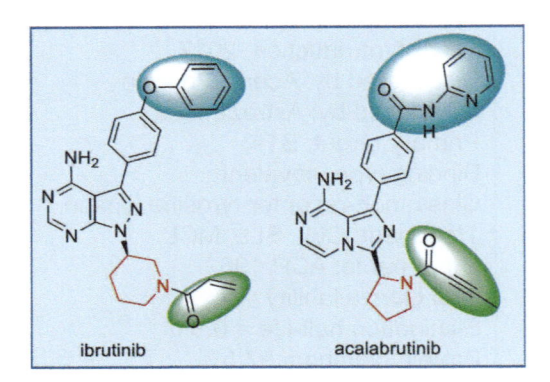

Figure 1. Structures of ibrutinib and acalabrutinib.

In January 2014-just one year after it was founded, Acerta initiated a phase I/II trial to evaluate the safety and efficacy of orally administered acalabrutinib in patients with CLL, Richter's syndrome and prolymphocytic leukemia. The clinical data showed fewer side effects (such as atrial fibrillation) than ibrutinib. It was at this stage of the development that the two-year old Acerta was bought by AstraZeneca for $5 billion, a modest price compared with AbbVie's $21 billion acquisition of Pharmacyclics to gain Ibrutinib, which occurred nine months earlier. In November 2019, the FDA approved acalabrutinib for the treatment of adult patients with CLL or small lymphocytic leukemia (SLL) based on the positive phase III data. Acalabrutinib generated approximately $1 billion in sales for 2021.

In December 2019-seven years after it sold BTK inhibitor, acalabrutinib, Merck added to its pipeline a phase II, reversible BTK inhibitor, nemtabrutinib (ARQ531) (Fig. 2), from its acquisition of ArQule for $2.7 billion.

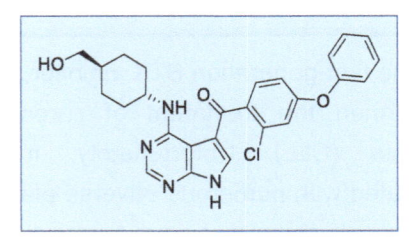

Figure 2. Structure of nemtabrutinib (ARQ531).

Table 1. Selectivity vs. kinases with active cysteine corresponding to Cys481 in BTK[8,9]

kinase	ibrutinib		acalabrutinib		zanubrutinib		tirabrutinib	
	IC_{50}	selectivity	IC_{50}	selectivity	IC_{50}	selectivity	IC_{50}	selectivity
BTK	0.47 nM	1.0	2.5 nM	1.0	0.30 nM	1.0	6.8 nM	1.0
BLK	0.17 nM	0.36	1 µM	410	1.13 nM	1.2	0.30 µM	44
BMX	0.86 nM	1.8	36 nM	14	0.62 nM	0.7[a]	6 nM	0.88
EGFR	3.8 nM	8.1	7.5 µM	3000	2.6 nM	8.7	3.0 µM	440
HER2	8.6 nM	18	0.61 µM	246	0.53 µM	1767	7.3 µM	1080
HER4	2.0 nM	4.3	12 nM	4.8	1.58 nM	1.7[a]	0.77 µM	113
ITK	55 nM	117	>20 µM	>8000	56 nM	187	>20 µM	>2940
JAK3	18 nM	38	>20 µM	>8000	0.58 µM	1933	0.55 µM	810
TXK	1.9 nM	4.0	170 nM	68	2.95 nM	3.2[a]	92 nM	14
TEC	3.2 nM	6.8	37 nM	15	2.0 nM	6.7	48 nM	7.1

[a]Selectivity calculated based on BTK IC_{50} = 0.92 nM.[9]

Binding Mode

The co-crystal structure of acalabrutinib–BTK complex has not been reported. A binding model[10] indicates that the 8-NH and N7 of the amino imidazopyrazine ring H-bonds with main chain atoms of Glu475 and Met477 in the hinge region, respectively. The 2-pyridylbenzamide engages in two H-bonds with main chain atoms of Ser538 and Asp539 (Fig. 3). The electrophilic butynamide covalently binds to the thiol of Cys481 residue at the edge of the pocket.

PK Properties and Metabolism[11]

Acalabrutinib has an absolute bioavailability of 25% and an elimination half-life of 0.9 h. Because of its better oral bioavailability but much shorter half-life than ibrutinib, acalabrutinib is taken at lower dosage (100 mg) than ibrutinib (420 mg) but twice a day (vs. QD for ibrutinib) (Table 2).

Figure 3. Putative acalabrutinib–BTK interactions.

The major circulating metabolite in the plasma is the pyrrolidine ring-opened product **1** (Fig. 4). This metabolite also exhibited covalent BTK inhibition with potency 2-fold lower than acalabrutinib, and its kinase selectivity profile is similar to acalabrutinib. This active metabolite has a half-life of 6.9 h, much longer than the parent drug. However, the extent of its contribution to on-target covalent inhibition of BTK in humans remains to be established.

Table 2. PK properties of BTK inhibitors[12]

PK	ibrutinib	acala-brutinib	zanu-brutinib	tira-brutinib
t_{max}	1–2 h	0.75 h	2 h	2.9 h
V_d/F	10000 L	101 L	881 L	
$t_{1/2}$	4–6 h	0.9 h	3.3 h	3.6 h
F	4%	25%	15%	
CL	62 L/h	159 L/h	182 L/h	
dose	QD 420 mg CLL/SLL/ WM 500 mg MCL/MZL	BID 100 mg CLL/SL L/MCL)	BID 160 mg MCL	QD 480 mg PCNSL

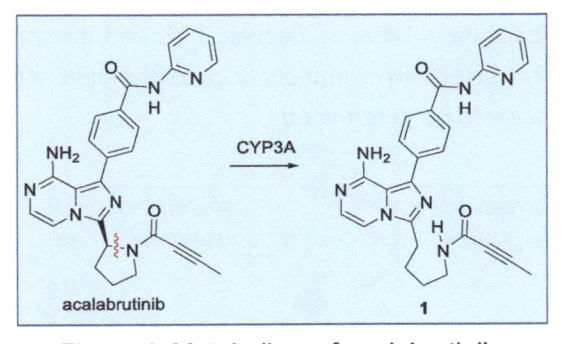

Figure 4. Metabolism of acalabrutinib.

1. H. Y. Estupiñán, A. Berglöf, R. Zain, C. Smith, C. Comparative analysis of BTK inhibitors and mechanisms underlying adverse effects. *Front. Cell Dev. Biol*. **2021**, *9*, 630942.
2. J. Liu, C. Chen, D. Wang, J. Zhang, T. Zhang. Emerging small-molecule inhibitors of the Bruton's tyrosine kinase (BTK): Current development. *Euro. J. Med. Chem*. **2021**, *217*, 113329.
3. M. Armstrong. Uncovering biopharma's hidden treasure. *Evaluate Vamtage*. October 14, 2019.
4. A. Ward, A. Massoudi. AstraZeneca adds 55% Acerta Pharma stake to swath of deals. Financial Times. Decemer 17, 2015.
5. Global University Venturing. AstraZeneca to inject $2.5 bn in Acerta. December 18, 2015.
6. J. Plieth. Astra signals a late run on BTK inhibition. *Evaluate Vamtage*. December 14, 2015.
7. Acalabrutinib - Acerta Pharma/AstraZeneca. AdisInsight.
8. R. Sakowicz, J. Y. Feng. Biochemical characterization of tirabrutinib and other irreversible inhibitors of Bruton's tyrosine kinase reveals differences in on - and off - target inhibition. *Biochim. Biophys. Acta Gen. Subj*. **2020**, *1864*, 129531.
9. Y. Guo, Y. Liu, N. Hu, D. Yu, C. Zhou, G. Shi, B. Zhang, M. Wei, J. Liu, L. Luo, Z. Tang, H. Song, Y. Guo, X. Liu, D. Su, S. Zhang, X. Song, X. Zhou, Y. Hong, S. Chen, Z. Wang. Discovery of zanubrutinib (BGB-3111), a novel, potent, and selective covalent inhibitor of Bruton's tyrosine kinase. *J. Med. Chem*. **2019**, *62*, 7923–7940.
10. T. Barf, T. Covey, R. Izumi, M. van de Kar, M. Gulrajani, B. van Lith, M. van Hoek, E. de Zwart, D. Mittag, D. Demont, S. Verkaik, F. Krantz, P. G. Pearson, R. Ulrich, A. Kaptein. Acalabrutinib (ACP-196): A covalent Bruton tyrosine kinase inhibitor with a differentiated selectivity and in vivo potency profile. *J. Pharmacol. Exp. Ther*. **2017**, *363*, 240–252.
11. T. Podoll, P. G. Pearson, J. Evarts, T. Ingallinera, E. Bibikova, H. Sun, M. Gohdes, K. Cardinal, M. Sanghvi, J. G. Slatter. Bioavailability, biotransformation, and excretion of the covalent Bruton tyrosine kinase inhibitor acalabrutinib in rats, dogs, and humans. *Drug Metab. Dispos*. **2019**, *47*, 145–154.
12. T. Wen, J. Wang, Y. Shi, H. Qian, P. Liu. Inhibitors targeting Bruton's tyrosine kinase in cancers: drug development advances. *Leukemia*. **2021**, *35*, 312–332.

3.3 Zanubrutinib (Brukinsa™) (10)

Structural Formula	Space-filling Model	Brief Information
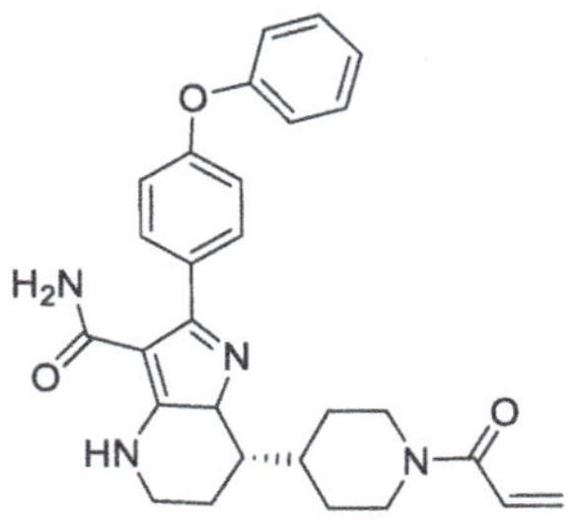 $C_{28}H_{30}N_4O_3$; MW = 470 clogP = 3.8; tPSA = 97	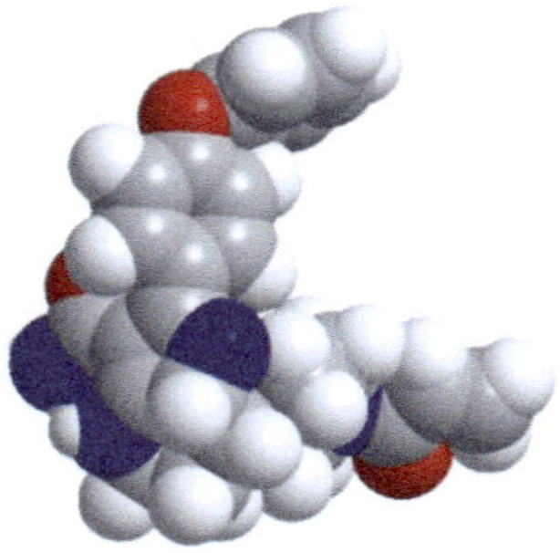	Year of discovery: 2013 Year of introduction: 2021 Discovered by: BeiGene Developed by: BeiGene Primary target: BTK Binding type: covalent Class: non-receptor tyrosine kinase Treatment: MCL, MZL, WM Other name: BGB-3111 Oral bioavailability = 15% Elimination half-life = 3.3 h Protein binding = 94%

● = C ○ = H ● = O ● = N ● = S ● = F ● = Cl ● = Br ● = I

Zanubrutinib, a second-generation BTK inhibitor discovered and developed by BeiGene in China, has been approved by the FDA (in 2019) for the treatment of chronic lymphocytic leukemia (CLL) as well as certain other indications. Zanubrutinib has lower toxicity and better efficacy in comparison with ibrutinib. It is in direct competition with AstraZeneca's acalabrutinib for the $12 billion blood cancer market currently dominated by the first-in-class BTK inhibitor ibrutinib.

Discovery of Zanubrutinib[1]

The discovery of zanubrutinib started with ibrutinib as the lead structure and the 4-aminopyrazolopyrimidine motif as the donor/acceptor hinger binder. The aminopyrimidine ring **A** was simulated by an intramolecular hydrogen bonded structure **B,** a 5-aminopyrazole-4-carboxamide scaffold with similar shape and hinge-binding elements (Fig. 1). Indeed, both carboxamide (R)-**1** and (S)-**1** maintained comparable activity and kinase selectivity to ibrutinib (Fig. 2). Although this bioisosteric replacement created a new series of pyrazole-based BTK inhibitors, the additional primary amino group increased the total of hydrogen bond donors to 4, thereby reducing the membrane permeability and oral absorption. To reduce the number of hydrogen bond donors, the 4-amino group attached to the pyrazole ring was connected to the proximal piperidine to give tetrahydropyrazolopyrimidine **2** (Fig. 2). The carboxamide subunit was retained to preserve its critical binding to the hinge. Further optimization of the linker between the Michael acceptor warhead and the bicyclic core led to zanubrutinib, which fortunately manifested improved selectivity over ibrutinib. For example, it exhibited 187-, 1933-, and 1800-fold selectivity, respectively, against ITK, JAK3, and HER2, which contain cysteines in the ATP-binding site corresponding to cysteine 481 in BTK. Zanubrutinib at 1 µM also exhibited low inhibition in SRC family of kinases such as SRC (29%), LYN (35%), FYN (50%), and LCK (73%, IC_{50} = 187 nM) which are associated with increased bleeding risk due to platelet activation. However, its overall selectivity is still less favorable than acalabrutinib,[2] another second-generation BTK inhibitor (see Table 1, Section 3.2), and the impact of the selectivity variation to potential side effects remains to be determined.

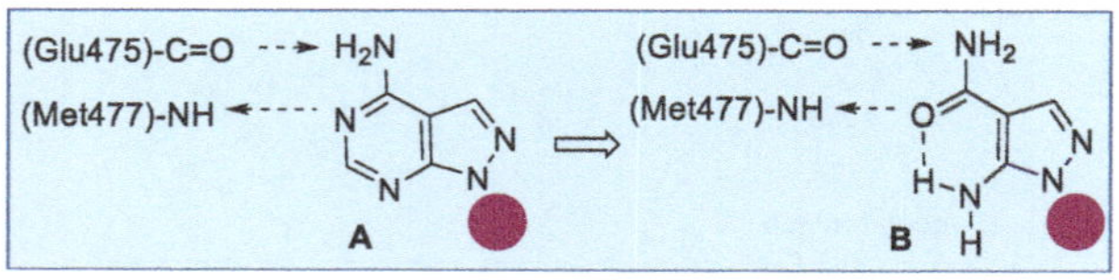

Figure 1. Bioisostere of aminopyrimidine.

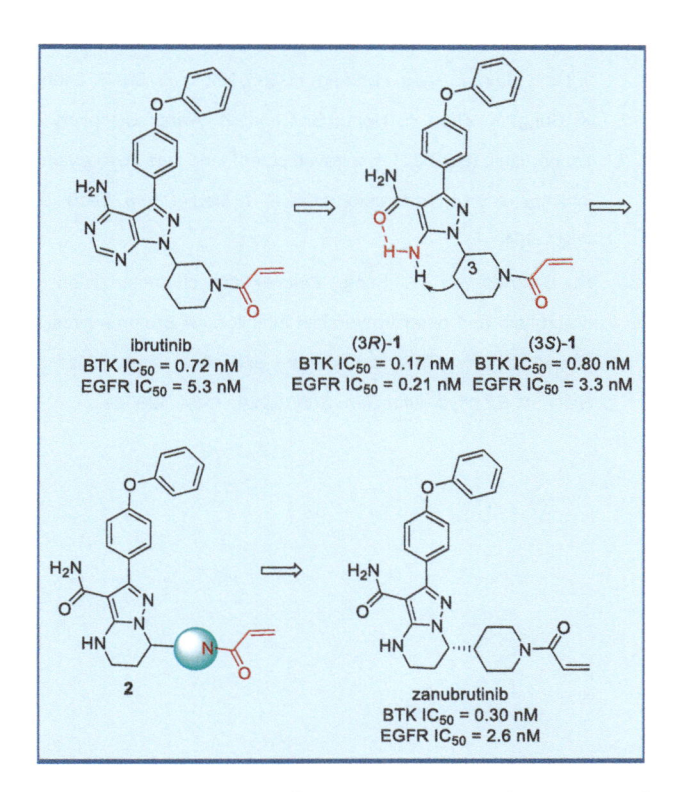

Figure 2. Summary of the chemical optimization of zanubrutinib.

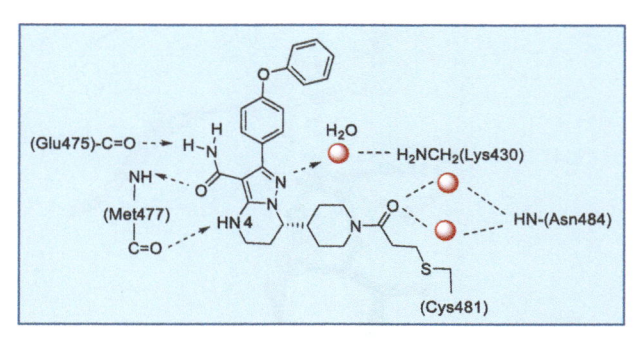

Figure 3. Summary of zanubrutinib -BTK interactions based on co-crystal structure.

Binding Mode[1]

The electron density map corresponding to zanubrutinib and Cys481 showed covalent linkage with Cys481 (Fig. 4). Zanubrutinib forms three critical hydrogen bonds with hinge residues Glu475 and Met477 (Fig. 5). As compared with the co-crystal structure of ibrutinib with BTK, there is additional hydrogen bond between the backbone carbonyl oxygen of Met477 and the 4-NH (Fig. 3). The terminal phenyl group engages in a T-shape π–π stacking with Phe540, and the pyrazolyl nitrogen interacts with Lys430 via a water bridge. The warhead carbonyl also interacts indirectly with amide NH of Asn484 via two water molecules. Both zanubrutinib and ibrutinib have a piperidinyl linker; however, it adopts two different binding modes in complex with BTK as shown in Figure 6. In addition, a single crystal X-ray structure of zanubrutinib showed a classic intramolecular H-bond between carboxamide oxygen and the 4-NH, which confirmed the bioisosteric mimicry of the aminopyrimidine ring (Fig. 1).

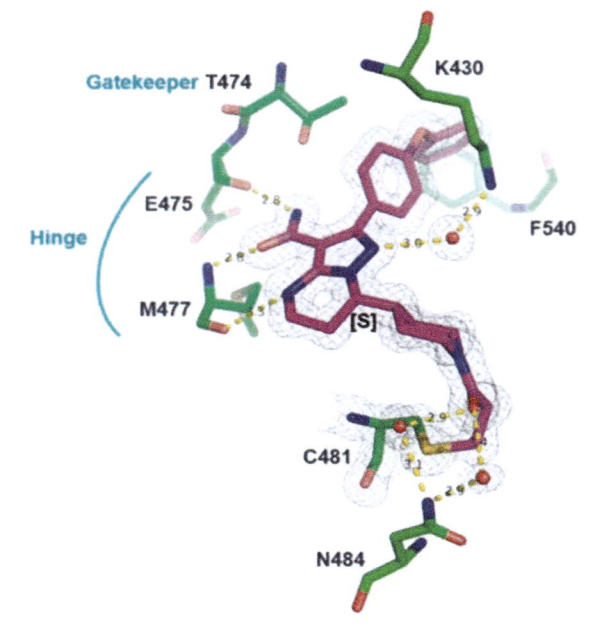

Figure 4. Electron density map corresponding to zanubrutinib and Cys481. Adapted with permission from *J. Med. Chem.* **2019**, *62*, 7923–7940.

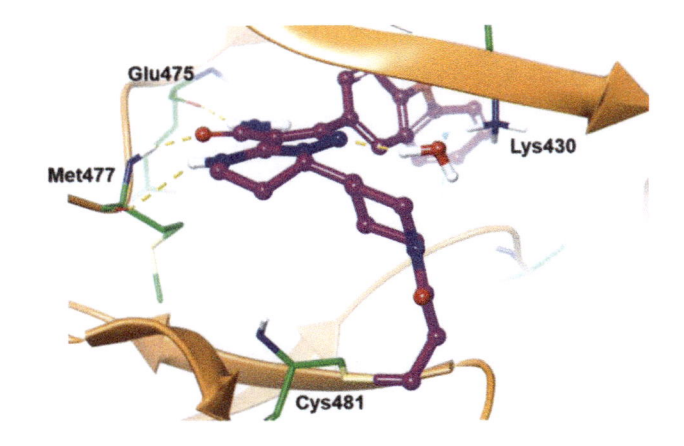

Figure 5. Co-crystal structure of zanubrutinib with BTK (PDB ID: 6J6M).

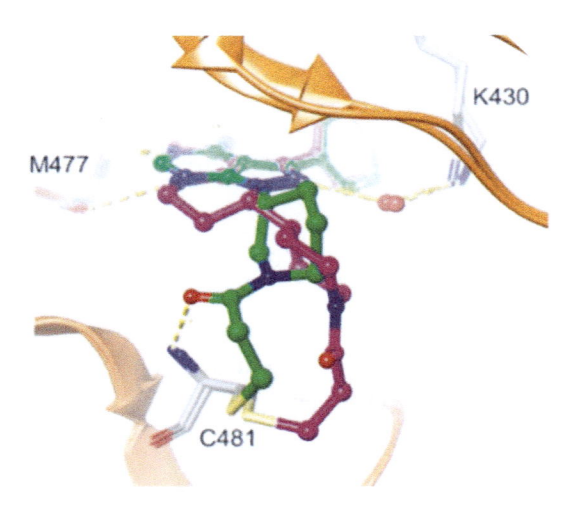

Figure 6. Overlay of Ibrutinib (green) with zanubrutinib (magenta) in complex with BTK. Adapted with permission from *J. Med. Chem.* **2019**, *62*, 7923–7940.

PK Properties and Metabolism[2]

Zanubrutinib showed a mean terminal elimination half-life of approximately 2–4 h (160 or 320 mg, QD) and an estimated oral bioavailability of 15%, relative to 3.9% (fasting state) for ibrutinib (see Table 2, acalabrutinib chapter). For MCL indications, it is taken at lower dosage (160 mg) but twice daily (vs. 420 mg, QD for ibrutinib).

Zanubrutinib is primarily eliminated hepatically via CYP3A4, but its metabolites have not been characterized.

1. Y. Guo, Y. Liu, N. Hu, D. Yu, C. Zhou, G. Shi, B. Zhang, M. Wei, J. Liu, L. Luo, Z. Tang, H. Song, Y. Guo, X. Liu, D. Su, S. Zhang, X. Song, X. Zhou, Y. Hong, S. Chen, Z. Wang. Discovery of zanubrutinib (BGB-3111), a novel, potent, and selective covalent inhibitor of Bruton's tyrosine kinase. *J. Med. Chem.* **2019**, *62*, 7923–7940.

2. R. Sakowicz, J. Y. Feng. Biochemical characterization of tirabrutinib and other irreversible inhibitors of Bruton's tyrosine kinase reveals differences in on - and off - target inhibition. *Biochim. Biophys. Acta Gen. Subj.* **2020**, *1864*, 129531.

3.4   Tirabrutinib (Velexbru™) (11)

Structural Formula	Space-filling Model	Brief Information
$C_{25}H_{22}N_6O_3$; MW = 454 clogP = 3.4; tPSA = 104		Year of discovery: 2012 Year of introduction: 2020 (Japan) Discovered by: Ono Pharmaceutical Developed by: Ono Pharmaceutical Primary target: BTK Binding type: covalent Class: non-receptor tyrosine kinase Treatment: PCNSL Other name: ONO-4059 Oral bioavailability = not reported Elimination half-life = 6.5–8 h Protein binding = 91%

= C   = H   = O   = N   = S   = F   = Cl   = Br   = I

In March 2020, tirabrutinib was approved in Japan for the treatment of primary central nervous system lymphoma (PCNSL).

Tirabrutinib shows excellent selectivity against all kinases with cysteines in the ATP-binding site corresponding to Cys481 in BTK with the exception of BMX, TXK and TEC; however, it displays weaker BTK inhibition than ibrutinib (IC_{50} = 6.8 nM vs. 0.47 nM) (see Table 1 in the Section 3.2).[1,2]

1. H. S. Walter, S. A. Rule, M. J. Dyer, L. Karlin, C. Jones, B. Cazin, P. Quittet, N. Shah, C. V. Hutchinson, H. Honda, K. Duffy, J. V. Jamieson, N. Courtenay-Luck, T. Yoshizawa, J. Sharpe, T. Ohno, S. Abe, A. Nishimura, G. Cartron, G. Salles. A phase 1 clinical trial of the selective BTK inhibitor ONO/GS-4059 in relapsed and refractory mature B-cell malignancies. *Blood*, **2016**, *127*, 411–419.

2. R. Sakowicz, J. Y. Feng. Biochemical characterization of tirabrutinib and other irreversible inhibitors of Bruton's tyrosine kinase reveals differences in on - and off - target inhibition. *Biochim. Biophys. Acta Gen. Subj.* **2020**, *1864*, 129531.

3.5 Orelabrutinib (Innobruka™) (12)

Structural Formula	Space-filling Model	Brief Information
$C_{26}H_{25}N_3O_3$; MW = 427 clogP = 3.8; tPSA = 85		Year of discovery: 2014 Year of introduction: 2020 (China) Discovered by: InnoCare Developed by: InnoCare Primary target: BTK Binding type: covalent Class: non-receptor tyrosine kinase Treatment: CLL, SLL, MCL Other name: ICP-022 Oral bioavailability = not reported Elimination half-life = 4 h Protein binding = not reported

● = C   ● = H   ● = O   ● = N   ● = S   ● = F   ● = Cl   ● = Br   ● = I

In December 2020, orelabrutinib was approved by the China National Medical Products Administration for the treatment of patients with relapsed/refractory MCL, CLL, and SLL. This compound is also being evaluated for the treatment of multiple sclerosis.

Orelabrutinib, which has a unique pyridylcarboxamide hinge binder, showed an IC_{50} value of 1.6 nM in the BTK enzymatic assay. When tested against a panel of 456 kinases at 1 µM, orelabrutinib only inhibited BTK (>90%) while ibrutinib also targeted several other kinases including EGFR, TEC and BMX. Orelabrutinib has a favorable half-life of 4 h in healthy volunteers.[1,2] It is administered orally at a daily dose of 150 mg.

1. S. Dhillon. Orelabrutinib: First Approval. *Drugs*. **2021**, *81*, 503–507.
2. B. Zhang, R. Zhao, R. Liang, Y. Gao, R. Liu, X. Chen, Z. Lu, Z. Wang, L. Yu, S. Shakib, J. Cui. Abstract CT132: Orelabrutinib, a potent and selective Bruton's tyrosine kinase inhibitor with superior safety profile and excellent PK/PD properties. AACR Annual Meeting 2020; April 27-28, 2020 and June 22-24, 2020; Philadelphia, PA.

Chapter 4. EGFR/HER Family Inhibitors

Approximately 30% of non-small-cell lung cancer (NSCLC) patients carry epidermal growth factor receptor (EGFR) mutations. Gefitinib is the first EGFR targeted therapy for these NSCLC patients carrying common mutations (e.g., EGFR exon 19 deletions or exon 21 (L858R) substitutions). Other follow-on first-generation EGFR TKIs include erlotinib and icotinib (Fig. 1).

Afatinib, the first second-generation irreversible pan-EGFR (HER) family inhibitor, has a broader inhibitory profile and greater potency than the first-generation EGFR TKIs against both wild-type EGFR and EGFR common mutations. It has also been approved for nonresistant EGFR mutations (e.g., S768I, L861Q, or G719X). As the first FDA-approved irreversible covalent kinase inhibitor, afatinib represents an important milestone in the discovery of antikinase therapies. Dacomitinib is a follow-on covalent EGFR inhibitor.

Despite their efficacy for the EGFR common mutations, the first- and second-generation EGFR TKIs lose activity against the gatekeeper T790M mutation, an acquired mutation that confers drug resistance. This mutation has been mitigated by osimeritinib, an irreversible third-generation TKI. Thanks to its superior safety and efficacy in patients with EGFR-mutant NSCLC regardless of T790M mutation status, osimertinib has quickly become one of the most successful kinase inhibitors with sales of $5 billion in 2021 and $5.4 billion in 2022.

Unfortunately, even osimertinib showed poor responses to exon 20 insertion mutations, the third most common EGFR mutations (after common mutations and T790M mutations). Fortunately, adding a CO_2Pr^i group to the pyrimidine hinge binder led to mobocertinib with better activity for these mutants than osimertinib.

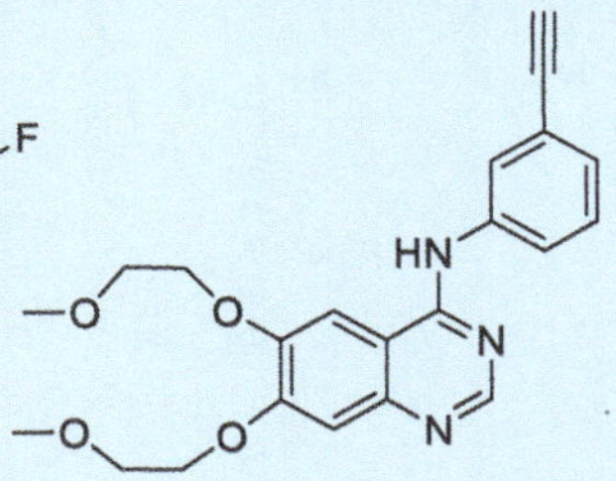

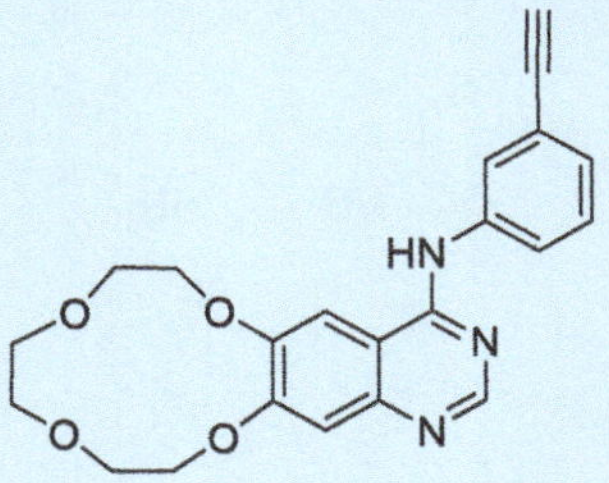

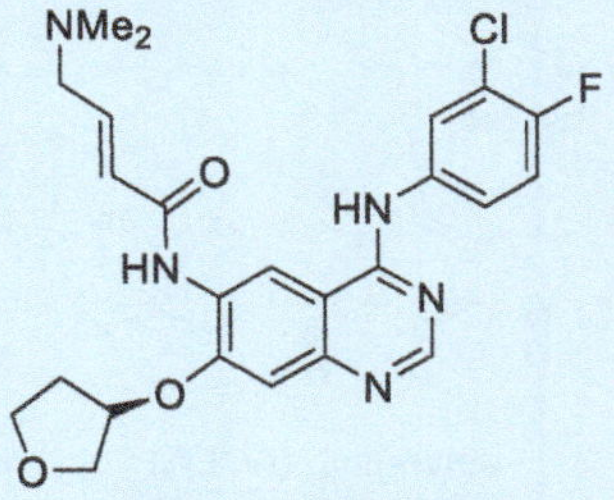

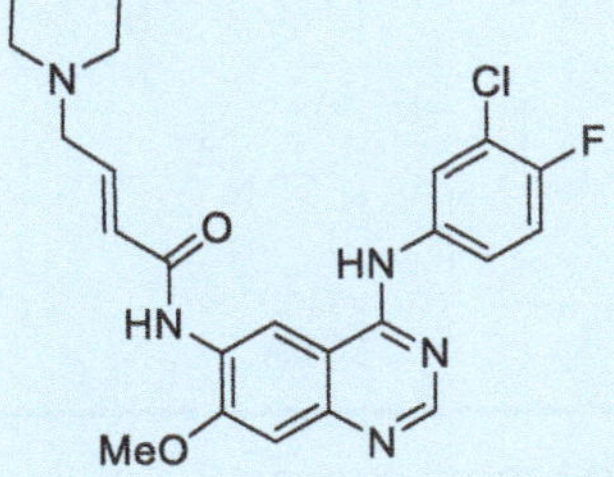

Recently, a new mutation (C797S) has emerged in EGFR exon 20 after the use of osimertinib, prompting the development of fourth-generation TKIs.[1,2]

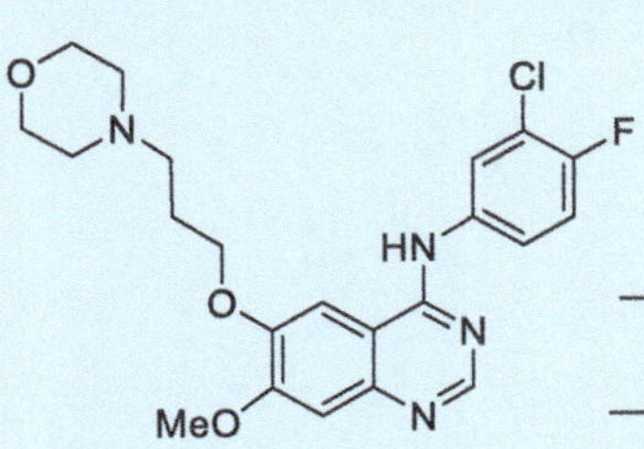

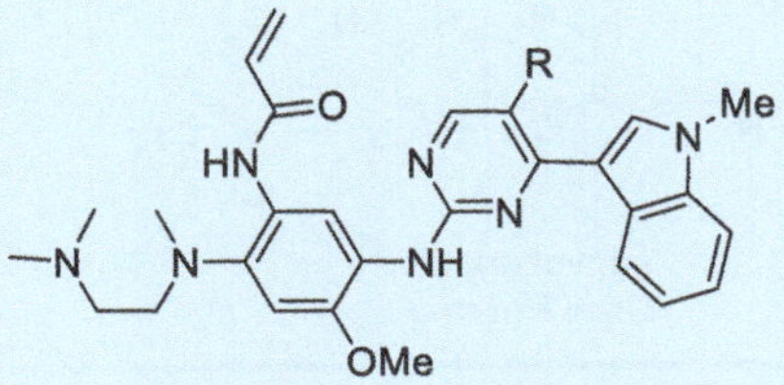

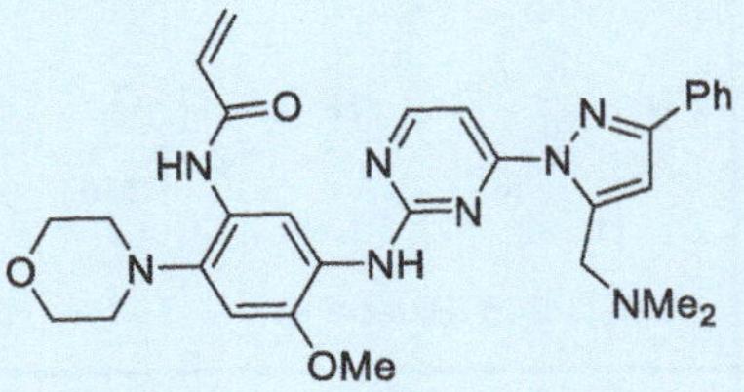

Figure 1. Approved EGFR inhibitors for NSCLC.

Molecules Engineered Against Oncogenic Proteins and Cancer, First Edition. E. J. Corey and Yong-Jin Wu.
© 2024 John Wiley & Sons Inc. Published 2024 by John Wiley & Sons Inc.

The EGFR (HER) family consists of four members: EGFR, HER2, HER3, and HER4. Of those, HER2 is overexpressed in ~15% to 20% of breast tumors, and these (HER2+) tumors can be treated with lapatinib, the first dual EGFR/HER2 inhibitor. Subsequently, more potent irreversible dual inhibitors, neratinib and pyrotinib were brought to market. The newest addition to the class is tucatinib which exhibits higher HER2 selectivity over EFGR than lapatinib or neratinib (which are nearly equipotent inhibitors of HER2 and EGFR). These inhibitors have become powerful weapons in the fight against HER2+ breast cancer.

This chapter describes the research that led to the discoveries of the EGFR/HER inhibitors **13–22** shown in Figures 1 and 2.

Several newer EGFR/HER candidates are shown in Figure 3.[3]

Figure 2. Approved HER2/EGFR inhibitors for HER2+ breast cancer.

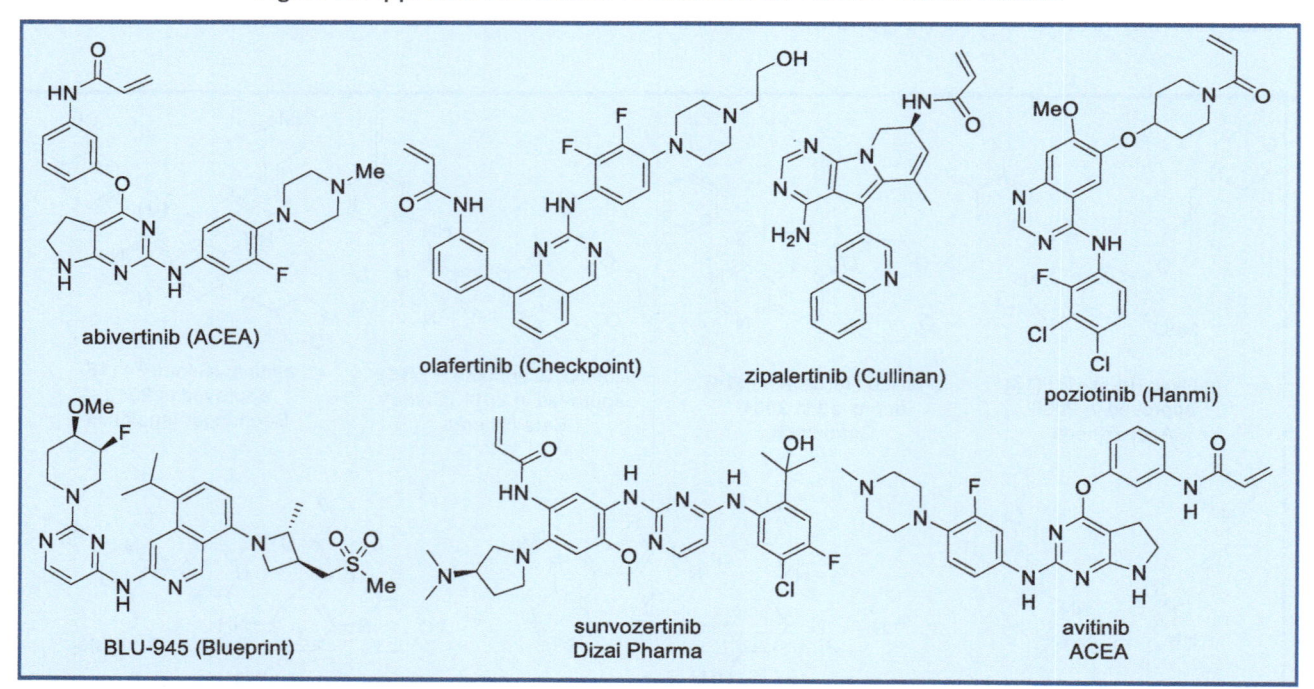

Figure 3. EGFR/HER2 inhibitors in phase II/III trials for NSCLC and breast cancer.

1. J. He, Z. Zhou, X. Sun, Z. Yang, P. Zheng, S. Xu, W. Zhu. The new opportunities in medicinal chemistry of fourth-generation EGFR inhibitors to overcome C797S mutation. *Euro. J. Med. Chem.* **2021**, *210*, 112995.

2. K. Shi, G. Wang, J. Pei, J. Zhang, J. Wang, L. Ouyang, Y. Wang, W. Li. Emerging strategies to overcome resistance to third-generation EGFR inhibitors. *J. Hematol. Oncol.* **2022**, *5*, 94.

3. clinicaltrials.gov.

4.1 Gefitinib (Iressa™) (13)

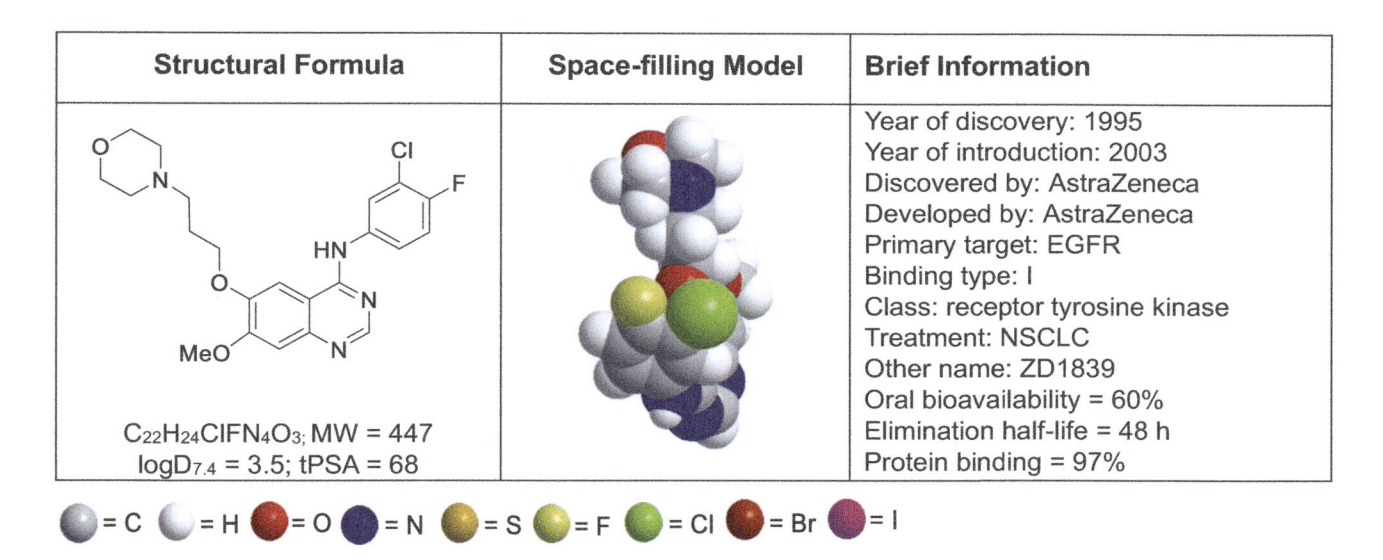

Structural Formula	Space-filling Model	Brief Information
$C_{22}H_{24}ClFN_4O_3$; MW = 447 $logD_{7.4}$ = 3.5; tPSA = 68		Year of discovery: 1995 Year of introduction: 2003 Discovered by: AstraZeneca Developed by: AstraZeneca Primary target: EGFR Binding type: I Class: receptor tyrosine kinase Treatment: NSCLC Other name: ZD1839 Oral bioavailability = 60% Elimination half-life = 48 h Protein binding = 97%

● = C   ○ = H   ● = O   ● = N   ● = S   ● = F   ● = Cl   ● = Br   ● = I

Gefitinib is the first epidermal growth factor receptor (EGFR)-targeted therapy approved in 2003 for the treatment of advanced non-small-cell lung cancer (NSCLC),[1] and in 2015, for the treatment of NSCLC patients whose tumors have EGFR exon 19 deletions or exon 21 (L858R) substitution mutations as detected by an FDA-approved test. Gefitinib is currently available in over 64 countries.

Lung cancer is the third most common cancer in the United States, with approximately 236,740 new cases and 154,050 deaths each year. NSCLC is the most common type of lung cancer, accounting for 82% of all lung cancer diagnoses.

In a vast majority of NSCLC tumors, EGFR is overexpressed, triggering EGFR pathway overactivation, which results in cellular proliferation, differentiation, and survival in the lung. Gefitinib is the first selective EGFR inhibitor to demonstrate clinical efficacy by a blockage of EGFR signaling in target cells[2] and is currently used as the first-line treatment for NSCLC patients harboring activating EGFR mutations through (1) in-frame deletions within exon 19 (E19 dels) or a leucine to arginine substitution (L858R) in E21 (90%) (Fig. 1); and (2) exon 20 insertions (5%).[3] These activating mutations are highly sensitive to first-generation EGFR TKIs, such as gefitinib and erlotinib. Treatment with these agents in EGFR-mutant NSCLC patients results in dramatically high response rates and prolonged progression-free survival relative to conventional standard chemotherapy. The discovery of the first-generation TKIs, gefitinib and erlotinib, to target the EGFR-activating mutations marked the dawn of a new era of personalized treatment for NSCLC.

Unfortunately, despite initial and often dramatic responses of gefitinib treatment, nearly all patients will eventually develop acquired resistance after 10–14 months primarily due to a secondary T790M gatekeeper mutation (Fig. 1). Second-generation irreversible EGFR TKIs were developed to overcome the T790M mutation, beyond the common activating EGFR mutations.[4]

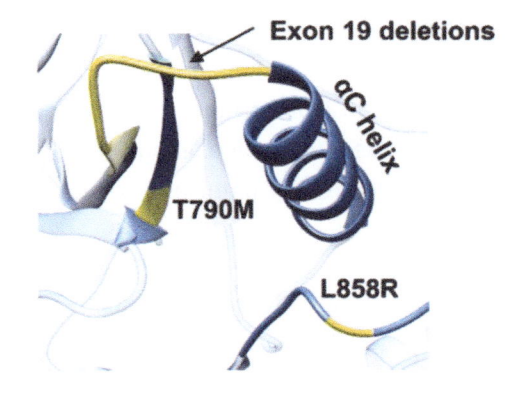

Figure 1. Main-chain location of the three common EGFR kinase somatic mutations (yellow) relative to the αC helix. Adapted from *PloS One*, **2019**,*14*(9), e0222814.

Gefitinib and erlotinib are the first two EGFR-targeted TKIs, and they share a 4-anilinoquinazoline scaffold (Fig. 2). Due to the same mechanism of action and structural similarity, no significant difference between the effectiveness of the two drugs was observed according to one comparative study. Their adverse effects were also similar except several dermal side effects. However, gefitinib was more cost-effective than erlotinib on the basis of pharmacoeconomic analysis.[5]

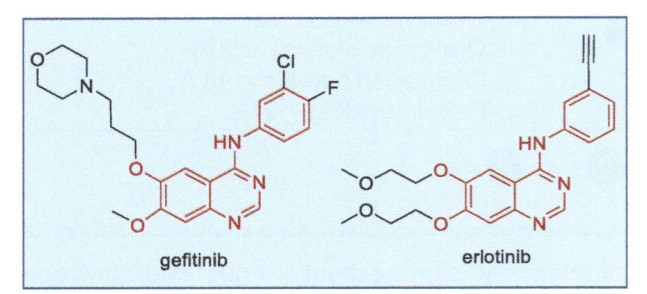

Figure 2. Structures of gefitinib and erlotinib.

EGFR Signaling

EGFR is a member of the HER (also known as ERBB) family of receptors, a subfamily of four closely related receptor tyrosine kinases (RTKs): EGFR (HER1, ERBB1), HER2 (ERBB2), HER3 (ERBB3), and HER4 (ERBB4). It can be activated by binding of its specific ligands, EGF and TGFα. Upon activation by EGF or TGFα in the extracellular domain, EGFR undergoes a transition from an inactive monomeric form to an active homodimer (Fig. 3). This dimerization activates the tyrosine kinase activity in the intracellular domain. As a result, receptor autophosphorylation of EGFR at specific sites occurs, which leads to the initiation of signal-transduction cascades, including RAS/MARK, AKT/PI3K, and JAK/STAT pathways, to promote cell proliferation and survival.[2]

EGFR is expressed on the surface of some normal cells and is involved in cell growth. However, elevated levels of EGFR kinase occur frequently in cancers such as NSCLC, metastatic colorectal cancer, glioblastoma, head and neck cancer, pancreatic cancer, and breast cancer where they drive unregulated growth. Interruption of EGFR signaling by inhibiting intracellular tyrosine kinase activity with ATP-competitive inhibitors can blunt the activity of the EGFR and prevent the growth of EGFR-expressing tumor cells.

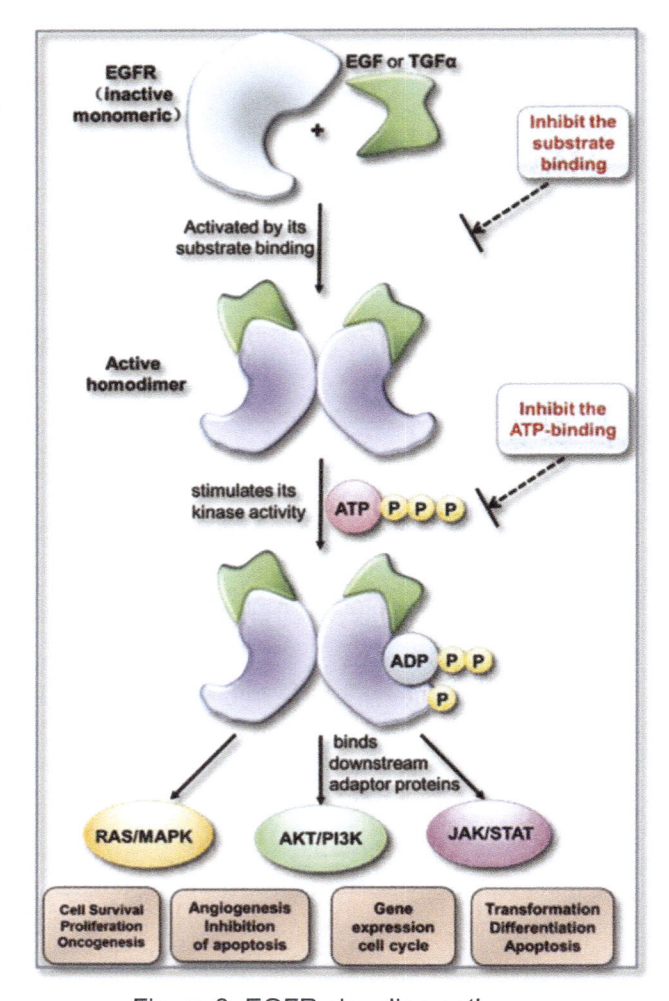

Figure 3. EGFR signaling pathway.

EGFR Activating Mutations in NSCLC[6–11]

In normal cells, EGFR activation is tightly controlled via ligand binding to initiate downstream signaling, but tumor cells can circumvent this requirement by mutation. Specific mutations within the EGFR kinase domain are common in NSCLC, with L858R, E19 dels and exon 20 insertions accounting for 41%, 44%, and 5% of all EGFR mutations, respectively. Variants of E19 dels differ in the length of the deleted amino acid sequence, with the Δ^{746}ELREA^{750} being the most common subtype. Overall, substitutions, deletions and insertions cause structural changes in the local structural features of the kinase domain and alter EGFR substrate specificity. Consequently, the mutated EGFR undergoes both ligand- and dimer-independent autophosphorylation, thus constitutively activating several intracellular pathways to result in proliferation and survival of lung cancer cells (Fig. 4). As such, EGFR inhibition represents an attractive approach to target NSCLC with specific EGFR mutations.

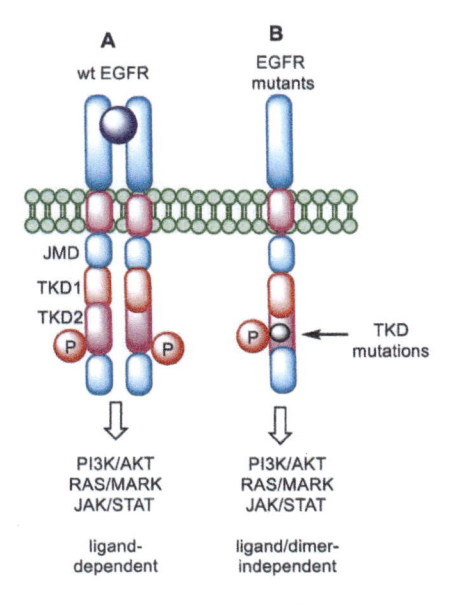

Figure 4. A: ligand-dependent EGFR activation in wild-type (wt); B: ligand- and dimer-independent EGFR activation in various mutants. TKD: tyrosine kinase domain.

The crystal structures of wild-type and L858R mutant can provide some structural insights into constitutive activation of activating mutations. Comparison of the structure of the inactive, wild-type EGFR in complex with lapatinib (Fig. 5) with that of the active, L858R mutant in complex with gefitinib (Fig. 6) reveals that the mutations destabilize the inactive conformation and subsequently lock the mutant in the active state. In the inactive state (Fig. 5), a helical turn (shown in orange), formed from the N-terminal lobe (N-lobe), displaces the αC helix, which is rotated away from the active site (αC-out conformation). Leu858 (shown in red), which is located adjacent to the highly conserved DFG motif in a part of the activation loop, penetrates through a cluster of hydrophobic residues (shown in yellow) from the N-lobe. Substitution of this leucine with a much larger, charged arginine disrupts these favorable hydrophobic interactions. In the active conformation (Fig. 6), the regulatory αC helix is turned inward to the active site with the activation loop (orange) being rearranged. Under the new setting, the hydrophobic cluster no longer exists, and Arg858 (red) is well accommodated. The structural differences between wild-type and mutant EGFR may be responsible for the oncogenic activation of EGFR due to the L858R mutation. In the mutated form, the activation loop is kept in the "DFG-in" active conformation, resulting in constitutively active mutant protein.

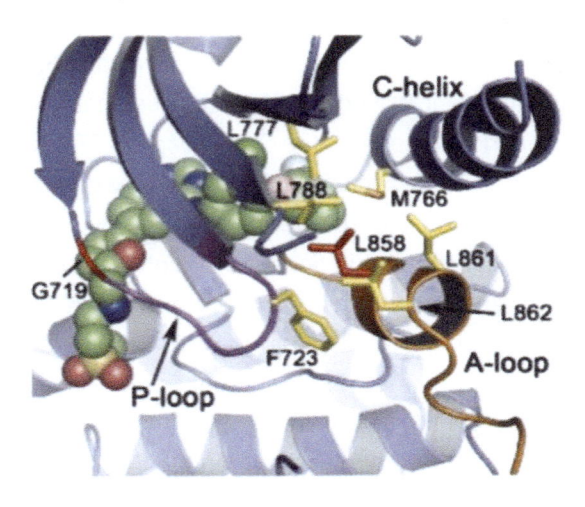

Figure 5. The structure of the inactive, wild-type EGFR in complex with lapatinib. Adapted with permission from *Cancer Cell.* **2007**, *11*, 217–227.

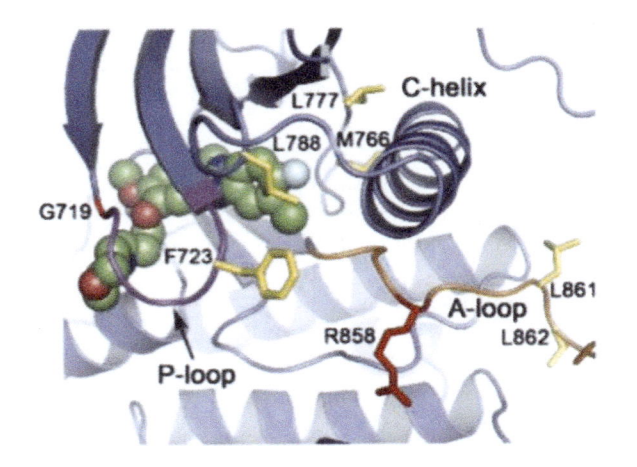

Figure 6. The structure of the active, L858R mutant in complex with gefitinib. Adapted with permission from *Cancer Cell.* **2007**, *11*, 217–227.

The activating mutations are highly responsive to first-generation TKIs such as gefitinib and erlotinib for two reasons. First, the mutant kinases exhibit much stronger affinity than the wt EGFR. For example, gefitinib binds 20-fold more tightly to the L858R mutant than to the wt enzyme. Second, the deletion and L858R mutations dramatically reduce the ATP-binding affinity. Both effects act together to result in the high potency of gefitinib and erlotinib against activating mutants.[7]

T790M Mutations in NSCLC[8]

Although lung cancers caused by activating mutations in EGFR are initially responsive to TKIs, their efficacy is short-lived due to resistance acquired by a second mutation, the gatekeeper T790M mutation in EGFR exon 20. This mutation accounts for about half of all resistance to gefitinib and erlotinib. The T790M mutation was thought to cause steric hindrance and impair the binding of the ATP-binding site of TKIs such as gefitinib. However, gefitinib binds with the T790M-mutant EGFR kinase with K_D of 4.6 nM, nearly as tightly as the L858R mutant (K_D = 2.4 nM) and tighter than the WT kinase (K_D = ~35 nM). Furthermore, the reversible inhibitors are well accommodated in the co-crystal structures of the active and inactive conformations of the EGFR T790M mutants. Taken together, these findings suggest that the gatekeeper-induced steric hindrance may not be the primary mechanism of resistance.

The T790M-mediated acquired resistance has been attributed to the increased ATP affinity which reduces the potency of any ATP-competitive kinase inhibitor. For example, T790M mutation increases the ATP affinity of the oncogenic L858R mutant by more than an order of magnitude. In essence, the activating mutations reduce the ATP affinity, rendering these mutants more responsive to TKI. In contrast, the T790M secondary mutation brings ATP affinity back to the level of the WT kinase. Because reversible TKIs such as gefitinib compete with ATP for binding to the kinase active site, the enhanced ATP affinity may diminish the inhibitor potency.[7]

Discovery of Gefitinib[12]

Compound **1**, an early lead, potently inhibited EGFR in biochemical assays (IC_{50} = 5 nM) and also EGF-stimulated human tumor cell growth (IC_{50} = 50 nM). Compound **1** showed oral activity against human tumor xenografts in mice, but it suffered from rapid clearance with a $t_{1/2}$ of 1 h. The short $t_{1/2}$ results from two major routes of metabolism: (1) oxidation of the methyl group to give the benzyl alcohol **2**; and (2)

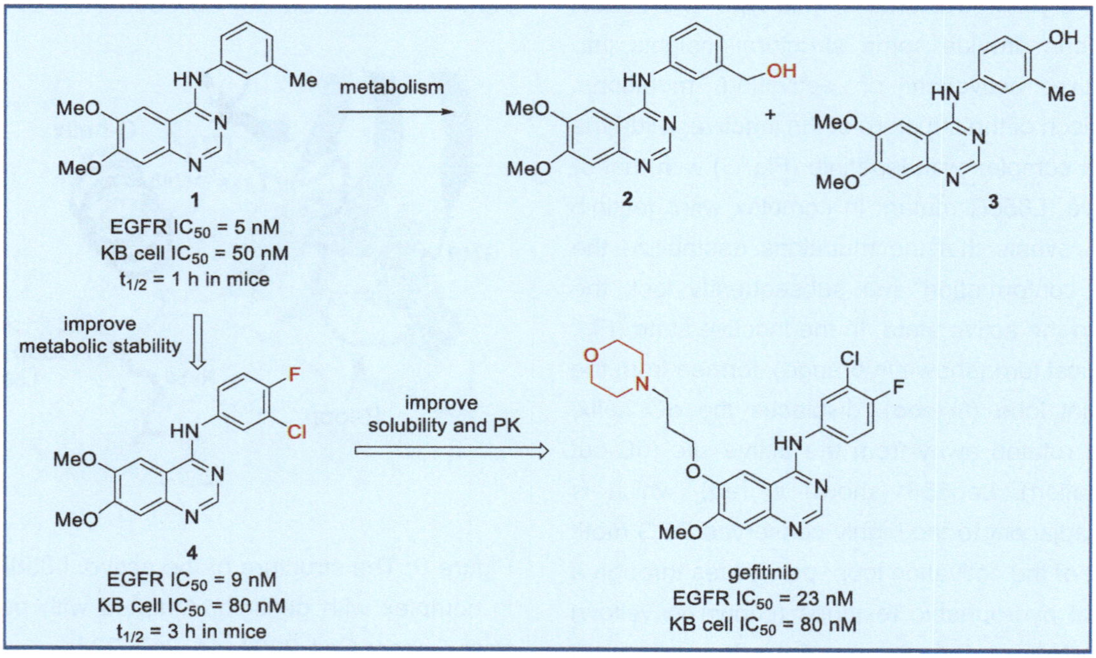

Figure 7. Structural optimization of gefitinib.

oxidation at the para position of the aniline moiety to produce the phenol metabolite **3** (Fig. 7). These two routes of metabolism were blocked by substitution with chlorine in place of the methyl group and introduction of a fluorine at the para position of the aniline to give **4**. Although **4** slightly lost potency in vitro, relative to **1** (EGFR-TK enzyme test, IC_{50} = 9 nM; cellular IC_{50} = 80 nM), it nevertheless showed reduced clearance ($t_{1/2}$ = 3 h) and improved oral efficacy. Next, various basic groups were introduced into the alkoxy side chains to reduce lipophilicity and improve solubility, leading to gefitinib with favorable PK properties in humans (60% oral bioavailability and $t_{1/2}$ = 60 h) compatible with once-daily oral dosing (200 mg).

Gefitinib is a potent and selective EGFR inhibitor with IC_{50} of 23 nM in biochemical assays, vs. >3.7 µM for the related HER-family member HER2. It is also highly selective against other kinases. The high potency and selectivity were also observed in the cellular assays.[13]

Binding Mode[6]

The key interactions of gefitinib with EGFR as determined from X-ray crystallographic data (Fig. 8) are summarized in Figure 9. It binds to the ATP-binding pocket of the active form of EGFR, with the quinazoline N1 in the rear of the ATP-binding pocket, where it hydrogen bonds with the main chain amide of Met793 in the hinge region. Unlike most kinase inhibitors, gefitinib forms only a single hydrogen bond to the hinge. The 3-chloro-4-fluoro aniline substituent extends into the back of the hydrophobic pocket which serves as the ATP-binding cleft. The ortho-fluoro activated chlorine atom attracts the backbone carbonyl oxygen atom of Leu788 within the active site of EGFR and forms a halogen bond (Fig. 9). The methoxy group in the 7 position of the quinazoline is in van der Waals contact with Gly796, while the 6-propylmorpholino group extends into solvent. As a type I inhibitor, gefitinib does not occupy the back cleft.

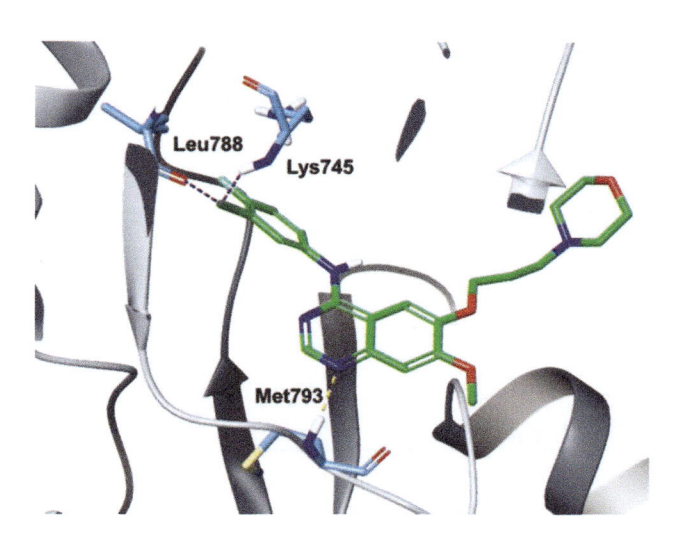

Figure 8. C-crystal structure of gefitinib–EGFR (PDB ID: 2ITY).

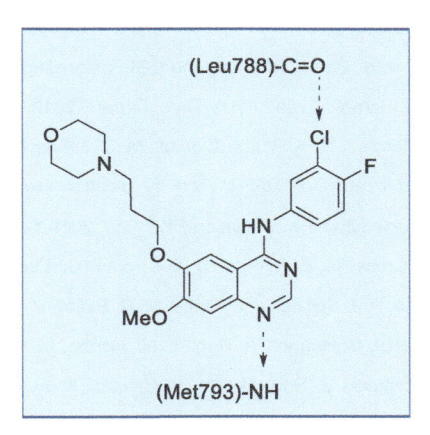

Figure 9. Summary of EGFR–gefitinib interactions based on X-ray co-structural data.

EGFR TKIs Built upon Gefitinib

Gefitinib contains a 4-anilinoquinazoline *core*, which is retained in virtually all second- and third-generation EGFR TKI's (Fig. 10). In essence, the discovery of gefitinib as a molecularly targeted approach ushered in a new beginning of EGFR TKIs for NSCLC treatment.

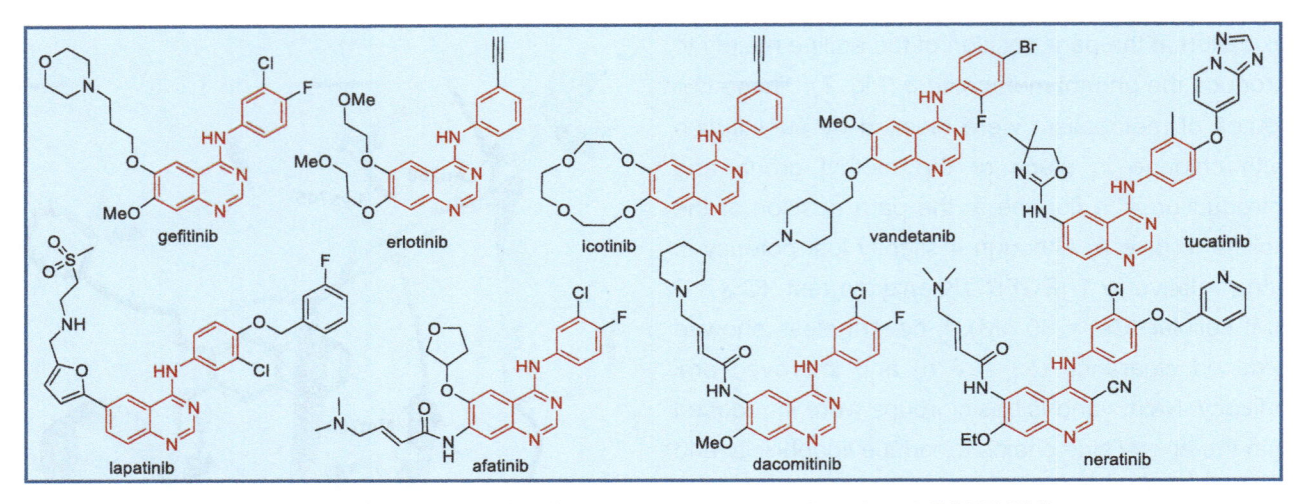

Figure 10. Structures of gefitinib and its structurally related EGFR TKIs.

1. R. S. Herbst, M. Fukuoka, J. Baselga. Gefitinib–a novel targeted approach to treating cancer. *Nat. Rev. Cancer.* **2004**, *12*, 956–965.

2. P. Seshacharyulu, M. P. Ponnusamy, D. Haridas, M. Jain, A. Ganti, S. K. Batra. Targeting the EGFR signaling pathway in cancer therapy. *Expert Opin. Ther. Targets.* **2012**, *16*, 15–31.

3. M. Z. Tamirat, M. Koivu, K. Elenius, M. S. Johnson. Structural characterization of EGFR exon 19 deletion mutation using molecular dynamics simulation. *PloS One*, **2019**, *14*, e0222814.

4. D. A. Cross, S. E. Ashton, S. Ghiorghiu, C. Eberlein, C. A. Nebhan, P. J. Spitzler, J. P. Orme, M. R. Finlay, R. A. Ward, M. J. Mellor, G. Hughes, A. Rahi, V. N. Jacobs, M. Red Brewer, E. Ichihara, J. Sun, H. Jin, P. Ballard, K. Al-Kadhimi, R. Rowlinson, W. Pao. AZD9291, an irreversible EGFR TKI, overcomes T790M-mediated resistance to EGFR inhibitors in lung cancer. *Cancer Discov.* **2014**, *4*, 1046–1061.

5. P. Thomas, B. Vincent, C. George, J. M. Joshua, K. Pavithran, M. Vijayan. A comparative study on erlotinib & gefitinib therapy in non-small cell lung carcinoma patients. *Indian J. Med. Res.* **2019**, *150*, 67–72.

6. C. H. Yun, T. J. Boggon, Y. Li, M. S. Woo, H. Greulich, M. Meyerson, M. J. Eck. Structures of lung cancer-derived EGFR mutants and inhibitor complexes: mechanism of activation and insights into differential inhibitor sensitivity. *Cancer Cell.* **2007**, *11*, 217–227.

7. C. C. Valley, D. J. Arndt-Jovin, N. Karedla, M. P. Steinkamp, A. I. Chizhik, W. S. Hlavacek, B.. S. Wilson, K. A. Lidke, D. S. Lidke. Enhanced dimerization drives ligand-independent activity of mutant epidermal growth factor receptor in lung cancer. *Mol. Biol. Cell.* **2015**, *26*, 4087–4099.

8. C. H. Yun, K. E. Mengwasser, A. V. Toms, M. S. Woo, H. Greulich K. K. Wong, M. Meyerson, M. J. Eck. The T790M mutation in EGFR kinase causes drug resistance by increasing the affinity for ATP. *PNAS.* **2008**, *105*, 2070–2075.

9. X. Zhang, J. Gureasko, K. Shen, P. A. Cole, J. Kuriyan. An allosteric mechanism for activation of the kinase domain of epidermal growth factor receptor. *Cell.* **2006**, *125*, 1137–1149.

10. K. S. Gajiwala, J. Feng, R. Ferre, K. Ryan, O. Brodsky, S. Weinrich, J. C. Kath, A. Stewart. Insights into the aberrant activity of mutant EGFR kinase domain and drug recognition. *Structure.* **2013**, *21*, 209–219.

11. R. Sordella, D. W. Bell, D. A. Haber, J. Settleman. Gefitinib-sensitizing EGFR mutations in lung cancer activate anti-apoptotic pathways. *Science.* **2004**, *305*, 1163–1167.

12. A. J. Barker, K. H. Gibson, W. Grundy, A. A. Godfrey, J. J. Barlow, M. P. Healy, J. R. Woodburn, S. E. Ashton, B. J. Curry, L. Scarlett, L. Henthorn, L. Richards. Studies leading to the identification of ZD1839 (Iressa™): An orally active, selective epidermal growth factor receptor tyrosine kinase inhibitor targeted to the treatment of cancer. *Bioorg. Med. Chem. Lett.* **2001**, *11*, 1911–1914.

13. A. E. Wakeling, S. P. Guy, J. R. Woodburn, S. E. Ashton, B. J. Curry, A. J. Barker, K. H. Gibson. GEFITINIB (Gefitinib): An orally active inhibitor of epidermal growth factor signaling with potential for cancer therapy. *Cancer Res.* **2002**, *62*, 5749–5754.

4.2 Erlotinib (Tarceva™) (14)

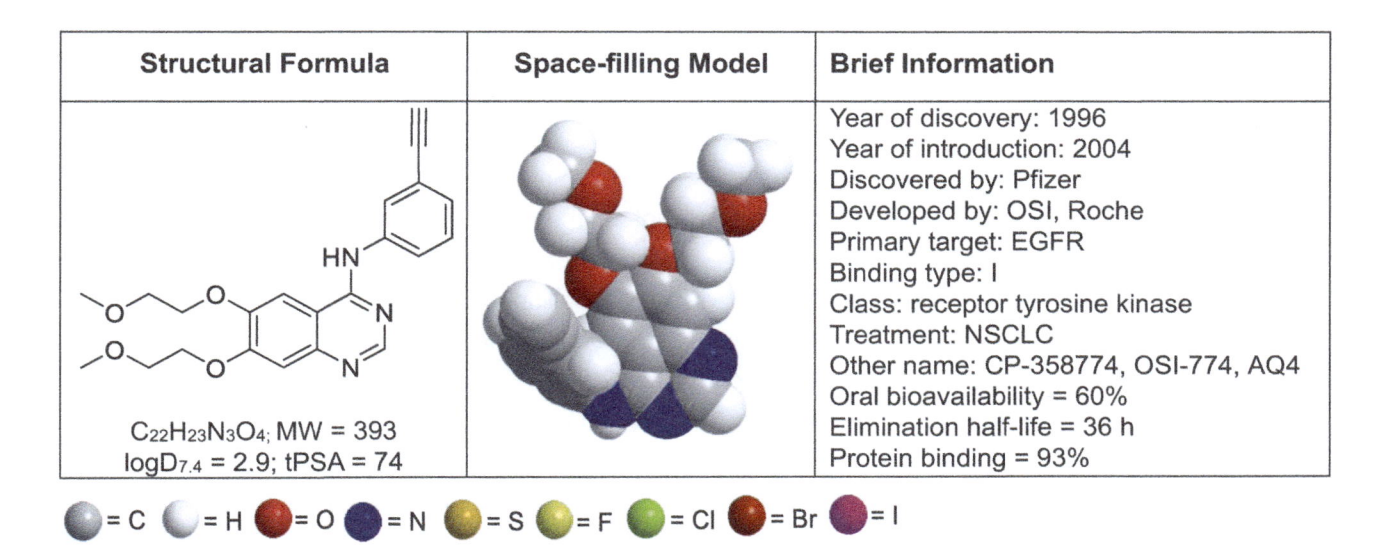

Structural Formula	Space-filling Model	Brief Information
$C_{22}H_{23}N_3O_4$; MW = 393 $logD_{7.4}$ = 2.9; tPSA = 74		Year of discovery: 1996 Year of introduction: 2004 Discovered by: Pfizer Developed by: OSI, Roche Primary target: EGFR Binding type: I Class: receptor tyrosine kinase Treatment: NSCLC Other name: CP-358774, OSI-774, AQ4 Oral bioavailability = 60% Elimination half-life = 36 h Protein binding = 93%

● = C ○ = H ● = O ● = N ● = S ● = F ● = Cl ● = Br ● = I

Erlotinib is the second quinazoline-based reversible epidermal growth factor receptor (EGFR)-targeted therapy first approved in 2004 for the treatment of patients with advanced or metastatic non-small-cell lung cancer (NSCLC) as a second or greater line treatment, after failure of at least one prior chemotherapy regimen. In 2005, it gained approval as a combination therapy with gemcitabine for the first-line treatment of patients with locally advanced, unresectable, or metastatic pancreatic cancer. In 2010, it was approved as a maintenance treatment of NSCLC patients whose disease had not progressed after four cycles of platinum-based first-line chemotherapy. In 2013, it gained further approval as a first-line therapy for metastatic NSCLC patients whose tumors had EGFR exon 19 deletions or exon 21 (L858R) substitution mutations as detected by an FDA-approved test.

The antineoplastic properties of erlotinib derive mainly from inhibition of EGFR with IC$_{50}$ values in the low-nanomolar range in both biochemical and cellular assays. Competing with ATP, erlotinib reversibly binds to the intracellular catalytic domain of EGFR tyrosine kinase, thereby reversibly inhibiting EGFR phosphorylation and blocking the signal transduction events and tumorigenic effects associated with EGFR activation.[1,2]

Erlotinib has been widely used in the treatment of NSCLC harboring activating mutations, and it is on the World Health Organization's List of Essential Medicines. Unfortunately, essentially all NSCLC patients with EGFR-activating mutations develop resistance to the first-generation TKIs such as erlotinib and gefitinib with a median duration of 10–13 months. The most common resistance mechanism, which occurs in 50–60% of patients, involves the development of the exon 20 T790M gatekeeper mutation. This mutation results in the replacement of threonine with the larger methionine near the ATP-binding pocket.[3] The gatekeeper mutation has been overcome by the second- and third-generation EGFR TKIs.

Erlotinib was discovered by Pfizer but developed by OSI Pharmaceuticals in partnership with Genentech and Roche. OSI, originally known as Oncogene Science, was formed in 1983 with specialty in molecular targeted therapies. In 1995, OSI formed alliance with Pfizer to identify EGFR inhibitors as anticancer agents. In 1996, Pfizer took advantage of a gap in Zeneca's EGFR TKI patent application and came up with an alkyne version of gefitinib, CP-358774, later known as erlotinib, which was developed as a treatment for solid tumors by the joint venture. However, in June 2000, Pfizer merged with Warner–Lambert, whose pharmaceutical division also had an irreversible EGFR inhibitor later known as dacomitinib (Fig. 1) (discovered in collaboration with the University of Auckland) in development. In order to meet Federal Trade Commission antitrust requirements for the merger, Pfizer kept dacomitinib but granted all developmental and marketing rights of erlotinib to OSI. This divestiture of the erlotinib portfolio, in effect, gave OSI a royalty-free, cashless license to the drug. More specifically, OSI's ownership of erlotinib increased from 6% to 100% without any extra cost. However, OSI was a small biotechnology company with limited financial prowess and expertise in clinical trials; therefore, it opted to co-develop and co-market erlotinib with Roche and Genentech, and the rest is history.

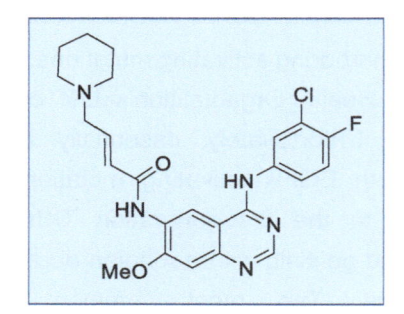

Figure 1. Structure of dacomitinib.

In contrast to the success of erlotinib, dacomitinib, Pfizer's once preferred EFGR inhibitor, did not meet expectations: it underperformed twice against erlotinib in the clinical trials in early 2014. In one, it failed to improve progression-free survival compared with erlotinib, and in the other, also vs. erlotinib, it demonstrated no significant increase in overall survival. Thus, Pfizer went back to conduct additional clinical trials, while erlotinib, which it gave away for free, achieved sales of $1.5 billion in the same year. Four years later – in 2018, the twice-failed dacomitinib was finally approved as the first second-generation irreversible EGFR TKI for NSCLC. However, its efficacy is still just comparable to erlotinib in the EGFR mutated patients, and its sales, expected to reach 182 million in 2022, are only a fraction of erlotinib's peak sales.

Discovery of Erlotinib

On October 20, 1993, Zeneca disclosed a series of quinazoline-based EGFR TKIs with a generic formula of **1**, including **2** as a specific example (Fig. 2) (EP0566226A1, priority date: January 20, 1992), which stimulated significant interests in the area. Because this application had not claimed C≡CH as an aniline substituent, Pfizer quickly exploited this gap and focused on the alkynylanilinoquinazolines, leading to erlotinib with an overall profile comparable to gefitinib.[4] Thanks to the C≡CH group, Pfizer's was able to obtain a granted patent for erlotinib (US005747498A),[5] even though it was filed on May 28, 1996 – three years after Zeneca's original patent was first published.

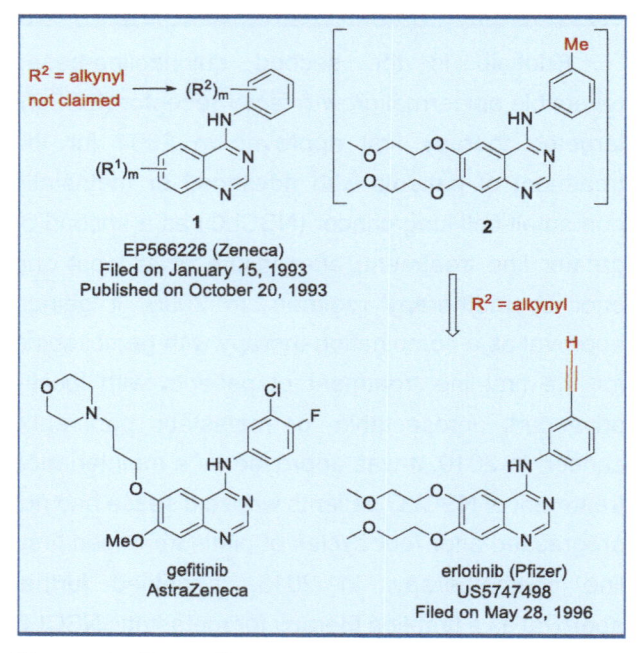

Figure 2. From Zeneca's compound **2** to Pfizer's erlotinib.

Pfizer's patent on erlotinib was not challenged by Zeneca but rather by Mylan, a global generic and specialty pharmaceuticals company, in 2012. Mylan argued: (1) erlotinib was obvious to compound **2** because they differed only at 3' of the aniline ring: C≡CH vs. methyl; and (2) a skilled medicinal chemist seeking new EGFR TKIs would be motivated to select **2** as the lead compound and end up with erlotinib. However, there was no credible evidence to support a seasoned medicinal chemist would focus on a potential problematic C≡CH group. In contrast, medicinal chemists have been advised to avoid this group because of its potential to form reactive metabolites, leading to inactivation of critical liver enzymes. In fact, very few drug molecules contain the C≡CH group, e.g., the contraceptives desogestrel and ethinyl estradiol, and the Parkinson's disease medications selegiline and rasagiline) (Fig. 3), all of which are used in very low dosages (<1.5 vs. 150 mg, once daily for erlotinib). Another supporting evidence is that C≡CH resulted in superior potency to the ethyl group. In the end, the court believed that replacement of the methyl with C≡CH was not obvious, and Mylan failed to invalidate Pfizer's erlotinib patent.[6]

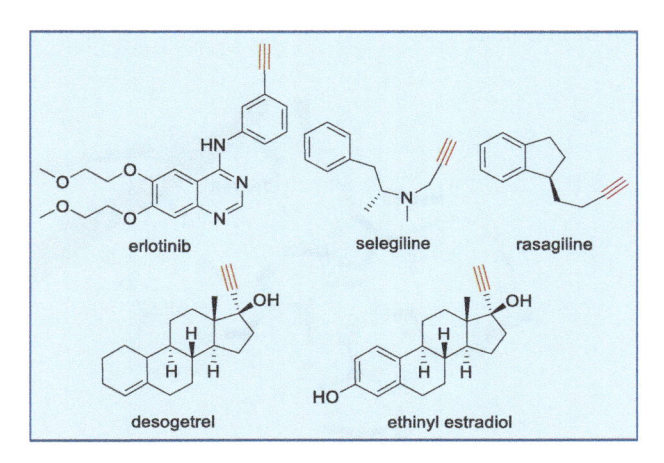

Figure 3. Drugs containing the C≡CH group.

C≡CH as a Bioisostere of Chlorine[7,8]

The ortho-fluoro activated chlorine atom in gefitinib attracts the backbone carbonyl oxygen atom of Leu788 within the active site of EGFR. Both the aromatic fluorine activated C−Cl and aryl C≡CH show electron affinity at the tip of these groups. The polarized CH moiety of the C≡CH group in erlotinib can act as a weak hydrogen bond donor to the carbonyl oxygen of Leu788 of the EGFR protein as shown from their co-crystal structures with EGFR (Fig. 4).

Metabolism of Erlotinib[9,10]

Studies in a single oral dose of [^{14}C]erlotinib hydrochloride in healthy male volunteers revealed three major biotransformation pathways: (1) O-demethylation of the side chains followed byoxidation to the carboxylic acid **5a** (29.4% of dose); (2) oxidation of the acetylene moiety to the carboxylic acid **5b** (21.0%); and (3) hydroxylation of the aromatic ring to **6** (9.6%) (Fig. 5). Formation of carboxylic acids **5a/b** involves CYP3A4-mediated metabolism to produce two reactive metabolites (oxirenes **3a/b** and ketenes **4a/b**) which can alkylate the heme group and/or protein, resulting in the mechanism-based inactivation of CYP3A4. Metabolite **6** contains an electron-rich aminophenol subunit which can form an electrophilic quinone-imine intermediate **7** (a similar intermediate is also formed from gefitinib), a known mechanism-based inactivator of CYP3A4. In addition to mechanism-based inactivation of CYP3A4, these reactive intermediates may also contribute to hepatotoxicity. Despite the potential formation of oxirene and ketene intermediates, C≡CH does not seem to result in any significant toxicity likely because they are quickly converted into stable carboxylic acid metabolites **5a/b**. In fact, erlotinib at an oral dose of 150 mg daily has been shown to be as safe as gefitinib at an oral dose of 250 mg daily in the treatment of NSCLC patients. Gefitinib and erlotinib also show comparable efficacy in the clinic.

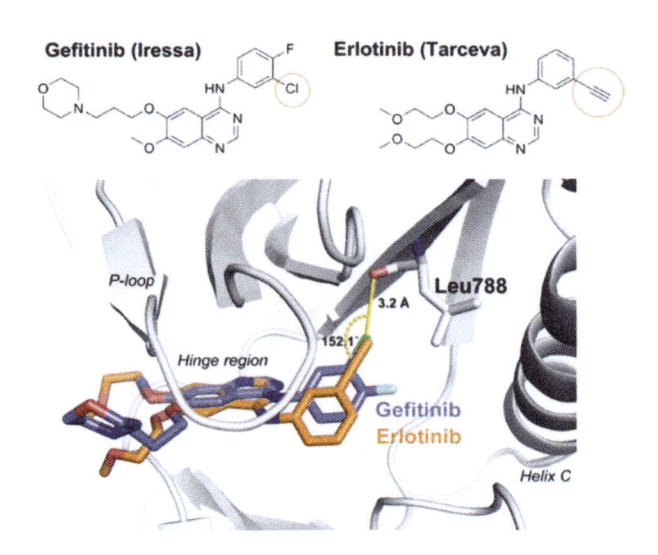

Figure 4. Co-crystal structure of gefitinib bound to EGFR in an overlay with the binding mode of erlotinib. The geometry of the Cl···O halogen bond is highlighted in yellow. Adapted with permission from *J. Med. Chem.* **2020**, *63*, 5625–5663.

Figure 5. Metabolism of erlotinib.

Binding Mode[11,12]

Erlotinib binds to the ATP-binding pocket of active EGFR as type I inhibitor (Fig. 7a, while it behaves as a type I1/2 inhibitor in binding to the inactive DFG-in and C helix-out conformation (Fig. 7b). In both co-crystal structures, the quinazoline N1 hydrogen bonds with the amide nitrogen of Met769 in the hinge region, while the N3 is interacting indirectly with the side chain hydroxyl of Thr766 via a bound water molecule (Figs. 6, 7).

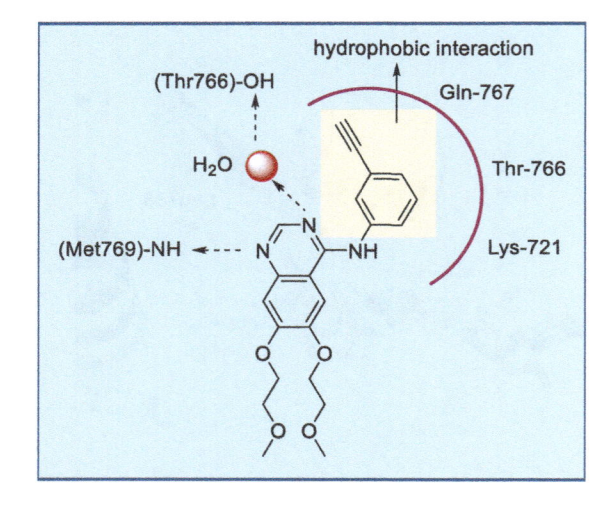

Figure 6. Summary of EGFR–erlotinib interactions based on X-ray structural data.

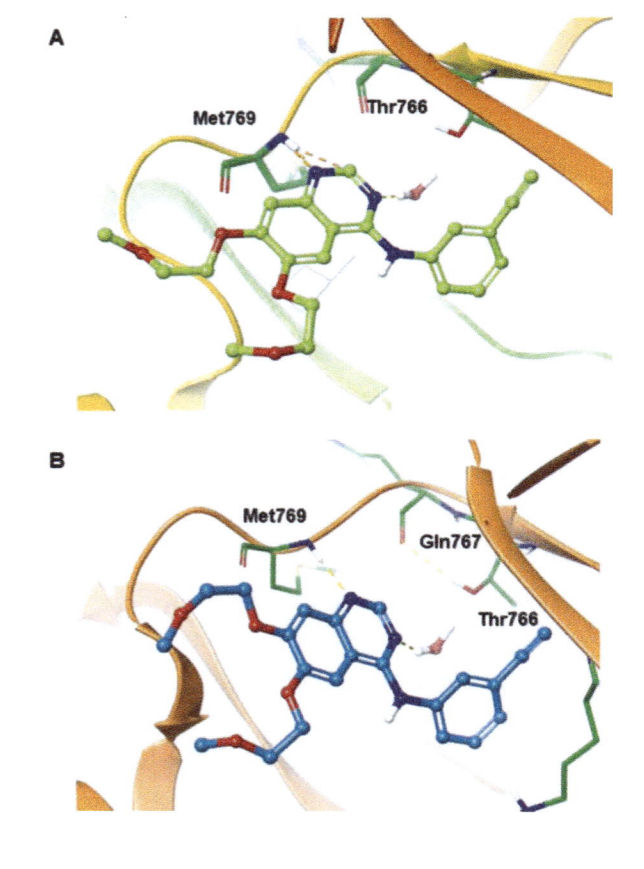

Figure 7. Co-crystal structures of erlotinib with EGFR: (A) active conformation (type I) (PDB ID: 1M17); (B) inactive conformation (type I1/2) (PDB ID: 4HJO).

Perspective

Pfizer boldly utilized the conventionally "undesirable" C≡CH to exploit a gap in Zeneca's patent, leading to erlotinib. Ironically, Pfizer unknowingly gave it away to OSI who turned it into one of the most successful TKIs. After all, the unique C≡CH provides favorable PK properties and causes no significant adverse effects, which enabled Pfizer to secure a granted patent. However, such a useful oncology drug may not be discovered today as contemporary medicinal chemists are routinely reminded to stay away from the C≡CH group.[8] Thus, it is still necessary to keep the C≡CH and other "undesirable" functionalities in the "toolbox" because drug metabolism depends on the entire molecule, not just one individual functionality.

1. K. N. Ganjoo, H. Wakelee. Review of erlotinib in the treatment of advanced non-small cell lung cancer. *Biol. Target Ther.* **2007**, *1*, 335–346.
2. J. Dowell, J. D. Minna, P. Kirkpatrick. Fresh from the pipeline: erlotinib hydrochloride. *Nature Rev. Drug Discov.* **2004**, *4*, 13–14.
3. Z. Tang, R.. Du, S. Jiang, C. Wu, D. S. Barkauskas, J. Richey, J. Molter, M. Lam, C. Flask, S. Gerson, A. Dowlati, L. Liu, Z. Lee, B. Halmos, Y. Wang, J. A. Kern, P. C. Ma. Dual MET–EGFR combinatorial inhibition against T790M-EGFR-mediated erlotinib-resistant lung cancer. *Brit. J. Cancer.* **2008**, *99*, 911–922.
4. M. Burotto, E. E. Manasanch, J. Wilkerson, T. Fojo. Gefitinib and erlotinib in metastatic non-small cell lung cancer: a meta-analysis of toxicity and efficacy of randomized clinical trials. *The oncologist.* **2015**, *20*, 400–410.
5. L. D. Arnold, R. C. Schnur. Alkynyl and azido-substituted 4-anilinoquinazolines. US Patent 5747498A.
6. S. Beeser. Mylan's obviousness challenge of erlotinib compound patent fails – US District Judge. May 03, 2012. https://www.aitkenklee.com/mylans-obviousness-challenge-of-erlotinib-compound-patent-fails-us-district-judge.
7. R. Wilcken, M. O. Zimmermann, M. R. Bauer, T. J. Rutherford, A. R. Fersht, A. C. Joerger, F. M. Boeckler. Experimental and theoretical evaluation of the ethynyl moiety as a halogen bioisostere. *ACS Chem. Biol.* **2015**, *10*, 2725–2732.
8. T. Talele. Acetylene Group, Friend or Foe in Medicinal Chemistry. *J. Med. Chem.* **2020**, *63*, 5625–5663.
9. H. Zhao, S. Li, Z. Yang, Y. Peng, X. Chen, J. Zheng. Identification of ketene-reactive intermediate of erlotinib possibly responsible for inactivation of P450 enzymes. *Drug Metab. Dispos.* **2018**, *46*, 442–450.
10. X. Li, T. M. Kamenecka, M. D. Cameron. Cytochrome P450-mediated bioactivation of the epidermal growth factor receptor inhibitor erlotinib to a reactive electrophile. *Drug Metab Dispos.* **2010**, *38*, 1238–45.
11. J. H. Park, Y. Liu, M. A. Lemmon, R. Radhakrishnan. Erlotinib binds both inactive and active conformations of the EGFR tyrosine kinase domain. *Biochem. J.* **2012**, *448*, 417–423.
12. J. Stamos, M. X. Sliwkowski, C. Eigenbrot. Structure of the epidermal growth factor receptor kinase domain alone and in complex with a 4-anilinoquinazoline inhibitor. *J. Biol. Chem.* **2002**, *277*, 46265–46272.

4.3 Icotinib (Conmana™) (15)

Structural Formula	Space-filling Model	Brief Information
$C_{22}H_{21}N_3O_4$; MW = 391 clogP = 4.2; tPSA = 73		Year of discovery: 2002 Year of introduction: 2011 (China) Discovered by: Beta Pharma Developed by: Zhejiang Beta Pharma Primary targets: EGFR Binding type: I Class: receptor tyrosine kinase Treatment: NSCLC Other name: BPI-2009 Oral bioavailability = 52% Elimination half-life = 5.5 h

● = C   ○ = H   ● = O   ● = N   ● = S   ● = F   ● = Cl   ● = Br   ● = I

Icotinib is a first-generation quinazoline-based reversible EGFR (epidermal growth factor receptor)-targeted therapy approved by the China Food and Drug Administration (CFDA) in 2011 for the second- or third-line therapy of metastatic non-small-cell lung cancer (NSCLC) and in 2014 for the first-line treatment of EGFR-mutant NSCLC patients. Since its approval, icotinib has gained over a third of the market share in lung cancer therapies in China with peak sales of 1.55 billion RMB ($250 million) in 2019.

Icotinib, once touted as China's first home-grown anticancer drug, was actually discovered by Beta Pharma (BP) (formerly Beta Chemicals) which was founded in 1996 in New Haven, CT to provide custom synthesis for pharmaceutical companies. In 2002, while working on various custom intermediate projects, BP exploited a gap in Pfizer's patent of erlotinib[1] by cyclizing its 6,7-bis(2-methoxyethoxy) substituents to form a crown ether derivative, BPI-2009, which became icotinib (Fig. 1).[2] However, BP was a tiny company of ~10 chemists with no financial prowess; therefore, it joined with several investors in China to form a joint venture – Zhejiang Beta Pharma Co. Ltd. (ZBP), a Chinese corporation to develop, test, and market icotinib in China. BP contributed its patent rights to icotinib to the joint venture and received in exchange a 45% interest in ZBP.

ZBP brought icotinib from early development to market launch in China in 8 years, ~2 years faster than in the USA and some other countries. In addition, drug development in China costs substantially less than that in the USA: icotinib's total development expenses are estimated in the range of $20–30 million.[3] These factors along with easy access to a huge NSCLC population have enabled ZBP to offer a more affordable price than its competitors: gefitinib and erlotinib – the two other first-generation EGFR TKIs.

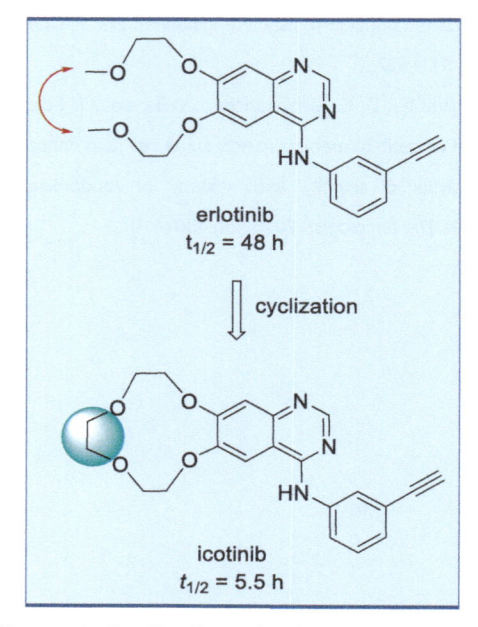

Figure 1. Cyclization of erlotinib to icotinib.

Icotinib was shown to have comparable efficacy to gefitinib for advanced NSCLC, and a marginally lower incidence of drug-related adverse events.[4,5] However, it has a much shorter half-life (5.5 h vs. 48 h for gefitinib and 36 h for erlotinib) due to extensive metabolism at the 12-crown-4 ether moiety (ring-opening and further oxidation); therefore, it needs to be taken three times a day (125 mg dose) as opposed to once daily with erlotinib (150 mg) and gefitinib (250 mg). Despite these disadvantages, it has taken a substantial market share from its competitors in China. For example, from the first quarter of 2013 to the second quarter of 2016, the market share of the purchasing volumes increased from 13% to 40% for icotinib, while that of erlotinib and gefitinib decreased from 35% to 15% and from 52% to 45%, respectively. The success of icotinib has been largely attributed to its significantly lower price: the costs per daily dose of gefitinib, erlotinib and icotinib were $87, $109, and $72, respectively, in the first quarter of 2013.[6]

Since 2016, the Chinese government has implemented a central drug price negotiation policy to make expensive medicines more affordable, and only medicines with successfully negotiated prices can be covered by national health insurance. These negotiations led to the daily costs of all three EGFR-TKIs around $30 by the fourth quarter of 2017, and subsequently, the market shares of gefitinib, erlotinib and icotinib changed to 57%, 8% and 35%, respectively, with gefitinib dominating again in 2018 due to its brand name and once-daily dosing administration.[6]

Although icotinib is a me-too cancer drug, it has played an important role in decreasing prices of gefitinib and erlotinib by creating competition in China. More importantly, the success of icotinib has helped jumpstart the Chinese anticancer drug industry.[3,6]

Icotinib binds reversibly to the ATP-binding site of the EGFR protein, preventing completion of the signal transduction cascade.[7] However, no co-crystal structure of icotinib in complex with EGFR has been disclosed. Because of its close structural similarity to erlotinib, it is highly likely that both inhibitors bind in a similar mode (Fig. 2).

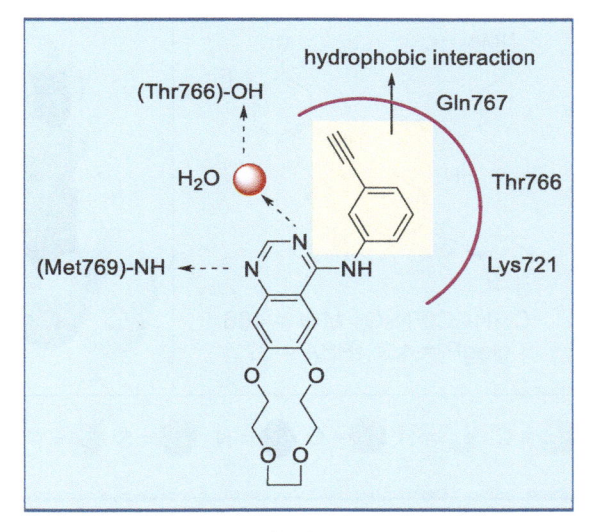

Figure 2. Putative EGFR–icotinib interactions.

1. L. D. Arnold, R. C. Schnur. Alkynyl and azido-substituted 4-anilinoquinazolines. US Patent 5747498A.
2. D. Zhang, G. Xie, C. Davis, Z. Cheng, H. Chen, Y. X. Wang, Z. Z. Cheng. Fused quinazoline derivatives useful as tyrosine kinase inhibitors. US Patent 7078409 B2, **2006**.
3. D. R. Camidge. Icotinib: Kick-starting the Chinese anticancer drug industry. *Lancet. Oncol.* **2013**, *14*, 913–914.
4. Y. Shi, L. Zhang, X. Liu, C. Zhou, L. Zhang, S. Zhang, D. Wang, Q. Li, S. Qin, C. Hu, Y. Zhang, J. Chen, Y. Cheng, J. Feng, H. Zhang, Y. Song, Y. L. Wu, N. Xu, J. Zhou, R. Luo, Y. Sun. Icotinib versus gefitinib in previously treated advanced non-small-cell lung cancer (ICOGEN): A randomised, double-blind phase 3 non-inferiority trial. *Lancet. Oncol.* **2013**, *14*, 953–961.
5. M. Burotto, E. E. Manasanch, J. Wilkerson, T. Fojo. Gefitinib and erlotinib in metastatic non-small cell lung cancer: A meta-analysis of toxicity and efficacy of randomized clinical trials. *Oncologist.* **2015**, *20*, 400–410.
6. Z. Luo, B. Gyawali, S. Han, L. Shi, X. Guan, A. K. Wagner. Can locally developed me-too drugs aid price negotiation? An example of cancer therapies from China. *Semin. Oncol.* **2021**, *48*, 141–144.
7. Y. S. Guan, Q. He, M. Li. Icotinib: Activity and clinical application in Chinese patients with lung cancer. *Expert Opin. Pharmacother.* **2014**, *15*, 717–728.

4.4 Afatinib (Gilotrif™) (16)

Structural Formula	Space-filling Model	Brief Information
$C_{24}H_{25}ClFN_5O_3$; MW = 486 clogP = 4.3; tPSA = 87		Year of discovery: 2000 Year of introduction: 2013 Discovered by: Boehringer Ingelheim Developed by: Boehringer Ingelheim Primary targets: EGFR Binding type: covalent Class: receptor tyrosine kinase Treatment: NSCLC Other name: BIBW2992 Oral bioavailability = not reported Elimination half-life = 37 h Protein binding = 95%

● = C   ○ = H   ● = O   ● = N   ● = S   ● = F   ● = Cl   ● = Br   ● = I

Afatinib is the first second-generation irreversible pan-EGFR (HER) family inhibitor initially approved in 2013 for the treatment of patients with metastatic NSCLC whose tumors have EGFR exon 19 deletions or exon 21 (L858R) substitution mutations as detected by an FDA-approved test, and in 2016 for metastatic, squamous NSCLC progressing after platinum-based chemotherapy. In 2018, afatinib was further approved for first-line treatment of patients with metastatic NSCLC harboring nonresistant EGFR mutations including S768I, L861Q, or G719X. Afatinib represents an important milestone in kinase inhibitor development because it is the first irreversible EGF (HER) family inhibitor approved as a frontline therapy for NSCLC patients with EGFR mutations.[1,2] It is on the World Health Organization's List of Essential Medicines.

The superiority of afatinib to the first-generation EGFR TKIs remains to be further defined. In two separate comparison studies,[3,4] afatinib showed no greater efficacy than gefitinib or erlotinib in the first-line treatment of EGFR-mutant NSCLC, but it was more effective than erlotinib as second-line treatment of patients with advanced squamous cell carcinoma (SCC). However, in several other investigations,[5] afatinib showed significant improvements over gefitinib in the first-line treatment of patients with EGFR-mutant NSCLC. Afatinib also proved efficacious in patients with tumors harboring certain uncommon EGFR mutations, and like gefitinib and erlotinib, it was effective in patients with brain metastases, which is common in NSCLC, especially in EGFR-mutant cancers.

Although EGFR-TKIs have been shown to be highly active against EGFR-mutant NSCLC, resistance to EGFR-TKIs seems inevitable primarily due to a secondary T790M gatekeeper mutation (see Section 4.1). Unfortunately, second-generation irreversible EGFR TKIs such as afatinib fail to overcome EGFR T790M-mediated resistance in patients. In the case of afatinib, the concentrations at which afatinib inhibits lung cancer cells harboring EGFR T790M are not achievable in humans due to dose limiting adverse effects such as skin rash and diarrhea.[6]

Both afatinib and erlotinib caused more frequent rash and other skin toxicities, as well as diarrhea, relative to gefitinib. The main cause of skin toxicity is the targeting effect on wild-type EGFR.

Even though afatinib contains a metabolically sensitive tetrahydrofuranyl moiety, it nonetheless exhibits an elimination half-life of 37 h which allows with once-daily administration of just 40 mg.

Discovery of Afatinib

Gefitinib is a potent, reversible EGFR inhibitor,

which was converted to an irreversible inhibitor afatinib by replacing: (1) 7-methoxy with (tetrahydrofuranyl)oxy; and (2) morpholinoethoxy group at the kinase solvent front with a water-solubilizing Michael acceptor designed to attach covalently to a cysteine residue (Cys773 in EGFR, Cys805 in HER2 and Cys803 in HER4) located in the ATP-binding pocket of these receptors (Fig. 1). The basic dimethylamino group on the Michael acceptor serves to facilitate deprotonation of the SH group of Cys797 and thus accelerate the conjugate addition which inactivates EGFR.[7]

Afatinib is an irreversible, covalent pan-human EGFR (HER) family inhibitor which contains a reactive acrylamide to react with a nearby cysteine SH group of the HER receptor family (Fig. 2). Covalent inhibitors have several potential advantages: (1) increased selectivity in the kinome; (2) longer duration of action; (3) potential for improved therapeutic index due to slower off rates and a lower effective dose; and (4) the potential to minimize drug resistance. However, covalent inhibitors are associated with off-target liabilities. More specifically, there are more than 10 kinases in addition to EGFR that have a reactive cysteine residue in the position equivalent to Cys797 in EGFR, and therefore, it is imperative to achieve adequate selectivity against these kinases.

Afatinib is more active than gefitinib and erlotinib against wild-type EGFR, L858R EGFR or L858R/T790M EGFR in both biochemical (Fig. 1)[6] and cellular assays (Table 1).[8] However, despite its improved activity against gatekeeper mutations T790M (Table 1), unfortunately, afatinib was found to be ineffective at the maximum tolerated dose against lung cancer harboring EGFR T790M positive mutations. So far, osimertinib is the only third-generation EGFR TKI that can overcome EGFR T790M-mediated resistance in patients.

Afatinib is also a highly potent inhibitor of HER2 (IC_{50} = 14 nM) and HER4 (IC_{50} = 1 nM).[6] In addition, afatinib strongly inhibits the proliferation of cancer cell lines driven by multiple HER2 receptor aberrations at concentrations below 100 nM. Because of afatinib's additional activity against HER2, it is still being investigated for breast cancer as well as other EGFR and HER2-driven cancers.

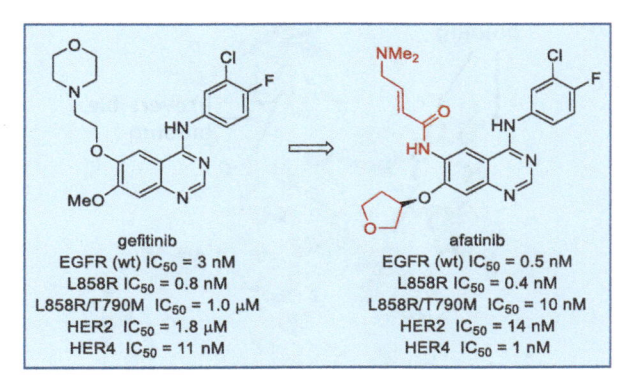

Figure 1. From gefitinib to afatinib.

Table 1. Cellular IC_{50}'s of TKIs[8]

inhibitor	erlotinib	afatinib	osmeritinib
exon 19del	23 nM	0.2 nM	12 nM
L858R	39 nM	0.2 nM	9 nM
exon 19del +T790M	1.6 µM	141 nM	3 nM
L858R +T790M	>10 µM	196 nM	13 nM
wild-type	1.0 µM	31 nM	0.94 µM

Binding Mode[9]

The protein complexes with afatinib shows a bi-lobal architecture characteristic of the protein kinase superfamily (Figs. 2, 3). A rather long hydrogen bond (3.3 Å) is formed between the amide nitrogen of Met793 at the hinge region and the quinazoline core of the inhibitor (Fig. 3). Most importantly, a covalent bond forms between Cys797 at the edge of the active site and the Michael-acceptor group of afatinib with a C–S bond length of 1.82 Å. Interestingly, two molecules of afatinib and one of water are observed to complex with T790M EGFR in the X-ray structure.

The key interactions of afatinib with T790M EGFR based on the X-ray crystal structure is summarized in Fig. 4.

The formation of the covalent bond allows these irreversible inhibitors to achieve greater occupancy of the ATP site as well as higher selectivity for the EGFR family tyrosine kinases relative to the reversible inhibitors. Unfortunately, afatinib and other second-generation irreversible EGFR inhibitors are far from ideal agents because of side effects and insufficient activity against the gatekeeper mutations.

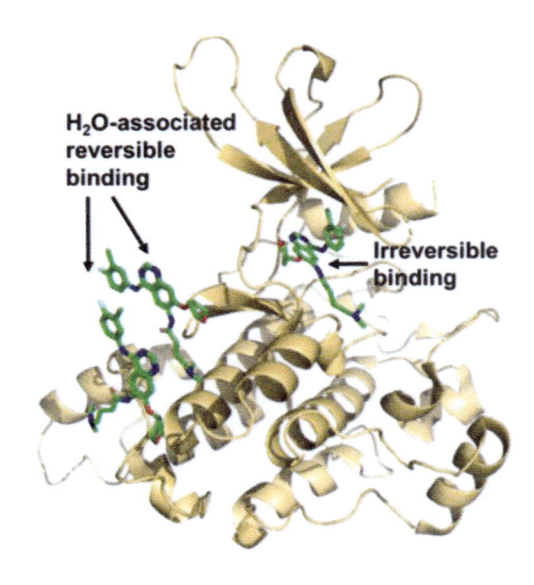

Figure 2. Co-crystal structure of afatinib with T790M EGFR.

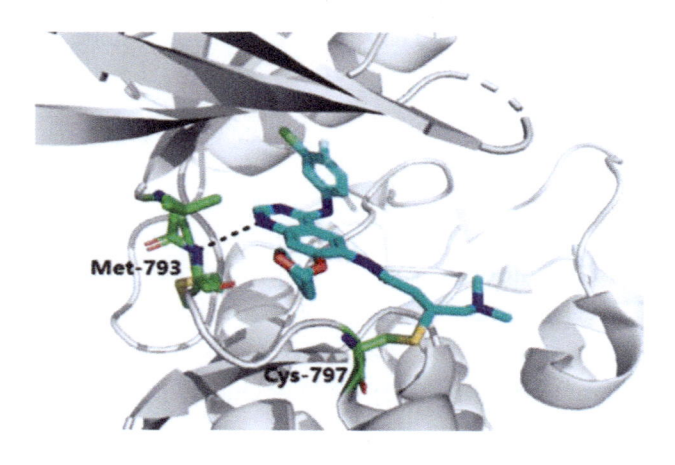

Figure 3. Co-crystal structure of afatinib with T790M EGFR (PDB ID: 4G5J).

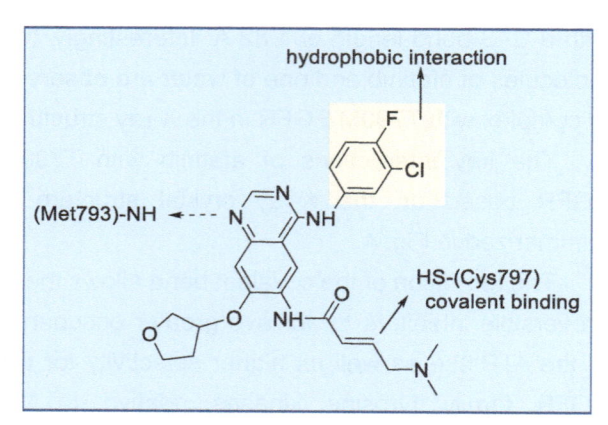

Figure 4. Summary of T790M EGFR–afatinib interactions based on X-ray co-crystal structural data.

1. R. T. Dungo, G. M. Keating. Afatinib: First global approval. *Drugs*. **2013**, *73*, 1503–1515.

2. G. M. Keating. Afatinib: A review of its use in the treatment of advanced non-small cell lung cancer. *Drugs*. **2014**, *74*, 207–221.

3. P. Krawczyk, D. M. Kowalski, R. Ramlau, E. Kalinka-Warzocha, K. Winiarczyk, K. Stencel, T. Powrózek, K. Reszka, K. Wojas-Krawczyk, M. Bryl, M. Wójcik-Superczyńska, M. Głogowski, A. Barinow-Wojewódzki, J. Milanowski, M. Krzakowski. Comparison of the effectiveness of erlotinib, gefitinib, and afatinib for treatment of non-small cell lung cancer in patients with common and rare *EGFR* gene mutations. *Oncol. Lett.* **2017**, *13*, 4433–4444.

4. Z. Yang, A. Hackshaw, Q. Feng, X. Fu, Y. Zhang, C. Mao, T. Tang. Comparison of gefitinib, erlotinib and afatinib in non-small cell lung cancer: A meta-analysis. *Int. J. Cancer*. **2017**, *140*, 2805–2819.

5. R. D. Harvey, V. R. Adams, T. Beardslee, P. Medina. Afatinib for the treatment of EGFR mutation-positive NSCLC: A review of clinical findings. *J. Oncol. Pharm. Pract*. **2020**, *26*, 1461–1474.

6. T. Hirano, H. Yasuda, T. Tani, J. Hamamoto, A. Oashi, K. Ishioka, D. Arai, S. Nukaga, M. Miyawaki, I. Kawada, K. Naoki, D. B. Costa, S. S. Kobayashi, T. Betsuyaku, K. Soejima. In vitro modeling to determine mutation specificity of EGFR tyrosine kinase inhibitors against clinically relevant EGFR mutants in non-small-cell lung cancer. *Oncotarget*. **2015**, *17*, 38789–803.

7. H. R. Tsou, E. G. Overbeek-Klumpers, W. A. Hallett, M. F. Reich, M. B. Floyd, B. D. Johnson, R. S. Michalak, R. Nilakantan, R. C. Discafani, J. Golas, S. K. Rabindran, R. Shen, X. Shi, Y. F. Wang, J. Upeslacis, A. Wissner. Optimization of 6,7-disubstituted-4-(arylamino)quinoline-3-carbonitriles as orally active, irreversible inhibitors of human epidermal growth factor receptor-2 kinase activity. *J. Med. Chem*. **2005**, *48*, 1107–1131.

8. D. Li, L. Ambrogio, T. Shimamura, S. Kubo, M. Takahashi, L. R. Chirieac, R. F. Padera, G. I. Shapiro, A. Baum, F. Himmelsbach, W. J. Rettig, M. Meyerson, F. Solca, H. Greulich, K. Wong. BIBW2992, an irreversible EGFR/HER2 inhibitor highly effective in preclinical lung cancer models. *Oncogene*. **2008**, *27*, 4702–4711.

9. F. Solca, G. Dahl, A. Zoephel, G. Bader, M. Sanderson, C. Klein, O. Kraemer, F. Himmelsbach, E. Haaksma, G. R. Adolf. Target binding properties and cellular activity of afatinib (BIBW 2992), an irreversible ErbB family blocker. *J. Oncol. Pharm. Pract*. **2012**, *343*, 342–350.

4.5 Dacomitinib (Vizimpro™) (17)

Structural Formula	Space-filling Model	Brief Information
$C_{24}H_{25}N_5O_2ClF$; MW = 470 clogP = 5.7; tPSA = 78		Year of discovery: 2004 Year of introduction: 2018 Discovered by: Pizer Developed by: Pfizer Primary targets: EGFR/HER2-4 Binding type: covalent Class: receptor tyrosine kinase Treatment: NSCLC with EGFR alterations Other name: PF-00299804 Oral bioavailability = 80% Elimination half-life = 70 h Protein binding = 98%

● = C   ○ = H   ● = O   ● = N   ● = S   ● = F   ● = Cl   ● = Br   ● = I

Dacomitinib is the second FDA-approved irreversible pan-EGFR (HER) family inhibitor (five years after afatinib) indicated for the treatment of metastatic non-small cell lung cancer (NSCLC) with EGFR exon 19 deletion or exon 21 L858R substitution mutations as detected by an FDA-approved test.[1,2]

Dacomitinib was originally developed to target EGFR-activating mutations and T790M mutants. However, as a monotherapy, it failed to overcome T790M-mediated resistance in the clinic because it has only modest potency (IC_{50} = 100–500 nM) and the concentrations required to suppress T790M activity are not achievable in humans[2] due to dose-limiting toxicity caused by its potent inhibition of wild-type EGFR (Table 1).[3]

Discovery of Dacomitinib [4–7]

Dacomitinib was derived from gefitinib, a potent, reversible EGFR inhibitor, by replacing the morpholinoethoxy group of the kinase solvent front with a water-solubilizing Michael acceptor designed to attach covalently to a cysteine residue (Cys773 in EGFR and Cys805 in HER2) located in the ATP-binding pocket of these receptors (Fig. 1). The piperidine ring at the end of the Michael acceptor is incorporated to accelerate the conjugative addition. Of special note is the quinoline analog, EKB-569, lacks HER2 activity despite remarkable structural similarity, whereas neratinib, a pan-EGFR inhibitor for breast cancer, requires a terminal pyridin-3-ylmethoxy group to gain HER2 potency.[7–10]

gefitinib
EGFR IC_{50} = 0.51 µM
HER2 IC_{50} = 1.6 µM

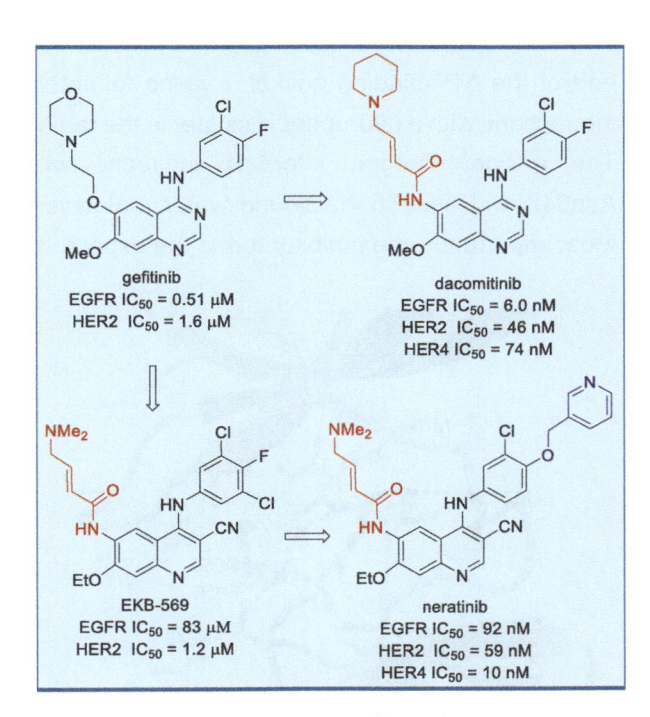

dacomitinib
EGFR IC_{50} = 6.0 nM
HER2 IC_{50} = 46 nM
HER4 IC_{50} = 74 nM

EKB-569
EGFR IC_{50} = 83 µM
HER2 IC_{50} = 1.2 µM

neratinib
EGFR IC_{50} = 92 nM
HER2 IC_{50} = 59 nM
HER4 IC_{50} = 10 nM

Figure 1. Structural optimization of dacomitinib vs. neratinib.

Dacomitinib inhibits EGFR, HER2, and HER4 with IC_{50} values of 6.0, 45.7, and 73.7 nM respectively, vs. 3.1, 343, and 476 nM for gefitinib and 0.56, 512, and 790 nM for erlotinib (Table 1). The in vitro IC_{50} of dacomitinib against the cell lines HCC827 (exon 19 del E746_A750), and H3255 (L858R substitution) is 2 nM and 7 nM, respectively, vs. 8 nM and 75 nM for gefitinib. Furthermore, dacomitinib shows modest potency with IC_{50} values in the 100–500 nM range for T790M, whereas gefitinib lacks activity for this mutant.[5]

Dacomitinib is characterized by high oral bioavailability (80%) and long half-life (70 h), leading to a low oral dosage of only 45 mg tablet once per day, relative to 250 mg for gefitinib and 150 mg for erlotinib.

Table 1. Enzyme and cellular IC_{50}'s (nM) of TKIs[3]

inhibitor	EGFR (wt)	HER2 (wt)	HER4 (wt)	Del19 (cell)	L858R (cell)
gefitinib	3.1 nM	343 nM	476 nM	4.8 nM	26 nM
erlotinib	0.56 nM	512 nM	790 nM	4.9 nM	16 nM
afatinib	0.5 nM	14 nM	1 nM	0.9 nM	4 nM
dacomitinib	6.0 nM	45.7 nM	73.7 nM	<1 nM	2.6 nM
osimertinib	184 nM	116 nM	46-67 nM	1.1 nM	9 nM

Binding Mode[11]

In the crystal structure of dacomitinib in complex with the EGFR gatekeeper T790M mutant (Fig. 2), the kinase adopts an inactive conformation, with the αC helix displaced. The quinazoline ring occupies the ATP-binding site, with the ring nitrogen forming a single hydrogen bond with the amide NH of Met793 in the hinge. The fluorochloroaniline group approaches the deep end of the ATP-binding pocket, making favorable interactions with a hydrophobic cluster in the region. The aniline nitrogen interacts indirectly with Asn842 and Asp855 via bound water molecules. Most importantly, the inhibitor forms the expected covalent bond with Cys797 at the edge of the active site cleft and the butenamide Michael-acceptor group, making binding irreversible (Fig. 3).

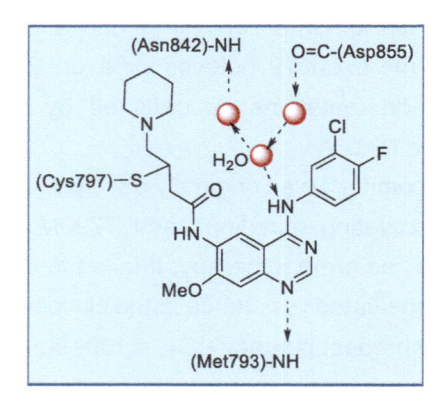

Figure 3. Summary of dacomitinib–EGFR T790M interactions based on the co-crystal structure.

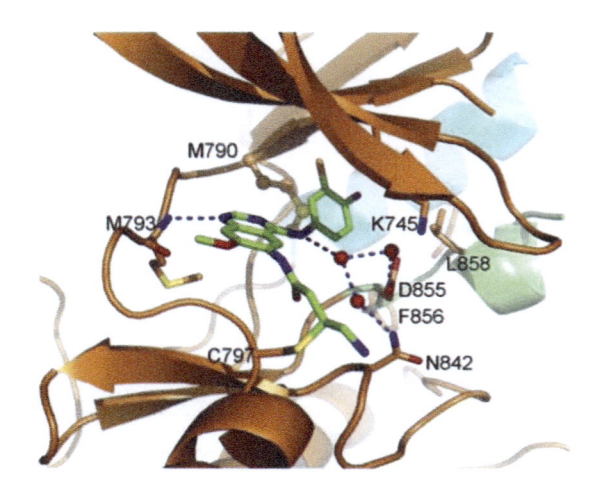

Figure 2. Co-crystal structure of dacomitinib in complex with the EGFR T790M mutant. Adapted with permission from *Structure*. **2013**, *21*, 209–219.

1. M. Shirley. Dacomitinib: First global approval. *Drugs*. **2018**, *78*, 1947–1953.

2. D. A. Cross, S. E. Ashton, S. Ghiorghiu, C. Eberlein, C. A. Nebhan, P. J. Spitzler, J. P. Orme, M. R. Finlay, R. A. Ward, M. J. Mellor, G. Hughes, A. Rahi, V. N. Jacobs, M. Red Brewer, E. Ichihara, J. Sun, H. Jin, P. Ballard, K. Al-Kadhimi, R. Rowlinson, W. Pao. AZD9291, an irreversible EGFR TKI, overcomes T790M-mediated resistance to EGFR inhibitors in lung cancer. *Cancer Discov*. **2014**, *4*, 1046–1061.

3. D. Lavacchi, F. Mazzoni, G. Giaccone. Clinical evaluation of dacomitinib for the treatment of metastatic non-small cell lung cancer (NSCLC): Current perspectives. *Drug Des. Devel. Ther*. **2019**, *13*, 3187–3198.

4. D. Lavacchi, F. Mazzoni, G. Giaccone. Clinical evaluation of dacomitinib for the treatment of metastatic non-small cell lung cancer (NSCLC): Current perspectives. *Drug Des. Devel. Ther.* **2019**, *13*, 3187–3198.

5. S. H. Ou, R. A. Soo. Dacomitinib in lung cancer: A "lost generation" EGFR tyrosine-kinase inhibitor from a bygone era? *Drug Des. Devel. Ther.* **2015**, *9*, 5641–5653.

6. J. A. Engelman, K. Zejnullahu, C. M. Gale, E. Lifshits, A. J. Gonzales, T. Shimamura, F. Zhao, P. W. Vincent, G. N. Naumov, J. E. Bradner, I. W. Althaus, L. Gandhi, G. I. Shapiro, J. M. Nelson, J. V. Heymach, M. Meyerson, K. K. Wong, P. A. Jänne. PF00299804, an irreversible pan-ERBB inhibitor, is effective in lungCancer models with EGFR and ERBB2 mutations that are resistantto gefitinib. *Cancer Res.* **2007**, *67*, 11924–11932.

7. H. R. Tsou, E. G. Overbeek-Klumpers, W. A. Hallett, M. F. Reich, M. B. Floyd, B. D. Johnson, R. S. Michalak, R. Nilakantan, R. C. Discafani, J. Golas, S. K. Rabindran, R. Shen, X. Shi, Y. F. Wang, J. Upeslacis, A. Wissner. Optimization of 6,7-disubstituted-4-(arylamino)quinoline-3-carbonitriles as orally active, irreversible inhibitors of human epidermal growth factor receptor-2 kinase activity. *J. Med. Chem.* **2005**, *48*, 1107–1131.

8. A. Wissner, E. Overbeek, M. F. Reich, M. B. Floyd, B. D. Johnson, N. Mamuya, E. C. Rosfjord, C. Discafani, R. Davis, X. Shi, S. K. Rabindran, B. C. Gruber, F. Ye, W. A. Hallett, R. Nilakantan, R. Shen, Y. F. Wang, L. M. Greenberger, H. R. Tsou. Synthesis and structure-activity relationships of 6,7-disubstituted 4-anilinoquinoline-3-carbonitriles. The design of an orally active, irreversible inhibitor of the tyrosine kinase activity of the epidermal growth factor receptor (EGFR) and the human epidermal growth factor receptor-2 (HER-2). *J. Med. Chem.* **2003**, *46*, 49–63.

9. A. Wissner, M. B. Floyd, S. K. Rabindran, R. Nilakantan, L. M. Greenberger, R. Shen, Y. F. Wang, H. R. Tsou. Syntheses and EGFR and HER-2 kinase inhibitory activities of 4-anilinoquinoline-3-carbonitriles: Analogues of three important 4-anilinoquinazolines currently undergoing clinical evaluation as therapeutic antitumor agents. *Bioorg. Med. Chem. Lett.* **2002**, *12*, 2893–2897.

10. A. Wissner, D. M. Berger, D. H. Boschelli, M. B. Floyd, Jr., L. M. Greenberger, B. C. Gruber, B. D. Johnson, N. Mamuya, R. Nilakantan, M. F. Reich, R. Shen, H. R. Tsou, E. Upeslacis, Y. F. Wang, B. Wu, F. Ye, N. Zhang. 4-Anilino-6,7-dialkoxyquinoline-3-carbonitrile inhibitors of epidermal growth factor receptor kinase and their bioisosteric relationship to the 4-anilino-6,7-dialkoxyquinazoline inhibitors. *J. Med. Chem.* **2000**, *43*, 3244.

11. K. S. Gajiwala, J. Feng, R. Ferre, K. Ryan, O. Brodsky, S. Weinrich, J. C. Kath, A. Stewart. Insights into the aberrant activity of mutant EGFR kinase domain and drug recognition. *Structure.* **2013**, *21*, 209–219.

4.6 Osimertinib (Tagrisso™) (18) and Analogs

Structural Formula	Space-filling Model	Brief Information
$C_{29}H_{33}N_7O_2$; MW = 500 $logD_{7.4}$ = 3.4; tPSA = 85		Year of discovery: 2011 Year of introduction: 2015 Discovered by: AstraZeneca Developed by: AstraZeneca Primary target: EGFR Binding type: covalent Class: receptor tyrosine kinase Treatment: NSCLC Other name: AZD9291 Oral bioavailability = 70% Elimination half-life = 48 h Protein binding = 94.7%

● = C   ○ = H   ● = O   ● = N   ● = S   ● = F   ● = Cl   ● = Br   ● = I

Osimeritinib is an irreversible third-generation epidermal growth factor receptor (EGFR) inhibitor approved first in 2015 as a second-line treatment option for patients with advanced EGFR-T790M mutation-positive non-small cell lung cancer (NSCLC) and then in 2018, for the first-line treatment of patients with metastatic NSCLC whose tumors have EGFR exon 19 deletions or exon 21 L858R mutations, as detected by an FDA-approved test. In 2020, osimeritinib was further approved for the adjuvant treatment of adult patients with early-stage EGFR mutated NSCLC after tumor resection with curative intent.[1]

Osimertinib has also been shown to result in the complete remission of multiple brain metastases of EGFR-mutated NSCLC without radiation therapy when used as the first-line treatment. In addition, radiation therapy for brain metastases can be deferred or even withheld especially when osimertinib is used as first-line treatment, because complete CNS remission can be occasionally achieved as early as one month even in patients with as many as 20 lesions. Therefore, first-line treatment with osimertinib is superior to conventional therapy for EGFR-mutated NSCLC with CNS metastases, which is common in EGFR-mutant cancers.[2]

Osimeritinib remains the only FDA-approved therapy for NSCLC patients who acquire EGFR T790M mutation in response to first-generation EGFR TKI treatments. Because of its superior safety and efficacy in patients with EGFR-mutant NSCLC regardless of T790M mutation status, osimertinib has quickly become a drug of choice for oncologists. Its sales rocketed to $5 billion in 2021 and $5.4 billion in 2022, only five years after launch, and it still has significant growth potential as an adjuvant, which would be given as an additional treatment to lower the risk of cancer recurrence.

Despite its broad coverage of T790M-resistant mutants, osimertinib exhibits only limited activity against resistant EGFR exon 20 insertion mutations. Interestingly, just a single substitution on the pyrimidine ring led to mobocertinib (Fig. 1), a next generation of EGFR TKI (see Section 4.7).

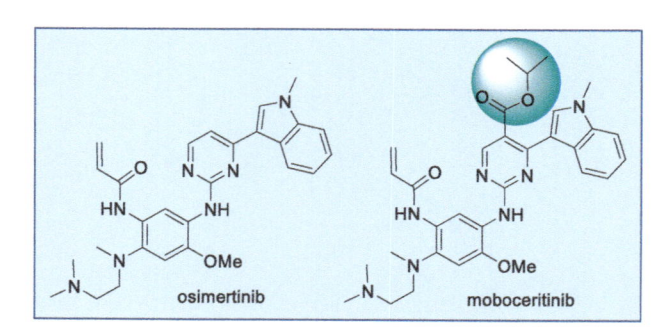

Figure 1. Osimertinib vs. moboceritinib.

Irreversible Inhibitors for T790M Acquired Resistance[3–12]

Lung cancers caused by activating mutations in EGFR are initially responsive to first- and second-generation tyrosine kinase inhibitors (TKIs), but the efficacy is often short-lived primarily due to structural changes that result from the gatekeeper T790M mutation in EGFR exon 20 located in the ATP-binding cleft. This mutation is frequently found in NSCLC carrying a TKI-sensitive EGFR activating mutation (e.g., in-frame deletion in exon 19 or an L858R point mutation in exon 21, see Section 4.1 for a detailed discussion on these mutations) that have developed resistance after TKI treatment. The first- and second-generation EGFR TKIs (e.g., gefitinib, erlotinib, and afatinib) are highly effective against one of the activating EGFR mutations, with objective response rates of 50–70%. However, the acquired resistance due to the EGFR T790M mutation is found in approximately 50–60% of resistant cases.

The T790M mutant exhibits tyrosine phosphorylation levels similar to wild-type (wt) EGFR, whereas the T790M/L858R double mutation significantly increases phosphorylation, relative to the L858R mutation alone. Thus, a combination of T790M and activating EGFR kinase domain mutation results in a substantial increase of its catalytic phosphorylating activity, producing a more potent kinase.

The T790M mutation was initially thought to cause steric hindrance and impair the binding of the ATP-binding site of TKIs such as gefitinib. However, the T790M mutant maintains high affinity to gefitinib. For example, the T790M mutant EGFR kinase binds gefitinib with a K_D of 4.6 nM, an affinity similar to that with L858R mutant (K_D = 2.4 nM). In contrast, introduction of the T790M mutation enhances the ATP affinity of the oncogenic L858R mutant by more than an order of magnitude, and the increased ATP affinity is largely responsible for the T790M mutation conferred drug resistance.

The increased ATP-binding of the oncogenic L858R/T790M double mutant has less pronounced impact on the irreversible inhibitors because they form covalent binding. Once bound covalently, they permanently block the active binding site and no longer compete with ATP. Thus, the covalent, irreversible inhibitors such as osimertinib are well suited to combat T790M resistance mechanism.

EGFR Wild-type Selectivity[13,14]

Second-generation TKIs incorporate a Michael acceptor to covalently target Cys797 in EGFR at the lip of the ATP-binding site. As this cysteine is present in all relevant forms of EGFR, these irreversible TKIs not only increase activity against the activating and resistant mutations but also against WT EGFR (Table 1). Inhibition of wt EGFR is not related to efficacy but rather causes adverse effects such as skin rash and diarrhea. Because of these dose limiting adverse effects, second-generation irreversible TKIs such as afatinib are unable to achieve the concentrations required to inhibit lung cancer cells harboring EGFR T790M in humans; therefore, they have not provided any clinical benefit in T790M positive patients, despite their in vitro activity against T790M mutations. This failure has stimulated the search for third-generation 790M mutant-selective EGFR TKIs that spare wt EGFR.

Discovery of Osimertinib[13–18]

The indole/N-(2-methoxyphenyl)pyrimidinamine scaffold of osimertinib originated from the screening hit compound **1** (Fig. 2), a potent reversible inhibitor of the L858R-T790M double mutant (DM) (IC_{50} = 9 nM in the biochemical assay) with 88-fold selectivity over wt EGFR. Unfortunately, in the double-mutant cellular assay, **1** showed an IC_{50} value of 0.77 µM, ~90-fold drop from the biochemical data. This discrepancy is likely driven by the high ATP affinity of this DM enzyme and the high ATP concentration present in the cellular environment, a primary resistance mechanism to ATP competitive EGFR inhibitors such as gefitinib. One approach of enhancing cellular activity is to convert a reversible inhibitor into an irreversible, non-ATP competitive inhibitor to partially overcome the effect of the increased affinity of ATP for the mutant kinase. Indeed, incorporation of an acrylamide warhead to C1 of lead compound **1** and removal of the piperazine substituent at C2 (to give **2**) increased cellular activity to 22 nM against DM. Compound **2** was also highly active against EGFR sensitizing/activating mutant (AM) exon del19 (IC$_{50}$ = 29 nM) while maintaining 25-fold WT selectivity. To mitigate the high lipophilicity of **2**, the acrylamide terminal methylene was substituted with a basic CH$_2$NMe$_2$ group to give **3** without compromising activity and selectivity. The CH$_2$NMe$_2$, which is present in several second-generation irreversible EGFR TKIs such as afatinib, also serves to facilitate deprotonation of the SH group of Cys797 and thus accelerate the conjugate addition to inactivate the mutant protein. To further reduce lipophilicity and improve aqueous solubility, the basic functionality was relocated from C1 to the C2 kinase solvent-front area as in the original lead compound **1** (which contains a piperazine at C2), giving rise to **4** with drastically improved DM cellular activity and wt selectivity (IC$_{50}$ = 0.2 nM; selectivity = 55). However, **4** lacked oral bioavailability (F = 1% in rats) and also potently inhibited IGF1R (IC$_{50}$ = 7 nM), potentially contributing dose limiting toxicity. These issues were addressed by: (1) removal of the chlorine substituent; and (2) N-methylation of the indole free NH to give osimertinib with 32-fold margin for wt EGFR and favorable PK properties (F = 70% in humans; $t_{1/2}$ = 40 h) compatible with once-daily administration (80 mg tablet).

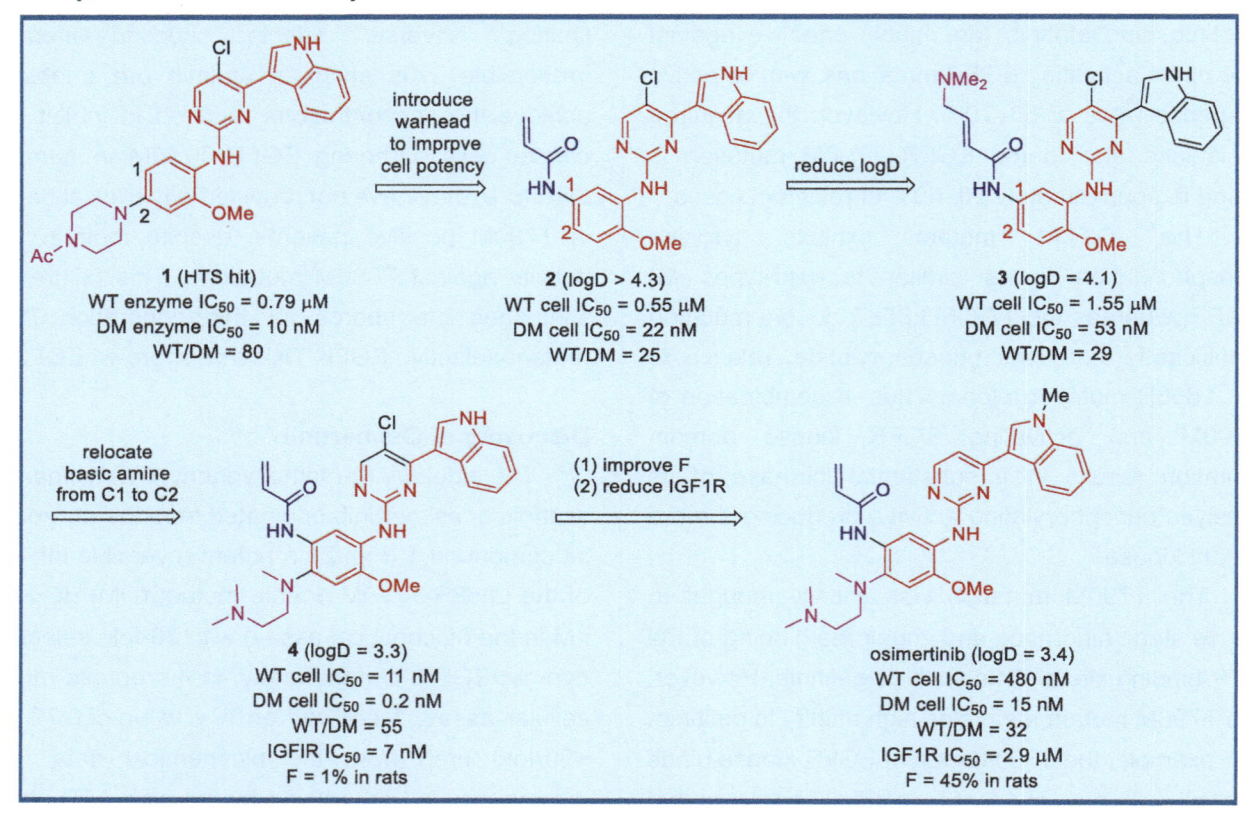

Figure 2. Structural optimization of osimeritinib.

Table 1. Cellular IC$_{50}$'s of EGFR TKIs[14]

inhibitor	IC$_{50}$ (nM)			WT/L858R-T790M margin
	L858R-T790M	E19del	WT	
gefitinib	3300	8.7	62	0.02
erlotinib	7600	5.9	77	0.01
afatinib	23	5.7	12	0.53
dacomitinib	42	0.63	11	0.27
osimertinib	15	17	480	32
olmutinib	10	9.2	2225	222
aumolertinib	3.3	4.1	596	181

Osimertinib differs from afatinib (a second-generation TKI) mainly in two aspects: (1) the core structure evolved from an anilinoquinazoline motif to an anilinopyrimidine motif; and (2) the main hydrophobic segment was changed from a halogenated phenyl ring to an indole ring (Fig. 3). The distinct anilinopyrimidine core together with the indole headgroup makes osimertinib a highly EGFR mutant-selective inhibitor over wt (wt IC_{50} = 480 nM, AM IC_{50} = 17 nM, DM IC_{50} = 15 nM) (Table 1), thus overcoming the dose limiting toxicity seen with the earlier generation of EGFR TKIs.[14]

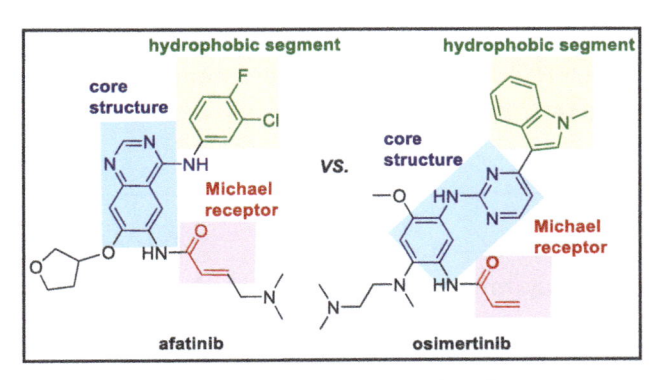

Figure 3. Osimertinib vs. afatinib.

Aumolertinib (Ameile™)[19–21]

A major human metabolite of osimertinib is the N-demethylated free NH indole **5**[22] (Fig. 4) which is a more potent inhibitor of DM (IC_{50} = 2 nM) and also wt EGFR (IC_{50} = 33 nM) than the parent drug, resulting in lower wt selectivity (WT/DM = 16 vs. 32 for osimertinib).[14] In addition, metabolite **5** showed ~11-fold more potent inhibition of IGF1R.[14]

The CYP3A4-mediated N-demethylation was largely overcome by replacement of methyl with the more oxidation-resistant cyclopropyl group to give aumolertinib (Fig. 4) which also showed 5-fold enhanced potency and wt selectivity (WT/DM = 181 vs. 32 for osimertinib). Furthermore, the cyclopropyl group increases the blood–brain barrier permeability, which is beneficial for the treatment of EGFR-mutated NSCLC with CNS metastases. Interestingly, despite these improvements, aumolertinib is still administered at higher dose (110 vs. 80 mg for osimertinib, once daily).

Aumolertinib (formerly known as almonertinib), which was discovered by Shanghai Hansoh Pharma, is a me-too third-generation EGFR TKI with higher selectivity of both EGFR sensitizing and T790M mutants over wt EGFR. It was approved in 2020 by the National Medical Products Administration (NMPA) of China for the second-line treatment for patients with emergent EGFR T790M-resistant mutations. There have been no head-to-head clinical trials of aumolertinib vs. osimertinib, but a comparison of the clinical trial data reveals that the clinical outcomes are comparable, with a slightly lower incidence of WT EGFR-mediated toxicities such as rash and diarrhea for aumolertinib.[23]

Copycat drugs such as aumolertinib involve much less cost and risk and thus can be priced below the original version, thus undercutting established drugs such as osimertinib. As described in the Section 4.3, the launch of icotinib (a cyclized version of erolitinib) in China at a reduced price in China, drove down the prices of the then-established first-generation TKIs, erlotinib and gefitinib. It is possible that aumolertinib might bring a similar cost reduction vs. osimertinib not only in China but more broadly if improved efficacy is demonstrative.

Hansoh Pharma has partnered with EQRx to expand global access to aumolertinib. Their initial plan to bring it swiftly to the US market was dropped after the US FDA declined to approve drugs such as Eli Lilly/Innovent Biologics' PD-1 inhibitor sintilimab whose application was based primarily on data from China. EQRx has just initiated a US clinical trial for aumolertinib.

Furmonertinib (Ivesa™)

Furmonertinib (formerly known as alflutinib), is another me-too third-generation EGFR TKI which has a pyridyl ring replacing the phenyl ring of osimertinib and a trifluoroethyl group replacing the methyl group (Fig. 4).[24] This TKI, which was discovered and developed by Shanghai Allist Pharma, was approved in November 2019 by the NMPA of China for the second-line treatment of

NSCLC with EGFR sensitive mutation and EGFR T790M drug-resistant mutation, and more recently as a first-line treatment for classical EGFR mutant NSCLC. In June 2022, furmonertinib was granted the Fast Track designation by the FDA in patients with advanced or metastatic NSCLC with activating EGFR or HER2 mutations, including exon 20 insertion mutations.

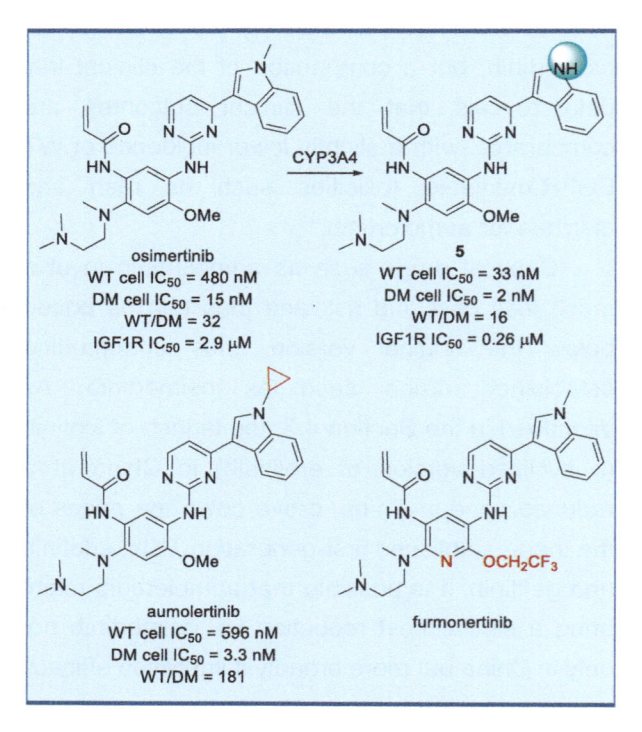

Figure 4. Metabolism of osimertinib and structures of aumolertinib and furmonetinib.

Binding Mode of Osimertinib[25]

A co-crystal structure of osimertinib with EGFR (Fig. 5) did not show a covalent bond between the acrylamide moiety and the protein side chain Cys 797 even though the wt protein was definitely covalently modified as confirmed by mass spectrometry. The co-crystal structure showed that there is an alternative reversible binding mode that is generated during in vitro crystallization.

Osimertinib binds on the outer edge of the ATP-binding pocket of the active kinase conformation. The pyrimidine N4 hydrogen bonds to the amide NH of Met793 in the hinge region, while the acrylamide carbonyl oxygen hydrogen bonds to the amide NH of Cys797 (Figs. 5, 6).

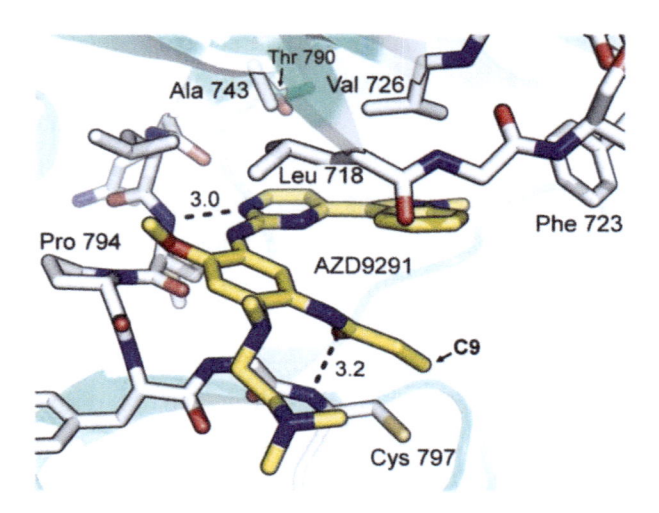

Figure 5. Co-crystal structure of osimertinib with EGFR. Adapted with permission from *J. Struct. Biol.* **2015**, *192*, 539–544.

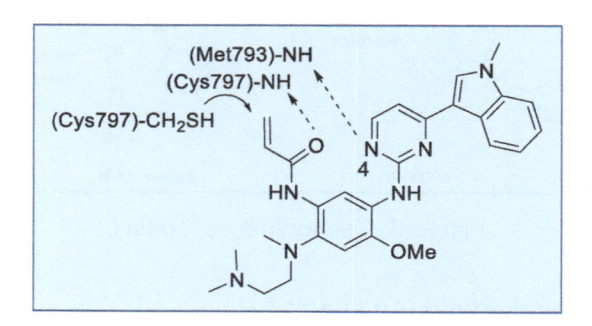

Figure 6. Summary of osimertinib–EGFR interactions.

1. S. L. Greig. Osimertinib: First global approval. *Drugs.* **2016**, *76*, 263–273.
2. Ameku K, Higa M. Complete remission of multiple brain metastases in a patient with EGFR-mutated non-small-cell lung cancer treated with first-line osimertinib without radiotherapy. *Case Rep. Oncol. Med.* **2020**, 9076168.
3. K. Suda, R. Onozato, Y. Yatabe, T. Mitsudomi. EGFR T790M mutation: A double role in lung cancer cell survival?. *J. thorac. Oncol.* **2009**, *4*, 1–4.
4. H. Patel, R. Pawara, A. Ansari, S. Surana. Recent updates on third generation EGFR inhibitors and emergence of fourth generation EGFR inhibitors to combat C797S resistance. *Eur. J. Med. Chem.* **2017**, *142*, 32–47.
5. S. D. Cesco, J. Kurian, C. Dufresne, A. K. Mittermaier, N. Moitessier. Covalent inhibitors design and discovery. *Eur. J. Med. Chem.* **2017**, *138*, 96–114.
6. L. H. Wang, J. Y. Zhao, Y. Yao, C. Y. Wang, J. B. Zhang, X. H. Shu, X. L. Sun, Y. X. Li, K. X. Liu, H. Yuan, X. D. Ma. Covalent

binding design strategy: A prospective method for discovery of potent targeted anticancer agents. *Eur. J. Med. Chem.* **2017**, *142*, 493–505.

7. C. E. Steuer, F. R. Khuri, S. S. Ramalingam. The next generation of epidermal growth factor receptor tyrosine kinase inhibitors in the treatment of lung cancer. *Cancer* **2015**, *121*, 8, 1–6.

8. R. Minari, P. Bordi, M. Tiseo. Third-generation epidermal growth factor receptor-tyrosine kinase inhibitors in T790M-positive non-small cell lung cancer: review on emerged mechanisms of resistance. *Transl. Lung Cancer Res.* **2016**, *5*, 695–708.

9. S. H. Wang, S. D. Cang, D. L. Liu. Third-generation inhibitors targeting EGFR T790M mutation in advanced non-small cell lung cancer. *J. Hematol. Oncol.* **2016**, *9*, 34–40.

10. S. N. Milik, D. S. Lasheen, R. A.T. Serya, K. A. M. Abouzid. How to train your inhibitor: Design strategies to overcome resistance to epidermal growth factor receptor inhibitors. *Eur. J. Med. Chem.* **2017**, *142*, 131–151.

11. B. C. Liao, C. C. Lin, J. C. H. Yang. Second and third-generation epidermal growth factor receptor tyrosine kinase inhibitors in advanced nonsmall cell lung cancer. *Curr. Opin. Oncol.* **2015**, *27*, 94–101.

12. H. Cheng, S. K. Nair, B. W. Murray. Recent progress on third generation covalent EGFR inhibitors. *Bioorg. Med. Chem. Lett.* **2016**, *26*, 1861–1868.

13. R. A. Ward, M. J. Anderton, S. Ashton, P. A. Bethel, M. Box, S. Butterworth, N. Colclough, C. G. Chorley, C. Chuaqui, D. A. Cross, L. A. Dakin, J. E. Debreczeni, C. Eberlein, M. R. Finlay, G. B. Hill, T. C. Klinowska, C. Lane, S. Martin, J. P. Orme, M. J. Waring. Structure- and reactivity-based development of covalent inhibitors of the activating and gatekeeper mutant forms of the epidermal growth factor receptor (EGFR). *J. Med. Chem.* **2013**, *56*, 7025–7048.

14. M. R. Finlay, M. Anderton, S. Ashton, P. Ballard, P. A. Bethel, M. R. Box, R. H. Bradbury, S. J. Brown, S. Butterworth, A. Campbell, C. Chorley, N. Colclough, D. A. E. Cross, G. S. Currie, M. Grist, L. Hassall, G. B. Hill, D. James, M. James, P. Kemmitt, T. Klinowska, G. Lamont, S. G. Lamont, N. Martin, H. L. McFarland, M. J. Mellor, J. P. Orme, D. Perkins, P. Perkins, G. Richmond, P. Smith, R. A. Ward, M. J. Waring, D. Whittaker, S. Wells, G. L. Wrigley. Discovery of a potent and selective EGFR inhibitor (AZD9291) of both sensitizing and T790M resistance mutations that spares the wild type form of the receptor. *J. Med. Chem.* **2014**, *57*, 8249–8267.

15. S. Butterworth, D. Cross, M. Finlay, R. Ward, M. J. Waring. The structure-guided discovery of osimertinib: the first U.S. FDA approved mutant selective inhibitor of EGFR T790M. *Medchemcomm.* **2017**, *8*, 820–822.

16. A. Yver. Osimertinib (AZD9291)–A science driven, collaborative approach to rapid drug design and development. *Ann. Oncol.* **2016**, *27*, 1165–1170.

17. S. Butterworth, M. R. Finlay, R. A. Ward, H. M. Redfearn. 2-(2,4,5-substituted-anilino)pyrimidine compounds. US 9732058 B2, **2017**.

18. D. A. Cross, S. E. Ashton, S. Ghiorghiu, C. Eberlein, C. A. Nebhan, P. J. Spitzler, J. P. Orme, M. R. V. Finlay, R. A. Ward, M. J. Mellor, G. Hughes, A. Rahi, V. N. Jacobs, M. R.Brewer, E. Ichihara, J.Sun, H. L. Jin, P. Ballard, K. A. Kadhimi, R. Rowlinson, T. Klinowska, G. H. P. Richmond, M. Cantarini, D. W. Kim, M. R. Ranson, W. Pao. AZD9291, an irreversible EGFR TKI, overcomes T790M-mediated resistance to EGFR inhibitors in lung cancer. *Cancer Discov.* **2014**, *4*, 1046–1061.

19. M. Shirley, S. J. Keam. Aumolertinib: A review in non-small cell lung cancer. *Drugs.* **2022**, *82*, 577–584.

20. S. Wang, S. Cang, D. Liu. Third-generation inhibitors targeting EGFR T790M mutation in advanced non-small cell lung cancer. *J. Hematol. Oncol.* **2016**, *9*, 34.

21. J. Wang, L. Wu. An evaluation of aumolertinib for the treatment of EGFR T790M mutation-positive non-small cell lung cancer. *Expert Opin. Pharmacother.* **2022**, *23*, 647–652.

22. P. A. Dickinson, M. V. Cantarini, J. Collier, P. Frewer, S. Martin, K. Pickup, P. Ballard. Metabolic disposition of osimertinib in rats, dogs, and humans: Insights into a drug designed to bind covalently to a cysteine residue of epidermal growth factor receptor. *Drug Metab. Dispos.* **2016**, *44*, 1201–1212.

23. C. Aggarwal, B. Gyawali. Aumolertinib in EGFR-mutant lung cancer: Will the promise of cost disruption ease access? *J. Clin. Oncol.* **2022**, JCO2200903. Advance online publication. https://doi.org/10.1200/JCO.22.00903

24. E. D. Deeks. Furmonertinib: First approval. *Drugs.* **2021**, *81*, 1775–1780.

25. Y. Yosaatmadja, S. Silva, J. M. Dickson, A. V. Patterson, J. B. Smaill, J. U. Flanagan, M. J. McKeage, C. J. Squire. Binding mode of the breakthrough inhibitor AZD9291 to epidermal growth factor receptor revealed. *J. Struct. Biol.* **2015**, *192*, 539–544.

4.7 Mobocertinib (Exkivity™) (19)

Structural Formula	Space-filling Model	Brief Information
$C_{32}H_{39}N_7O_4$; MW = 585 clogP = 5.5; tPSA = 111		Year of discovery: 2013 Year of introduction: 2021 Discovered by: Ariad Developed by: Takeda Primary target: EGFR, HER2-4 Binding type: covalent Class: receptor tyrosine kinase Treatment: NSCLC with EGFR alterations Other name: AP32788, TAK788 Oral bioavailability = 37% Elimination half-life = 18 h Protein binding = not reported

◯ = C ◯ = H ●= O ●= N ●= S ●= F ●= Cl ●= Br ●= I

Mobocertinib received an accelerated approval in 2021 for the treatment of adult patients with locally advanced or metastatic non-small cell lung cancer (NSCLC) with EGFR exon 20 insertion (EGFRex20ins) mutations as detected by an FDA-approved test, whose disease has progressed on or after platinum-based chemotherapy.[1,2]

Mobocertinib is a third-generation tyrosine kinase inhibitor (TKI) that overcomes the drug resistance of NSCLC tumors harboring exon 20 insertions in EGFR and HER2, and it is the first-in-class targeted therapy for the treatment of these specific tumors.

Mobocertinib is derived from osimertinib, a second-generation TKI which exhibits only limited activity against several resistant EGFRex20ins mutants. The two inhibitors differ structurally only in that the pyrimidine ring of moboceritinib carries an added (rationally chosen) isopropyl ester group (Fig. 1) which produces expanded coverage of EGFRex20ins mutations as well as improved selectivity over wt EGFR. The successful elaboration of osimeritinib to moboceritinib via a single substitution illustrates that a next-generation TKI could be just one substituent away from the existing TKI.

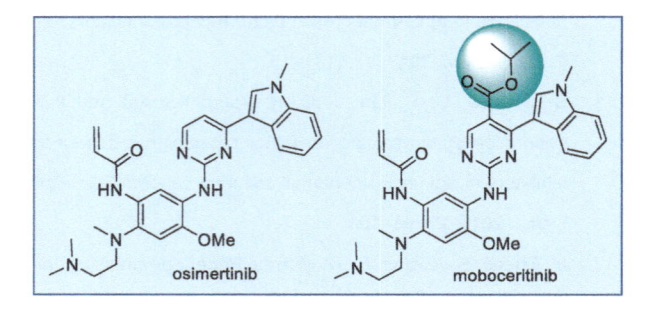

Figure 1. Structures of osimertinib and moboceritinib.

EGFRex20ins Mutations in NSCLC[3-6]

The EGFRex20ins mutations result in a cluster of mutants with amino acids inserted predominantly between amino acid 767 and 774 near the C-terminus of the αC helix, leading to constitutive activation of EGFR. These insertions, which account for approximately 4–12% of EGFR-mutant NSCLC patients, are particularly prevalent in nonsmokers and Asian population. Unfortunately, an overwhelming majority of these mutations are resistant to first- and second-generation EGFR TKIs except osimertinib which may be effective against some resistant mutants. The exon 20 insertion mutations also occur within the residues 775–783 of HER2, accounting for 3% of NSCLC patients carrying HER2 alterations. Together, EGFR and HER2 exon 20 mutations represent ~4% of all NSCLC patients. HER2 TKIs (e.g., afatinib, lapatinib, neratinib, and dacomitinib) exhibit only limited potency against HER2-mutant tumors. Against the backdrop of TKI resistance of NSCLC tumors harboring exon 20 insertions in EGFR and HER2, mobocertinib was developed as a third-generation TKI.

From Osimertinib to Mobocertinib[4,7]

Superimposition of crystal structures of wt EGFR with EGFRex20ins NPG reveals significant similarity between the two proteins in the ATP-binding site where no amino acid substitutions occur.[4] Although the EGFRex20ins NPG does shift the αC-helix to the active conformation of EGFR, this shift is still far away from the ATP-binding site (data not shown). As such, it has been challenging to identify selective TKIs over wt. Nevertheless, the subtle structural variations between the two proteins in the αC-helix region have been fruitfully explored for selectivity. A docking model of osimertinib in complex with the EGFRex20ins NPG mutant (data not shown) showed an unoccupied pocket (close to the αC-helix) formed primarily by the gatekeeper residue Met790, the catalytic lysine Lys745, and the DFG triad preceding Thr854. This pocket can be accessed by a substituent attached to the pyrimidine hinge binder, and consequently, an isopropyl ester group was incorporated onto osimeritinib (to give mobocertinib)[4,7] to interact with the gatekeeper residue within this selectivity pocket and to take advantage of the conformational nuances between EGFRex20ins mutants and wt EGFR. It is this isopropyl ester group that gives rise to the improved activity for the EGFRex20ins mutants over osimertinib.

The acrylamide Michael-acceptor of mobocertinib was incorporated to allow covalent linkage with Cys797 in EGFR, as with other second-generation irreversible TKIs (e.g., afatinib and osimertinib). In general, the irreversible inhibitors are more potent than reversible analogs and produce more sustained kinase inhibition. However, covalent inhibitors suffer from off-target liabilities as there are at least 10 kinases in addition to EGFR that have a reactive cysteine residue in the position equivalent to Cys797.

Mobocertinib was designed to inhibit oncogenic variants containing mutations in exon 20 with selectivity over the wt protein. The IC_{50} in wt EGFR cells was about 35 nM, while the average IC_{50} for cells bearing various exon 20 insertions was about 10 nM. In NSCLC cell lines, mobocertinib was highly active against common EGFR-activating mutations (exon 19 deletions and L858R) (IC_{50} = 1.3–4.0 nM) or with a gatekeeper T790M mutant (IC_{50} = 9.8 nM), as well as selective for wt EGFR (IC_{50} = 35 nM). In Ba/F3 cells, mobocertinib displayed activity against 14 EGFR mutations with IC_{50} values ranging from 2.7 to 22.5 nM, vs. 34.5 nM for wt EGFR. More specifically, mobocertinib inhibited all five variants of EGFRex20ins mutations with IC_{50} ranging from 4.3 nM to 22.5 nM.

Mobocertinib also irreversibly targets HER2 exon 20 insertion mutations; however, its specific function on HER2 exon 20 insertion–mutant NSCLC remains to be determined.[6,8]

Binding Mode[3,7]

In the co-crystal structure of mobocertinib in complex with EGFRT790M+V948R (Fig. 2), the aminopyrimidine scaffold occupies the ATP-binding site, forming two critical hydrogen bonds with Met793 of the hinge, and the isopropyl ester occupies a selectivity pocket in the vicinity of αC helix. The ester carbonyl group interacts with Gln791 backbone carbonyl via a bound water molecule. Most importantly, the inhibitor forms the expected covalent bond with Cys797 at the edge of the active site cleft and the acrylamide Michael-acceptor group, making binding irreversible (Fig. 4).

The co-crystal structure of wt EGFR kinase in complex with mobocertinib (Fig. 3) shows that the ester group is slightly tilted to enable a hydrogen bond between the Thr790 gatekeeper side chain hydroxyl group and the ester carbonyl oxygen (Fig. 5). However, comparison of the wt and mutant proteins with mobocertinib reveals that the binding pockets are similar despite differences in overall kinase conformation.

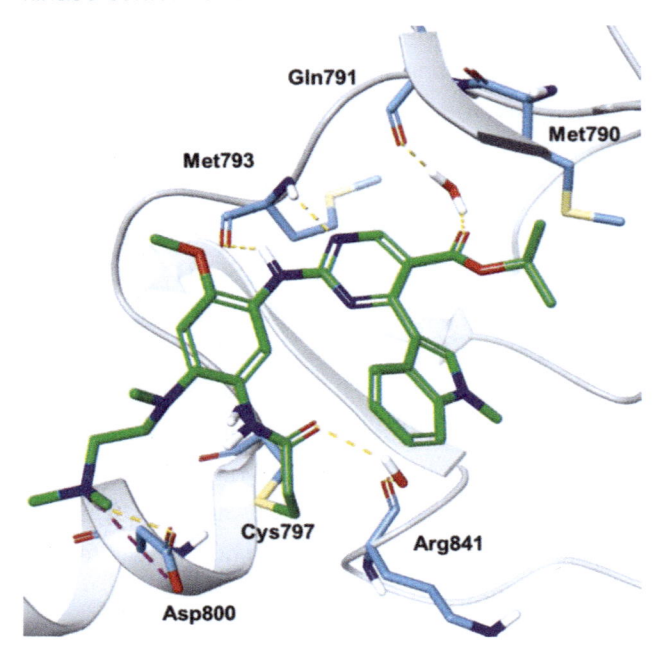

Figure 2. Co-crystal structure of mobocertinib–EGFR-T790M+V948R (PDB ID: 7A6K).

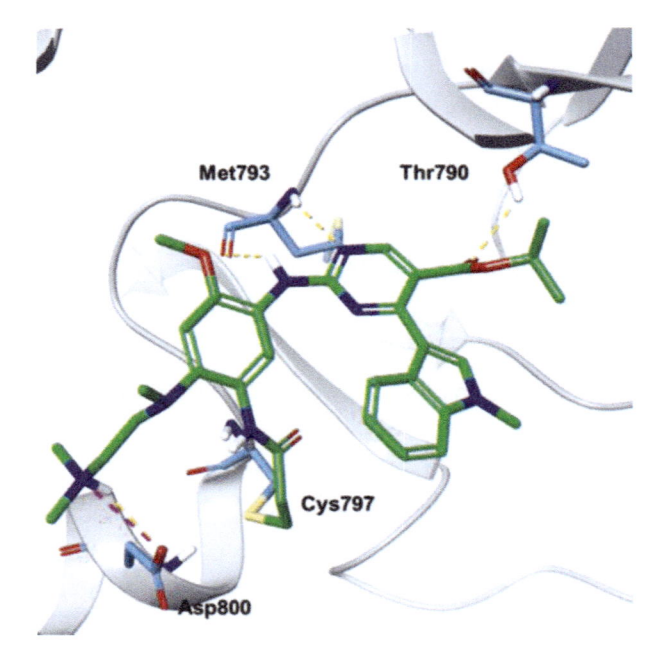

Figure 3. Co-crystal structure of mobocertinib–wt EGFR (PDB ID: 7B85).

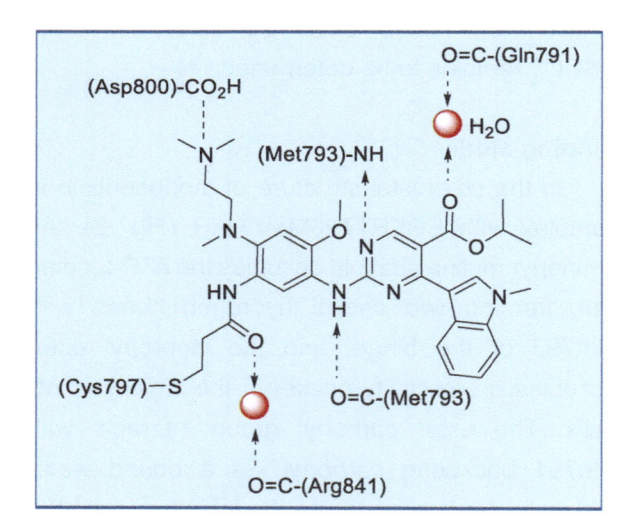

Figure 4. Summary of mobocertinib–EGFR T790M+V948R interactions.

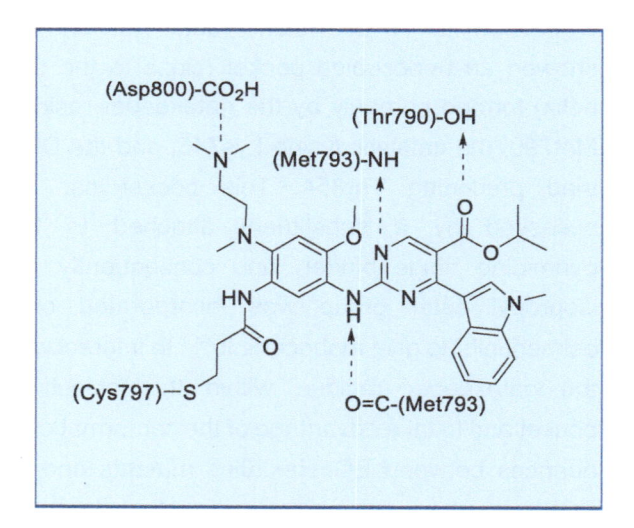

Figure 5. Summary of mobocertinib–wt EGFR interactions.

Metabolism of Moboceritinib[4]

Moboceritinib undergoes CYP450-mediated metabolism to give two major N-demethylated metabolites, AP32914 and AP32960 (Fig. 6), whose IC$_{50}$ values are within 2-fold of mobocertinib for both wt and mutant EGFR. It is likely that these metabolites also contribute to the pharmacologic activity of mobocertinib. Interestingly, the typically labile isopropyl ester appears to be resistant to esterase-induced hydrolysis.

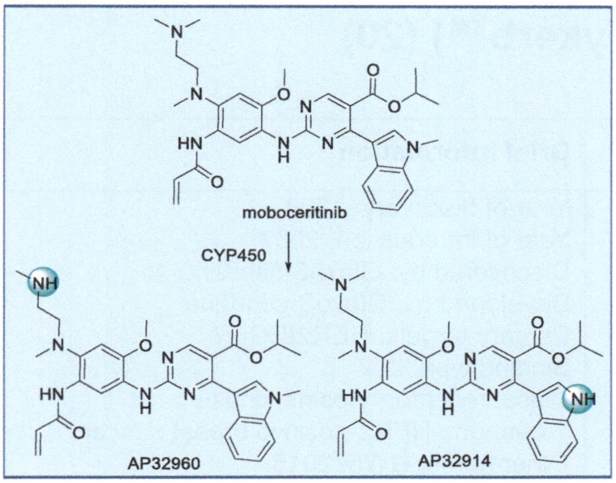

Figure 6. Metabolism of moboceritinib.

1. A. Markham. Mobocertinib: First approval. *Drugs*. **2021**, *81*, 2069–2074.
2. R. Roskoski., Jr. Properties of FDA-approved small molecule protein kinase inhibitors: A 2022 update. *Pharmacol. Res.* **2022**, *175*, 106037.
3. J. Lategahn, H. L. Tumbrink, C. Schultz-Fademrecht, A. Heimsoeth, L. Werr, J. Niggenaber, M. Keul, F. Parmaksiz, M. Baumann, S. Menninger, E. Zent, I. Landel, J. Weisner, K. Jeyakumar, L. Heyden, N. Russ, F. Müller, C. Lorenz, J. Brägelmann, I. Spille, D. Rauh. Insight into targeting exon20 insertion mutations of the epidermal growth factor receptor with wild type-sparing inhibitors. *J. Med. Chem.* **2022**, *65*, 6643–6655.
4. F. Gonzalvez, S. Vincent, T. E. Baker, A. E. Gould, S. Li, S. D. Wardwell, S. Nadworny, Y. Ning, S. Zhang, W. S. Huang, Y. Hu, F. Li, M. T. Greenfield, S. G. Zech, B. Das, N. I. Narasimhan, T. Clackson, D. Dalgarno, W. C. Shakespeare, M. Fitzgerald, V. M. Rivera. Mobocertinib (TAK-788): A targeted inhibitor of *EGFR* exon 20 insertion mutants in non-small cell lung cancer. *Cancer Discov.* **2021**, *11*, 1672–1687.
5. S. S. Zhang, V. W. Zhu. Spotlight on mobocertinib (TAK-788) in NSCLC with *EGFR* exon 20 insertion mutations. *Lung Cancer.* **2021**, *12*, 61–65.
6. H. Han, S. Li, T. Chen, M. Fitzgerald, S. Liu, C. Peng, K. H. Tang, S. Cao, J. Chouitar, J. Wu, D. Peng, J. Deng, Z. Gao, T. E. Baker, F. Li, H. Zhang, Y. Pan, H. Ding, H. Hu, V. Pyon, K. K. Wong. Targeting HER2 exon 20 insertion-mutant lung adenocarcinoma with a novel tyrosine kinase inhibitor mobocertinib. *Cancer Res.* **2021**, *81*, 5311–5324.
7. J. P. Robichaux, Y. Y. Elamin, Z. Tan, B. W. Carter, S. Zhang, S. Liu, S. Li, T. Chen, A. Poteete, A. Estrada-Bernal, A. T. Le, A. Truini, M. B. Nilsson, H. Sun, E. Roarty, S. B. Goldberg, J. R. Brahmer, M. Altan, C. Lu, V. Papadimitrakopoulou, J. V. Heymach. Mechanisms and clinical activity of an EGFR and HER2 exon 20-selective kinase inhibitor in non-small cell lung cancer. *Nature Med.* **2018**, *24*, 638–646.

4.8 Lapatinib (Tykerb™) (20)

Structural Formula	Space-filling Model	Brief Information
$C_{29}H_{26}N_4O_4FClS$; MW = 581 clogP = 5.8; tPSA = 101		Year of discovery: 1998 Year of introduction: 2007 Discovered by: GlaxoSmithKline Developed by: GlaxoSmithKline Primary targets: HER2/EGFR Binding type: I1/2 Class: receptor tyrosine kinase Treatment: HER2-positive breast cancer Other name: GW572016 Oral bioavailability = not reported Elimination half-life = 24 h Protein binding = 99%

● = C   ○ = H   ● = O   ● = N   ● = S   ● = F   ● = Cl   ● = Br   ● = I

Lapatinib, a dual inhibitor of EGFR (also known as HER1) and HER2 receptor tyrosine kinases, was approved in 2007 for the treatment of patients with advanced metastatic breast cancer whose tumors overexpress HER2 and who have received prior therapy, including anthracycline, taxane, and trastuzumab. It is the first targeted, once-daily oral treatment option for this patient population. In 2010, it was also granted accelerated approval as a combination therapy with letrozole as a first-line, all-oral treatment for women with metastatic breast cancer.

Breast cancer is the most common type of cancer in the United States, affecting a total of more than 3.8 million women and 264,000 new cases every year. The average risk of a woman in the United States developing breast cancer sometime in her lifetime is ~13%, and ~42,000 women die each year. The 20–25% of breast cancers that overexpress HER2 (HER2-positive breast cancers) tend to be more aggressive than other types of breast cancer.

Lapatinib works against HER2-positive breast cancers by blocking both EGFR and HER2 which can cause uncontrolled cell growth. It inhibits both proteins with nanomolar potency as well as >300-fold selectivity over other kinases.[1–3]

Lapatinib has poor aqueous solubility (7 μg/mL) and high protein binding (>99%) due to high lipophilicity (clogP = 5.8), both of which contribute to poor oral bioavailability. This property leads to an unusually high daily dosage of 1250 mg.

Lapatinib was discovered through extensive structural optimization of the 6-furanylquinazoline chemotype.[2]

EGFR Signaling in Cancer[4,5]

Epidermal growth factor receptor (EGFR) is a member of the HER (also known as ERBB) family of receptors, a subfamily of four closely related receptor tyrosine kinases (RTKs): EGFR (HER1, ERBB1), HER2 (ERBB2), HER3 (ERBB3) and HER4 (ERBB4). It is expressed on the surface of some normal cells and is involved in cell growth. However, elevated levels of EGFR kinase occur frequently in cancers such as non-small-cell lung cancer (NSCLC), metastatic colorectal cancer, glioblastoma, head and neck cancer, pancreatic cancer, and breast cancer where they drive unregulated growth.

EGFR is a canonical RTK that spans the cell membrane with the ligand-binding site outside and the phosphorylation site inside of the cell (Fig. 1). Binding to the extracellular region of the EGFR with a specific set of ligands (e.g., EGF, TGF-α)

triggers either homodimer formation or heterodimerization with other HER family members. Subsequently, dimerization of EGFR activates its cytoplasmic tyrosine kinase domain which then triggers further signal transduction, including via the PI3K/AKT/mTOR and RAS/RAF/MEK/ERK pathways (Fig. 1). The EGFR signaling pathway is one of the most important pathways that regulate growth, survival, proliferation, and differentiation in mammalian cells, and aberrant activation of EGFR signaling contributes to diverse cancers. Consequently, interruption of EGFR signaling, either by blocking the extracellular region of EGFR or by inhibiting intracellular tyrosine kinase activity can prevent the growth of EGFR-expressing tumors with clinical benefit. Indeed, EGFR has become a well-established target for both monoclonal antibodies and specific tyrosine kinase inhibitors. Monoclonal antibodies attach to target tumor cells by binding to cell-surface antigens whereas small-molecule inhibitors of TKIs function intracellularly.

Increasing knowledge of the structure and function of the EGFR subfamily of tyrosine kinases and of their role in the initiation and progression of various cancers has provided the impetus for a substantial research effort aimed at developing new anticancer therapies that target specific components of the EGFR signal transduction pathway.

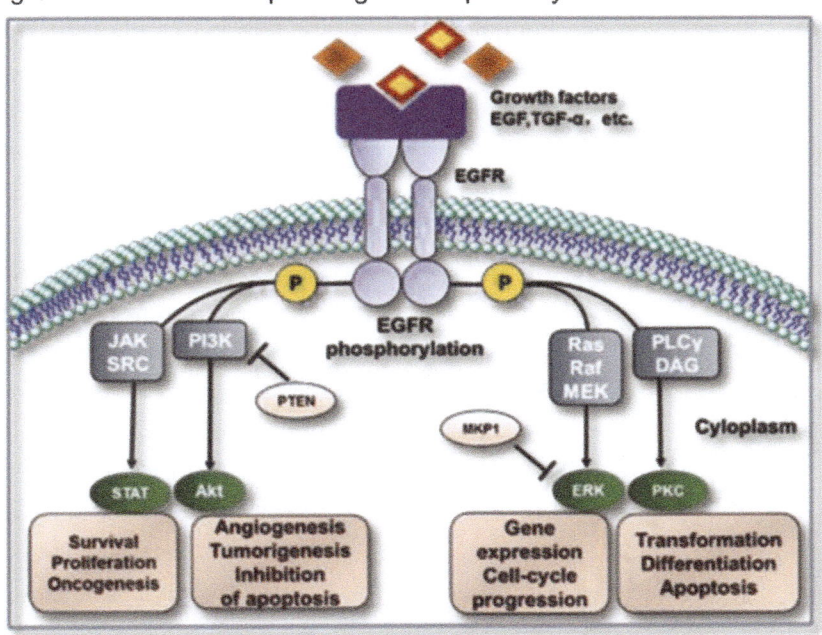

Figure 1. Ligand-dependent signaling pathway of EGFR.

In addition to cancer, hyperactivation of EGFR has also been implicated in various neurodegenerative disorders, including Alzheimer's disease. It is possible that CNS penetrant anti-cancer EGFR kinase inhibitors may be of value for the treatment of certain CNS disorders.[6]

HER2 Signaling in Breast Cancer[7–9]

HER2 is another member of the HER family of RTK receptors; however, its extracellular domain lacks a ligand-binding site and has no identifiable cell surface ligand, unlike the other HER family receptors. It exists in a constitutively active conformation and can undergo ligand-independent dimerization with other HER receptors or homodimerization when it is present in high concentrations, such as in cancer. The most active and tumor-promoting combination is thought to be the HER2/HER3 dimer. Receptor dimerization and activation of the HER2 tyrosine kinase domain initiate downstream-signaling cascades similar to EGFR, resulting in uncontrolled cellular proliferation and survival of breast cancer cells.

HER2 kinase has been shown to potentiate EGFR signaling by increasing the binding affinity with its ligand EGF. Furthermore, selective EGFR inhibitors can reduce HER2 signaling and growth of breast cancer cells that express high levels of HER2. Thus, dual EGFR/HER2 inhibition may be more effective than targeting each separately, and lapatinib seems to accomplish exactly that.

Binding Mode[10]

Lapatinib is a type I1/2 inhibitor that binds to a DFG-in, αC helix-out inactive conformation of EGFR (Fig. 2). The quinazoline N1 forms a hydrogen bond with the amide NH group of Met793 of the hinge (Fig. 3). Although the aniline nitrogen is not involved in any direct hydrogen-bonding interactions with the protein, it serves to increase the basicity of quinazoline N1. The proximal methylsulfonylethylamino group, which functions to increase aqueous solubility of the drug, extends toward the solvent exposed region.

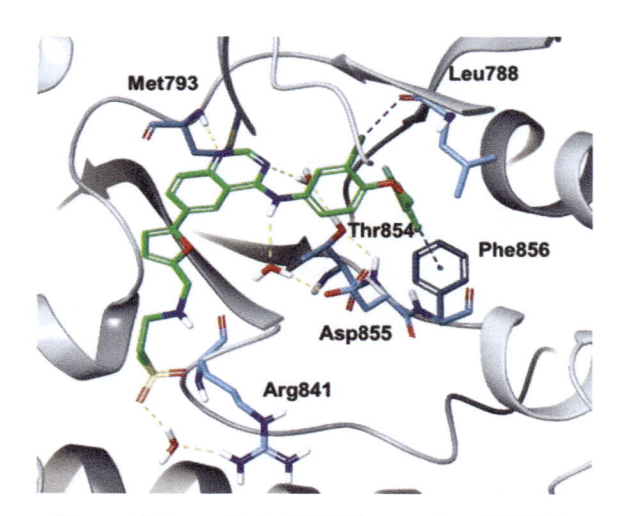

Figure 2. Lapatinib–EGFR complex (1XKK).

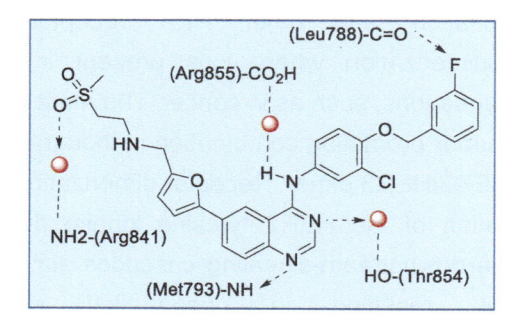

Figure 3. Lapatinib–EGFR interactions.

1. D. W. Rusnak, K. Lackey, K. Affleck, E. R. Wood, B. R. Keith, D. M. Murray, W. B. Knight, R. J. Mullin, T. M. Gilmer. The effects of the novel, reversible epidermalGrowth factor receptor/ErbB-2 tyrosine kinase inhibitor, GW2016, on the growth of human normal and tumor-derived cell lines in vitro and in vivo. *Mol. Cancer Ther.* **2001**, *1*, 85–94.

2. K. G. Petrov, Y. M. Zhang, M. Carter, G. S. Cockerill, S. Dickerson, C. A. Gauthier, Y. Guo, R. A. Mook, D. W. Rusnak, A. L. Walker, E. R. Wood, K. E. Lackey. Optimization and SAR for dual ErbB-1/ErbB-2 tyrosine kinase inhibition in the 6-furanylquinazoline series. *Bioorg. Med. Chem. Lett.* **2006**, *16*, 4686–4691.

3. D. Bilancia, G. Rosati, A. Dinota, D. Germano, R. Romano, L. Manzione. Lapatinib in breast cancer. *Ann. Oncol.* **2007**, *18* (Suppl. 6), vi26–vi30.

4. M. Sebastian, A. Schmittel, M. Reck. First-line treatment of EGFR-mutated nonsmall cell lung cancer: Critical review on study methodology. *Eur. Respir. Rev.* **2014**, *23*, 92–105.

5. K. Oda, Y. Matsuoka, A. Funahashi, H. Kitano. A comprehensive pathway map of epidermal growth factor receptor signaling. *Mol. Syst. Biol.* **2005**, *1*, 2005.0010.

6. H. M. Mansour, H. M. Fawzy, A. S. El-Khatib, M. Khattab. Potential repositioning of anti-cancer EGFR inhibitors in Alzheimer's disease: Current perspectives and challenging prospects. *Neuroscience.* **2021**, *469*, 191–196.

7. J. C. Xuhong, X. W. Qi, Y. Zhang, J. Jiang. Mechanism, safety and efficacy of three tyrosine kinase inhibitors lapatinib, neratinib and pyrotinib in HER2-positive breast cancer. *Amer. J. Cancer Res.* **2019**, *9*, 2103–2119.

8. Z. Mitri, T, Constantine, R. O'Regan. The HER2 receptor in breast cancer: Pathophysiology, clinical use, and new advances in therapy. *Chemother. Res. Pract.* **2012**, 743193.

9. F. Le Du, G. Diéras, G. Curigliano. The role of tyrosine kinase inhibitors in the treatment of HER2+ metastatic breast cancer. *Eur. J. Cancer.* **2021**, *154*, 175–189.

10. K. Aertgeerts, R. Skene, J. Yano, B. C. Sang, H. Zou, G. Snell, A. Jennings, K. Iwamoto, N. Habuka, A. Hirokawa, T. Ishikawa, T. Tanaka, H. Miki, Y. Ohta, S. Sogabe. Structural analysis of the mechanism of inhibition and allosteric activation of the kinase domain of HER2 protein. *J. Biol. Chem.* **2011**, *286*, 18756–18765.

4.9 Tucatinib (Tukysa™) (21)

Structural Formula	Space-filling Model	Brief Information
$C_{26}H_{24}N_8O_2$; MW = 480 clogP = 5.2; tPSA = 108		Year of discovery: 2006 Year of introduction: 2020 Discovered by: Array BioPharma Developed by: Array BioPharma Primary targets: HER2 Binding type: I1/2 Class: receptor tyrosine kinase Treatment: HER2-positive breast cancer Other names: ONT-380, ARRY-380 Oral bioavailability = not reported Elimination half-life = 5.4 h Protein binding = 97%

= C = H = O = N = S = F = Cl = Br = I

Tucatinib, a highly selective HER2 inhibitor, was approved in 2020 as a combination therapy with trastuzumab and capecitabine for the treatment of patients with advanced unresectable or metastatic HER2-positive breast cancer, including patients with brain metastases, who have received one or more prior anti-HER2-based regimens in the metastatic setting.[1,2]

The high HER2 selectivity of tucatinib over EFGR offers a significant clinical advantage over the other two aminoquinazoline kinase drugs, lapatinib and neratinib, which are nearly equipotent inhibitor of HER2 and EGFR. Blockage of the latter is associated with gastrointestinal and dermatologic adverse events.

Combination of tucatinib with trastuzumab and capecitabine resulted in more than 60% reduction in the risk of central nervous system progression or death in both patients with previously treated HER2-positive metastatic breast cancer who have active or stable brain metastases. Tucatinib is the first tyrosine kinase inhibitor to demonstrate prolongation of overall survival in patients with HER2-positive metastatic breast cancer with brain metastases in the clinical trials.[3] It is now the property of Seattle Genetics.

Discovery of Tucatinib[4]

Both EGFR and HER2 belong to the HER (ERBB) family of receptors and share a high degree of structural and functional homology. Moreover, the ATP-binding pockets of both receptors differ by just two amino acids, only one of which is involved in the binding with an ATP-competitive inhibitor (Cys775 in EGFR vs. Ser783 in HER2).[4] Therefore, it is challenging to identify selective HER2 inhibitors over EGFR. Against this backdrop, tucatinib has been developed as the most selective HER2 inhibitor drug for HER2-positive breast cancers.

The lead compound **1** showed good HER2 potency in the biochemical and cellular assays as well as 7-fold selectivity over EGFR; however, it suffered from poor pharmacokinetic properties. For example, it underwent facile metabolic oxidation to give the pyridine *N*-oxide **2**, which lacked membrane permeability. Interestingly, **2** was found to have high HER2 selectivity (33-fold over EGFR). To circumvent the metabolic stability issue, the problematic pyridine was replaced with a bicyclic triazolopyridine to give **3** as a potent HER2 inhibitor with suboptimal selectivity and PK properties. Further structural optimization led to tucatinib with a balanced profile of HER2 selectivity, potency and PK parameters (Fig. 1).

Tucatinib demonstrates high selectivity for HER2 compared with EGFR, unlike the HER2-targeted TKIs, lapatinib and neratinib. More specifically, tucatinib is ~65- and ~500-fold selective for HER2 vs. EGFR in the biochemical and cellular assays, respectively, vs. no differentiation for lapatinib and neratinib (Tables 1, 2).[4,5] In addition, tucatinib displays reduced potency in a HER4 biochemical assay relative to HER2, with an IC_{50} of 310 nM. Tucatinib also shows high selectivity against a panel of 223 various kinases.

The exact binding mode of tucatinib with HER2 has not yet been established due to lack of co-crystal structure with HER2. However, because it shares the same aminoquinazoline hinge-binding scaffold as lapatinib as well as other structural features (Fig. 2), it is highly likely that they bind similarly to HER2.

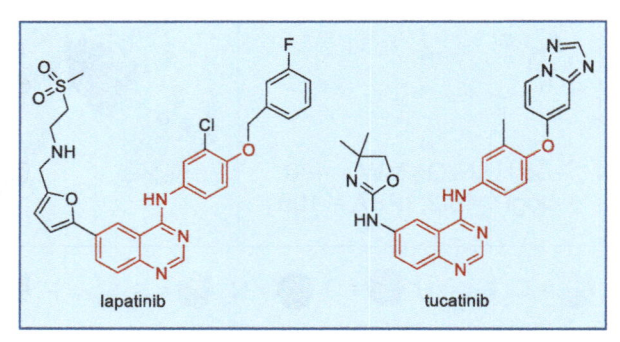

Figure 2. Structures of lapatinib and tucatinib.

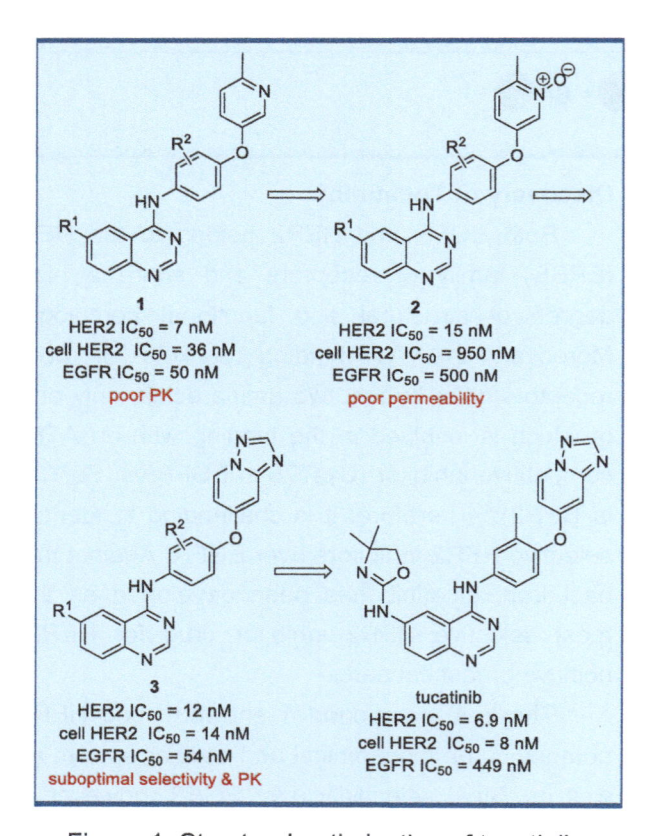

Figure 1. Structural optimization of tucatinib.

Table 1. Enzyme IC_{50}'s (nM) of HER2 inhibitors[4,5]

kinase	tucatinib	lapatinib	neratinib
HER2	6.9 nM	109 nM	5.6 nM
EGFR	449 nM	48 nM	1.8 nM

Table 2. Cellular IC_{50}'s (nM) of HER2 inhibitors[4,5]

kinase	tucatinib	lapatinib	neratinib
HER2	8 nM	49 nM	7 nM
EGFR	4000 nM	31 nM	8 nM

Binding Mode

1. M. Shah, S. Wedam, J. Cheng, M. H. Fiero, M. Xia, F. Li, J. Fan, X. Zhang, J. Yu, P. Song, W. Chen, T. K. Ricks, X. H. Chen, K. B. Goldberg, Y. Gong, W. F. Pierce, S. Tang, M. R. Theoret, R. Pazdur, L. Amiri-Kordestani, J. A. Beaver. FDA approval summary: Tucatinib for the treatment of patients with advanced or metastatic HER2-positive breast cancer. *Clin. Res.* **2021**, *27*, 1220–1226.

2. A. Lee A. Tucatinib: First Approval. *Drugs.* **2020**, *80*, 1033–1038.

3. N. U. Lin, V. Borges, C. Anders, R. K. Murthy, E. Paplomata, E. Hamilton, S. Hurvitz, S. Loi, A. Okines, V. Abramson, P. L. Bedard, M. Oliveira, V. Mueller, A. Zelnak, M. P. DiGiovanna, T. Bachelot, A. J. Chien, R. O'Regan, A. Wardley, A. Conlin, E. P. Winer. Intracranial efficacy and survival with tucatinib plus trastuzumab and capecitabine for previously treated HER2-positive breast cancer with brain metastases in the HER2CLIMB trial. *J. Clin. Oncol.* **2020**, *38*, 2610–2619.

4. S. L. Moulder, T. Baetz, V. Borges, S. K. Chia, E. Barrett, J. Garrus, K. Guthrie, C. L. Kass, E. Laird; J. Lyssikatos, F. Marmsater, E. Wallace. Abstract A143: ARRY-380, a selective HER2 inhibitor: From drug design to clinical evaluation. *Mol Cancer Ther.* **2011**, *10* (Suppl. 11), A143.

5. A. Kulukian, P. Lee, J. Taylor, R. Rosler, P. De Vries, D. Watson, A. Forero-Torres, S. Peterson. Preclinical activity of HER2-selective tyrosine kinase inhibitor tucatinib as a single agent or in combination with trastuzumab or docetaxel in solid tumor models. *Mol. Cancer Ther.* **2020**, *19*, 976–987.

4.10 Neratinib (Nerlynx™) (22)

Structural Formula	Space-filling Model	Brief Information
$C_{30}H_{29}N_6O_3Cl$; MW = 557 clogP = 5.0; tPSA = 111		Year of discovery: 2005 Year of introduction: 2017 Discovered by: Wyeth Developed by: Puma Biotechnology Primary targets: HER2/EGFR Binding type: I1/2 Class: receptor tyrosine kinase Treatment: HER2-positive breast cancer Other name: HKI272 Oral bioavailability = not reported Elimination half-life = 14.6 h Protein binding = 99%

● = C ○ = H ● = O ● = N ● = S ● = F ● = Cl ● = Br ● = I

Neratinib, a dual irreversible inhibitor of EGFR (HER1) and HER2 receptors, was approved in 2017 as a single agent, for the extended adjuvant treatment of adult patients with early stage HER2-positive breast cancer, and in 2020, as a combination therapy with capecitabine for the treatment of adult patients with advanced or metastatic HER2-positive breast cancer who have received two or more prior anti-HER2-based regimens in the metastatic setting.

Neratinib irreversibly blocks the signaling pathways of both EGFR and HER2 which can cause uncontrolled cell growth in breast cancers. It is also active against cancer cells possessing both L858R and T790M mutations.[1]

Discovery of Neratinib[2-5]

Gefitinib is a potent, reversible EGFR inhibitor, which was converted to an irreversible inhibitor EKB-569 by replacing its morpholinoethoxy group at the kinase solvent front with a water-solubilizing Michael acceptor designed to attach covalently to a cysteine residue (Cys773 in EGFR and Cys805 in HER2) located in the ATP-binding pocket of these receptors. The polar dimethylamino group at the end of the Michael acceptor serves as an intramolecular base catalyst via a cyclic five-membered transition state to trigger the conjugative addition. This irreversible inhibitor elicited potent antitumor oral activity in the EGFR-dependent tumor models; however, it was less effective in the HER2-dependent tumor models due to its modest in vitro HER2 activity (HER2 IC$_{50}$ = 1.2 μM). Comparison of EKB-569 with lapatinib, the first dual EGFR/HER2 inhibitor approved for HER2-positive breast cancer, shows that EKB-569 lacks the terminal fluorobenzyloxy substituent which confers HER2 potency. Thus, a pyridin-3-ylmethoxy group, effectively a bioisostere of the fluorobenzyloxy group, was attached to EKB-569 to give neratinib which proved to have a balanced EGFR/HER2 inhibition profile (Fig. 1)

Binding Mode[6]

In the crystal structure of neratinib in complex with the EGFR T790M/L858R double mutant (Fig. 2), the kinase adopts an inactive conformation, with the αC-helix displaced. The enlarged hydrophobic pocket created by the outward rotation of the αC-helix appears to be induced by the bulky aniline substituent (pyridin-3-ylmethoxy) in neratinib. The quinoline N1 forms a single hydrogen bond with the amide NH of Met793 of the hinge region. The 2-pyridinyl group interacts with the hydrophobic residues in the enlarged pocket. Most importantly, the inhibitor forms the expected covalent bond between Cys-797 at the edge of the active site cleft and the crotonamide Michael-acceptor group, making binding irreversible (Fig. 3).

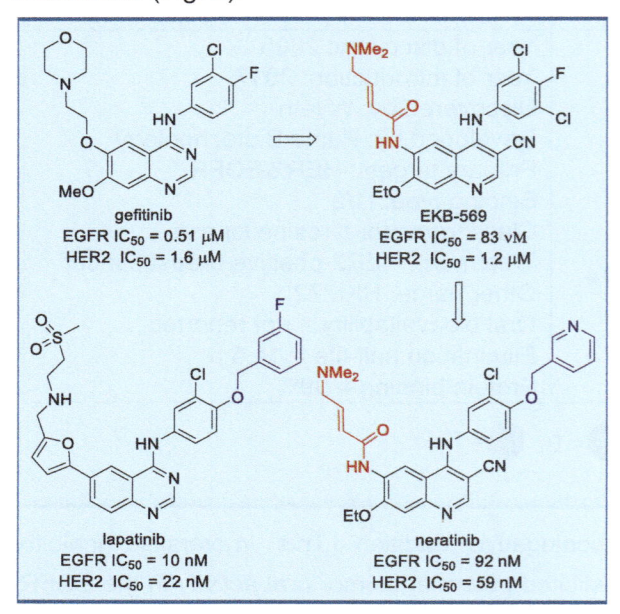

Figure 1. Structural optimization of neratinib.

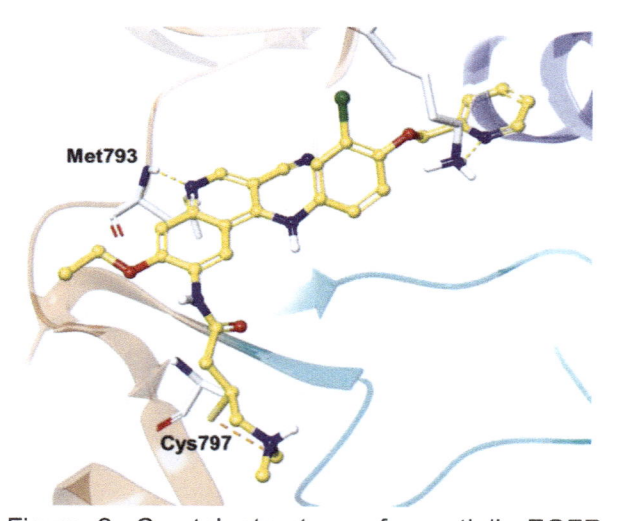

Figure 2. Crystal structure of neratinib–EGFR T790M/L858R mutant (PDB ID: 3W2Q).

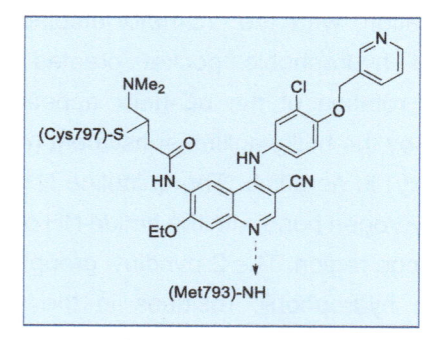

Figure 3. Neratinib–EGFR interactions.

1. E. D. Deeks. Neratinib: First global approval. *Drugs*. **2017**, *7*, 1695–1704.

2. H. R. Tsou, E. G. Overbeek-Klumpers, W. A. Hallett, M. F. Reich, M. B. Floyd, B. D. Johnson, R. S. Michalak, R. Nilakantan, R. C. Discafani, J. Golas, S. K. Rabindran, R. Shen, X. Shi, Y. F. Wang, J. Upeslacis, A. Wissner. Optimization of 6,7-disubstituted-4-(arylamino)quinoline-3-carbonitriles as orally active, irreversible inhibitors of human epidermal growth factor receptor-2 kinase activity. *J. Med. Chem*. **2005**, *48*, 1107–1131.

3. A. Wissner, E. Overbeek, M. F. Reich, M. B. Floyd, B. D. Johnson, N. Mamuya, E. C. Rosfjord, C. Discafani, R. Davis, X. Shi, S. K. Rabindran, B. C. Gruber, F. Ye, W. A. Hallett, R. Nilakantan, R. Shen, Y. F. Wang, L. M. Greenberger, H. R. Tsou. Synthesis and structure-activity relationships of 6,7-disubstituted 4-anilinoquinoline-3-carbonitriles. The design of an orally active, irreversible inhibitor of the tyrosine kinase activity of the epidermal growth factor receptor (EGFR) and the human epidermal growth factor receptor-2 (HER-2). *J. Med. Chem*. **2003**, *46*, 49–63.

4. A. Wissner, M. B. Floyd, S. K. Rabindran, R. Nilakantan, L. M> Greenberger, R. Shen, Y. F. Wang, H. R. Tsou. Syntheses and EGFR and HER-2 kinase inhibitory activities of 4-anilinoquinoline-3-carbonitriles: Analogues of three important 4-anilinoquinazolines currently undergoing clinical evaluation as therapeutic antitumor agents. *Bioorg. Med. Chem. Lett*. **2002**, *12*, 2893–2897.

5. A. Wissner, D. M. Berger, D. H. Boschelli, M. B. Floyd,Jr.; L. M. Greenberger, B. C. Gruber, B. D. Johnson, N. Mamuya, R. Nilakantan, M. F. Reich, R. Shen, H. R. Tsou, E. Upeslacis, Y. F. Wang, B. Wu, F. Ye, N. Zhang. 4-Anilino-6,7-dialkoxyquinoline-3-carbonitrile inhibitors of epidermal growth factor receptor kinase and their bioisosteric relationship to the 4-anilino-6,7-dialkoxyquinazoline inhibitors. *J. Med. Chem*. **2000**, *43*, 3244.

6. C. H. Yun, K. E. Mengwasser, A. V. Koms, M. Woo, H. Greulich, K. K. Wong, M. Meyerson, M. J. Eck. The T790M mutation in EGFR kinase causes drug resistance by Increasing the affinity for ATP. *PNAS*. **2008**, *105*, 2070–2075.

Chapter 5. VEGFR/Multikinase Inhibitors

Vascular endothelial growth factor (VEGF) is overexpressed in a variety of cancers, including colorectal, gastric, pancreatic, breast, prostate, lung, and skin (melanoma). It can induce endothelial cell proliferation and migration leading to new blood vessels (angiogenesis), which are essential for tumor oxygenation and growth. Because of its critical role in tumor angiogenesis, the VEGF/VEGFR pathway has been targeted extensively in the field of oncology, leading on the one hand to the injectable bevacizumab (a humanized immunoglobulin G monoclonal antibody) as the first FDA-approved anti-angiogenic drug in 2004, and on the other hand to the oral anti-angiogenic drugs sorafenib and sunitinib which were approved in 2005–2006 for the treatment of renal cell carcinoma (RCC) and other cancer types. The discovery of agents that block VEGF/VEGFR-mediated signaling pathways and subsequent angiogenesis has revolutionized the treatment of metastatic RCC.[1,2]

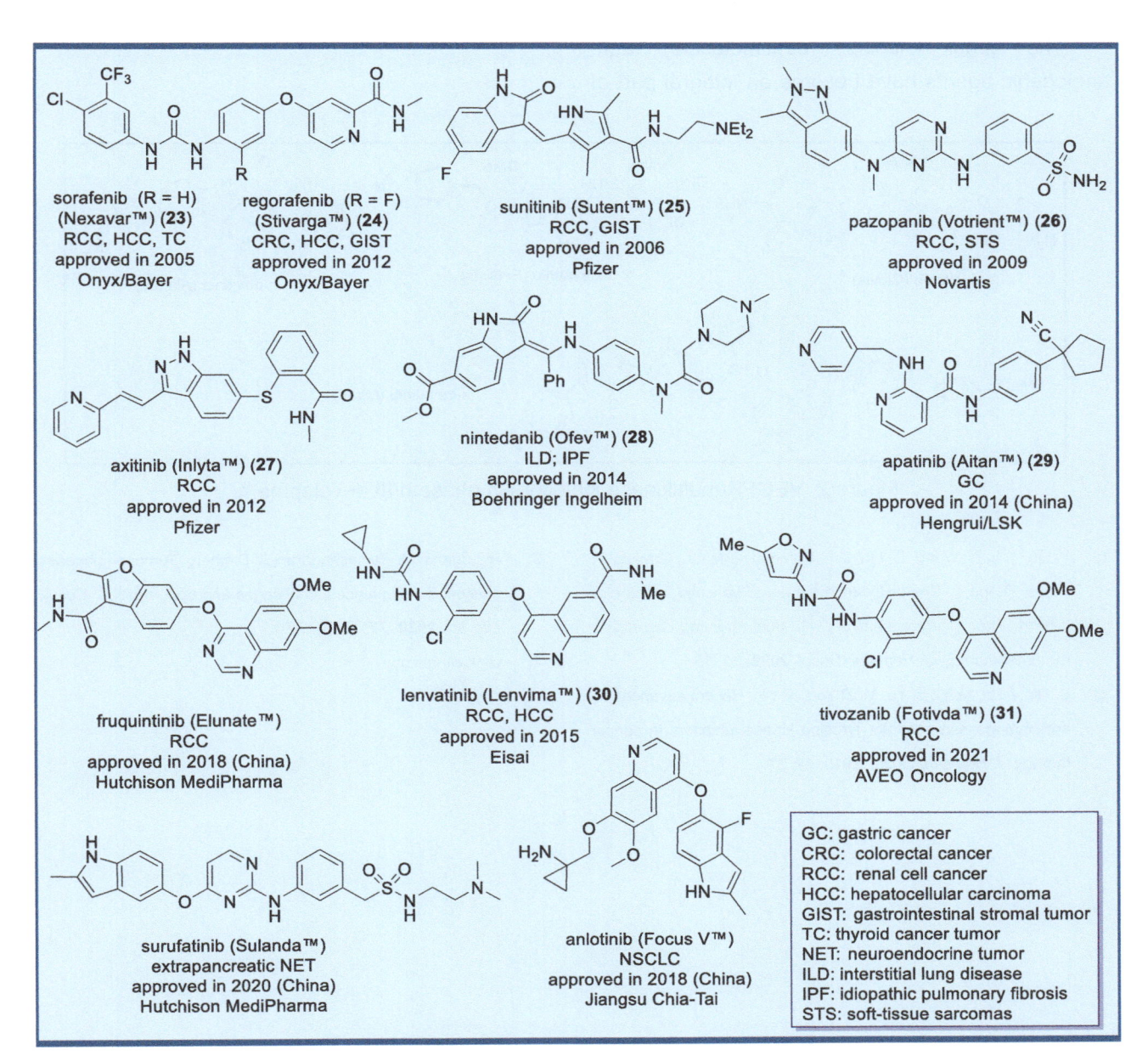

Figure 1. Approved VEGFR/multikinase inhibitors.

The angiogenic inhibitors sunitinib and sorafenib also target a number of different kinases (in addition to VEGFRs), which are involved in several signaling pathways. It is thought that inhibitors of multiple kinases that block VEGFR signaling together with other signaling pathways such as PDGFR might be more effective than an inhibitor that targets only VEGFR. Due to the multiple targets inhibited by these inhibitors, effects in different tumor types are likely to be mediated through a variety of mechanisms, resulting in potent anti-proliferative effects. However, the multikinase inhibitors frequently cause off-target side effects.[3]

The small-molecule (centimolecule) anti-angiogenic agents have become an integral part of the arsenal against RCC, HCC, and thyroid cancer (Fig. 1). Sunitinib and sorafenib reached respective peak sales of $1.2 and $1.0 billion in 2014, while bevacizumab (Avastin™) generated even larger sales – $7.1 billion in 2014.

This chapter describes the process by which VEGFR inhibitors **23–31** (Fig. 1) were discovered.

Research on antikinases for the VEGF/VEGFR pathway remains highly active as evidenced by numerous clinical trials of new candidates for RCC and other cancers, four of which are shown in Figure 2.[4]

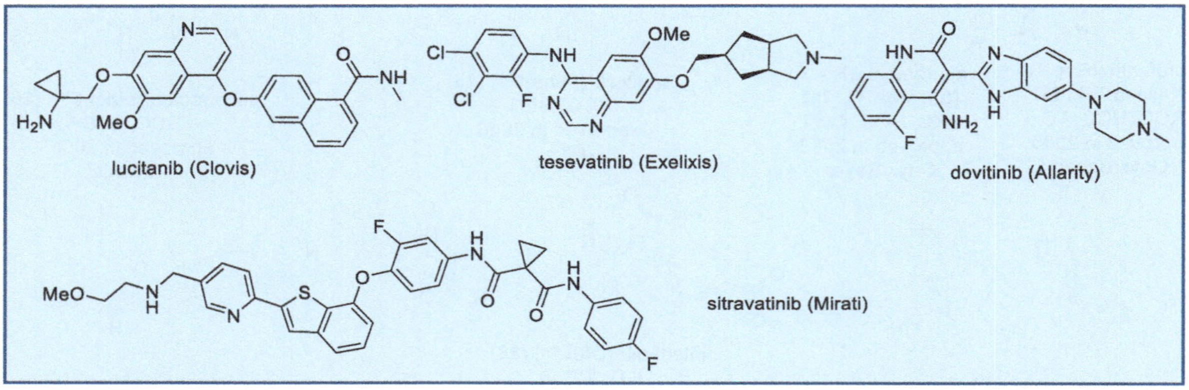

Figure 2. VEGFR/multikinase inhibitors in phase II/III development.

1. Y. Liu, Y. Li, Y. Wang, C. Lin, D. Zhang, J. Chen, L. Ouyang, F. Wu, J. Zhang, L. Chen. Recent progress on vascular endothelial growth factor receptor inhibitors with dual targeting capabilities for tumor therapy. *J. Hematol. Oncol.* **2022**, *15*, 89.
2. S. Qin, A. Li, M. Yi, S. Yu, M. Zhang, K. Wu. Recent advances on anti-angiogenesis receptor tyrosine kinase inhibitors in cancer therapy. *J. Hematol. Oncol.* **2019**, *12*, 27.
3. R. Lugano, M. Ramachandran, A. Dimberg. Tumor angiogenesis: causes, consequences, challenges and opportunities. *Cell Mol. Life Sci.* **2020**, *77*, 1745–1770.
4. clinicaltrials.gov.

5.1 Sorafenib (Nexavar™) (23)

Structural Formula	Space-filling Model	Brief Information
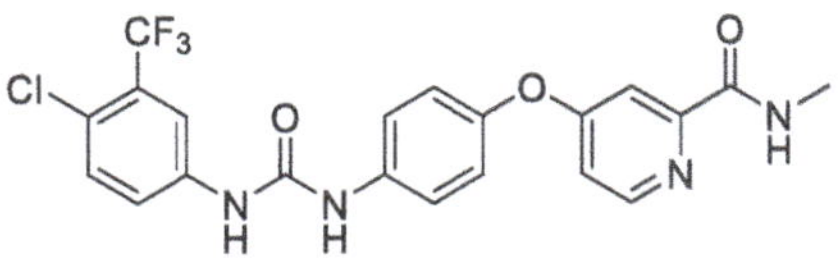 $C_{21}H_{16}ClF_3N_4O_3$; MW = 464 clogP = 5.5; tPSA = 92	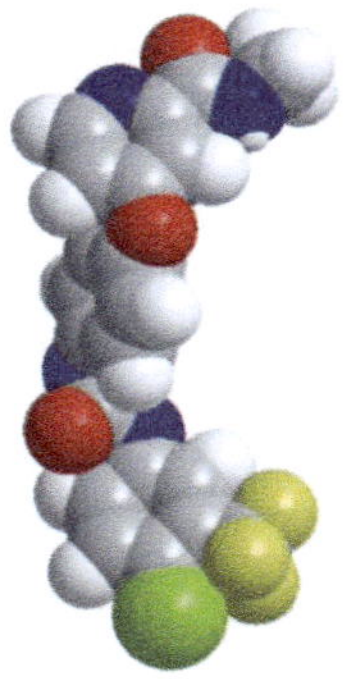	Year of discovery: 1999 Year of introduction: 2005 Discovered by: Onyx/Bayer Developed by: Bayer Primary target: VEGFR/multikinase Binding type: II Class: receptor tyrosine kinase Treatment: RCC, HCC, thyroid cancer Other name: BAY43-9006 Oral bioavailability = not reported Elimination half-life = 25–48 h Protein binding = 99.7%

● = C ○ = H ● = O ● = N ● = S ● = F ● = Cl ● = Br ● = I

Sorafenib is a multikinase inhibitor approved first in 2005 for the treatment of patients with advanced renal cell carcinoma (RCC), or kidney cancer, and in 2007, for the treatment of patients with unresectable hepatocellular carcinoma (HCC), or liver cancer. Its label was further expanded in 2013 for the treatment of late-stage (metastatic) differentiated thyroid cancer.[1–4]

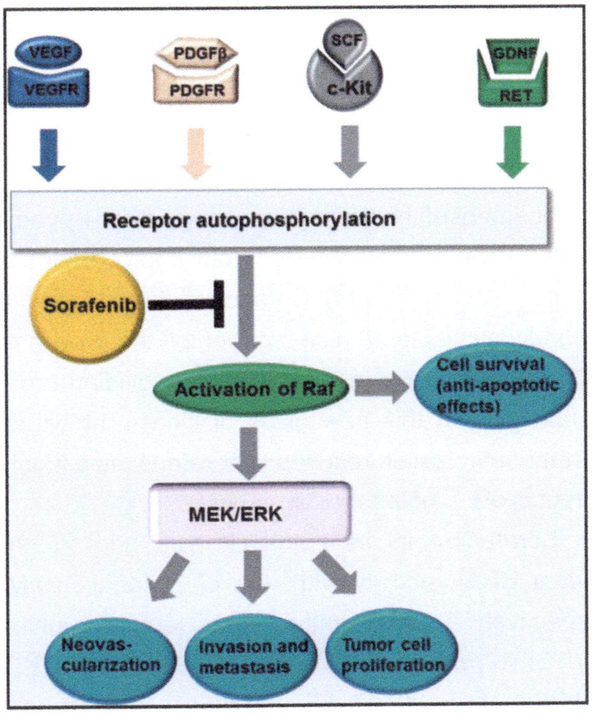

Sorafenib inhibits tumor cell proliferation and angiogenesis by targeting multiple serine/threonine and tyrosine kinases (RAF1, BRAF, VEGFR 1/2/3, PDGFR, KIT, FLT3, FGFR1, and RET) in multiple oncogenic signaling pathways (Fig. 1).[1,2] It was initially identified as a RAF kinase inhibitor and subsequently shown to inhibit several RTKs including vascular endothelial growth factor receptors (VEGFRs).

VEGF/VEGFR in Angiogenesis and Cancer[5–11]

Angiogenesis is the process which forms new blood vessels. Like normal tissues in the body, tumors need to develop new blood vessels to supply oxygen and nutrients in order for them to grow beyond 2 mm in diameter. Without adequate vascular support, tumors may become necrotic or even apoptotic. As a result, the tumor cells secrete vascular endothelial growth factors (VEGFs) to reach the nearby blood vessels in which VEGFRs are expressed on the surface of endothelial cells. Binding of VEGFs to their receptors (VEGFRs) leads to the dimerization of VEGFRs and activation of downstream signaling cascades primarily via the RAS/MEK and PI3K/AKT pathways (Figs. 1, 2), inducing endothelial cell proliferation and migration to form new vessels. With the new vessels sprouting and growing towards the tumor, the tumor enlarges,

Figure 1. Mode of action of sorafenib.

and the cancer cells penetrate nearby tissue, expanding throughout the body, and forming new colonies of cancer cells, a process known as metastases.

Angiogenesis is regulated by the VEGF/VEGFR signaling. The VEGF family consists of at least 7 members: VEGF-A, VEGF-B, VEGF-C, VEGF-D, VEGF-E, and placental growth factor 1 and 2 (PIGF-1 and PIGF-2, respectively). Within the VEGF family, VEGF-A is the most effective inducer of blood vessel growth.

VEGFs signal through three receptor tyrosine kinases (RTKs), VEGFR-1, VEGFR-2, and VEGFR-3 predominantly expressed by endothelial cells. VEGF-B, PIGF-1, and PIGF-2 specifically bind to VEGFR-1, whereas VEGF-C and VEGF-D bind to VEGFR-3. Although VEGF-A interacts with both VEGFR-1 and VEGFR-2 to induce endothelial cell proliferation and migration and to form new vessels, VEGFR-2 is thought to be more involved in the angiogenic processes than VEGFR-1, and therefore, VEGF-A/VEGFR2 pathway is a more attractive antiangiogenic target.

Endothelial cells are normally quiescent but can be induced to sprout and initiate angiogenesis by VEGF pro-angiogenic factors. Overexpression of VEGF has been found in various cancers, including colorectal carcinoma, gastric carcinoma, pancreatic carcinoma, breast cancer, prostate cancer, lung cancer, and melanoma. Because of its critical role in tumor angiogenesis, the VEGF/VEGFR pathway has attracted significant attention in the field of oncology.

The essential role of angiogenesis in tumor growth was first proposed by Judah Folkman in 1971.[11] Since then, a number of angiogenic inhibitors have been developed to prevent new vessel formation and result in tumor dormancy,[5,6] leading to bevacizumab as the first FDA-approved antiangiogenic drug in 2004. Bevacizumab is a humanized immunoglobulin G monoclonal antibody that binds to VEGF with high specificity, thereby blocking VEGF-mediated signaling pathways and thus angiogenesis. In 2005, the first small-molecule anti-angiogenic agent, sorafenib, was approved for the treatment of RCC. It is a multi-kinase inhibitor (MKI) which effectively blocks the action of several hyperactivated receptor kinases including BRAF and VEGFR. A MKI may provide better efficacy in the clinic than a highly selective single kinase inhibitor because tumor progression usually involves the action of multiple kinases rather than just one, although adverse effects can be problematic. Other licensed anti-angiogenic multi-kinase inhibitors include cabozantinib, lenvatinib, regorafenib, nintedanib, sunitinib, and pazopanib (Fig. 2).

Discovery of Sorafenib[12–18]

At the onset of the medicinal chemistry program that led to sorafenib, there were no marketed kinase inhibitors in the field of oncology, but the two first-generation targeted anticancer agents – imatinib (Gleevec™, Novartis) and erlotinib (Tarceva™, Genentech), which target BCR-ABL and EGFR, respectively – had begun to show clinical benefit.

The ~11-year journey of sorafenib from bench to bedside started in 1995 with HTS screening of about 200,000 compounds for RAF1 kinase inhibitory activity. This effort led to 3-thienyl urea **1** (Fig. 3) as a lead compound with a cRAF1 IC_{50} of 17 µM, which was improved by 10-fold with a single 4-methyl substitution on the phenyl ring to generate compound **2**. A combinatorial library synthesis was then carried out on the urea analogs with simultaneous variations on both the thiophene heterocycle and phenyl substituents, giving rise to the 3-amino-isoxazole **3** with an IC_{50} of 1.1 µM. Replacement of the distal phenyl ring with a 4-pyridyl moiety gave **4** that showed 5-fold improvement in potency (IC_{50} = 230 nM). Compound **4** was found to be orally available in mice, and inhibit the growth of HCT116 xenografts in vivo, thereby providing proof of principle for this new class of kinase inhibitors. Further optimization resulted in sorafenib as a highly potent cRAF1 inhibitor (IC_{50} = 6 nM).

Sorafenib inhibits both wild-type and V599E mutant BRAF activity with IC_{50} of 22 and 38 nM, respectively. Subsequently, it was found to inhibit mVEGFR2, mVEGFR3, mPDGFRβ, FLT3, and KIT with IC_{50} of 15, 20, 57, 58, and 68 nM, respectively. Sorafenib weakly inhibited FGFR-1 with IC_{50} of 580 nM, and it was inactive against ERK-1, MEK-1, EGFR, HER-2, IGFR-1, MET, PKB, PKA, CDK1/cyclin B, PKCα, PKCγ, and PIM-1.[18]

Oral sorafenib inhibited tumor growth and angiogenesis in a wide variety of preclinical cancer models, including human breast, colon, ovarian, thyroid, and pancreatic carcinomas, melanoma, and RCC, HCC, and NSCLC. Due to the multiple targets inhibited by sorafenib, effects in different tumor types are likely to be mediated through a variety of mechanisms. Although it may be difficult or even impossible to determine the precise contribution of individual targets to each tumor type, evaluation of preclinical data may help elucidate the contributions of given mechanisms of action of sorafenib in different tumor types.

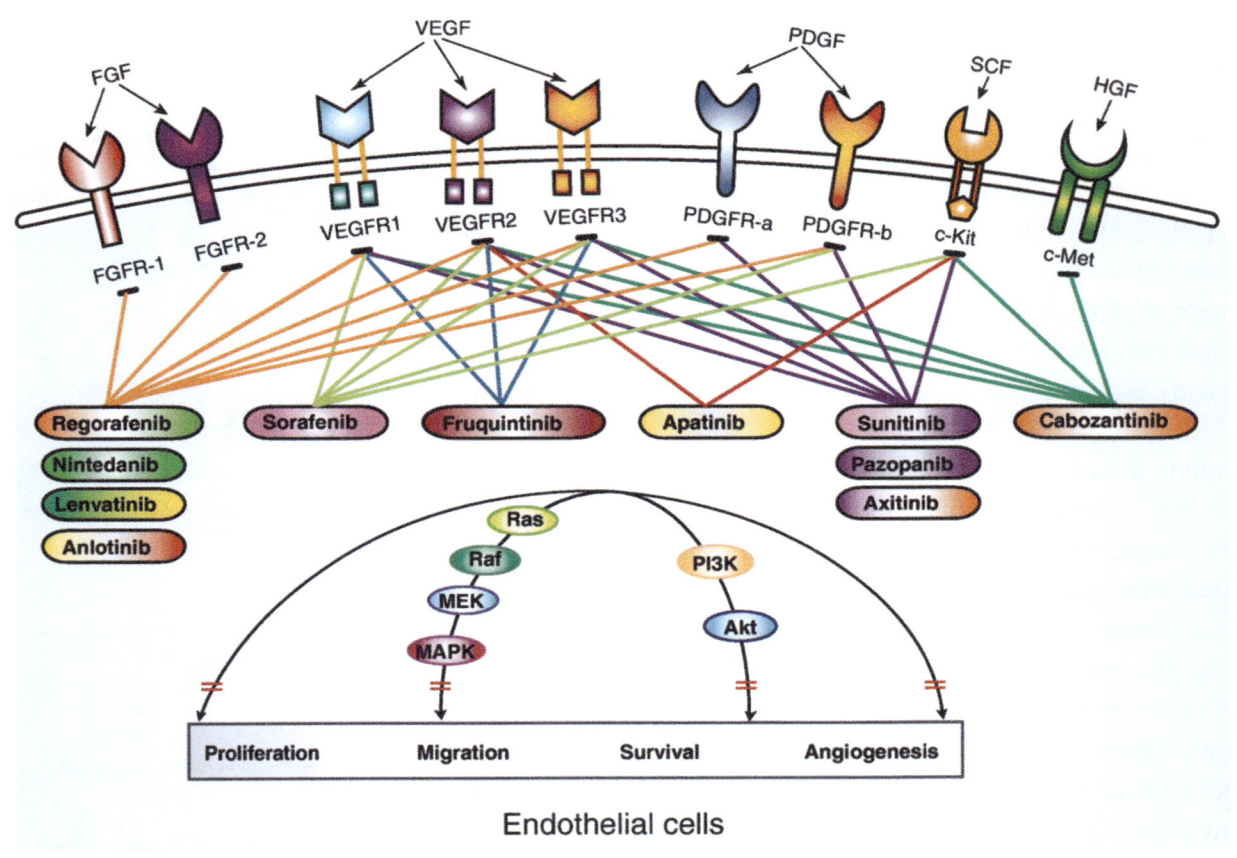

Figure 2. Main targets of anti-angiogenic TKIs. Adapted from *J. Hematol. Oncol.* **2019**, *12*, 27.

Figure 3. Structural optimization of sorafenib.

Sorafenib has a well-characterized tolerability and safety profile, with strategies available to prevent and manage adverse effects such as hand–foot skin reactions. With its novel mechanisms of action, the favorable safety profile of sorafenib in single-agent trials has led to its evaluation in combination with other anticancer therapies, including cytotoxic chemotherapy, immunological agents, and anti-angiogenic agents in a broad range of tumor types. It is anticipated that its favorable tolerability profile will make sorafenib suitable for longer-term administration, either as a monotherapy or in combination with other therapies.

It is noteworthy that sorafenib is currently the only systemic agent, which is approved for use in hepatocellular carcinoma HCC,[19,20] based on the results of the pivotal clinical trials which showed significantly longer median overall survival (OS) and time to radiological progression (TTP) with this agent.

Binding Mode[14]

The complex of sorafenib with a nonphosphorylated VEGFR2 construct comprising the catalytic and juxtamembrane (JM) domain shows the activation loop adopts a DFG-out position (Figs. 4, 5). The net effect of sorafenib's interactions with the kinase is to stabilize the DFG motif in an inactive conformation. The lipophilic trifluoromethyl phenyl ring at the opposite end of the molecule inserts into a hydrophobic pocket formed between the α-C and α-E helices and amino-terminal regions of the DFG motif and catalytic loop (Figs. 4, 5).[6] The urea functionality forms two crucial hydrogen bonds, one with the backbone aspartate of the DFG loop, and the other with the glutamate side chain of the α-C helix. In the auto-inhibited state, the DFG-out segment, particularly Phe1047, blocks ATP-binding. When the JM rearranges out of the regulatory domain pocket (RDP), then the DFG is unlocked and able to switch to the open activated state. Autophosphorylation, which activates the kinase presumably through stabilization of an active conformation, was shown to occur on the JM and then the DFG-containing activation loop sequentially.

Sorafenib fills the binding channel, making an important H-bond to Asp1046, two H-bonds to the side chain carboxylate oxygen of Glu885 and two H-bonds to Cys919 (Figs. 4–6). The selectivity of sorafenib derives from the formation of these four hydrogen bonds and attractive van der Waals (dispersion) interactions (Fig. 6).

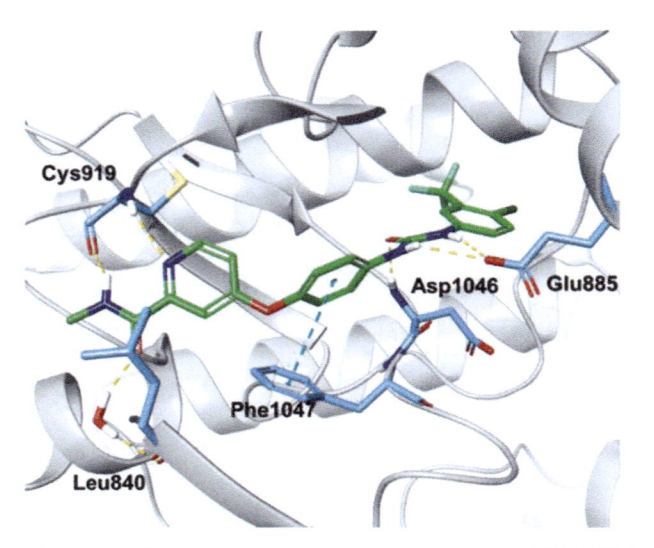

Figure 4. Co-crystal structure of sorafenib–VEGFR2 (juxtamembrane and kinase domains) (PDB ID: 4ASD).

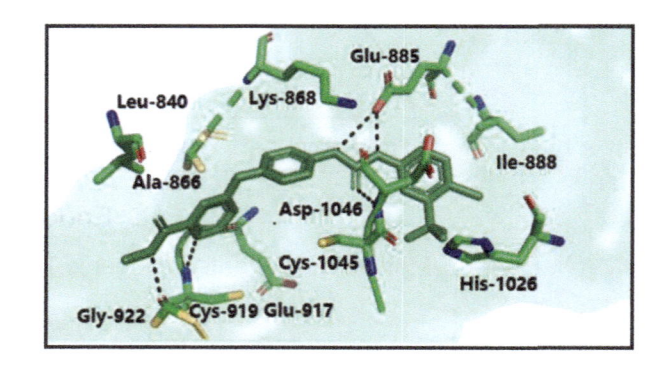

Figure 5. Co-crystal structure of sorafenib–VEGFR2 kinase domain (PDB ID: 4ASD).

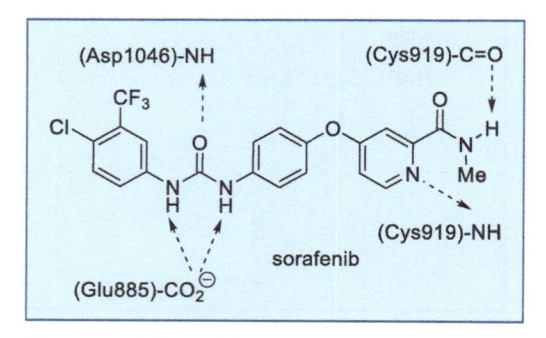

Figure 6. VEGFR2–sorafenib interactions.

1. S. Wilhelm, C. Carter, M. Lynch, T. Lowinger, J. Dumas, R. A. Smith, B. Schwartz, R. Simantov, S. Kelley. Discovery and development of sorafenib: A multikinase inhibitor for treating cancer. *Nat. Rev. Drug Discov.* **2006**, *5*, 835–844.
2. S. M. Wilhelm, L. Adnane, P. Newell, A. Villanueva, J. M. Llovet, M. Lynch. Preclinical overview of sorafenib, a multikinase inhibitor that targets both Raf and VEGF and PDGF receptor tyrosine kinase signaling. *Mol. Cancer Ther.* **2008**, *7*, 3129–3140.
3. B. Escudier, T. Eisen, W. M. Stadler, C. Szczylik, S. Oudard, M. Siebels, S. Negrier, C. Chevreau, E. Solska, A. A. Desai, F. Rolland, T. Demkow, T. E. Hutson, M. Gore, S. Freeman, B. Schwartz, M. Shan, R. Simantov, R. M. Bukowski. Sorafenib in advanced clear-cell renal-cell carcinoma. *N. Engl. J. Med.* **2007**, *356*, 125–134.
4. J. M. Llovet, S. Ricci, V. Mazzaferro, P. Hilgard, E. Gane, J.-F. Blanc, A. C. de Oliveira, A. Santoro, J.-L. Raoul, A. Forner, M. Schwartz, C. Porta, S. Zeuzem, L. Bolondi, T. F. Greten, P. R. Galle, J.-F. Seitz, I. Borbath, D. Häussinger, T. Giannaris, M. H. Shan, M. Moscovici, D. Voliotis, J. Bruix. Sorafenib in advanced hepatocellular carcinoma. *N. Engl. J. Med.* **2008**, *359*, 378–390.
5. Y. Liu, Y. Li, Y. Wang, C. Lin, D. Zhang, J. Chen, L. Ouyang, F. Wu, J. Zhang, L. Chen. Recent progress on vascular endothelial growth factor receptor inhibitors with dual targeting capabilities for tumor therapy. *J. Hematol. Oncol.* **2022**, *15*, 89.
6. S. Qin, A. Li, M. Yi, S. Yu, M. Zhang, K. Wu. Recent advances on anti-angiogenesis receptor tyrosine kinase inhibitors in cancer therapy. *J. Hematol. Oncol.* **2019**, *12*, 27.
7. L. Liu, Y. Cao, C. Chen, X. Zhang, A. McNabola, D. Wilkie, S. Wilhelm, M. Lynch, C. Carter. Sorafenib blocks the RAF/MEK/ERK pathway, inhibits tumor angiogenesis, and induces tumor cell apoptosis in hepatocellular carcinoma model PLC/PRF/5. *Cancer Res.* **2006**, *66*, 11851–11858.
8. M. Shibuya. Vascular endothelial growth factor (VEGF) and Its receptor (VEGFR) signaling in angiogenesis: A Crucial target for anti- and pro-angiogenic therapies. *Genes Cancer.* **2011**, *2*, 1097–1105.
9. G. Niu, X. Chen. Vascular endothelial growth factor as an anti-angiogenic target for cancer therapy. *Curr. Drug Targets.* **2010**, *11*, 1000–1017.
10. R. Lugano, M. Ramachandran, A. Dimberg. Tumor angiogenesis: Causes, consequences, challenges and opportunities. *Cell. Mol. Life Sci.* **2020**, *77*, 1745–1770.
11. J. Folkman. Tumor angiogenesis: Therapeutic implications. *N. Engl. J. Med.* **1971**, *285*, 1182–1186.
12. R. A. Smith, J. Barbosa, C. L. Blum, M.A. Bobko, Y. V. Caringal, R. Dally, J. S. Johnson, M. E. Katz, N. Kennure, J. Kingery-Wood, W. Lee, T. B. Lowinger, J. Lyons, V. Marsh, D. H. Rogers, S. Swartz, T. Walling, H. Wilda. Discovery of heterocyclic ureas as a new class of Raf kinase inhibitors: Identification of a second generation lead by a combinatorial chemistry approach. *Bioorg. Med. Chem. Lett.* **2001**, *11*, 2775–2778.
13. T. B. Lowinger, B. Riedl, J. Dumas, R. A. Smith. Design and discovery of small molecules targeting raf-1 kinase. *Curr. Pharm. Des.* **2002**, *8*, 2269–2278.
14. M. McTigue, B. W. Murray, J. H. Chen, Y. L. Deng, J. Solowiej, R. S. Kania. Molecular conformations, interactions, and properties associated with drug efficiency and clinical performance among VEGFR TK inhibitors. *PNAS.* **2012**, *109*, 18281–18289.
15. B. Riedl, J. Dumas, U. Khire, T. B. Lowinger, W. J. Scott, R. A. Smith, J. E. Wood, M. K. Monahan, R. Natero, J. Renick, R. N. Sibley. ω-carboxyaryl substituted diphenyl ureas as raf kinase inhibitors. US 7235576, **2007**.
16. B. Riedl, J. Dumas, U. Khire, T. B. Lowinger, W. J. Scott, R. A. Smith, J. E. Wood, M. K. Monahan, R. Natero, J. Renick, R. N. Sibley. ω-carboxyaryl substituted diphenyl ureas as P38 kinase inhibitors. US 7897623, **2011**.
17. J. Dumas, W. J. Scott, J. Elting, H. H. Makdad. Arylureas with angiogenesis inhibiting activity. US 8618141, **2013**.
18. S. M. Wilhelm, C. Carter, L. Y. Tang, D. Wilkie, A. McNabola, H. Rong, C. Chen, X. M. Zhang, P. Vincent, M. McHugh, Y. C. Cao, J. Shujath, S. Gawlak, D Eveleigh, B. Rowley, L. Liu, L. Adnane, M. Lynch, D. Auclair, I.Taylor, R. Gedrich, A. Voznesensky, B. Riedl, L. E. Post, G. Bollag, P. A. Trail. BAY 43-9006 exhibits broad spectrum oral antitumor activity and targets the RAF/MEK/ERK pathway and receptor tyrosine kinases involved in tumor progression and angiogenesis. *Cancer Res.* **2004**, *64*, 7099–7109.
19. G. M. Keating, A. Santoro. Sorafenib: A review of its use in advanced hepatocellular carcinoma. *Drugs.* **2009**, *69*, 223–240.
20. M. Keating. Sorafenib: A Review in Hepatocellular Carcinoma. *Targ. Oncol.* **2017**, *12*, 243–253.

5.2 Regorafenib (Stivarga™) (24)

Structural Formula	Space-filling Model	Brief Information
$C_{21}H_{17}ClF_4N_4O_3$; MW = 482 clogP = 5.2; tPSA = 92		Year of discovery: 2003 Year of introduction: 2012 Discovered by: Bayer and Onyx Developed by: Bayer and Onyx Primary target: VEGFR/multikinase Binding type: II Class: receptor tyrosine kinase Treatment: colorectal cancer, GIST, HCC Other name: BAY73-4506 Oral bioavailability = not reported Elimination half-life = 26-28 h Protein binding = 99.5%

● = C ○ = H ● = O ● = N ● = S ● = F ● = Cl ● = Br ● = I

Regorafenib is another multi-kinase inhibitor approved first in 2012 for advanced colorectal cancer, then in 2013, for advanced gastrointestinal stromal tumors (GIST); and more recently in 2017 for hepatocellular carcinoma (HCC or liver cancer). It is the first drug approved for the treatment of HCC in patients who have progressed during or after sorafenib therapy.[1–6]

Regorafenib is a follow-up to sorafenib co-developed by Bayer and Onyx.[7,8] It is identical to sorafenib with the exception of addition of a fluorine atom ortho to the urea at the central phenyl group (Fig. 1).[9,10] Biochemically, this fluorine results in a ~20-fold[10] increase in potency against VEFGR2 relative to sorafenib. One explanation for such a boost in potency is that the electron-withdrawing fluorine increases the acidity of the urea hydrogens, thereby resulting in stronger H-bonds to the side chain carboxylate oxygen of Glu885 (Fig. 2). In addition, the intramolecular interaction of the F atom with the adjacent NH of regorafenib "masks" one of the urea H-bond donors. This improves membrane permeability and cellular activity. Regorafenib also potently inhibits VEGFR1-3, PDGFR, FGFR and BRAF, including BRAFV600E, KIT and RET with IC_{50}'s at nanomolar concentrations. Thanks to its increased potency against VEGFR2, PDGFRβ and other kinases, regorafenib is orally dosed at 160 mg (4 × 40 mg tablets, once daily), much lower than 400 mg (2 × 200 mg tablets, twice daily) for the des-fluorine analog, sorafenib.

The co-crystal structure of regorafenib bound to any protein kinase has been reported. However, because of its high structural similarities to sorafenib, both inhibitors likely share the same binding modes in complex with VEGFR2. More specifically, regorafenib fills the binding channel of VEGFR2, making an important H-bond to Asp1046, two H-bonds to the side chain carboxylate oxygen of Glu885 and two H-bonds to Cys919 (Fig. 2).

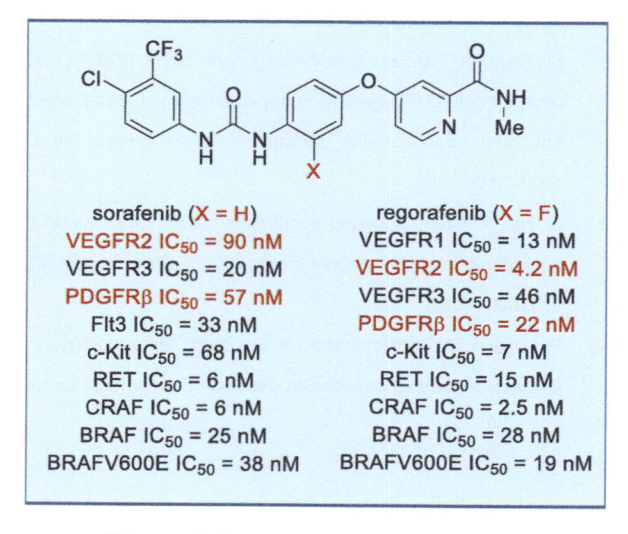

sorafenib (X = H)
VEGFR2 IC_{50} = 90 nM
VEGFR3 IC_{50} = 20 nM
PDGFRβ IC_{50} = 57 nM
Flt3 IC_{50} = 33 nM
c-Kit IC_{50} = 68 nM
RET IC_{50} = 6 nM
CRAF IC_{50} = 6 nM
BRAF IC_{50} = 25 nM
BRAFV600E IC_{50} = 38 nM

regorafenib (X = F)
VEGFR1 IC_{50} = 13 nM
VEGFR2 IC_{50} = 4.2 nM
VEGFR3 IC_{50} = 46 nM
PDGFRβ IC_{50} = 22 nM
c-Kit IC_{50} = 7 nM
RET IC_{50} = 15 nM
CRAF IC_{50} = 2.5 nM
BRAF IC_{50} = 28 nM
BRAFV600E IC_{50} = 19 nM

Figure 1. Regorafenib vs. sorafenib.

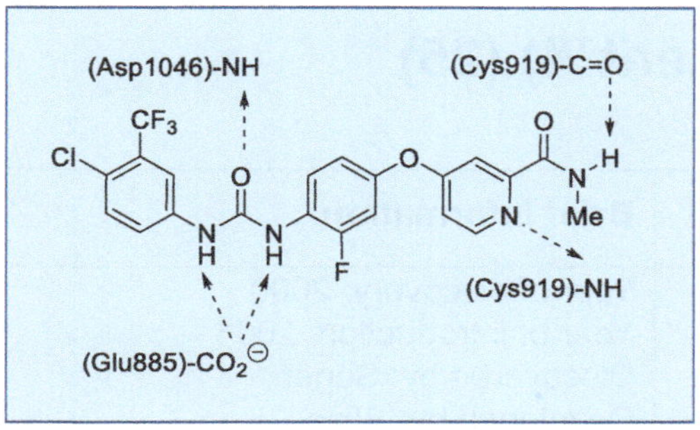

Figure 2. Putative interactions of regorafenib–VEGFR2 based on X-ray data of sorafenib–VEGFR2.

1. S. M. Wilhelm, J. Dumas, L. Adnane, M. Lynch, C. A. Carter, G. Schütz, K. H. Thierauch, D. Zopf. Regorafenib (BAY 73-4506): a new oral multikinase inhibitor of angiogenic, stromal and oncogenic receptor tyrosine kinases with potent preclinical antitumor activity. *Int. J. Cancer*, **2011**, *129*, 245–255.
2. J. Gotfrit, M. Vickers, S. Sud, T. Asmis, C. Cripps, R. Goel, T. Hsu, D. Jonker, R. Goodwin. Real-life treatment of metastatic colorectal cancer with regorafenib: A single-centre review. *Curr. Oncol.* **2017**, *24*, 234–239.
3. R. S. Finn. Review of regorafenib for the treatment of hepatocellular carcinoma. *Hepatology*. **2017**, *13*, 492 495.
4. B. Gyawali, V. Prasad. Me too drugs with limited benefits – The tale of regorafenib for HCC. *Nat. Rev. Clin. Oncol.* **2017**, *14*, 653–654.
5. A. G. Duffy, T. F. Greten. Regorafenib as second-line therapy in hepatocellular carcinoma. *Lancet.* **2017**, *389*, 56–66.
6. J. Bruix, S. K. Qin, P. Merle, A. Granito, Y. H. Huang, G. B. M. Pracht, O. Yokosuka, O. Rosmorduc, V. Breder, R. Gerolami, G. Masi, P. J. Ross, T. Q. Song, J.-P. Bronowicki, I. O. Hourmand, M. Kudo, A.-L. Cheng, J. M. Llovet, R. S. Finn, M. A. LeBerre, A. Baumhauer, G. Meinhardt, G. H. Han. Regorafenib for patients with hepatocellular carcinoma who progressed on sorafenib treatment (RESORCE): A randomised, double-blind, placebo-controlled, phase 3 trial. *Lancet.* **2017**, *389*, 56–66.
7. J. Stiehl, W. Heilmann, M. Lögers, J. Rehse, M. Gottfried, S. Wichmann. Process for the preparation of 4-{4-[({[4-chloro-3-(trifluoromethyl)-phenyl]amino]-3-flurphenoxy-*N*-ethylpyridie-carboxamide, its salts and monohydrate. US Patent 9458107 B2, **2016**.
8. S. Boyer, J. Dumas, B. Riedl, S. Wilhelm. Fluro substituted omega-carboxyaryl diphenyl urea for the treatment and prevention of diseases and conditions. US Patent 8637553B2 **2014**.
9. P. A. Harris. Inhibitors of vascular endothelial growth factor receptor. *Top. Med. Chem.* **2018**, *28*, 105–140.
10. G. Goel. Evolution of regorafenib from bench to bedside in colorectal cancer: Is it an attractive option or merely a "me too" drug? *Cancer Manag. Res.* **2018**, *10*, 425 437.

5.3 Sunitinib (Sutent™) (25)

Structural Formula	Space-filling Model	Brief Information
$C_{22}H_{27}FN_4O_2$; MW = 398 clogP = 3.0; tPSA = 73		Year of discovery: 2000 Year of introduction: 2006 Discovered by: Sugen Developed by: Pfizer Primary targets: VEGFR/multikinase Binding type: II (VEGFR2) Class: receptor tyrosine kinase Treatment: RCC, GIST Other name: SU-11248 Oral bioavailability = not reported Elimination half-life = 70 h Protein binding = 95%

● = C   ○ = H   ● = O   ● = N   ● = S   ● = F   ● = Cl   ● = Br   ● = I

Sunitinib is a multikinase inhibitor approved first in 2006 for gastrointestinal stromal tumors (GIST) and advanced renal cell carcinoma (RCC), and then in 2011 for patients with progressive neuroendocrine cancerous tumors located in the pancreas that cannot be removed by surgery or that have spread to other parts of the body (metastatic). Its indications were further expanded in 2017 to adjuvant treatment of adult patients at high risk of recurrent RCC.

Sunitinib is one of the earliest small-molecule antiangiogenic agents that prevent solid tumors from growing blood vessels so that they become starved of oxygen and nutrients required for growth. However, all VEGF angiogenic inhibitors are associated with a class-wide cardiovascular toxicity, especially hypertension, which requires monitoring of blood pressure during treatment and can, in most cases, be effectively controlled with standard antihypertensive therapy.[1-4]

Background

In the research field of tumor progression, it has long been known that neoplasms are often accompanied by an increased and unique vascularity; however, the molecular mechanisms responsible for these vascular changes in tumor tissue were not well understood and remained elusive for over 50 years. Advances in molecular biology and DNA cloning techniques in the middle 1980s through the early 1990s enabled the discovery and functional characterization of key mutated genes involved in the progression of human cancer. The identification of oncogenes, especially RTK families, provided new understanding of the encoded proteins critical to signal transduction pathways and the mechanisms promoting tumor progression. In 1971, Judah Folkman introduced the concept of neoangiogenesis as a critical mechanism by which neoplasms secrete a "tumor angiogenic factor or TAF" in order to recruit vascular supply and gain access to nutrients and oxygen.[5] Efforts to identify TAF resulted in the cloning and identification of two receptor tyrosine kinases VEGFR1 and VEGFR2 in 1992.[6,7]

VEGF was found to be overexpressed in a variety of tumor types and identified as an independent prognostic factor. VEGF expression is induced in cells under hypoxic conditions by hypoxia-inducible factor (HIF).[7] All members of the VEGF family stimulate cellular responses by binding to tyrosine kinase receptors (VEGFRs) on the cell surface, causing them to dimerize and become activated through transphosphorylation (Fig. 1) (for an in-depth discussion on the role of VEGF/VEGFR in angiogenesis and cancer, see Section 5.1).[4,5] With the discovery of VEGF, intensive research was undertaken to develop monoclonal antibody VEGF binders as well as small-molecule VEGFR inhibitors to prevent new vessel formation and induce tumor dormancy,[6-8] leading to the approval of bevacizumab in 2004, and sorafenib and sunitinib in 2005 and 2006, respectively. As with sorafenib, sunitinib inhibits multiple kinases,[1,2] including VEGFR1, VEGFR2, VEGFR3, PGFRα/β, and KIT (Fig. 1).[9]

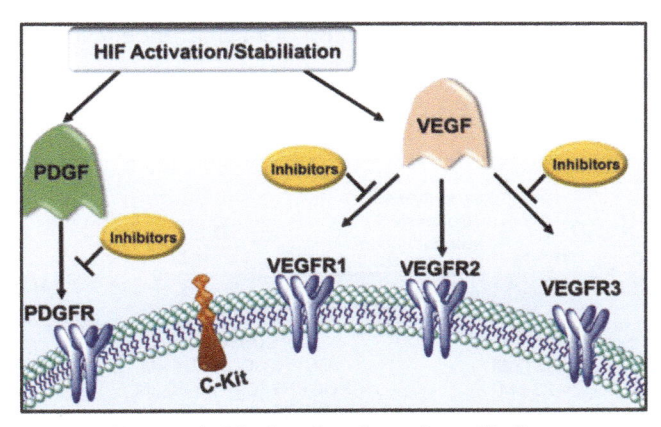

Figure 1. Mode of action of sunitinib.

Sunitinib simultaneously targets both tumor and surrounding supportive cells. Thus, it can interact with the complex multi-molecular lesions that drive tumor growth and survival. Specific single-target inhibitors are unable to achieve a similar effect. Another advantage of the multi-targeting sunitinib is that resistance is less likely to arise. A disadvantage of using multi-targeted sunitinib is that it might be difficult to determine which particular kinase inhibition(s) results in an antitumor effect. Therefore, individualized targeted therapy is required to resolve this issue.

Antitumor activity of sunitinib was observed in numerous tumor types, including RCC, GIST, NETs, NSCLC, HCC, sarcoma other than GIST, thyroid cancer, and melanoma. Although sunitinib shows promise as a monotherapy for these previously treated patients, add-on therapy with this multikinase inhibitor has been increasingly adopted to avoid drug resistance. More recently, it was reported that sunitinib quantitatively and qualitatively modified Tregs and repressed regulatory T cells to overcome immunotolerance in a murine model of hepatocellular cancer.[10] This suggests the potential of sunitinib as a therapeutic immune activator for HCC control.[1,2]

Discovery of Sunitinib[11-17]

In 1988, Brian Druker at Oregon Health and Sciences University and Nick Lydon of the former Ciba-Geigy (which after merger with Sandoz became Novartis in 1996) started a collaboration using targeted chemotherapies to treat CML, which ushered in the era of kinase inhibitors as precision medicines. Three year later – in 1991, Sugen was founded by two eminent signal transduction scientists Joseph Schlessinger (the "S" in Sugen) and Axel Ullrich (the "U" in Sugen) with the mission of developing protein kinase inhibitors. At that time, targeting kinases was considered a foolhardy approach, since most pharmaceutical companies viewed them undruggable, reminiscent of how work on KRAS was viewed only 10 years ago. After all, there are 518 protein kinases in the human body, and yet so many different kinases are expressed at various levels in individual cells. Thus, it was thought very challenging, if not impossible, to achieve a desirable level of selectivity for a particular kinase. Additionally, it was believed that a multikinase inhibitor would elicit too many off-target side effects to become a viable drug. Despite the overwhelming skepticism towards the feasibility of kinase inhibitors, Sugen scientists forged ahead with research targeting the ATP-binding site with small molecules.

While Genentech was focusing on monoclonal antibodies that bind to VEGFs to prevent activation of their receptors VEGFRs, Sugen was exploring small-molecule VEGFR inhibitors to block the VEGF/VEGFR signaling pathway and subsequently attenuate blood vessel formation required for tumor growth.

The medicinal chemistry program at Sugen started in 1994 with the random screening of a commercial compound library collection using whole-cell ligand-dependent autophosphorylation assays.[12] These efforts led to the identification of the 3-substituted (Z)-indolin-2-one **1** (Fig. 2) as a potent and selective inhibitor of VEGFR2. Although **1** is overwhelmingly favored through stabilization via an intramolecular hydrogen bond, some isomerization to 2 (a nonselective inhibitor of RTKs) occurs in solution,[18] complicating its further development as a single agent. Incorporation of 3′,5′-dimethyl substituents onto the pyrrole ring, which keeps the olefin geometry in the Z configuration, gave SU5416 with only slightly reduced potency against PDGF (IC_{50} = 20.26 µM) and VEGFR2 (IC_{50} = 1.04 µM). It became the first VEGFR2 inhibitor to enter into the clinic for cancer therapy in 1997 – one year ahead of imatinib from Novartis. SU5416 was dosed through IV infusion rather than oral administration due to its poor oral bioavailability which resulted from its poor solubility and metabolic instability (oxidation of the two methyl groups).[19] However, SU5416 failed to show significant efficacy, and its clinical development was halted in 2002. The lack of robust activity of SU5416 was attributed to the incomplete inhibition of its intended targets, a result of inadequate systemic exposure for therapeutic efficacy possibly due to very short half-life (<90 min) and 50–60% auto-induction of its clearance following chronic dosing.[20,21]

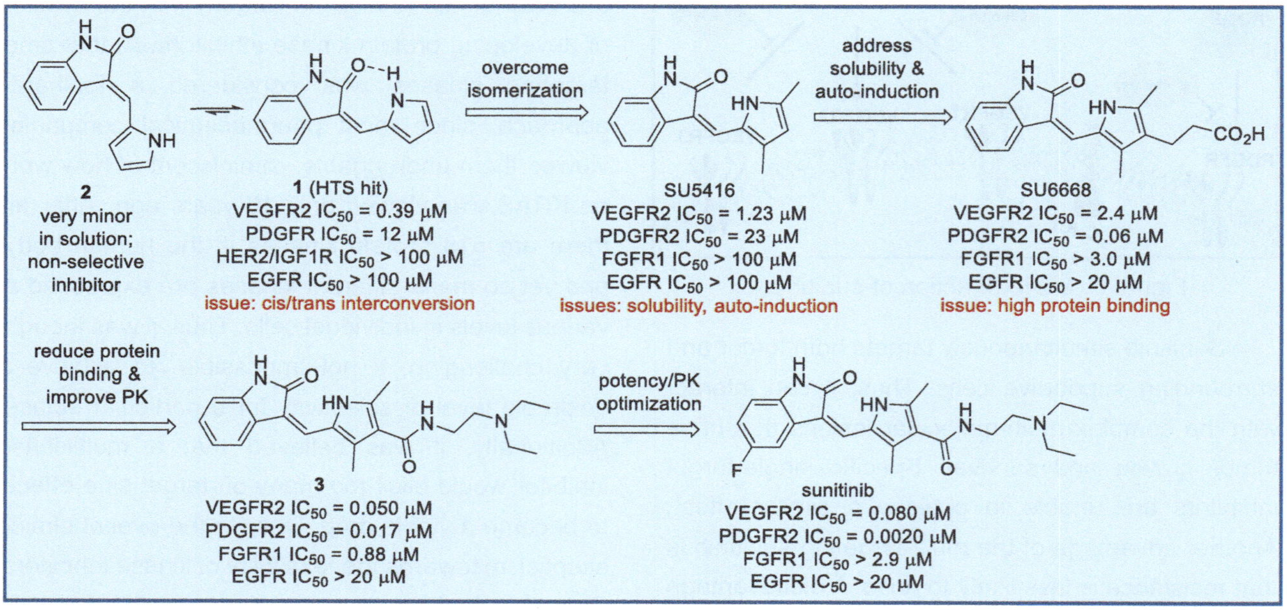

Figure 2. Structural optimization of sunitinib.

Consequently, second-generation VEGF inhibitors were sought to improve solubility.[13] Introduction of a water-solubilizing $-(CH_2)_3CO_2H$ side chain at the C4 position of the pyrrole gave SU6668 (Fig. 2) as a potent and selective inhibitor of PDGFRα (IC_{50} = 60 nM) with modest VEGFR2 activity (IC_{50} = 2.4 µM). The improved solubility and oral bioavailability enabled oral administration in clinical trials. However, due to its high-binding affinity with albumin, its cellular potency against PDGF-induced proliferation was substantially compromised by the presence of serum protein. For example, in the PDGF-induced bromodeoxyuridine incorporation cellular assay, its IC_{50} values shifted upward from 0.3 µM (no bovine serum albumin (BSA)) to 9.5 µM (0.1% BSA); 16 µM (0.5% BSA); and 36 µM (1% BSA).[11] In addition, SU6668 is an acidic drug with increased affinity for plasma proteins and lowered ability to enter the extravascular compartments of the body where some solid tumors reside. Hence, it exhibits low volume of distribution and poor tumor penetration. These factors along with its short plasma half-life (2–4 h) contributed to a non-robust clinical response of SU6668,[22,23] which was subsequently replaced with a third-generation drug candidate, sunitinib.

The third-generation VEGF inhibitors were designed to overcome the high affinity of SU6668 for plasma proteins, largely attributable to the carboxylic acid residue.[11] Replacement of −(CH$_2$)$_3$CO$_2$H with a basic water-solubilizing −C(O)(CH$_2$)$_2$NEt$_2$ group gave carboxamide **3** (Fig. 2) which was 30-fold more potent against VEGFR2 and PDGFRβ in biochemical assays and significantly more soluble under both acidic and neutral conditions. Further 5-fluorine substitution of the oxindole core led to SU11248, later known as sunitinib, which showed a 10-fold further enhancement in PDGFRβ inhibition and better metabolic stability. It is a potent ATP-competitive inhibitor of VEGFR2 and PDGFRβ with K_i of 9 nM and 8 nM, respectively, displaying >10-fold higher selectivity for VEGFR2 and PDGFR than FGFR-1, EGFR, CDK2, Met, IGFR-1, ABL, and SRC. Most importantly, its cellular potency was only marginally affected by the presence of serum protein (IC$_{50}$ < 7 nM at 0.1% and 0.3% BSA, 8 nM at 1% and 5% BSA in the PDGF-induced assay).[11] Thus, in the presence of a high level of serum, sunitinib remained highly potent against PDGFRβ, whereas SU6668 lost its activity. In addition, sunitinib and its active metabolite SU12662 (the *N*-desethyl derivative of sunitinib) showed prolonged half-lives of ~40 and 80 h, respectively, in heathy volunteers and cancer patients.[24] Furthermore, sunitinib is well absorbed, and its bioavailability is unaffected by food.[25] Because of these favorable PK properties, it elicited a much more sustained tumor response than SU6668. Sunitinib was approved in 2006 for GIST and RCC, following only one month after Bayer's sorafenib which was initially identified as a RAF inhibitor and subsequently shown to inhibit several VEGFRs (see Section 5.1).

Although Sugen was a pioneering signal transduction company, its early clinical compounds showed disappointing clinical outcomes due to poor PK properties, and as a result, it had been constantly struggling to raise sufficient funding to operate as an independent company. Ultimately, Sugen was bought by Pharmacia in 1999 for only $645 million. Then Pharmacia was acquired by Pfizer in 2003. Three months after the mega-merge, Pfizer management shut down Sugen.[26-28] Several scientists were then moved from Sugen in San Francisco to Pfizer in San Diego and contributed to the discoveries of crizotinib and lorlatinib (see Sections 9.1 and 9.5). Some of the original Sugen group later started a new company called Turning Point (TP) Therapeutics, which was acquired by Bristol-Myers Squibb in 2022 for $4.1 billion, primarily for repotrectinib (see Section 15.3), a phase III macrocyclic multikinase inhibitor for the treatment of patients with ROS1-positive metastatic NSCLC.

Binding Mode[29]

Sunitinib binds the DFG-out conformation of VEGFR2 at the front cleft and gate area (type II) (Fig. 3, 4). It occupies the ATP-binding site, forming three hydrogen bonds: (1) between the oxindole NH and the amide carbonyl of Glu-917 of the hinge, (2) the oxindole carbonyl and the amide NH of Cys919, and (3) the pyrrole NH and the amide carbonyl of Cys919. Several hydrophobic interactions with VEGFR also contribute to the potency (Fig. 4).

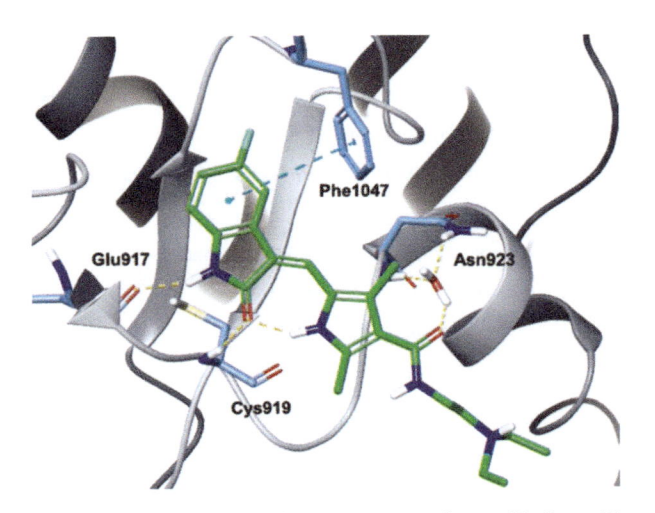

Figure 3. Co-crystal structure of sunitinib with VEGFR2 (PDB ID: 4AGD).

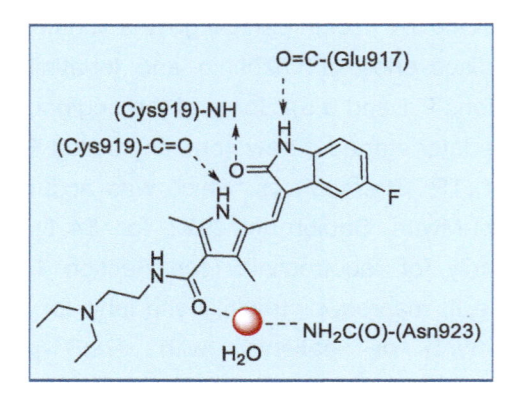

Figure 4. Summary of sunitinib–VEGFR2 interactions based on the co-crystal structural data.

1. B. Carlisle, N. Demko, G. Freeman, A. Hakala, N. MacKinnon, T. Ramsay, S. Hey, A. J. London, J. Kimmelman. Benefit, risk, and outcomes in drug development: A systematic review of sunitinib. *J. Natl. Cancer Inst.*, **2016**, *108*, djv292.
2. E. Nassif, C. Thibault, Y. Vano, L. Fournier, L. Mauge, V. Verkarre, M. Olivier. Timsit. Sunitinib in kidney cancer: 10 years of experience and development. *Expert Rev. Anticancer Ther.* **2017**, *17*, 129–142.
3. S. Faivre, G. Demetri, W. Sargent, E. Raymond. Molecular basis for sunitinib efficacy and future clinical development. *Nat. Rev. Drug Discov.* **2007**, *6*, 734–745.
4. X. L. Zhu, K. Stergiopoulos, S. H. Wu. Risk of hypertension and renal dysfunction with an angiogenesis inhibitor sunitinib: Systematic review and meta-analysis. *Acta Oncol.*, **2009**, *48*, 9–17.
5. J. Folkman. Tumor angiogenesis: Therapeutic implications. *N. Engl. J. Med.* **1971**, *285*, 1182–1186.
6. Y. Liu, Y. Li, Y. Wang, C. Lin, D. Zhang, J. Chen, L. Ouyang, F. Wu, J. Zhang, L. Chen. Recent progress on vascular endothelial growth factor receptor inhibitors with dual targeting capabilities for tumor therapy. *J. Hematol. Oncol.* **2022**, *15*, 89.
7. S. Qin, A. Li, M. Yi, S. Yu, M. Zhang, K. Wu. Recent advances on anti-angiogenesis receptor tyrosine kinase inhibitors in cancer therapy. *J. Hematol. Oncol.* **2019**, *12*, 27.
8. J. A. Forsythe, B.-H. Jiang, N. V. Iyer, F. Agani, S. W. Leung, R. D. Koos, G .L. Semenza. Activation of vascular endothelial growth factor gene transcription by hypoxia-inducible factor 1. *Mol. Cell. Bio.* **1996**, *16*, 4604–4613.
9. J. T. Hartmann, L. Kanz. Sunitinib and periodic hair depigmentation due to temporary c-KIT inhibition. *Arch. Dermatol.* **2008**, *144*, 1525–1526.
10. D. Liu, G. Li, D. M. Avella, E.T. Kimchi, J. T. Kaifi, M. P. Rubinstein, E. R. Camp, D. C. Rockey, T. D. Schell, K. F. Staveley-O'Carroll. Sunitinib represses regulatory T cells to overcome immunotolerance in a murine model of hepatocellular cancer. *Oncoimmunol.* **2018**, *7*, e1372079.
11. L. Sun, C. Liang, S. Shirazian, Y. Zhou, T. Miller, J. Cui, J. Y. Fukuda, J. Y. Chu, A. Nematalla, X.Y. Wang, H. Chen, A. Sistla, T. C. Luu, F. Tang, J. Wei, C. Tang. Discovery of 5-[5-fluoro-2-oxo-1,2-dihydroindol-(3Z)-ylidenemethyl]-2,4-dimethyl-1*H*-pyrrole-3-carboxylic acid (2-diethylaminoethyl)amide, a novel tyrosine kinase inhibitor targeting vascular endothelial and platelet-derived growth factor receptor tyrosine kinase. *J. Med. Chem.*, **2003**, *46*, 1116–1119.
12. L. Sun, N. Tran, T. Tang, H. App, P. Hirth, G. McMahon, C. Tang. Synthesis and biological evaluations of 3-substituted indolin-2-ones: A novel class of tyrosine kinase inhibitors that exhibit selectivity towards particular receptor tyrosine kinases. *J. Med. Chem.* **1998**, *41*, 2588–2603.
13. L. Sun, N. Tran, C. Liang, F. Tang, A. Rice, R. Schreck, K. Waltz, L. K. Shawver, G. McMahon, C. Tang. Design, synthesis, and evaluations of substituted 3-[(3- or 4-carboxyethylpyrrol-2-ylmethylidenyl]indolin-2-ones as inhibitors of VEGF, FGF, and PDGF receptor tyrosine kinases. *J. Med. Chem.* **1999**, *42*, 5120–5130.
14. R. S. Li, J. A. Stafford. Chapter 1. Discovery and development of sunitinib (SU11248): A multitarget tyrosine kinase inhibitor of tumor growth, survival, and angiogenesis. *Kinase Inhibitor Drugs*, Editors: R. S. Li; J. A. Stafford. *John Wiley & Sons, Inc.,* **2009**, 1–39.
15. P. C. Tang, T. A. Miller, X. Y. Li, L. Sun, C. C. Wei, S. Shirazian, C. X. Liang, T. Vojkovsky, A. S. Nematalla, M.Hawley. Pyrrole substituted 2-indolinone protein kinase inhibitors. U S Patent 7119090, **2003**.
16. P. C. Tang, T. A. Miller, X. Y. Li, L. Sun, C. C. Wei, S. Shirazian, C. X. Liang, T. Vojkovsky, A. S. Nematalla, M.Hawley. Pyrrole substituted 2-indolinone protein kinase inhibitors. U S Patent 7125905, **2006**.
17. T. A. T. Fong, L. K. Shawver, L. Sun, Cho Tang, Harald App, T. Jeff Powell, Young H. Kim, R. Schreck, X. Y. Wang, W. Risau, A. Ullrich, K. P. Hirth, G. McMahon. SU5416 Is a potent and selective inhibitor of the vascular endothelial growth factor receptor (Flk-1/KDR) that inhibits tyrosine kinase catalysis, tumor vascularization, and growth of multiple tumor types. *Cancer Res.* **1999**, *59*, 99–106.
18. Heather H. Zhang, L. Sun. Former Sugen scientists. Persnonal communication.

19. L. Antonian, H. Zhang, C. Yang, G. Wagner, L. K. Shawver, M. Shet, B. Ogilvie, A. Madan, A. Parkinson. Biotransformation of the anti-angiogenic compound SU5416. *Drug Metab. Dispos.* **2000**, *28*, 1505–1512.

20. J. V. Heymach, J. Desai, J. Manola, D. W. Davis, D. J. McConkey, D. Harmon, D. P. Ryan, G. Goss, T. Quigley, A. D. Van den Abbeele, S. G. Silverman, S. Connors, J. Folkman, C. D. Fletcher, G. D. Demetri. Phase II study of the antiangiogenic agent SU5416 in patients with advanced soft tissue sarcomas. *Clin. Cancer Res.* **2004**, *10*, 5732–5740.

21. M. W. Kieran, J. G. Supko, D. Wallace, R. Fruscio, T. Y. Poussaint, P. Phillips, I. Pollack, R. Packer, J. M. Boyett, S. Blaney, A. Banerjee, R. Geyer, H. Friedman, S. Goldman, L. E. Kun, T. Macdonald. Pediatric Brain Tumor Consortium (2009). Phase I study of SU5416, a small molecule inhibitor of the vascular endothelial growth factor receptor (VEGFR) in refractory pediatric central nervous system tumors. *Pediatr. Blood Cancer.* **2009**, *52*, 169–176.

22. A. D. Laird, P. Vajkoczy, L. K. Shawver, A. Thurnher, C. Liang, M. Mohammadi, J. Schlessinger, A. Ullrich, S. R. Hubbard, R. A. Blake, T. A. Fong, L. M. Strawn, L. Sun, C. Tang, R. Hawtin, C. Tang, N. Shenoy, K. P. Hirth, G. McMahon, C. Herrington. SU6668 is a potent antiangiogenic and antitumor agent that induces regression of established tumors. *Cancer Res.* **2000**, *60*, 4152–4160.

23. B. C. Kuenen, G. Giaccone, R. Ruijter, A. Kok, C. Schalkwijk, K. Hoekman, H. M. Pinedo. Dose-finding study of the multitargeted tyrosine kinase inhibitor SU6668 in patients with advanced malignancies. *Clin. Cancer Res.* **2005**, *11*, 6240–6246.

24. B. E. Houk, C. L. Bello, D. Kang, M. Amantea. A population pharmacokinetic meta-analysis of sunitinib malate (SU11248) and its primary metabolite (SU12662) in healthy volunteers and oncology patients. *Clin. Cancer Res.* **2009**, *15*, 2497–2506.

25. B. E. Houk, C. L. Bello, B. Poland, L. S. Rosen, G. D. Demetri, R. J. Motzer. Relationship between exposure to sunitinib and efficacy and tolerability endpoints in patients with cancer: results of a pharmacokinetic/pharmacodynamic meta-analysis. *Cancer Chemother. Pharmacol.* **2010**, *66*, 357–371.

26. K. Garber. Research retreat: Pfizer eliminates Sugen, shrinks cancer infrastructure. *J. Nat. Cancer Inst.* **2003**, *95*, 1036–1038.

27. K. Garber. Sugen falls as casualty of Pfizer-Pharmacia merger. *Nat. Biotechnol.* **2003**, *21*, 722–723.

28. A. Levitzki. The closure of Sugen. *Nat. Biotechnol.* **2003**, *21*, 969.

29. M. Mctigue, B. W. Murray, J. H. Chen, Y. L. Deng, J. Solowiej, R. S. Kania. Molecular conformations, interactions, and properties associated with drug efficiency and clinical performance among VEGFR TK inhibitors. *PNAS.* **2012**, *109*, 18281–18289.

5.4 Pazopanib (Votrient™) (26)

Structural Formula	Space-filling Model	Brief Information
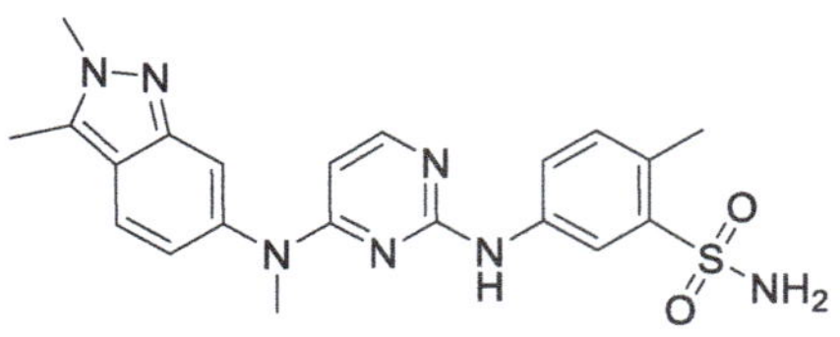 $C_{21}H_{23}SN_7O_2$; MW = 437 clogP = 3.8; tPSA = 116		Year of discovery: 2000 Year of introduction: 2009 Discovered by: GSK Developed by: Novartis/GSK Primary targets: VEGFR/multikinase Binding type: II (VEGFR2) Class: receptor tyrosine kinase Treatment: RCC, STS Other name: GW786034 Oral bioavailability = 14–39% Elimination half-life = 31 h Protein binding > 99.9%

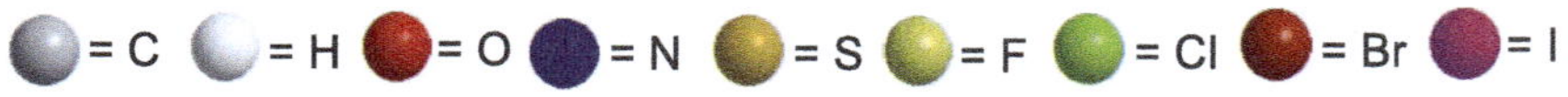

● = C ● = H ● = O ● = N ● = S ● = F ● = Cl ● = Br ● = I

Pazopanib is a multikinase inhibitor approved first in 2009 for advanced renal cell carcinoma (RCC) and then in 2012, for advanced soft tissue sarcoma (STS).

Pazopanib predominantly inhibits VEGFR1-3, PDGFRα/β, and the stem cell factor receptor KIT, resulting in inhibition of tumor angiogenesis, cell growth, and survival.[1] Patients treated with pazopanib have been observed to benefit from a reduction in tumor blood flow, increased tumor apoptosis, inhibition of tumor growth, and tumor hypoxia.[1-3]

Pazopanib is also active against chemotherapy-resistant non-adipocytic STS. It causes less myelosuppression, hand–foot syndrome, mucositis/stomatitis, dysgeusia, and fatigue than sunitinib, but more loss of liver function.

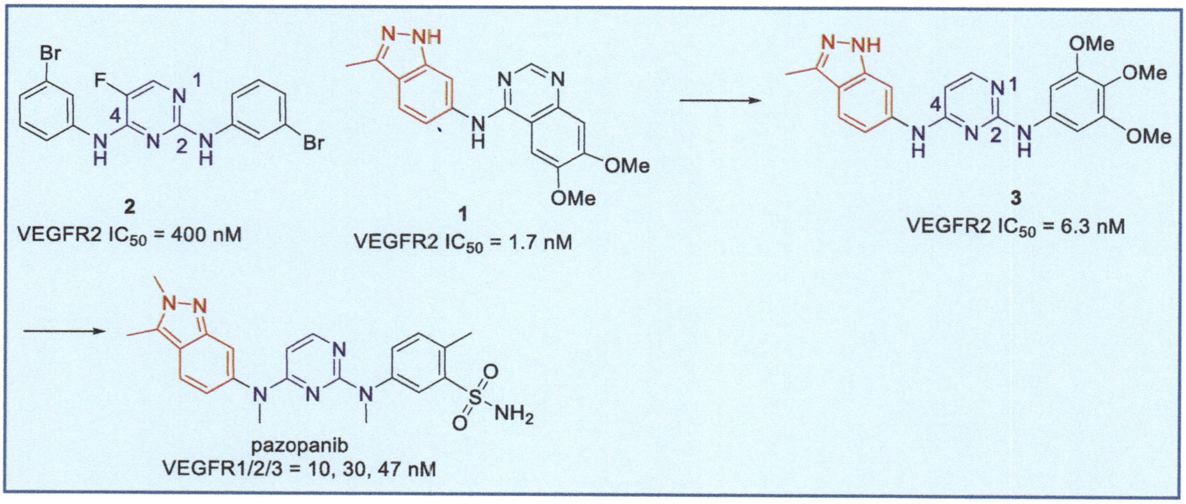

2
VEGFR2 IC$_{50}$ = 400 nM

1
VEGFR2 IC$_{50}$ = 1.7 nM

3
VEGFR2 IC$_{50}$ = 6.3 nM

pazopanib
VEGFR1/2/3 = 10, 30, 47 nM

Figure 1. Structural optimization of pazopanib.

Pazopanib contains an N-linked indazole headgroup and amino pyrimidine hinge binder, which originated from HTS hits pyrimidine **1** and quinazoline **2**, respectively (Fig. 1). Combination of **1** with **2**, assisted by computational modelling, gave a hybrid compound **3**, which was further optimized to give pazopanib.[4,5] It showed potent inhibition of all the human VEGFR receptors with an IC_{50}'s of 10, 30, and 47 nM for VEGFR-1, -2, and -3, respectively. Moderate activity was also seen with the closely related tyrosine receptor kinases PDGFR, KIT, FGF-R1, and c-FMS with IC_{50}'s of 84, 74, 140, and 146 nM, respectively. In cellular assays, in addition to inhibiting the VEGF-induced proliferation of HUVECs, pazopanib potently blocked VEGF-induced phosphorylation of VEGFR-2 in HUVEC cells with an IC_{50} of ~8 nM.[4]

Pazopanib differs from sorafenib and sunitinib by having a dimethylated indazole headgroup instead of a carboxylate group. Consequently, it is unique among these TKIs by forming one H-bond with DFG-Asp1046 rather than Glu885 (data not shown), based on the X-ray co-crystal structure with nonphosphorylated plus JM (juxtamembrane) VEGFR2[6] (which was not uploaded to PDB because of insufficient electron density for the assignment of specific JM residues).[7] The indazole subunit of pazopanib penetrates into the channel of the JM position in the complex, making a direct H-bond to the amide NH of Asp1046, a kinase backbone pivot point, and stabilizing the DFG-out conformation (type II inhibitor).[6] The greater structural rigidity of pazopanib vs. sunitinib leads to greater selectivity for VEGFR. The key interactions of pazopanib with VEGFR2 based on the co-crystal structure are summarized in Figure 2.

1. P. L. McCorma. Pazopanib: A review of its use in the management of advanced renal cell carcinoma. *Drugs.* **2014**, *74*, 1111–1125.
2. E. D. Deeks. Pazopanib in advanced soft tissue sarcoma. *Drugs.* **2012**, *72*, 2129–2140.
3. J. Verweij, S. Sleijfer. Pazopanib, a new therapy for metastatic soft tissue sarcoma. *Expert Opin. Pharmacother.* **2013**, *14*, 929–935.
4. P. A. Harris, A. Boloor, M. Cheung, R. Kumar, R. M. Crosby, R. G. Davis-Ward, A. H. Epperly, K. W. Hinkle, R. N. Hunter III, J. H. Johnson, V. B. Knick, C. P. Laudeman, D. K. Luttrell, R. A. Mook, R. T. Nolte, S. K. Rudolph, J. R. Szewczyk, A. T. Truesdale, J. M. Veal, L. Wang, J. A. Stafford. Discovery of 5-[[4-[(2,3-dimethyl-2H-indazol-6-yl)methylamino]-2-pyrimidinyl]amino]-2-methyl-benzenesulfonamide (Pazopanib), a novel and potent vascular endothelial growth factor receptor inhibitor. *J. Med. Chem.* **2008**, *51*, 4632–4640.
5. A. Boloor, M. Cheung, R. Davis, P. A. Harris, K. Hinkle, R. A. Mook, Jr., J. A. Stafford, J. M. Veal, C, Assigne. Pyrimidineamines as angiogenesis modulators. US7262203, **2000**.
6. M. Mctigue, B. W. Murray, J. H. Chen, Y. L. Deng, J. Solowiej, R. S. Kania. Molecular conformations, interactions, and properties associated with drug efficiency and clinical performance among VEGFR TK inhibitors. *PNAS.* **2012**, *109*, 18281–18289.
7. C. C. Ayala-Aguilera, T. Valero, A. Lorente-Macías, D. J. Baillache, S. Croke, A. nciti-Broceta. Small molecule kinase inhibitor drugs (1995-2021): Medical indication, pharmacology, and synthesis. *J. Med. Chem.* **2022**, *65*, 1047–1131.

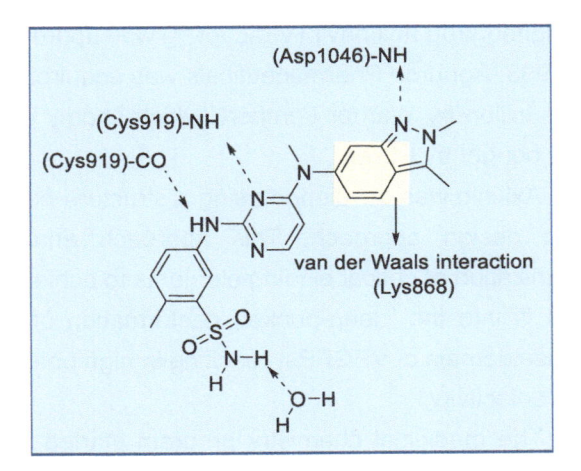

Figure 2. Pazopanib–VEGFR2 interactions.

5.5 Axitinib (Inlyta™) (27)

Structural Formula	Space-filling Model	Brief Information
$C_{22}H_{18}SN_4O$; MW = 386 clogP = 3.3; tPSA = 66		Year of discovery: 1999 Year of introduction: 2012 Discovered by: Agouron Developed by: Pfizer Primary targets: VEGFR/PDGFR Binding type: II (VEGFR2) Class: receptor tyrosine kinase Treatment: RCC Other name: AG-013736 Oral bioavailability = 58% Elimination half-life = 2.5–6.1 h Protein binding = 99%

● = C ○ = H ● = O ● = N ● = S ● = F ● = Cl ● = Br ● = I

Axitinib was first approved in 2012 as a second-line treatment for advanced kidney cancer (renal cell carcinoma, RCC)[1,2] and in 2019 in combination with pembrolizumab or avelumab as first-line treatment for advanced RCC.[3]

Axitinib is different from sorafenib and sunitinib in having a unique indazole hinge binder (vs. pyrimidine for pazopanib and oxindole-pyrrole for sunitinib). It also contains an uncommon sulfide moiety, which was incorporated to induce a DFG-out binding conformation of VEGFR2 to improve selectivity. Because the sulfide is readily oxidized to give an inactive sulfoxide as a major metabolite,[4] the half-life in humans is only 2.5–6.1 h and, as a result, axitinib needs to be administered twice daily (5 mg tablet).

The primary mechanism of action of axitinib is thought to be VEGFR1-3, KIT, and PDGFR inhibition and prevention of angiogenesis. In cellular assays, it inhibits VEGFR1-3 autophosphorylation at picomolar concentrations, consistent with its potent and highly selective activity against these receptors in biochemical assays. Axitinib also inhibits platelet-derived growth factor receptors α and β (PDGFRα/β) and KIT at nanomolar concentrations (Fig. 1). Importantly, axitinib displays only marginal inhibition against "off-target" protein kinases and no significant activity in a broad protein kinase screen.

In vitro studies showed that axitinib did not significantly inhibit other receptor kinases, including colony-stimulating factor CSF-1R, FLT-3, FGFR-1, RET, EGFR, and Met.

Discovery of Axitinib[5–8]

Axitinib was discovered at Agouron, first known as Agouron Institute, a non-profit research organization founded in 1978 with a mission to apply X-ray crystallography of proteins and computer graphics-assisted design to drug discovery. In 1984, it was transformed into a commercial entity, Agouron Pharmaceuticals, to exploit the commercial side of such rational drug design. In 1997, its first rationally designed drug nelfinavir (Viracept™) was approved. In 1998, Agouron Pharmaceuticals was acquired for $2.1 billion by Warner Lambert, which shortly after was bought by Pfizer.

Axitinib was developed using a structure-based drug design approach. This approach enabled optimization of critical binding elements to achieve a tight fit into the "deep-pocket" conformation of the kinase domain of VEGFRs, resulting in high potency and selectivity.

The medicinal chemistry program started with the high-throughput screening of the Agouron exploratory compound libraries which identified pyrazole **1** (Fig. 1) with modest potency (K_i = 5.1 µM).

In an effort to rigidify the inhibitor for stronger binding, the vinyl pyrazole core was cyclized to form an indazole core, and the styrene subunit was truncated to give a pharmacophore indazole 2 without compromising potency (K_i = 2.3 µM). Adding back the styrene subunit to 2 led to the full length indazole 3 as a highly potent VEGFR2 inhibitor (K_i = 0.3 nM). However, 3 suffered from poor kinase selectivity and low oral bioavailability in mouse. The broad-spectrum kinase activity of 3 was attributed to its binding to the active DFG-in conformation as shown in its co-crystal structure with VEGFR2. In contrast, other HTS hits which bound the DFG-out inactive conformation exhibited a superior kinase selectivity profile, and this prompted the design of DFG-out binding elements into this chemotype. Replacement of the 2-methoxyphenol at the 6-position of the indazole ring in 2 with a phenyl thiol functionality led to another pharmacophore indazole 4 which successfully targeted the DFG-out conformation.

Compound 4 largely maintained VEGFR activity (K_i = 6.6 µM for 4 vs. 2.3 µM for 2), while eliminating the metabolic liabilities of the phenolic hydroxyl group and the methoxy group which are present in 2. Once again, adding back the styrene subunit gave 5 which restored most VEGFR2 inhibitory activity (K_i = 17 nM). Despite a substantial drop in potency relative to 3 (K_i = 0.3 nM vs. 17 nM for 3), compound 5 showed superior kinase selectivity due to the phenyl sulfide-favored binding to the DFG-out conformation, Subsequently, the terminal styrene phenyl was replaced with a more polar 2-pyridyl group to allow hydrogen-bonding with a crystallographic water. Lastly, an ortho-substituted carboxamide was installed to form the two key H-bonds critical for potency to the DFG-out conformation. Taken together, these changes resulted in axitinib with picomolar potency and good oral bioavailability in humans (58%).

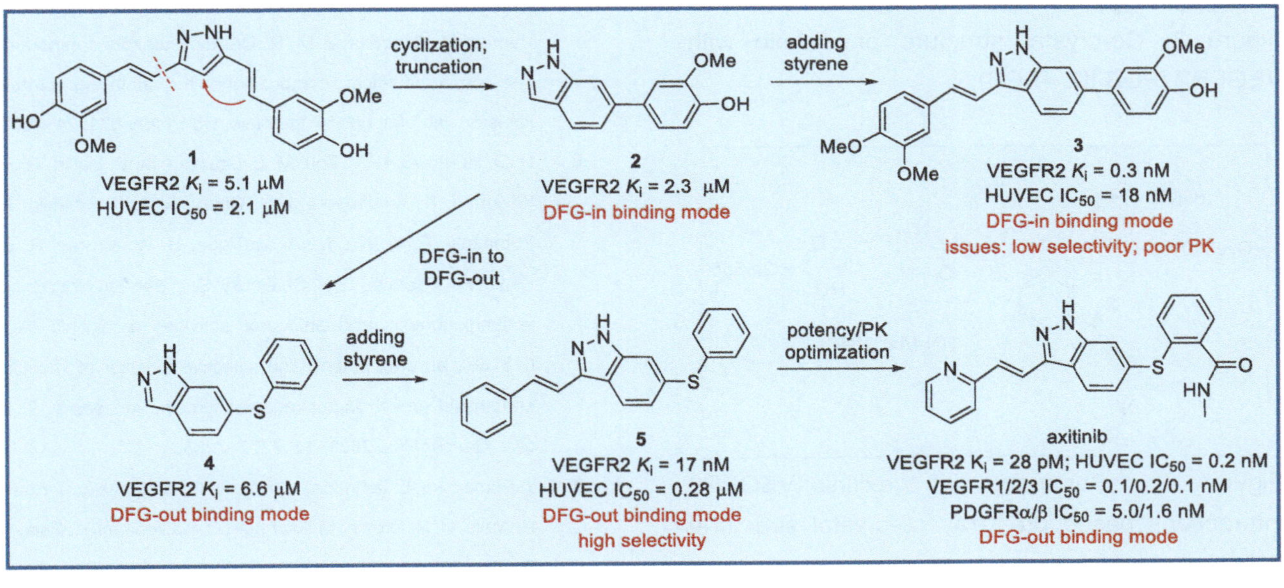

Figure 1. Structural optimization of axitinib.

Binding Mode[9]

In the co-crystal structure of axitinib bound to VEGFR2 DFG-out inactive conformation (Fig. 2), the indazole scaffold forms two hydrogen bonds with the hinge: one between the indazole NH and the backbone carbonyl of Glu917; and the other between the indazole nitrogen the amide NH backbone of Cys919. The styryl group penetrates through the narrow tunnel and extends out toward the solvent front. The phenyl sulfide is positioned slightly higher and deeper into the back pocket as compared to other type II inhibitors. The carboxamide forms one H-bond to the NH backbone of Asp1046 and a second H-bond to the carboxylate side chain of Glu885 (Fig. 3). The indazole head group substantially complements the full length of the channel, contributing to high affinity with both polar charge stabilization and hydrophobic interactions.

It has also been shown that axitinib binds (in a different conformation from the VEGFR2 binding) to the BCR-ABL fusion protein, specifically inhibiting the drug-resistant T315I mutant isoform.[10]

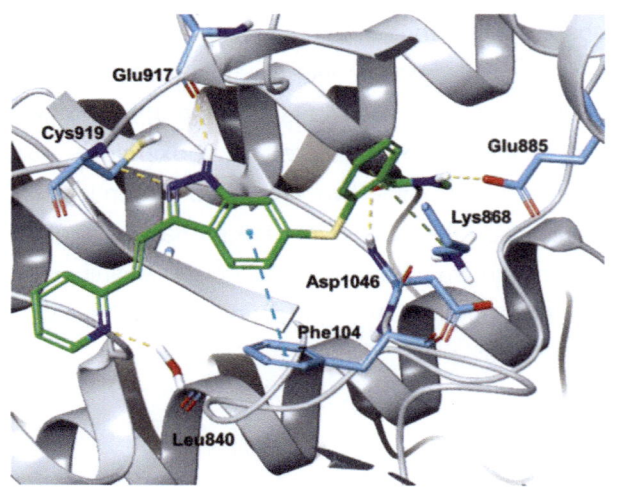

Figure 2. Co-crystal structure of axitinib with VEGFR2 (PDB ID: 4AG8).

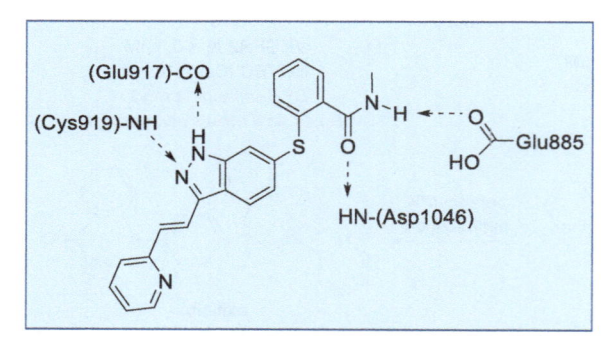

Figure 3. Summary of axitinib–VEGFR2 interactions based on X-ray co-crystal structural data.

1. B. Escudier, M. Gore. Axitinib for the Management of metastatic renal cell carcinoma. *Drugs R. D.* **2011**, *11*, 113–126.
2. D. Hu-Lowe, M. Hallin, R. Feeley, H. Zou, D. Rewolinski, G. Wickman, E. Chen, Y. Kim, S. Riney, J. Reed, D. Heller, B. Simmons, R. Kania, M. McTigue, M. Niesman, S. Gregory, D. Shalinsky, S. Bender. Characterization of potency and activity of the VEGF/PDGF receptor tyrosine kinase inhibitor AG013736. *Proc. Am. Assoc. Cancer Res.* **2002**, *43*, A5356.
3. M. B. Atkins, E. R. Plimack, I. Puzanov, M. N. Fishman, D. F. McDermott, D. C. Cho, U. Vaishampayan, S. George, T. E. Olencki, J. C. Tarazi, B. Rosbrook, K. C. Fernandez, M. Lechuga, T. K. Choueiri. Axitinib in combination with pembrolizumab in patients with advanced renal cell cancer: A non-randomised, open-label, dose-finding, and dose-expansion phase 1b trial. *Lancet. Oncol.* **2018**, *19*, 405–415.
4. B. J. Smith, Y. Pithavala, H. Z. Bu, P. Kang, B. Hee, A. J. Deese, W. F. Pool, K. J. Klamerus, E. Y. Wu, D. K. Dalvie. Pharmacokinetics, metabolism, and excretion of [^{14}C]axitinib, a vascular endothelial growth factor receptor tyrosine kinase inhibitor, in humans. *Drug Metab. Dispos.* **2014**, *42*, 918–931.
5. P. A. Harris. Inhibitors Inhibitors of vascular endothelial growth factor receptor. *Top. Med. Chem.* **2018**, *28*, 105–140.
6. R. S. Kania. Structure-based design and characterization of axitinib. In: R. Li, J. A. Stafford (eds) Kinase inhibitor drugs. Wiley, Hoboken, **2009**, p167.
7. R. S. Kania, S. L. Bender, A. J. Borchardt, S. J. Cripps, Y. Hua, M. D. Johnson, T. O. Johnson, H. T. Lulu, C. L. Palmer, S. H. Reich, A. M. T. Russell, M. Teng, C. Thomas, M. D. Varney, M. B. Wallace, M. R. Collins. Indazole compounds and pharmaceutical compositions for inhibiting protein kinases, and methods for their use. US Patent 534524, **2003**.
8. D. D. Hu-Lowe, H. Y. Zou, M. L. Grazzini, M. E. Hallin, G. R. Wickman, K. Amundson, J. H. Chen, D. A. Rewolinski, S. Yamazaki, E. Y. Wu, M. A. McTigue, B. W. Murray, R. S. Kania, P. O'Connor, D. R. Shalinsky, S. L. Bender. Nonclinical antiangiogenesis and antitumor activities of axitinib (AG-013736), an oral, potent, and selective inhibitor of vascular endothelial growth factor receptor tyrosine kinases 1, 2, 3. *Clin. Cancer Res.* **2008**, *14*, 7272–7283.
9. T. Pemovska, E. Johnson, M. Kontro, G. A. Repasky, J. Chen, P. Wells, C. N. Cronin, M. McTigue, O. Kallioniemi, K. Porkka, B. W. Murray, K. Wennerberg. Axitinib effectively inhibits BCR-ABL1 (T315I) with a distinct binding conformation. *Nature.* **2015**, *519*, 102–105.
10. M. Mctigue, B. W. Murray, J. H. Chen, Y. L. Deng, J. Solowiej, R. S. Kania. Molecular conformations, interactions, and properties associated with drug efficiency and clinical performance among VEGFR TK inhibitors. *PNAS.* **2012**, *109*, 18281–18289.

5.6 Nintedanib (Ofev™) (28)

Structural Formula	Space-filling Model	Brief Information
$C_{31}H_{33}N_5O_4$; MW = 539 clogP = 3.0; tPSA = 94		Year of discovery: 2003 Year of introduction: 2014 Discovered by: Boehringer Ingelheim Developed by: Boehringer Ingelheim Primary targets: PDGFR; VEGFR; FGFR Binding type: II Class: receptor tyrosine kinase Treatment: ILD, IPF Other name: BIBF1120 Oral bioavailability = 4.7% Elimination half-life = 10–15 h Protein binding = 98%

● = C   ○ = H   ● = O   ● = N   ● = S   ● = F   ● = Cl   ● = Br   ● = I

Nintedanib was approved in 2014 for idiopathic pulmonary fibrosis (IPF), and in 2020 for chronic fibrosing (scarring) interstitial lung diseases (ILDs) with a progressive phenotype (trait).[1]

IPF is the most common type of the idiopathic ILDs, and it is an uncommon disorder characterized by progressive loss of lung function, dyspnea, and cough, affecting ~100,000 people in the United States.

IPF patients show increased levels of fibrocytes in the lungs and in the circulation. Fibrocytes are circulating precursors for fibroblasts.[2] Fibrocytes express FGFR2 (fibroblast growth factor receptor) and VEGFR1 (vascular endothelial growth factor receptor), while fibroblasts express FGFR2, and more abundantly PDGFRα/β (platelet-derived growth factor receptor). Upon binding to their specific growth factors, these cell surface receptors undergo autophosphorylation to activate subsequent downstream pro-fibrotic and proliferative pathways and cause IPF symptoms.

Nintedanib inhibits the proliferation of lung fibroblasts induced by fibrocytes by blocking signaling pathways activated by growth factor receptors on fibroblasts, including VEGFRs and PDGFRs.

VEGF, PDGF and FGFR in IPF[3–5]

In the process of pulmonary fibrosis, fibroblasts and myofibroblasts become activated by growth factors secreted by the airway epithelium after the inflammatory damage. The fibroblasts change to myofibroblasts or proliferate, depending on the growth factors. This event leads to areas of fibroblastic foci, the leading edge of fibrotic destruction of the lung. The pivotal growth factors for the fibrotic process include VEGF, FGF, and PDGF. However, the specific contribution of VEGF/VEGFR signaling to IPF remains to be determined.

PDGFs are dimeric ligands which bind to homodimers or heterodimers of PDGFRα/β. These receptors consist of five extracellular immunoglobulin-like domains, a single-span transmembrane domain, and an intracellular tyrosine kinase domain. Ligand binding brings about autophosphorylations of the receptors and subsequent downstream signaling via RAS-RAF-MEK-ERK and PI3K-AKT-mTOR (Fig. 1). PDGF, a potent mitogen for fibroblasts, plays a critical role in the expansion of myofibroblasts by stimulating proliferation, migration, and survival. Excessive PDGF causes myofibroblasts to deposit additional connective tissue products in the interstitial space, leading to a distorted alveolar architecture. In addition, PDGF was found to be significantly elevated in the alveolar macrophage cells from the lungs of IPF patients. The significance of PDGFR was further demonstrated by PDGFR-specific tyrosine kinase inhibitors such as imatinib, a PDGFR and BCR-ABL inhibitor, which reduced bleomycin-induced lung fibrosis in mice. Despite its efficacy in animal models of pulmonary fibrosis, imatinib failed to improve survival or lung function in IPF patents, suggesting that PDGFR inhibitors alone may not be sufficient to treat IPF.[4] Therefore, inhibition of PDGFR along with other growth factor receptors such as FGFR is necessary for anti-fibrotic treatment.

FGFR shares a canonical receptor tyrosine kinase (RTK) architecture, with a split intracellular kinase domain. Binding to the FGF ligand triggers downstream signaling cascades similar to those of PDGFR (Fig. 1). FGF and FGFR were shown to be overexpressed in the pathogenic regions of IPF, which is consistent with enhanced FGF and FGFR signaling. The excessive PGFR activity promotes fibroblast migration and causes the scarring of the lungs. Indeed, eradication of FGF signaling reduced pulmonary fibrosis and improved survival in bleomycin-treated mice. Thus, it appears likely that constitutive FGF/FGFR signaling contributes to pulmonary fibrotic diseases.[5]

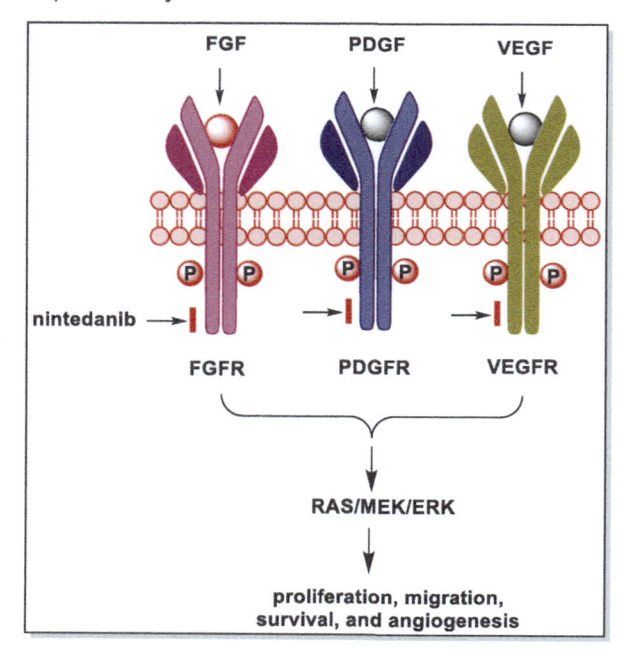

Figure 1. FGFR/PDGFR/VEGFR pathways.

Discovery of Nintedanib

Evaluation of compounds from a CDK4 kinase inhibitor program at Boehringer Ingelheim in a set of kinase selectivity assays identified indolinone **1** as a moderately active VEGFR2 inhibitor (IC_{50} = 763 nM). Interestingly, this compound was devoid of any CDK4 activity and highly selective against a panel of other kinases (e.g., IGF1R, InsR, CDK1, CDK2, EGFR, HER2, PIK1: IC_{50} > 10 µM). Because of its favorable selectivity profile, compound **1** was utilized as a lead compound for a VEGFR inhibitor oncology program. Subsequent optimization of both potency and pharmacokinetic properties led to nintedanib as a multikinase inhibitor (Fig. 2).[6,7] It was initially developed as an anticancer drug because it inhibits tumor angiogenesis, but later repurposed as an anti-fibrotic agent.

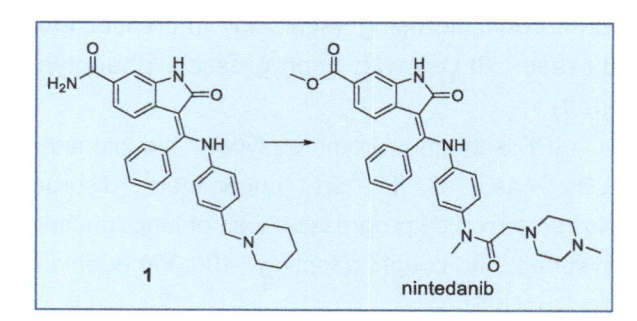

Figure 2. Structural optimization of nintedanib.

Nintedanib is an inhibitor of the receptor tyrosine kinases PDGFRα/β with IC_{50} values of 59 and 65 nM, respectively. It also inhibits FGFR1-4 with IC_{50} values of 69, 37, 108, and 610 nM, respectively, and VEGFR1-3 with IC_{50} values of 34, 21 and 13 nM, respectively. In addition, nintedanib is a potent inhibitor of FLT-3 (FMS-like tyrosine kinase-3) (IC_{50} = 26 nM) and RET (rearranged during transfection) (IC_{50} = 0.67 nM).[1,3]

Nintedanib has poor solubility in the intestinal tract environment and poor metabolic stability due to ester hydrolysis, both of which contribute to low oral bioavailability (~5%).[8]

Binding Mode

In the co-crystal structure of nintedanib in complex with VEGFR2 (Fig. 3), the indolinone core binds to the hinge, forming two hydrogen bonds: the NH group of the indolinone with the amide oxygen of Glu917, and the carbonyl oxygen of the indolinone with the amide NH of Cys919. The methyl ester carbonyl oxygen hydrogen bonds with the side chain amine of Lys868. The external solvent exposed N-methyl amine of piperazine potentially engages in ionic interactions with the carboxylic acid reside of Glu850 (Fig. 4).[9]

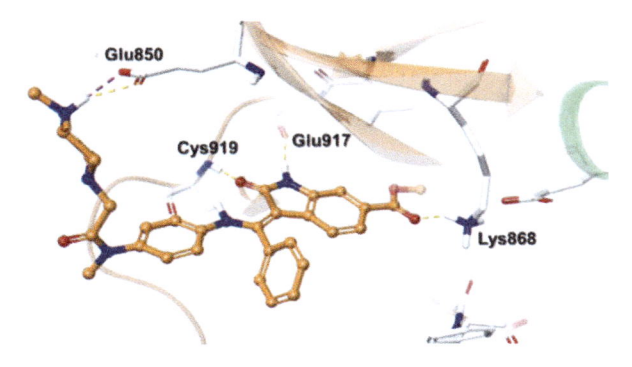

Figure 3. Co-crystal structure of nintedanib–VEGFR2 (PDB ID: 3C7Q).

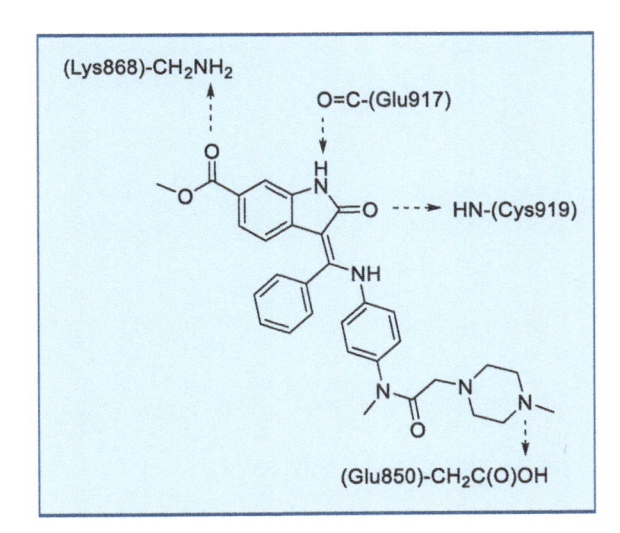

Figure 4. Nintedanib–VEGFR2 interactions.

An X-ray co-crystal structure of RET kinase domain–nintedanib complex has also been determined (Fig. 5). It binds to the RET kinase via three crucial hydrogen bond interactions similar to those observed in its complex with VEGFR2 (Fig. 6).[10]

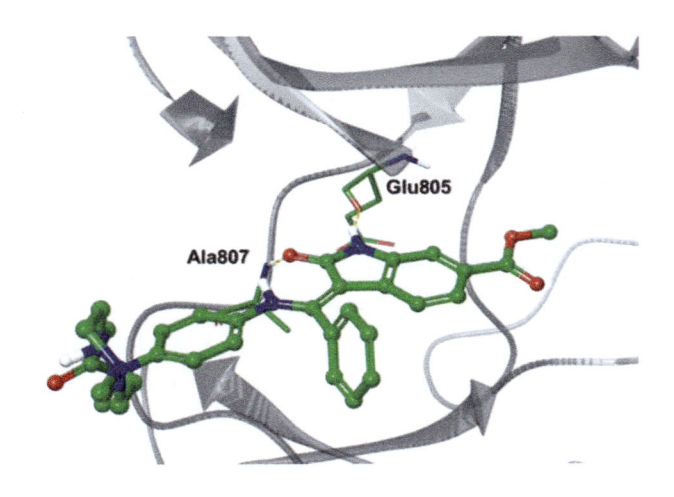

Figure 5. Co-crystal structure of nintedanib–RET (PDB ID: 6NEC).

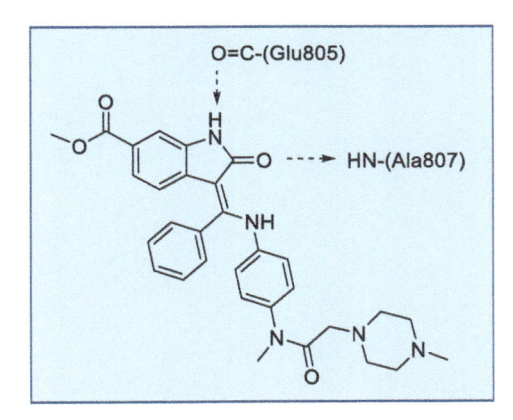

Figure 6. Summary of nintedanib–RET interactions.

1. P. L. McCormack. Nintedanib: first global approval. *Drugs.* **2015**, *75*, 129–139.

2. P. Heukels, J. van Hulst, M. van Nimwegen, C. E. Boorsma, B. N. Melgert, L. M. van den Toorn, K. Boomars, M. S. Wijsenbeek, H. Hoogsteden, J. H. von der Thüsen, R. W. Hendriks, M. Kool, B. van den Blink. Fibrocytes are increased in lung and peripheral blood of patients with idiopathic pulmonary fibrosis. *Respir. Res.* **2018**, *19*, 90.

3. L. Wollin, E. Wex, A. Pautsch, G. Schnapp, K. E. Hostettler, S. Stowasser, M. Kolb. Mode of action of nintedanib in the treatment of idiopathic pulmonary fibrosis. *Eur. Respir. J.* **2015**, *45*, 1434–1445.

4. N. I. Chaudhary, G. J. Roth, F. Hilberg, J. Müller-Quernheim, A. Prasse, G. Zissel, A. Schnapp, J. E. Park. Inhibition of PDGF, VEGF and FGF signalling attenuates fibrosis. *The Eur. Respir. J.* **2007**, *29*, 976–985.

5. B. MacKenzie, M. Korfei, I. Henneke, Z. Sibinska, X. Tian, S. Hezel, S. Dilai, R. Wasnick, B. Schneider, J. Wilhelm, W. Klepetko, W. Seeger, R. Schermuly, A. Günther, S. Bellusci. Increased FGF1-FGFRc expression in idiopathic pulmonary fibrosis. *Respir. Res.* **2015**, *16*, 83.

6. G. J. Roth, A. Heckel, F. Colbatzky, S. Handschuh, J. Kley, T. Lehmann-Lintz, R. Lotz, U. Tontsch-Grunt, R. Walter, F. Hilberg. Design, Synthesis, and Evaluation of Indolinones as TripleAngiokinase Inhibitors and the discovery of a highly specific 6-methoxycarbonyl-substituted indolinone (BIBF 1120). *J. Med. Chem.* **2009**, 52, 4466–4480.

7. G. J. Roth, R. Binder, F. Colbatzky, C. Dallinger, R. Schlenker-Herceg, F. Hilberg, S. L. Wollin, R. Kaiser. Nintedanib: from discovery to the clinic. *J. Med. Chem.* **2015**, *58*, 1053–1063.

8. H. Liu, K. Du, D. Li, Y. Du J. Xi, Y. Xu, Y. Shen, T. Jiang, T. J. Webster. A high bioavailability and sustained-release nano-delivery system for nintedanib based on electrospray technology. *Int. J. Nanomedicine.* **2018**, *13*, 8379–8393.

9. R. Roskoski, Jr. Classification of small molecule protein kinase inhibitors based upon the structures of their drug-enzyme complexes. *Pharmacol. Res.* **2016**, *103*, 26–48.

10. S. S. Terzyan, T. Shen, X. Liu, Q. Huang, P. Teng, M. Zhou, F. Hilberg, J. Cai, B. Mooers, J. Wu. Structural basis of resistance of mutant RET protein-tyrosine kinase to its inhibitors nintedanib and vandetanib. *J. Biol. Chem.* **2019**, *294*, 10428–10437.

5.7 Apatinib (Aitan™) (29)

Structural Formula	Space-filling Model	Brief Information
$C_{24}H_{23}N_5O$; MW = 397 clogP = 3.7; tPSA = 89		Year of discovery: 2004 Year of introduction: 2014 (China) Discovered by: Advenchen Laboratories Developed by: Hengrui/LSK/HLB Primary target: VEGFR/multikinase Binding type: unknown Class: receptor tyrosine kinase Treatment: gastric cancer Other name: YN968D1, Rivoceranib Oral bioavailability = not reported Elimination half-life = 9 h Protein binding = not reported

● = C   ● = H   ● = O   ● = N   ● = S   ● = F   ● = Cl   ● = Br   ● = I

Apatinib was approved by the China Food and Drug Administration (CFDA) in 2014 for patients with late-stage gastric cancer in China. Clinical development for other types of cancer such as colorectal cancer, liver cancer (hepatocellular carcinoma), and soft tissue sarcoma is ongoing in several countries worldwide.[1,2]

Apatinib selectively binds to and inhibits vascular endothelial growth factor receptor 2 (VEGFR2), thereby blocking VEGF-stimulated endothelial cell migration and proliferation, resulting in a decrease in tumor microvessel density.

Apatinib has a strong affinity for VEGFR2 (IC_{50} = 1), which is 90-fold that of sorafenib (IC_{50} = 90 nM). It is less effective against KIT (IC_{50} = 429 nM), RET (IC_{50} = 13 nM), and c-SRC (IC_{50} = 53 nM) and does not inhibit EGFR, HER-2, or FGFR1 (IC_{50} >10 µM).[3,4]

Apatinib differs from sorafenib-type VEGFR inhibitors due to lack of a urea functionality, which may account for its distinct kinase selectivity profile.

1. M. Gou, H. Si, Y. Zhang, N. Qian, Z. Wang, W. W. Shi, G. H. Dai. Efficacy and safety of apatinib in patients with previously treated c-metastatic colorectal cancer: A real-world retrospective study. *Sci. Rep.* **2018**, *8*, 4602–4068.
2. L. J. Scott. Apatinib: A review in advanced gastric cancer and other advanced cancers. *Drugs*, **2018**, *12*, 1–12.
3. S. Tian, H. Quan, C. Xie, H. Guo, F. Lü, Y. Xu, J. Li, L. Lou. YN968D1 is a novel and selective inhibitor of vascular endothelial growth factor receptor-2 tyrosine kinase with potent activity in vitro and in vivo. *Cancer Sci.* **2011**, *10*, 1374–1380.
4. P. A. Harris. Inhibitors of vascular endothelial growth factor receptor. *Top. Med. Chem.* **2018**, *28*, 105–140.

5.8 Lenvatinib (Lenvima™) (30)

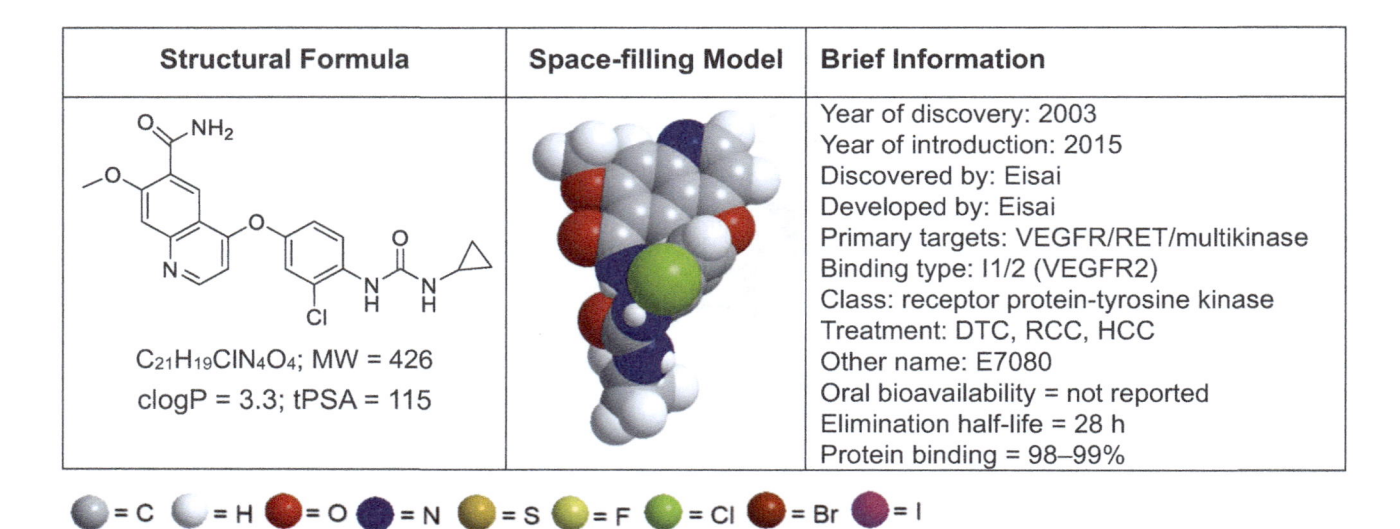

Structural Formula	Space-filling Model	Brief Information
$C_{21}H_{19}ClN_4O_4$; MW = 426 clogP = 3.3; tPSA = 115		Year of discovery: 2003 Year of introduction: 2015 Discovered by: Eisai Developed by: Eisai Primary targets: VEGFR/RET/multikinase Binding type: I1/2 (VEGFR2) Class: receptor protein-tyrosine kinase Treatment: DTC, RCC, HCC Other name: E7080 Oral bioavailability = not reported Elimination half-life = 28 h Protein binding = 98–99%

● = C   ○ = H   ● = O   ● = N   ● = S   ● = F   ● = Cl   ● = Br   ● = I

Lenvatinib was first approved in 2015 for the treatment of patients with locally recurrent or metastatic, progressive, radioactive iodine-refractory differentiated thyroid cancer (DTC). Subsequently, it won three additional approvals: (1) in combination therapy with everolimus for the treatment of patients with advanced renal cell carcinoma (RCC) following one prior antiangiogenic therapy; (2) for the first-line treatment of patients with unresectable hepatocellular carcinoma (HCC); and (3) in combination with pembrolizumab, for the treatment of patients with advanced endometrial carcinoma that is not microsatellite instability-high or mismatch repair deficient, who have disease progression following prior systemic therapy and are not candidates for curative surgery or radiation.[1]

Lenvatinib inhibits tumor cell proliferation and angiogenesis by targeting multiple serine/threonine and tyrosine kinases (VEGFR1-3, FGFR1-4, PDGFR, and KIT) in multiple oncogenic signaling pathways.[2] It inhibits VEGFR1-3 with IC_{50}'s of 4.7, 3.0, and 2.3 nM, in biochemical assays, respectively. The other angiogenic factors sensitive to Lenvatinib are: RET (IC_{50} = 6.4 nM), KIT (IC_{50} = 85 nM), FGFR1–4 (IC_{50} = 61, 27, 52, and 43 nM) and PDGFRα (IC_{50} = 29 nM) (Fig. 1).[3] The efficacy against thyroid cancers likely results from a combination of VEGFR anti-angiogenic and RET antitumorigenic pathway inhibition.[4]

Lenvatinib was originally identified as an antiangiogenic agent that inhibited vascular endothelial growth factor receptor (VEGFR1-3), fibroblast growth factor receptor (FGFR1-4), platelet-derived growth factor receptor α (PDGFRα), stem cell factor receptor (KIT), and rearranged during transfection (RET).[5,6] These transmembrane tyrosine kinase receptors, when activated by their respective ligands, enable transduction of extracellular signals to the cytoplasm through several pathways including the PI3K/AKT/mTOR pathway. A somatic or germline mutation in the DNA coding for one of these TKRs or in one of the downstream components can cause a constitutive activation of the intracellular signaling, leading to uncontrolled cell proliferation, and dedifferentiation. Lenvatinib blocks the signal transduction pathways through these TRKs by inhibiting their kinase activities, thus shutting off or slowing down cancer cell proliferation.

Lenvatinib is a close analog of sorafenib, and essentially, replacing Cl, CF_3-benzene and pyridine of sorafenib with cyclopropane and quinoline, respectively, gives lenvatinib (Fig. 1).[7] These two changes led to ~30-fold and ~10-fold enhancement in VEGFR2/3 inhibition, respectively.[5,6] One contributory factor to the increased potency could be the electron-withdrawing chlorine ortho to the urea moiety, which increases the acidity of the urea hydrogens, thereby resulting in stronger H-bonds to the side chain carboxylate oxygen of Glu885 (Fig. 2). Thanks to the enhanced potency against VEGFRs, lenvatinib is dosed at 24 mg (orally once daily), much lower than 400 mg (2 × 200 mg tablets, twice daily) for sorafenib.

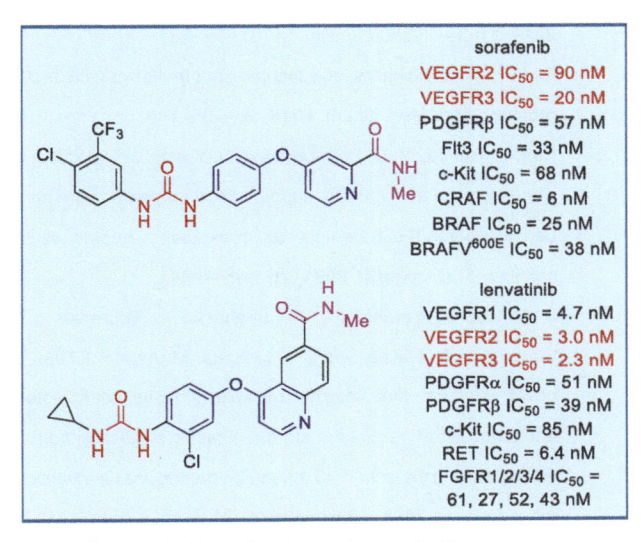

Figure 1. Sorafenib vs. lenvatinib.

As expected from their similar structures, lenvatinib and cabozantinib show comparable multikinase inhibition profiles (Fig. 2), and both have been approved for similar cancer treatments. Cabozantinib is used as a monotherapy, while lenvatinib can be combined with everolimus or pembrolizumab for various indications.

As with other MKIs, the use of lenvatinib is limited by a variety of adverse events, which have been attributed to its activity against other receptor tyrosine kinases, such as EGFR (diarrhea and dermatologic toxicities) and VEGFR (hypertension). These off-target side effects often necessitate dose reduction, which may result in poor suppression of the RET signaling pathway.

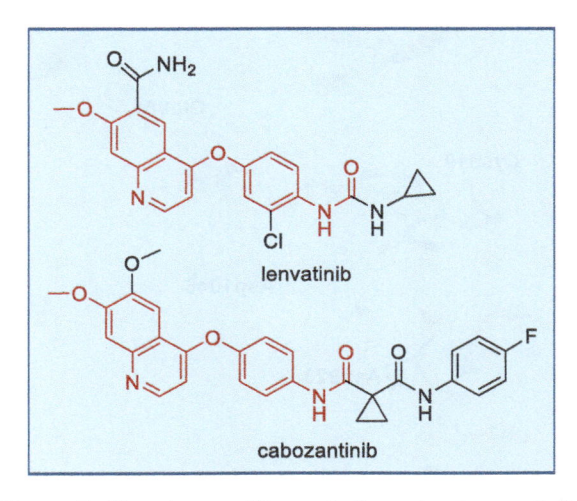

Figure 2. Structures of lenvatinib and cabozantinib.

Binding Mode[8,9]

As shown in the co-crystal structure with VEGFR2, lenvatinib binds to both the ATP-binding site and the neighboring allosteric region of VEGFR2 in the DFG-in conformation, a distinct mode of interaction classified as type I1/2 inhibitor (Fig. 3).[10,12] The quinoline ring of lenvatinib completely occupies the adenine binding site, forming a crucial hydrogen bond between N1 and the amide NH of Cys919 in the hinge region. The two urea NH groups hydrogen bond with the carboxylate group of Glu885, while the urea carbonyl forms a hydrogen bond with the amide NH of Asp1046. In addition, the 6-carbamoyl group attached to the quinoline ring engages in two hydrophilic interactions with the main and side chain atoms of Asn923 via two bound water molecules (Fig. 3). In addition, a CH−π interaction occurs between the cyclopropane ring and the benzene ring of Phe1047 (DFG motif) in the neighboring allosteric region (Fig. 4). The cyclopropane ring is 3.43 Å away from the centroid of the benzene ring, and their planes are oriented at a 105.6° angle. This interaction contributes to a DFG-in, αC helix-out binding conformation. The key hydrogen-bonding interactions of the lenvatinib–VEGFR2 complex are summarized in Fig. 5.

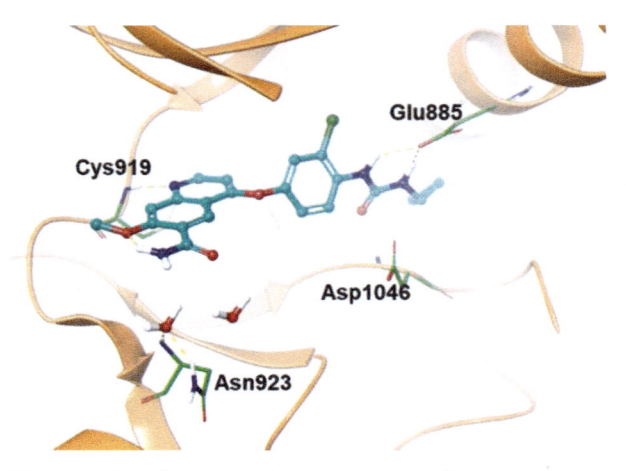

Figure 3. Co-crystal structure of lenvatinib–VEGFR2 (PDB ID: 3WZD).

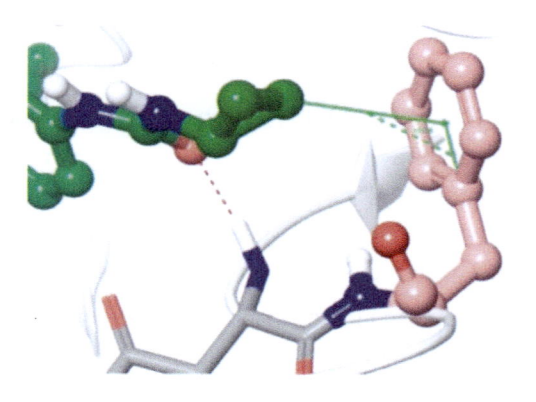

Figure 4. C-H/π interaction between the cyclopropane ring and the benzene ring of Phe1047 (DFG motif). Adapted with permission from *ACS Med. Chem. Lett.* **2015**, *6*, 89–94.

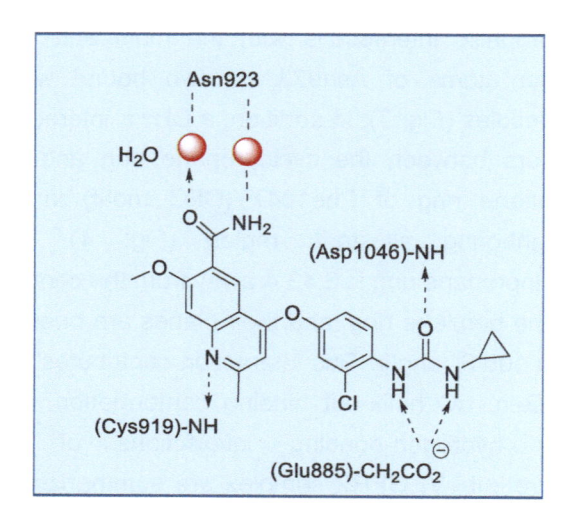

Figure 5. Summary of lenvatinib–VEGFR2 interactions based on their co-crystal structure.

1. L. J. Scott. Lenvatinib: First global approval. *Drugs.* **2015**, *75*, 553–560.

2. J. Matsui, Y. Yamamoto, Y. Funahashi, A. Tsuruoka, T. Watanabe, T. Wakabayashi, T. Uenaka. M. Asada. E7080, a novel inhibitor that targets multiple kinases, has potent antitumor activities against stem cell factor producing human small cell lung cancer H146, based on angiogenesis inhibition. *Int. J. Cancer,* **2008**, *122*, 664–671.

3. M. Capozzi, C. De Divitiis, A. Ottaiano, C. von Arx, S. Scala, F. Tatangelo, P. Delrio, S. Tafuto. Lenvatinib, a molecule with versatile application: from preclinical evidence to future development in anti-cancer treatment. *Cancer Manag. Res.* **2019**, *11*, 3847–3860.

4. P. A. Harris. Inhibitors of vascular endothelial growth factor receptor. *Top. Med. Chem.* **2018**, *28*, 105–140.

5. Nair, K. Reece, M. B. Donoghue, W. V. Yuan, L. Rodriguez, P. Keegan, R. Pazdur. FDA supplemental approval summary: Lenvatinib for the treatment of unresectable hepatocellular carcinoma. *Oncologist.* **2021**, *26*, e484–e491.

6. J. Matsui, Y. Yamamoto, Y. Funahashi, A. Tsuruoka, T. Watanabe, T. Wakabayashi, T. Uenaka, M. Asada. E7080, a novel inhibitor that targets multiple kinases, has potent antitumor activities against stem cell factor producing human small cell lung cancer H146, based on angiogenesis inhibition. *Int. J. Cancer.* **2008**, *122*, 664–671.

7. Y. Funahashi, A. Tsuruoka, M. Matsukura, T. Haneda, Y. Fukuda, J. Kamata, K. Takahashi, T. Matsushima, K. Miyazaki, K. I. Nomoto, T. Watanbe, H. Obaishi, A.Yamaguchi, S. Suzuki, K. Nakamura, F. Mimura, Y. Yamamoto, J. Matsui, K. Matsui, T. Yoshiba, Y. Szuki, I. Arimoto. Nitrogen-containing aromatic derivatives. US Patent 7253286B2.

8. K. Okamoto, M. Ikemori-Kawada, A. Jestel, K. von König, Y. Funahashi, T. Matsushima, A. Tsuruoka, A. Inoue, J. Matsui. Distinct binding mode of multikinase inhibitor lenvatinib revealed by biochemical characterization. *ACS Med. Chem. Lett.* **2015**, *6*, 89–94.

9. R. Roskoski, Jr, A. Sadeghi-Nejad. Role of RET protein-tyrosine kinase inhibitors in the treatment RET-driven thyroid and lung cancers. *Pharmacol. Res.* **2018**, *128*, 1–17.

5.9 Tivozanib (Fotivda™) (31)

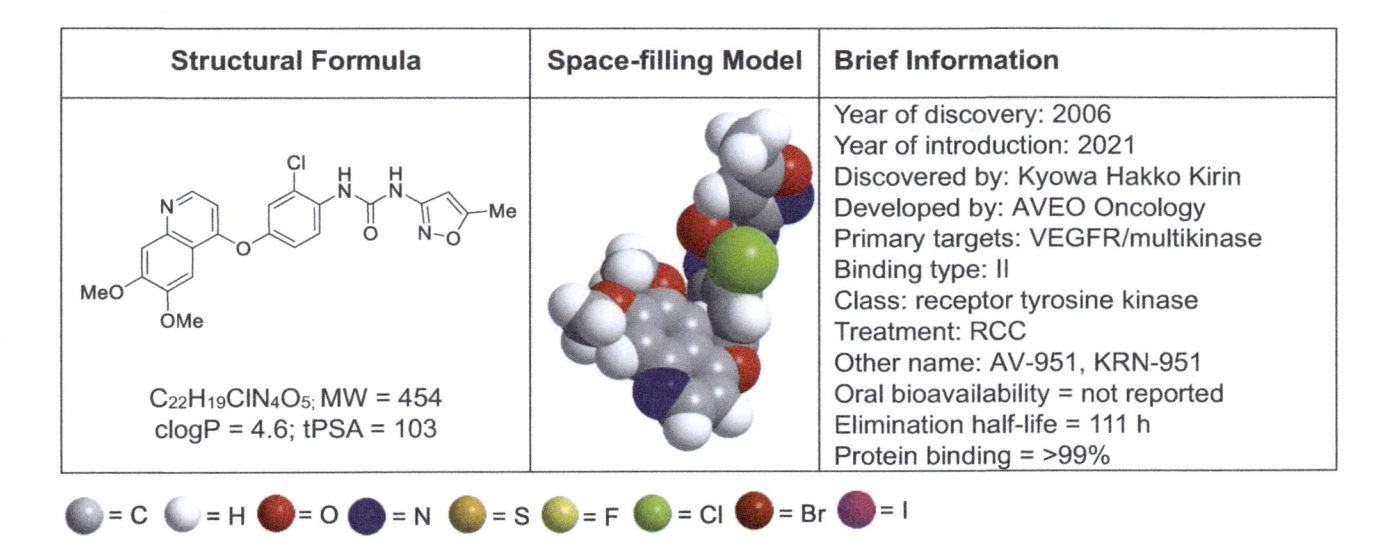

Structural Formula	Brief Information
$C_{22}H_{19}ClN_4O_5$; MW = 454 clogP = 4.6; tPSA = 103	Year of discovery: 2006 Year of introduction: 2021 Discovered by: Kyowa Hakko Kirin Developed by: AVEO Oncology Primary targets: VEGFR/multikinase Binding type: II Class: receptor tyrosine kinase Treatment: RCC Other name: AV-951, KRN-951 Oral bioavailability = not reported Elimination half-life = 111 h Protein binding = >99%

Tivozanib is a highly potent multi-kinase inhibitor approved by the FDA in 2021 for adult patients with relapsed or refractory advanced renal cell carcinoma (RCC) following two or more prior systemic therapies.

Tivozanib inhibits tumor cell proliferation and angiogenesis by targeting multiple serine/threonine and tyrosine kinases (VEGFR 1/2/3, PDGFR, and KIT) in multiple oncogenic signaling pathways (Fig. 1).[1–3]

Tivozanib is a close analog of sorafenib, and they were derived from lead compound **1** originally identified in a RAF inhibitor program (Fig. 1).[4] Essentially, replacing isoxazole moiety of **1** with a substituted phenyl gave sorafenib, while switching the pyridyl of **1** with 6,7-dimethoxyquinoline led to tivozanib.[5] There are profound differences in activity and PK properties despite the subtle structural variations between the two inhibitors. For example, tivozanib is more potent than sorafenib by 560- and 58-fold against VEGFR2 in biochemical[6] and cellular assays,[7] respectively. Most notably, tivozanib exhibits a half-life of 111 h, vs. 25–48 h for sorafenib. Because of its superb potency and extremely long half-life, tivozanib is dosed orally at only 0.89 mg (once a day), significantly lower than 400 mg (2 × 200 mg tablets, twice daily) for sorafenib.

Tivozanib binds to VEGFR2 very similarly to sorafenib due to their structural similarity (Figs. 2, 3). It forms four hydrogen bonds: two hydrogen bonds between the urea NH with Glu885; one hydrogen bond between the urea C=O with Asp1046; and the nitrogen atom in the quinoline ring with Cys919.

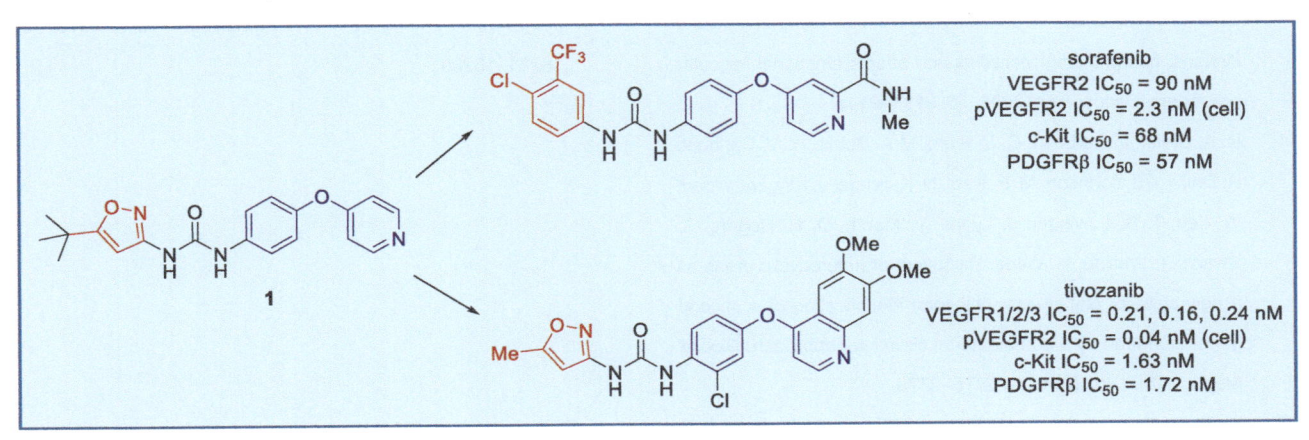

Figure 1. Structural optimization of tivozanib vs. sorafenib.

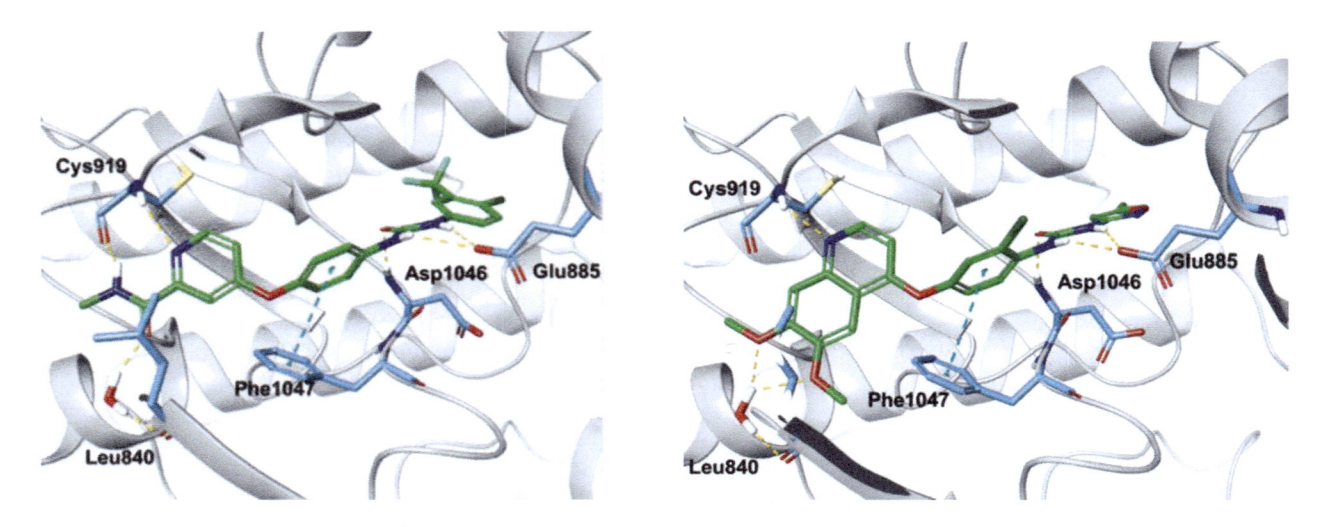

Figure 2. Co-crystal structures of VEGFR2 with sorafenib (left) (4ASD) and tivozanib (right) (PDBID: 4ASE).

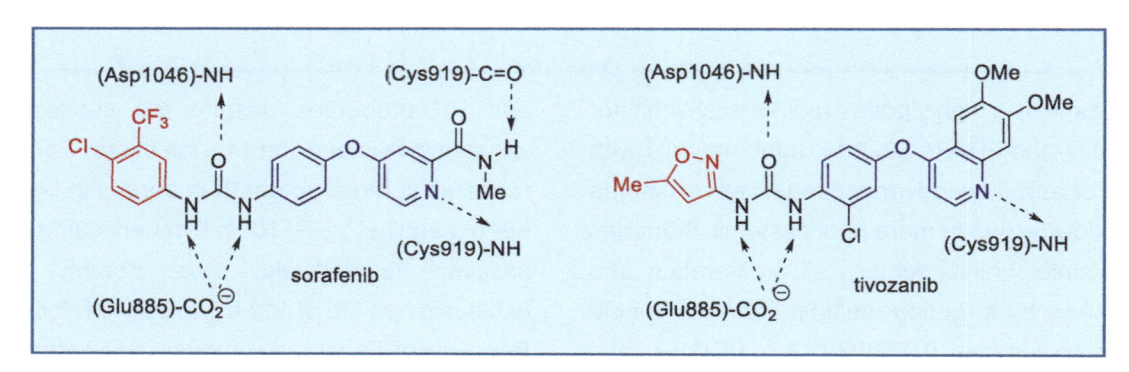

Figure 3. Summary of interactions between VEGFR2 and tivozanib vs. sorafenib based on X-ray co-structural data.

1. S. Kim. Tivozanib: First Global Approval. *Drugs.* **2017**, *77*,
2. M. Hepgur, S. Sadeghi, T. B. Dorff, D.I. Quinn. Tivozanib in the treatment of renal cell carcinoma. *Biologics.* **2013**, *7*, 139–148.
3. K. Nakamura, E. Taguchi, T. Miura, A.Yamamoto, K. Takahashi, F. Bichat, N. Guilbaud, K. Hasegawa, K. Kubo, Y. Fujiwara, R. Suzuki, K. Kubo, M. Shibuya, T. Isoe. KRN951, A highly potent inhibitor of vascular endothelial growth factor receptor tyrosine kinases, has antitumor activities and affects functional vascular properties. *Cancer Res.* **2006**, *66*, 9134–9142.
4. R. A. Smith, J. Barbosa, C. L. Blum, M.A. Bobko, Y. V. Caringal, R. Dally, J. S. Johnson, M. E. Katz, N. Kennure, J. Kingery-Wood, W. Lee, T. B. Lowinger, J. Lyons, V. Marsh, D. H. Rogers, S. Swartz, T. Walling, H. Wilda. Discovery of heterocyclic ureas as a new class of Raf kinase inhibitors: identification of a second generation lead by a combinatorial chemistry approach. *Bioorg. Med. Chem. Lett.* **2001**, *11*, 2775–2778.
5. S. Ramachandra, W. R. Bishop, L. Masat, C. B. Huang, T. Takeuchi, S. Kantak, C. Y. Huang. Fully human anti-VEGF antibodies and methods of using. US Patent 8216571B2, **2012**.
6. P. A. Harris. Inhibitors of vascular endothelial growth factor receptor. *Top. Med. Chem.* **2018**, *28*, 105–140.
7. M. McTiguea, B. W. Murrayb, J. H. Chen, Y. L. Deng, J. Solowiej, R. S. Kania. Molecular conformations, interactions, and properties associated with drug efficiency and clinical performance among VEGFR TK inhibitors. *PNAS.* **2012**, *109*, 18281–18289.

Chapter 6. CDK4/6 Inhibitors

Cyclins and cyclin-dependent kinases (CDKs) control cell cycle progression through four distinct phases (G_1, S, G_2, and M) and thereby regulate cell division and tumor growth. The critical G_1-to-S cell cycle checkpoint is strictly controlled by the cyclin D1, CDK4, and CDK6. Cyclin D1 is overexpressed in up to 50% of breast cancer, especially hormone receptor–positive (HR+) breast cancer, making the G_1-to-S checkpoint an ideal therapeutic target. CDK4/6 inhibitors target specifically at this checkpoint and block cell cycle progression through this checkpoint, thus arresting tumor cells in the G_1 phase and suppressing their growth.

CDK4/6 inhibitors (palbociclib, ribociclib, and abemaciclib) have significantly prolonged progression-free survival in HR+ breast cancer when used in combination with endocrine therapy by targeting the cell cycle machinery and overcoming some aspects of endocrine resistance. Because of their well-established efficacy, they have become the standard of care in the treatment of HR+, HER2− metastatic breast cancer.[1–3]

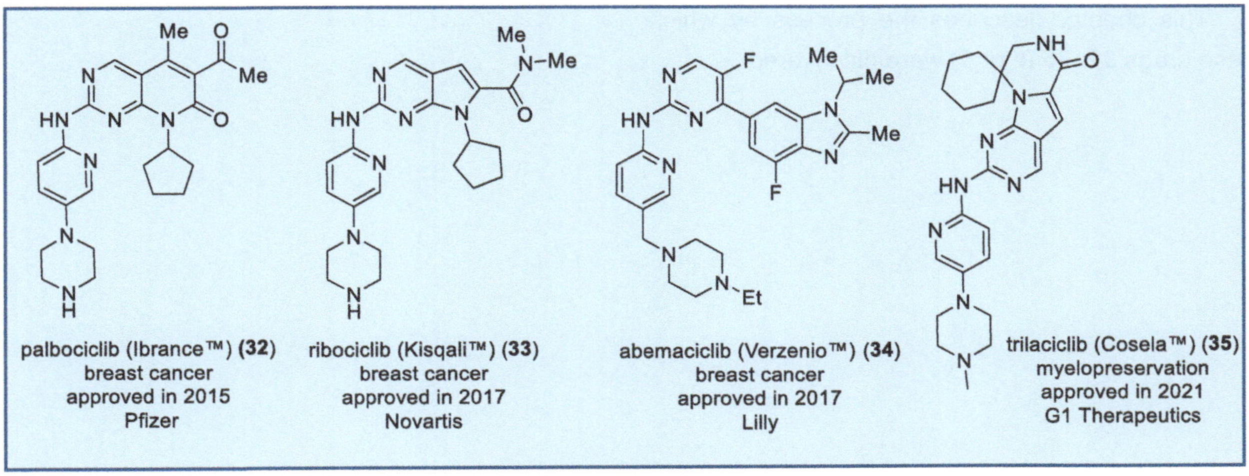

Figure 1. Approved CDK4/6 inhibitors.

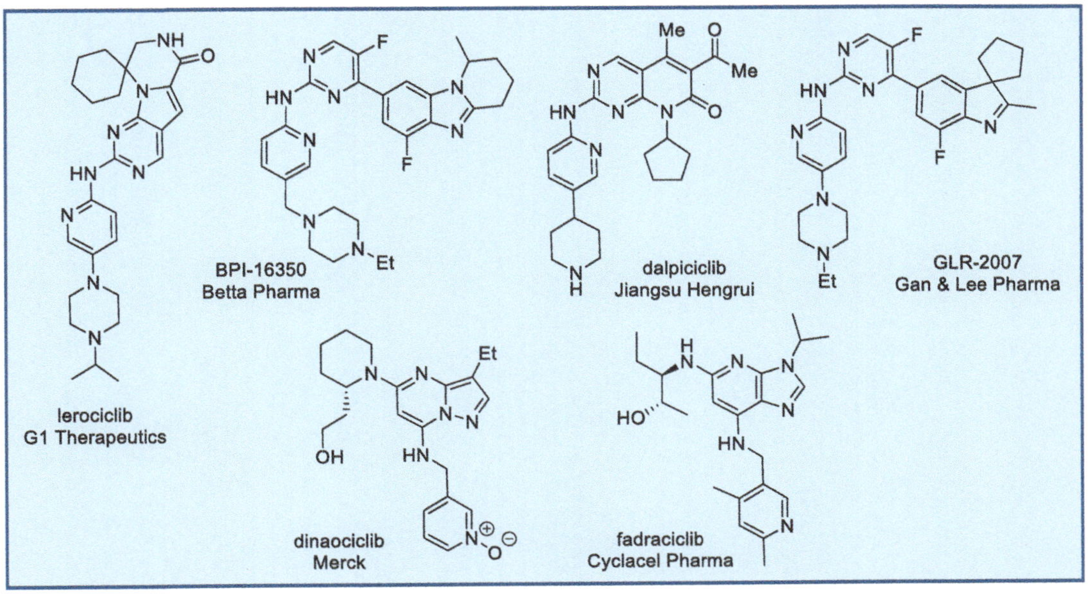

Figure 2. CDK inhibitors in clinical trials for breast cancer and NSCLC.

Molecules Engineered Against Oncogenic Proteins and Cancer, First Edition. E. J. Corey and Yong-Jin Wu.
© 2024 John Wiley & Sons Inc. Published 2024 by John Wiley & Sons Inc.

Palbociclib generated peak sales of $5.4 billion in 2021, second only to the leading BTK inhibitor, ibrutinib (within the kinase drug class), which earned $9.7 billion in 2020. Abemaciclib and ribociclib had sales of $2.5 and $1.2 billion, respectively, in 2022.

Unfortunately, breast cancer cells can acquire resistance to the existing CDK4/6 inhibitor drugs primarily by producing higher amounts of CDK4/6.[4] This prompted the development of next-generation of inhibitors with improved potency and selectivity, some of which are currently in clinical trials (Fig. 2).[5]

Palbociclib, ribociclib, and abemaciclib are used for the treatment of breast cancer, whereas trilaciclib is approved for myelosuppression.

This chapter describes the process by which these drugs **32–35** (Fig. 1) were discovered.

1. A. Ammazzalorso, M. Agamennone, B. De Filippis, M. Fantacuzzi. Development of CDK4/6 inhibitors: A five years update. *Molecules*. **2021**, 26, 1488.
2. Z. Shi, L. Tian, T. Qiang, J. Li, Y. Xing, X. Ren, C. Liu, C.Liang, From structure modification to drug launch: A systematic review of the ongoing development of cyclin-dependent kinase inhibitors for multiple cancer therapy. *J. Med. Chem.* **2022**, *65*, 6390–6418.
3. M. A. George, S. Qureshi, C. Omene, D. L. Toppmeyer, S. Ganesan. Clinical and Pharmacologic Differences of CDK4/6 Inhibitors in Breast Cancer. *Front. Oncol.* **2021**, *11*, 693104.
4. M. Álvarez-Fernández, M. Malumbres. Mechanisms of sensitivity and resistance to CDK4/6 inhibition. *Cancer cell.* **2020**, *37*, 514–529.
5. clinicaltrials.gov.

6.1 Palbociclib (Ibrance™) (32)

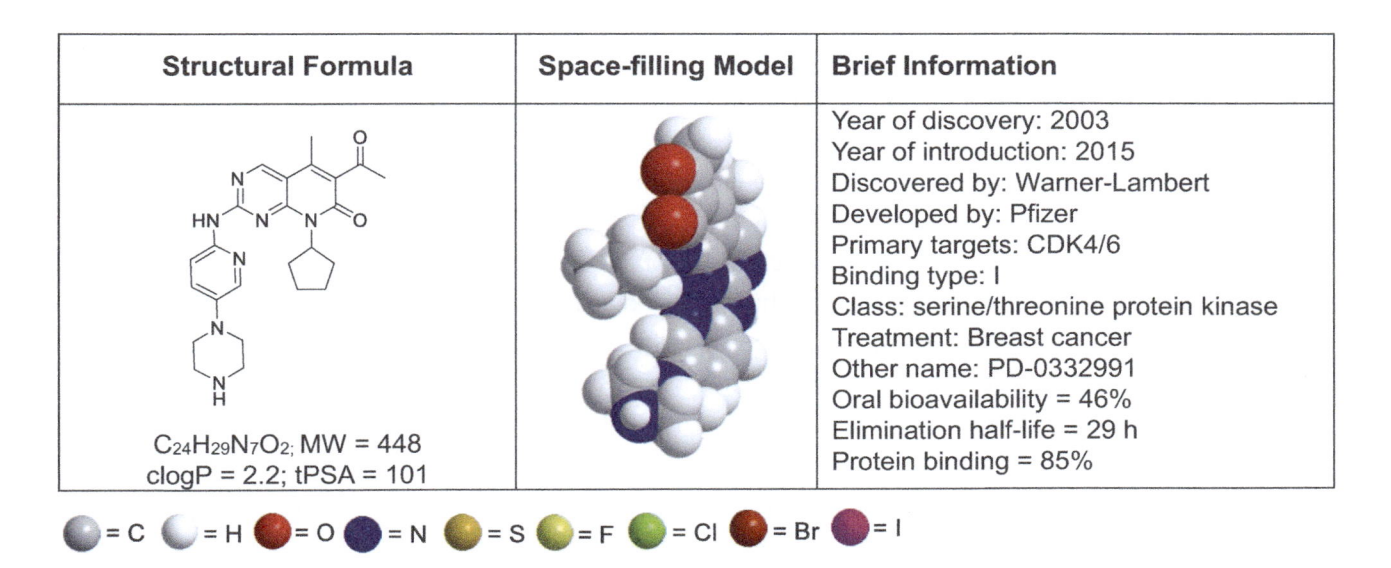

Structural Formula	Space-filling Model	Brief Information
$C_{24}H_{29}N_7O_2$; MW = 448 clogP = 2.2; tPSA = 101		Year of discovery: 2003 Year of introduction: 2015 Discovered by: Warner-Lambert Developed by: Pfizer Primary targets: CDK4/6 Binding type: I Class: serine/threonine protein kinase Treatment: Breast cancer Other name: PD-0332991 Oral bioavailability = 46% Elimination half-life = 29 h Protein binding = 85%

● = C ○ = H ● = O ● = N ● = S ● = F ● = Cl ● = Br ● = I

Palbociclib is a first-in-class CDK4/6 (cyclin-dependent kinase) inhibitor which was approved: (1) in 2015, as a combination therapy with letrozole for postmenopausal women with advanced breast cancer; (2) in 2016, in combination with fulvestrant for the treatment of women with estrogenic hormone receptor-positive (HR+), human epidermal growth factor receptor-2-negative (HER2−) advanced or metastatic breast cancer; (3) in 2017, as the first-line treatment of women with HR+, HER2− metastatic breast cancer; and (4) in 2019, for the treatment of men with HR+, HER2− metastatic breast cancer.[1–4]

Breast cancer is the most common type of cancer in the United States, affecting a total of more than 3.8 million women and 264,000 new cases every year. The average risk of a woman in the United States developing breast cancer sometime in her lifetime is ~13%, and ~42,000 women die each year. Although the overall five-year survival has approached to 90%, the five-year survival rate remains only 25% in metastatic or advanced breast cancer. The most common subtype, ER+/HER2−, which accounts for ~73% of all breast cancers, are traditionally treated with the antiestrogens tamoxifen or letrozole which target the hormone receptor pathway. Combination therapy of CDK4/6 inhibitors such as palbociclib with endocrine therapy represents a major advance in the management of HR+/HER2− advanced, or metastatic breast cancer.[1]

Palbociclib selectively inhibits cyclin D-CDK4/6 complex activity, thus preventing the phosphorylation of retinoblastoma (Rb) protein and subsequently blocking tumor cell cycle progression from G_1 into S phase. This arrests tumor cells in the G_1 phase and suppresses their growth.[1]

Palbociclib travelled an unusually long road over 14 years from a preclinical candidate in 2001 to its launch in 2015.[5,6] Remarkably, although palbociclib was almost relegated to the drug development graveyard due to its failure to show robust antitumor effect in the phase I trials in 2004, it was resurrected for breast cancer therapy by investigators outside Pfizer, notably Richard Finn and Dennis Slamon at UCLA's Jonsson Cancer Center in 2007.[5,6]

Palbociclib was the first-to-market, two years ahead of its rivals: Novartis' ribociclib and Lilly's abemaciclib. Because of its efficacy in delaying disease progression as well as its favorable safety and tolerability profile, palbociclib quickly became the drug of choice for treatment of breast cancer. Subsequently, its sales climbed to $5.4 billion in 2020 and 2021,[7] second only to AbbVie's ibrutinib (within kinase drug class), a BTK inhibitor for CLL, which racked up $9.7 billion in sales in 2020. However, its sales ranking dropped from No.13 in

2020 to No. 20 in 2021 primarily due to competition with the two newer CDK4/6 inhibitors, abemaciclib and ribociclib (see Fig. 5 for structures). Abemaciclib's sales reached $1.35 billion in 2021, a 48% increase from 2020, while ribociclib's hit $936 million, a jump of 36%. The two competitors are possibly superior because palbociclib has not been proven overall to provide survival benefit in metastatic breast cancer patients. Novartis' ribociclib is the only drug in the class with consistently (three clinical trials) proven overall survival benefit by one year in HR+/HER2− metastatic breast cancer. With three survival wins for ribociclib in clinical trials, unmatched by palbociclib, Novartis has launched an ambitious head-to-head trial dubbed Harmonia for HR+/HER2− advanced breast cancer patients. However, robociclib is the only one of the three CDK4/6 inhibitors that carries a warning for risk of QT prolongation which signals problematic heart rhythm.

Although palbociclib contains an uncommon methyl ketone subunit, which usually is prone to reduction, in palbociclib, this subunit appears to be metabolically resistant as shown by the favorable PK properties (F = 46%; $t_{1/2}$ = 29 h), compatible with once-daily oral administration (125 mg tablet).[8] The methyl ketone contributes to PK as well as to its binding affinity as the ketone oxygen hydrogen bonds to the amide NH of Asp163 of the CDK4/6 proteins (see binding mode).

All three marketed CDK4/6 inhibitors may cause rare but severe inflammation of the lungs in certain patients.

CDK4/6 in Cell Cycle and Breast Cancer[9–12]

Tissue homeostasis is strictly controlled by cell division and death, through which cells precisely manage tissue damage and repair thus avoiding over-proliferation to cause cancer. The cell division cycle (or simply, cell cycle) consists of four distinct phases: (1) G_1 phase (containing G_0 and the restriction point or checkpoint R); (2) S phase (synthesis); (3) G_2 phase (preparation for mitosis); and (4) M phase (mitosis) (Fig. 1). The majority of differentiated cells in adult tissues are parked in the G_0 state, ready to enter the cell cycle. Cells within the G_0 phase can be there either temporarily (quiescence) or permanently due to aging or deterioration (senescence). The quiescent cells can re-enter the cell cycle upon receiving mitogenic stimuli, such as from growth factors or hormones. The cell cycle relies on a series of checkpoints to block the cells from progressing into the next phase of the cycle. In particular, the cyclin proteins bind to and activate their kinase partners (the CDKs) to drive the progression through the cell cycle. More specifically, a cyclin is synthesized and then binds to a CDK, forming an active CDK/cyclin kinase complex that phosphorylates substrates to initiate the first cell cycle transition. After the transition has completed, the cyclin is degraded, thereby deactivating the kinase. Then, a second cyclin is synthesized and forms another active CDK/cyclin kinase complex that phosphorylates other substrate to trigger the second transition. Each CDK/cyclin pair acts on specific substrates, resulting in the activation of different signaling pathways that control different transitions.[12]

In the early G_1 phase, endogenous or exogenous mitogenic stimuli induce expression of cyclin D (D1, D2 and D3), and the activated cyclin D forms complexes with CDK4 and CDK6, known as the cyclin D-CDK4/6 complex. This active complex phosphorylates the C-terminal region of Rb to give the hyperphosphorylated Rb protein, which releases pRb from its association with transcription factor E2F, allowing the cell cycle to proceed. The partially detached E2F facilitates expression of cyclin E, which activates CDK2. The activated CDK2 hyperphosphorylates pRb, fully releasing E2F suppression and triggering cells to move from G_1 to S phase, in which DNA synthesis starts (Fig. 1).

The cell cycle is a fundamental biological process that is highly controlled by the actions of a series of CDKs, and its dysregulation causes excessive cell division, tumor growth and cancer. Abnormalities of the cyclin/CDK/Rb pathway are frequent in breast cancer, in which CDK4/6 have been identified as key drivers of proliferation in HR+ breast cancer. For example, amplification and overexpression of CCND1 oncogene, encoding cyclin D1 protein, is common in breast cancer, more specifically within luminal A (29%), luminal B (58%), and HER2 enriched (38%) subtypes, Similarly, CDK4 gene amplification has been identified in 14% of luminal A, 25% of luminal B, and 24% of HER2 enriched tumors.[11] Thus, inhibition of CDK4/6 has the potential to overcome all of these mechanisms. Although, theoretically, a cell cycle inhibitor can only prevent a tumor from growing, CDK4/6 inhibitors have been shown to shrink human breast and lung tumors. The exact mechanism remains to be determined. It is likely that the inhibitor induces cancer cells into a senescence G_0 state with a permanent inability to divide. Such senescent cells may ultimately be removed by immune cells, resulting in tumor shrinkage. Another possibility is that the prolonged cell cycle arrest by the inhibitor makes the tumor cells more sensitive to other cancer drugs.[5,6]

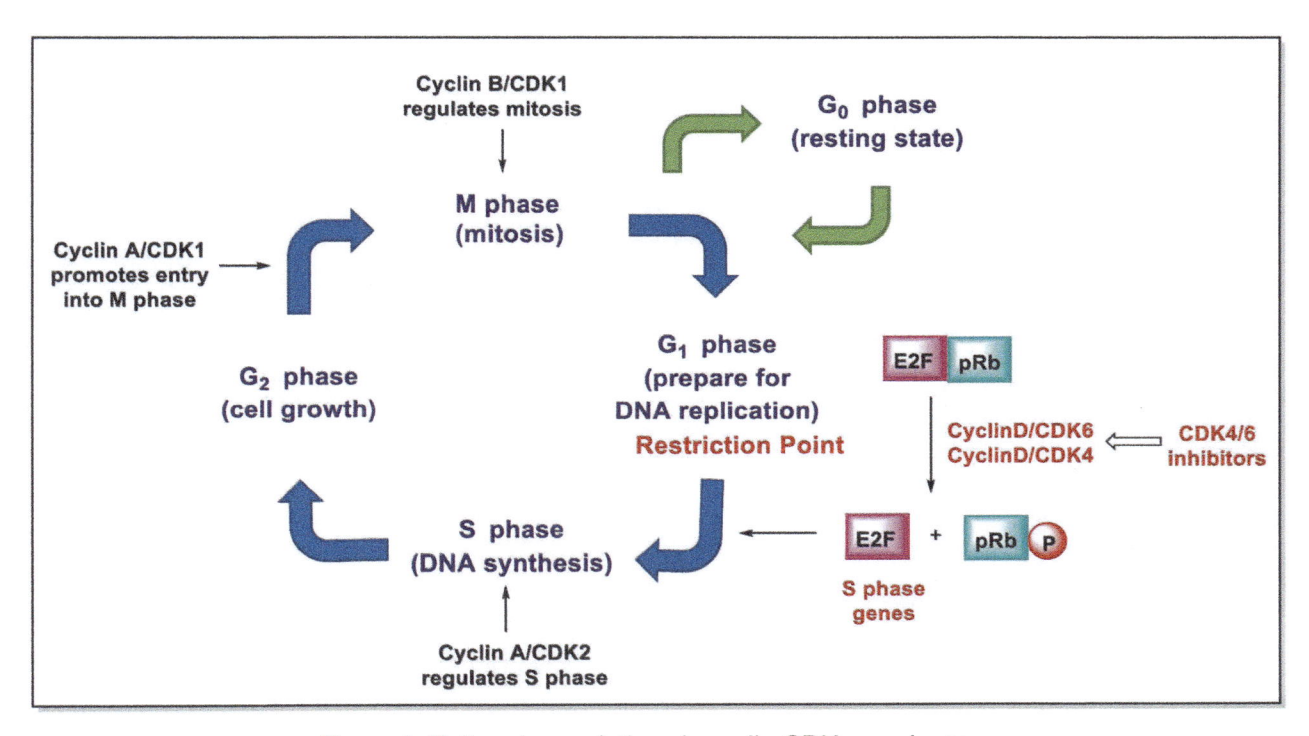

Figure 1. Cell cycle regulation via cyclin-CDK complexes.

Discovery of Palbociclib[13–15]

Research in the 1970s and 1980s revealed the key molecular regulators of the cell cycle, which led to the 2001 Nobel Prize in physiology or medicine for Leland H. Hartwell, Tim Hunt, and Sir Paul M. Nurse. By 1995, CDK4/6 had been identified as a viable target for cancer treatment, and scientists at Parke-Davis in Ann Arbor, Michigan, started to explore selective CDK4 inhibitors in collaboration with Onyx Pharmaceuticals, which provided CDK4 and other cell cycle proteins for various selectivity assays. SAR for selective inhibition of CDK4 was enabled by testing compounds against a small panel of four enzymes: (1) CDK4/D1 (Cyclin D1); (2) CDK2/A (Cyclin A); (3) FGFR (fibroblast growth factor receptor); and (4) PGFR (platelet-derived growth factor receptor).

The starting point for selective CDK4/D inhibitors was pyridopyrimidinone **1a** which contained a 2-aminophenyl substituent at C2. This compound was a nonselective CDK4/D inhibitor with an IC_{50} value of 210 nM (Fig. 2). A comparison of **1a** with its 2-aminopyridyl counterpart **1b** revealed a striking difference in selectivity: the pyridyl analogue **1b** substantially favoring CDK4/D inhibition over CDK2/A. Incorporation of a piperazine moiety onto **1a** gave **2a** as a highly potent inhibitor of CDK4/D (IC_{50} = 2 nM). Like **1a**, **2a** was also nonselective and it inhibited CDK2/A (IC$_{50}$ = 43 nM) and FGFR (IC$_{50}$ = 80 nM). Replacement of the aniline side chain of **2a** with aminopyridyl led to **2b** with excellent selectivity for CDK4/D versus other CDKs, FGFR, and PDGFR, while largely maintaining CDK4/D potency. The remarkable selectivity of 2-aminopyridyl analogs **1b/2b** vs. their 2-aminophenyl counterparts **1a/2a** established the pyrimidine−NH−pyridine motif as a key element for CDK4 isoform inhibition and prompted a comprehensive exploration of pyridopyrimidinones with 2-aminopyridine side chains at the C2 position, leading to PD-0332991, later known as palbociclib, as a potent and highly selective CDK4/6 inhibitor (IC$_{50}$: CDK4/D1 = 11 nM, CDK4/D3 = 9 nM and CDK6/D2 = 15 nM). It also exhibited no activity against a panel of 36 additional protein kinases. Because of its favorable in vitro and in vivo profile as well as good PK properties in rats (F = 56%), palbociclib became a preclinical candidate in 2001.

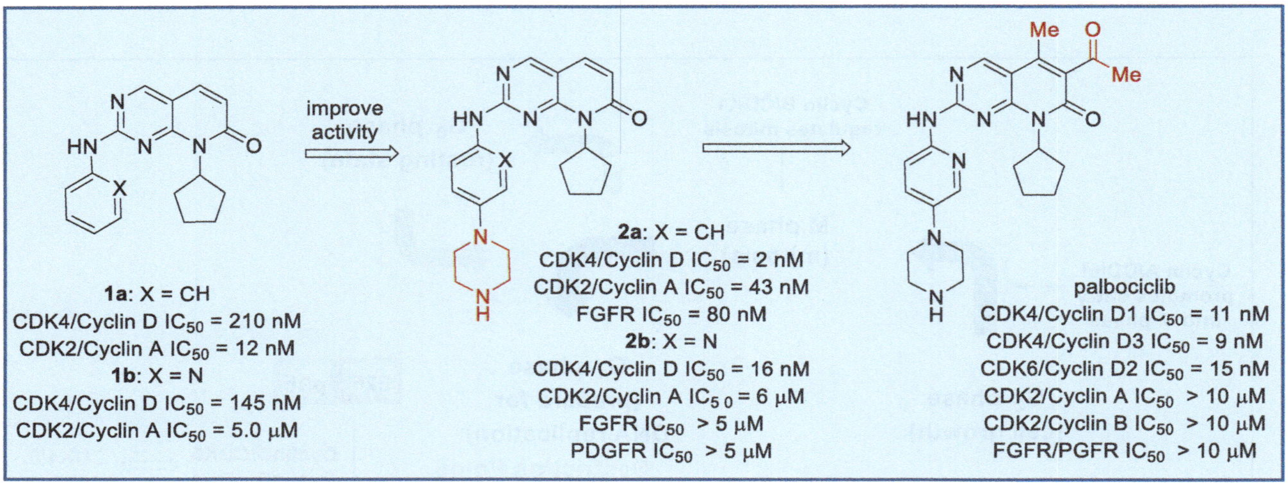

Figure 2. Structural optimization of palbociclib.

Development of Palbociclib[5,6]

It took Parke-Davis 6 years to identify palbociclib as a preclinical candidate, but its further journey to bedside took even longer – 14 years. In 2000, Pfizer acquired Warner-Lambert (which owned Parke-Davis drug discovery unit in Ann Arbor, MI) primarily for the cholesterol drug Lipitor, and palbociclib was not given high priority for development. In 2003, Pfizer acquired Swedish-American drug giant Pharmacia, which added multiple preclinical drug candidates to Pfizer's pipeline, delaying the clinical development of some drug candidates including palbociclib. In the same year, Pfizer management closed all research in Ann Arbor. In 2004, Pfizer finally launched a phase I dose-finding study in 33 patients with Rb-intact advanced solid tumors and non-Hodgkin lymphoma, which failed to show much antitumor effect. Unfortunately, there was no biomarker at the time to monitor the alterations in the CDK4/6 proteins so it was difficult to predict who might benefit from the drug. As a result, the project was shelved for another five years. In May 2007, Dennis Slamon, a UCLA oncologist whose earlier research had contributed to the breast cancer drug Herceptin, evaluated palbociclib against a large panel of breast cancer cell lines. To his surprise, it is most active towards the ER+ cell lines, a significant finding as more than 60% of all breast cancers are ER+.[16] Subsequently, he and Richard Finn (another UCLA oncologist) tested palbociclib in combination with letrozole, a standard antiestrogen, in a dozen women with ER+ metastatic breast cancer, three of whom showed at least 30% tumor shrinkage.[17] These results rejuvenated the Pfizer palbociclib project – the rest is history.

The long timeline for development of Pfizer's CDK4/6 inhibitor palbociclib enabled Lilly and Novartis to catch up with abemaciclib and ribociclib.

Although Pfizer moved slowly in the clinical development of palbociclib, it made fast progress with dacomitinib (an irreversible EGFR inhibitor), another drug candidate discovered at Parke-Davis. Under the Federal Trade Commission antitrust requirements for the merger with Warner–Lambert, Pfizer kept dacomitinib but gave away its own EGFR inhibitor (discovered in Groton, CT), erlotinib, to OSI for free (see Section 4.2). After failing twice in the clinic, dacomitinib was finally approved in 2018 for EGFR mutated cancers, but its sales in 2022 (<176 million) are dwarfed by erlotinib's peak sales (1.8 billion in 2014).

Binding Mode[18–20]

Figure 3 shows the ATP-binding site of CDK6 (in brown) in complex with palbociclib aligned with that of CDK4 (in cyan). It binds in the ATP-binding pocket of the kinases, forming three hydrogen bonds: (1) between the carbonyl group of Val101 of the hinge and the 2-NH group attached to C-2 of the pyridopyrimidine core; (2) between the N3 nitrogen of the core and the amide NH group of Val101; and (3) between the DFG-Asp163 amide NH group and the 6-acetyl oxygen (Fig. 4). The 5-methyl and 6-acetyl groups occupy hydrophobic pocket in the gate area. The piperazinylpyridinylamino substituent at the 2-position of the pyrido[2,3-d]pyrimidinone core borders the solvent front, while the cyclopentyl group is located where the ATP ribose is expected to reside.

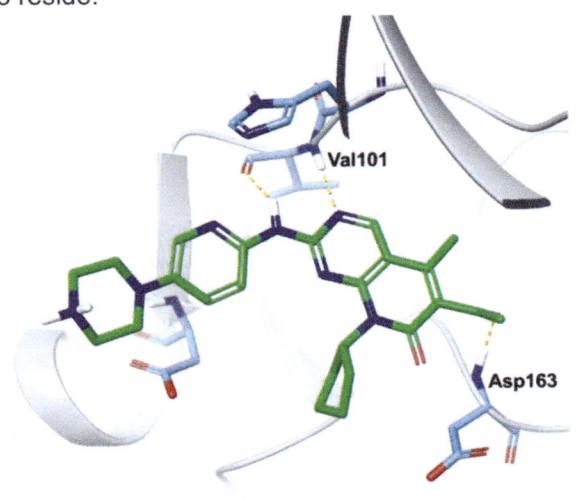

Figure 3. Palbociclib–CDK6 complex (PDB: 5L2I).

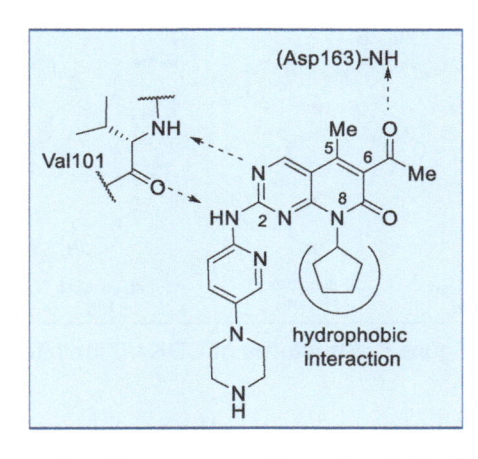

Figure 4. Summary of palbociclib–CDK4/6 interactions.

Origin of Selectivity[22,23]

All three CDK4/6 selective inhibitors, palbociclib, ribociclib, and abemaciclib, share the common pyrimidine–NH–pyridine motif (Fig. 5).[21] This motif may be responsible for their CDK4/6 isoform selectivity and other kinase selectivity as revealed by the co-crystal structure of abemaciclib in complex with CDK6 (Fig. 6)[22] which shows an ordered water molecule bridging the imidazole of hinge residue His100 and the pyridinyl nitrogen (Figs. 6, 7). The electron density for this water was not observed in the co-crystal structures of palbociclib and ribociclib possibly due to limited resolution. However, space is still available for a water molecule at this position, and such a bridging interaction with His100 may result in high kinase selectivity, because this histidine residue is found in only 8 kinases based on sequence alignment (total of 442 kinases used for alignment). Moreover, within the CDK family, only CDK4/6 contain His100. The critical contribution of the pyridinyl nitrogen atom to the selectivity is consistent with the SAR shown in Figure 2: pyridyl analogs **1b/2b** showed excellent selectivity, while removal of the pyridyl nitrogen gave nonselective analogs **1a/2a**.

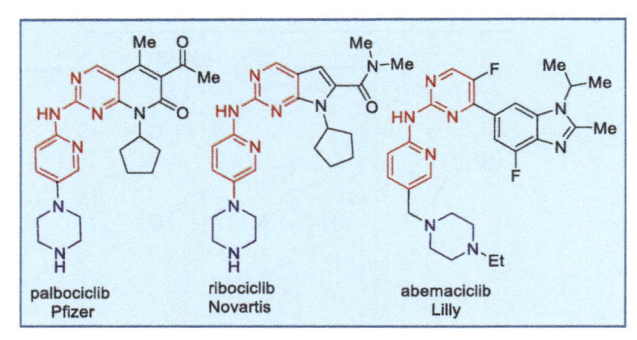

Figure 5. Structures of CDK4/6 inhibitors.

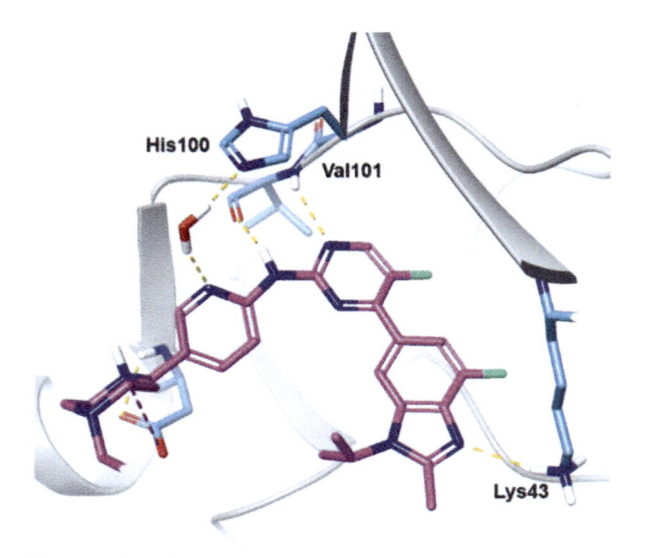

Figure 6. Co-crystal structure of abemaciclib–CDK6 (PDB ID: 5L2S).

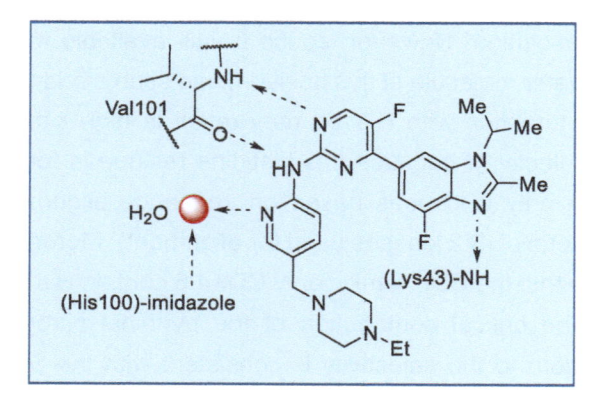

Figure 7. Summary of abemaciclib–CDK6 interactions.

Perspective

A classic SAR study from Parke-Davis identified the pyrimidine–NH–pyridine motif as a pharmacophore for CDK4/6 isoform selectivity, and subsequent SAR studies gave palbociclib. The key finding by Dennis Slamon that ER+ cell lines are very sensitive to palbociclib led to resurrection of CDK4/6 inhibition for breast cancer. The 14-year path of palbociclib from a preclinical candidate to a marketed drug is reminiscent of the path to midostaurin: a twice failed drug ultimately repurposed for FLT3 mutant AML (see Section 18.1). The discovery of selective CDK4/6 inhibitors led by palbociclib ushered in a new beginning of combination therapies for HR+/HER− breast cancer.

1. J. McCain. First-in-Class CDK4/6 Inhibitor palbociclib could usher in a new wave of combination therapies for HR+, HER2- breast cancer. *P & T.* **2015**, *40*, 511–520.

2. J. Lu. Palbociclib: A first-in-class CDK4/CDK6 inhibitor for the treatment of hormone-receptor positive advanced breast cancer. *J. Hematol. Oncol.* **2015**, *8*, 98.

3. K. A. Cadoo, A. Gucalp, T. A. Traina. Palbociclib: an evidence-based review of its potential in the treatment of breast cancer. *Breast Cancer.* **2014**, *6*, 123–133.

4. B. O'Leary, R. S. Finn, N. C. Turner. Treating cancer with selective CDK4/6 inhibitors. *Nat. Rev. Clin. Oncol.* **2016**, *13*, 417–430.

5. K. Garber. The cancer drug that almost wasn't. *Science.* **2014**, *345*(6199), 865–867.

6. K. Garber. Breakthrough. Six years after Pfizer left town, a powerful cancer drug discovered here is on the fast track. Ann Arbor Observer. June 29, 2014.

7. K. Dunleavy. The top 20 drugs by worldwide sales. *Fierce Pharma.* May 31, 2022.

8. B. B. Chavan, S. Tiwari, R. D. Nimbalkar, P. R. Garg, M. Talluri. In vitro and in vivo metabolic investigation of the Palbociclib by UHPLC-Q-TOF/MS/MS and in silico toxicity studies of its metabolites. *J. pharma. Biomed. Anal.* **2018**, *157*, 59–74.

9. C. J. Sherr, D. Beach, G. I. Shapiro. Targeting CDK4 and CDK6: From discovery to therapy. *Cancer Discov.* **2016**, *6*, 353–367.

10. F. Schettini, I. De Santo, C. G. Rea, P. De Placido, L. Formisano, M. Giuliano, G. Arpino, M. De Laurentiis, F. Puglisi, S. De Placido, L. Del Mastro. CDK 4/6 inhibitors as single agent in advanced solid tumors. *Front. Oncol.* **2018**, *8*, 608.

11. M. Piezzo, S. Cocco, R. Caputo, D. Cianniell, G. D. Gioia, V. D. Lauro, G. Fusco, C. Martinelli, F. Nuzzo, M. Pensabene, M. De Laurentiis. Targeting cell cycle in breast cancer: CDK4/6 inhibitors. *Int. J. Mol. Sci.* **2020**, *21*, 6479.

12. R. Uzbekov, C. Prigent. A journey through time on the

discovery of cell cycle regulation. *Cells.* **2022**, *11*(4), 704.

13. P. L.Toogood, P. J. Harvey, J. T. Repine, D. J. Sheehan, S. N. VanderWel, H. Zhou, P. R. Keller, D. J. McNamara, D. Sherry, T. Zhu, J. Brodfuehrer, C. Choi, M. R. Barvian, D. W. Fry. Discovery of a potent and selective inhibitor of cyclin-dependent kinase 4/6. *J. Med. Chem.* **2005**, *48*, 2388–2406.

14. S. N. VanderWel, P. J. Harvey, D. J. McNamara, J. T. Repine, P. R. Keller, J. Quin, R. J. Booth, W. R. Elliott, E. M. Dobrusin, D. W. Fry, P. L. Toogood. Pyrido[2,3-*d*]pyrimidin-7-ones as specific inhibitors of cyclin-dependent kinase 4. *J. Med. Chem.* **2005**, 48, 2371–2387.

15. D. W. Fry, P. J. Harvey, P. R. Keller, W. L. Elliott, M. A. Meade, E. Trachet, M. Albassam, X. X. Zheng, W. R. Leopold, W. R. Pryer, P. L. Toogood. Specific inhibition of cyclin-dependent kinase 4/6 by PD 0332991 and associated antitumor activity in human tumor xenografts. *Mol. Cancer Ther.* **2004**, *3*, 1427–1438.

16. R. S. Finn, J. P. Crown, I. Lang, K. Boer, I. M. Bondarenko, S. O. Kulyk, J. Ettl, R. Patel, T. Pinter, M. Schmidt, Y. Shparyk, A. R. Thummala, N. L. Voytko, C. Fowst, X. Huang, S. T. Kim, S. Randolph, D. J. Slamon. The cyclin-dependent kinase 4/6 inhibitor palbociclib in combination with letrozole versus letrozole alone as first-line treatment of oestrogen receptor-positive, HER2-negative, advanced breast cancer (PALOMA-1/TRIO-18): a randomised phase 2 study. *Lancet Oncol.* **2015**, *16*, 25–35.

17. S. Finn, J. Dering, D. Conklin, O. Kalous, D. J. Cohen, A. J. Desai, C. Ginther, M. Atefi, I. Chen, C. Fowst, G. Los, D. J. Slamon. PD 0332991, a selective cyclin D kinase 4/6 inhibitor, preferentially inhibits proliferation of luminal estrogen receptor-positive human breast cancer cell lines in vitro. *Breast Cancer Res.* **2009**, *11*, R77.

18. S. Tadesse, M. Yu, L. B. Mekonnen, F. Lam, S. Islam, K. Tomusange, M. H. Rahaman, B. Noll, S. K. Basnet, T. Teo, H. Albrecht, R. Milne, S. Wang. Highly potent, selective, and orally bioavailable 4-thiazol-N-(pyridin-2-yl)pyrimidin-2-amine cyclin-dependent kinases 4 and 6iInhibitors as anticancer drug candidates: Design, synthesis, and evaluation. *J. Med. Chem.* **2017**, *60*, 1892–1915.

19. H. S. Lu, U. S. Gahmen. Toward understanding the structural basis of cyclin-dependent kinase 6 specific inhibition. *J. Med. Chem.* **2006**, *49*, 3826–3831.

20. Roskoski, R. Cyclin-dependent protein kinase inhibitors Including palbociclib as anticancer drugs. *Pharmacol. Res.* **2016**, *107*, 249−275.

21. Y. S. Cho, M. Borland, C. Brain, C. H. Chen, H. Cheng, R. Chopra, K. Chung, J. Groarke, G. He, Y. Hou, S. Kim, S. Kovats, Y. Lu, M. O'Reilly, J. Shen, T. Smith, G. Trakshel. 4-(Pyrazol-4-yl)-pyrimidines as selective inhibitors of cyclin-dependent kinase 4/6. *J. Med. Chem.* **2010**, *53*, 7938–7957.

22. P. Chen, N. V. Lee, W. Y. Hu, M. R. Xu, R. A. Ferre, H. Lam, S. Bergqvist, J. Solowiej, W. Diehl, Y. A. He, X. Yu, A. Nagata, T. VanArsdale, B. W. Murray. Spectrum and Degree of CDK drug interactions predicts clinical performance. *Mol. Cancer Ther.* **2016**, *15*, 2273–2281.

23. A. Ammazzalorso, M. Agamennone, B. De Filippis, M. Fantacuzzi. Development of CDK4/6 inhibitors: A five years update. *Molecules.* **2021**, *26*, 1488.

6.2 Ribociclib (Kisqali™) (33)

Structural Formula	Space-filling Model	Brief Information
$C_{23}H_{30}N_8O$; MW = 435 clogP = 1.8; tPSA = 87		Year of discovery: 2011 Year of introduction: 2017 Discovered by: Astex, Novartis Developed by: Novartis Primary targets: CDK4/6 Binding type: I1/2 Class: serine/threonine protein kinase Treatment: Breast cancer Other name: LEE011 Oral bioavailability = 66% Elimination half-life = 32 h Protein binding = 70%

● = C ○ = H ● = O ● = N ● = S ● = F ● = Cl ● = Br ● = I

Ribociclib was first approved in 2017 as the second CDK4/6 (cyclin-dependent kinase) inhibitor in combination with aromatase inhibitor as initial endocrine-based therapy for the treatment of postmenopausal women with hormone receptor positive (HR+), human epidermal growth factor receptor-2 negative (HER2−) advanced or metastatic breast cancer. In 2018, its label was expanded to pre/perimenopausal women with HR+/HER2− advanced or metastatic breast cancer.[1–7]

Ribociclib differs from palbociclib only in the A ring (pyrrole vs. pyridinone) and the C6 substitution (carboxamide vs. acetyl) (Fig. 1), and they show very similar binding poses in complex with CDK6.[8] These minor structural variations give rise to significant differences in pharmacological properties. First, ribociclib inhibits CDK4/D1 more potently than CDK6/D1 complexes (IC_{50} = 10 nM and 39 nM, respectively),[9] whereas palbociclib is equally potent on both isoforms (CDK4/D1, IC_{50} = 11 nM; CDK6/D1, IC_{50} = 16 nM).[10] Second, palbociclib inhibits pRb phosphorylation (Ser780, Ser807) more potently than ribociclib (IC_{50}'s: 9–21 nM vs. 31–89 nM) in cellular assays, and it also has superior antiproliferative activity in breast cancer cell lines.[11] Third, ribociclib carries a warning for risk of QT prolongation – abnormal heart rhythm, while palbociclib does not. Fourth, although both inhibitors display comparable PK properties, they differ in recommended dosage (600 mg, once daily for ribociclib, vs. 125 mg for palbociclib). Finally, ribociclib has shown one-year survival benefit in three clinical trials in HR+/HER2− metastatic breast cancer, whereas such evidence has not been presented for palbociclib. Thanks to these survival data from clinical trials, ribociclib sales jumped 36% from $687 million in 2020 to $936 million in 2021. In an effort to provide further evidence of its efficacy, Novartis has launched an ambitious head-to-head trial dubbed Harmonia for HR+/HER2− advanced breast cancer patients.

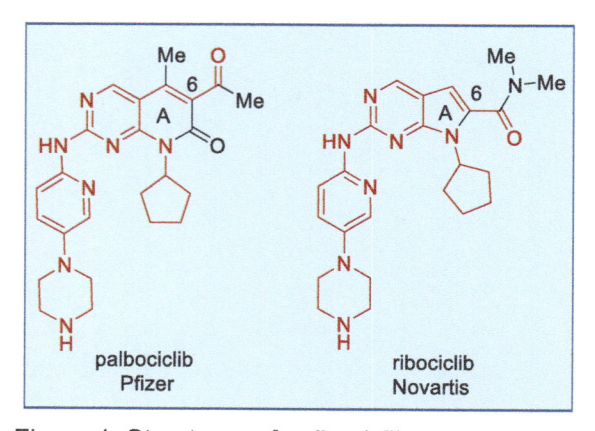

Figure 1. Structures of palbociclib and ribociclib.

The co-crystal structure of ribociclib in

complex with CDK6 shows the inhibitor is bound to the ATP-binding pocket of the inactive DFG-in, αC-helix-out configuration.[8,11] Ribociclib interacts with the protein primarily via three hydrogen bonds: (1) between the carbonyl group of Val101 (the third hinge residue) and the 2-amino group attached to the pyrrolopyrimidine core; (2) between the N3 nitrogen of the core and the amide NH group of Val101; and (3) between the DFG-Asp163 amide NH group and the carboxamide oxygen (Figs. 2, 3). The amino-pyrrolopyrimidine sits within the adenine pocket and gate area while the cyclopentyl group occupies the front pocket. Overall, the binding mode of ribociclib is very similar to that of palbociclib (Figs. 2, 3).

The pyridyl nitrogen of ribociclib likely interacts with His100 imidazole ring via a bound water molecule (Fig. 2), and this distinct network of hydrogen bonding interactions may account for its CDK4/6 isoform selectivity and other kinase selectivity (see Section 6.1 for details).[11]

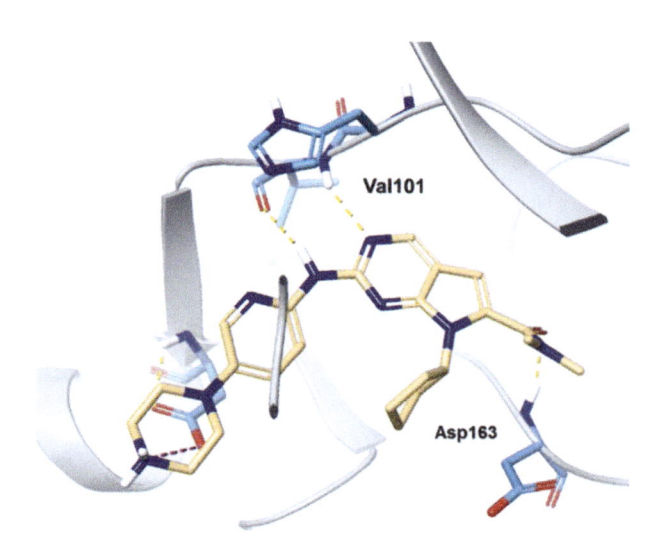

Figure 2. Co-crystal structure of ribociclib–CDK6 (PDB ID: 5L2T).

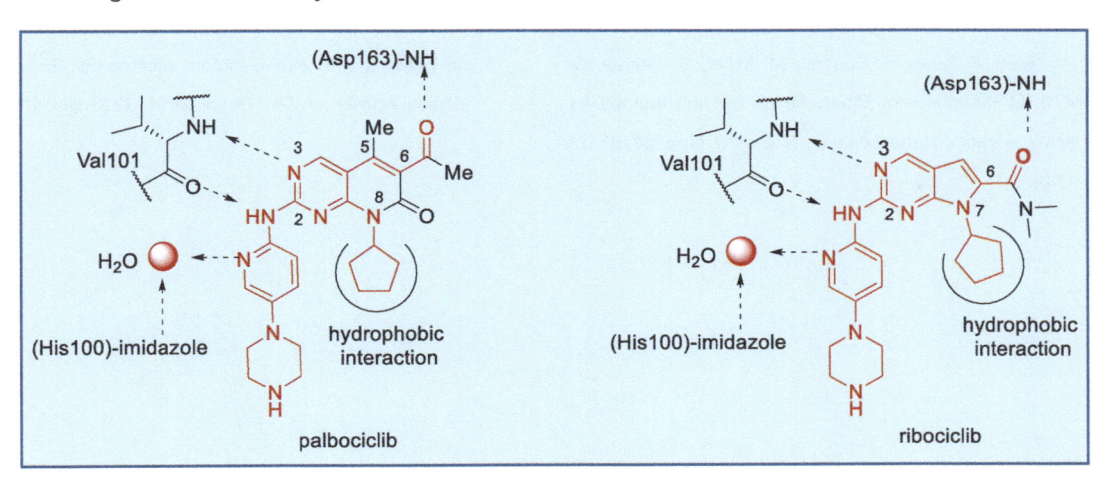

Figure 3. Summary of CDK6 interactions with palbociclib (left) and ribociclib (right). Interaction of pyridyl nitrogen with His100 via a bound water is highly likely but was not observed in the co-crystal structure possibly due to limit of resolution.[11]

When used in combination with other drugs such as an ALK or an MEK inhibitor, ribociclib has been shown to have a synergistic effect, resulting in improved responses.[12] This is likely a result of tumor strains that evolve to compensate for the blocked signaling pathways. Simply blocking one pathway in cancer tumorigenesis can sometimes result in "tumor compensation", where the tumor compensates for the blocked signaling pathway by utilizing other proliferative pathways to survive. By blocking several pathways at once, it is thought that the tumor is less able to evolve into an anti-tumor response. Utilizing ribociclib in combination with other agents has been shown to reduce the development of resistance to the single agent.

1. Y. Syed. Ribociclib: First global approval. *Drugs.* **2017**, *77*, 799–807.
2. R. Bartsch. Ribociclib: a valuable addition to treatment options in breast cancer? *Esmo Open* **2017**, *2*, e000246–e000249.

3. D. Tripathy, A. Bardia, W. R. Sellers. Ribociclib (LEE011): mechanism of action and clinical impact of this selective cyclin-dependent kinase 4/6 inhibitor in various solid tumors. *Clin. Cancer Res.* **2017**, *23*, 3251–3262.

4. M. Piezzo, S. Cocco, R. Caputo, D. Cianniell, G. D. Gioia, V. D. Lauro, G. Fusco, C. Martinelli, F. Nuzzo, M. Pensabene, M. De Laurentiis. Targeting cell cycle in breast cancer: CDK4/6 inhibitors. *Int. J. Mol. Sci.* **2020**, *21*, 6479

5. J. R. Infante, P. A. Cassier, J. F. Gerecitano, P. O. Witteveen, R. Chugh, V. Ribrag, A. Chakraborty, A.Matano, J. R. Dobson, A. S. Crystal, S. Parasuraman, G. I. Shapiro. A phase I study of the cyclin-dependent kinase4/6Inhibitor Ribociclib (LEE011) in patients with advanced solid tumors and lymphomas. *Clin. Cancer Res.* **2016**, *22*, 5696–5705.

6. G. N. Hortobagyi, S. M. Stemmer, H. A. Burris, Y. S. Yap, G. S. Sonke, S. P. Shimon, M. Campone, K. L. Blackwell, F. André, E. P. Winer, W. Janni, S. Verma, P. Conte, C. L. Arteaga, D. A. Cameron, K. Petrakova, L. L. Hart, C. Villanueva, A. Chan, E. Jakobsen, A. Nusch, O. Burdaeva, E. M. Grischke, E. Alba, E. Wist, N. Marschner, A. M. Favret, D. Yardley, T. Bachelot, L. M. Tseng, S. Blau, F. Xuan, F. Souami, M. Miller, C. Germa, S. Hirawat, J. O'Shaughnessy. Ribociclib as first-line therapy for HR-positive, advanced breast cancer. *N. Engl. J. Med.* **2016**, *375*, 1738–1748.

7. C. T. Brain, M. J. Sung, B. Lagu. Pyrrolopyrimidine compounds and their uses. US Patent 8324225B2, **2012**.

8. R. Roskoski. Properties of FDA-approved small molecule protein kinase inhibitors. *Pharmacol. Res.* **2019**, *144*, 19–50.

9. S. P. Corona, D. Generali. Abemaciclib: a CDK4/6 inhibitor for the treatment of HR+/HER2- advanced breast cancer. *Drug Des. Devel. Ther.* **2018**, *12*, 321–330.

10. P. L.Toogood, P. J. Harvey, J. T. Repine, D. J. Sheehan, S. N. VanderWel, H. Zhou, P. R. Keller, D. J. McNamara, D. Sherry, T. Zhu, J. Brodfuehrer, C. Choi, M. R. Barvian, D. W. Fry. Discovery of a potent and selective inhibitor of cyclin-dependent kinase 4/6. *J. Med. Chem.* **2005**, *48*, 2388–2406.

11. P. Chen, N. V. Lee, W. Y. Hu, M. R. Xu, R. A. Ferre, H. Lam, S. Bergqvist, J. Solowiej, W. Diehl, Y. A. He, X. Yu, A. Nagata, T. VanArsdale, B. W. Murray. Spectrum and degree of CDK drug interactions predicts clinical performance. *Mol. Cancer Ther.* **2016**, *15*, 2273–2281.

12. S. J. Alan, K. Muaiad, J. K. Martijn, P. M. Andrew, S. Gary, F. Catherine, M. Alessandro, B. Suraj, P. Sudha, K. Kevin. A phase 1b/2 study of LEE011 in combination with binimetinib (MEK162) in patients with NRAS-mutant melanoma: Early encouraging clinical activity". *J. Clin. Oncol.* **2014**, *32* (Suppl. 15), 9009–9009.

6.3 Abemaciclib (Verzenio™) (34)

Structural Formula	Space-filling Model	Brief Information
$C_{27}H_{32}N_8F_2$; MW = 506 clogP = 4.8; tPSA = 71		Year of discovery: 2008 Year of introduction: 2017 Discovered by: Eli Lilly Developed by: Eli Lilly Primary targets: CDK4/6 Binding type: I1/2 Class: serine/threonine protein kinase Treatment: Breast cancer Other name: LY2835219 Oral bioavailability = 46% Elimination half-life = 18 h Protein binding = 95–98%

● = C   ○ = H   ● = O   ● = N   ● = S   ● = F   ● = Cl   ● = Br   ● = I

Abemaciclib, the third CDK4/6 (cyclin-dependent kinase) inhibitor (following palbociclib and ribociclib), was first approved in 2017, in combination with fulvestrant or as a monotherapy for the treatment of women with HR+/HER2− advanced or metastatic breast cancer following endocrine therapy and prior chemotherapy, and then in 2018, as the first-line treatment of metastatic breast cancer, and further in 2021, in combination with endocrine therapy (tamoxifen or an aromatase inhibitor) for the adjuvant treatment of adult patients with HR+/HER2−, node-positive, early breast cancer (EBC) at high risk of recurrence and a Ki-67 percentage score of ≥20% as determined by an FDA-approved test. So far, abemaciclib is the only CDK4/6 Inhibitor approved for this group of high risk EBC patients.[1–5]

All three marketed CDK4/6 selective inhibitors (ribociclib, palbociclib, and abemaciclib, Fig. 2) share the common pyrimidine–NH–pyridine scaffold, which is critical for CDK4/6 isoform selectivity. Palbociclib and ribociclib are closer in structure than abemaciclib which is a more flexible molecule. Its distinct structure leads to several unique pharmacological and PK properties. First, abemaciclib shows the highest potency for the CDK4/cyclin D1 complex, with an IC_{50} value 5-fold lower than the other two inhibitors in biochemical assays (Table 1).[1] It is also the most potent CDK4/6 inhibitor in the pRb phosphorylation (Ser780, Ser807) cellular assays as well as in breast cancer cell proliferation assays.[6] Palbociclib and ribociclib are highly selective for CDK4/6 relative to other human protein kinases, whereas abemaciclib exhibits more complex pharmacology, including potent CDK9 inhibition (Table 1) and other kinase activities of the DYRK, PIM, HIPK, and CaMK kinase families (27 human protein kinases showed K_i < 10 nM).[6-8] Unlike palbociclib and ribociclib, abemaciclib also induces in vivo inhibition of CDK1, CDK2, CDK5, CDK9, CDK14, CDKs16-18, GSK3a/b, CAMKIIg/d, and PIM1 kinases.[6–8] The selectivity variations can be attributed to their subtle yet critical differences in binding interactions.[7,8] More specifically, the co-crystal structure of abemaciclib with CDK6 shows a hydrogen bond between the invariant catalytic residue Lys43 and benzimidazole N3 nitrogen (Fig. 1), which is not present in the other two inhibitors.

The decreased kinase selectivity of abemaciclib may be a consequence of its relatively greater conformational flexibility which arises from the possibility of free rotation about the C–C bond between the pyrimidine and benzimidazole subunits. In addition, whereas albociclib's acetyl and adjacent methyl, and ribociclib's dimethylamino group, all pack against the back wall of the ATP-binding pocket, the much larger substituents of palbociclib and ribociclib may not be able to fit in other kinases (Fig. 1), which could also contribute to the lower kinase selectivity observed with abemaciclib.

Palbociclib ($t_{1/2}$ = 29 h) and ribociclib (32 h) have a longer half-life than abemaciclib (18 h) and hence are administered once daily, whereas abemaciclib requires twice-daily administration.[8] The relative short half-life of abemaciclib likely arises from its metabolic instability and high clearance, which are consistent with its high lipophilicity (clogP = 4.8, vs. 1.8 for ribociclib and 2.2 for palbociclib).

Figure 1 (right) shows the ATP-binding site of CDK6 in complex with abemaciclib. It binds in the ATP-binding pocket of the inactive configuration, forming three hydrogen bonds: (1) between the carbonyl group of Val101 of the hinge and the 2-NH group attached to C2 of the pyrimidine core; (2) between the N3 nitrogen of the core and the amide NH group of Val101; and (3) between the amide NH of Lys43 and benzimidazole N3 nitrogen (see Fig. 2). The aminopyridine occupies the adenine pocket of the cleft and the benzimidazole penetrates through the front pocket.

The co-crystal structure also shows an ordered water molecule bridging the imidazole of hinge residue His100 and the pyridinyl nitrogen.[7] The electron density for this water was not observed in the co-crystal structures of palbociclib and ribociclib possibly due to limited resolution (Fig. 2, middle and left). However, space is still available for a water molecule at this position, and such a bridging interaction with His100 could explain the high kinase selectivity of these CDK4/6 inhibitors, because this histidine residue is found in only eight kinases based on sequence alignment (total of 442 kinases used for alignment). Moreover, within the CDK family, only CDK4/6 contain His100.

Table 1. IC_{50}'s of CDK/Cyclin complexes[1]

CDK complex	palbociclib	ribociclib	abemaciclib
CDK4/D1 IC_{50}	11 nM	10 nM	2 nM
CDK6/D1 IC_{50}	16 nM	39 nM	10 nM
CDK1/B IC_{50}	>10 µM	>10 µM	1.6 µM
CDK2/A/E IC_{50}	>70 µM	>10 µM	504 nM
CDK5/p25 IC_{50}	>40 µM	>10 µM	355 nM
CDK9/T1 K_i	NR	NR	57 nM

NR: not reported.

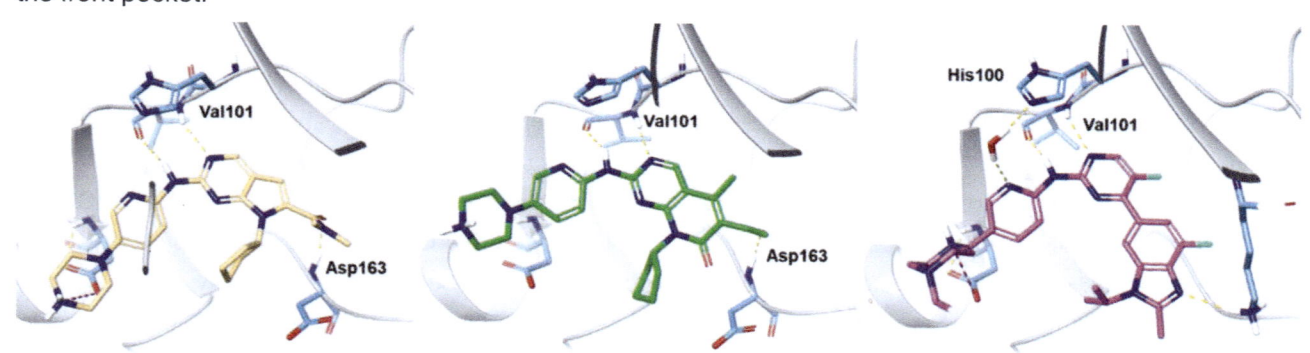

Figure 1. CDK6 complexes of ribociclib (left) (5L2T), palbociclib (middle) (5L2I) and abemaciclib (right) (5L2S).

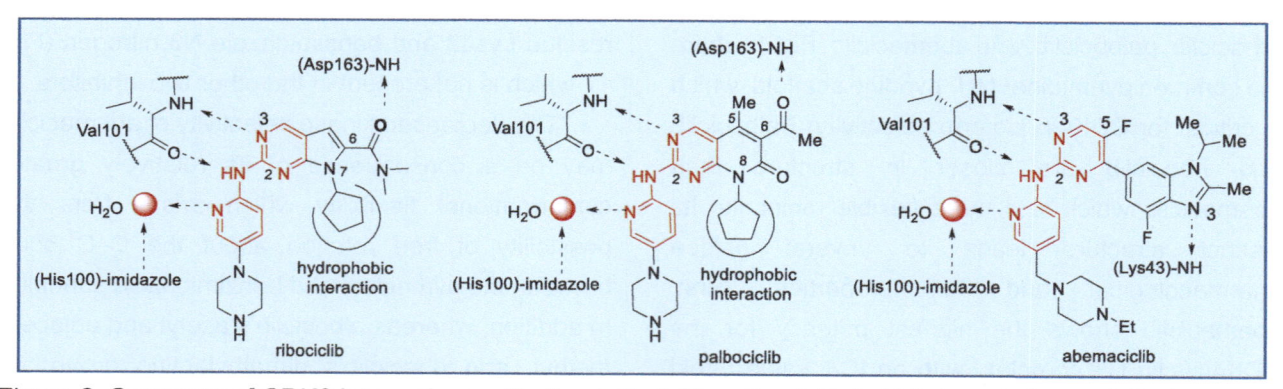

Figure 2. Summary of CDK6 interactions with ribociclib (left), palbociclib (middle), and abemaciclib (right).

1. S. P. Corona, D. Generali. Abemaciclib: A CDK4/6 inhibitor for the treatment of HR+/HER2- advanced breast cancer. *Drug Des. Devel. Ther.* **2018**, *12*, 21–330.
2. D. A. Coates, L. M. Gelbert, J. M. Knobeloch, A. D. D. Magana, A. D. P. Gonzalez, M. F. D. P. Catalina, M. C. G. Paredes, E. M. M. D. L. Nava, M. D. M. O. Finger, J. A. M. Perez, A. I. M. Herranz, C. P. Martinez, C. S. Martinez. Protein kinase inhibitors. US Patent 7855211, **2010**.
3. M. P. Goetz, M. Toi, M. Campone, J. Sohn, S. P. Shimon, J. Huober, I. H. Park, O. Trédan, S. C. Chen, L. Manso, O. C. Freedman, G. G. Jaliffe, T. Forrester, M. Frenzel, S. Barriga, I. C. Smith, N, Bourayou, A. D. Leo. MONARCH 3: Abemaciclib As initial therapy for advanced breast cancer. *J. Clin. Oncol.* **2017**, *35*, 3638–3646.
4. L. M. Gelbert, S. Cai, X. Lin, C. S. Martinez, M. D. Prado, M. J. Lallena, R. Torres, R. T. Ajamie, G. N. Wishart, R. S. Flack, B. L. Neubauer, J. Young, E. M. Chan, P. Iversen, D. Cronier, E. Kreklau, A. D. Dios. Preclinical characterization of the CDK4/6 inhibitor LY2835219: In-vivo cell cycle-dependent/independent anti-tumor activities alone/in combination with gemcitabine. *Invest. New. Drugs.* **2014**, *32*, 825–837.
5. P. Sidaway. Breast Cancer: Abemaciclib effective in combination with aromatase inhibition. *Nat. Rev. Clin. Oncol.* **2017**, 14, 714.
6. A. Patnaik, L. S. Rosen, S. M. Tolaney, A. W. Tolcher, J. W. Goldman, L. Gandhi, K. P. Papadopoulos, M. Beeram, D. W. Rasco, J. F. Hilton, A. Nasir, R. P. Beckmann, A. E. Schade, A. D. Fulford, T. S. Nguyen, R. Martinez, P. Kulanthaivel, L. Q. Li, M. Frenzel, D. M. Cronier, E. M. Chan, K. T. Flaherty, P. Y. Wen, G. I. Shapiro. Efficacy and safety of abemaciclib, an inhibitor of CDK4 and CDK6, for patients with breast cancer, non–small cell lung cancer, and other solid tumors. *Cancer. Discov.* **2016**, *6*, 740–753.
7. P. Chen, N. V. Lee, W. Y. Hu, M. R. Xu, R. A. Ferre, H. Lam, S. Bergqvist, J. Solowiej, W. Diehl, Y. A. He, X. Yu, A. Nagata, T. Van Arsdale, B. W. Murray. Spectrum and degree of CDK drug interactions predicts clinical performance. *Mol. Cancer Ther.* **2016**, *15*, 2273–2281.
8. M. A. George, S. Qureshi, C. Omene, D. L. Toppmeyer, S. Ganesan. Clinical and pharmacologic differences of CDK4/6 inhibitors in breast cancer. *Front. Oncol.* **2021**, *11*, 693104.

6.4 Trilaciclib (Cosela™) (35)

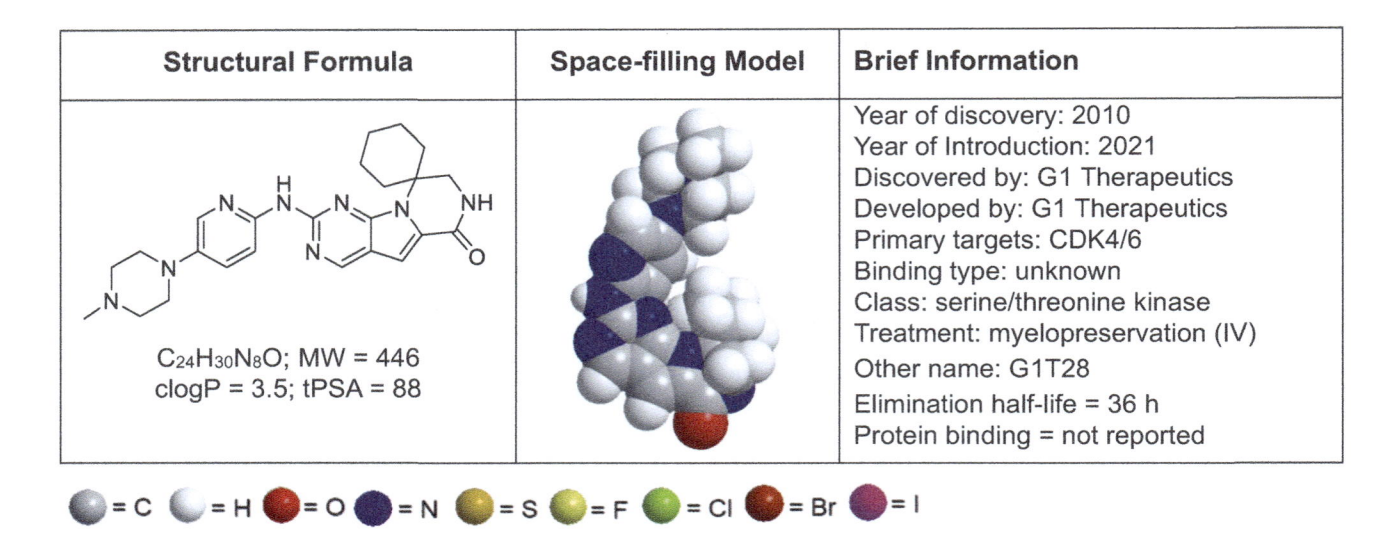

Structural Formula	Space-filling Model	Brief Information
$C_{24}H_{30}N_8O$; MW = 446 clogP = 3.5; tPSA = 88		Year of discovery: 2010 Year of Introduction: 2021 Discovered by: G1 Therapeutics Developed by: G1 Therapeutics Primary targets: CDK4/6 Binding type: unknown Class: serine/threonine kinase Treatment: myelopreservation (IV) Other name: G1T28 Elimination half-life = 36 h Protein binding = not reported

● = C   ○ = H   ● = O   ● = N   ● = S   ● = F   ● = Cl   ● = Br   ● = I

Trilaciclib (administered by injection) has been introduced as a treatment to reduce chemotherapy-induced bone marrow suppression in adults receiving cytotoxic drugs for metastatic small cell lung cancer. It is thought to protect bone marrow cells from damage caused by chemotherapy by inhibiting CDK4/6 (cyclin-dependent kinase 4/6). While several other CDK4/6 inhibitors are approved to treat breast cancer, trilaciclib is a first-in-class myeloprotective agent.[1]

Cytotoxic chemotherapy drugs kill fast-growing cancer cells, but they cannot discriminate between cancer cells and fast-growing normal healthy cells, such as bone marrow stem cells. These cells divide quickly to replenish relatively short-lived blood cells (lifespan = 100 days for red blood cells; one week for platelets; and a few hours to a few days for white blood cells). Damage of stem cells in the bone marrow by chemotherapy results in low blood counts, which can be treated with blood cell growth factors and blood transfusions. However, the best approach to treat low blood counts is to prevent them before they occur. There have been long-standing efforts to develop agents that enable chemotherapy drugs to target only the cancer cells and not the healthy bone marrow stem cells. It is possible that trilaciclib, a CDK4/6 inhibitor, may help with that.[2]

Bone marrow stem cells use CDK4/6 to move from the G1 phase (where the cell is preparing to divide) to the S phase where DNA synthesis starts, and inhibition of CDK4/6 causes stem cells to temporarily stay at the G1 phase. In other words, the stem cells stop moving on to the S phase (DNA synthesis) or the G2 phase (growth), and certainly not to the M phase (mitosis). As a result, stem cells are protected from being damaged or even destroyed as chemotherapy only affects cells that are in other phases: the S, the G2, and the M phase. In the meantime, the tumor cells continue to transition to the S and M phases to divide, and they are killed by chemotherapy, causing the tumor to shrink (Fig. 1). By preventing the stem cells from injury during chemotherapy, trilaciclib allows the patients to complete their course of treatment on time, and it avoids the expensive treatment of potential myelosuppressive adverse effects. So far, trilaciclib does not appear to reduce the efficacy of chemotherapy.[3,4]

In biochemical assays, trilaciclib reversibly inhibited CDK4/cyclin D1 and CDK6/cyclin D3 with an IC_{50} of 1 and 4 nM, respectively. It also displayed high selectivity against other CDK/cyclin complex family members and other protein kinases.[3]

There is no reported co-crystal structure of trilaciclib in complex with CDK4/6, and its binding mode remains unknown.

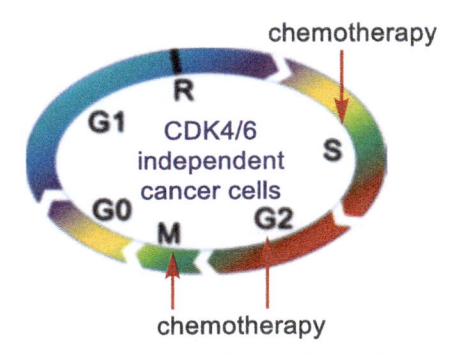

Figure 1. Mechanism of CDK4/6 inhibitors in protecting stem cells from chemotherapy-induced damage.

1. S. Dhillon. Trilaciclib: First approval. *Drugs*. **2021**, *81*, 867–874.
2. C. Wang. CDK4/6 Kinase inhibitors: limiting chemotherapy Toxicity.https://oncobites.blog/2020/06/10/cdk4-6-kinase-inhibitors-limiting-chemotherapy-toxicity..
3. J. E. Bisi, J. A. Sorrentino, P. J. Roberts, F. X. Tavares, J. C. Strum. Preclinical characterization of G1T28: A novel CDK4/6 inhibitor for reduction of chemotherapy-induced myelosuppression. *Mol. Cancer Ther.* **2016**, *15*, 783–793.
4. A. Ammazzalorso, M. Agamennone, B. De Filippis, M. Fantacuzzi. Development of CDK4/6 inhibitors: A five years update. *Molecules*. **2021**, *26*, 1488.

Chapter 7. JAK Inhibitors

The JAK/STAT signal transduction pathway, mediated by four Janus kinases (JAK) and seven signal transducer and activator of transcription (STAT) transcription factors, has been implicated in the pathology of autoimmune, allergic, and inflammatory diseases. Blockage of this pathway using JAK inhibitors with varying isoform selectivity has been explored by more than 15 pharmaceutical companies, resulting in a number of potent JAK inhibitors, 10 of which have been developed and approved as useful drugs for various indications (Figs. 1, 2)[1–7] Pfizer's tofacitinib was the first JAK inhibitor to enter clinical trials for renal transplantation, but Incyte's ruxolitinib was the first to win FDA approval in 2011 for the treatment of patients with myelofibrosis, a rare blood disease caused by a JAK2 V617F mutation that leads to myeloproliferative neoplasms (MPN). More than eight years later, two more selective JAK2 inhibitors (fedratinib and pacritinib) were approved for the same indication. Tofacitinib is the first-in-class pan-JAK inhibitor approved for rheumatoid arthritis (RA), psoriatic arthritis (PA), polyarticular course juvenile idiopathic arthritis (JIA) and ulcerative colitis (UC). In addition to tofacitinib, Pfizer also launched abrocitinib for atopic dermatitis (AD). Baricitinib is approved for RA and atopic dermatitis in Europe. In addition, it has recently received emergency approval in combination with the antiviral remdesivir for the treatment of COVID-19.

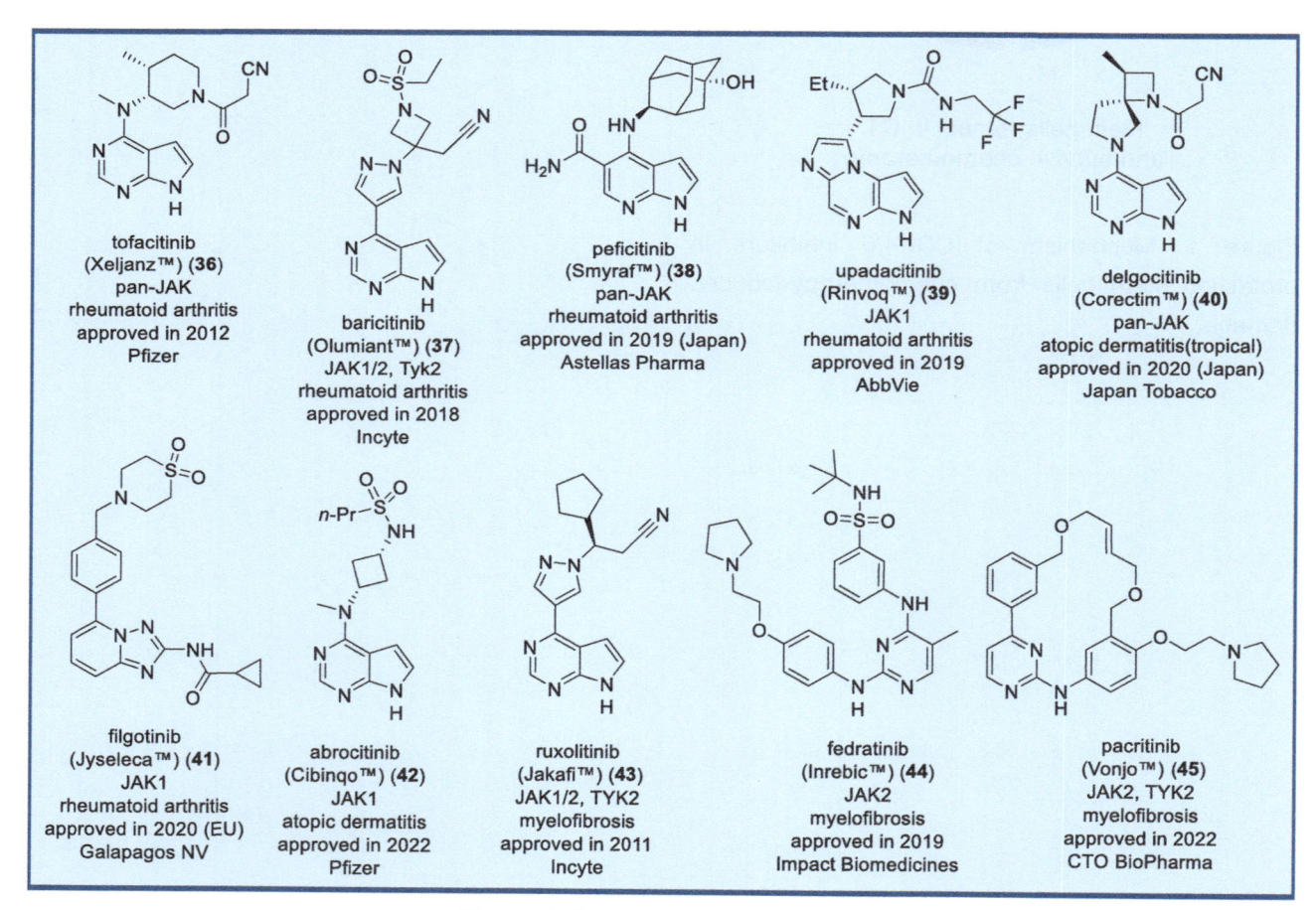

Figure 1. Approved JAK inhibitors.

While being proven to help ease joint pain and swelling, certain JAK inhibitors have also been associated with an excess risk of serious heart-related events, cancer, blood clots, and death. For this reason, tofacitinib, baricitinib, and upadacitinib carry a black box warning of such side effects. These potential liabilities may be linked to insufficient isoform specificity (Table 1), likely due to the high structural homology between family members.

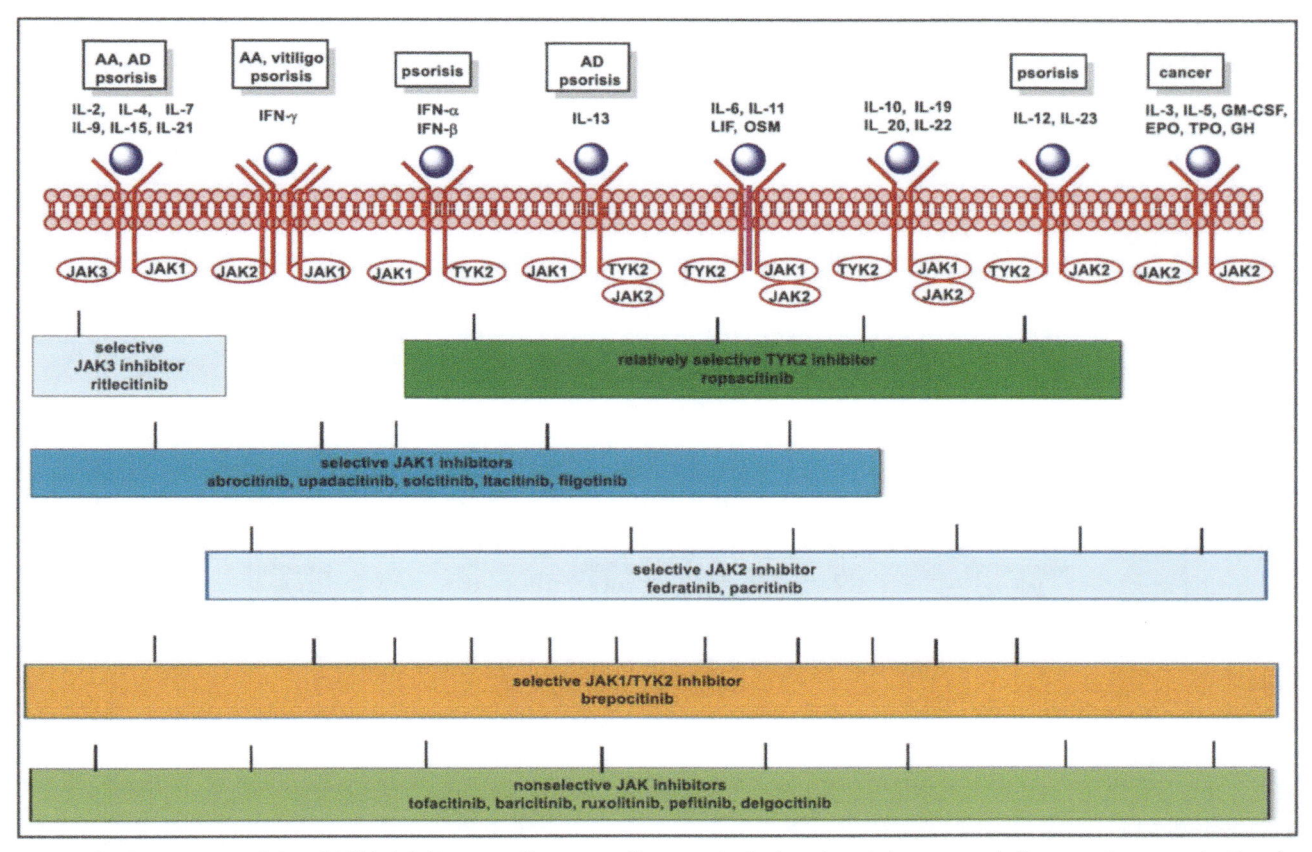

Figure 2. Summary of the JAK inhibitors as therapeutic agents to treat autoimmune, inflammatory, and allergic diseases as well as cancer. AA: alopecia areata; AD: atopic dermatitis.

Table 1. IC$_{50}$ values of JAK inhibitors[8]

inhibitor	JAK1	JAK2	JAK3	TYK2 (JH1)
tofacitinib	15 nM	77 nM	55 nM	489 nM
ruxolitinib (INCB018424)	6 nM	9 nM	487 nM	30 nM
baricitinib (LY3009104)	4 nM	7 nM	787 nM	61 nM
peficitinib	4 nM	5 nM	0.7 nM	5 nM
abrocitinib (PF-04965842)	29 nM	803 nM	>15 µM	1.3 µM
upadacitinib (ABT-494)	47 nM	120 nM	2.3 µM	4.7 µM
delgocitinib (JTE-0521)	2.8 nM	26 nM	13 nM	58 nM
filgotinib (GLPG0634)	363 nM	2.4 µM	>10 µM	2.6 µM
fedratinib (TG101348)	105 nM	3 nM	1.0 µM	405 nM
ritlecitinib (PF-06651600)	>10 µM	>10 µM	33 nM	>10 µM
brepocitinib (PF-06700841)	17 nM	77 nM	6.5 µM	23 nM
ropsacitinib (PF-06826647)	383 nM	74 nM	>10 µM	17 nM

There are several novel JAK inhibitors in advanced development, including ritlecitinib, brepocitinib and ropsacitinib (Fig. 3). Ritlecitinib acts as a dual inhibitor of JAK3 and the tyrosine-protein kinase TEC. Brepocitinib is a TYK2 (JH1)/JAK1 dual inhibitor, whereas ropsacitinib is a TYK2 JH2 inhibitor with modest selectivity over JAK2. Considering the broad and sophisticated roles of JAKs and the interdependency of many cytokine receptors, the unique profile of these newer candidates might enable a larger role in the treatment of a variety of inflammatory diseases. The clinical results with these compounds will clarify the relationship between target selectivity, clinical efficacy and safety, and guide the discovery of next-generation JAK inhibitors with optimized isoform specificity.

JAK inhibitors have been increasingly used to treat inflammatory and autoimmune diseases, and this chapter details the research that led the discoveries of JAK inhibitors **36–48** (Figs. 1, 3).

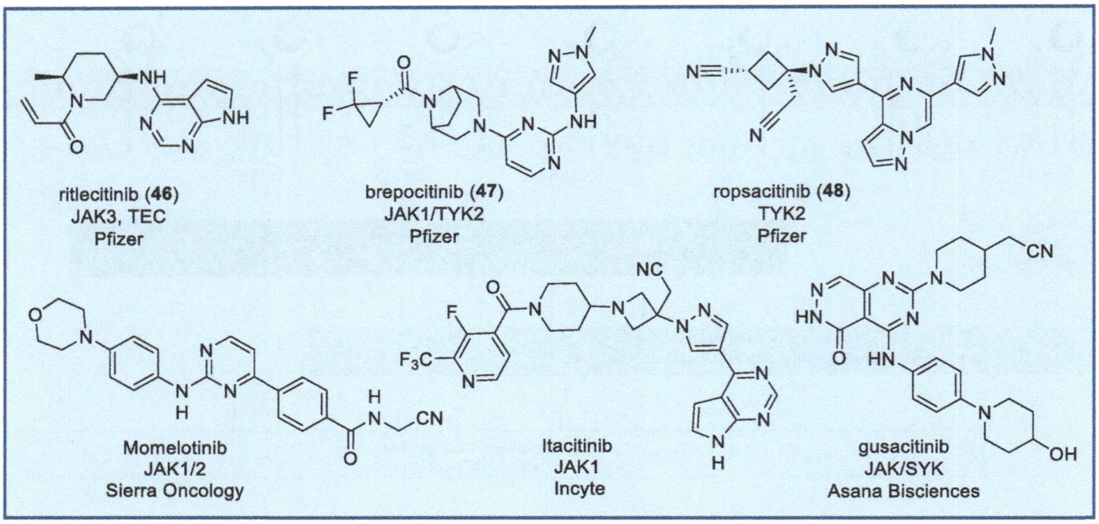

Figure 3. JAK inhibitors in phases II/III clinical trials.

1. Y. Tanaka, Y. Luo, J. O'Shea, S. Nakayamada. Janus kinase-targeting therapies in rheumatology: A mechanisms-based approach. *Nat. Rev. Rheumatol.* **2022**, *18*, 133–145.

2. M. Gadina, D. A. Chisolm, R. L. Philips, I. B. McInness, P. S. Changelian, J. J. O'Shea. Translating JAKs to Jakinibs. *J. Immunol.* **2020**, *204*, 2011–2020.

3. J. E. Pope, J. Gotlib, C. N. Harrison. Current and future status of JAK inhibitors. *Lancet.* **2021**, *398*, 803–816.

4. F. R. Spinelli, F. Meylan, J. J. O'Shea, M. Gadina. JAK inhibitors: Ten years after. *Eur. J. Immunol.* **2021**, *51*, 1615–1627.

5. J. J. O'Shea, M. Gadina. Selective Janus kinase inhibitors come of age. *Nat. Rev. Rheumatol.* **2019**, *15*, 74–75.

6. A. Cinats, E. Heck, L. Robertson. Janus Kinase inhibitors: A review of their emerging applications in dermatology. *Skin Ther. Lett.* **2018**, *23*, 5–9.

7. G. E Fragoulis, I. B McInnes, S. Siebert. JAK-inhibitors. New players in the field of immune-mediated diseases, beyond rheumatoid arthritis. *Rheumatol.* **2019**, *58*, i43–i54.

8. S. T. Wrobleski, R. Moslin, S. Lin, Y. Zhang, S. Spergel, J. Kempson, J. S. Tokarski, J. Strnad, A. Zupa-Fernandez, L. Cheng, D. Shuster, K. Gillooly, X. Yang, E. Heimrich, K. W. McIntyre, C. Chaudhry, J. Khan, M. Ruzanov, J. Tredup, D. Mulligan, D. S. Weinstein. Highly selective inhibition of tyrosine kinase 2 (TYK2) for the treatment of autoimmune diseases: Discovery of the allosteric inhibitor BMS-986165. *J. Med. Chem.* **2019**, *62*, 8973–8995.

7.1 Tofacitinib (Xeljanz™) (36)

Structural Formula	Space-filling Model	Brief Information
$C_{16}H_{20}N_6O$; MW = 312 clogP = 1.6; tPSA = 84		Year of discovery: 2000 Year of introduction: 2012 Discovered by: Pfizer Developed by: Pfizer Primary targets: pan-JAK Binding type: II Class: non-receptor tyrosine kinase Treatment: RA, PA, JIA, UC Other name: CP-690550 Oral bioavailability = 74% Elimination half-life = 3.2 h Protein binding = 40%

● = C ○ = H ● = O ● = N ● = S ● = F ● = Cl ● = Br ● = I

Tofacitinib is the first-in-class pan-JAK inhibitor approved for rheumatoid arthritis (RA), psoriatic arthritis (PA), polyarticular course juvenile idiopathic arthritis (JIA), and ulcerative colitis (UC). It works by blocking the JAK/STAT signal transduction pathway, which is implicated in the pathology of autoimmune, allergic, and inflammatory diseases. The development of molecular JAK antikinases is now a major area of research on the control of autoimmune diseases and also certain malignancies.

JAK/STAT Pathway

The Janus kinase (JAK) family consists of four intracellular, non-receptor tyrosine kinases: JAK1, JAK2, JAK3, and tyrosine kinase 2 (TYK2). TYK2 is the black sheep of the family not named as JAK4 because it was discovered first before the unique structure and function of other members of the JAK family were fully characterized. The name of JAK originated from acronym "Just Another Kinase," but ironically, JAKs were shown to be remarkably different from other classes of protein tyrosine kinases (PTKs) due to the presence of one true kinase domain JH1 and one pseudokinase domain JH2, which acts to negatively regulate the kinase activity of the JH1. To emphasize the two faces of JAKs, these kinases were renamed as "Janus kinases" after the two-headed mythical Roman god Janus.

JAK proteins generate signals from the binding of cytokines to specific membrane receptors which serve as transducers and activators of transcription (STAT). The canonical JAK/STAT pathway (Fig. 1), which is initiated by the binding of cytokines (e.g. interferons and interleukins) to the extracellular portions of their cognate receptors, causes JAKs to phosphorylate each other at tyrosine residues in "activation loops" (internal transphosphorylation). The activated JAKs phosphorylate tyrosine residues on the tails of the receptors, producing docking sites for SH2 domain-containing signaling proteins such as STATs. Once recruited to the receptors, STATs become phosphorylated by JAKs, and the phosphorylated STATs dissociate from receptor complex, dimerize, and then translocate to the nucleus where they bind to specific DNA sequences and regulate transcription of a variety of target genes. The JAK/STAT pathway is crucial for the downstream effects of cytokines as regulators of immunity, inflammation and hematopoiesis.

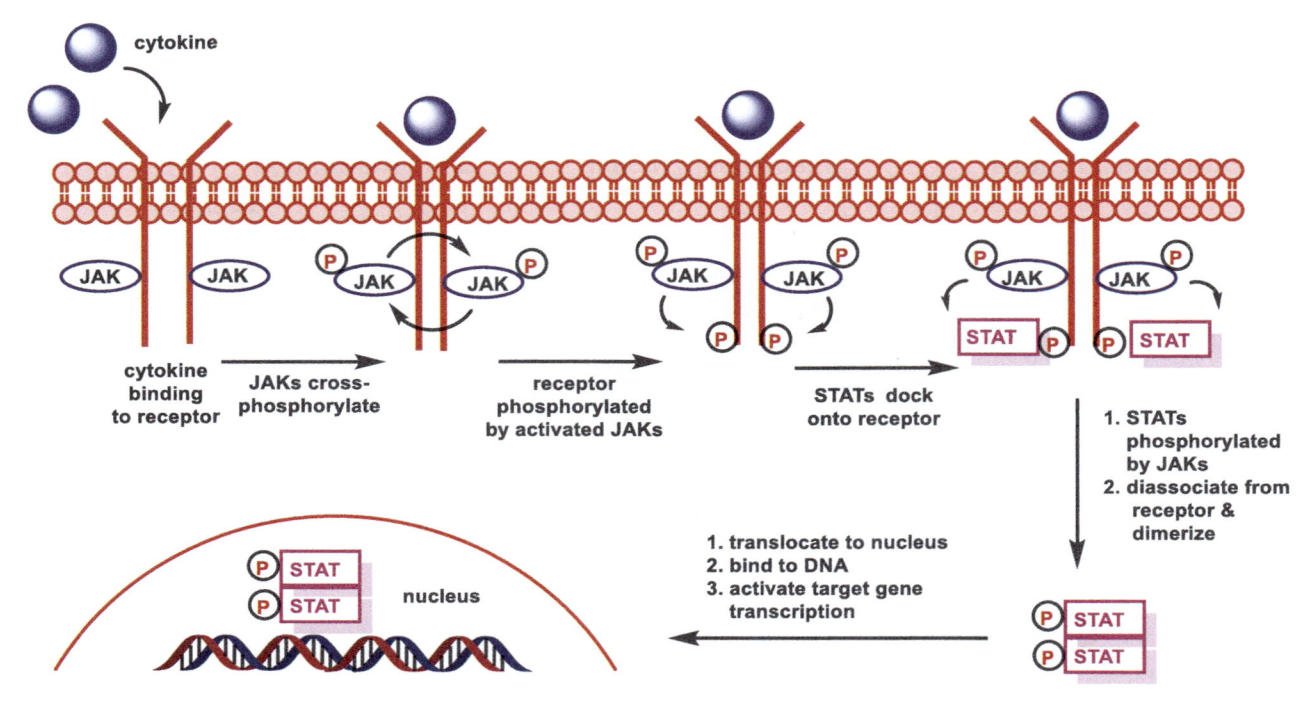

Figure 1. JAK/STAT pathway.

The biological significance of JAK/STAT signaling pathway depends on the pairing of a given cytokine receptor with two or three JAKs as determined by the subunit composition of the cytokine receptor. For example, JAK3, which associates only with the common γ-subunit, always pairs with JAK1, and this pairing regulates the signaling of six known γ-common cytokines IL-2, IL-4, IL7, IL-9, IL-15, and IL-21, all of which are predominantly linked to adaptive immune activities. However, JAK1 pairs with JAK2 and TYK2 in various combinations, which control the signaling of a variety of cytokine including several proinflammatory cytokines associated with the innate immune response, such as IL-6 and the type I interferons. Interestingly, JAK2 is the sole JAK to couple with itself, and this self-pairing modulates the signaling of several cytokines and growth factors, e.g., IL-3, IL-5, granulocyte macrophage colony-stimulating factor (GM-CSF), erythropoietin (EPO), and thrombopoietin (TPO).

JAK inhibitors transiently occupy the ATP-binding pocket of JAK enzymes, thus obstructing their kinase function. Inhibition of either one or both JAK monomers associated with specific cytokines has been shown to be sufficient to block cytokine signaling implicated in the pathogenesis of certain diseases. However, modulation of other cytokines may also contribute to unwanted side effects. Therefore, for a given therapeutic indication, it may be critical to strike an appropriate balance in JAK isoform selectivity, which has proved to be both rewarding and challenging for drug development. The intensive search for JAK inhibitors with adequate selectivity has culminated in the discovery of 12 marketed drugs to treat autoimmune disorders and various cancers.

Discovery of Tofacitinib

The seven-year journey from target identification to tofacitinib from 1993 to 2000 started with the joint venture between NIH led by Dr. John O'Shea and Pfizer.[1] High-throughput screening of Pfizer compound library (~400,000 compounds) against JAK3 catalytic domain identified pyrrolo[2,3-d]pyrimidine 1 (Fig. 2) with modest inhibitory activity (JAK3 IC_{50} = 210 nM).[2] In order to improve potency and reduce lipophilicity of the lead compound, a high-speed analog synthesis of 4-chloropyrrolopyrimidine 3 with N-methyl-cycloalkylamine 4 was carried out, and library synthesis led to the 2′,5′-dimethyl substituted cyclohexane derivative **2** (mixture of diastereomers) with 10-fold improvement in potency over the original lead **1**. The stereochemical complexity associated with the two methyl groups of the cyclohexane ring was simplified when the commercially available, enantiomerically pure carvone natural products were incorporated onto the pyrrolo[2,3-*d*]pyrimidine core, an exercise that identified the milestone analog **5** as a highly potent JAK3 inhibitor (IC_{50} = 2 nM). However, this compound showed only modest cellular activity (IC_{50} = 50 nM) in the inhibition of IL-2 dependent T cell blast proliferation presumably due to the high protein binding, which can be attributed to its lipophilicity (clogP = 4.8). Thus, a nitrogen atom was added the cyclohexane ring to form the piperidine head group, a bioisosteric replacement which served to reduce clogP, mitigate the diastereochemical issue, and allow straightforward amide coupling reactions to afford a variety of amide analogs. One of them was CP-690550, later known as tofacitinib, the first JAK inhibitor to be tested in the clinic. Tofacitinib exhibited single-digit nanomolar "PAN" inhibitory IC_{50} profile against JAK1, JAK2, and JAK3 and much reduced activity towards TYK2 (IC_{50} = 34 nM) based on the solution-phase caliper assays. Tofacitinib was approved for the treatment of rheumatoid arthritis in 2012, joint pain and swelling, psoriatic arthritis in 2017, and ulcerative colitis in 2018. However, severe side effects observed during clinical studies prevented the FDA from approving more efficacious higher doses for the treatment of rheumatoid arthritis (RA). Tofacitinib also carries a black box warning for an increased risk of pulmonary embolism and death associated with the 10 mg BID dose. The adverse effects may be linked to the lack of specificity or JAK selectivity.

Binding Mode

The binding mode of tofacitinib with JAKs is exemplified by JAK3 (Figs. 3, 4).[3] The pyrrolopyrimidine ring core is positioned against the hinge region where it forms two hydrogen bonds with the amide carbonyl oxygen of Glu 903 (2.8 Å) and amide NH of Leu905 (3.1 Å). The piperidine ring is oriented in the hydrophobic pocket of the JAK3 binding cavity, engaging hydrophobic interactions with Met902, Val836, Cys909, Leu956, and Ala966. The *N*-methyl group of the piperidine ring is located in an underlying pocket at the lower rim of the binding cavity of JAK3. This pocket is postulated to contribute to the overall kinome selectivity profile of tofacitinib. The terminal cyano group, which is situated against the glycine-rich loop, forms polar interactions with main chain atoms spanning this flexible loop, particularly Gly831, Gly834, Ser835 and Val836. For this reason, replacement of the cyano with cyclopropyl and trifluoromethyl reduces JAK3 inhibitory activity by 6-fold.

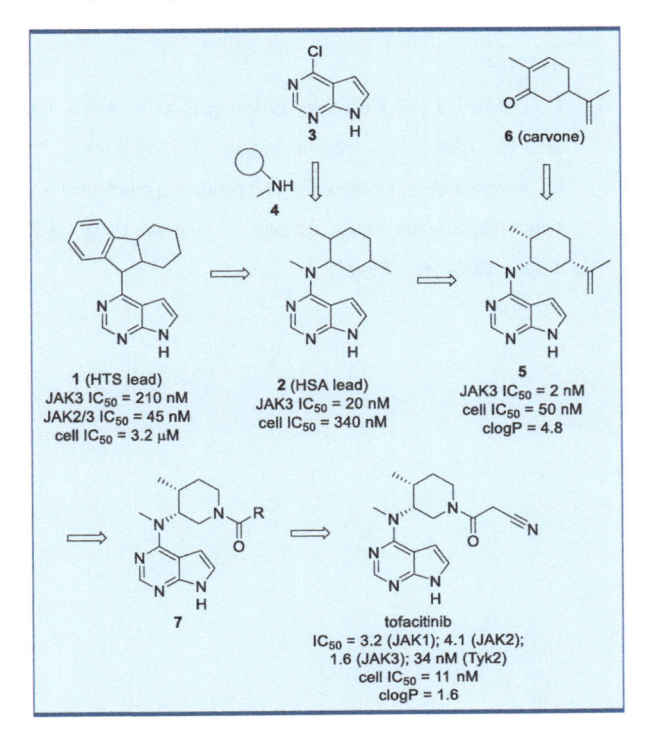

Figure 2. Chemical optimization of tofacitinib.

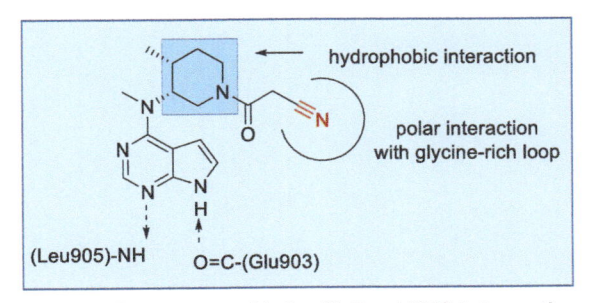

Figure 3. Summary of tofacitinib–JAK3 interactions.

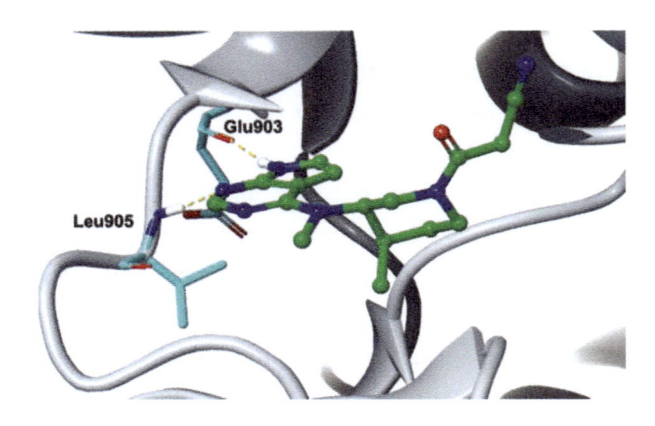

Figure 4. Co-crystal structure of tofacitinib with JAK3 (PDB ID: 3LXN).

Pharmacokinetics

Tofacitinib has good solubility (>4 mg/mL in water for the crystalline citrate salt),[2] which contributes to its excellent oral bioavailability (74%) in healthy volunteers.[4,5] However, tofacitinib exhibits relatively short elimination half-life of 3.2 hours, and its protein binding is very low (approximately 40%, predominantly to albumin). Tofacitinib undergoes major hepatic clearance (70% of total clearance) and minor renal clearance (30%). The metabolism of tofacitinib proceeds primarily in the liver by CYP3A4 (major) and CYP2C19 (minor), resulting in oxidation of the pyrrolopyrimidine and piperidine rings, oxidation of the piperidine ring side-chain, N-demethylation, and glucuronidation.

As opposed to the longer half-life of biologics, small-molecule JAK inhibitors such as tofacitinib typically display much shorter half-lives with once or twice-daily dosing regimens, which may be inconvenient to some patients. However, the shorter half-life ensures faster recovery to the full ability to fight infections which may occur with any immune suppressing medication.

1. J. LaMattina. A brief history of tofacitinib. Forbes. May 10, 2012.
2. M. E. Flanagan, T. A. Blumenkopf, W. H. Brissette, M. F. Brown, J. M. Casavant, C. Shang-Poa, J. L. Doty, E. A. Elliott, M. B. Fisher, M. Hines, C. Kent, E. M. Kudlacz, B. M. Lillie, K. S. Magnuson, S. P. McCurdy, M. J. Munchhof, B. D. Perry, P. S. Sawyer, T. J. Strelevitz, C. Subramanyam, J. Sun, P. S. Changelian. Discovery of CP-690,550: A potent and selective Janus kinase (JAK) inhibitor for the treatment of autoimmune diseases and organ transplant rejection. *J. Med. Chem.* **2010**, *53*, 8468–8484.
3. J. E. Chrencik, A. Patny, I. K. Leung, B. Korniski, T. L. Emmons, T. Hall, R. A. Weinberg, J. A. Gormley, J. M. Williams, J. E. Day, J. L. Hirsch, J. R. Kiefer, J. W. Leone, H. D. Fischer, C. D. Sommers, H. C. Huang, E. J. Jacobsen, R. E. Tenbrink, A. G. Tomasselli, T. E. Benson. Structural and thermodynamic characterization of the TYK2 and JAK3 kinase domains in complex with CP-690550 and CMP-6. *J. Mol. Biol.* **2010**, *400*, 413–433.
4. D. J. Cada, Kendra Demaris, T. L. Levien, D. E. Baker. Formulary drug reviews tofacitinib. *Hosp. Pharm.* **2013**, *48*, 413–424.
5. M. E. Dowty, J. Lin, T. F. Ryder, W. Wang, G. S. Walker, A. Vaz, G. L. Chan, S. Krishnaswami, C. Prakash. The pharmacokinetics, metabolism, and clearance mechanisms of tofacitinib, a janus kinase inhibitor, in humans. *Drug Metab. Dispos.* **2014**, *42*, 759–773.

7.2 Baricitinib (Olumiant™) (37)

Structural Formula	Space-filling Model	Brief Information
$C_{16}H_{17}N_7O_2S$; MW = 371 clogP = 0.41; tPSA = 114		Year of discovery: 2010 Year of introduction: 2018 Discovered by: Incyte Developed by: Incyte/Lilly Primary targets: JAK1/2 Binding type: II Class: non-receptor tyrosine kinase Treatment: rheumatoid arthritis Other name: INCB28050, LY3009104 Oral bioavailability = 79% Elimination half-life = 12.5 h Protein binding = 50%

● = C   ○ = H   ● = O   ● = N   ● = S   ● = F   ● = Cl   ● = Br   ● = I

Baricitinib is a close analog of ruxolitinib (44) (Fig. 1), and both are potent selective JAK1 and JAK2 inhibitors.[1] However, in contrast to ruxolitinib which gained FDA approval to treat myelofibrosis in 2011, baricitinib was approved by the FDA in 2018 for the treatment of moderate-to-severe active rheumatoid arthritis (as the second JAK inhibitor after tofacitinib).

The pyrrolopyrimidine hinge binder of baricitinib in JAK2 occupies an almost identical position to that of ruxolitinib in the ATP site (Figs. 2, 3), and one of the sulfone oxygen atoms forms a hydrogen bond with amide NH of the Lys857.[2]

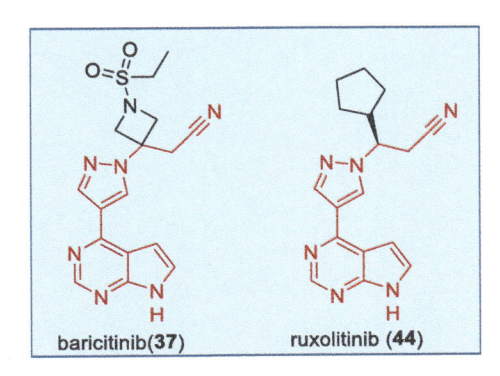

Figure 1. Baricitinib vs. ruxolitinib.

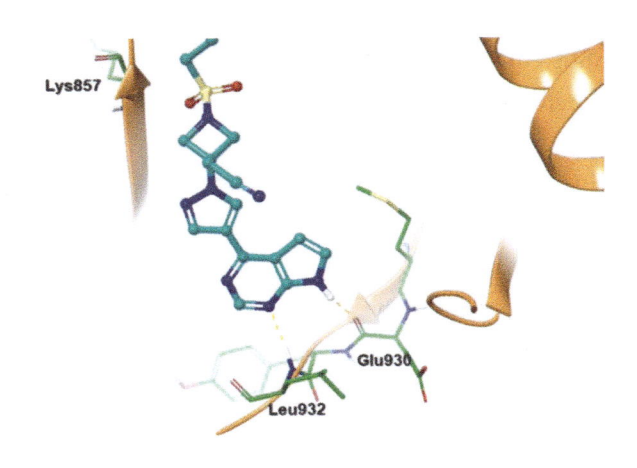

Figure 2. Co-crystal structure of baricitinib with JAK2 (PDB ID: 6VN8).

Table 1. IC$_{50}$'s of baricitinib/ruxolitinib[1]

kinase	baricitinib	ruxolilitinib
JAK1	5.9 nM	2.3 nM
JAK2	5.7 nM	2.8 nM
JAK3	>300 nM	428 nM
TYK2	53 nM	19 nM
CHK2	>1 µM	>1 µM
MET	>10 µM	>10 µM

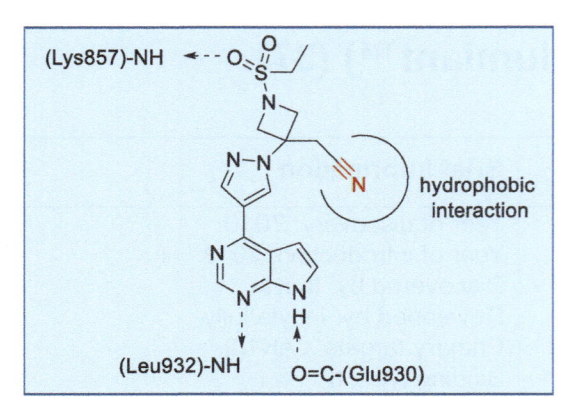

Figure 3. Summary of baricitinib–JAK3 interactions based on X-ray co-crystal structure.

Despite structural similarity, baricitinib differs from ruxolitinib in metabolic stability.[3,4] While ruxolitinib is metabolized primarily by CYP3A4 in the liver, barcitinib resists oxidation and is cleared by the kidney. Therefore, it is not recommended for use in patients with severe renal impairment.

Despite its poor aqueous solubility, baricitinib still exhibits good oral bioavailability (80%) presumably due to its good metabolic stability and prolonged dissolution rate. The elimination half-life is approximately 12 h in patients with rheumatoid arthritis. The good oral bioavailability and long half-life support once-daily regimen (2 mg tablet).

1. J. S. Fridman, P. A. Scherle, R. Collins, T. C. Burn, Y. Li, J. Li, M. B. Covington, B. Thomas, P. Collier, M. F. Favata, X. Wen, J. Shi, R. McGee, P. J. Haley, S. Shepard, J. D. Rodgers, S. Yeleswaram, G. Hollis, R. C. Newton, B. Metcalf, K. Vaddi. Selective inhibition of JAK1 and JAK2 is efficacious in rodent models of arthritis: Preclinical characterization of INCB028050. *J. Immunol.* **2010**, *184*, 5298–5307.

2. R. R. Davis, B. Li, S. Y. Yun, A. Chan, P. Nareddy, S. Gunawan, M. Ayaz, H. R. Lawrence, G. W. Reuther, N. J. Lawrence, E. Schönbrunn. Structural insights into JAK2 inhibition by ruxolitinib, fedratinib, and derivatives thereof. *J. Med. Chem.* **2021**, *64*, 2228–2241..

3. J. G. Shi, X. Chen, F. Lee, T. Emm, P. A. Scherle, Y. Lo, N. Punwani, W. V. Williams, S. Yeleswaram. The pharmacokinetics, pharmacodynamics, and safety of baricitinib, an oral JAK 1/2 inhibitor, in healthy volunteers. *J. Clin. Pharmacol.* **2014**, *54*, 1354–1361.

4. https://www.drugs.com/monograph/baricitinib.html.

7.3 Peficitinib (Smyraf™) (38)

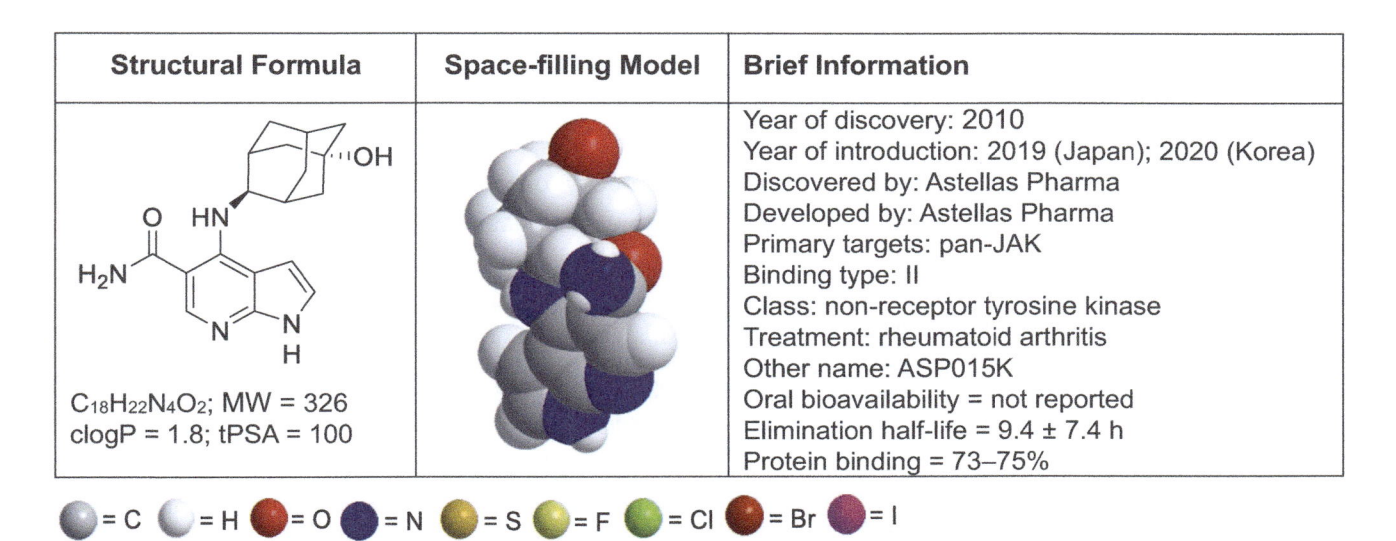

Structural Formula	Space-filling Model	Brief Information
$C_{18}H_{22}N_4O_2$; MW = 326; clogP = 1.8; tPSA = 100		Year of discovery: 2010 Year of introduction: 2019 (Japan); 2020 (Korea) Discovered by: Astellas Pharma Developed by: Astellas Pharma Primary targets: pan-JAK Binding type: II Class: non-receptor tyrosine kinase Treatment: rheumatoid arthritis Other name: ASP015K Oral bioavailability = not reported Elimination half-life = 9.4 ± 7.4 h Protein binding = 73–75%

● = C   ○ = H   ● = O   ● = N   ● = S   ● = F   ● = Cl   ● = Br   ● = I

Peficitinib is a pan-JAK inhibitor approved only in Japan and Korea for the treatment of rheumatoid arthritis.

Discovery of Peficitinib[1-3]

As described in the Section 7.1, the lead compound 2 to tofacitinib was identified through high-speed analog synthesis of 4-chloro-7H-pyrrolo[2,3-d]pyrimidine 1 with various amines.[4] Since the N3 of the pyrrolo[2,3-d]pyrimidine is not involved in the hinge binding, this nitrogen was eliminated to give 4-chloro-1H-pyrrolo[2,3-b]pyridine 3 (Fig. 1), which was surveyed with a variety of amines, leading to 4 with modest JAK3 potency. Molecular docking of 4 with JAK3 showed ample space in the C5 region; thus, a carbamoyl group was introduced to give 5. However, this compound showed no improvement in activity. To reduce steric repulsion between C4 amino moiety and the C5 carbamoyl group, the N-methyl group was removed to afford 6, a milestone compound with an over 100-fold increase in JAK3 inhibition (IC$_{50}$ = 14 nM) over 5. The improved potency may be attributed to the internal hydrogen bonding of the C4 NH and C5 carbamoyl oxygen to form a six-membered ring as shown in the co-crystal structures with JAKs, thus resulting a favorable conformation for enhanced hinge binding. Addition of a cis methyl to the cyclohexane ring adjacent to the amino group further enhanced potency by 3-fold (7, IC$_{50}$ = 5.1 nM, cell IC$_{50}$ = 86 nM). Incubation of 7 in rat liver microsomes (RLM) showed unacceptably high intrinsic clearance (CL$_{int}$) value of more than 1000 mL/min/kg primarily due to CYP mediated hydroxylation of the methyl group to form CH$_2$OH. Thus, the cyclohexane ring was replaced with larger bridged ring systems to fill the hydrophobic cavity of JAK3, and the 2-adamantane derivative was then identified to have superior potency (JAK3 IC$_{50}$ = 2.1 nM). Nevertheless, this compound still suffered from high clearance in RLM (CL$_{int}$ > 1000 mL/min/kg) as metabolic instability remained, which can be attributed to its high lipophilicity (clogP = 5.1). Thus, a polar hydroxyl group was incorporated in the adamantane core leading to peficitinib with clogP of 3.3, which contributed to a much lower CL$_{int}$ (124 mL/min/kg) and also increased potency due to interaction with the main chain atom either directly or via a water molecule (vide infra). As with tofacitinib, pelitinib is a potent pan-JAK inhibitor with IC$_{50}$ values of 3.9 nM (JAK1), 5.0 nM (JAK2), 0.71 nM (JAK3), and 4.8 nM (TYK2), exhibiting modest JAK3 selectivity. Peficitinib has been approved in Japan in and Korea for the treatment of rheumatoid arthritis.

The selection of hydroxyadamantane to reduce lipophilicity and improve metabolic stability was previously utilized in the discovery of two DPP-4 inhibitors, saxagliptin and vildagliptin, for the treatment of type II diabetes (Fig. 2).[5]

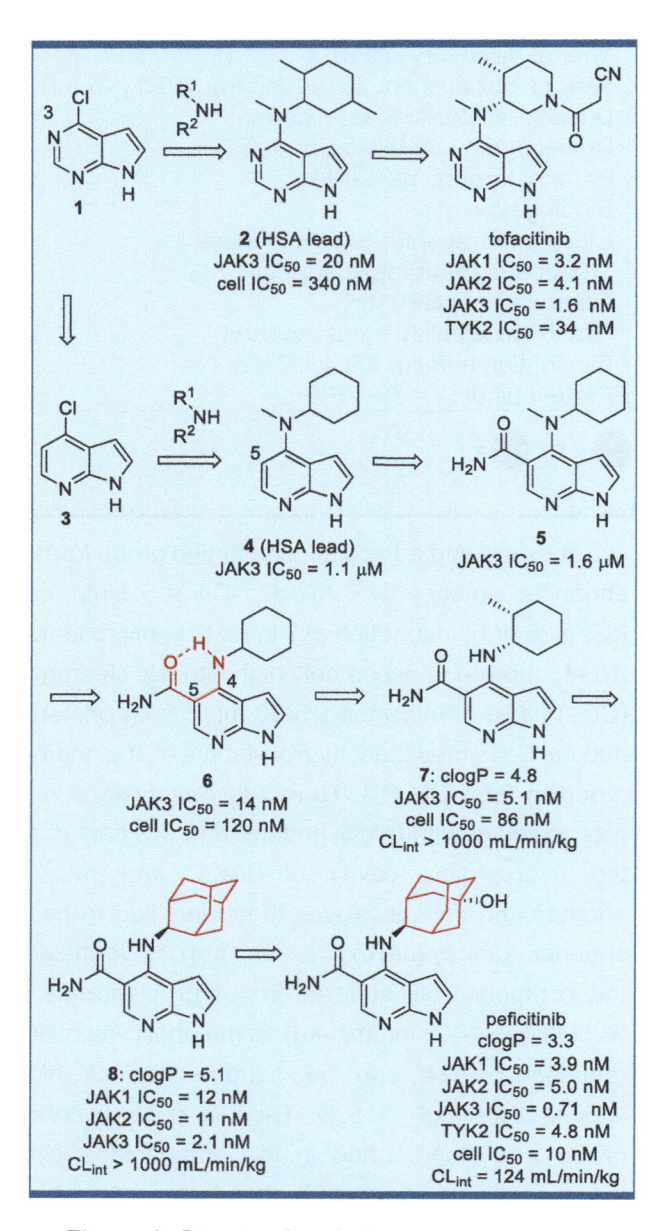

Figure 1. Structural optimization of peficitinib.

Figure 2. Hydroxyl-adamantane based DPP-4 inhibitors.

Binding Mode[1]

In the co-crystal structure with JAK1 (Fig. 4), the pyrrolopyridine ring forms two direct hydrogen bonds with amide carbonyl oxygen of Glu957 (3.1 Å) and amide NH of Leu959 (3.2 Å) (Fig. 3). The carbamoyl NH also forms a hydrogen bond with the amide carbonyl oxygen of Leu959 (3.3 Å). The carbamoyl group interacts indirectly with Pro960, Glu966 and Ser963 through a network of hydrogen bonds via two water molecules. The carbamoyl oxygen forms an intramolecular hydrogen bond with the C4 amino group of the pyrrolopyridine ring. The lipophilic adamantyl group situated in the large cavity engages in hydrophobic interactions with residues of Val889. The hydroxyl group inside the pocket forms a hydrogen bond with side chain of Asn1008 (Fig. 4).

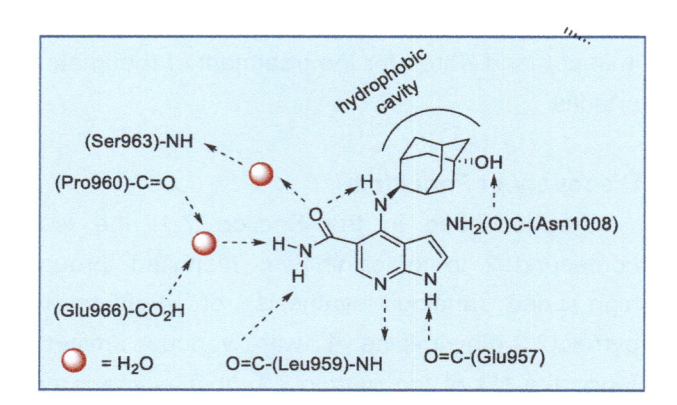

Figure 3. Peficitinib–JAK1 interactions based on X-ray co-crystal structure.

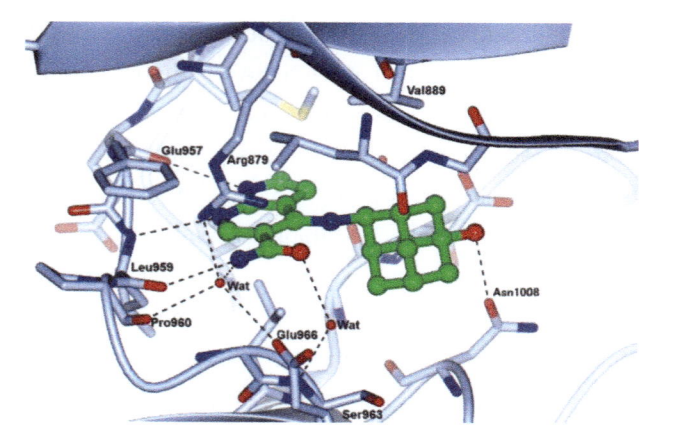

Figure 4. Crystal structure of JAK1 in a complex with peficitinib. Adapted with permission from *Bioorg. Med. Chem.* **2018**, *26*, 4971–4983.

The binding mode of peficitinib to the hinge region of JAK2 (Fig. 6) differs from that of JAK1 (Fig. 4): peficitinib is flipped in JAK2 as compared to JAK1 complex. More specifically, the carbamoyl is oriented inside the hinge pocket, while the pyrrolopyridine ring extends outside the pocket. As a result, the carbamoyl NH and the N7 atom of the pyrrolopyridine core form hydrogen bonds with the main chain atoms of Glu930 (2.8 Å) and Leu932 (3.2 Å), respectively (Fig. 5). The N1 atom of the pyrrolopyridine ring forms a weak hydrogen bond with the amide carbonyl oxygen of Leu932 (3.5 Å). The bound water molecules were not observed in the co-crystal structure with JAK2, possibly due to insufficient resolution. However, the co-crystal structures with both JAK2 and JAK3 showed conserved coordinates of residues around these water molecules. It is conceivable that both the carbamoyl and adamantyl hydroxyl groups would interact with JAK2 via water molecules in an analogous manner to JAK3.

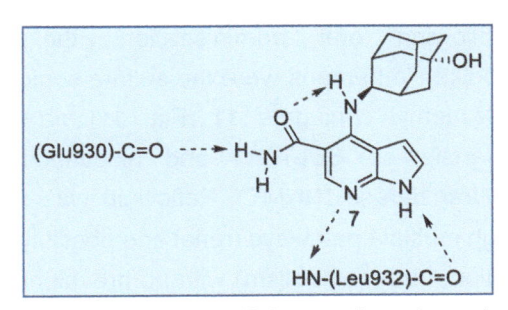

Figure 5. Peficitinib–JAK2 interactions based on X-ray co-crystal structure.

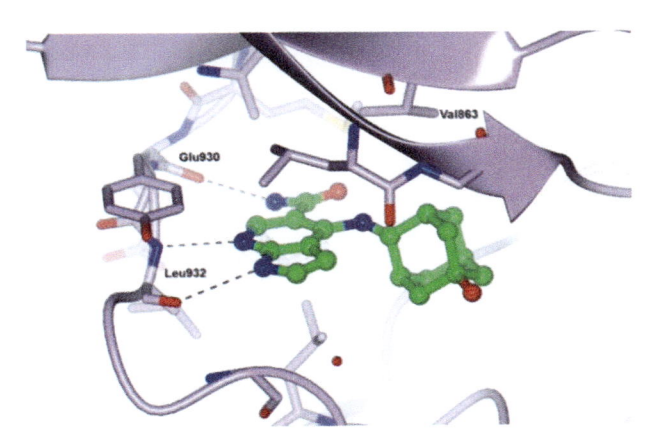

Figure 6. Crystal structure of JAK2 in complex with peficitinib. Adapted with permission from *Bioorg. Med. Chem.* **2018**, *26*, 4971–4983.

When bound to JAK3 (Fig. 8), the pyrrolopyridine ring forms two hydrogen bonds with the main chain atoms of Leu905 (3.4 and 3.1 Å), and the carbamoyl group forms a hydrogen bond with the main chain atom of Glu903 (3.4 Å) in the hinge region. The carbamoyl oxygen interacts indirectly with the main chain atom of Asp967 via a water molecule, and it also forms an intramolecular hydrogen bond with the C4 amino group of the pyrrolopyridine core (Fig. 7), as observed in the complex with JAK1 structure. The hydroxyl group of the adamantyl moiety is involved in a network of hydrogen bond interactions with the main chain atom of Arg953 via a water molecule, which may account for the superior JAK3 inhibitory activity relative to other isoforms.

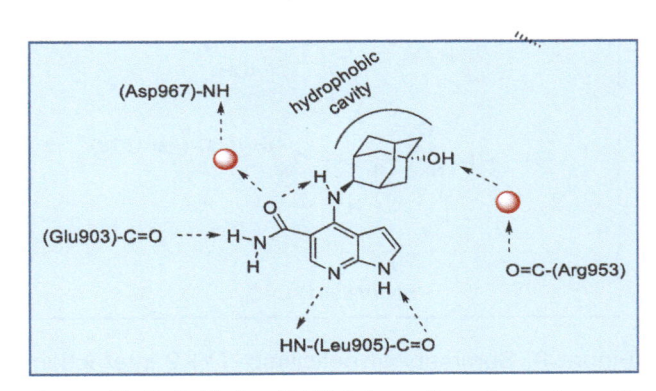

Figure 7. Peficitinib–JAK3 interactions based on X-ray co-crystal structure.

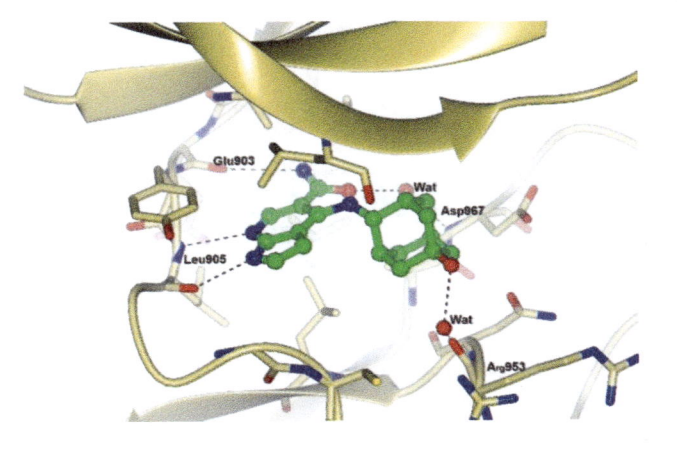

Figure 8. Crystal structure of JAK3 in a complex with peficitinib. Adapted with permission from *Bioorg. Med. Chem.* **2018**, *26*, 4971–4983.

In the complex of peficitinib with TYK2 (Fig. 10),

the pyrrolopyridine core forms two hydrogen bonds with the main chain atoms of Val981 (2.9 and 2.8 Å), and the carbamoyl group forms a hydrogen bond with the main chain atom of Glu979 (3.1 Å) in the hinge region. The carbamoyl oxygen forms a hydrogen bond with a water molecule which binds to the main chain atom of Asp1041 (Fig. 9), similar to that observed for the JAK3 complex. The adamantyl hydroxyl group forms hydrogen bonds with the side chains of Asp1041 and Asn1028. Overall, the binding mode of the pyrrolopyridine ring to the hinge region in TYK2 is similar to that in JAK2 and JAK3, while the binding mode of the adamantyl moiety is similar to that in the JAK1 complex.

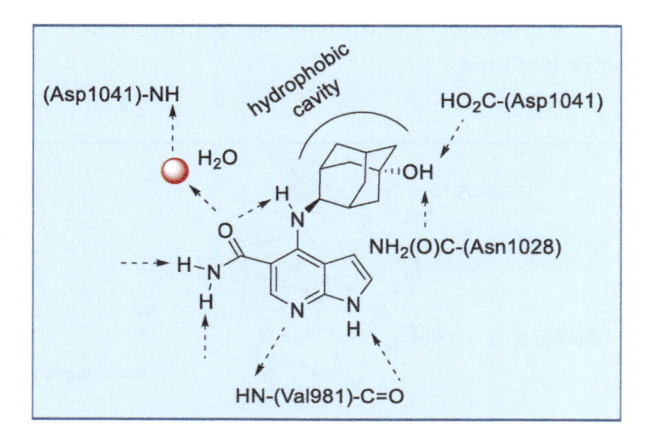

Figure 9. Summary of peficitinib–TYK2 interactions based on X-ray co-crystal structure.

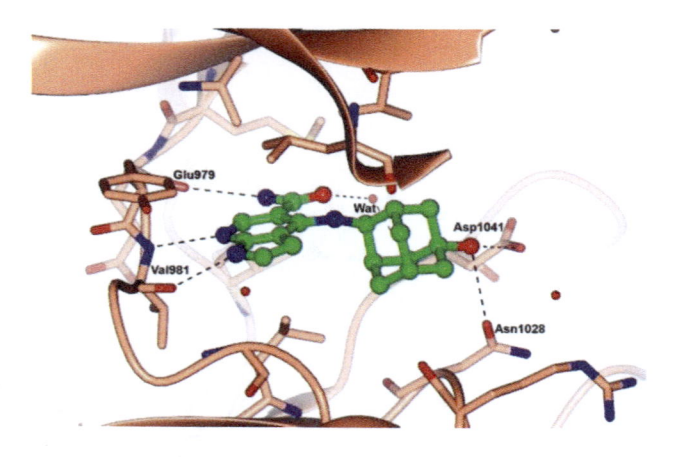

Figure 10. Crystal structure of TYK2 in a complex with peficitinib. Adapted with permission from *Bioorg. Med. Chem.* **2018**, *26*, 4971–4983.

The adamantyl hydroxyl interacts with the main chain atom either directly or indirectly via a water molecule, and this interaction may be responsible for the increase in inhibitory activity observed in JAK1, JAK2, and JAK3. The impact of the hydroxyl group on TYK2 remains to be determined as the TYK2 data for **8** have not reported in the literature.

The disposition of 1*H*-pyrrolo[2,3-*b*]pyridine-5-carboxamide scaffold in the hinge region of the ATP-binding site in JAK1 differs from that of the other JAKs, an observation that has been rationalized in terms of location of unfavorable water molecules using WaterMap analysis of the co-crystal structures with all the isoforms.

PK Properties and Metabolism[6–8]

Peficitinib exhibited low aqueous solubility (≤0.1 mg/mL at pH 7) and modest cell permeability. It also showed moderate oral bioavailability in rats (46%) and monkeys (19%). In healthy volunteers, a single oral dose of 150 mg peficitinib showed an elimination half-life of 9.4 h and high clearance (CL/F = 85.7 L/h, 78% of human hepatic blood flow = 109 L/h for 70 kg body weight).

Following oral administration, the major metabolites in humans were the sulfate conjugate **9** and *N*-methyl conjugate **11** (Fig. 11) via human sulfotransferase SULT2A1 and nicotinamide *N*-methyltransferase (NNMT). Peficitinib was cleared through multiple pathways (renal and possibly biliary excretion, and metabolism) with no prevalent single mechanism.[9]

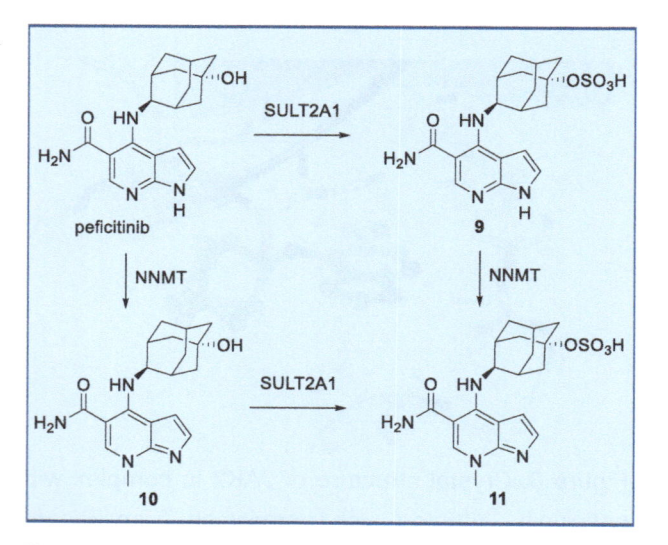

Figure 11. Metabolism of peficitinib in humans.

1. H. Hamaguchi, Y. Amano, A. Moritomo, S. Shirakami, Y. Nakajima, K. Nakai, N. Nomura, M. Ito, Y. Higashi, T. Inoue. Discovery and structural characterization of peficitinib (ASP015K) as a novel and potent JAK inhibitor. *Bioorg. Med. Chem.* **2018**, *26*, 4971–4983.

2. Y. Nakajima, T. Tojo, M. Morita, K. Hatanaka, S. Shirakami, A. Tanaka, H. Sasaki, K. Nakai, K. Mukoyoshi, H. Hamaguchi, F. Takahashi, A. Moritomo, Y. Higashi, T. Inoue. Synthesis and evaluation of 1H-pyrrolo[2,3-b]pyridine derivatives as novel immunomodulators targeting Janus kinase 3. *Chem. Pharm. Bull.* **2015**, *63*, 341–353.

3. Y. Nakajima, T. Inoue, K. Nakai, K. Mukoyoshi, H. Hamaguchi, K. Hatanaka, H. Sasaki, A. Tanaka, F. Takahashi, S. Kunikawa, H. Usuda, A. Moritomo, Y. Higashi, M. Inami, S. Shirakami. Synthesis and evaluation of novel 1H-pyrrolo[2,3-b]pyridine-5-carboxamide derivatives as potent and orally efficacious immunomodulators targeting JAK3. *Bioorg. Med. Chem.* **2015**, *23*, 4871–4883.

4. M. E. Flanagan, T. A. Blumenkopf, W. H. Brissette, M. F. Brown, J. M. Casavant, C. Shang-Poa, J. L. Doty, E. A. Elliott, M. B. Fisher, M. Hines, C. Kent, E. M. Kudlacz, B. M. Lillie, K. S. Magnuson, S. P. McCurdy, M. J. Munchhof, B. D. Perry, P. S. Sawyer, T. J. Strelevitz, C. Subramanyam, J. Sun, P. S. Changelian. Discovery of CP-690,550: a potent and selective Janus kinase (JAK) inhibitor for the treatment of autoimmune diseases and organ transplant rejection. *J. Med. Chem.* **2010**, *53*, 8468–8484.

5. D. J. Augeri, J. A. Robl, D. A. Betebenner, D. R. Magnin, A. Khanna, J. G. Robertson, A. Wang, L. M. Simpkins, P. Taunk, Q. Huang, S. P. Han, B. Abboa-Offei, M. Cap, L. Xin, T. Tao, E. Tozzo, G. E. Welzel, D. M. Egan, J. Marcinkeviciene, S. Y. Chang, L. G. Haman. Discovery and preclinical profile of Saxagliptin (BMS-477118): A highly potent, long-acting, orally active dipeptidyl peptidase IV inhibitor for the treatment of type 2 diabetes. *J. Med. Chem.* **2005**, *48*, 5025–5037.

6. M. Shibata, J. Toyoshima, Y. Kaneko, K. Oda, T. Kiyota, A. Kambayashi, T. Nishimura. The bioequivalence of two peficitinib formulations, and the effect of food on the pharmacokinetics of peficitinib: Two-way crossover studies of a single dose of 150 mg peficitinib in healthy volunteers. *Clin. Pharmacol. Drug Dev.* **2021**, *10*, 283–290.

7. D. Miyatake, T. Shibata, M. Shibata, Y. Kaneko, K. Oda, T. Nishimura, M. Katashima, H. Sekino, K. Furihata, A. Urae. Pharmacokinetics and Safety of a Single Oral Dose of Peficitinib (ASP015K) in Japanese Subjects with Normal and Impaired Renal Function. *Clin. Drug Investig.* **2020**, *40*, 149–159.

8. Report on the Deliberation Results – PMDA. February 29, 2019. https://www.pmda.go.jp/files/000233074.pdf.

9. K. Oda, Y. J. Cao, T. Sawamoto, N. Nakada, O. isniku, Y. Nagasaka, K. Y. Sohda. Human mass balance, metabolite profile and identification of metabolic enzymes of [^{14}C]ASP015K, a novel oral janus kinase inhibitor. *Xenobiotica.* **2015**, *45*, 887–902.

7.4 Upadacitinib (Rinvoq™) (39)

Structural Formula	Space-filling Model	Brief Information
$C_{17}H_{19}F_3N_6O$; MW = 380 clogP = 2.3; tPSA = 72		Year of discovery: 2013 Year of introduction: 2019 Discovered by: AbbVie Developed by: AbbVie Primary targets: JAK1 Binding type: II Class: non-receptor tyrosine kinase Treatment: psoriatic arthritis, rheumatoid arthritis Other name: ABT-494 Oral bioavailability = not reported Elimination half-life = 6–16 h Protein binding = 52%

● = C ● = H ● = O ● = N ● = S ● = F ● = Cl ● = Br ● = I

Upadacitinib is a selective JAK1 inhibitor approved in 2019 for the treatment of adults with moderately to severely active rheumatoid arthritis who have had an inadequate response or intolerance to methotrexate.

Discovery of Upadacitinib[1–3]

The overwhelming majority of kinase hinge-binding scaffolds incorporate monocyclic or bicyclic heterocyclic cores.[4] In contrast, upadacitinib, a JAK1-selective inhibitor, possesses a structurally unique tricyclic pyrrolopyrazine motif, formed by fusing a third ring to the bicyclic hinge binder of tofacitinib. The initial tricyclic pyrrolopyrazine analog **1**, which contained the piperidine amide moiety of tofacitinib (albeit without the methyl group), showed no improvement in potency or selectivity as compared with tofacitinib (Fig. 1). Evidently the tricyclic ring system caused a distortion that displaced the cyano group to an unfavorable location. The lead pyrrolidine analog **2** was even less active than the piperdine counterpart **1**, but SAR studies varying the terminal group (amide to urea) led to upadacitinib which exhibited IC$_{50}$ of 45 nM vs. JAK1 and 109 nM vs. JAK2, > 40-fold selectivity over JAK3 (IC$_{50}$ = 2.1 µM) and 100-fold selectivity over TYK2 (IC$_{50}$ = 4.7 µM) as compared to JAK1. In the cellular assays, upadacitinib was 42-fold selective for JAK1 (IC$_{50}$ = 14 nM) as compared to JAK2 (IC$_{50}$ = 0.59 µM) (Table 1).

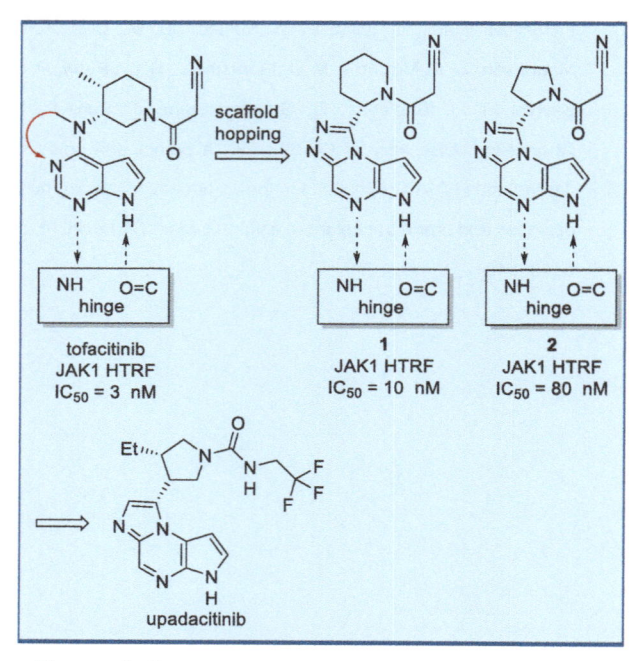

Figure 1. Structural optimization of upadacitinib.

Table 1. JAK1 selectivity data of upadacitinib.[3]

assay	enzyme	cell
JAK1 IC$_{50}$	47 nM	14 nM
JAK2 IC$_{50}$/JAK1 IC$_{50}$	2.5	42
JAK3 IC$_{50}$/JAK1 IC$_{50}$	49	133
TYK2 IC$_{50}$/JAK1 IC$_{50}$	100	194

Binding Mode[3]

No co-crystal structures of upadacitinib with any of the JAKs have been reported. Based on the structural similarity of upadacitinib with tofacitinib, it is possible that both inhibitors bind to JAK1 in a similar fashion, with the pyrrolopyrazine core packed against the hinge region and with two hydrogen bonds between the main chain atoms of the hinge (Figs. 2, 3). The pyrrolidine ring is situated in the hydrophobic pocket of the JAK1 binding cavity, engaging hydrophobic interactions with hydrophobic residues. The terminal trifluoromethyl group, which sits against the glycine-rich loop, can interact with main-chain atoms. Modelling studies suggested that these interactions in JAK1 are different from those in JAK2. It was postulated that the canonical glycine-rich loop adopted a "closed" backbone conformation in JAK1, as opposed to JAK2, due to amino acid sequence differences in the region. Figure 3 shows a model of upadacitinib in JAK1 overlaid with the protein backbone of an X-ray structure of JAK2. The trifluoroethyl group just fits in the tight space under the glycine-rich loop of JAK1 with stabilizing van der Waals interactions.

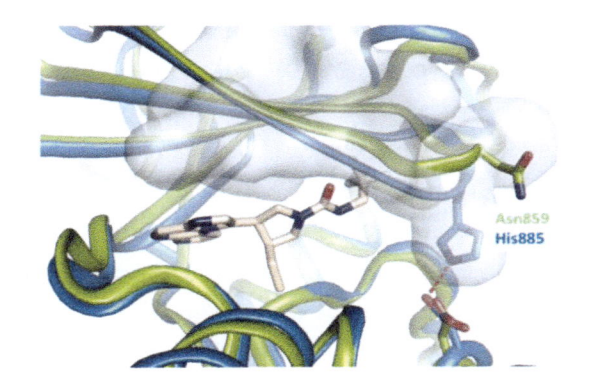

Figure 3. Upadacitinib modeled in the crystal structure of JAK1 (blue) and JAK2 (green). Adapted from *BMC rheumatol.* **2018**, *2*, 23.

PK Properties and Metabolism[5,6]

Upadacitinib exhibits an elimination half-life between 6 and 16 h, supporting once-daily oral dosing. In a human radio-labelled study, 79% of the total plasma radioactivity was accounted for as the parent drug, and 13% of the total plasma radioactivity accounted for the main inactive metabolite **5**, which was formed through CYP3A4-mediated oxidation to the carboxylic acid **3**, followed by glucuronidation with glucuronic acid (**4**) (Fig. 4).

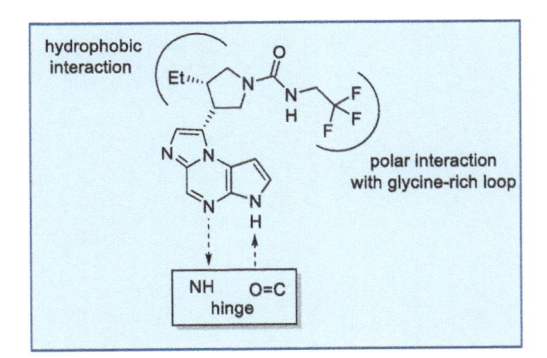

Figure 2. Proposed upadacitinib–JAK1 interactions.

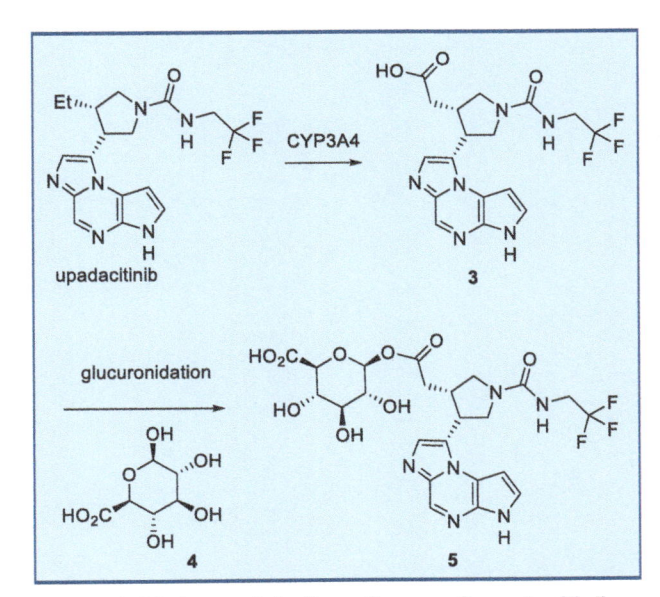

Figure 4. Major metabolic pathway of upadacitinib.

1. S. Van Epps, B. Fiamengo, J. Edmunds, A. Ericsson, K. Frank, M. Friedman, D. George, J. George, E. Goedken, B. Kotecki, G. Martinez, P. Merta, M. Morytko, S. Shekhar, B. Skinner, K. Stewart, J. Voss, G. Wallace, L. Wang, N. Wishart. Design and synthesis of tricyclic cores for kinase inhibition. *Bioorg. Med. Chem. Lett.* **2013**, *23*, 693–698.

2. M. Friedman, K. E. Frank, A. Aguirre, M. A. Argiriadi, H. Davis, J. J. Edmunds, D. M. George, J. S. George, E. Goedken, B. Fiamengo, D. Hyland, B. Li, A. Murtaza, M. Morytko, G. Somal, K. Stewart, E. Tarcsa, S. Van Epps, J. Voss, L. Wang, N. Wishart. Structure activity optimization of 6H-pyrrolo[2,3-e][1,2,4]triazolo[4,3-a]pyrazines as Jak1 kinase inhibitors. *Bioorg. Med. Chem. Lett.* **2015**, *25*, 4399–4404.

3. J. M. Parmentier, J. Voss, C. Graff, A. Schwartz, M. Argiriadi, M. Friedman, H. S. Camp, R. J. Padley, J. S. George, D. Hyland, M. Rosebraugh, N. Wishart, L. Olson, A. J. Long. In vitro and in vivo characterization of the JAK1 selectivity of upadacitinib (ABT-494) *BMC rheumatol.* **2018**, *2*, 23.

4. L. Xing, J. Klug-Mcleod, B. Rai, E. A. Lunney. Kinase hinge binding scaffolds and their hydrogen bond patterns. *Bioorg. Med. Chem.* **2015**, *23*, 6520–6527.

5. M. F. Mohamed, B. Klünder, A. A. Othman. Clinical pharmacokinetics of upadacitinib: Review of data relevant to the rheumatoid arthritis indication. *Clin. Pharmacokinet.* **2010**, *59*, 531–544.

6. Clinical pharmacology and biopharmaceutics review(s). https://www.accessdata.fda.gov/drugsatfda_docs/nda/2019/211675Orig1s000ClinPharmR.pdf.

7.5 Delgocitinib (Corectim™) (40)

Structural Formula	Space-filling Model	Brief Information
$C_{16}H_{18}N_6O$; MW=: 310 logD (pH 7) = 0.6; tPSA = 84		Year of discovery: 2010 Year of introduction: 2020 (Japan) Discovered by: Japan Tobacco Developed by: Japan Tobacco/LEO Pharma Primary targets: pan-JAK Binding type: II Class: non-receptor tyrosine kinase Treatment: atopic dermatitis (topical ointment) Other name: JET-052 Oral bioavailability = not reported Elimination half-life = not reported Protein binding = 22–29%

● = C ○ = H ● = O ● = N ● = S ● = F ● = Cl ● = Br ● = I

Delgocitinib is a pan-JAK inhibitor which inhibits all members of the Janus Kinase (JAK) family. It is being developed for the treatment of autoimmune disorders and hypersensitivity, including inflammatory skin conditions. Topical delgocitinib 0.5% ointment received its first approval in Japan in 2020 for the treatment of adults with atopic dermatitis. The FDA has granted delgocitinib cream fast track designation for chronic hand eczema. An oral formulation of delgocitinib is being investigated in Japan for the treatment of autoimmune disorders and hypersensitivity.[1,2]

Discovery of Delgocitinib

The lead compound to delgocitinib was tofacitinib, a potent pan-JAK inhibitor. To maintain the overall pan-JAK inhibition profile of tofacitinib, the pyrrolopyrimidine hinge binder and the cyanoacetyl group were kept fixed while the central aminopiperidine linker was replaced by a variety of spirocyclic scaffolds to maintain comparable binding patterns (Fig. 1).[3] This effort identified the three dimensional diazaspiro[3.4]octane scaffold as the optimal linker, and further SAR studies led to delgocitinib, which demonstrates better selectivity against JAK3 compared to tofacitinib, and its selectivity for the JAK family over LCK is also improved. Delgocitinib exhibits no significant inhibition of non-JAK kinases under 1 µM except ROCK2.

tofacitinib
JAK1 IC_{50} = 2.9 nM
JAK2 IC_{50} = 1.2 nM
JAK3 IC_{50} = 1.1 nM
TYK2 IC_{50} = 4.2 nM
LCK IC_{50} = 0.44 µM

diazaspirocycle derivatives

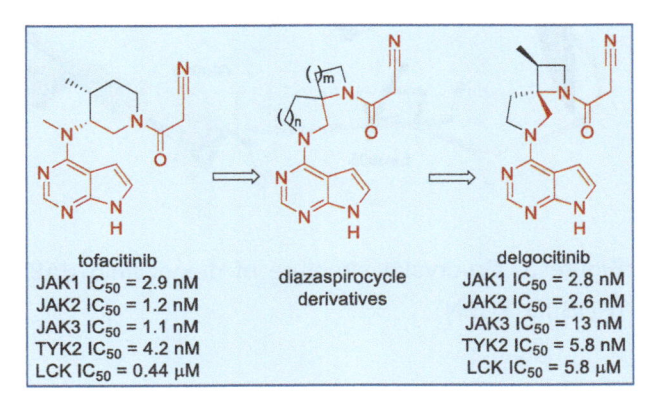

delgocitinib
JAK1 IC_{50} = 2.8 nM
JAK2 IC_{50} = 2.6 nM
JAK3 IC_{50} = 13 nM
TYK2 IC_{50} = 5.8 nM
LCK IC_{50} = 5.8 µM

Figure 1. Summary of the chemical optimization of delgocitinib.

Binding Mode

Delgocitinib shares very similar binding patterns to tofacitinib in complex with JAK3 (Figs. 2 and 3).[3] The pyrrolopyrimidine ring forms hydrogen bond interactions with JAK3 hinge region of Glu903 (2.6 Å) and Leu905 (2.9 Å). The diazaspirooctane moiety occupies the lipophilic pocket. The cyanoacetyl group interacts with the glycine-rich loop, making orthogonal dipolar interactions with main chain atoms (Gly829−Lys830 and Gly834−Ser835).

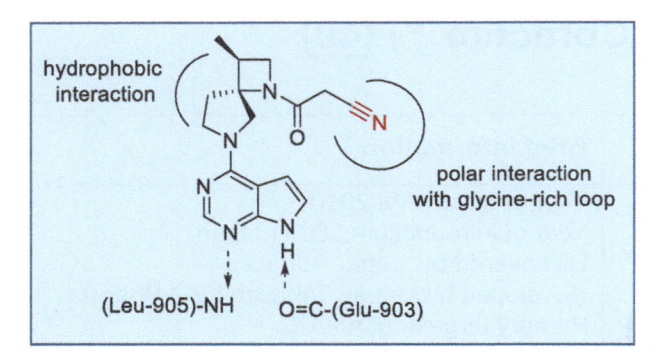

Figure 2. Summary of delgocitinib–JAK3 interactions.

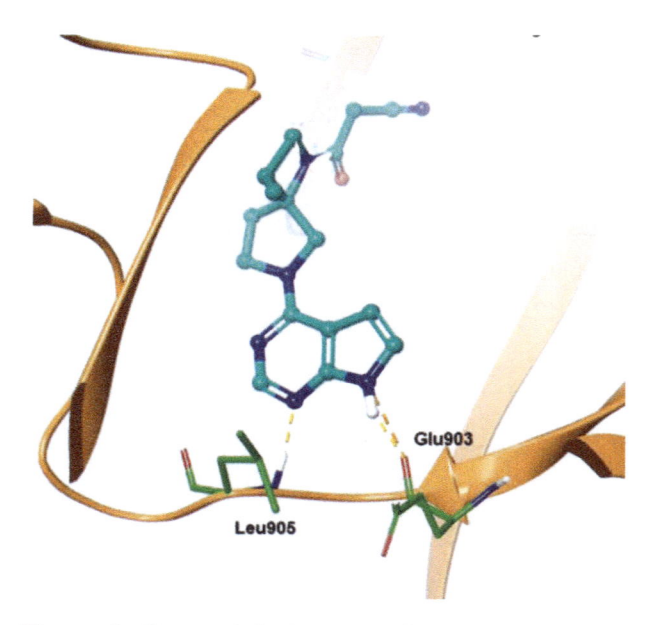

Figure 3. Co-crystal structure of delgocitinib–JAK3 (PDB ID: 7C3N).

1. S. Dhillon. Delgocitinib: First approval. *Drugs*. **2020**, *80*, 609–615.
2. R. Chovatiya, A. S. Paller. JAK inhibitors in the treatment of atopic dermatitis. *J. Allergy Clin. Immunol*. **2021**, *148*, 927–940.
3. M. L. Vazquez, N. Kaila, J. W. Strohbach, J. D. Trzupek M. F. Brown, M. E. Flanagan, M. J. Mitton-Fry, T. A. Johnson, R. E. TenBrink, E. P. Arnold, A. Basak, S. E. Heasley, S. Kwon, J. Langille, M. D. Parikh, S. H. Griffin, J. M. Casavant, B. A. Duclos, A. E. Fenwick, T. M. Harris, R. Unwalla. Identification of N-{cis-3-[methyl(7H-pyrrolo[2,3-d]pyrimidin-4-yl)amino]cyclobutyl}-propane-1-sulfonamide (PF-04965842): A selective JAK1 clinical candidate for the treatment of autoimmune diseases. *J. Med. Chem*. **2018**, *61*,1130–1152.

PK Properties

Delgocitinib is metabolically stable in both liver microsome and hepatocytes across species (human, rat, dog, and monkey), and it is also not metabolized in human skin microsomes. It has favorable oral bioavailability in rats (78%) and dogs (124%).[3]

Following repeat topical application to the affected area twice daily with a maximum dose of 5 g per application, plasma levels of delgocitinib were detected in 12%, 16%, 14%, and 12% of the patients at week 4, 12, 28, and 52, respectively.[1]

7.6 Filgotinib (Jyseleca™) (41)

Structural Formula	Space-filling Model	Brief Information
$C_{21}H_{23}N_5O_3S$; MW = 425 clogP = 1.8; tPSA = 94		Year of discovery: 2009 Year of introduction: 2020 (EU, Japan, UK) Discovered by: Galapagos NV Developed by: Galapagos NV Primary target: JAK1 Binding type: II Class: non-receptor tyrosine kinase Treatment: ulcerative colitis, rheumatoid arthritis Other name: GLPG0634 Oral bioavailability = not reported Elimination half-life = 6–11 h Protein binding = 32%

● = C ○ = H ● = O ● = N ● = S ● = F ● = Cl ● = Br ● = I

Filgotinib is a JAK1 selective inhibitor approved in EU and Japan for the treatment of inflammatory autoimmune diseases, including inflammatory arthritis and inflammatory bowel disease.

Discovery of Filgotinib[1]

High-throughput screening of the BioFocus kinase-focused library collection against the kinase domain of JAK1 in an in vitro biochemical assay identified the triazolo[1,5-a]pyridine **1** (Fig. 1) with encouraging JAK1 inhibitory activity (IC_{50} = 70 nM). Although this lead was attractive because of its low molecular weight and modest JAK1 selectivity; it suffered from poor aqueous solubility, high protein binding, and poor metabolic stability (only 6% of the parent compound remaining after 60 min incubation in human liver microsomes (HLM). Replacement of methoxy by the polar, moderately basic, morpholinomethyl group maintained JAK1 inhibitory activity, while improving metabolic stability ($t_{1/2}$ >90 min in HLM), membrane permeability, and aqueous solubility as well as reducing plasma protein binding (PPB = 70% in rat). To further improve the metabolic stability of **2** in rats (RLM = 38% remaining after 60 min incubation in rat liver microsomes (RLM)), the morpholine oxygen was replaced by SO_2 forming a thiopiperidine dioxide, a transformation that culminated in the discovery of filgotinib with 9-fold improvement in JAK1 potency and 2- to 9-fold in JAK1 isoform selectivity. The enhanced potency can be attributed to the polar interactions of the polar SO_2 group with the glycine-rich loop as revealed in the co-crystal structure with JAK2. Filgotinib also outperformed the morpholine analog **2** in terms of metabolic stability in RLM (65% vs. 38% remaining for **2**) as well as plasma protein binding (48% vs. 71%).

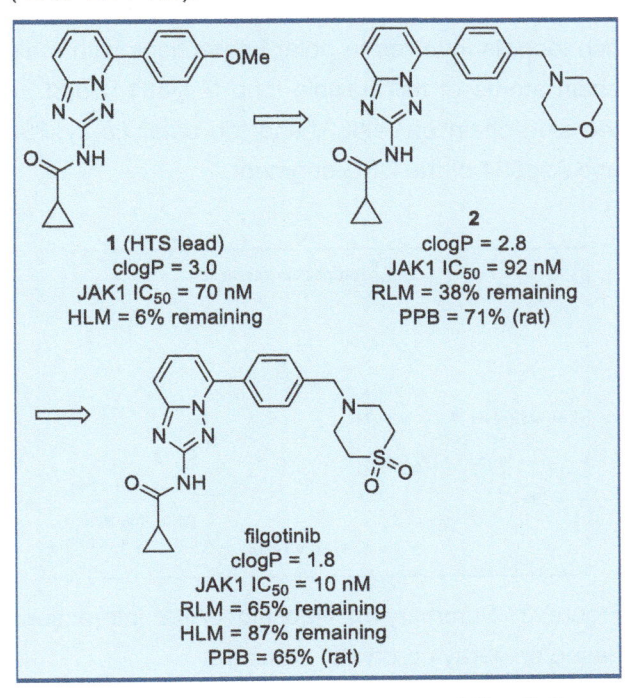

Figure 1. Structural optimization of filgotinib.

Filgotinib exhibited only modest selectivity vs. JAK2 in the biochemical assay (JAK1 IC_{50}/JAK2 IC_{50} = 2.8); however, it displayed a 28-fold selectivity for JAK1- over JAK2-dependent signaling in human whole blood, which is more relevant to in vivo setting. In the same assay, tofacitinib and baricitinib showed 10- and 3-fold selectivity, respectively (Table 1).

Filgotinib has been recently approved for the treatment of ulcerative colitis and rheumatoid arthritis in several countries outside the United States.

Table 1. JAK1 selectivity data of filgotinib vs. tofacitinib[1]

inhibitor	filgotinib	tofacitinib
JAK1 IC_{50}	10 nM	0.8 nM
JAK2 IC_{50}/JAK1 IC_{50}	2.8	0.8
JAK3 IC_{50}/JAK1 IC_{50}	81	0.4
TYK2 IC_{50}/JAK1 IC_{50}	12	42

Binding Mode[1]

In the co-crystal structure of filgotinib with JAK2 (Figs. 3, 4), the triazolopyridine ring is oriented against the hinge, and the exocyclic NH and the triazolo nitrogen N3 forms two crucial hydrogen bonds with the amide carbonyl oxygen and the NH of Leu932 (Fig. 2). The phenyl ring fits within a hydrophobic pocket. The terminal thiomorpholine dioxide group, which is positioned against the glycine rich loop, is involved in polar interactions with main chain atoms of this flexible loop (Gly861, Ser862), the side chain of Val863, and the catalytic Lys882 and Asp994 of the DFG segment.

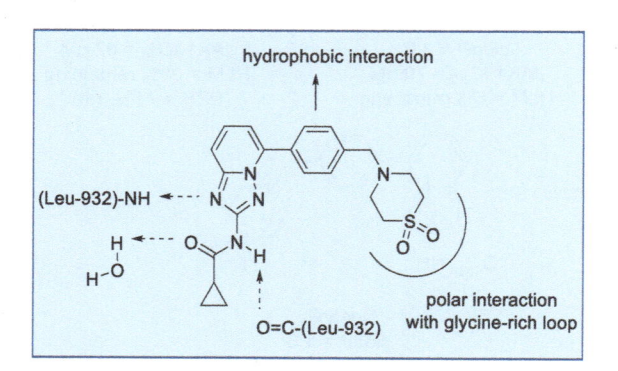

Figure 2. Summary of filgotinib–JAK2 interactions based on X-ray co-crystal structure.

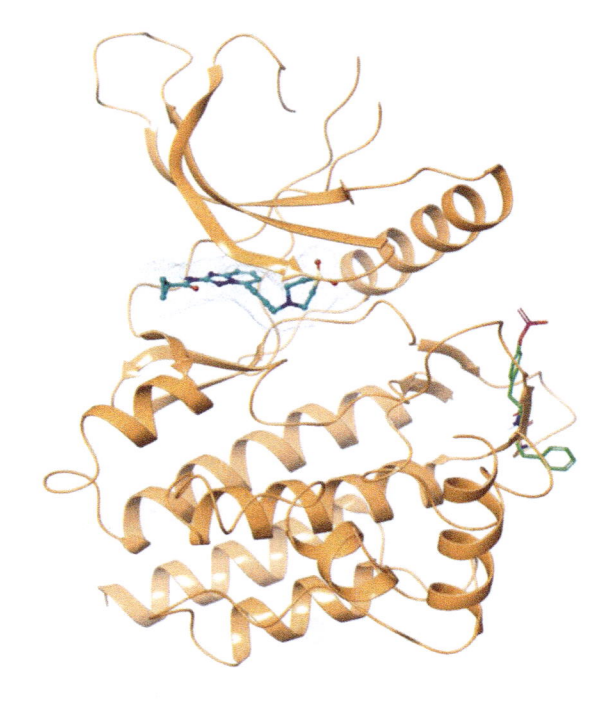

Figure 3. Co-crystal structure of filgotinib with JAK2 (PDB ID: 4P7E).

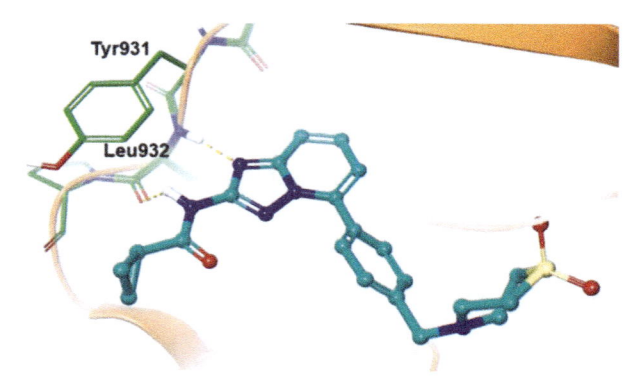

Figure 4. Detailed view of the X-ray crystal structure of filgotinib bound to the ATP-binding site of JAK2 (PDB ID: 4P7E).

PK Properties and Metabolism[2,3]

Filgotinib has favorable metabolic stability in vitro (65%, 45%, and 87% of compound remaining in rat, dog, and human liver microsomes, respectively, after 60 min incubation; and a half-life of more than 200 min in hepatocytes for all species). Filgotinib is a substrate of P-gp as revealed by an efflux ratio of 16 from the Caco-2 assay. In spite of the high efflux ratio, it still showed 45% and 67% oral bioavailability in rats and dogs, respectively, presumably due to its good solubility (0.18 mg/mL in water), high membrane permeability and good metabolic stability. It exhibited low plasma protein bindings across species (48% in rats, 29% in dogs, and 32% in humans).

The elimination half-lives of filgotinib were 6 and 11 hours in the Japanese and Caucasian volunteers, respectively. In both trials, the active metabolite **3** (Fig. 5) showed plasma concentrations far beyond those of filgotinib with half-life values ranging from 17 to 20 hours, which supports once a day oral administration.

The active metabolite **3** was formed by hydrolysis of the amide functionality to release the free amine. This metabolite is less potent than filgotinib but still maintained good JAK1 selectivity. This metabolite is believed to contribute to the overall therapeutic effects along with the parent drug.

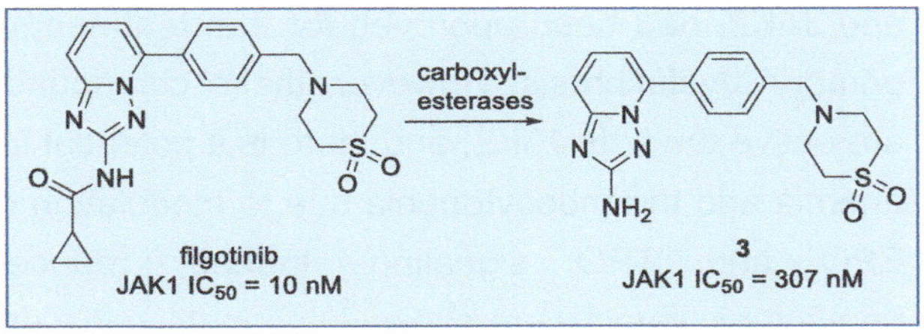

1. C. J. Menet, S. R. Fletcher, G. Van Lommen, R. Geney, J. Blanc, K. Smits, N. Jouannigot, P. Deprez, E. M. van der Aar, P. Clement-Lacroix, L. Lepescheux, R. Galien, B. Vayssiere, L. Nelles, T. Christophe, R. Brys, M. Uhring, F. Ciesielski, L. Van Rompaey. Triazolopyridines as selective JAK1 inhibitors: from hit identification to GLPG0634. *J. Med. Chem.* **2014**, *57*, 9323–9342.

2. L. Van Rompaey, R. Galien, E. M. van der Aar, P. Clement-Lacroix, L. Nelles, B. Smets, L. Lepescheux, T. Christophe, K. Conrath, N. Vandeghinste, B. Vayssiere, S. De Vos, S. Fletcher, R. Brys, G. van 't Klooster, J. H. Feyen, C. Menet. Preclinical characterization of GLPG0634, a selective inhibitor of JAK1, for the treatment of inflammatory diseases. *J. Immunol.* **2013**, *191*, 3568–3577.

3. F. Namour, P. M. Diderichsen, E. Cox, B. Vayssière, A. Van der Aa, C. Tasset, G. Van't Klooster. Pharmacokinetics and pharmacokinetic/pharmacodynamic modeling of filgotinib (GLPG0634), a selective JAK1 inhibitor, in support of phase IIB dose selection. *Clin. Pharmacokinet.* **2015**, *54*, 859–874.

Figure 5. Metabolic pathway of filgotinib.

7.7 Abrocitinib (Cibinqo™) (42)

Structural Formula	Space-filling Model	Brief Information
$C_{12}H_{17}N_5O_2S$; MW = 295 logD (pH7.4) = 1.9; tPSA = 86		Year of discovery: 2014 Year of introduction: 2022 Discovered by: Pfizer Developed by: Pfizer Primary target: JAK1 Binding type: II Class: non-receptor tyrosine kinase Treatment: atopic dermatitis Other name: PF-04965842 Oral bioavailability = 60% Elimination half-life = 5 h Protein binding = 50%

● = C ○ = H ● = O ● = N ● = S ● = F ● = Cl ● = Br ● = I

Abrocitinib, a selective JAK1 inhibitor, was approved in 2022 for the treatment of adults with refractory, moderate-to-severe atopic dermatitis (AD) whose disease is not adequately controlled with other systemic drug products or when use of those therapies is inadvisable.

Atopic dermatitis is a chronic inflammatory skin disease characterized by intense itch, dry skin, and skin pain. Systemic immunosuppressants, such as cyclosporine, methotrexate, and azathioprine, are frequently prescribed for the treatment of moderate-to-severe atopic dermatitis which is nonsusceptible to topical therapy or oral antihistamines. However, none of these therapies is suitable for long-term use because of adverse effects. In 2017, dupilumab (Dupixent™), an anti-IL-4 receptor alpha monoclonal antibody, was approved by the FDA to treat adult patients with moderate-to-severe atopic dermatitis which cannot be managed with topical prescriptions. Dupilumab is generally effective, but either biweekly or monthly subcutaneous injections can be inconvenient for some patients. Against the backdrop of dupilumab, abrocitinib, an orally bioavailable selective JAK1 inhibitor, has been developed for atopic dermatitis.[1]

Previously, tofacitinib, a JAK1/JAK3 inhibitor with moderate activity on JAK2, and baricitinib, a JAK1/JAK2 inhibitor, had been approved to treat patients with rheumatoid arthritis. Ruxolitinib, a close analog of baricitinib and also a dual inhibitor of JAK1 and JAK2, had been approved for the treatment of primary myelofibrosis. However, these compounds are active towards JAK2, and there is a potential for anemia and thrombocytopenia due to modulation of EPO and TPO signaling. Indeed, reduced hemoglobin levels were shown in patients who were administered with baricitinib and tofacitinib.[2]

Abrocitinib is an orally bioavailable, selective JAK1 inhibitor to treat moderate-to-severe atopic dermatitis. Abrocitinib preferentially blocks cytokine signaling involving JAK1 and is selective against signaling pathways using dual JAK2 or JAK2/TYK2. Selective inhibition of JAK1 can regulate multiple cytokine pathways implicated in the pathology of atopic dermatitis, such as IL-4 and IL-13, while mitigating potential risk associated with JAK2 inhibition.[2]

Discovery of Abrocitinib

The discovery of abrocitinib started with tofacitinib, a potent pan-JAK inhibitor which exhibits excellent overall kinome selectivity presumably due to the pyrrolopyrimidine hinge binder.[2] To gauge the JAK1 selectivity, the N-methyl-7H-pyrrolo[2,3-d]pyrimidin-4-amine scaffold was kept intact while the piperidine group was replaced with a variety of cycloalkylamines in which the terminal amine was converted to a variety of amide and sulfonamide derivatives (Fig. 1). This work led to a series of *cis*-1,3-cyclobutane diamine-linked sulfonamides with both excellent JAK1 potency and more importantly excellent selectivity within the JAK family. One of these sulfonamides was abrocitinib which is approximately 28-fold selective over JAK2, >340-fold over JAK3 and 43-fold over TYK2, and it has even higher selectivity over the broader kinome.

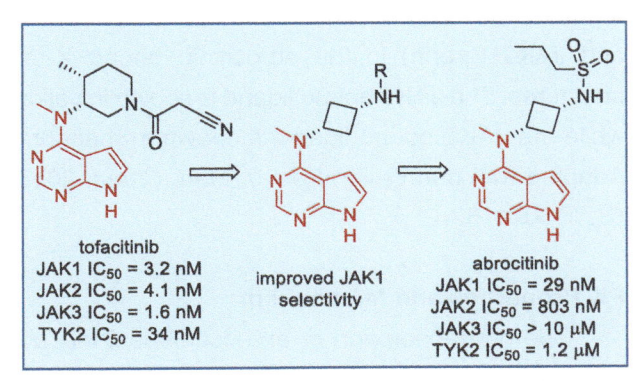

Figure 1. Summary of the chemical optimization of abrocitinib.

Binding Mode

The co-crystal structure of abrocitinib in complex with JAK1 (Fig. 2) showed hydrogen bond interactions of the pyrrolopyrimidine ring with hinge residues Leu959, Glu957, and the sulfonamide NH with Asn1008 and Arg1007. A bound water molecule is hydrogen bonded to N-3 of the hinge binder and also forms hydrogen bond with Glu966 residue (Fig. 3).

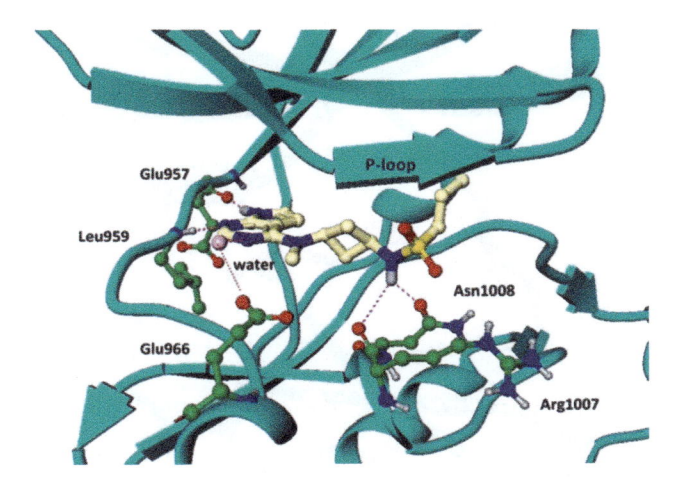

Figure 3. X-ray crystal structure of abrocitinib in complex with JAK1. Adapted with permission from *J. Med. Chem.* **2018**, *61*, 1130–1152.

The co-crystal structure of abrocitinib with JAK2 (Fig. 5) is similar to that with JAK1 (Fig. 3). The bicyclic pyrrolopyrimidine core forms two hydrogen bond interactions with hinge residues Leu932 and Glu930. The N-3 of the hinge binder interacts with a bound water molecule (light pink), which is also hydrogen bonded to Asp939 through a network of water molecules (not shown in Fig. 5 for clarity). Another bound water molecule engages in hydrogen bond interactions with the sulfur atom of abrocitinib and the NH backbone of the Lys857 residue in the P-loop. The sulfonamide NH is involved in hydrogen bond interactions with Asn981 and Arg980 (Fig. 4).

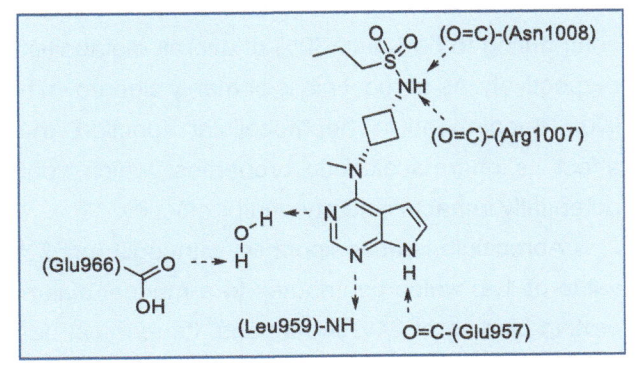

Figure 2. Summary of abrocitinib–JAK1 interactions based on X-ray co-crystal structure.

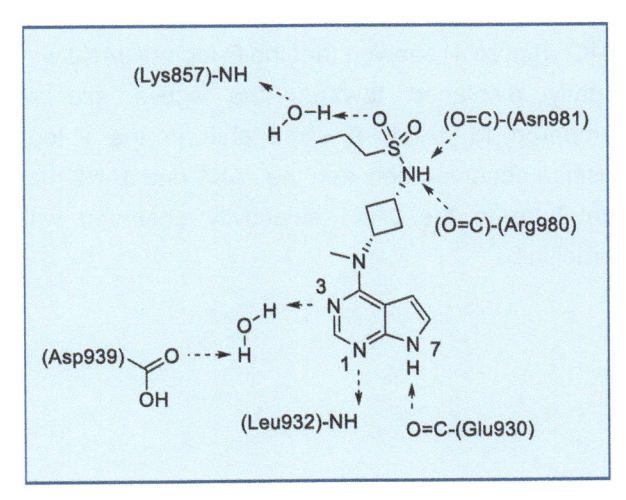

Figure 4. Summary of abrocitinib–JAK2 interactions based on X-ray co-crystal structure.

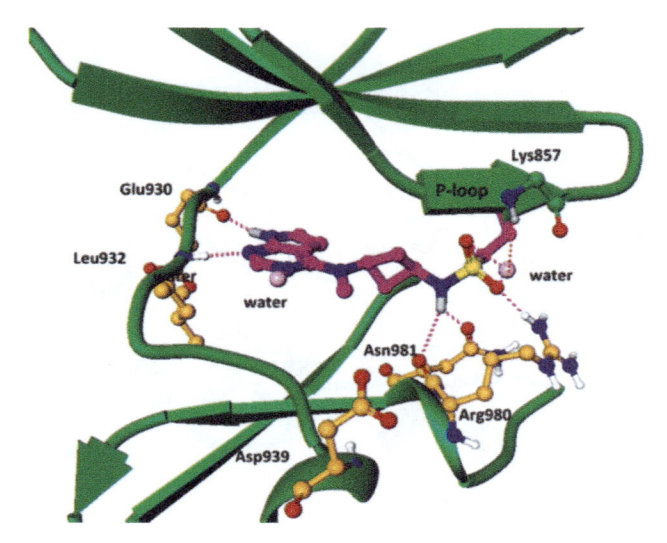

Figure 5. X-ray crystal structure of abrocitinib (in magenta) in complex with JAK2. Adapted with permission from *J. Med. Chem.* **2018**, *61*, 1130–1152.

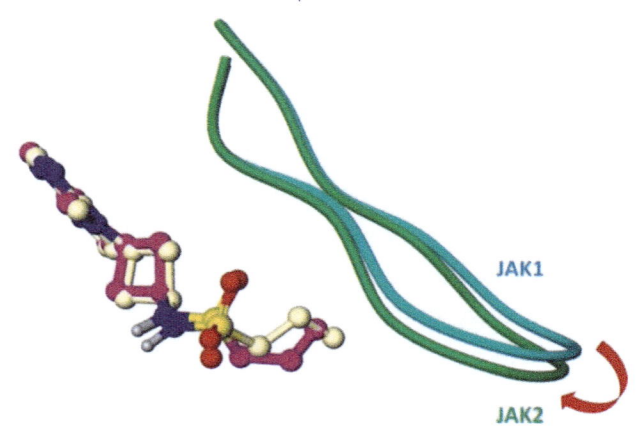

Figure 6. Orientations of the P-loops in JAK1 (cyan) and JAK2 (green) in the abrocitinib bound X-ray structures. The JAK1 bound ligand is shown in yellow, while the JAK2 bound ligand is shown in magenta. Adapted with permission from *J. Med. Chem.* **2018**, *61*, 1130–1152.

Abrocitinib binds to both JAK1 and JAK2 in a similar fashion, but the minor differences in the hydrogen bond patterns and in the disposition of the terminal propyl group in the P-loop area could explain the JAK1 selectivity (Fig. 6). The pyrrolopyrimidine core participates in two key hinge-binding interactions: (1) between the pyrrole NH and the amide carbonyl oxygen of the backbone Glu957(JAK1)/Glu930(JAK2); and (2) between the pyrimidine N1 nitrogen and the amide NH of the backbone Leu959(JAK1)/Leu932(JAK2). The terminal *n*-propyl group is oriented differently toward the P-loop region. An overlay of the co-crystal structures of abrocitinib in complex with JAK1 and JAK2 (Figure 4) showed that the P-loop of JAK2 was slightly displaced towards the active site as compared to JAK1. A slight shift in the P-loop position observed between the JAK1 and JAK2 may contribute to the JAK1 selectivity observed with abrocitinib.

PK Properties and Metabolism

The oral absorption of abrocotinib is 91%, but its oral bioavailability is moderate (60%) due to hepatic metabolism.[3,4] Human PK studies with oral administration of [^{14}C]-abrocitinib showed that the parent drug was the most abundant circulating species (26%), along with 3 oxidative metabolites: PF-06471658 (11%), PF-07055087 (12%), and PF-07054874 (14%) (Fig. 7).[5] The two hydroxyl compounds are active metabolites, with comparable JAK1 selectivity profiles to abrocitinib, whereas the pyrrolidinone pyrimidine metabolite lacks in kinase activity. Based on the in vitro cytochrome P450 phenotyping studies in human hepatocytes, CYP2C19 and CYP2C9 are the major CYP isoforms involved in the oxidative metabolism of abrocitinib, contributing to 53% and 30% of overall metabolism, respectively. As abrocotinib is primarily cleared in the liver, impairment in hepatocellular function may affect its pharmacokinetic properties, which could potentially impact its safety and/or efficacy.

Abrocitinib is quite lipophilic with logD (pH 7.4) value of 1.9, which contributes to a modest plasma protein binding of 64% in humans. Its terminal half-life of 5 h along with 60% oral bioavailability appears to be adequate for once-daily, oral treatment of moderate-to-severe atopic dermatitis.

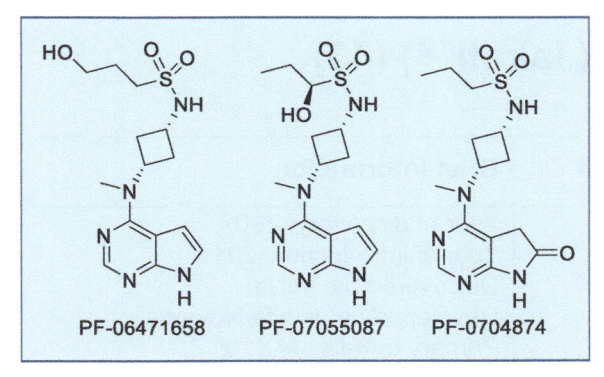

Figure 7. Three major metabolites of abrocitinib in human plasma.

1. R. Chovatiya, A. S. Paller. JAK inhibitors in the treatment of atopic dermatitis. *J. Allergy Clin. Immunol.* **2021**, *148*, 927–940.
2. M. L. Vazquez, N. Kaila, J. W. Strohbach, J. D. Trzupek M. F. Brown, M. E. Flanagan, M. J. Mitton-Fry, T. A. Johnson, R. E. TenBrink, E. P. Arnold, A. Basak, S. E. Heasley, S. Kwon, J. Langille, M. D. Parikh, S. H. Griffin, J. M. Casavant, B. A. Duclos, A. E. Fenwick, T. M. Harris, R. Unwalla. Identification of N-{cis-3-[methyl(7H-pyrrolo[2,3-d]pyrimidin-4-yl)amino]cyclobutyl}-propane-1-sulfonamide (PF-04965842): A selective JAK1 clinical candidate for the treatment of autoimmune diseases. *J. Med. Chem.* **2018**, *61*, 1130–1152.
3. E. Q. Wang, V. Le, M. O'Gorman, S. Tripathy, M. E. Dowty, L. Wang, B. K. Malhotra. Effects of hepatic impairment on the pharmacokinetics of abrocitinib and its metabolites. *J. Clin. Pharmacol.* **2021**, *61*, 1311–1323.
4. https://www.ema.europa.eu/en/documents/assessment-report/cibinqo-epar-public-assessment-report_en.pdf.
5. S. Tripathy, D. Wentzel, X. K. Wan, O. & Kavetska. Validation of enantioseparation and quantitation of an active metabolite of abrocitinib in human plasma. *Bioanalysis.* **2021**, *13*, 1477–1486.

7.8 Ruxolitinib (Jakafi™) (43)

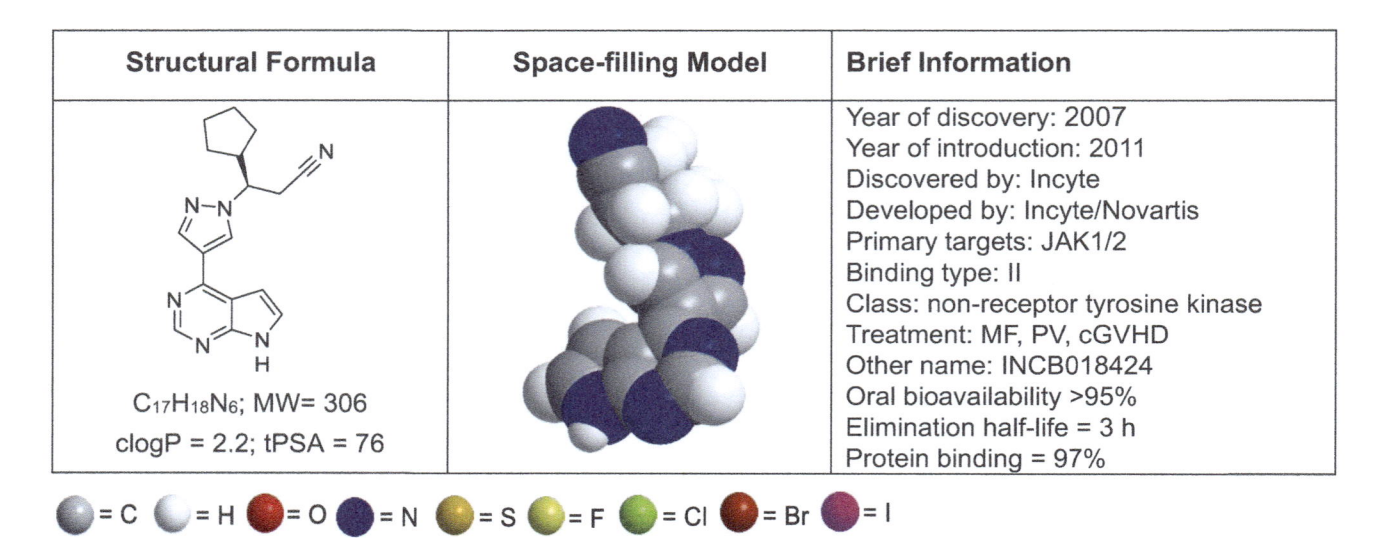

Structural Formula	Space-filling Model	Brief Information
$C_{17}H_{18}N_6$; MW= 306 clogP = 2.2; tPSA = 76		Year of discovery: 2007 Year of introduction: 2011 Discovered by: Incyte Developed by: Incyte/Novartis Primary targets: JAK1/2 Binding type: II Class: non-receptor tyrosine kinase Treatment: MF, PV, cGVHD Other name: INCB018424 Oral bioavailability >95% Elimination half-life = 3 h Protein binding = 97%

● = C ● = H ● = O ● = N ● = S ● = F ● = Cl ● = Br ● = I

Ruxolitinib is a JAK1/2 inhibitor which was approved for the treatment of myelofibrosis, a malignancy of bone marrow that results in diminished production of blood cells. JAK2, as the other JAKs, consists of a pseudokinase domain (JH2) and a catalytic tyrosine kinase domain (JH1). The JH2 domain has only weak kinase activity *per se*, but it serves two purposes: (1) it inhibits the tyrosine kinase activity of JH1 by interacting with its activation loop; and (2) it enables full activation upon cytokine receptor binding. When no cytokines (e.g., erythropoietin, EPO) bind to their receptors, JAK2 is inactive as the kinase activity of the JH1 domain is blocked by the JH2 domain (Fig. 1A). Binding of EPO to its receptor brings about conformational changes in the receptor and in the bound JAK2, leading to a separation of the JH2 inhibitory domain from JH1 to make JAK2 activated (Fig. 1B). However, a single somatic V617F mutation in the JH2 domain of JAK2 prohibits JH2 from inhibiting JH1, thus keeping the kinase active even in the absence of any cytokine binding (Fig. 1C).[1]

The JAK2V617F mutation has been observed in 90% of patients with polycythemia vera (PV) and 50% of patients with essential thrombocythemia (ET) and primary myelofibrosis (MF). The discovery of JAK2V617F as one of the causes for myeloproliferative neoplasms (MPNs) has stimulated significant interest in the development of selective JAK2 inhibitors to treat MF, ET, and PV. In common with most JAK inhibitors, JAK2 inhibitors such as ruxolitinib target the ATP-binding site of the catalytic domain (JH1) and not the pseudokinase domain (JH2), thus inhibiting both mutated and wild-type kinases. These drugs work by inhibiting neoplastic cell proliferation and downregulating cytokine signaling through proinflammatory cytokine receptors. JAK2 inhibitors have proved to be valuable for MPN patients with and without the JAK2V617F mutation.[2]

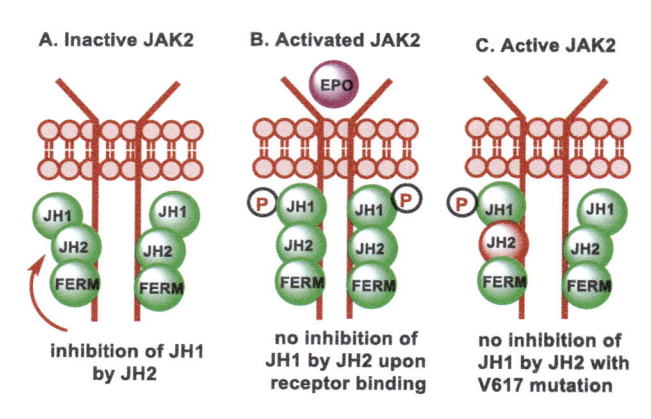

Figure 1. Three different states of JAK2. A: Inactive JAK2; B: Activated JAK2; and C: Active JAK2.

Ruxolitinib was the first JAK inhibitor to reach the market, and the first US FDA-approved medication for myelofibrosis in 2011. Roxolitinib exhibits good potency against JAK1 and JAK2 with IC$_{50}$ values of 3.3 and 2.8 nM, respectively (Table 1), and it displayed 6-fold selectivity against TYK2 and 130-fold against JAK3. Since JAK3 plays a central role in the control of lymphopoiesis, the lack of JAK3 activity spares damage to lymphopoiesis. Roxolitinib also demonstrated excellent selectivity against CHK2 and MET (Table 1) as well as a commercial panel of 26 additional kinases.[3]

Table 1. Kinase selectivity of ruxolitinib[3]

kinase	IC$_{50}$
JAK1	3.3 nM
JAK2	2.8 nM
JAK3	428 nM
TYK2	19 nM
CHK2	>1 µM
MET	>10 µM

In the co-crystal complex with JAK2 (Fig. 3),[4,5] ruxolitinib is situated deep inside the ATP site, forming two hydrogen bond interactions between the pyrrolopyrimidine moiety and amide carbonyl oxygen of Glu930 and amide NH of Leu932 of the hinge region (Fig. 2). The inhibitor is also involved in several van der Waals hydrophobic interactions with surrounding residues, notably the P-loop (Leu855 and Gly856) and the DFG motif (Asp994). The hydrophobic and flexible gatekeeper residue Met929 also improves ruxolitinib binding via multiple van der Waals attractions.

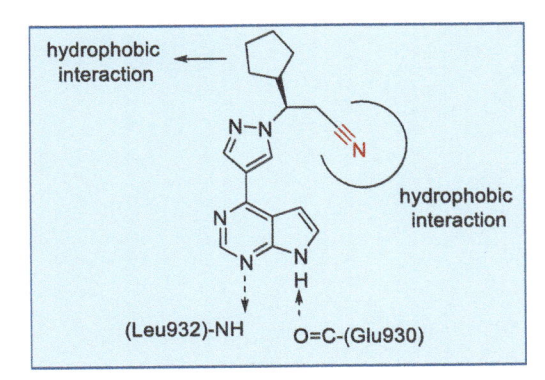

Figure 2. Ruxolitinib–JAK3 interactions.

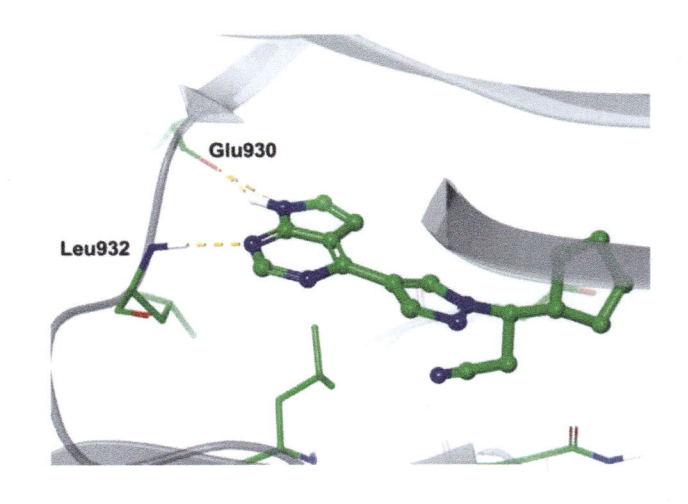

Figure 3. Co-crystal structure of ruxolitinib with JAK2 (PDB ID: 6VGL).

Ruxolitinib and tofacitinib share the same pyrrolopyrimidine hinge binder but differ primarily in the middle linkage: heterocycle pyrazole vs. weakly tertiary amine (Fig. 4). The more lipophilic pyrazole of ruxolitinib increases clogP to 2.2 (vs. 1.6 for tofacitinib), resulting in much higher protein binding (97% bound to plasma proteins, vs. 40% for tofacitinib).

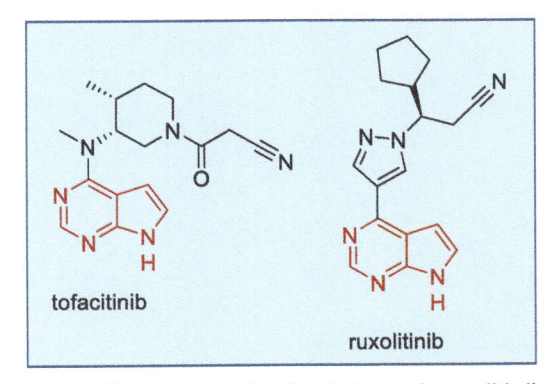

Figure 4. Structures of tofacitinib and ruxolitinib.

Ruxolitinib is metabolized primarily by CYP3A4 in the liver, forming the hydroxylated cyclopentyl derivative **1** as the major active metabolite. Ruxolitinib has excellent oral bioavailability (>95%), but elimination half-life is relatively short (approximately 3 h for the parent compound, 5.8 h for ruxolitinib plus other metabolites).[6] To overcome the metabolic instability of the cyclopentyl moiety, the Concert Pharmaceuticals has developed deuruxolitinib (Fig. 5).

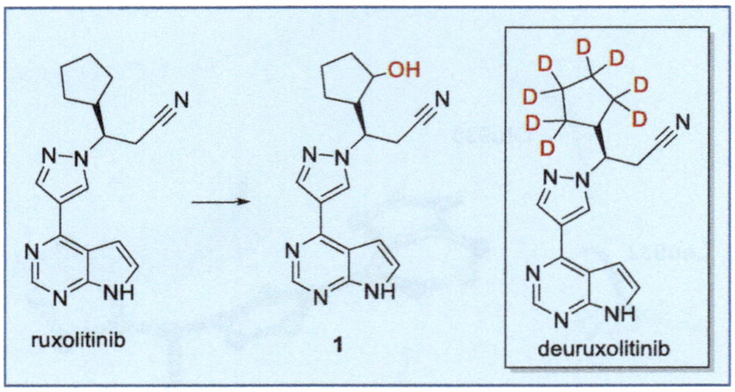

Figure 5. Major metabolism of ruxolitinib in humans.

1. M. Bennett, D. F. Stroncek, D. F. Recent advances in the bcr-abl negative chronic myeloproliferative diseases. *J. Trans. Med.* **2006**, *4*, 41.
2. F. P. Santos, S. Verstovsek. JAK2 inhibitors for myelofibrosis: why are they effective in patients with and without JAK2V617F mutation?. *Anti-cancer Agents Med. Chem.* **2012**, *12*, 1098–1109.
3. A. Quinta's-Cardama, K. Vaddi, P. Liu, T. Manshouri, J. Li, P. A. Scherle, E. Caulder, X. Wen, Y. Li, P. Waeltz, M. Rupar, T. Burn, Y. Lo, J. Kelley, M. Covington, S. Shepard, J. Rodgers, P. Haley, H. Kantarjian, J. S. Fridman, S. Verstovsek. Preclinical characterization of the selective JAK1/2 inhibitor INCB018424: Therapeutic implications for the treatment of myeloproliferative neoplasms. *Blood.* **2010**, *115*, 3109–3117
4. T. Zhou, S. Georgeon, R. Moser, D. J. Moore, A. Caflisch, O. Hantschel. Specificity and mechanism-of-action of the JAK2 tyrosine kinase inhibitors ruxolitinib and SAR302503 (TG101348). *Leukemia.* **2014**, *28*, 404–407.
5. R. R. Davis, B. Li, S. Y. Yun, A. Chan, P. Nareddy, S. Gunawan, M. Ayaz, H. R. Lawrence, G. W. Reuther, N. J. Lawrence, E. Schönbrunn. Structural insights into JAK2 inhibition by ruxolitinib, fedratinib, and derivatives thereof. *J. Med. Chem.* **2021**, *64*, 2228–2241.
6. A. D. Shilling, F. M. Nedza, T. Emm, S. Diamond, E. McKeever, N. Punwani, W. Williams, A. Arvanitis, L. G. Galya, M. Li, S. Shepard, J. Rodgers, T. Y. Yue, S. Yeleswaram. Metabolism, excretion, and pharmacokinetics of [14C]INCB018424, a selective Janus tyrosine kinase 1/2 inhibitor, in humans. *Drug Metab. Dispos.* **2010**, *38*, 2023–2031.

7.9 Fedratinib (Inrebic™) (44)

Structural Formula	Space-filling Model	Brief Information
$C_{27}H_{36}N_6O_3S$; MW = 524 clogP = 6.2; tPSA = 107		Year of discovery: 2010 Year of introduction: 2019 Discovered by: TargeGen Developed by: Impact Biomedicines/Celgene Primary target: JAK2 Binding type: II Class: non-receptor tyrosine kinase Treatment: myeloproliferative diseases Other name: SAR302503, TG101348 Oral bioavailability = 18–37% Elimination half-life = 2–3 days Protein binding = 91–96%

◉ = C ○ = H ● = O ● = N ● = S ● = F ● = Cl ● = Br ● = I

Following the discovery of a JAK2 mutation, V617F, in myeloproliferative neoplasms in 2005, ruxolitinib, a dual JAK1/2 inhibitor, was found to be effective for the treatment of myelofibrosis (MF) and eventually approved in the United States in 2011. Eight years later, fedratinib, a highly selective JAK2 inhibitor, became the second drug approved for MF, more specifically for the treatment of adult patients with intermediate-risk level 2 or high-risk primary or secondary MF or as second-line therapy for patients previously treated with ruxolitinib. However, fedratinib bears a black box warning for encephalopathy, likely due to its concurrent reduction of thiamine uptake. Therefore, fedratinib should not be started in patients with thiamine deficiency, and thiamine levels need to be monitored and maintained during treatment. It is noteworthy that the encephalopathy concern contributed to a prolonged clinical halt and delayed approval of the drug by several years.

In cell-free kinase activity assays, fedratinib exhibited an IC_{50} value of 3 nM for both wild-type JAK2 and JAK2V617F, which is 35-fold selective vs. JAK1, >300-fold vs. JAK3, and >100-fold vs. TYK2 (Table 1).[1] In addition, fedratinib showed little or no resistance against 211 ruxolitinib-resistant JAK2 variants in vitro studies. The lack of genetic resistance has been rationalized by dual binding of fedratinib in the kinase domain at both the ATP and peptide-substrate binding sites (Fig. 1).[2]

Table 1. IC_{50} data of fedratinib vs. ruxolitinib[3]

inhibitor	fedratinib	ruxolitinib
JAK1	105 nM	6 nM
JAK2	3 nM	9 nM
JAK3	1002 nM	487 nM
TYK2	405 nM	30 nM

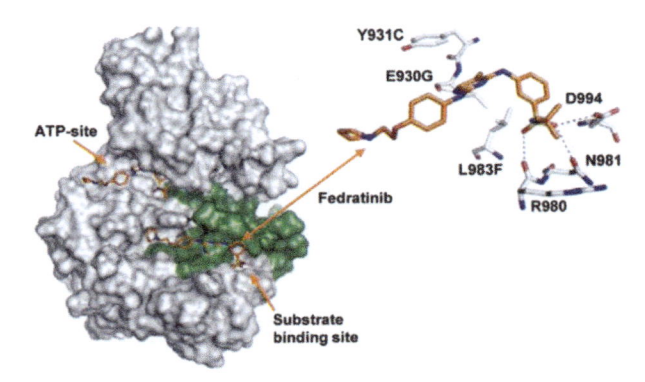

Figure 1. Dual-binding sites of fedratinib at the JAK2 ATP and peptide-substrate binding sites. Adapted from *Leukemia*. **2021**, *35*, 1–17.

In the co-crystal structure of fedratinib with JAK2 (Fig. 3), the aminopyrimidine ring is oriented against the hinge, and the exocyclic NH at C2 and the pyrimidine N1 form two hydrogen bonds with the amide oxygen and the NH of Leu932 in the hinge (Fig. 2). Both the pyrimidine and the sulfonamide-substituted phenyl rings penetrate hydrophobic pockets, and the sulfonamide forms polar interactions with P-loop residues. The 1-(2-phenoxyethyl)pyrrolidine subunit projects towards the solvent-exposed front of the binding pocket.[4]

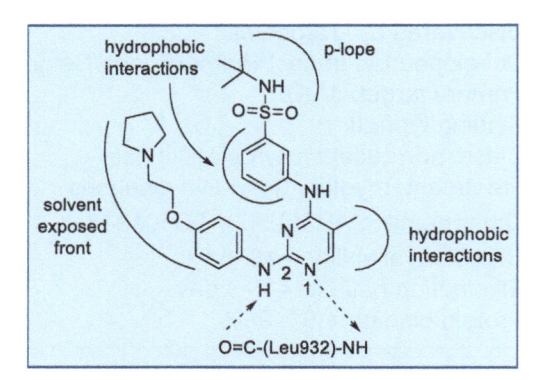

Figure 2. Fedratinib –JAK2 interactions.

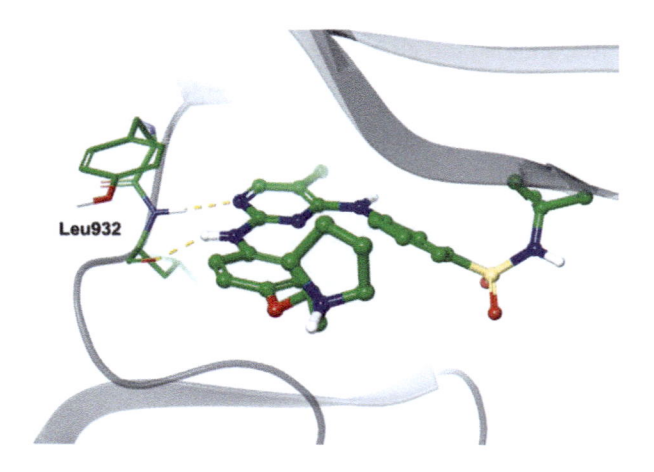

Figure 3. Co-crystal structure of fedratinib with JAK2 (PDB ID: 6VNE).

Fedratinib pharmacokinetics are characterized by prolonged duration of action: effective half-life of 41 hours, terminal half-life of approximately 114 hours, and elimination half-life of 2–3 days, which makes it suitable for once-daily dosing. In contrast, ruxolitinib has a short terminal half-life of approximately 3 h, suggesting that twice-daily administration of ruxolitinib is more suitable than once-daily dosing.[5]

Following oral administration, fedratinib is mainly biooxidized by CYP3A4 to give two major circulating metabolites in the plasma: the pyrrolidone derivative **1** and the *N*-butyric acid **2** (Fig. 4).[1]

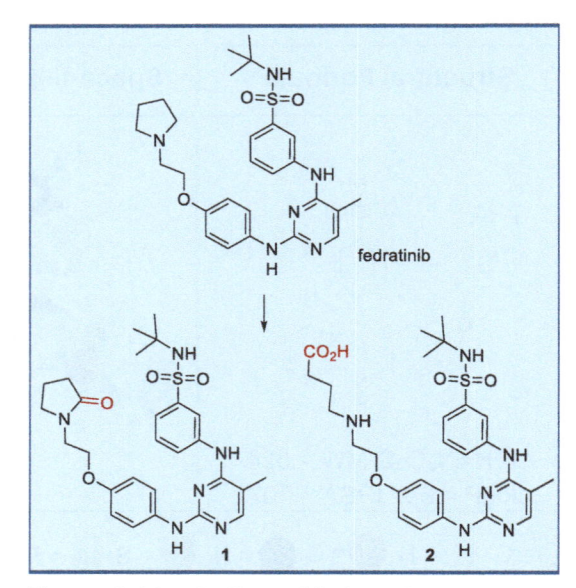

Figure 4. Metabolism of fedratinib in humans.

1. M. Talpaz, J. J. Kiladjian. Fedratinib, a newly approved treatment for patients with myeloproliferative neoplasm-associated myelofibrosis. *Leukemia.* **2021**, *35*, 1–17.

2. M. Kesarwani, E. Huber, Z. Kincaid C. R. Evelyn, J. Biesiada, M. Rance, M. B. Thapa, N. P. Shah, J. Meller, Y. Zheng, M. Azam. Targeting substrate-site in Jak2 kinase prevents emergence of genetic resistance. *Sci. Rep.* **2015**, *5*, 14538.

3. J. S. Tokarski, A. Zupa-Fernandez, J. A. Tredup, K. Pike, C. Chang, D. Xie, L. Cheng, D. Pedicord, J. Muckelbauer, S. R. Johnson, S. Wu, S. C. Edavettal, Y. Hong, M. R. Witmer, L. L. Elkin, Y. Blat, W. J. Pitts, D. S. Weinstein, J. R. Burke. Tyrosine kinase 2-mediated signal transduction in T lymphocytes is blocked by pharmacological stabilization of Its pseudokinase domain. *J. Biol. Chem.* **2015**, *290*, 11061–11074.

4. R. R. Davis, B. Li, S. Yun, A. Chan, P. Nareddy, S. Gunawan, M. Ayaz, H. R. Lawrence, G. W. Reuther, N. J. Lawrence, E. Schönbrunn. Structural insights into JAK2 inhibition by ruxolitinib, fedratinib, and derivatives thereof. *J. Med. Chem.* **2021**, *64*, 2228–2241.

5. C. N. Harrison, N. Schaap, A. M. Vannucchi, J. J. Kiladjian, E. Jourdan, R. T. Silver, H. C. Schouten, F. Passamonti, S. Zweegman, M. Talpaz, S. Verstovsek, S. Rose, J. Shen, T. Berry, C. Brownstein, R. A. Mesa. Fedratinib in patients with myelofibrosis previously treated with ruxolitinib: An updated analysis of the JAKARTA2 study using stringent criteria for ruxolitinib failure. *Am. J Hematol.* **2020**, *95*, 594–603.

7.10 Pacritinib (Vonjo™) (45)

Structural Formula	Space-filling Model	Brief Information
$C_{28}H_{32}O_3N_4$; MW = 472 clogP = 4.2; tPSA = 68		Year of discovery: 2005 Year of introduction: 2022 Discovered by: Capricorn Developed by: CTI BioPharma Primary targets: JAK2/FLT3 Binding type: I Class: non-receptor tyrosine kinase (JAK3) Treatment: myelofibrosis Other name: SB1518 Oral bioavailability = not reported Elimination half-life = 40 h Protein binding = 99%

● = C ○ = H ● = O ● = N ● = S ● = F ● = Cl ● = Br ● = I

Pacritinib received accelerated approval in 2022 for the treatment of adults with intermediate or high-risk primary or secondary (post-polycythemia vera or post-essential thrombocythemia) myelofibrosis with a platelet count below 50 × 10⁹/L, which affects 26% of all myelofibrosis patients (~20,000 in the United States).

Myelofibrosis is a rare hematological malignancy characterized by splenomegaly and debilitating symptoms. Currently, there are two first-line therapy options for myelofibrosis: JAK inhibitors ruxolitinib (Incyte), and fedratinib (Biomedicine). BioPharma's pacritinib became the third JAK inhibitor to reach the myelofibrosis market. It is indicated for the treatment of patients with cytopenic myelofibrosis who are ineligible for the other two JAK inhibitors. It brings hope to this critically underserved patient population, which previously had no approved treatment options.

Unlike several other JAK2 inhibitors, pacritinib exhibits good selectivity (~60-fold) for JAK2 over JAK1, while also inhibiting JAK3 with about 25-fold lower potency than JAK2 (Table 1). Pacritinib exhibits equipotent activity against FLT3 (IC$_{50}$ = 22 nM) and mutant FLT3-D835Y (IC$_{50}$ = 6 nM) at the same potency range as JAK2 (JAK2 IC$_{50}$ = 23 nM).[1] Because of its FLT3 activity, pacritinib is undergoing clinic trials for the treatment of acute myeloid leukemia (AML).

Table 1. IC$_{50}$'s (nM) of JAK inhibitors.

kinase	JAK1	JAK2	JAK3	TYK2	FLT3
pacritinib[1]	1280	23	520	50	22
ruxotinib[2]	6	9	487	30	NA
fedratinib[3]	105	3	1002	405	15

Pacritinib originated from compound **1**, which was identified through screening of a Capricorn library. This lead compound showed broad kinase inhibition with modest cellular potency; however, the ubiquitous pyrimidine-based motif has been claimed in numerous issued patents and patent applications. To develop novel analogs with strong proprietary position, a series of small-molecule macrocycles was prepared using ring-closing metathesis (RCM) reaction (Fig. 1). This effort led to pacritinib as a potent JAK2/FLT3 inhibitor.[4,5]

The development of pacritinib suffered a serious setback in 2016 due to a full clinical hold which resulted from concerns over excess deaths and cardiac and hemorrhagic complications in the preliminary clinical trials. After a careful review of the data by the FDA, the hold was revoked in 2017, allowing further use. It now appears that a relatively high dose of pacritinib (200 mg, twice daily) elicits favorable efficacy without reducing platelet count in the high risk myelofibrosis population suffering from cytopenic myelofibrosis.[6]

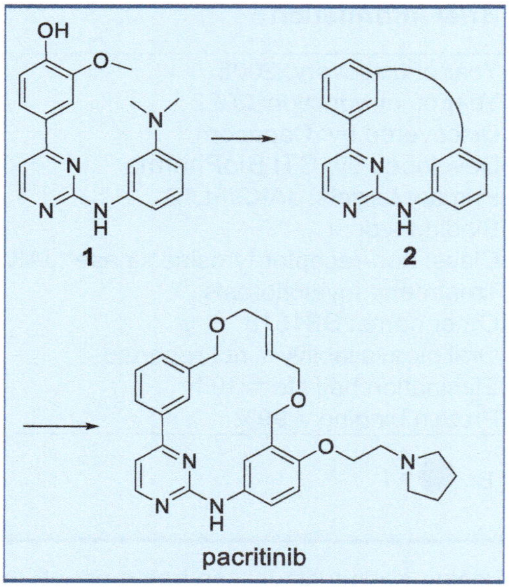

Figure 1. Structural optimization of pacritinib.

Pacritinib is a type I JAK2 inhibitor, preferentially binding to an activated form of JAK2. However, its exact binding mode has not yet been determined due to lack of X-ray co-crystal structures with and of the kinases.[4]

1. S. Hart, K. C. Goh, V. Novotny-Diermayr, Y. C. Tan, B. Madan, C. Amalini, L. C. Ong, B. Kheng, A. Cheong, J. Zhou, W. J. Chng, J. M. Wood. Pacritinib (SB1518), a JAK2/FLT3 inhibitor for the treatment of acute myeloid leukemia. *Blood Cancer J.* **2011**, *1*, e44.

2. S. T. Wrobleski, R. Moslin, S. Lin, Y. Zhang, S. Spergel, J. Kempson, J. S. Tokarski, J. Strnad, A. Zupa-Fernandez, L. Cheng, D. Shuster, K. Gillooly, X. Yang, E. Heimrich, K. W. McIntyre, C. Chaudhry, J. Khan, M. Ruzanov, J. Tredup, D. Mulligan, D. S. Weinstein. Highly selective inhibition of tyrosine kinase 2 (TYK2) for the treatment of autoimmune diseases: Discovery of the allosteric inhibitor BMS-986165. *J. Med. Chem.* **2019**, *62*, 8973–8995.

3. M. Talpaz, J. J. Kiladjian. Fedratinib, a newly approved treatment for patients with myeloproliferative neoplasm-associated myelofibrosis. *Leukemia.* **2021**, *35*, 1–17.

4. A. D. William, A. C. Lee, S. Blanchard, A. Poulsen, E. L. Teo, H. Nagaraj, E. Tan, D. Chen, M. Williams, E. T. Sun, K. C. Goh, W. C. Ong, S. K. Goh, S. Hart, R. Jayaraman, M. K. Pasha, K. Ethirajulu, J. W. Wood, B. W. Dymock. Discovery of the macrocycle 11-(2-pyrrolidin-1-yl-ethoxy)-14,19-dioxa-5,7,26-triaza-tetracyclo[19.3.1.1(2,6).1(8,12)]heptacosa-1(25),2(26),3,5,8,10,12(27),16,21,23-decaene (SB1518), a potent Janus kinase 2/fms-like tyrosine kinase-3 (JAK2/FLT3) inhibitor for the treatment of myelofibrosis and lymphoma. *J. Med. Chem.* **2011**, *54*, 4638–4658.

5. J. Y. Jeon, Q. Zhao, D. R. Buelow, M. Phelps, A. R. Walker, A. S. Mims, S. Vasu, G. Behbehani, J. Blachly, W. Blum, R. B. Klisovic, J. C. Byrd, R. Garzon, S. D. Baker, B. Bhatnagar. (Preclinical activity and a pilot phase I study of pacritinib, an oral JAK2/FLT3 inhibitor, and chemotherapy in FLT3-ITD-positive AML. *Investig. New Drugs.* **2020**, *38*, 340–349.

6. A. B. Duenas-Perez, A. J. Mead. Clinical potential of pacritinib in the treatment of myelofibrosis. *Ther. Adv. Hematol.* **2015**, *6*, 186–201.

7.11 Ritlecitinib (46)

Structural Formula	Space-filling Model	Brief Information
$C_{15}H_{19}N_5O$; MW = 285 clogP = 2.3; tPSA = 70		Year of discovery: 2013 Status: phase III Discovered by: Pfizer Developed by: Pfizer Primary targets: JAK3/TEC Binding type: covalent Class: non-receptor tyrosine kinase Treatment: alopecia areata (phase III) Other name: PF-06651600 Protein binding = 14%

● = C ○ = H ● = O ● = N ● = S ● = F ● = Cl ● = Br ● = I

Ritlecitinib, a dual JAK3/TEC inhibitor, had been granted FDA breakthrough status for the treatment of alopecia areata (autoimmune-induced hair loss) in September 2018. Promising phase 3 trial results in alopecia areata were reported by Pfizer in August 2021, placing it in competition with Lilly/Incyte's Olumiant. It is also in clinical studies for the treatment of vitiligo, rheumatoid arthritis, Crohn's disease and ulcerative colitis.

Background

JAK3 plays an important role in cytokine signaling associated with either T-cell proliferation, differentiation, or development. It is also crucial in the development of B-cells and NK-cells. Inhibitors of JAK3 can potentially act as powerful immunosuppressants for the treatment of autoimmune diseases such as rheumatoid arthritis and psoriasis.[1] Highly selective inhibition of JAK3 may minimize adverse effects and improve tolerability. However, the design of selective JAK3 inhibitors has been very challenging because only two unique residues of JAK3 could be targeted for selectivity. In addition, the JAK family shows only subtle conformational differences among isoforms. Fortunately, JAK3 has a unique cysteine residue at the gatekeeper position, which is replaced by a serine residue in the other JAK isoforms. This residue, Cys909 in human JAK3, is reminiscent of the cysteine residues in epidermal growth factor receptor (EGFR) and Bruton's tyrosine kinase (BTK) that have been successfully utilized in the design of useful covalent kinase inhibitors. The irreversible inhibition of JAK3 by covalent tagging can be advantageous because of the prolonged pharmacodynamic effect after target inactivation, provided potential liabilities such as off-target reactions can be avoided.[2]

Discovery of Ritlecitinib[2,3]

The design of a covalent JAK3 inhibitor to intercept Cys909 was inspired by the co-crystal structure of tofacitinib (Fig. 1), in which the amide substituent of the piperdine projected away from Cys909.[2] If an electrophilic warhead is attached to the piperdine nitrogen, the pyrimidyl amine N–C bond needs to undergo rotation by approximately 180°C to bring the electrophilic warhead into close proximity with Cys909. Attachment of an acryloyl group at the piperidine nitrogen led to compound **1**, which exhibited only modest inhibition of JAK3 and also poor JAK isoform selectivity, indicating no covalent interactions. The co-crystal structure of **1** with JAK3 showed that the piperidine subunit was not correctly positioned, although the pyrrolopyrimidine core was located in the hinge active site. The required binding mode could be accessed by removal of both methyl groups, the N-methyl and the methyl attached to the piperdine ring to produce a low-energy conformation having the reactive acrylamide in close proximity to Cys909. Indeed, the desmethyl **2** formed a covalent adduct with JAK3 (Fig. 2), resulting in a significant improvement in JAK3 potency (IC$_{50}$ = 56 nM, [ATP] = 1 μM) along with excellent selectivity against other JAK isoforms (Fig. 3).

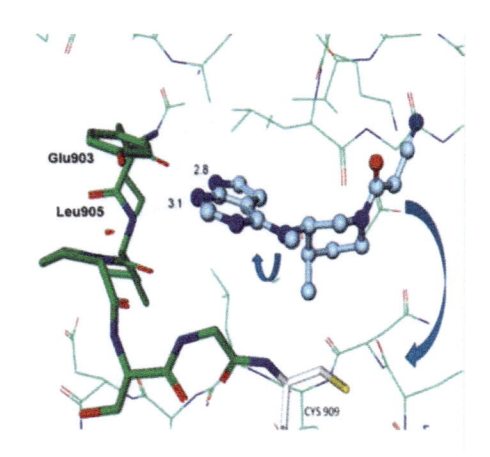

Figure 1. Crystal structure of tofacitinib–JAK3 illustrating rotation of the N–C axis would project amide close to Cys909. Adapted with permission from *J. Med. Chem.* **2017**, *60*, 1971–1993.

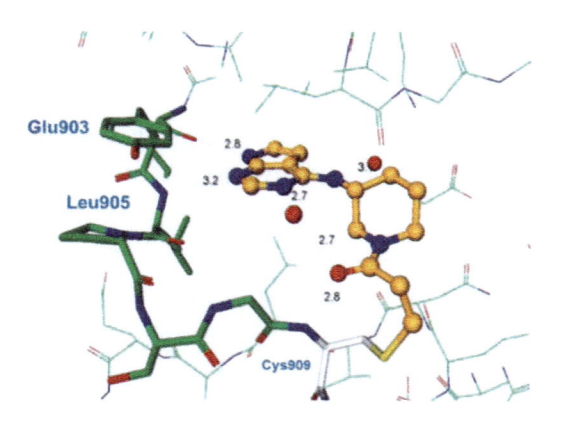

Figure 2. Crystal structure of **2** in JAK3 with a covalent bond to Cys909. Adapted with permission from *J. Med. Chem.* **2017**, *60*, 1971–1993.

Compound **2** exhibited favorable metabolic stability in rat liver microsomes, but its PK properties in rats were moderate to poor presumably due to glutathione-*S*-transferase (GST) mediated glutathione (GSH) conjugate addition to the acrylamide moiety. When tested in a whole blood stability assay, an in vitro surrogate measure of whole body GST mediated clearance, compound **2** showed $t_{1/2}$ values of 98 and 332 min in rat and human whole blood, respectively. Deploying substituents such as methyl to the acrylamide olefin moiety resulted in a substantial reduction in GST-mediated clearance, but unfortunately JAK3 activity was substantially compromised. Fortunately, the introduction of a methyl group in proximity to the acryloyl side chain on the piperdine ring to modulate GST binding proved effective and led to the discovery of ritlecitinib, which showed superior whole blood stability ($t_{1/2}$ >360 min in human blood vs. 332 min for **2**, 161 min in rat blood vs. 98 min for **2**). Incubation with metabolically viable human hepatocytes showed 98.4% of radiolabeled ritlecitinib remained free, indicating that only a small percentage of the daily dose would undergo covalent conjugation with proteins in hepatocytes.[3] Interestingly, this remote methyl substitution also resulted in 2-fold improvement in both JAK3 potency (JAK3 IC$_{50}$ = 33 vs. 56 nM for **2**) and passive membrane permeability in the Ralph Russ canine kidney cell (RRCK) assay (P$_c$ = 15 × 10^{-6} cm/s vs. 8 × 10^{-6} cm/s for **2**).[2]

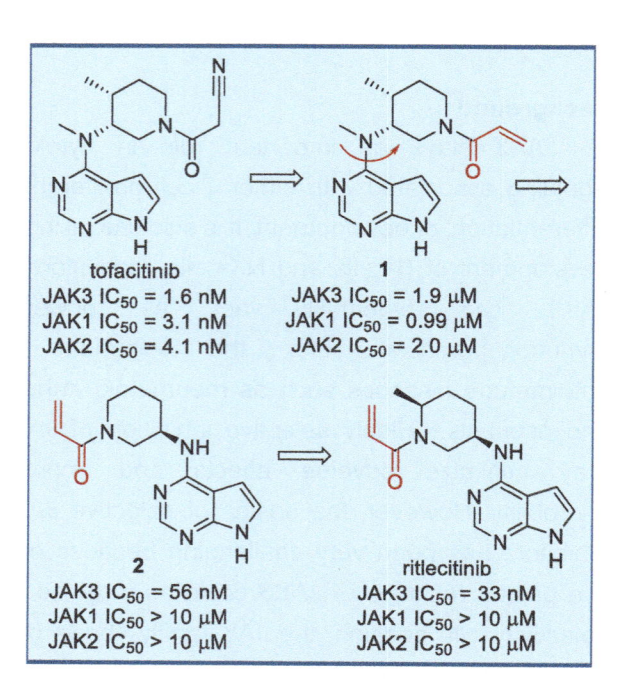

Figure 3. Summary of the chemical optimization of ritlecitinib.

Ritlecitinib achieved its high JAK3 selectivity via covalent interaction with a unique cysteine residue

(Cys909) in the catalytic domain. However, there are 10 other kinases in the kinome bearing a cysteine at the equivalent position of Cys909 to JAK3. Five of those kinases are within the TEC kinase family, which were also inhibited by ritlecitinib.[4] On the basis of preclinical data, the TEC inhibition also contributed to inhibition of the cytolytic function of CD8+ T cells and NK cells. Thus, the dual inhibition JAK3 and the TEC kinase family may contribute to favorable therapeutic effects.

Other potent, selective JAK3 covalent inhibitors include aryl aminoacrylamide **3**[2] and piperidine analog RB1 (Fig. 4).[5] Compound **3** was less favorable than aliphatic aminoacrylamides such as ritlecitinib due to a greater susceptibility to glutathione addition. RB1 showed good oral bioavailability (73%) and long half-life (15 h) in rats, and it showed efficacy in a mouse collagen-induced arthritis model. However, no further development of RB1 has been reported.

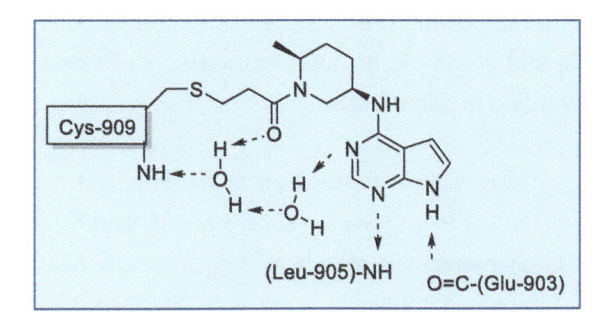

Figure 5. Summary of ritlecitinib–JAK3 interactions based on X-ray co-crystal structure.

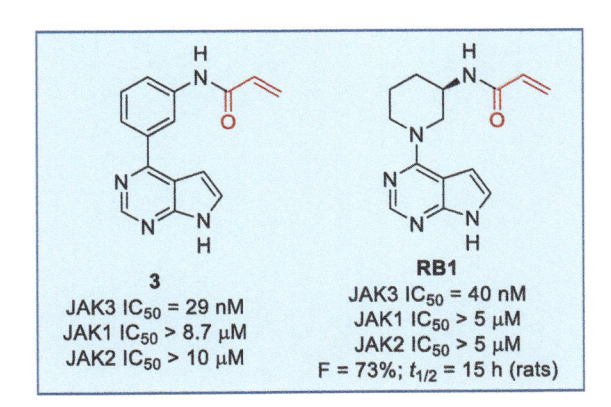

Figure 4. Two other JAK3 covalent inhibitors.

Binding Mode[2,3]

Ritlecitinib forms two direct hydrogen bonds between the pyrrolopyrimidine moiety and amide carbonyl oxygen of Glu903 and amide NH of Leu905 of the hinge region (Figs 5, 6).[3] The thiol of Cys909 adds to acrylamide unsaturated double bond to result in a covalent bond between the inhibitor and JAK3. The N3 nitrogen interacts with the amide carbonyl oxygen and the NH of the Cys909 through a network of hydrogen bonds via two well-ordered water molecules.

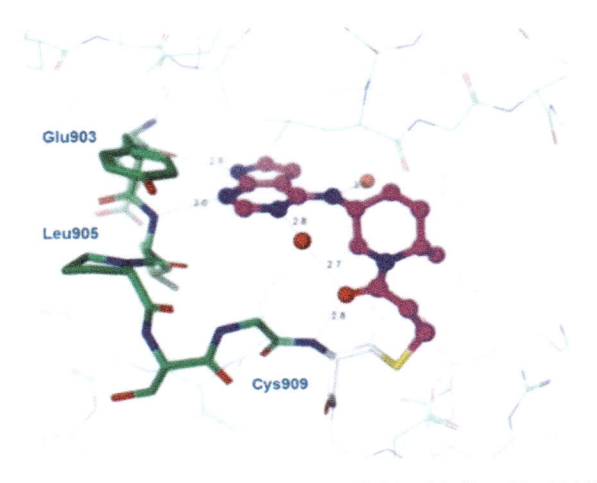

Figure 6. Co-crystal structure of ritlecitinib with JAK3. The hinge residues (JAK3 residues 903–908) are shown as green thick sticks, with Cys909 as white sticks. Adapted with permission from *ACS Chem. Biol.* **2016**, *11*, 3442–3451.

PK Properties and Metabolism[2]

Ritlecitinib exhibited low clearance in rats and dogs, which contributed to its good oral bioavailability (85% in rats and 109% in dogs). Its lower oral bioavailability in monkeys (56%) primarily resulted from its fast clearance. The pharmacokinetic properties are directly linked to its stability in hepatocytes and whole blood, demonstrating the power of in vitro assays to predict in vivo pharmacokinetic parameters. The good oral bioavailability obtained from rats and dogs was a combination of high passive permeability, high aqueous solubility (>2 mg/mL), and low hepatic clearance. Ritlecitinib underwent CYP3A4 mediated oxidation of the acrylamide and piperdine ring as well as glutathione conjugation with the acrylamide.

Allometric scaling predicted an oral bioavailability of 90% and a half-life of 2 h in humans, which need to be verified in clinic trials.

1. L. Tan, K. Akahane, R. McNally, K. M. Reyskens, S, B. Ficarro, S. Liu, G. S. Herter-Sprie, S. Koyama, M. J. Pattison, K. Labella, L. Johannessen, E. A. Akbay, K. K. Wong, D. A. Frank, J. A. Marto, T. A. Look, J. S. Arthur, M. J. Eck, N. S. Gray. Development of selective covalent janus kinase 3 inhibitors. *J. Med. Chem.* **2015**, *58*, 6589–6606.
2. A. Thorarensen, M. E. Dowty, M. E. Banker, B. Juba, J. Jussif, T. Lin, F. Vincent, R. M. Czerwinski, A. Casimiro-Garcia, R. Unwalla, J. I. Trujillo, S. Liang, P. Balbo, Y. Che, A. M. Gilbert, M. F. Brown, M. Hayward, J. Montgomery, L. Leung, X. Yang, J.B. Telliez. Design of a janus kinase 3 (JAK3) specific inhibitor 1-((2S,5R)-5-((7H-pyrrolo[2,3-d]pyrimidin-4-yl)amino)-2-methylpiperidin-1-yl)prop-2-en-1-one (PF-06651600) allowing for the interrogation of JAK3 signaling in humans. *J. Med. Chem.* **2017**, *60*, 1971–1993.
3. J. B. Telliez, M. E. Dowty, L. Wang, J. Jussif, T. Lin, L. Li, E. Moy, P. Balbo, W. Li, Y. Zhao, K. Crouse, C. Dickinson, P. Symanowicz, M. Hegen, M. E. Banker, F. Vincent, R. Unwalla, S. iang, A. M. Gilbert, M. F. Brown, A. Thorarensen. Discovery of a JAK3-selective inhibitor: Functional differentiation of JAK3-selective inhibition over pan-JAK or JAK1-selective inhibition. *ACS Chem. Biol.* **2016**, *11*, 3442–3451.
4. H. Xu, M. I. Jesson, U. I. Seneviratne, T. H. Lin, M. N. Sharif, L. Xue, C. Nguyen, R. A. Everley, J. I. Trujillo, D. S. Johnson, G. R. Point, A. Thorarensen, I. Kilty, J. B. Telliez. PF-06651600, a dual JAK3/TEC family kinase inhibitor. *ACS Chem. Biol.* **2019**, *14*, 1235–1242.
5. H. Pei, L. He, M. Shao, Z. Yang, Y. Ran, D. Li, Y. Zhou, M. Tang, T. Wang, Y. Gong, X. Chen, S. Yang, M. Xiang, L. Chen. Discovery of a highly selective JAK3 inhibitor for the treatment of rheumatoid arthritis. *Sci. Rep.* **2018**, *8*, 5273.

7.12 Brepocitinib (47)

Structural Formula	Space-filling Model	Brief Information
$C_{18}H_{21}F_2N_7O$; MW = 389 logD = 1.6 (pH 7.4); tPSA = 72		Year of discovery: 2015 Current status: phase II Discovered by: Pfizer Outlicensed by: Pfizer in 2021 Primary targets: JAK1/TYK2 Binding type: II Class: non-receptor tyrosine kinase Treatment: immunological disorders Other name: PF-06700841 Oral bioavailability = not reported Elimination half-life = 3.8–7.5 h Protein binding = 39%

● = C   ○ = H   ● = O   ● = N   ● = S   ● = F   ● = Cl   ● = Br   ● = I

Brepocitinib is a dual TYK2/JAK1 inhibitor that had been in phase II clinic trials by Pfizer for a variety of immunological disorders. On November 2, 2021, Pfizer announced a decision to outlicense brepocitinib and ropsacitinib (a relatively selective TYK2 inhibitor) to another company.

Discovery of Brepocitinib[1]

The lead compound for a Pfizer TYK2/JAK1 dual inhibitor program started with compound **1**, which was selective against JAK2 and JAK3 when tested under the pseudophysiological conditions with [ATP] = 1 μM (Fig. 1). Under these conditions, **1** showed weak inhibition of TYK2 (TYK2 IC$_{50}$ = 774 nM), and weaker activity against JAK1, JAK2, and JAK3 (JAK1 IC$_{50}$ = 3.7 μM, JAK2 and JAK3 IC$_{50}$'s > 10 μM). Comprehensive SAR studies were undertaken primarily in three regions of the lead compound: the bridged piperazine moiety, the solvent exposed aromatic motif, and the amide substituent that interacts with the glycine-rich P-loop of the kinase, leading to brepocitinib with a balanced TYK2 and JAK1 inhibition profile (TYK2 and JAK1 IC$_{50}$ = 23 and 17 nM, respectively). This compound is modestly selective against JAK2 (IC$_{50}$ = 77 nM) but highly selective against JAK3 (IC$_{50}$ = 6.5 μM) (Table 1).

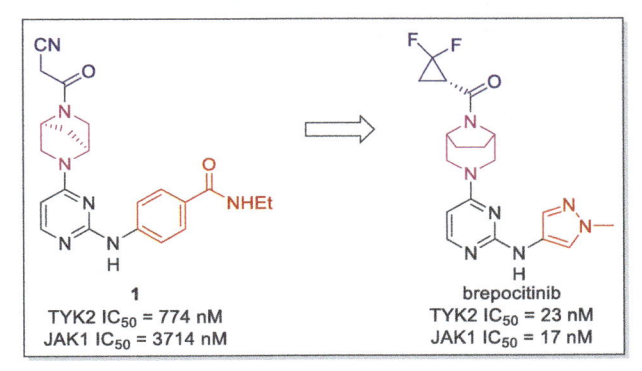

Figure 1. Structural optimization of brepocitinib.

Table 1. Selectivity of brepocitinib vs. tofacitinib[1]

inhibitor	brepocitinib	tofacitinib
JAK1 IC$_{50}$	17 nM	15 nM
JAK2 IC$_{50}$	77 nM	77 nM
JAK3 IC$_{50}$	6.5 μM	55 nM
TYK2 IC$_{50}$	23 nM	0.49 μM

Binding Mode[1,2]

In the complex with TYK2 (Figs. 2, 4), brepocitinib is packed against the hinge, forming two classic donor and acceptor H-bonds between the aminopyrimidine and the amide NH and amide carbonyl oxygen of Val981. The ethylene bridge of the [3.2.1]-diazabicycle fits into the lower hydrophobic portion of the binding site, and the difluoromethylene of the cyclopropyl amide extends into the P-loop of the ATP-binding site.

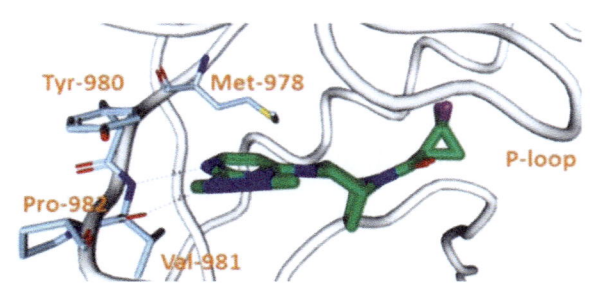

Figure 2. Co-crystal structure of brepocitinib–TYK2.

Brepocitinib binds to JAK1 in approximately the same manner as TYK2 (Figs. 3, 4), consistent with the comparable IC$_{50}$'s for both isoforms (IC$_{50}$ = 23 nM for TYK2 ; 17 nM for JAK1).

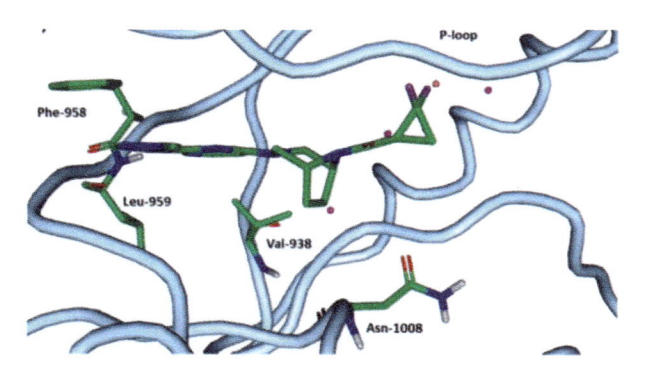

Figure 3. Co-crystal structure of brepocitinib with JAK1. Adapted with permission from *J. Med. Chem.* **2018**, *61*, 8597–8612 (Figs. 2, 3).

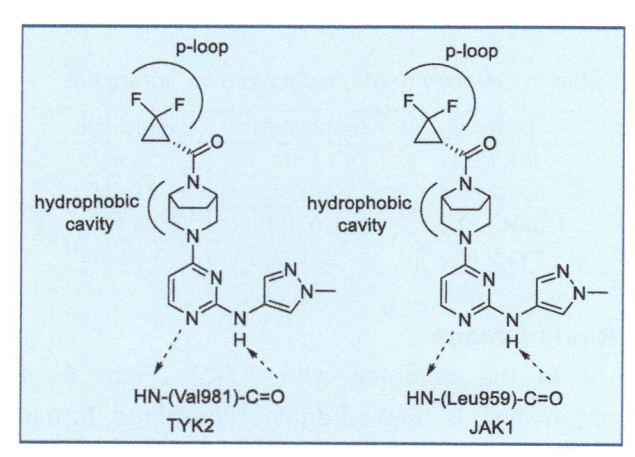

Figure 4. Interactions of brepocitinib with TYK2 (left) and JAK1 (right).

Brepocitinib exhibited comparable potency against JAK1 and TYK2, whereas baricitinib, ruxolitinib, and tofacitinib showed substantially reduced TYK2 inhibition (Table 2). This can be rationalized by two different TYK2 binding modes (Fig. 5).[2] Examination of the back pockets of all JAK isoforms revealed Ile960 in TYK2 and a valine in the other isoforms. This unique amino acid variation may account for different activity of TYK2 from other isoforms. Overlaying of the pyrrolopyridimidine hinge binder in baricitinib, ruxolitinib and tofacitinib shows a typical two-point hydrogen bond donor and acceptor interaction with main chain atoms of Glu979 and Val981 in the hinge region (Fig. 6). However, this binding arrangement causes steric repulsion between the pyrrolopyridimidine ring and the larger side chain of Ile960 in TYK2, resulting in a decrease in TYK2 inhibition. In contrast, the aminopyrimidine hinge-binding group of brepocitinib does not interact with the carbonyl oxygen of Glu969; instead, it forms two direct hydrogen bonds with main chain atoms of Val981, which resolves the clash with TYK2 Ile960. As a result brepocitinib has good TYK2 potency.

Table 2. Selectivity data of brepocitinib vs. other JAK inhibitors[1,2]

inhibitor	JAK1	TYK2
brepocitinib	17 nM	23 nM
baricitinib	4 nM	61 nM
ruxolitinib	6 nM	30 nM
tofacitinib	23 nM	0.49 μM

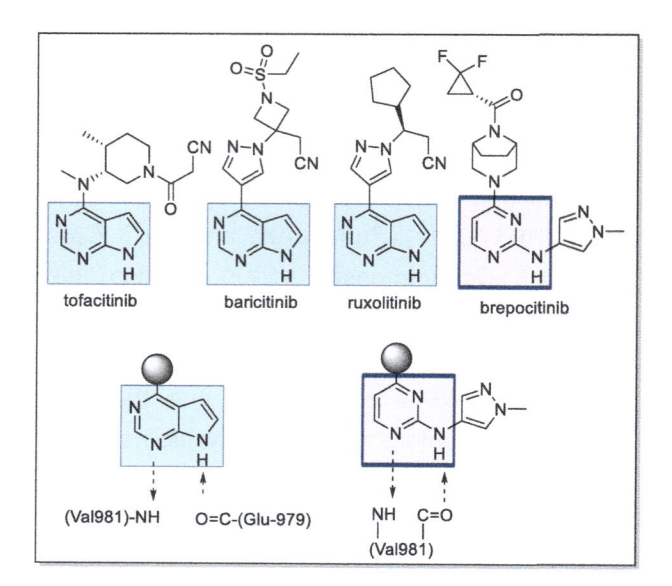

Figure 5. Two different TYK2 hinge-binding modes.

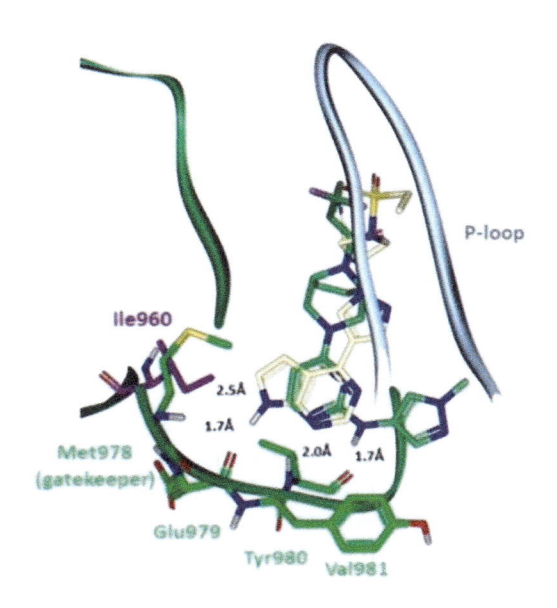

Figure 6. Two different TYK2 binding modes: modeled baricitinib overlayed with brepocitinib-TYK2 complex. Adapted with permission from *J. Med. Chem.* **2020**, *63*, 13561−13577.

PK Properties and Metabolism[1,3,4]

The *p*-toluenesulfonic acid salt of brepocitinib exhibits good aqueous solubility (4.84 mg/mL at pH 7.64; >7 mg/mL in simulated gastric fluids). This compound also displayed high passive membrane permeability (mean P_{APP} = 18.8 × 10^{-6} cm/s as determined using Ralph Russ canine kidney cells (RRCK)). The high oral bioavailability of brepocitinib obtained from rats (83%) is consistent with its high passive permeability and good solubility.

The elimination half-life of brepocitinib ranged from 3.8 to 7.5 h after a single oral dose and from 4.9 to 10.7 h after multiple-dose administration. It was eliminated from the body by CYP450 mediated hepatic metabolism (84%) (mainly via CYP3A4) and renal clearance (16%). Oxidation of the *N*-methyl pyrazole (**2**) is the major metabolic pathway, followed by N-demethylation (**3**) and N-dealkylation with loss of pyrazole (**4**) (Fig. 7).

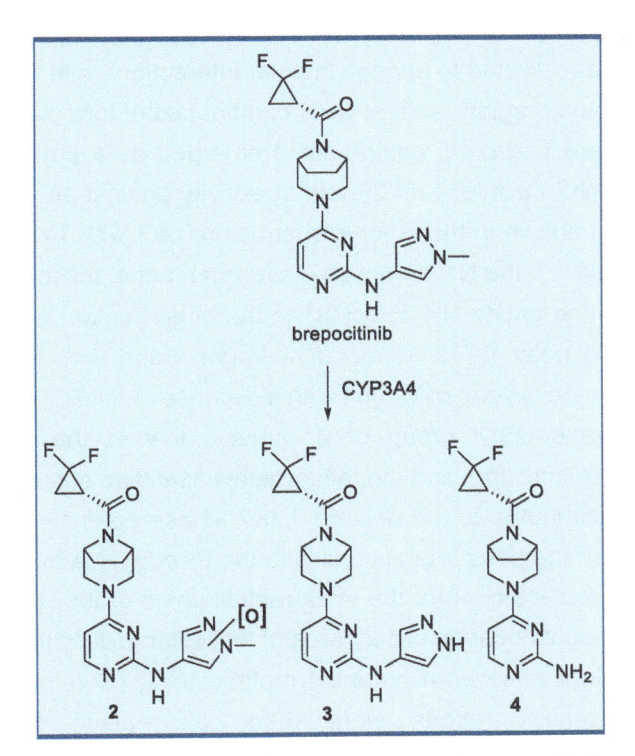

Figure 7. Major metabolic pathways of brepocitinib.

1. A. Fensome, C. M. Ambler, E. Arnold, M. E. Banker, M. F. Brown, J. Chrencik, J. D. Clark, M. E. Dowty, I. V. Efremov, A. Flick, B. S. Gerstenberger, A. Gopalsamy, M. M. Hayward, M. Hegen, B. D. Hollingshead, J. Jussif, J. D. Knafels, D. C. Limburg, D. Lin, T. H. Lin, L. Zhang. Dual inhibition of TYK2 and JAK1 for the treatment of autoimmune diseases: Discovery of ((S)-2,2-difluorocyclopropyl)((1R,5S)-3-(2-((1-methyl-1H-pyrazol-4-yl)amino)pyrimidin-4-yl)-3,8-diazabicyclo[3.2.1]octan-8-yl)methanone (PF-06700841). *J. Med. Chem.* **2018**, *61*, 8597–8612.

2. B. S. Gerstenberger, C. Ambler, E. P. Arnold, M. E. Banker, M. F. Brown, J. D. Clark, A. Dermenci, M. E. Dowty, A. Fensome, S. Fish, M. M. Hayward, M. Hegen, B. D. Hollingshead, J. D. Knafels, D. W. Lin, D. R. Owen, E. Saiah, R. Sharma, F. F. Vajdos, S. W. Wright. Discovery of tyrosine Kinase 2 (TYK2) inhibitor (PF-06826647) for the treatment of autoimmune diseases. *J. Med. Chem.* **2020**, *63*, 13561–13577.

3. C. Banfield, M. Scaramozza, W. Zhang, E. Kieras, K. M. Page, A. Fensome, M. Vincent, M. E. Dowty, K. Goteti, P. J. Winkle. The safety, tolerability, pharmacokinetics, and pharmacodynamics of a TYK2/JAK1 inhibitor (PF-06700841) in healthy subjects and patients with plaque psoriasis. *J. Clin. Pharmacol.* **2018**, *58*, 434–447.

4. P. Lefevre, N. Casteele. Clinical pharmacology of Janus kinase inhibitors in inflammatory bowel disease. *J. Crohns colitis.* **2010**, *14* (Suppl. 2), S725–S736.

7.13 Ropsacitinib (48)

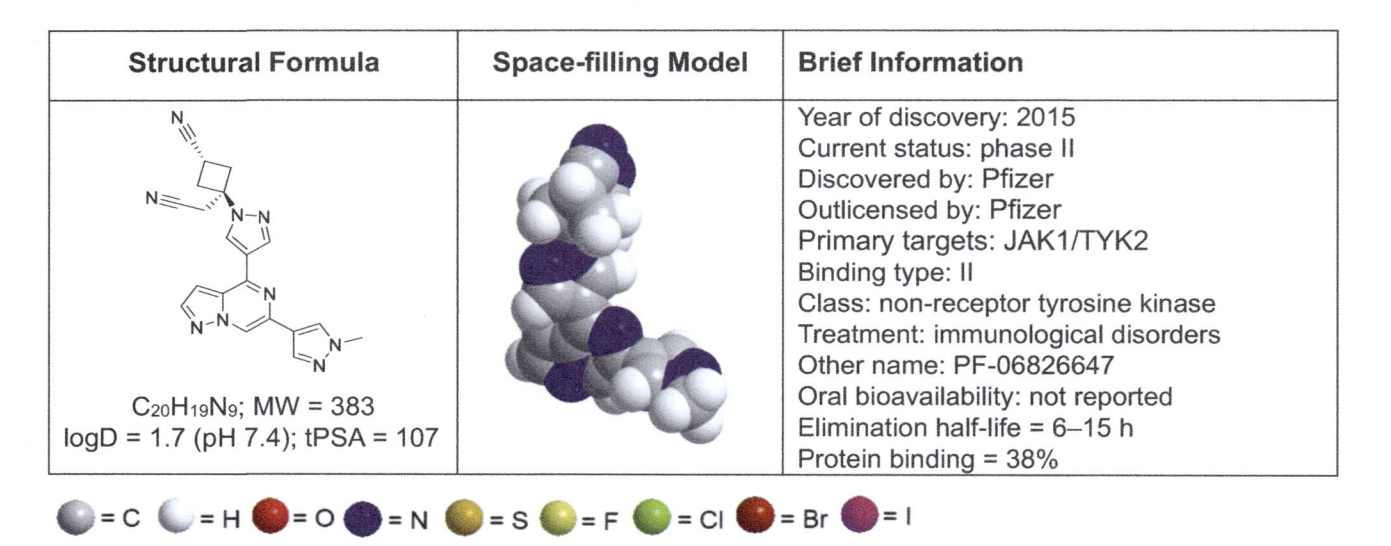

Structural Formula	Space-filling Model	Brief Information
$C_{20}H_{19}N_9$; MW = 383 logD = 1.7 (pH 7.4); tPSA = 107		Year of discovery: 2015 Current status: phase II Discovered by: Pfizer Outlicensed by: Pfizer Primary targets: JAK1/TYK2 Binding type: II Class: non-receptor tyrosine kinase Treatment: immunological disorders Other name: PF-06826647 Oral bioavailability: not reported Elimination half-life = 6–15 h Protein binding = 38%

● = C   ● = H   ● = O   ● = N   ● = S   ● = F   ● = Cl   ● = Br   ● = I

Ropsacitinib is a relatively selective TYK2 inhibitor that had been in phase II clinic trials by Pfizer for a variety of immunological disorders. On November 2, 2021, Pfizer announced a decision to outlicense it to another company.

Discovery of Ropsacitinib[1]

The discovery of ropsacitinib, a dual JAK1/TYK2 inhibitor, was guided by previous research and molecular modeling on brepocitinib that revealed distinct hinge binding to the main chain atoms of Val981.[2] This binding mode evades the steric clash with the larger side chain of Ile960, which is unique to the back pocket of TYK2 (equivalent to a valine for other isoforms). In contrast, the two-point donor/acceptor pattern of the hinge binding with Glu979 and Val981 as shown in baricitinib causes clash with the Ile960 residue, leading to reduced activity against TYK2 (Fig. 1). Thus, a hinge binder that does not interact with the amide oxygen of Glu969 would overcome the steric repulsion with the bulky residue of TYK2 Ile960. The aminopyrimidine core as in brepocitinib meets these criteria; however, neither TYK2 selectivity over JAK1 nor subnanomolar potency could be achieved with this hinge binder. To identify a new lead for selective and potent TYK2 inhibitors, several [5,6]-fused aromatic rings systems with a hydrogen bond acceptor and donor both to Val981 as shown in **1** were surveyed.

A methylpyrazole was chosen as the solvent-front group as in brepocitinib due to its steric bulk, which, in combination with the unsubstituted five-membered ring of the bicyclic system, would lock the hinge binder into the desired binding mode. A substituted trifluoroethylazetidinyl pyrazole group was selected to engage in polar interactions with the P-loop region. After several compounds of formula **2** were prepared, compound **3** emerged as a potent TYK2 inhibitor with 36-fold selectivity against JAK1. As shown in the co-crystal structure of **3** with TYK2 (Fig. 2), the N1 serves as a hydrogen bond acceptor to the amide NH of Val981 in the hinge region, and the polar C7−H forms a hydrogen bond with the amide oxygen of Val981 with a distance of 2.4 Å. The cyanomethyl group of **3** projects toward the C-terminal lobe, and the trifluoroethyl azetidine subunit contributes to the desired TYK2 kinase potency by forming polar interactions with the P-loop. However, a replacement for the trifluoroethylamino group was sought because of concerns of testicular toxicity that might arise from potential metabolites.[3] Therefore, extensive efforts were made to replace the trifluoroethylamino group of the azitidine ring, eventually leading to ropsacitinib which exhibited high JAK1 and JAK3 selectivity (>160-fold) and modest JAK2 selectivity (12-fold).

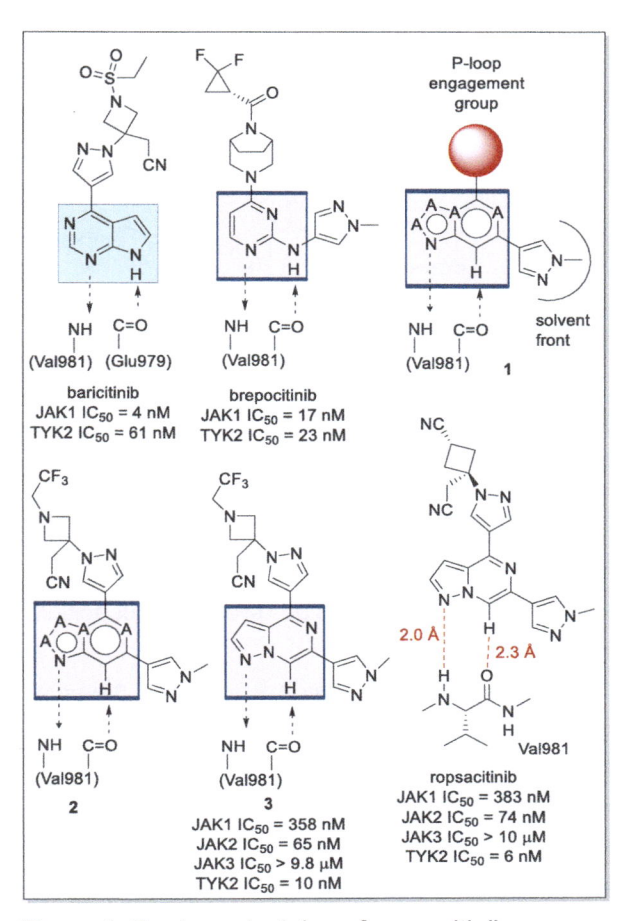

Figure 1. Design principles of ropsacitinib.

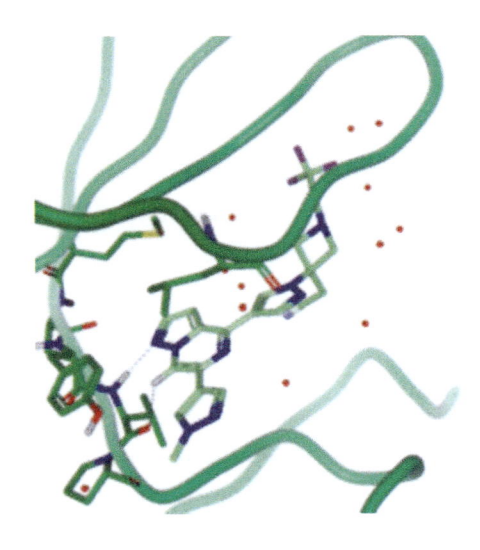

Figure 2. Co-crystal structure of **3** with TYK2. Adapted with permission from *J. Med. Chem.* **2020**, *63*, 13561–13577.

Binding Mode[1]

The co-crystal structure of ropsacitinib with TYK2 (Figs. 4, 5) shows the hinge-binding mode with N1 serving as a hydrogen bond donor to the amide NH of Val981 in the TYK2 hinge region, and the distance between C7−H and the carbonyl oxygen of Val981 is 2.3 Å, within van der Waals radii. The *cis* cyclobutane cyanomethyl subunit is oriented below and toward the tip of the P-loop. The potent TYK2 potency of ropsacitinib results from the optimized interactions with both the hinge and the P-loop (Fig. 3).

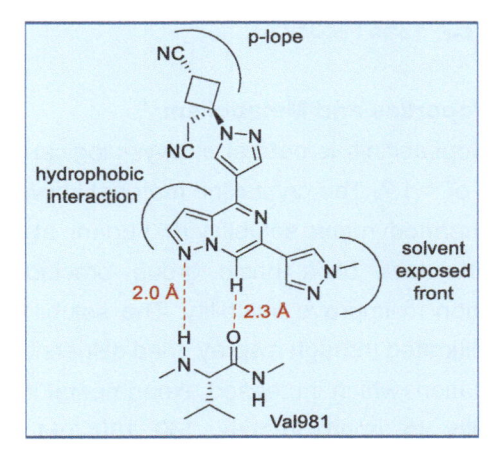

Figure 3. Summary of ropsacitinib–TYK2 interactions based on the co-crystal structure data.

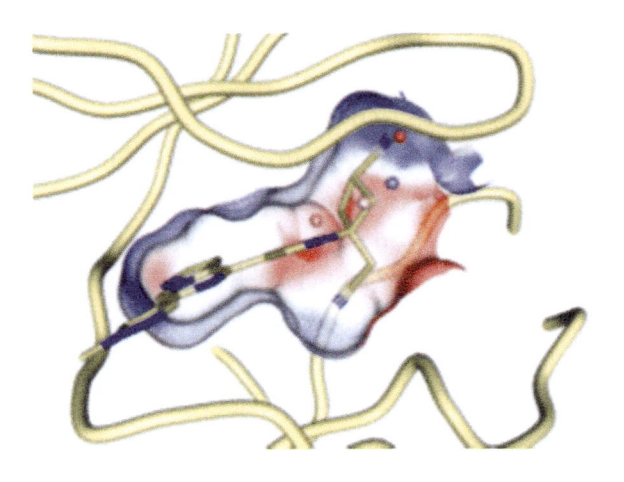

Figure 4. Co-crystal structure of ropsacitinib with TYK2. Adapted with permission from *J. Med. Chem.* **2020**, *63*, 13561–13577.

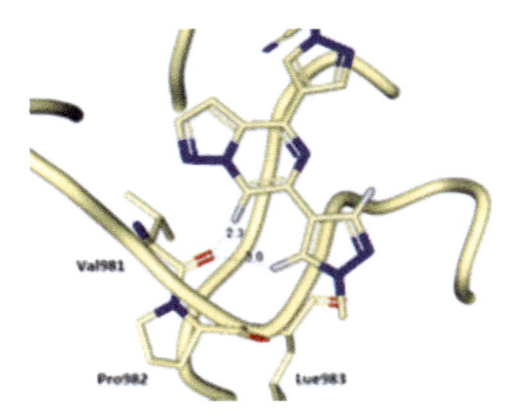

Figure 5. Top view of the hinge and solvent exposed region of the co-crystal structure of ropsacitinib with TYK2. Adapted with permission from *J. Med. Chem.* **2020**, *63*, 13561–13577.

PK Properties and Metabolism[4,5]

Ropsacitinib is neutral at physiological pH with a pKa of < 1.7. The crystalline material showed very poor thermodynamic solubility (0.3 μg/mL at pH 7.4), and the lack of a basic group precluded salt formation to improve solubility. The solubility issue was mitigated through a spray dried dispersion (SDD) formulation, which increased experimental intestinal solubility to approximately 160−180 μM (60−70 μg/mL). Ropsacitinib displayed high passive membrane permeability in MDCK-LE cells (mean P_{app} = 16.7 × 10^{-6} cm/s) but is a substrate for MDR1 (efflux ratio > 6) and BCRP (efflux ratio > 2) efflux transport. Following oral administration of the crystalline (non SDD) material to rats at 3 and 30 mpk, ropsacitinib showed an oral bioavailability of 15% and 8% respectively. The SDD formulation increased oral bioavailability to 58% and 63% at 3 and 30 mpk, respectively. The oral bioavailability in humans was expected to be dose-dependent due to poor solubility.

In healthy volunteers, the terminal elimination half-life of ropsacitinib in a single ascending dose (SAD) study is also dose-dependent: for example, approximately 6 hours at 10 mg dose and 15 hours at 400 mg dose.

Ropsacitinib underwent hepatic clearance in humans primarily through the CYP450-mediated (via CYP1A2, CYP2D6, and CYP3A) metabolism (Fig. 6), resulting in N-demethylation of the pyrazole (**6**), hydroxylation of the pyrazole (**5**), and addition of oxygen (e.g., **4**) and loss of the nitrile group (e.g., **7**).

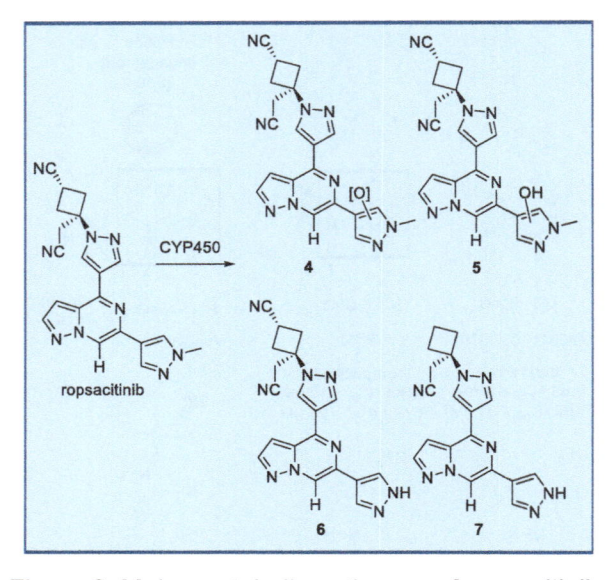

Figure 6. Major metabolic pathways of ropsacitinib.

1. B. S. Gerstenberger, C. Ambler, E. P. Arnold, M. E. Banker, M. F. Brown, J. D. Clark, A. Dermenci, M. E. Dowty, A. Fensome, S. Fish, M. M. Hayward, M. Hegen, B. D. Hollingshead, J. D. Knafels, D. W. Lin, D. R. Owen, E. Saiah, R. Sharma, F. F. Vajdos, S. W. Wright. Discovery of tyrosine kinase 2 (TYK2) inhibitor (PF-06826647) for the treatment of autoimmune diseases. *J. Med. Chem.* **2020**, *63*, 13561–13577.

2. A. Fensome, C. M. Ambler, E. Arnold, M. E. Banker, M. F. Brown, J. Chrencik, J. D. Clark, M. E. Dowty, I. V. Efremov, A. Flick, B. S. Gerstenberger, A. Gopalsamy, M. M. Hayward, M. Hegen, B. D. Hollingshead, J. Jussif, J. D. Knafels, D. C. Limburg, D. Lin, T. H. Lin, L. Zhang. Dualiinhibition of TYK2 and JAK1 for the treatment of autoimmune diseases: Discovery of ((S)-2,2-difluorocyclopropyl)((1R,5S)-3-(2-((1-methyl-1H-pyrazol-4-yl)amino)pyrimidin-4-yl)-3,8-diazabicyclo[3.2.1]octan-8-yl)methanone (PF-06700841). *J. Med. Chem.* **2018**, *61*, 8597–8612.

3. M. E. Dowty, G. Hu, F. Hua, F. B. Shilliday, H. V. Dowty. Drug design structural alert: Formation of trifluoroacetaldehyde through N-dealkylation is linked to testicular lesions in rat. *Int. J. Toxicol.* **2011**, *30*, 546−550.

4. P. Lefevre, P., N. Vande Casteele. Clinical Pharmacology of janus kinase inhibitors in inflammatory bowel disease. *J. Crohns colitis*. **2010**, *14* (Suppl. 2), S725–S736.

5. https://clinicaltrials.gov/ProvidedDocs/61/NCT03210961/Prot_0 00.pdf.

Chapter 8. Allosteric TYK2 Inhibitors

Psoriasis affects more than 7.5 million Americans aged 20 years or older, with plaque psoriasis being the most common type. For years, psoriasis has been treated with injectable biologic drugs (e.g., TNF inhibitors, IL-17 inhibitors, and IL-23 inhibitors). Pharmaceutical companies have long sought to replace them with orally bioavailable small-molecule therapies. While some JAK inhibitors have been evaluated in clinical studies, none have been granted an indication for the treatment of plaque psoriasis.[1,2] While being proven effective at treating several inflammatory diseases, their use has been restrained due to a black box warning of serious heart-related events, cancer, blood clots and death. These potential liabilities may be linked to insufficient isoform specificity, likely due to the high structural homology between JAK family members.

The JAK family is characterized by the presence of a canonical catalytic kinase domain and a pseudokinase domain, which have been termed JAK homology 1 (JH1) and JAK homology 2 (JH2), respectively. Because the catalytic domains of all JAK isoforms exhibit a high degree of homology, the discovery of highly selective JH1 inhibitors has been daunting. In fact, none of the FDA-approved JAK inhibitor drugs, all of which target JH1, are selective inhibitors against a particular JAK isoform. More recently, ropsacitinib has emerged as a potent and relatively selective TYK2 JH1 inhibitor (IC_{50} = 6 nM); however, its selectivity vs. JAK2 is only modest (12-fold). The formidable challenge of achieving high selectivity of TYK2/JH1 inhibition has inspired an alternative strategy to selectively target TYK2-STAT signaling via TYK2 JH2 inhibition, i.e., allosteric inhibition (type IV inhibitor).

The TYK2 protein, constitutively expressed in immune cells, can undergo mutations and polymorphisms, resulting in aberrant IL-12-, IL-23- and IFNα/β-promoted TYK2-STAT signaling. This contributes to the pathogenesis of psoriasis. Selective TYK2 inhibitors modulate TYK2-mediated pathways, but spare those driven by JAKs; therefore, they may avoid JAK-associated liabilities (Fig. 1).[3]

Deucravacitinib is such an allosteric inhibitor approved in 2022 for the treatment of adults with moderate-to-severe plaque psoriasis who are candidates for systemic therapy or phototherapy.[4] Unlike JAK inhibitors, it does not carry a black box warning of serious adverse effects.

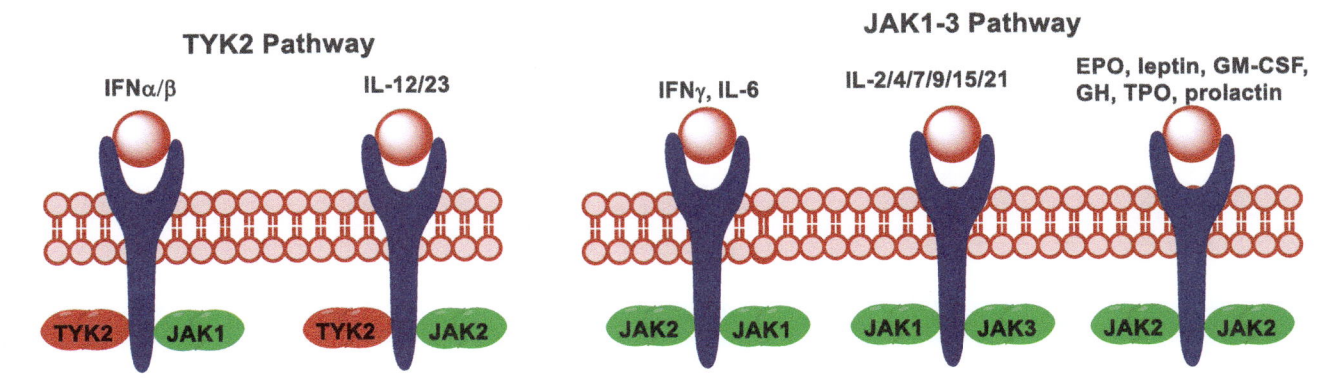

Figure 1. Cytokine responses in TYK2 and JAK pathways.

The research on allosteric TYK2 inhibitors remains highly competitive. For example, NDI-034858 (Fig. 2), a selective TYK2 (JH2) inhibitor, has been recently acquired by Takeda for $4 billion. This compound exhibited positive results in an ongoing phase 2b clinical trial for treating plaque psoriasis. VTX-958 (the structure has not been disclosed) is another allosteric inhibitor to enter into phase IIB clinical trials.

This chapter describes the process by which

deucravacitinib (Fig. 2) was discovered.

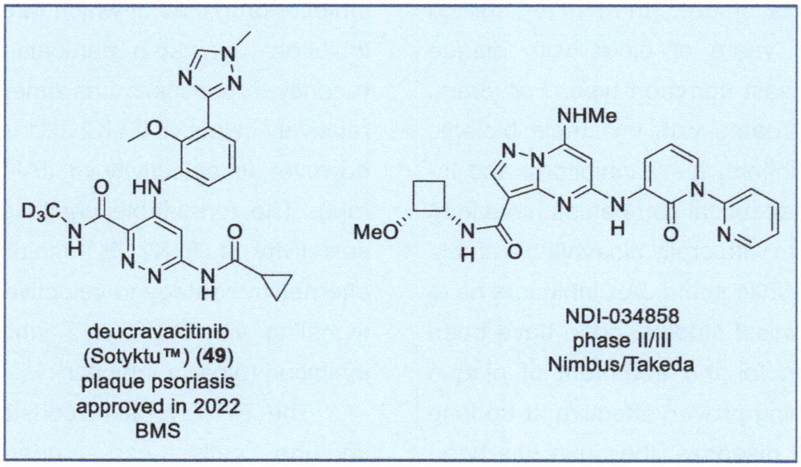

Figure 2. Allosteric TYK2 inhibitors.

1. Y. Tanaka, Y. Luo, J. O'Shea, S. Nakayamada. Janus kinase-targeting therapies in rheumatology: A mechanisms-based approach. **2022**, *Nat. Rev. Rheumatol.* **2022**, *18*, 133–145.
2. F. R. Spinelli, F. Meylan, J. J. O'Shea, M. Gadina. JAK inhibitors: Ten years after. *Eur. J. Immunol.* **2021**, *51*, 1615–1627.
3. M. J. Gooderham, H. C. Hong, I. V. Litvinov. Selective TYK2 inhibition in the treatment of moderate to severe chronic plaque psoriasis. *Skin Thera. Lett.* **2022**, *27*, 1–5.
4. S. T. Wrobleski, R. Moslin, S. Lin, Y. Zhang, S. Spergel, J. Kempson, J. S. Tokarski, J. Strnad, A. Zupa-Fernandez, L. Cheng, D. Shuster, K. Gillooly, X. Yang, E. Heimrich, K. W. McIntyre, C. Chaudhry, J. Khan, M. Ruzanov, J. Tredup, D. Mulligan, D. S. Weinstein. Highly selective inhibition of tyrosine kinase 2 (TYK2) for the treatment of autoimmune diseases: Discovery of the allosteric inhibitor BMS-986165. *J. Med. Chem.* **2019**, *62*, 8973–8995.

8.1 Deucravacitinib (Sotyktu™) (49)

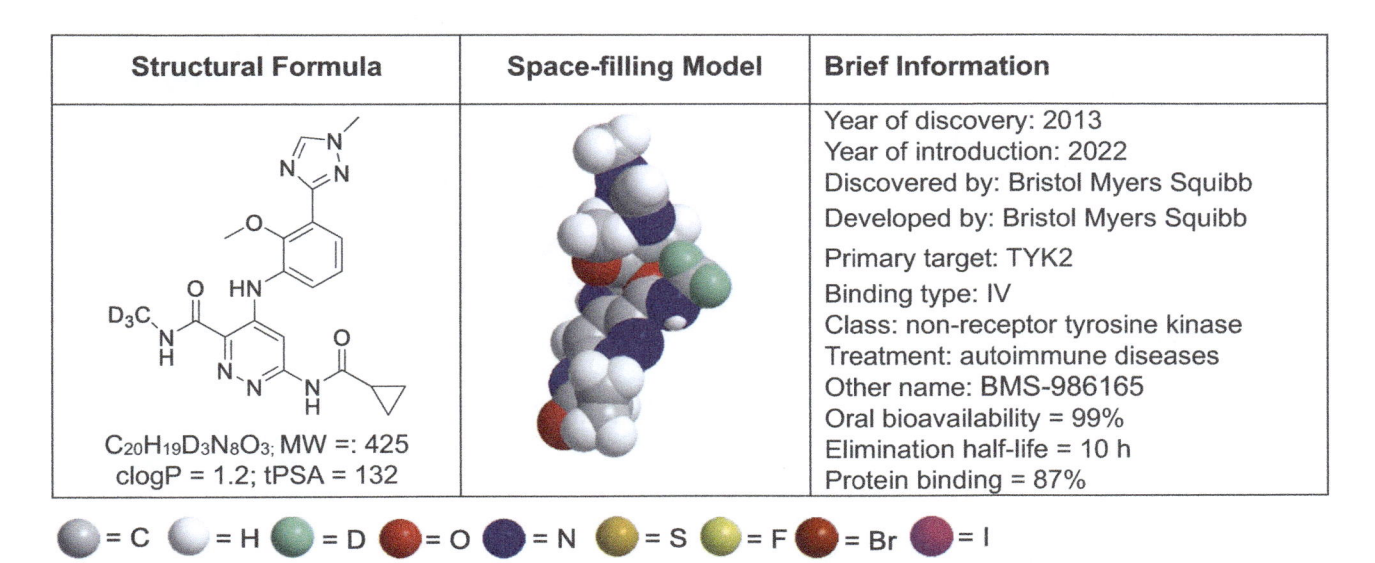

Structural Formula	Space-filling Model	Brief Information
$C_{20}H_{19}D_3N_8O_3$; MW =: 425 clogP = 1.2; tPSA = 132		Year of discovery: 2013 Year of introduction: 2022 Discovered by: Bristol Myers Squibb Developed by: Bristol Myers Squibb Primary target: TYK2 Binding type: IV Class: non-receptor tyrosine kinase Treatment: autoimmune diseases Other name: BMS-986165 Oral bioavailability = 99% Elimination half-life = 10 h Protein binding = 87%

● = C ○ = H ● = D ● = O ● = N ● = S ● = F ● = Br ● = I

Deucravacitinib, a highly selective allosteric TYK2 inhibitor, received its first approval from the FDA in 2022 for the treatment of adults with moderate-to-severe plaque psoriasis who are candidates for systemic therapy or phototherapy.

Background[1]

The allosteric TYK2 inhibitors selectively target TYK2/STAT signaling via TYK2 JH2 inhibition. TYK2 JH2 shows the overall fold of a typical catalytic domain, but there are a series of differences between individual residues of the TYK2 JH1 and JH2 domains at the binding pockets between the two domains, which could enable selective JH2 inhibition. Superposition of the crystal structures of TYK2 JH2, JAK1 JH2, JAK2 JH2, and JAK3 JH2 (data not shown) also reveals unique differences in the binding pockets of these pseudokinase domains, suggesting the potential for selective TYK2 JH2 inhibitors. In addition, mutation studies implicated that TYK2 JH2 may play an important role in the regulation of TYK2 function. Furthermore, TYK2 JH2 inhibitors were shown to inhibit TYK2-dependent downstream receptor signaling but not JAK1-, JAK2-, JAK3- dependent (but TYK2-independent) signaling. Taken together, these findings support the hypothesis that the JH2 domain regulates activation of the catalytic domain (JH1) via receptor-regulated inhibitory interactions. Hence, selective inhibition of TYK2 JH2 domain emerges as a new approach for the treatment of autoimmunity with reduced potential for isoform-related liabilities.

Discovery of Deucravacitinib[1–5]

The pursuit of selective TYK2 JH2 inhibitors started with a chemogenomics approach which combined kinome-wide inhibitory profiling of a compound library for cellular activity against an IL-23-stimulated transcriptional response in T lymphocytes. This screening identified a series of inhibitors which are bound to and stabilized the TYK2 JH2 domain to result in blockade of receptor-mediated activation of the adjacent catalytic domain (JH1). A co-crystal structure of the JH2 domain with BMS-066 (Fig. 1), a representative inhibitor, confirmed that the binding site is similar to the ATP-binding site in catalytic kinases. The crystal structure also showed a number of amino acid and structural differences from catalytic kinases, indicating that this domain is incapable of providing catalytic activity, which is consistent with a pseudokinase. In the complex, BMS-066 occupies the site typically bound by the adenine group of ATP in catalytic kinases. The imidazole basic nitrogen and the amino moiety form hydrogen bonds with Val690 of the hinge (Fig. 2). The amide NH of the methoxyacetamide group hydrogen bonds with the backbone carbonyl of the P-loop Leu595 as the group protrudes out toward solvent (Fig. 1). The pyridine nitrogen is also hydrogen bonded to a water molecule. Figure 3 summarizes the TYK2 JH2 domain structure with bound BMS-066, highlighting the residue differences between the JH2 domains of other JAK isoforms. The individual residue differences between the JH2 domain of TYK2 and those of other isoforms can be explored for the rational design of selective TYK2 JH2 inhibitors. Indeed, the unique Ala671 of TYK2 JH2 vs. the bulkier Val671 of other isoforms played an important role in the discovery of deucravacitinib (*vide infra*).

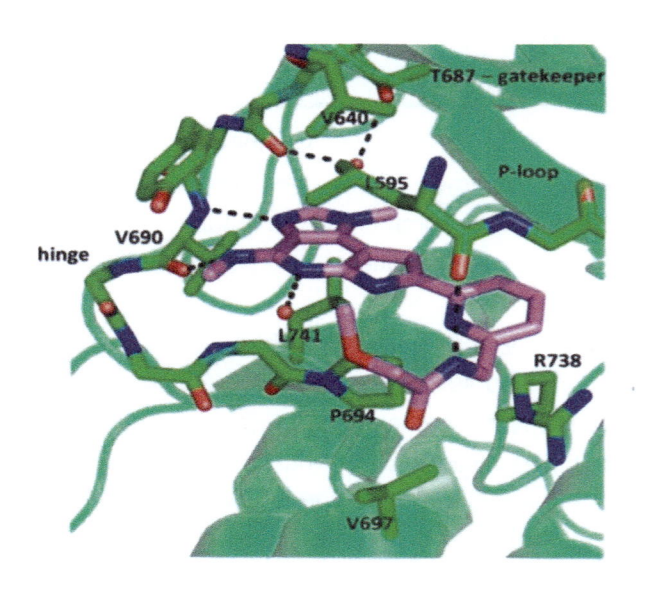

Figure 1. BMS-066–TYK2 JH2 complex. Adapted from *J. Biol, Chem.* **2015**, *290*, 11061–11074.

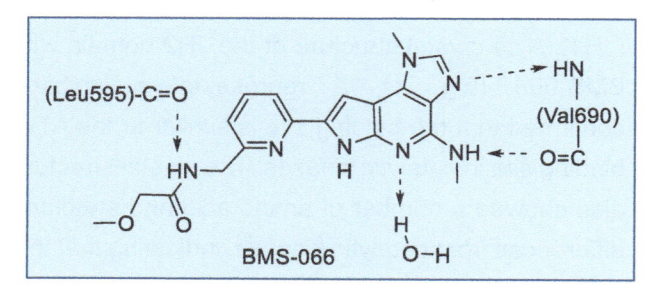

Figure 2. Key BMS-066–TYK2 JH2 interactions.

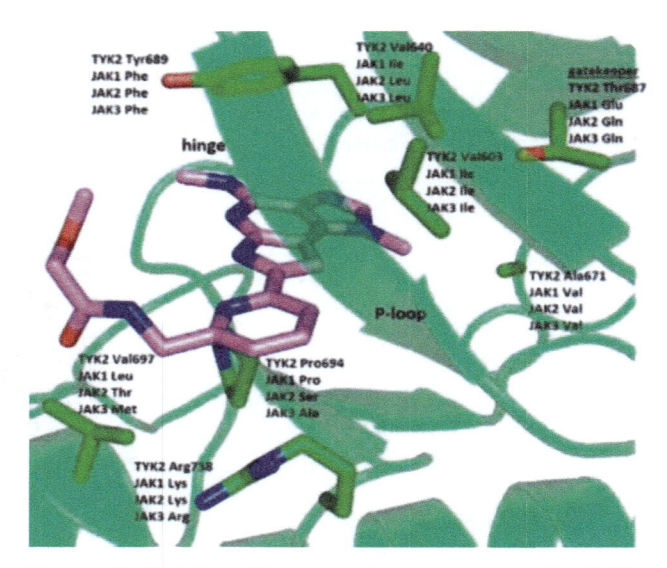

Figure 3. TYK2 JH2 co-crystal structure with BMS-066 highlighting the residue differences between the JH2's of other isoforms. Adapted from *J. Biol, Chem.* **2015**, *290*, 11061–11074.

Despite the insightful binding mode obtained from its co-crystal structure, BMS-066 was not further pursued because of its suboptimal activity (TYK2 JH2 IC$_{50}$ = 72 nM) and potent inhibition of IKKβ kinase (IC$_{50}$ = 9 nM).[6] To search for more potent TYK2 JH2-binding scaffolds, a high-throughput screening of the Bristol Myers Squibb compound collection was carried out using a scintillation proximity assay (SPA), which led to the identification of nicotinamide **1** as a potent TYK2 JH2 inhibitor (Fig. 4). However, this compound lacked selectivity over the catalytic (JH1) domains of the four JAK family members as well as the other kinases. Nevertheless, it showed high selectivity against both PDE4 and IKKβ, two off-targets that were involved with previously known chemotypes. In addition, compound **1** exhibited good metabolic stability in the in vitro assays. On the basis of its overall balanced profile, this compound was chosen for further selectivity exploration.

As part of the SAR efforts, the primary amide was converted to the *N*-methyl derivative **2** (Fig. 4), which surprisingly showed substantial improvement in JAK isoform as well as overall kinome selectivity. Furthermore, the breakthrough compound **2** was shown to be inactive against the JAK (including

TYK2) JH1 domains at 2 μM concentration, and it inhibited only 4 of 265 kinases tested by more than 50% at 1 μM concentration.

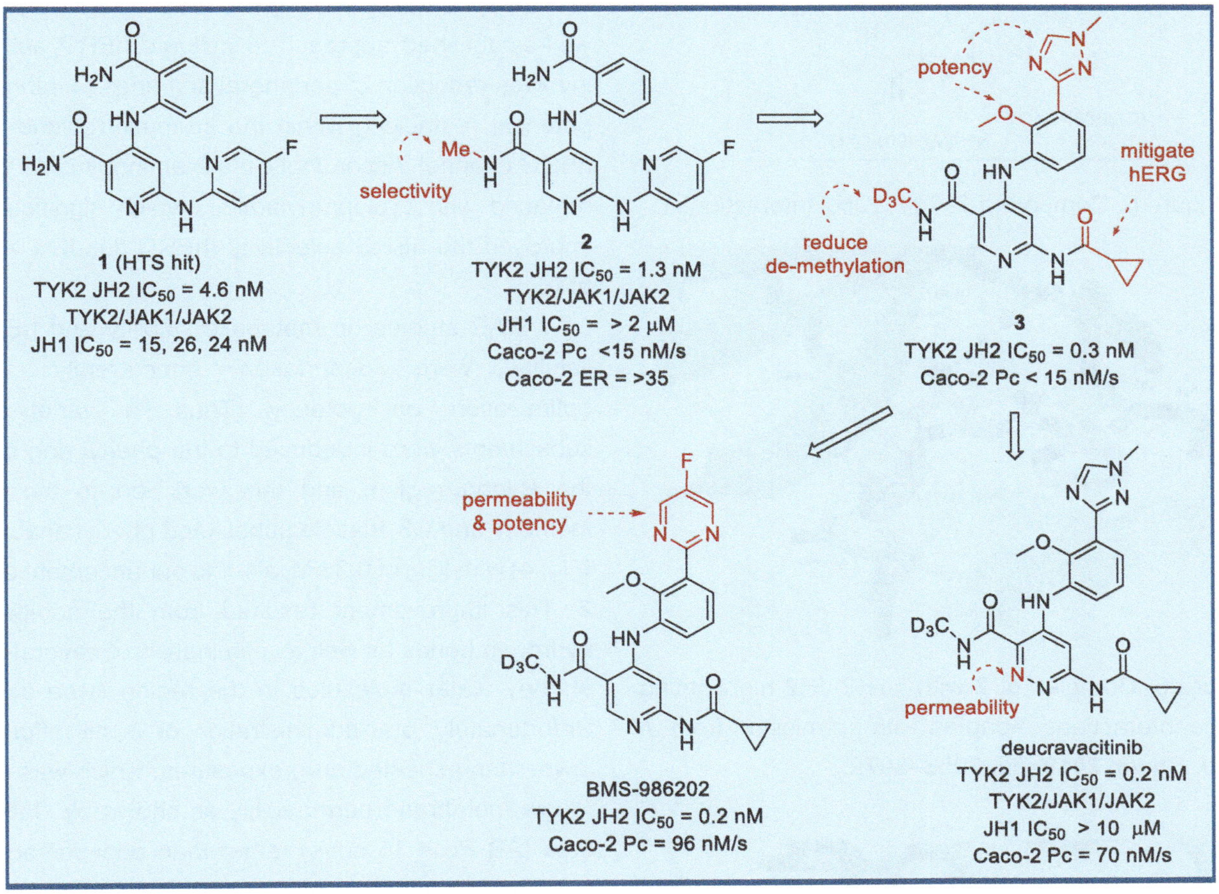

Figure 4. Summary of the chemical optimization of deucravacitinib and BMS-986202.

The profound effect of the *N*-methyl group on selectivity can be rationalized by examining its environment in the co-crystal structure of **2** with the TYK2 JH2 domain (Figs. 5–7). The aminopyridinecarboxamide is oriented against the hinge, and the C6 NH, pyridine nitrogen and the C3 carboxamide NH form three donor–acceptor–donor hydrogen bonds with backbone amides of Val690 and Glu688 in the hinge. An additional intramolecular hydrogen bond between the C3 carboxamide oxygen and the C4 amino hydrogen further stabilizes the bioactive conformation. All these hydrogen bonds together with other favorable hydrophobic interactions effectively tie the molecule in the exact place, protruding the *N*-methyl to approach the proximal Ala671. Of special note is that the alanine in this particular position, which is found in only nine other kinases, would be substituted with a much larger residue such as valine, leucine, or isoleucine in a majority of other kinases. Conceivably, the C3 primary amide of **1** can be tolerated by larger residues of other kinases, thus leading to very modest selectivity profile. In contrast, the limited space provided by an adjacent alanine ("alanine pocket") of TYK2 JH2 domain can only tightly fit the *N*-methyl of **2**, consistent with the high selectivity.

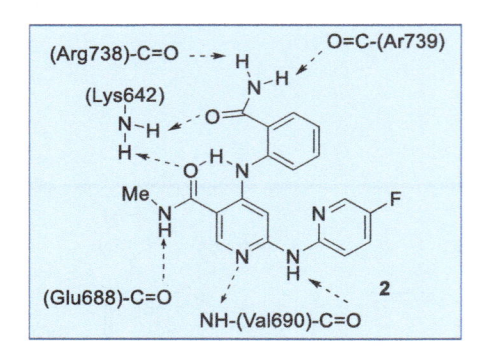

Figure 5. Compound **2**–TYK2 JH2 interactions.

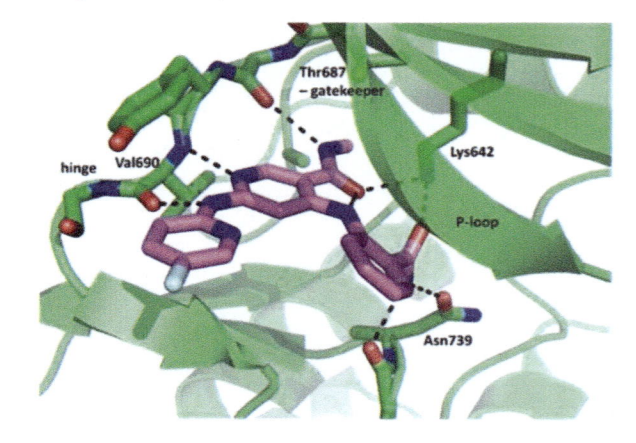

Figure 6. Complex of **2** with TYK2 JH2 highlighting hinge interactions. Adapted with permission from *J. Med. Chem.* **2019**, *62*, 8953–8972.

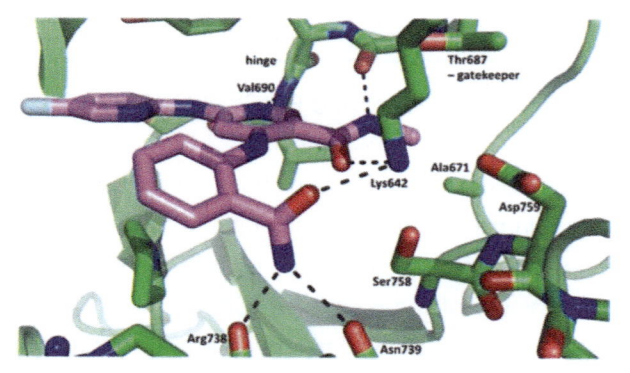

Figure 7. co-crystal structure of **2** with TYK2 JH2 highlighting interactions of C2′ amide and the Ala671 residue. Adapted with permission from *J. Med. Chem.* **2019**, *62*, 8953–8972.

Because of its good potency and selectivity, **2** was evaluated in oral pharmacokinetic studies in mice, which showed poor serum levels. The poor exposure was consistent with its low membrane permeability (Caco-2 AB Pc < 15 nm/s) and a high efflux ratio (ER = 35). Metabolic N-demethylation to form the primary amide **1** also contributed to the low serum level, and as such, the trideuteromethyl secondary amide was incorporated. The other potential liability associated with **2** was hERG inhibition, which frequently occurs with compounds consisting of multiple heteroaryls. A well-established approach to mitigate hERG activity involves reduction of peripheral aryl rings to interrupt potential π-stacking within the aromatic residue-rich hERG channel. Accordingly, the 2-aminopyridine was replaced with cyclopropylamide, which significantly improved the hERG selectivity (hERG flux IC$_{50}$ > 80 µM).

SAR studies on metabolic stability and hERG liability were undertaken concurrently with optimization on potency. Thus, a variety of substituents were introduced to the phenyl ring near the P-loop region, and this work led to the C2′ methoxy and C3′ triazole substituted phenyl analog **3** (Fig. 4) with IC$_{50}$ of 0.3 nM, a 4-fold enhancement over **2**. This improvement resulted from the additional hydrogen bonds as well as elimination of several low entropy water molecules in the region (*vide infra*). Unfortunately, oral administration of **3** still afforded lower than expected drug exposures, which was due to low membrane permeability as shown by Caco-2 data (AB Pc < 15 nm/s) rather than poor metabolic stability. To specifically address the permeability issue while maintaining the other favorable attributes, two bioisosteric substitutions proved to be beneficial: (1) pyridine with less basic pyridazine; and (2) triazole with 5-fluoropyrimidine. The former approach led to the discovery of deucravacitinib with more than 4-fold enhancement in permeability (AB Pc = 70 nm/s), which translated to a 24-fold improvement in C_{max} over its pyridine counterpart **3** in mouse oral pharmacokinetic studies, even though it was a P-gp substrate with an efflux ratio of 10 (BA Pc = 740 nm/s). The latter approach led to BMS-986202, which showed >6-fold improvement in permeability (Pc = 96 nm/s) and 42-fold improvement in C_{max} as compared to **3**.

Deucravacitinib shows specific, high affinity binding to TYK2 JH2 and low affinity in the canonical TYK2, JAK1, JAK2, and JAK3 JH1 assays (IC$_{50}$'s >10 µM). It is >1000-fold selective against a panel of 249 protein and lipid kinases and pseudokinases, with the exception of BMPR2 (IC_{50} = 193 nM) and JAK1 JH2 pseudokinase domain (IC_{50} = 1 nM). Despite its potent affinity for JAK1 JH2, deucravacitinib elicited low functional activity in a JAK1/JAK3-dependent IL-2 stimulated cellular assay.

BMS-986202 displays >10,000-fold selectivity for TYK2 JH2 over a diverse panel of 273 kinases and pseudokinases that include JAK family members. Like deucravacitinib, its high binding affinity to JAK1 JH2 (IC_{50} = 7.8 nM) did not translate to any functional activity in the cellular assay. BMS-986202 was under clinical trials for the treatment of moderate-to-severe psoriasis.

Binding Mode[5]

The hinge-binding mode of deucravacitinib is the same as **2**. The aminopyridazine carboxamide core forms three donor–acceptor–donor hydrogen bonds with backbone amides of Val690 and Glu688 in the hinge (Figs. 8, 9). The C2′ methoxy also forms a direct hydrogen bond interaction with the main chain atom of Lys642, which likely contributes to the observed increase in potency. The N-2 triazole nitrogen engages in a direct hydrogen bond with Arg738, and the triazole also serves to displace several energetically unfavorable water molecules, which contributes to the high binding affinity of deucravacitinib to the TYK2 JH2 domain.

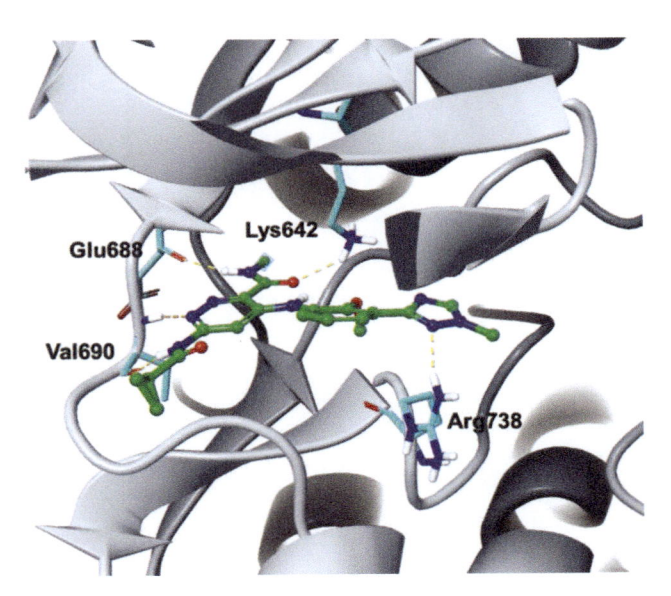

Figure 8. Deucravacitinib–TYK2 JH2 complex (PDB ID: 6NZP).

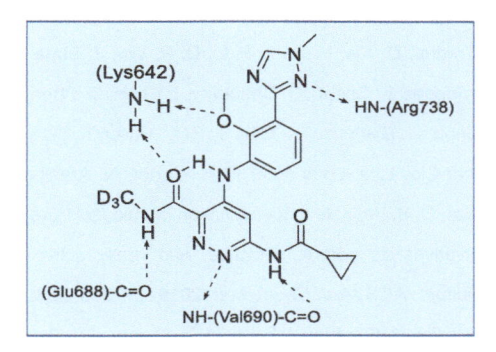

Figure 9. Summary of deucravacitinib–TYK2 JH2 interactions based on co-crystal structure.

PK Properties and Metabolism

The crystalline free base form of deucravacitinib exhibited poor aqueous solubility (5.2 μg/mL), which was still acceptable for preclinical studies. It showed moderate half-lives of 4–5 h across species (mouse, dog and monkey). Excellent exposures and high bioavailability (F > 85%) in mouse, dog, and monkey were obtained from oral pharmacokinetic studies at 10 mpk dose.[5]

Following oral administration of deucravacitinib, the major metabolite in human plasma was the cyclopropyl carboxamide hydrolytic cleavage product **4** (Fig. 10).[7]

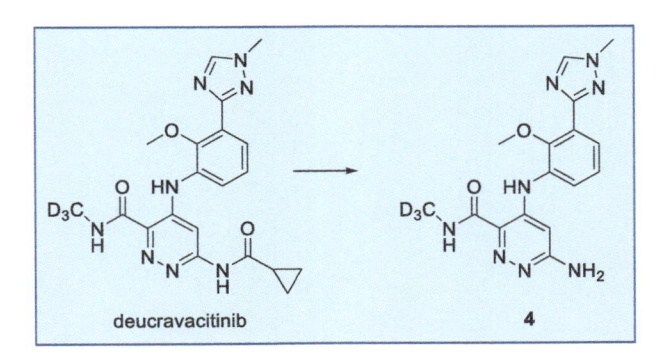

Figure 10. Major metabolism of deucravacitinib in humans.

1. J. S. Tokarski, A. Zupa-Fernandez, J. A. Tredup, K. Pike, C. Chang, D. Xie, L. Cheng, D. Pedicord, J. Muckelbauer, S. R. Johnson, S. Wu, S. C. Edavettal, Y. Hong, M. R. Witmer, L. L. Elkin, Y. Blat, W. J. Pitts, D. S. Weinstein, J. R. Burke. Tyrosine kinase 2-mediated signal transduction in T lymphocytes is

blocked by pharmacological stabilization of its pseudokinase domain. *J. Biol. Chem.* **2015**, *290*, 11061–11074.

2. C. Liu, J. Lin, R. Moslin, J. S. Tokarski, J. Muckelbauer, C. Chang, J. Tredup, D. Xie, H. Park, P. Li, D. R. Wu, J. Strnad, A. Zupa-Fernandez, L. Cheng, C. Chaudhry, J. Chen, C. Chen, H. Sun, P. Elzinga, C. D'arienzo, K. Gillooly, T. L. Taylor, K. W. McIntyre, L. Salter-Cid, L. J. Lombardo, P. H. Carter, N. Aranibar N, J. R. Burke, D. S. Weinstein. Identification of imidazo[1,2-*b*]pyridazine derivatives as potent, selective, and orally active Tyk2 JH2 inhibitors. *ACS Med. Chem. Lett.* **2019**, *21*, 383–388.

3. R. Moslin, D. Gardner, J. Santella, Y. Zhang, J. V. Duncia, C. Liu, J. Lin, J. S. Tokarski, J. Strnad, D. Pedicord, J. Chen, Y. Blat, A. Zupa-Fernandez, L. Cheng, H. Sun, C. Chaudhry, C. Huang, C. D'Arienzo, J. S. Sack, J. K. Muckelbauer, C. Chang, J. Tredup, D. Xie, N. Aranibar, J. R. Burke, P. H. Carter, D. S. Weinstein. Identification of imidazo[1,2-*b*]pyridazine TYK2 pseudokinase ligands as potent and selective allosteric inhibitors of TYK2 signalling. *Medchemcomm.* **2016**, *15*, 700–712.

4. S. T. Wrobleski, R. Moslin, S. Lin, Y. Zhang, S. Spergel, J. Kempson, J. S. Tokarski, J. Strnad, A. Zupa-Fernandez, L. Cheng, D. Shuster, K. Gillooly, X. Yang, E. Heimrich, K. W. McIntyre, C. Chaudhry, J. Khan, M. Ruzanov, J. Tredup, D. Mulligan, D. S. Weinstein. Highly selective inhibition of tyrosine kinase 2 (TYK2) for the treatment of autoimmune diseases: Discovery of the allosteric inhibitor BMS-986165. *J. Med. Chem.* **2019**, *62*, 8973–8995.

5. R. Moslin, Y. Zhang, S. T. Wrobleski, S. Lin, M. Mertzman, S. Spergel, J. S. Tokarski, J. Strnad, K. Gillooly, K. W. McIntyre, A. Zupa-Fernandez, L. Cheng, H. Sun, C. Chaudhry, C. Huang, C. D'Arienzo, E. Heimrich, X. Yang, J. K. Muckelbauer, C. Chang, D. S. Weinstein. Identification of *N*-methyl nicotinamide and *N*-methyl pyridazine-3-carboxamide pseudokinase domain ligands as highly selective allosteric inhibitors of tyrosine kinase 2 (TYK2). *J. Med. Chem.* **2019**, *62*, 8953–8972.

6. S. H. Watterson, C. M. Langevine, K. Van Kirk, J. Kempson, J. Guo, S. H. Spergel, J. Das, R. V. Moquin, A. Dyckman, D. Nirschl, K. Gregor, M. A. Pattoli, X. Yang, K. W. McIntyre, G. Yang, M. A. Galella, H. Booth-Lute, L. Chen, Z. Yang, D. Wang-Iverson, W. J. Pitts. Novel tricyclic inhibitors of IKK2: Discovery and SAR leading to the identification of 2-methoxy-N-((6-(1-methyl-4-(methylamino)-1,6-dihydroimidazo[4,5-d]pyrrolo[2,3-b]pyridin-7-yl)pyridin-2-yl)methyl)acetamide (BMS-066). *Bioorg. Med. Chem. Lett.* **2011**, *21*, 7006–7012.

7. M. Yao, X. Gu, J. Brailsford, J. C. Cortes, R. Iyer, W. Li, P177 - Comparative metabolism of [^{14}C]-BMS-986165 in mice, rats, monkey, and humans. *Drug Metab. Pharmacokinet.* **2020**, *35*, Suppl., S77.

Chapter 9. ALK/multikinase Inhibitors

Anaplastic lymphoma kinase (ALK) gene fusion results in oncogenic fusion proteins, which account for ~3–5% of genetic alterations in patients with non-small cell lung cancer (NSCLC). The most common ALK fusion involves echinoderm microtubule-associated protein like 4 (EML4) gene, producing the EML4-ALK hybrid gene as an important tumor driver in NSCLC. The resulting fusion proteins undergo aberrant dimerization or oligomerization, leading to constitutive ALK activation. This triggers the associated downstream cellular signaling pathways, such as RAS/MAPK, PI3K/AKT and JAK/STAT, resulting in dysregulated cellular proliferation and tumor survival. Because ALK+ lung cancers depend on ALK signaling for their survival, ALK inhibitors have been established as a useful targeted therapy in the treatment of advanced NSCLC.

Crizotinib is the first targeted therapy approved for ALK-rearranged NSCLC. It is a multikinase inhibitor initially identified as a MET inhibitor but later found to be a potent ALK and ROS1 inhibitor. Despite its substantial efficacy in patients with advanced ALK+ NSCLC, acquired resistance to crizotinib is common, typically occurring within one year of treatment. To overcome this resistance, more potent and more selective second-generation ALK TKIs were developed, including ceritinib, alectinib, and brigatinib (Fig. 1). Thanks to the improved brain penetration, these TKIs have also been shown to be more effective in the treatment of brain metastases in NSCLC, a common complication among patients with lung cancer. Subsequently, the second-generation TKIs have largely replaced crizotinib as the standard first-line therapy for patients with advanced ALK-rearranged NSCLC because of their superior efficacy to crizotinib (Fig. 1).[1]

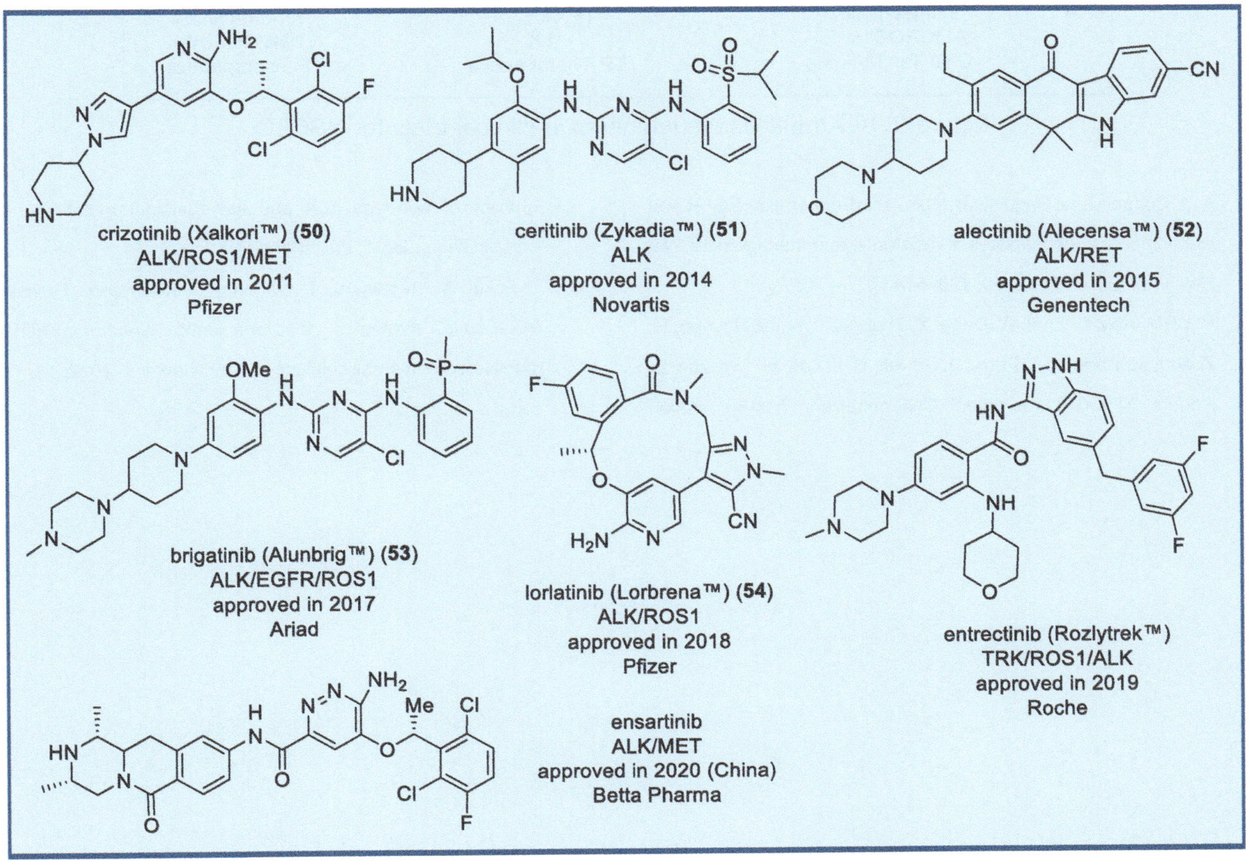

Figure 1. Approved ALK/multikinase inhibitors for NSCLC.

Molecules Engineered Against Oncogenic Proteins and Cancer, First Edition. E. J. Corey and Yong-Jin Wu.

Lorlatinib is a third-generation, macrocyclic, highly potent and selective ALK/ROS1 TKI, developed to overcome to acquired resistance to earlier-generation of ALK TKIs, particularly the gatekeeper mutation L1196M, as well as to achieve improved CNS penetration (due to its compact structure and desirable lipophilicity).

TPX-0131 (Fig. 2) and NVL-655 (whose structure has not been disclosed) are fourth-generation ALK TKIs which are currently being evaluated in clinical trials for ALK+ advanced NSCLC.[2,3] The former is a highly potent and brain-penetrant, compact macrocyclic TKI targeting both wild-type and a broad spectrum of ALK drug-resistant mutations (solvent-front, gatekeeper, hinge region, and compound mutations).

Although ALK pathology represents a niche market in NSCLC, accounting for roughly 3–5% of all cases, Roche's alectinib still brought in sales of $1.26 billion in 2020, while Pfizer's crizotinib delivered $524 million in sales in 2018.

This chapter describes the discovery process that led to ALK inhibitors **50**–**54**. The two TRK/ROS1/ALK inhibitors, entrectinib (Fig. 1) and repotrectinib (Fig. 2) are presented in Chapter 16 (TRK inhibitors).

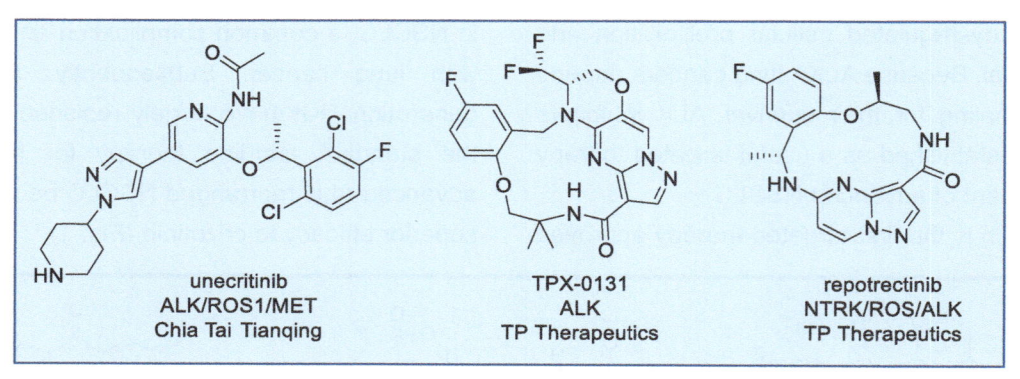

Figure 2. ALK/multikinase inhibitors in clinical trials for NSCLC.

1. A. J. Cooper, L. V. Sequist, J. J. Lin. Third-generation EGFR and ALK inhibitors: mechanisms of resistance and management. *Nat. Rev. Clin. Oncol.* **2022**, *19*, 499–514.
2. B. W. Murray, D. Zhai, W. Deng, X. Zhang, J. Ung, V. Nguyen, H. Zhang, M. Barrera, A. Parra, J. Cowell, D. J. Lee, H. Aloysius, E. Rogers. TPX-0131, a potent CNS-penetrant, next-generation inhibitor of wild-type ALK and ALK-resistant mutations. *Mol. Cancer Thera.* **2021**, *20*, 1499–1507.
3. T. Fukui, M. Tachihara, T. Nagano, K. Kobayashi. Review of therapeutic strategies for anaplastic lymphoma kinase-rearranged non-small cell lung cancer. *Cancers.* **2022**, *14*, 1184.

9.1 Crizotinib (Xalkori™) (50)

Structural Formula	Brief Information
$C_{21}H_{22}Cl_2FON_5$; FW = 449 logD = 1.96; tPSA = 66	Year of discovery: 2004 Year on Introduction: 2011 Discovered by: Pfizer Developed by: Pfizer Primary targets: ALK/ROS1/MET Binding type: I Class: receptor tyrosine kinase Treatment: NSCLC Other name: PF2341066 Oral bioavailability = 43% Elimination half-life = 42 h Protein binding = 90%

● = C ○ = H ● = O ● = N ● = S ● = F ● = Cl ● = Br ● = I

Crizotinib is the first-in-class anaplastic lymphoma kinase (ALK) inhibitor indicated for the treatment of patients with advanced non-small cell lung cancer (NSCLC) whose tumors are ALK-positive or ROS1-positive, and for the treatment of pediatric patients with relapsed or refractory systemic anaplastic large cell lymphoma (ALCL) that is ALK-positive.

Crizotinib selectively targets the activity of ALK alterations, including NPM1 (nucleophosmin)-ALK and EML4 (echinoderm microtubule-associated protein-like)-ALK fusions, and thereby inhibiting the growth of tumors driven by the increased ALK activity. The ALK rearrangements account for approximately 5–6% of NSCLC patients, with 9000 new cases per year in the United States (vs. 6700 for CML) and about 45,000 worldwide. The ALK fusions are also prevalent in more than 50% of all ALCL patients, with NMP-ALK being the most common variant.

Crizotinib also inhibits MET (mesenchymal–epithelial transition factor) and ROS1 (ROS proto-oncogene 1) tyrosine kinases. The ROS1 gene fusions have also been identified as oncogenic drivers in NSCLC, accounting for 1–2% of NSCLC cases.

Crizotinib is used as the first-line therapy in the treatment of advanced NSCLC harboring ALK or ROS1 fusions and has transformed the treatment of these cancer patients.

ALK Fusions in NSCLC and ALCL[1–5]

ALK is a canonical receptor-protein tyrosine kinase (RTK) which is activated through ligand binding with its extracellular domain of ALK. The binding leads to dimerization and autophosphorylation (at the intracellular tyrosine kinase domain) and consequent activation of downstream intracellular signaling pathways (Fig. 1, top right).

ALK is highly expressed in the neonatal brain and is involved in the development of brain and neurons. It was originally identified in 1994 in ALCL as a fusion partner of NPM-ALK resulting from a chromosomal translocation. Subsequently, ALK fusions have been associated with other types of cancers, such as NSCLC, neuroblastomas, squamous cell, colorectal, and breast carcinomas.[1]

ALK gene fusion occurs when the 3′-end of ALK is fused to the 5′-end of EML4 and NPM1 to form the EML4-ALK and ALK-NMP1-ALK fusion genes, respectively, through chromosomal rearrangements. Although the chromosomal breakpoint within the EML4 gene varies, all the EML4-ALK variants maintain the same cytoplasmic tyrosine kinase domain of ALK. While EML4-ALK is the most common ALK fusion in NSCLC, more than 19 different ALK fusion partners have been reported, including KIF5B, TFG, and KLC1.

In the process of ALK rearrangement, the ALK intracellular kinase domain combines with the coiled coil domain of the partner genes such as EML4 and NPM1 to form hybrid genes (Fig. 1, bottom left). The ALK kinase domain is maintained after fusion (despite the breakpoint) with no change in downstream intracellular kinase activity. However, the newly incorporated coiled-coil domain functions as a dimerization unit, which induces constitutive homodimerization, subsequent autophosphorylation and sustained activation of fusion kinases in a ligand-independent manner. The persistently active tyrosine kinases, EML4-ALK and NPM1-ALK, trigger multiple intracellular downstream signaling pathways, resulting in uncontrolled cellular proliferation and survival of NSCLC and ALCL cells, respectively (Fig. 1, bottom right).

ROS1 is another canonical RTK that is closely related to ALK. Like ALK, ROS1 has also been found to form hybrid genes with more than 26 fusion partners, some of which are already known to hybridize with ALK. All the fusion proteins retain the ROS1 kinase domain, while the fusion partners retain the dimerization domains, which result in ligand-independent, constitutive ROS1 tyrosine kinase activation.[6]

Crizotinib is a dual ALK and ROS1 inhibitor that competitively blocks the ATP-binding sites of these aberrantly active ALK and ROS1 fused kinases, and subsequently turns off their downstream activities.

The discovery of crizotinib represents a classic example of the X-ray structure guided truncation of inefficient scaffolding to reduce molecular weight and lipophilicity to increase oral bioavailability.

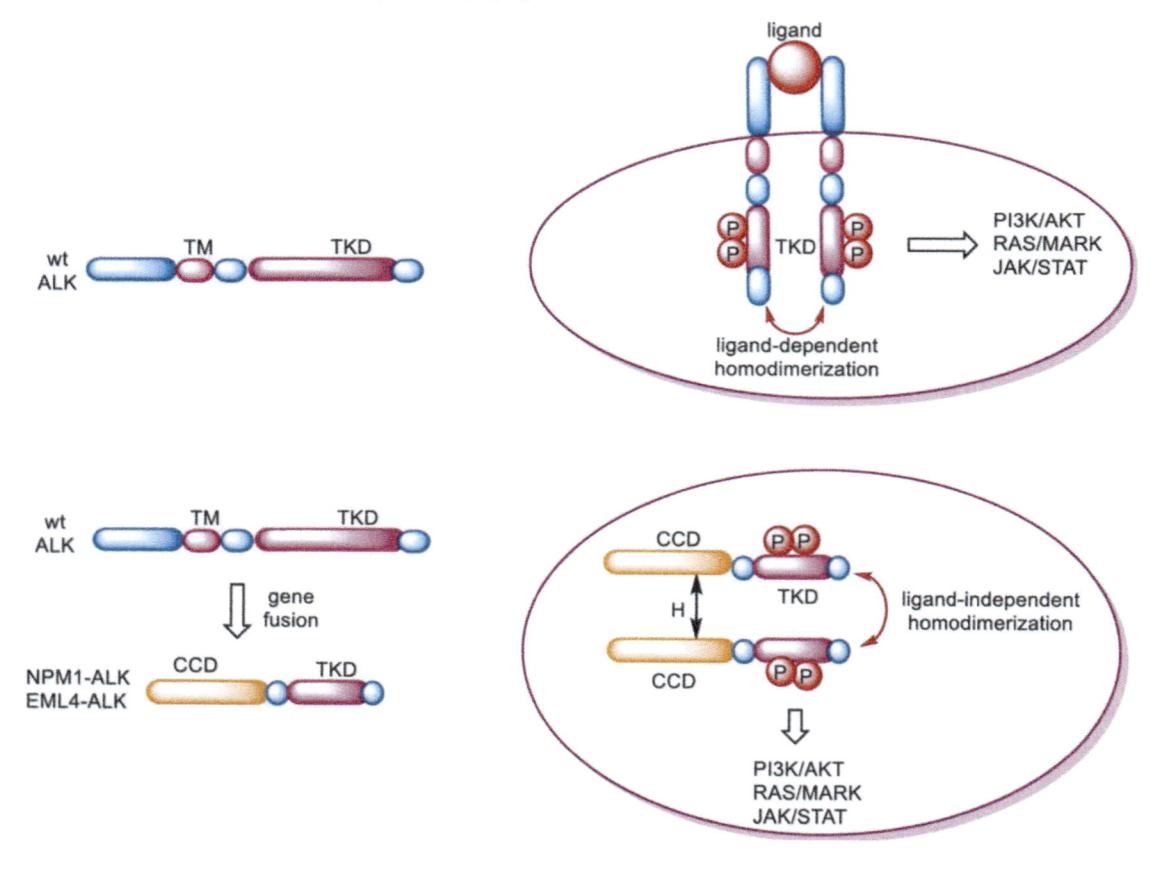

Figure 1. Top left: schematic representation of wild-type (wt) ALK; Top right: ligand-dependent ALK activation in wild-type; Bottom left: ALK fusion; Bottom right: ligand-independent ALK activation in NPM1-ALK and EML4-ALK. TM: transmembrane domain; TKD: tyrosine kinase domain; CCD: coiled-coil domain.

Discovery of Crizotinib[7]

Criziotinib was originally developed as a MET inhibitor. The lead compound, SU11274, was identified as a MET inhibitor with an IC_{50} of 10 nM in the biochemical assay, but it lacked in cellular activity. Optimization of SU11274 led to PHA-665752 with significantly improved cellular potency (cell IC_{50} = 9 nM). However, despite its potent and selective inhibition of MET autophosphorylation and related biological functions in both in vitro and in vivo assays, PHA-665752 suffered from poor oral bioavailability due to low solubility, high metabolic clearance and poor permeability, which limited its further development. To ameliorate oral bioavailability, smaller and less lipophilic inhibitors were designed using the X-ray crystal structure of PHA-665752 (Fig. 2).

The co-crystal structure of PHA-665752 with MET shows that the pyrrole NH forms one hydrogen bond with the carbonyl group of Pro1158 hinge and the oxindole carbonyl forms another hydrogen bond with the amide NH of Met1160 of the hinge (Fig. 2). The oxindole and pyrrole rings are oriented in a coplanar conformation due to an intramolecular hydrogen bond between the pyrrole NH and the oxindole carbonyl. The binding conformation is stabilized by the critical π–π stacking interaction between the benzyl group and Tyr1230 of the A-loop. This interaction is further reinforced by a hydrogen bond between the oxygen of the sulfonyl group and the backbone NH of Asp1222. However, the phenyl sulfone methylene linker takes an unnecessary U-turn to project the 2,6-dichlorophenyl group for the π–π stacking interaction with Tyr1230 (Fig. 2). The X-ray crystal structural analyses suggested that both the pyrrole-oxindole scaffold and the U-turn linker can be truncated to give a much smaller inhibitor without compromising critical interactions within ATP-binding site.

Bioisosteric replacement of the coplanar conformation of pyrrole-oxindole leads to the tricyclic pyridoindole **1**, and a C–N cleavage of the pyrrole ring affords aminopyridine **2** as a conceptionally new MET inhibitor (Fig. 3). The U-turn phenyl sulfone methylene linker was straightened out to a simple methyleneoxy to give **3**, which exhibited only marginally activity (K_i = 0.46 μM). Interestingly, introduction of a "magic" methyl group onto the ether linkage increased activity by 7-fold (**4**: K_i = 68 nM; cell IC_{50} = 140 nM) because the methyl group enables a stronger π–π stacking interaction of the benzyl group with Tyr1230 while contributing additional hydrophobic interactions. The potency was further boosted by a fluorine substitution of the dichlorophenyl ring (**5**: K_i = 12 nM; cell IC_{50} = 20 nM). To further reduce molecular weight and lipophilicity, the entire amidophenyl group was replaced with a variety of simple aryl and heteroaryl groups, and the methylpyrazole analog **7** emerged as a moderately potent inhibitor with substantially reduced molecular weight (FW = 380; K_i = 46 nM; cell IC_{50} = 44 nM). Nevertheless, the high lipophilicity (clogP = 4.7) contributed to its high metabolic instability, which was circumvented by replacing the methyl group with a water soluble piperdine moiety to give crizotinib as a highly potent MET inhibitor with 43% oral bioavailability. Subsequently, crizotinib was found to be a potent inhibitor of ALK (IC_{50} = 20 nM) and ROS1 (K_i < 0.025 nM).

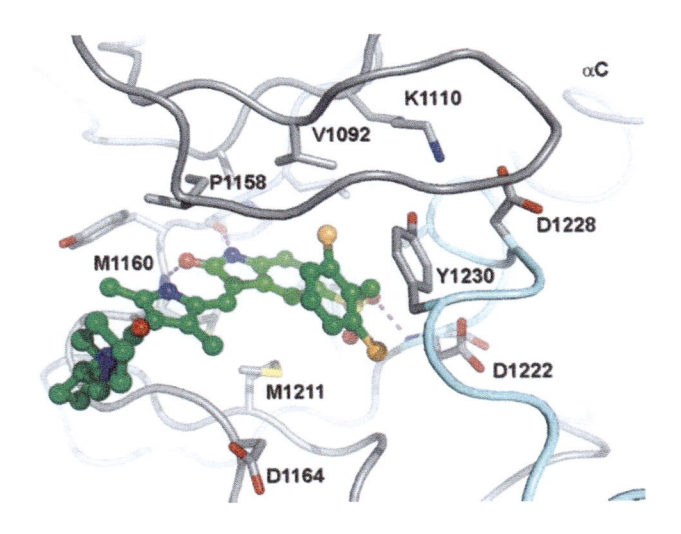

Figure 2. Co-crystal structure of PHA-665752 bound to the kinase domain of MET. Adapted with permission from *J. Med. Chem.* **2011**, *54*, 6342–6363.

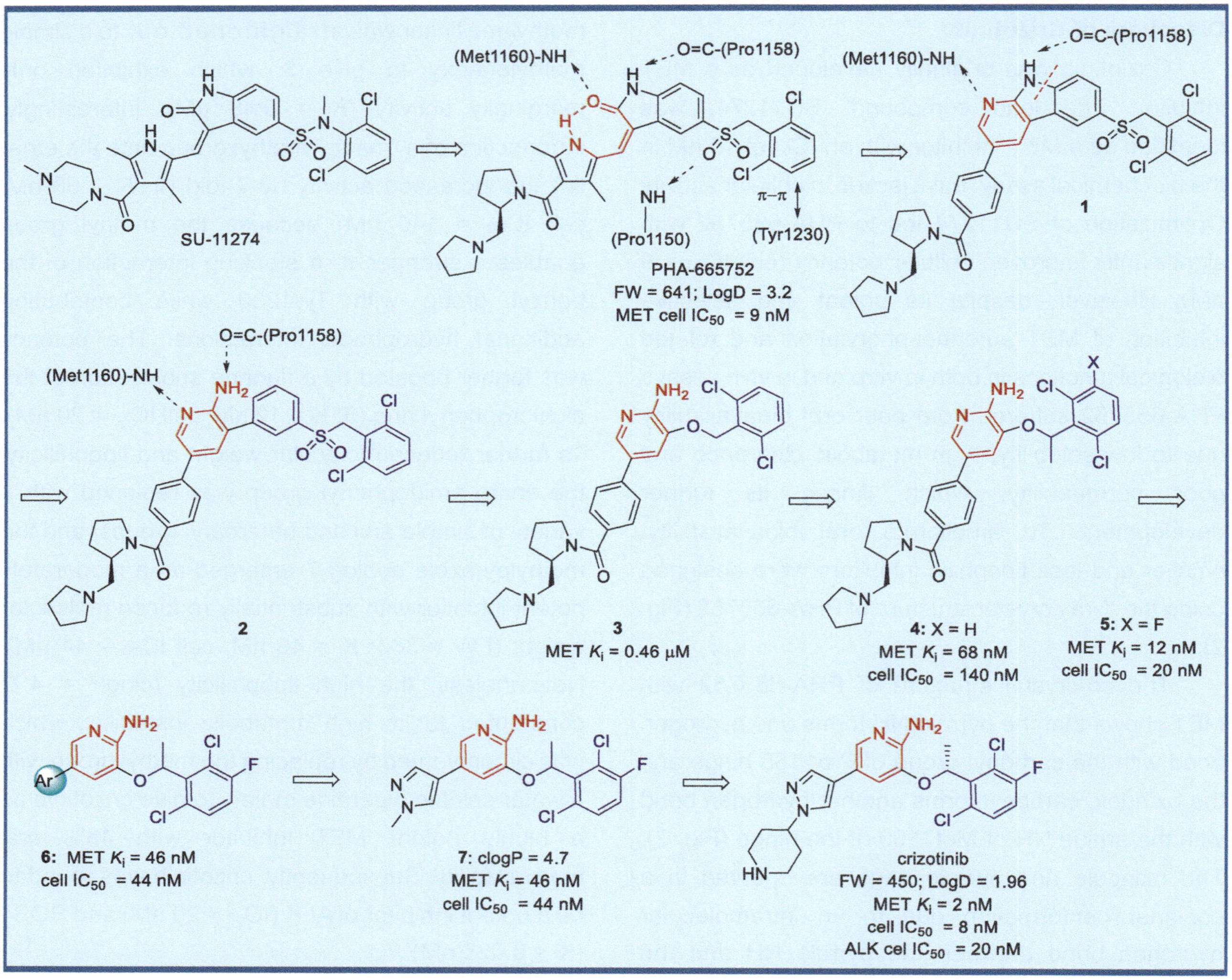

Figure 3. Structural optimization of crizotinib.

Crizotinib Resistance[4]

Despite initial robust efficacy in ALK-positive tumors, relapse and resistance to crizotinib occur in most ALK-rearranged patients within 12 months due to acquired secondary mutations (L1196M, G1269A, F1174L, S1206Y, 1151 T-ins, L1152R, C1156Y and G1202R) in the ALK kinase domain. The ALK gatekeeper mutation L1196M is the most common in NSCLC patients; therefore, significant efforts have gone into the search for potent L1196M inhibitors, leading to the development of second-generation ALK inhibitors including ceritinib, alectinib, lorlatinib and brigatinib that overcome the acquired secondary mutations.

Binding Mode

In the co-crystal structure of crizotinib with MET (Fig. 4),[7] the amino group of the aminopyridine forms one hydrogen bond with the amide carbonyl of Pro1158 hinge and the N1 nitrogen of the pyridine ring forms another hydrogen bond with the amide NH of Met1160 of the hinge (Fig. 5). The (R)-methyl group plays a critical role in orienting the 2,6-dichlorophenyl to engage a π–π stacking with Tyr1230. In addition, this methyl contributes to the high potency by making favorable hydrophobic interactions. The proximal piperidine ring, which is attached to the N1 position of the pyrazol-4-yl, is directed toward the solvent (Fig. 5).

A co-crystal structure of ALK kinase domain in complex with crizotinib reveals a conformation similar to crizotinib bound to MET (Figs. 6, 7).[7,8] However, the π–π stacking interaction between the dichlorophenyl and MET Tyr1230 is absent in the complex with ALK, and this may explain a weaker potency observed with ALK (IC_{50} = 20 nM vs. 8 nM for MET).

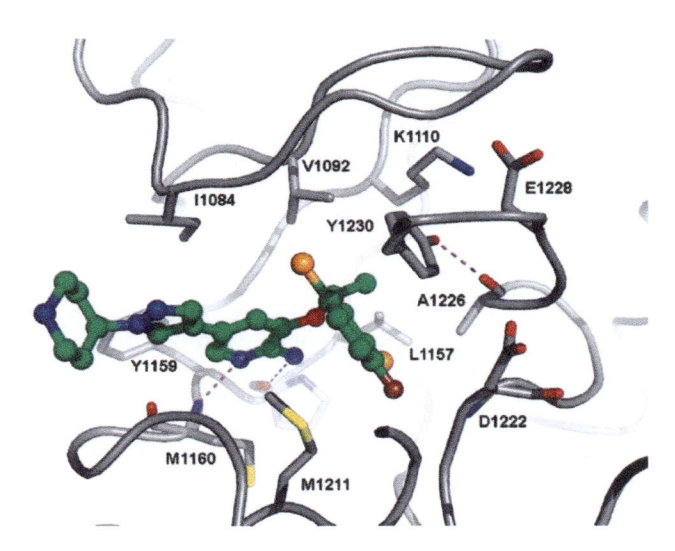

Figure 4. Co-crystal structure of crizotinib in complex with MET. Adapted with permission from *J. Med. Chem.* **2011**, *54*, 6342–6363.

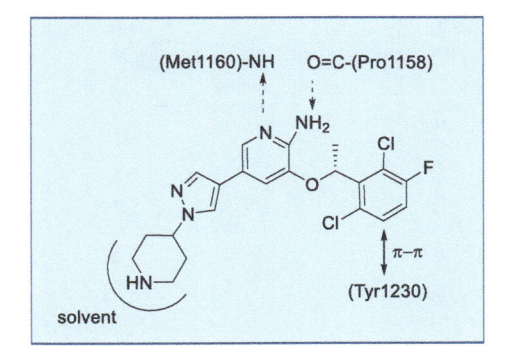

Figure 5. Interactions of MET–crizotinib.

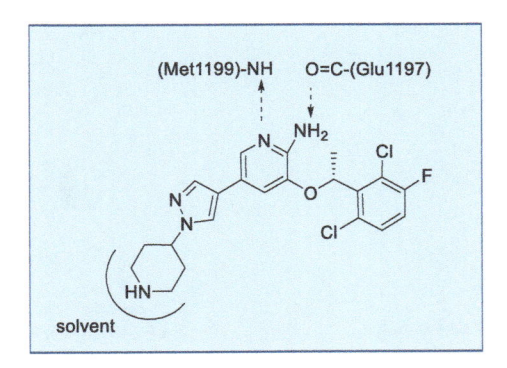

Figure 6. Interactions of ALK–crizotinib.

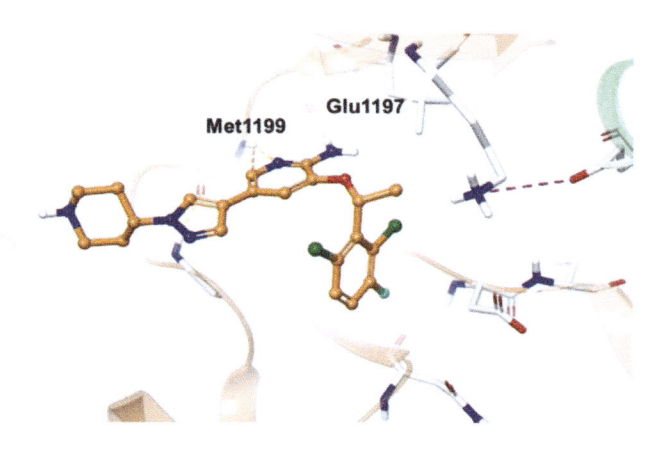

Figure 7. Crizotinib in complex with ALK (PDB ID: 2XP2).

1. B. Hallberg, R. H. Palmer. The role of the ALK receptor in cancer biology. *Ann. Oncol*, **2016**, *27* (Suppl. 3), iii4–iii15.
2. S. P. Ducray, K. Natarajan, G. D. Garland, S. D. Turner, G. Egger. The transcriptional roles of ALK fusion proteins in tumorigenesis. *Cancers*. **2019**, *11*, 1074.
3. A. T. Shaw, J. A. Engelman. ALK in lung cancer: past, present, and future. *J. Clin. Oncol.* **2013**, *31*, 1105–1111.
4. J. Wu, J. Savooji, D. Liu. Second- and third-generation ALK inhibitors for non-small cell lung cancer. *J. Hematol. Oncol.* **2016**, *9*, 19.
5. X. Du, Y. Shao, H. F. Qin, Y. H. Tai, H. J. Gao. ALK-rearrangement in non-small-cell lung cancer (NSCLC). *Thorac. Cancer.* **2018**, *9*, 423–430.
6. K. Sehgal, R. Patell, D. Rangachari, D. B. Costa. Targeting *ROS1* rearrangements in non-small cell lung cancer with crizotinib and other kinase inhibitors. *Transl. Cancer Res*. **2018**, *7* (Suppl. 7), S779–S786.
7. J. J. Cui, M. Tran-Dube, H. Shen, M. Nambu, P. P. Kung, M. Pairish, L. Jia, J. Meng, L. Funk, I. Botrous, M. Mctigue, N. Grodsky, K. Ryan, E. Padrique, G. Alton, S. Timofeevski, S. Yamazaki, Q. Li, H. Zou, J. Christensen, B. Mroczkowski, S. Bender, R. S. Kania, M. P. Edwards. Structure based drug design of crizotinib (PF-02341066), a potent and selective dual inhibitor of mesenchymal epithelial transition factor (c-MET) kinase and anaplastic lymphoma kinase (ALK). *J. Med. Chem.* **2011**, *54*, 6342–6363.
8. R. Roskoski, Jr. Classification of small molecule protein kinase inhibitors based upon the structures of their drug-enzyme complexes. *Pharmacol. Res.* **2016**, *103*, 26–48.

9.2 Ceritinib (Zykadia™) (51)

Structural Formula	Space-filling Model	Brief Information
$C_{28}H_{36}ClSO_3N_5$; FW = 558 logD = 6.4; tPSA = 104		Year of discovery: 2010 Year on Introduction: 2014 Discovered by: Novartis Developed by: Novartis Primary targets: ALK Binding type: I Class: receptor tyrosine kinase Treatment: NSCLC Other name: LDK378 Oral bioavailability = NA Elimination half-life = 41 h Protein binding = 97%

● = C   ○ = H   ● = O   ● = N   ● = S   ● = F   ● = Cl   ● = Br   ● = I

Ceritinib is a second-generation anaplastic lymphoma kinase (ALK) tyrosine kinase inhibitor which was granted accelerated approval in 2014 as a second-line treatment for patients with ALK-positive, late-stage (metastatic) non-small cell lung cancer (NSCLC) which has become resistant to crizotinib, the first approved ALK inhibitor. In 2017, its use was expanded to the first-line treatment of patients with metastatic ALK-positive NSCLC.[1]

Discovery of Ceritinib[2]

The lead Compound, TAE684,[3] is a highly potent ALK inhibitor whose kinase activity could be improved by replacing the methoxy group by isopropoxy giving analog **1** which had favorable overall kinase selectivity (Fig. 1). Unfortunately, both TAE684 and **1** were found to form reactive metabolites. For example, incubation in human liver microsomes of both compounds resulted in significant amount of reactive species which can be trapped with glutathione (GSH). Presumably, metabolic oxidation of the electron-rich benzene-1,4-diamine gives a highly reactive 1,4-diiminoquinone **3**, which reacts with GSH to form GSH adduct **4** (Fig. 2). The formation of reactive metabolites from both compounds prevented their further development due to potentially significant toxicological liabilities.

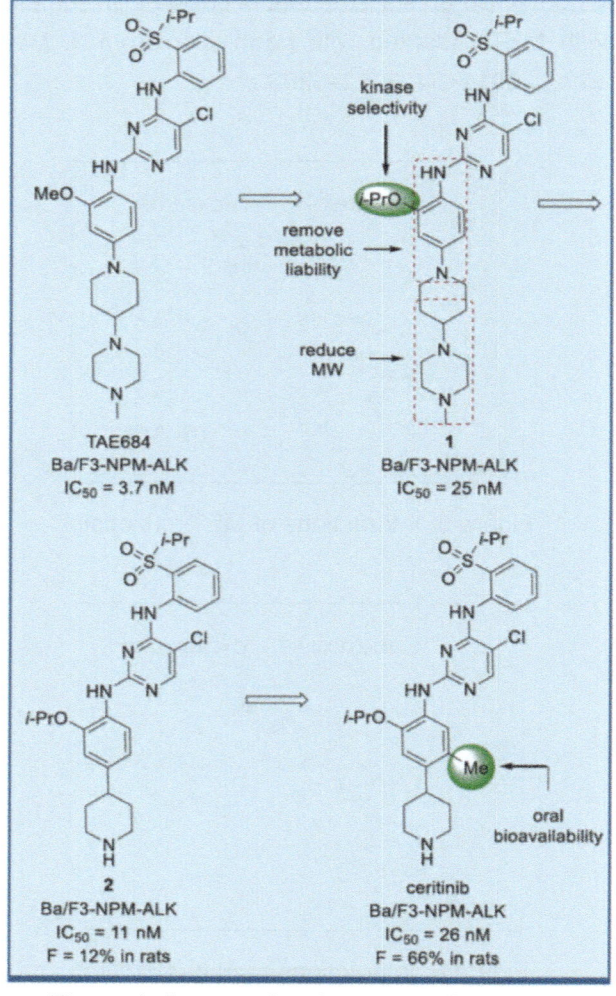

Figure 1. Structural optimization of ceritinib.

To overcome the reactive metabolite issue, the problematic 4-(piperidin-1-yl)aniline scaffold was changed to 4-(piperidin-4-yl)aniline, and the terminal piperazine was removed to reduce molecular weight, leading to **2** which was a potent and selective ALK inhibitor. Because **2** suffered from poor oral bioavailability in rats (12%), it was further modified by addition of methyl to the isopropoxyaniline ring to give ceritinib with improved (66%) oral bioavailability in rats.

Ceritinib exhibits potent inhibitory activity against EML4-ALK fusion mutations, including L1196M, G1269A, I1171T, and S1206Y, all of which are resistant to crizotinib. However, it is less active toward ALK with G1202R and F1174C mutations.

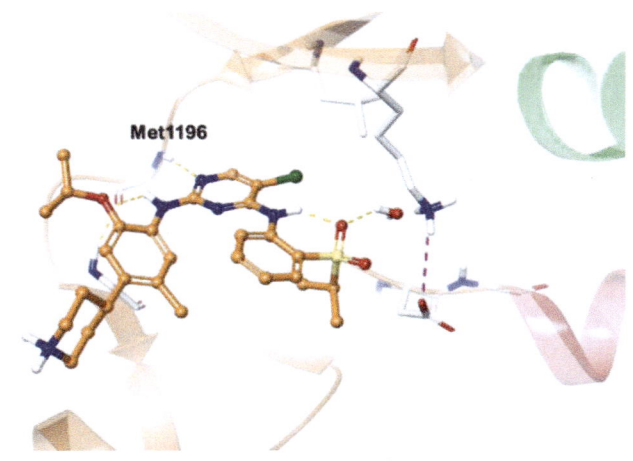

Figure 3. Co-crystal structure of ceritinib in complex with ALK (PDB ID: 4MKC).

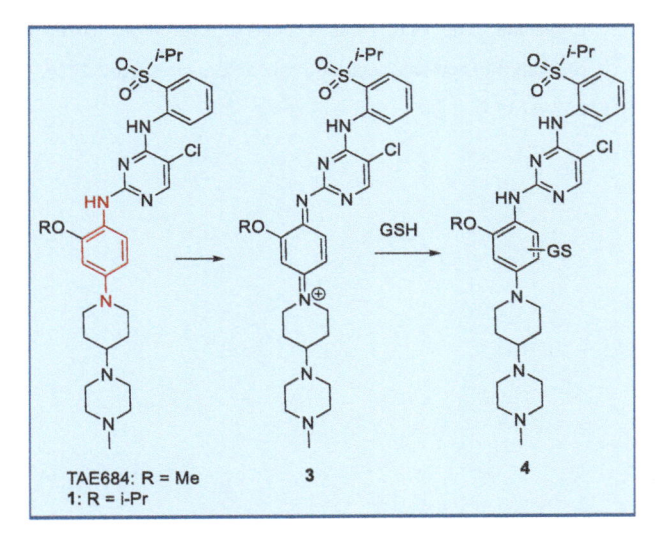

Figure 2. Presumed pathway for the formation of glutathione adduct **4**.

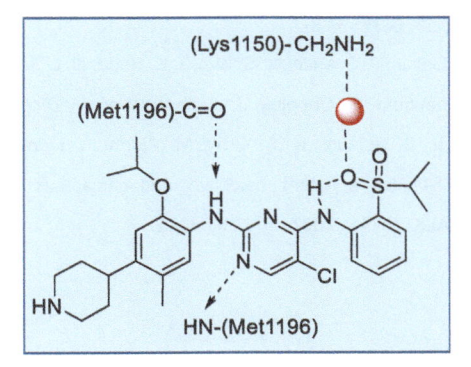

Figure 4. Summary of ceritinib–ALK interactions based on the co-crystal structure.

Binding Mode

Ceritinib is a type I inhibitor of ALK, and it binds similarly to crizotinib with the DFG-Asp in conformation (Fig. 3).[4-6] The 2,4-diaminopyrimidine core forms two hydrogen bonds with Met1199 of the ALK hinge (Fig. 4). One of the sulfonyl oxygens forms a 6-membered intramolecular hydrogen bond with the adjacent amine, and this oxygen also interacts with the side chain amine of Lys1150 via a water molecule. The terminal 2-isopropoxy-3-(piperidin-4-yl) phenyl group resides between the solvent and the ATP-binding pocket.

Ceritinib maintains high potency toward the most common G1269A and L1196M crizotinib-resistant mutants. Examination of both ceritinib– and crizotinib–ALK co-crystal structures (data not shown)[9] reveals that mutation of Gly to Ala in the G1269A mutant is unlikely to affect ceritinib binding, whereas the Ala methyl would cause a steric clash with the proximate dichlorofluorphenyl ring of crizotinib. The chlorine of the pyrimidine hinge binder of ceritinib, which approaches the gatekeeper Leu1196 side chain, engages in a hydrophobic interaction with the Leu side chain. In the L1196M mutant, a comparable interaction would occur with Met. However, introduction of a Met at the gatekeeper position 1196 would result in steric interference and unfavorable interactions with the 2-amino group of the pyridine hinge binder and the α-methyl group of crizotinib. The co-crystal structural data provides insights into the structural basis for the increased potency of ceritinib against both crizotinib resistant mutants.[6]

1. S. Dhillon, M. Clark. Ceritinib: first global approval. *Drugs.* **2014**, *74*, 1285–1291.
2. T. H. Marsilje, W. Pei, B. Chen, W. Lu, T. Uno, Y. Jin, T. Jiang, S., Kim, N. Li, M. Warmuth, Y. Sarkisova, F. Sun, A. Steffy, A. C. Pferdekamper, A. G. Li, S. B. Joseph, Y. Kim, B. Liu, T. Tuntland, X. Cui, P. Y. Michellys. Synthesis, structure-activity relationships, and in vivo efficacy of the novel potent and selective anaplastic lymphoma kinase (ALK) inhibitor 5-chloro-N2-(2-isopropoxy-5-methyl-4-(piperidin-4-yl)phenyl)-N4-(2-(isopropylsulfonyl)phenyl)pyrimidine-2,4-diamine (LDK378) currently in phase 1 and phase 2 clinical trials. *J. Med. Chem.* **2013**, *56*, 5675–5690.
3. A. V. Galkin, J. S. Melnick, S. Kim, T. L. Hood, N. Li, L. Li, G. Xia, R. Steensma, G. Chopiuk, J. Jiang, Y. Wan, P. Ding, Y. Liu, F. Sun, P. G. Schultz, N. S. Gray, M. Warmuth. Identification of NVP-TAE684, a potent, selective, and efficacious inhibitor of NPMALK. *PNAS.* **2007**, *104*, 270–275.
4. R. Roskoski, Jr. Classification of small molecule protein kinase inhibitors based upon the structures of their drug-enzyme complexes. *Pharmacol. Res.* **2016**, *103*, 26–48.
5. J. J. Cui, M. Tran-Dube, H. Shen, M. Nambu, P. P. Kung, M. Pairish, L. Jia, J. Meng, L. Funk, I. Botrous, M. Mctigue, N. Grodsky, K. Ryan, E. Padrique, G. Alton, S. Timofeevski, S. Yamazaki, Q. Li, H. Zou, J. Christensen, B. Mroczkowski, S. Bender, R. S. Kania, M. P. Edwards. Structure based drug design of crizotinib (PF-02341066), a potent and selective dual inhibitor of mesenchymal epithelial transition factor (c-MET) kinase and anaplastic lymphoma kinase (ALK). *J. Med. Chem.* **2011**, *54*, 6342–6363.
6. L. Friboulet, N. Li, R. Katayama, C. Lee, J. F. Gainor, A. S. Crystal, P. Y. Michellys, M. M. Awad, N. Yanagitani, S. Kim, A. C. Pferdekamper, J. Li, S. Kasibhatla, F. Sun, X. Sun, S. Hua, P. McNamara, S. Mahmood, E. L. Lockerman, N. Fujita, J. A. Engelman. The ALK inhibitor ceritinib overcomes crizotinib resistance in non-small cell lung cancer. *Cancer Discov.* **2014**, *4*, 662–673.

9.3 Alectinib (Alecensa™) (52)

Structural Formula	Space-filling Model	Brief Information
$C_{30}H_{34}O_2N_4$; MW = 482 clogP = 5.5; tPSA = 69		Year of discovery: 2008 Year of introduction: 2015 Discovered by: Chugai Developed by: Genentech/Roche Primary targets: ALK/RET Binding type: I Class: receptor tyrosine kinase Treatment: NSCLC Other names: CH5424802, RO5424802 Oral bioavailability = 37% Elimination half-life = 33 h Protein binding = >99%

● = C ○ = H ● = O ● = N ● = S ● = F ● = Cl ● = Br ● = I

Alectinib is a second-generation ALK inhibitor that received accelerated approval in 2015 for treatment of patients with ALK-positive metastatic non-small cell lung cancer (NSCLC) with disease progression or intolerance of crizotinib. In 2017, it was granted regular approval for first-line treatment of patients with metastatic ALK-positive NSCLC.

Alectinib is a highly selective ALK inhibitor with potent inhibitory activity against ALK mutations conferring resistance to crizotinib, including the gatekeeper mutation L1196M.[1,2] It also potently inhibits RET kinase (IC_{50} = 4.8 nM) and RET kinases with point mutations that were identified in medullary thyroid cancer (IC_{50} = 5.7 – 53 nM).[3]

Alectinib has a distinct benzo[b]carbazole hinge binder, vs. the aminopyridine and aminopyrimidine hinge binders commonly used in other ALK inhibitors, such as crizotinib and brigatinib.

Discovery of Alectinib[4,5]

The tetracyclic compound 1 was identified as a weak ALK inhibitor with an IC_{50} value of 1.3 μM from the Chugai kinase inhibitor library (Fig. 1). It has a unique tetracyclic scaffold with the ketone to form one hinge-binding interaction. Replacement of the 3-ethoxy with a chlorine furnished 2 with 2-fold lower ALK IC_{50} value than 1. The benzofuran scaffold was changed to the indole core to give 3 with 17-fold improved potency because the indole NH is involved in a hydrogen bond with the kinase. The electron donation from the indole nitrogen might reinforce the hinge binding with the ketone and contribute to the increased activity. The 3-cyano derivative 4 was 5-fold more potent than the 3-chloro counterpart 3 due to the additional interaction of the cyano group with the ALK protein. However, 4 suffered from high lipophilicity (clogP = 6.0) as well as poor metabolic stability due to oxidative dealkylation of diethylamino group. Both issues were mitigated by replacing the diethylamino)ethoxy side chain with a more water soluble 4-morpholinopiperidine to give 5 with much lower lipophilicity (clogP = 4.5). Finally, ethyl substitution at C9 led to alectinib with improved target selectivity, antiproliferative activity, and metabolic stability.

Binding Mode[6]

In the co-crystal structure with human ALK (Fig. 2), alectinib binds to the ATP site with the DFG-in mode, forming one hinge hydrogen bond between the carbonyl oxygen and the backbone NH of Met1199 (vs. two or three hinge hydrogen bonds observed with other ALK inhibitors, such as crizotinib, lorlatanib, brigatinib). This distinct hinge binding likely contributes to its high ALK selectivity. The indole NH hydrogen bonds to amide carbonyl of

Gly1269, and the cyano group participates in a complicated network of hydrogen bonds to the neighboring amino acids via the co-crystallized water or ethylene glycol molecules (Fig. 3).

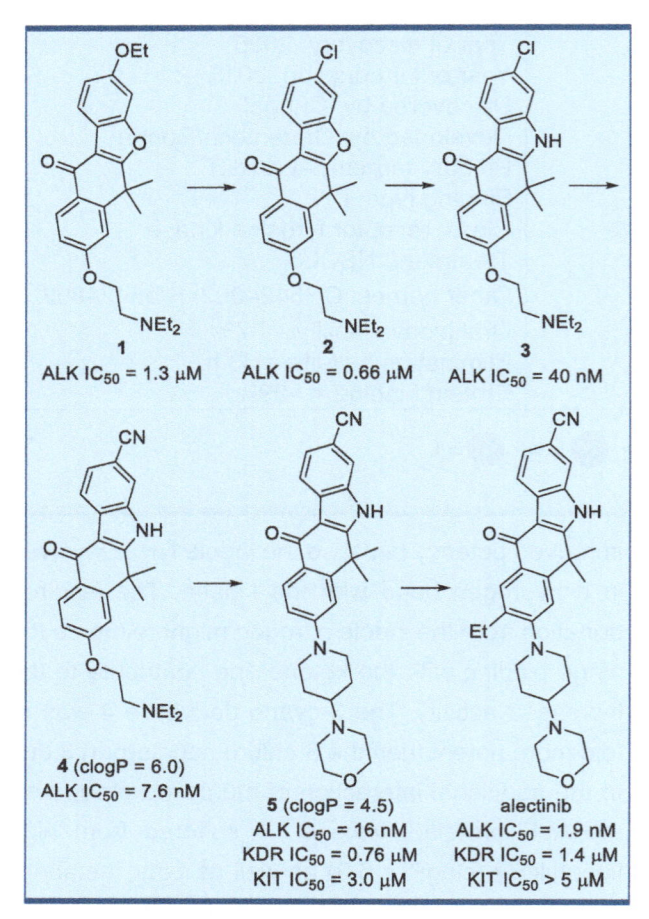

Figure 1. Structural optimization of alectinib.

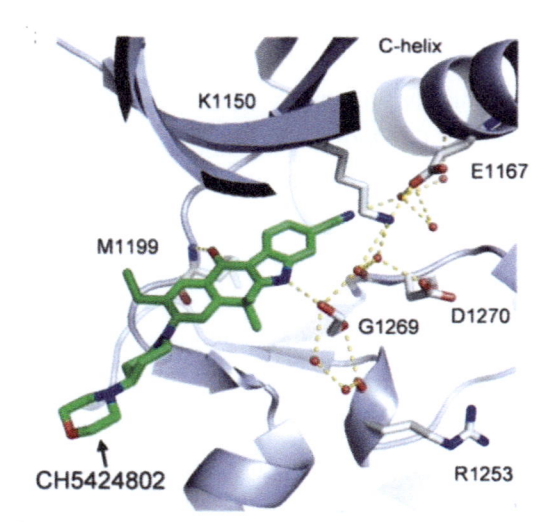

Figure 2. Co-crystal structure of alectinib–hALK (CH5424802 = alectinib). Adapted with permission from *Cancer Cell.* **2011**, *19*, 679–690.

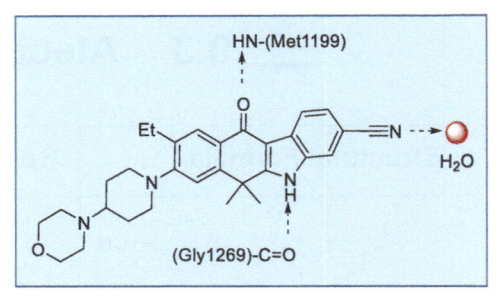

Figure 3. Summary of alectinib–ALK interactions based on the co-crystal structure.

1. Paik, J., Dhillon, S. Alectinib: A review in advanced, ALK-positive NSCLC. *Drugs* **2018**, *78*, 1247–1257.
2. M. Santarpia, G. Altavilla, R. Rosell. Alectinib: a selective, next-generation ALK inhibitor for treatment of ALK-rearranged non-small-cell lung cancer. *Expert Rev. Respir. Med.* **2015**, *9*, 255–268.
3. T. Kodama, T. Tsukaguchi, Y. Satoh, M. Yoshida, Y. Watanabe, O. Kondoh, H. Sakamoto. Alectinib shows potent antitumor activity against RET-rearranged non-small cell lung cancer. *Mol. Cancer Thera.* **2014**, *13*, 2910–2918.
4. K. Kinoshita, K. Asoh, N. Furuichi, T. Ito, H. Kawada, S. Hara, J. Ohwada, T. Miyagi, T. Kobayashi, K. Takanashi, T. Tsukaguchi, H. Sakamoto, T. Tsukuda, N. Oikawa. Design and synthesis of a highly selective, orally active and potent anaplastic lymphoma kinase inhibitor (CH5424802). *Bioorg. Med. Chem.* **2012**, *20*, 1271–1280.
5. K. Kinoshita, T. Kobayashi, K. Asoh, N. Furuichi, T. Ito, H. Kawada, S. Hara, J. Ohwada, K. Hattori, T. Miyagi, W. S. Hong, M. J. Park, K. Takanashi, T. Tsukaguchi, H. Sakamoto, T. Tsukuda, N. Oikawa. 9-Substituted 6,6-dimethyl-11-oxo-6,11-dihydro-5H-benzo[b]carbazoles as highly selective and potent anaplastic lymphoma kinase inhibitors. *J. Med. Chem.* **2011**, *54*, 6286–6294.
6. H. Sakamoto, T. Tsukaguchi, S. Hiroshima, T. Kodama, T. Kobayashi, T. A. Fukami, N. Oikawa, T. Tsukuda, N. Ishii, Y. Aoki. CH5424802, a selective ALK inhibitor capable of blocking the resistant gatekeeper mutant. *Cancer Cell.* **2011**, *19*, 679–690.

9.4 Brigatinib (Alunbrig™) (53)

Structural Formula	Space-filling Model	Brief Information
$C_{29}H_{39}ClPO_2N_7$; FW = 584 clogP = 1.7; tPSA = 85		Year of discovery: 2008 Year on Introduction: 2017 Discovered by: ARIAD Developed by: ARIAD Primary targets: ALK/EGFR Binding type: I Class: receptor tyrosine kinase Treatment: NSCLC Other name: AP26113 Oral bioavailability = not reported Elimination half-life = 25 h Protein binding = 90%

● = C　○ = H　● = O　● = N　● = S　● = F　● = Cl　● = Br　● = P

Brigatinib is a second-generation anaplastic lymphoma kinase (ALK) tyrosine kinase inhibitor that was granted the FDA-accelerated approval as a second-line treatment for patients with ALK-positive late-stage (metastatic) non-small cell lung cancer (NSCLC) which has become resistant to crizotinib, the first approved ALK inhibitor. In 2020, its use was expanded to the first-line treatment of patients with metastatic ALK-positive NSCLC.[1]

Brigatinib is highly active against a number of crizotinib-resistant mutants, including ALK L1196M, F1174C/V, I1171N, and G120R. It is even effective against ALK mutants that fail to respond to other second-generation ALK inhibitors such as alectinib and ceritinib. It is also a potent inhibitor of EGFR, and the mutant EGFR T790M.[2] Based on relative efficacy estimates, brigatinib may be more effective than ceritinib in terms of PFS (progression-free survival) and OS (overall survival) and alectinib in terms of OS in crizotinib-refractory ALK+ NSCLC patients.[3] Brigatinib is now marketed by Takeda which acquired Ariad for $3.66 billion.

Brigatinib is a rare phosphine oxide-containing drug, whose discovery provides an instructive example of bioisosteric replacement to revive a disposed drug candidate.

From TAE 684 to Brigatinib

TAE684 is a highly potent ALK inhibitor;[4] however, the formation of reactive metabolites prevented its further development due to potentially toxicological liabilities.[5] Because its reactive metabolites were believed to result from metabolic oxidation of the electron-rich benzene-1,4-diamine moiety, that subunit was replaced by the 4-(piperidin-4-yl)aniline subunit (Fig. 1). Additional structural optimizations led to ceritinib as a second-generation ALK inhibitor (Fig. 1)[5] (also see Section 9.2).

Remarkably, although TAE684 was relegated to the drug development graveyard due to reactive metabolite liabilities, brigatinib emerged to become the best-in-class ALK inhibitor for the treatment of crizotinib-resistant NSCLC cancers.[2,3,6] TAE684 and brigatinib are structurally identical except that the isopropylsulfonyl group of the former is replaced by an isosteric dimethylphosphine oxide (DMPO) of the latter. (One may wonder how many discarded drug candidates could be resurrected through judicious structural modifications.) Interestingly, reactive metabolite formation from brigitinib does not seem to cause serious problems, even though incubation in rat liver microsomes showed metabolite formation.[7] It remains unknown if and to what extent these reactive metabolites contribute to any side effects in humans.

The DMPO subunit of brigatinib plays a decisive role in its superior kinase selectivity profile because the basic oxygen of the P=O bond forms a 6-membered intramolecular hydrogen bond with the adjacent NH, which stabilizes and rigidifies the U-shaped bioactive conformation (Fig. 1). In the biochemical assays for IGF1R (insulin-like growth factor receptor) and the InsR (insulin receptor kinase) (both of which share high sequence homology with ALK), brigatinib exhibits 67- and 529-fold selectivity, vs. 2- and 13-fold, respectively, for TAE684.[6]

The rigidified and flattened conformation of the DMPO analog, relative to the corresponding isopropyl sulfone, due to the intramolecular hydrogen bond (masking the hydrogen bond donor NH) led not only to higher selectivity, but also substantial reductions in lipophilicity (logD = 3.0 for brigatinib vs. 4.1 for TAE684), tPSA (84 vs. 101) and plasma protein binding, as well as a marked increase in aqueous solubility (337 vs. 15 µg/mL) (Table 1).[6] Most importantly, the DMPO group proves to be indispensable in mitigating the reactive metabolite issue resulting from the electron-rich benzene-1,4-diamine scaffold, but its exact role remains to be determined.

Table 1. TAE684 vis-à-vis brigatinib[6]

properties	brigatinib	TAE684
ALK IC_{50}	0.37 nM	0.15 nM
IGF1R IC_{50}	24.9 nM	0.35 nM
InsR IC_{50}	196 nM	1.9 nM
Karpas cell IC_{50}	29 nM	14 nM
logD (pH 7.0)	3.0	4.1
tPSA	84	101
plasma protein binding	66% (human); 73% (rat); 64% (mouse)	99% (human, rat, mouse)
solubility (pH 7.4)	337 µg/mL	15 µg/mL
reactive metabolites	insignificant	significant

Binding Mode[6]

In the co-crystal structure with ALK, brigitinib adopts a U-shaped ligand conformation, and the 2-aminopyrimidine NH forms two hydrogen bonds with Met1199 of the ALK hinge (Figs. 2, 3). The methoxy group and DMPO aniline resides within proximity to the hinge and DFG sites, respectively (Fig. 4). The DMPO group is involved in an intramolecular hydrogen bond with the neighboring NH, which stabilizes the U-shaped conformation of brigatinib. Overlay of the two co-crystal structures of brigatinib and TAE684 reveals that the DMPO aniline in brigatinib is oriented toward the lower lobe of ALK compared with the sulfone aniline in TAE604 (Fig. 4). Consequently, the DMPO aniline is further away from the P-loop. The overall conformational variations of both inhibitors contribute to their differences in both potency and selectivity.

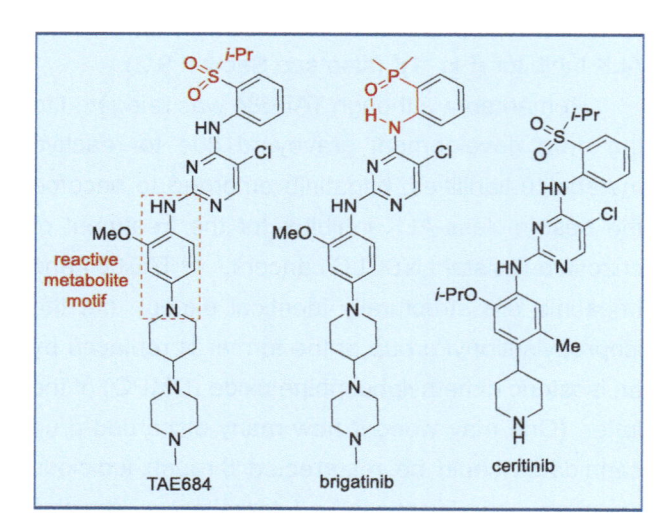

Figure 1. Structures of TAE684, brigatinib and ceritinib.

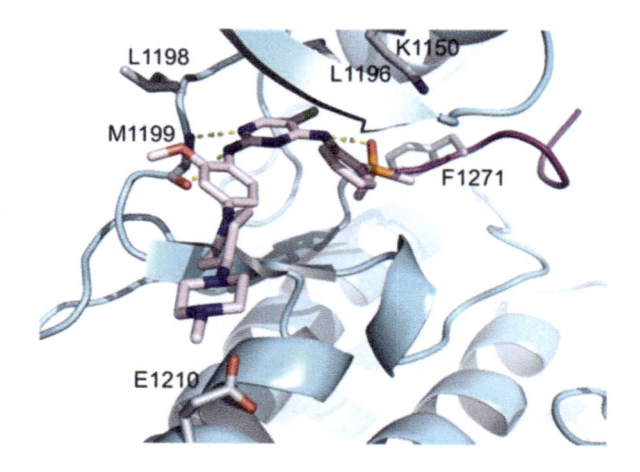

Figure 2. Co-crystal structure of brigatinib in complex with ALK. Adapted with permission from *J. Med. Chem.* **2016**, *59*, 4948–4964.

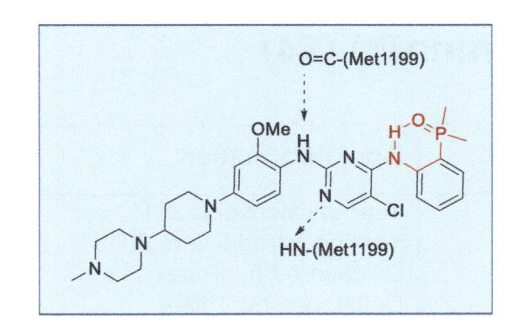

Figure 3. Ceritinib–ALK interaction.

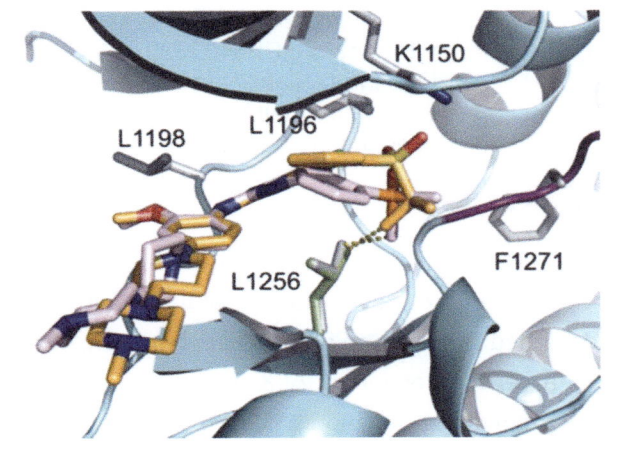

Figure 4. Overlay of brigatinib (pink) and TAE684 (orange) bound to wt ALK. Adapted with permission from *J. Med. Chem.* **2016**, *59*, 4948–4964.

1. A. Markham. Brigatinib: First global approval. *Drugs.* **2017**, *77*, 1131–1135.
2. S. Bedi, S. A. Khan, M. M. AbuKhader, P. Alam, N. A. Siddiqui, A. Husain. A comprehensive review on brigatinib – A wonder drug for targeted cancer therapy in non-small cell lung cancer. *Saudi Pharm.J.* **2018**, *26*, 755–763.
3. K. Reckamp, H. Lin, J. Huang, I. Proskorovsky, W. Reichmann, S. Krotneva, D. Kerstein, H. Huang, J. Lee. Comparative efficacy of brigatinib versus ceritinib and alectinib in patients with crizotinib-refractory anaplastic lymphoma kinase-positive non-small cell lung cancer. *Curr. Med. Res. Opion.* **2019**, *35* 569–576.
4. A. V. Galkin, J. S. Melnick, S. Kim, T. L. Hood, N. Li, L. Li, G. Xia, R. Steensma, G. Chopiuk, J. Jiang, Y. Wan, P. Ding, Y. Liu, F. Sun, P. G. Schultz, N. S. Gray, M. Warmuth. Identification of NVP-TAE684, a potent, selective, and efficacious inhibitor of NPMALK. *PNAS.* **2007**, *104*, 270–275.
5. T. H. Marsilje, W. Pei, B. Chen, W. Lu, T. Uno, Y. Jin, T. Jiang, S., Kim, N. Li, M. Warmuth, Y. Sarkisova, F. Sun, A. Steffy, A. C. Pferdekamper, A. G. Li, S. B. Joseph, Y. Kim, B. Liu, T. Tuntland, X. Cui, P. Y. Michellys. Synthesis, structure-activity relationships, and in vivo efficacy of the novel potent and selective anaplastic lymphoma kinase (ALK) inhibitor 5-chloro-N2-(2-isopropoxy-5-methyl-4-(piperidin-4-yl)phenyl)-N4-(2-(isopropylsulfonyl)phenyl)pyrimidine-2,4-diamine (LDK378) currently in phase 1 and phase 2 clinical trials. *J. Med. Chem.* **2013**, *56*, 5675–5690.
6. W. S. Huang, S. Liu, D. Zou, M. Thomas, Y. Wang, T. Zhou, J. Romero, A. Kohlmann, F. Li, J. Qi, L. Cai, T. Dwight, Y. Xu, R. Xu, D. Dodd, A. Toms, L. Parillon, X. Lu, R. Anjum, S. Zhang, F. Wang, J. Keats, S. D. Wardwell, Y. Ning, Q. Xu, L. E. Moran, Q. K. Mohemmad, H. G. Jang, T. Clackson, N. I. Narasimhan, V. M. Rivera, X. Zhu, D. Dalgarno, W. C. Shakespeare. Discovery of brigatinib (AP26113), a phosphine oxide-containing, potent, orally active inhibitor of anaplastic lymphoma kinase. *J. Med. Chem.* **2016**, *59*, 4948–4964.
7. A. A. Kadi, M. W. Attwa, H. W. Darwish. LC-ESI-MS/MS reveals the formation of reactive intermediates in brigatinib metabolism: elucidation of bioactivation pathways. *RSC. Adv.* **2018**, *8*, 1182–1190.

9.5 Lorlatinib (Lorbrena™) (54)

Structural Formula	Space-filling Model	Brief Information
$C_{21}H_{19}Cl_2FO_2N_6$; FW = 406 logD = 2.3; tPSA = 107		Year of discovery: 2011 Year on Introduction: 2018 Discovered by: Pfizer Developed by: Pfizer Primary targets: ALK/ROS1 Binding type: I Class: receptor tyrosine kinase Treatment: NSCLC Other name: PF6463922 Oral bioavailability = 81% Elimination half-life = 24 h Protein binding = 66%

● = C ● = H ● = O ● = N ● = S ● = F ● = Cl ● = Br ● = I

Lorlatinib is a third-generation anaplastic lymphoma kinase (ALK) and ROS1 (ROS proto-oncogene 1) inhibitor with high efficacy against multiple ALK-resistant mutations and good blood–brain barrier permeability. It was granted accelerated approval in 2018 for the second- or third-line treatment of ALK-positive metastatic non-small cell lung cancer (NSCLC). It also gained full FDA approval in 2021 for patients with metastatic NSCLC whose tumors are ALK-positive as detected by an FDA-approved test.[1]

Lorlatinib is a macrocyclic analog of crizotinib, the first targeted drug approved for the treatment of ALK gene rearranged NSCLC. Although crizotinib demonstrates initial robust efficacy in ALK+ tumors, it is less effective against the acquired secondary mutations (L1196M, G1269A, F1174L, S1206Y, 1151T-ins, L1152R, C1156Y and G1202R) which occur in the ALK kinase domain, with the gatekeeper mutation L1196M being the most common variant in NSCLC.[2] As the disease progresses during crizotinib treatment, lung cancer frequently spreads, or metastasizes, to the brain. It is estimated that 20–40% of NSCLC patients ultimately develop brain metastasis. Unfortunately, crizotinib has limited intracranial efficacy in treating brain metastases in ALK-positive NSCLC patients due to marginal brain penetration in humans (CSF/free plasma = 0.03).[3] Thus, the next generation of ALK inhibitors were designed to cross the blood–brain barrier as well as to target the secondary ALK mutations, particularly the gatekeeper mutation L1196M.[3,4]

Discovery of Lorlatinib[3]

Crizotinib has poor brain penetration in part due to the transporter-mediated efflux at the blood–brain barrier as shown by a MDR efflux ratio of 45 in the MDR1-MDCK permeability assay, which has been widely utilized to identify and characterize P-gp (P-glycoprotein) substrates and inhibitors (Fig. 1). The efflux ratio was reduced to 7.6 in the tertiary amide analog **1**, which lost activity presumably due to steric clash of the dimethyl substituted pyrazole and N,N-dimethyl amide in the preferred bound conformation. The co-crystal structures of similar analogs in complex with ALK reveals a U-shaped binding mode where the substituted fluorophenyl head group is in proximity to the heteroaromatic tail. Thus, macrocyclization connecting the pyrazole carbon to the amide N-methyl was carried out. This transformation resulted in substantially increased activity against crizotinib-resistant L1196M (**2**, K_i = 0.29 nM; cell IC_{50} = 14 nM) while curtailing the MDR efflux radio (from 7.6 to 4.2). However, compound **2** also exhibited potent inhibition of TRKB kinase, which is predominantly expressed in the nervous system and plays an important role in neuronal growth and development. Thus, SAR work was undertaken to mitigate the ALK selectivity margin.

The ATP-binding pocket of ALK consists of 27 residues including Leu1198, which is conserved in 26% of kinome at this position. As other kinases frequently have Phe or Tyr at this site, one could take advantage of this smaller Leu residue to gain selectivity against a majority of other kinases. For example, a larger group at this site may be acceptable to ALK Leu1198, but it may cause severe steric clash with TRKB Tyr635.

Overlay of the co-crystal structures of **2** in complex with ALK and TRKB (Fig. 2) reveals that the pyrazole *N*-methyl group of **2** in the TRKB complex is relatively far away from TRKB Tyr635 (3.2–4.1 Å from the three terminal Tyr atoms), and this explains why **2** is able to bind strongly to TRKB (Fig. 2). However, replacement of the methyl with a cyano group significantly increases TRKB selectivity to ~38-fold because the linear (and longer) cyano group of lorlatinib clashes with the TRKB Tyr635 (2.1–3.1 Å from the three terminal Tyr atoms) (structure not shown). The steric clash between the cyano and TRKB Tyr365 is unavoidable because of the rigidity of the macrocycle which prevents the cyano group from rotating away from the Tyr residue.

In biochemical and cellular assays, lorlatinib exhibits high potency against wild-type ALK and 13-resistant ALK mutants, and it is much more potent against ROS1 than ALK.

The relatively low molecular weight (406), desirable lipophilicity (logD = 2.3) and low MDR efflux ratio (1.5) of lorlatinib lead to favorable pharmacokinetic properties: good oral bioavailability (81%); prolonged elimination half-life (24 h); low plasma protein binding (66% in human plasma), and good brain penetration (brain tissue partition coefficient = 0.7 in mice).[5] These parameters support 100 mg once-daily dosing regimen, vs. 250 mg twice-daily administration for crizotinib. Because it readily penetrates the blood–brain barrier to achieve high CNS exposure, lorlatinib has demonstrated high efficacy for treating patients with CNS metastasis. So far, it appears that lorlatinib is the most effective ALK inhibitor against CNS metastasis among all ALK inhibitors.[6–8]

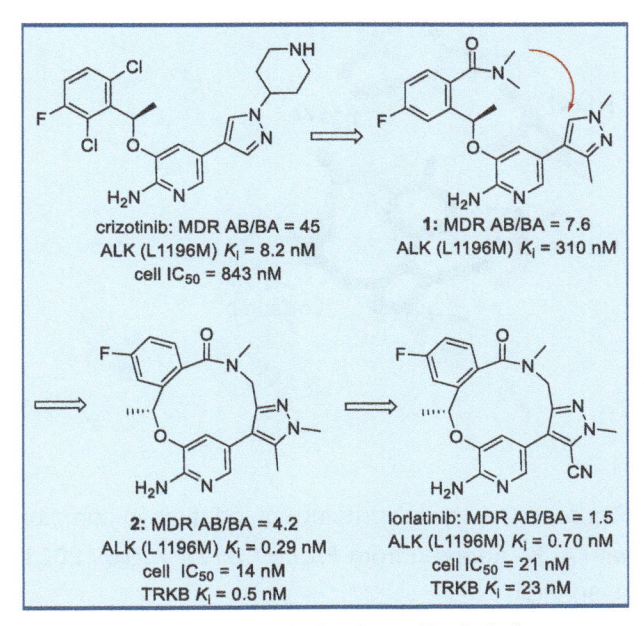

crizotinib: MDR AB/BA = 45
ALK (L1196M) K_i = 8.2 nM
cell IC_{50} = 843 nM

1: MDR AB/BA = 7.6
ALK (L1196M) K_i = 310 nM

2: MDR AB/BA = 4.2
ALK (L1196M) K_i = 0.29 nM
cell IC_{50} = 14 nM
TRKB K_i = 0.5 nM

lorlatinib: MDR AB/BA = 1.5
ALK (L1196M) K_i = 0.70 nM
cell IC_{50} = 21 nM
TRKB K_i = 23 nM

Figure 1. Structural optimization of lorlatinib.

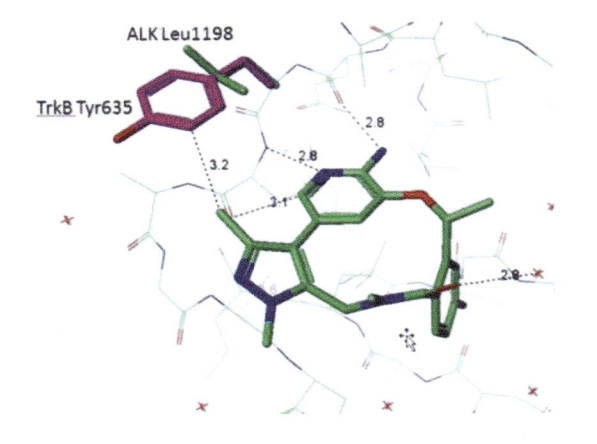

Figure 2. Macrocycle **2** in complex with ALK (green) and TRKB (purple) with selectivity residues highlighted. Adapted with permission from *J. Med. Chem.* **2014**, *57*, 4720–4744.

Binding Mode

As a macrocyclic analog of crizotinib, lorlatinib binds to the ALK kinase domain similarly to crizotinib[3,9–11] (Fig. 3). The amino group of the aminopyridine forms one hydrogen bond with the amide carbonyl of Glu1197 hinge and the N1 nitrogen of the pyridine ring forms another hydrogen bond with the amide NH of Met1199 of the hinge (Fig. 4).

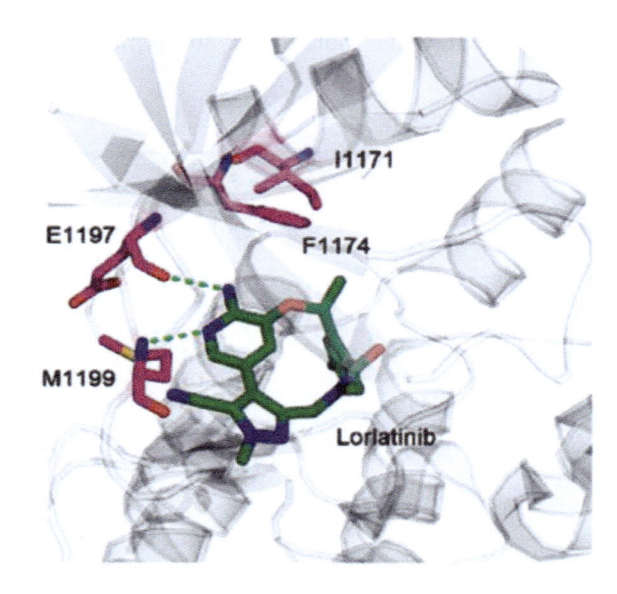

Figure 3. Co-crystal structure of lorlatinib in complex with ALK. Adapted from *Front. Cell Dev. Biol.* **2021**, *9*, 808864.

Figure 4. Summary of lorlatinib–ALK interactions based on the co-crystal structure.

The binding conformation of lorlatinib to ROS1 is also similar to that of crizotinib (data not shown).[12] The aminopyridine core makes two key hydrogen bonding interactions with hinge residues Glu2027 and Met2029. The macrocyclic amide carbonyl interacts with Lys1980 via a bound water molecule. In addition, the pyrazole nitrogen forms a water-mediated hydrogen bond to Asp2033 (Fig. 5).[12]

Although both crizotinib and lorlatinib display similar bound conformations, the latter shows greater potency for ROS1/ALK. This can be attributed to a loss of conformational entropy for the acyclic crizotinib to achieve the bioactive binding mode. In contrast, the rigid macrocylic lorlatinib, which is already in well-defined conformation, incurs little entropic penalties during binding.[13–15]

Lorlatinib is one of the only approved two macrocyclic PTK inhibitors. The other is pacritinib, a JAK1 inhibitor for myelofibrosis. Rigid macrocycles are often superior medicinal agents, e.g., simeprevir, an NS3/4A protease inhibitor for hepatitis C.[15]

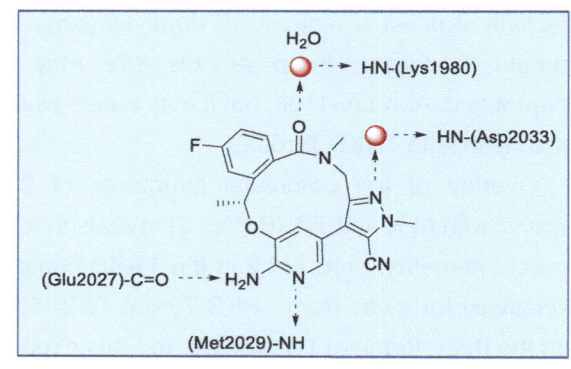

Figure 5. Summary of lorlatinib–ROS1 interactions based on the co-crystal structure.

1. Y. Y. Syed. Lorlatinib: First global approval. *Drugs.* **2019**, *79*, 93–98.
2. J. M. Hatcher, M. Bahcall, H. G. Choi, Y. Gao, T. Sim, R. George, P. A. Jänne, N. S. Gray. Discovery of inhibitors that overcome the G1202R anaplastic lymphoma kinase resistance mutation. *J. Med. Chem.* **2015**, *58*, 9296–9308.
3. T. W. Johnson, P. F. Richardson, S. Bailey, A. Brooun, B. J. Burke, M. R. Collins, J. J. Cui, J. G. Deal, Y. L. Deng, D. Dinh, L. D. Engstrom, M. He, J. Hoffman, R. L. Hoffman, Q. Huang, R. S. Kania, J. C. Kath, H. Lam, J. L. Lam, P. T. Le, M. P. Edwards. Discovery of (10R)-7-amino-12-fluoro-2,10,16-trimethyl-15-oxo-10,15,16,17-tetrahydro-2H-8,4-(metheno)pyrazolo[4,3-h][2,5,11]-benzoxadiazacyclotetradecine-3-carbonitrile (PF-06463922), a macrocyclic inhibitor of anaplastic lymphoma kinase (ALK) and c-ros oncogene 1 (ROS1) with preclinical brain exposure and broad-spectrum potency against ALK-resistant mutations. *J. Med. Chem.* **2014**, *57*, 4720–4744.
4. T. P. Heffron. Small molecule kinase Inhibitors for the treatment of brain cancer. *J. Med. Chem.* **2016**, *59*, 10030–10066.
5. W. Chen, J. Jin, Y. Shi, Y. Zhang, H. Zhou, G. Li. The underlying mechanisms of lorlatinib penetration across the blood-brain barrier and the distribution characteristics of lorlatinib in the brain. *Cancer Med.* **2020**, *9*, 4350–9.
6. K. Nakashima, Y. Demura, K. Kurokawa, T. Takeda, N. Jikuya, M. Oi, T. Tada, M. Akai, T. Ishizuka. Successful treatment with

lorlatinib in a patient with meningeal carcinomatosis of ALK-positive non-small cell lung cancer resistant to alectinib and brigatinib: A case report. *Medicine.* **2021**, *100*(39), e27385.

7. L. Wang, Z. Sheng, J. Zhang. Comparison of lorlatinib, alectinib and brigatinib in ALK inhibitor-naive/untreated ALK-positive advanced non-small-cell lung cancer: a systematic review and network meta-analysis. *J. Chemother.* **2022**, *34*, 87–96.

8. J. Wu, J. Savooji, D. Liu. Second- and third-generation ALK inhibitors for non-small cell lung cancer. *J. Hematol. Oncol.* **2016**, *9*, 19.

9. S. Liang, Q. Wang, X. Qi, Y. Liu, G. Li, S. Lu, L. Mou, X. Chen. Deciphering the mechanism of gilteritinib overcoming lorlatinib resistance to the double mutant I1171N/F1174I in anaplastic lymphoma kinase. *Front. Cell Dev. Biol.* **2021**, *9*, 808864.

10. J. J. Cui, M. Tran-Dube, H. Shen, M. Nambu, P. P. Kung, M. Pairish, L. Jia, J. Meng, L. Funk, I. Botrous, M. Mctigue, N. Grodsky, K. Ryan, E. Padrique, G. Alton, S. Timofeevski, S. Yamazaki, Q. Li, H. Zou, J. Christensen, B. Mroczkowski, S. Bender, R. S. Kania, M. P. Edwards. Structure based drug design of crizotinib (PF-02341066), a potent and selective dual inhibitor of mesenchymal epithelial transition factor (c-MET) kinase and anaplastic lymphoma kinase (ALK). *J. Med. Chem.* **2011**, *54*, 6342–6363.

11. R. Roskoski, Jr. Classification of small molecule protein kinase inhibitors based upon the structures of their drug-enzyme complexes. *Pharmacol. Res.* **2016**, *103*, 26–48.

12. H. Zou, Q. Li, L. D. Engstrom, M. West, V. Appleman, K. A. Wong, M. Mctigue, Y.-L. Deng, W. Liu, A. Brooun, S. Timofeevski, S. R. Mcdonnell, P. Jiang, M. D. Falk, P. B. Lappin, T. Affolter, T. Nichols, W. Hu, J. Lam, T. W. Johnson, T. Smeal, A. Charest, V. R. Fantin. PF-06463922 Is a potent and selective next-generation ROS1/ALK inhibitor capable of blocking crizotinib-resistant ROS1 mutations. *PNAS.* **2015**, *112*, 3493–3498.

13. A. Ali, C. Aydin, R. Gildemeister, K. P. Romano, H. Cao, A. Ozen, D. Soumana, A. Newton, C. J. Petropoulos, W. Huang, C. A. Schiffer. Evaluating the role of macrocycles in the susceptibility of hepatitis C virus NS3/4A protease inhibitors to drug resistance. *ACS Chem. Biol.* **2013**, *8*, 1469-78.

14. A. K. Yudin. Macrocycles: lessons from the distant past, recent developments, and future directions. *Chem. Sci.* **2015**, *6*, 30-49.

15. M. D. Cummings, S. Sekharan. Structure-based macrocycle design in small-molecule drug discovery and simple metrics to identify opportunities for macrocyclization of small-molecule ligands. *J. Med. Chem.* **2019**, *62*, 6843–6853.

Chapter 10. BRAF/Multikinase Inhibitors

Melanoma, the deadliest form of skin cancer, affects more than 1 million Americans annually. Although melanoma represents only 4% of all dermatological cancers, it accounts for 80% of all deaths. Fortunately, the treatment of patients with melanoma has been significantly advanced by the development of molecular-targeted therapies. For example, the BRAFV660E mutation is found in more than half of the patients diagnosed with cutaneous melanoma, making V660E a logical target. In 2011–2018, three BRAF kinase inhibitors were approved for use in the treatment of BRAF mutant melanoma, including vemurafenib, dabrafenib, and encorafenib (Fig. 1). More recently, dual inhibitors of BRAF and MEK (a downstream signaling target of BRAF in the MAPK pathway) showed synergistic benefit and therefore have been approved for the treatment of melanoma.[1,2]

Dabrafenib and encorafenib, which are more potent and more selective than the first-approved vemurafenib, can treat patients harboring BRAFV600E or BRAFV600K mutations, whereas vemurafenib is limited to the BRAFV600E-driven melanoma.[1,2]

Figure 2 shows six next-generation BRAF inhibitors currently in clinical trials. Two of these, tovorafenib and tinlorefenib, were shown to be brain penetrant.[3]

The development of BRAF-targeted therapies has revolutionized the treatment for BRAF-mutant metastatic melanoma. This chapter describes the discovery process that led to BRAF inhibitors **55–57**. Sorafenib (**23**) and regorafenib (**24**), which inhibit multikinases including BRAF600E and VEGFR, are discussed in Chapter 5 (VEGFR inhibitors).

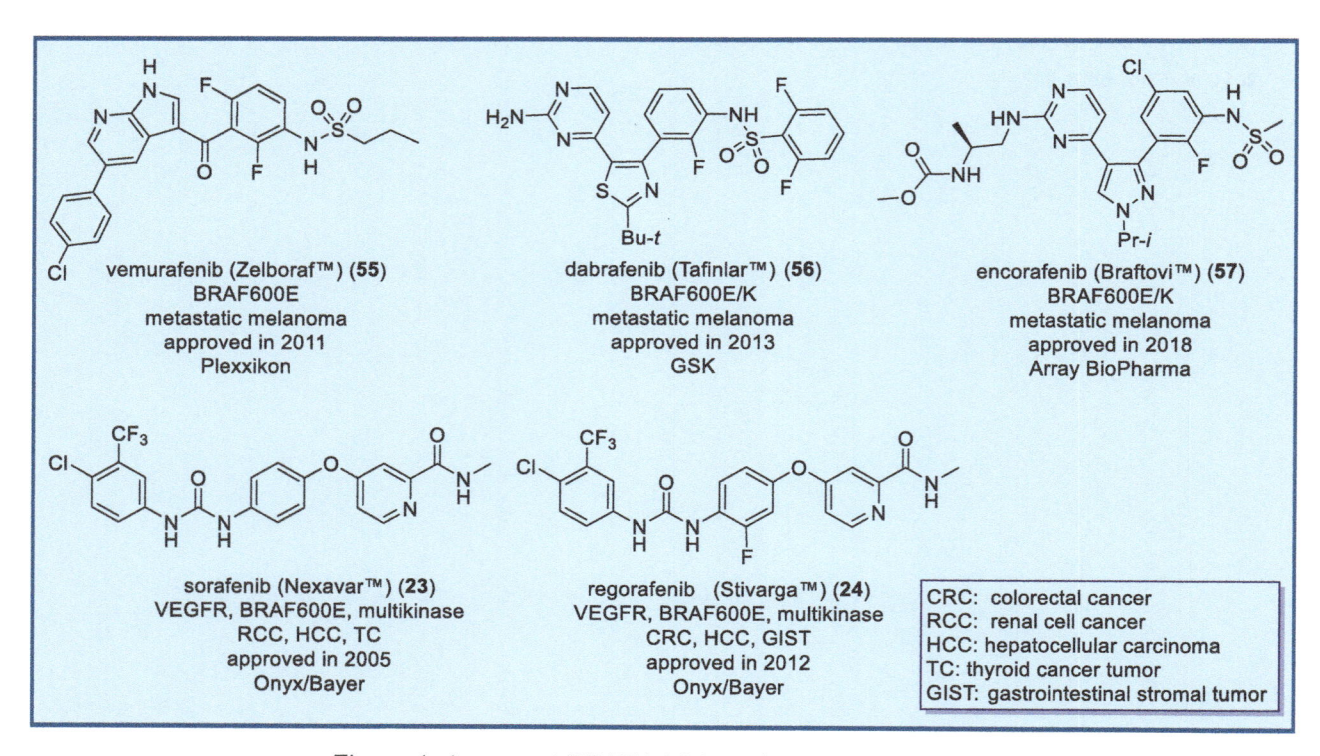

Figure 1. Approved BRAF inhibitors for various cancers.

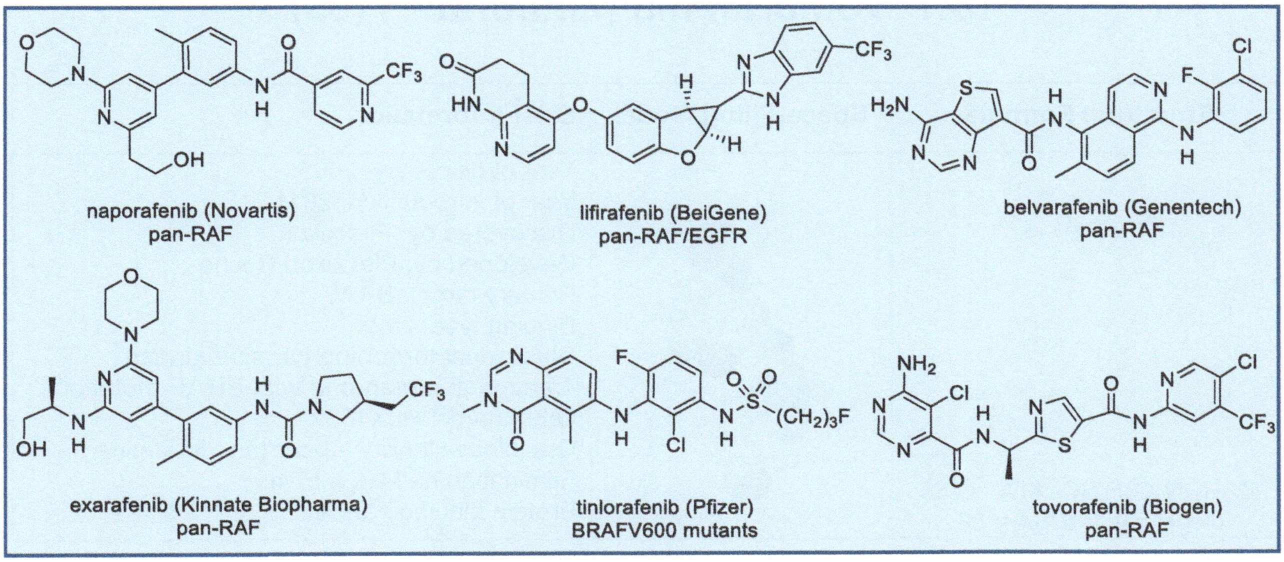

Figure 2. RAF inhibitors in clinical trials for solid tumors.

1. I. Proietti, N. Skroza, S. Michelini, A. Mambrin, V. Balduzzi, N. Bernardini, A. Marchesiello, E. Tolino, S. Volp, P. Maddalena, M. Di Fraia, G. Mangino, G. Romeo, C. Potenza. BRAF Inhibitors: Molecular Targeting and Immunomodulatory Actions. *Cancers*. **2020**, *12*, 1823.

2. V. Subbiah, C. Baik, J. M. Kirkwood. Clinical Development of BRAF plus MEK Inhibitor Combinations. *Trends Cancer*. **2020**, *6*, 797–810.

3. clinicaltrials.gov.

10.1 Vemurafenib (Zelboraf™) (55)

Structural Formula	Space-filling Model	Brief Information
$C_{23}H_{18}N_3O_3F_2SCl$; MW = 489 clogP = 4.1; tPSA = 88		Year of discovery: 2005 Year of introduction: 2011 Discovered by: Plexxikon Developed by: Plexxikon/Roche Primary target: BRAF Binding type: I1/2 Class: dual threonine/tyrosine kinase Treatment: melanoma with BRAF mutations Other name: PLX4032 Oral bioavailability = poor (not determined) Elimination half-life = 57 h Protein binding > 99%

● = C   ○ = H   ● = O   ● = N   ● = S   ● = F   ● = Cl   ● = Br   ● = I

Vemurafenib is a first-in-class inhibitor of oncogenic BRAF (BRAF proto-oncogene serine/threonine kinase), whose name derives from V600E mutated BRAF inhibitor.[1] It is used either alone or in combination with a MEK inhibitor (such as cobimetinib) for the treatment of patients with unresectable or metastatic melanoma driven by the BRAFV600E mutation as determined by the FDA-approved diagnostic test. This mutation is caused by transversion of thymine (T) to adenine (A) at nucleotide 1799 (T1799A), resulting in a substitution of valine (V) to glutamic acid (E) at position 600 of the amino acid sequence. The BRAFV660E mutation, found in about half of all cases of melanoma, is the deadliest and most aggressive form of skin cancer. The inhibition of mutated active BRAF V600E kinase by vermurafenib results in reduced signaling through the aberrant RAS/RAF/MEK/ERK pathway.

RAS/RAF/MEK/ERK Signaling Pathway[2–5]

The major RAS pathway which consists of rat sarcoma (RAS), rapidly accelerated fibrosarcoma (RAF), mitogen-activated extracellular signal-regulated kinase (MEK) and extracellular signal-related kinase 1/2 (ERK1/2), is commonly known as the RAS/RAF/MEK/ERK cascade (Fig. 1). Upon activation of the cell-membrane receptors, an adaptor protein GRB2 recruits guanine nucleotide exchange factor (GEF) such as Son of Sevenless (SOS) to the cell membrane. SOS is responsible for catalyzing the exchange of GDP for GTP on RAS, leading to RAS-GTP. The activated RAS interacts with RAF, triggering RAF activation through dimerization. The RAF proteins directly phosphorylate and activate MEK1/2, which in turn activate ERK1/2 proteins. The ERK1/2 phosphorylate diverse substrates involved in cell proliferation, survival, growth, metabolism, migration and differentiation.

RAS proteins serve as binary molecular switches cycling between the GTP-bound active state and the GDP-bound inactive state to transduce cell survival and growth signals. These proteins are normally tightly controlled, but in RAS-related cancers, the RAS switch is permanently turned "on", resulting in sustained cancer cell proliferation and thus tumor growth. These cancers can be caused by mutations in the RAS gene itself, or in upstream or downstream regulators (such as BRAF) (Fig. 2), which can drive the RAS pathway into a constitutively active, cancer-inducing state. Considerable effort has been directed towards the development of therapeutic agents that inhibit critical downstream pathway components, such as RAF and MEK kinases. Although RAF and MEK inhibitors both inhibit cellular proliferation, they exhibit different pharmacodynamic effects: MEK inhibitors prevent ERK phosphorylation regardless of cellular genotype, whereas RAF inhibitors only block ERK phosphorylation in RAF-mutated cells.

Selective RAF inhibitors including vemurafenib and dabrafenib have been shown to be effective in terms of tumor regression and overall survival benefit in BRAFV600E-driven melanoma. However, the clinical efficacy of these inhibitors requires near-complete inhibition of the RAS pathway.[6] Interestingly, these compounds also induce the paradoxical activation of the RAS pathway in RAF wild-type cells, which has been associated with cutaneous squamous cell carcinomas and keratoacanthomas during the RAF inhibitor treatment.[7] One strategy to combat paradoxical activation relies on a combination of BRAF and MEK inhibitors to target two steps in the RAS/RAF/MEK/ERK pathway (vertical inhibition). It has become standard of care in advanced-stage melanoma harboring the BRAFV600E mutation. Despite some increase in toxicity, the combination therapy dramatically improves response rates, progression-free survival and overall survival in patients with BRAF-mutant metastatic melanoma.

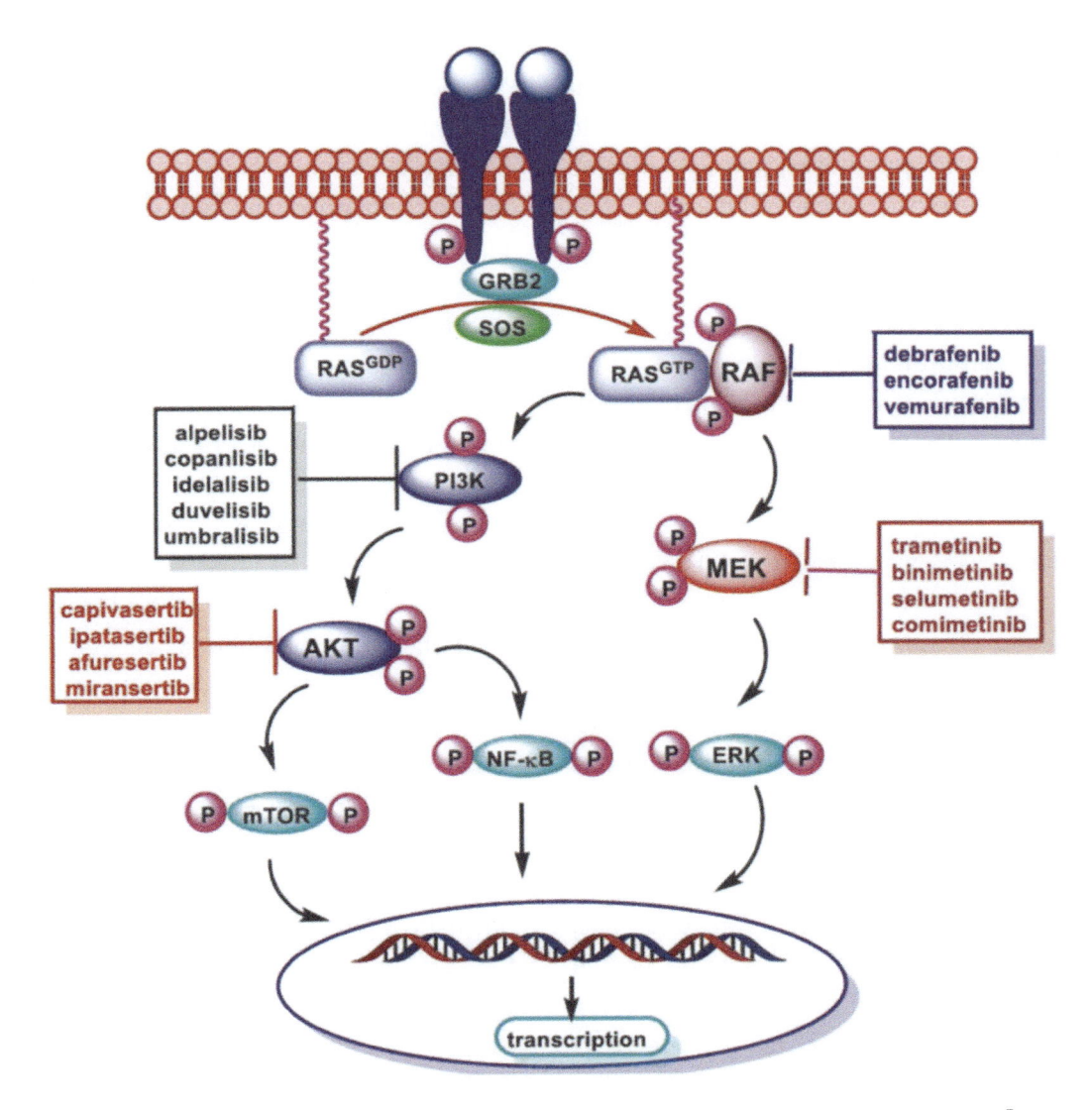

Figure 1. RAS/RAF/MEK/ERK and RAS/PI3K/AKT/mTOP signaling pathways.[5]

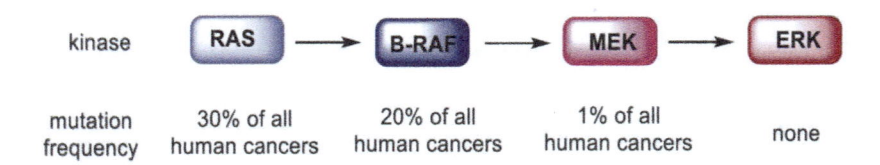

Figure 2. Genetic alterations of the RAS-dependent ERK1/2 pathway in cancer.[4]

RAS proteins also stimulate signaling through the PI3K phosphatidylinositol 3-kinase (PI3K) pathway which initiates a cascade that promotes cell growth and survival via AKT and mTOR as downstream signaling proteins (Fig. 1). However, there is extensive crosstalk between the PI3K and RAS/RAF/MEK signaling pathways, which provides compensatory signaling when either is inhibited. Therefore, therapeutic approaches to occlude either the MEK or PI3K pathway may not be sufficient to inhibit the growth of cancer cells, and to overcome resistance and improve efficacy. Therefore it is desirable to combine MEK and PI3K inhibitors that could simultaneously block both pathways.[8]

Discovery of Vemurafenib [1,9,10]

Vemurafenib was discovered through fragment-based approach using high throughput screening for kinase inhibition followed by crystallography of inhibitor-protein complexes (Fig. 3). To identify protein kinase ATP hinge binders, 20,000 small molecules (MW = 150–350) were screened at a concentration of 200 µM against five diversified and structurally well characterized kinases, resulting in 238 compounds with more than 30% inhibition vs. three of the five kinases. These compounds underwent high-throughput co-crystallization with recombinant kinase domains from several kinases, to yield over 100 structures of kinases co-crystallized with bound compounds. This process also eliminated some false positives generated in the initial screening at 200 µM. One of the co-crystal structures with PIM-1 revealed the binding of 7-azaindole 1 to the ATP-binding site with weak affinity (IC_{50} = 200 µM). To increase binding affinity, various substituents were introduced at C7 resulting in the 3-aminophenyl analog 2 with IC_{50} of 100 µM for PIM-1. The co-crystal structure of 2 with PIM-1 showed a single hydrogen bond between azaindole 1-NH and the hinge (data not shown).[1] Overlay of the azaindole scaffold, which contains a donor–acceptor motif, with the ATP-binding sites of many kinases indicated its capability to form two hydrogen bonds with the hinge. Indeed, the co-crystal structure of 3-benzyl analog 3 with the kinase domain of FGFR1 (data not shown)[1] showed two hydrogen bonds with the hinge, which is consistent with its enhanced binding affinity (IC_{50} = 1.9 µM). To improve potency and selectivity, a library of mono- and di-substituted azaindole analogs were synthesized, resulting in a series of difluorophenylsulfonamide-based compounds that potently and selectively inhibited the oncogenic BRAF. These compounds were co-crystallized with engineered forms of BRAFV600E and wild-type BRAF, and the co-crystal structural data guided further SAR studies. This effort identified compound 4 with IC_{50} of 13 nM against BRAFV600E as well as 12-fold selectivity vs. wild-type BRAF. Furthermore, this compound displayed high selectivity against a diverse panel of 70 other kinases. The co-crystal structure of 4 with BRAF confirmed the two hydrogen bond interactions of the azaindole to the hinge residues (data not shown).[1]

The pharmacokinetic properties of 4 were improved by replacing the 5-chlorine with 4-chlorophenyl to give vemurafenib. This compound exhibits comparable potency for BRAFV600E (IC_{50} = 31 nM) and CRAF-1 (IC_{50} = 48 nM) and is selective against many other kinases. Despite its modest (3-fold) preference for BRAFV600E to wild-type BRAF (IC_{50} = 30 vs. 100 nM) in the biochemical assays, vemurafenib manifested pronounced selectivity for BRAFV600E over wild-type BRAF in melanoma or colorectal cancer cell lines. However, vemurafenib appears to be less potent and less selective than dabrafenib, the second FDA-approved BRAF inhibitor (Table 1).

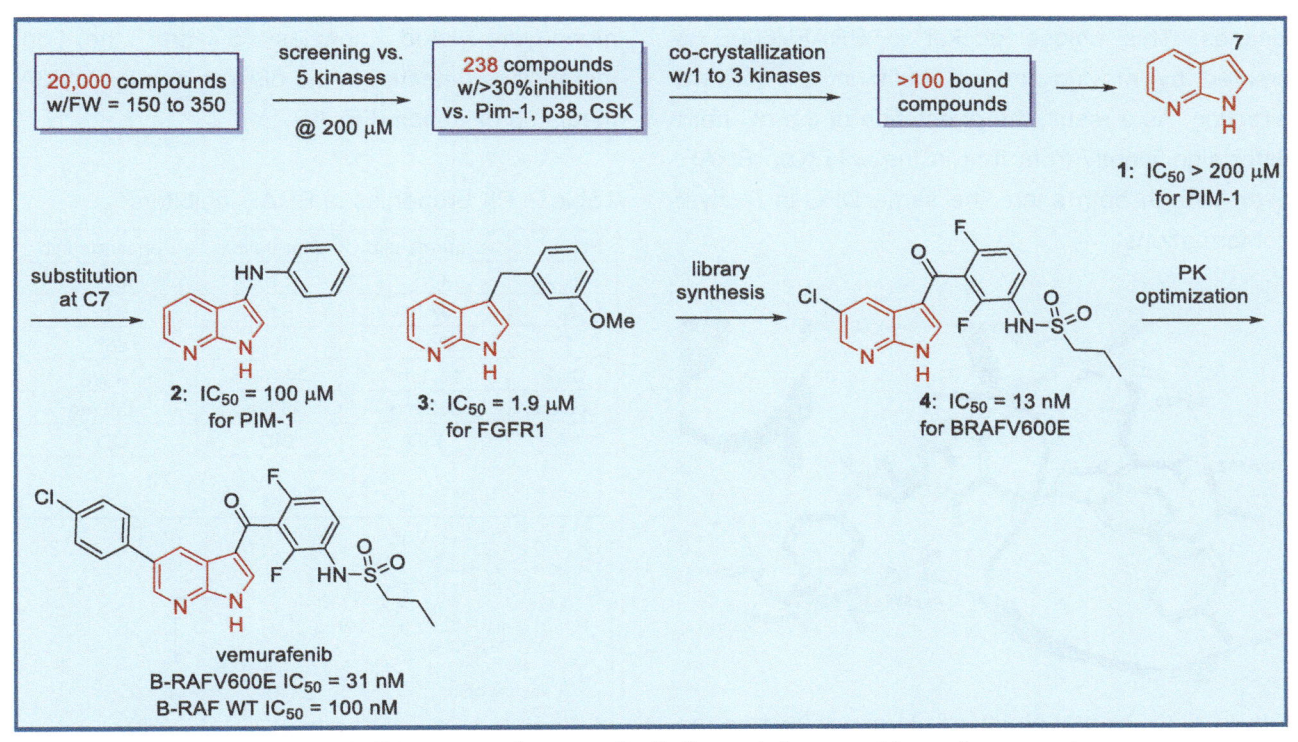

Figure 3. Discovery of vemurafenib through fragment-based approach.

Table 1. In vitro IC$_{50}$'s of RAF inhibitors[9,10]

kinase	vemurafetinib	Dabrafetinib	encorafetinib
BRAFV600E	31 nM	0.7 nM	0.35 nM
BRAF wt	100 nM	5.2 nM	0.47 nM
CRAF	48 nM	6.3 nM	0.3 nM

Binding Mode[1,9–11]

Vemurafenib underwent co-crystallization in the presence of the BRAFV600E protein to form a dimer with two protomers (data not shown).[1] Both protomers adopt the DFG-in (active) conformations, but interestingly, vemurafenib occupies the ATP-binding pocket in only one of the protomers. However, in the co-crystal structure of wild-type BRAF with compound **4** (a close analog of vemurafenib) (structure not shown), one protomer adopts the DFG-in (active) conformation, whereas the other adopts the DFG-out (inactive) conformation. Compound **4** binds to the protomer in active conformation with 100% occupancy, and the other protomer in inactive conformation with 60% occupancy. The structural differences between wild-type and mutant BRAF are responsible for the oncogenic activation of BRAF due to the V600E mutation. In the mutated form, the activation loop is kept in the DFG-in active conformation, resulting in constitutively active mutant protein.

As shown in the co-crystal structure of vemurafenib in the active site of BRAFV600E (Fig. 4), the azaindole is anchored to the hinge through two critical hydrogen bonds. More specifically, the N7 serves as hydrogen bond acceptor from the backbone amide of Cys532, while N1 serves as hydrogen bond donor to the backbone carbonyl oxygen of Gln530 (Fig. 5). The inhibitor occupies the adenine pocket with a DFG-in conformation, thereby enabling the formation of two hydrogen bonds between the sulfonamide moiety and Asp594 and Phe595 of the DFG sequence. The nitrogen atom of the sulfonamide moiety hydrogen bonds to the main chain NH group of Asp594, which suggests that this nitrogen atom is deprotonated (sulfonamide NH pK_a = 7.7, measured). One of the oxygen atoms of the sulfonamide hydrogen bonds to the backbone NH of Phe595 (Fig. 5). Both hydrogen bonds enable the propyl side chain to fit an interior pocket specific to the mutant BRAF protein. Because a similar pocket at this exact location is rare among other non-RAF kinases, access to the RAF-selective pocket plays an important role in the selectivity against non-RAF kinases. This unique pocket in BRAFV600E is created by moving the αC helix in an outward direction. As a result, the orientation of the αC-helix differ significantly from that in the wild-type BRAF, even though both share the same DFG-in (active) conformations.

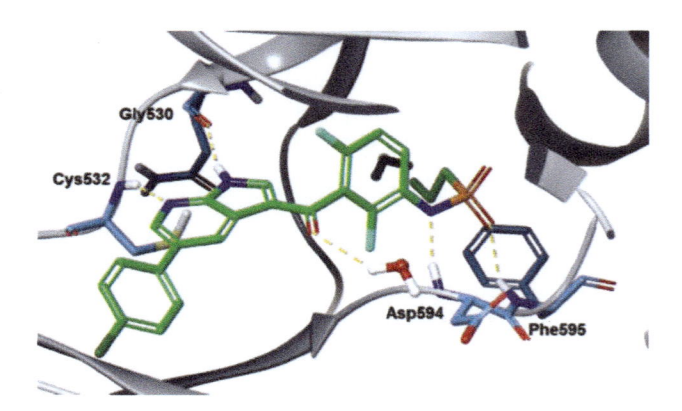

Figure 4. Co-crystal structure of vemurafenib in the active site of BRAFV600E (PDB ID: 3OG7).

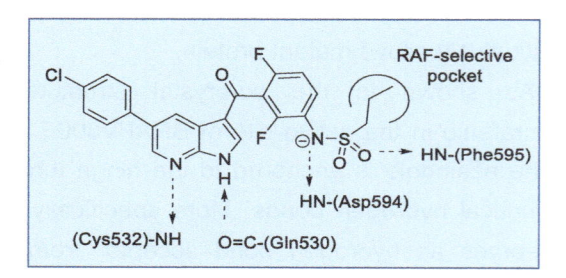

Figure 5. Summary of vemurafenib–BRAFV600E interactions based on an X-ray co-crystal structure.

PK Properties and Metabolism[12,13]

The recommended daily dose of vemurafenib is 1,920 mg (4 × 240 mg tablets, BID), the highest dosage among the three FDA-approved RAF inhibitors (Table 2). One contributing factor for such a high dose is the poor and variable oral bioavailability due to both low cell permeability and poor aqueous solubility. Nevertheless, vemurafenib is absorbed rapidly after a single oral dose of 960 mg, reaching a maximum drug concentration approximately 4 h after administration. It also exhibits long elimination half-life (57 h). It is cleared predominantly via the hepatic route.

Following oral administration, the parent drug predominates in the plasma, and the isomeric monohydroxylated species **5** are the only metabolites detected in the plasma due to CYP3A4 mediated oxidation (Fig. 6).

Table 2. PK properties of BRAF inhibitors[9]

PK	vemurafetinib	dabrafetinib	encorafetinib
t_{max}	4 h	2 h	2 h
$t_{1/2}$	57 h	8 h	6 h
F	Poor	95%	86%
CL/F	1.2 L/h	34.3 L/h	14 L/h
dosage	960 mg	150 mg	450 mg
dose	BID	BID	QD

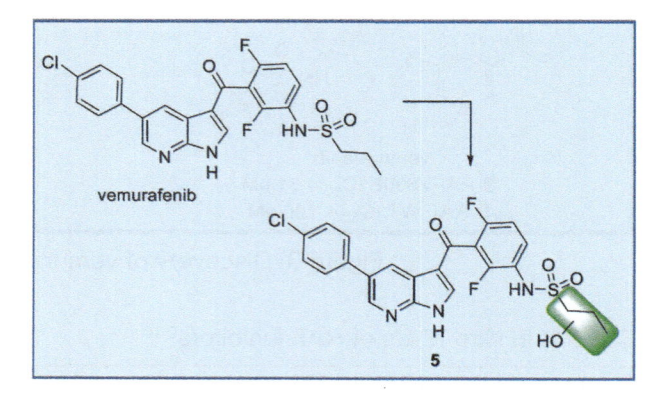

Figure 6. Major metabolic pathway of vemurafenib in humans.

1. G. Bollag, J. Tsai, J. Zhang, C Zhang, P. Ibrahim, K. Nolop, P. Hirth. Vemurafenib: the first drug approved for BRAF-mutant cancer. *Nat. Rev. Drug Discov.* **2012**, *11*, 873–886.
2. M. Trott. The ras pathway and cancer: regulation, challenges and the therapeutic progress. Technology Networks. May 04, 2021.
3. S. Okumura, P. A. Jänne. Molecular pathways: the basis for rational cobination using MEK inhibitors in KRAS-mutant cancers. *Clin. Cancer Res.* **2014**, *20*, 4193–4199.
4. C. Fremin, S. Meloche. From basic research to clinical development of MEK1/2 inhibitors for cancer therapy. *J. Hematol. Oncol.* **2010**, *3*, 8.
5. J. Han, Y. Liu, S. Yang, X. Wu, H. Li, Q. Wang. MEK inhibitors for the treatment of non-small cell lung cancer. *J. Hematol. Oncol.* **2021**, *14*, 1.
6. G. Bollag, P. Hirth, J. Tsai, J. Zhang, P. N. Ibrahim, H. Cho, W. Spevak, C. Zhang, Y. Zhang, G. Habets, E. A. Burton, B. Wong, G. Tsang, B. L. West, B. Powell, R. Shellooe, A. Marimuthu, H. Nguyen, K. Y. Zhang, D. R. Artis, K. Nolop. Clinical efficacy of a RAF inhibitor needs broad target blockade in BRAF-mutant melanoma. *Nature.* **2010**, *467*, 596–599.

7. C. Zhang, W. Spevak, Y. Zhang, E. A. Burton, Y. Ma, G. Habets, J. Zhang, J. Lin, T. Ewing, B. Matusow, G. Tsang, A. Marimuthu, H. Cho, G. Wu, H. Wang, D. Fong, S. Guyen, P. Shi, M. Womack, G. Nespi, G. Bollag. RAF inhibitors that evade paradoxical MAPK pathway activation. *Nature.* **2015**, *526*, 583–586.

8. S. Okumura, P. A. Jänne. Molecular pathways: the basis for rational combination using MEK inhibitors in KRAS-mutant cancers. *Clin. Cancer Res.* **2014**, *20*, 4193–4199.

9. J. Tsai, J. T. Lee, W. Wang, J. Zhang, H. Cho, S. Mamo, R. Bremer, S. Gillette, J. Kong, N. K. Haass, K. Sproesser, L. Li, K. S. Smalley, D. Fong, Y. L. Zhu, A. Marimuthu, H. Nguyen, B. Lam, J. Liu, I. Cheung, G. Bollag. Discovery of a selective inhibitor of oncogenic B-Raf kinase with potent antimelanoma activity. *PNAS.* **2008**, *105*, 3041–3046.

10. R. Roskoski R., Jr. Classification of small molecule protein kinase inhibitors based upon the structures of their drug-enzyme complexes. *Pharmacol. Res.* **2016**, *103*, 26–48.

11. C. C. Ayala-Aguilera, T. Valero, A. Lorente-Marcias, S.Croke, A. Unciti-Broceta. Small Molecule Kinase Inhibitor Drugs Synthesis (1995–2021). *J. Med. Chem.* **2022**, *65*, 1047–1131.

12. A. Kim, M. S. Cohen. The discovery of vemurafenib for the treatment of BRAF-mutated metastatic melanoma. *Expert Opin. Drug Discov.* **2016**, *1*, 907-16.

13. S. M. Goldinger, J. Rinderknecht, R. Dummer, F. P. Kuhn, Yang, L. Lee, R. C. Ayala, J. Racha, W. Geng, D. Moore, M. Liu, A. K. Joe, S. P. Bazan, J. F. Grippo. A single-dose mass balance and metabolite-profiling study of vemurafenib in patients with metastatic melanoma. *Pharmacol. Res. Perspect.* **2015**, *3*, e00113.

10.2 Dabrafenib (Tafinlar™) (56)

Structural Formula	Space-filling Model	Brief Information
$C_{23}H_{20}N_5O_2F_3S_2$; MW = 519 clogP = 4.6; tPSA = 109		Year of discovery: 2008 Year of introduction: 2013 Discovered by: GSK Developed by: GSK Primary target: BRAF Binding type: I1/2 Class: dual threonine/tyrosine kinase Treatment: melanoma with BRAF mutations Other name: GSK-2118436 Oral bioavailability = 95% Elimination half-life = 8 h Protein binding = 99.7%

● = C   ○ = H   ● = O   ● = N   ● = S   ● = F   ● = Cl   ● = Br   ● = I

Dabrafenib received accelerated FDA approval in 2013 to treat patients with unresectable or metastatic melanoma harboring the BRAFV600E mutation, as determined by the FDA approved diagnostic test. A combination therapy with trametinib (MEK inhibitor) was also approved in 2014 for the treatment of patients with unresectable or metastatic melanoma with BRAFV600E or BRAFV600K mutations. Dabrafenib follows vemurafenib as the second FDA-approved BRAFV600E inhibitor. It has been shown to be more potent and more selective.

Discovery of Dabrafenib [1,2]

The pathway of discovery started with the initial BRAF-active compound 1, which consists of an imidazopyridine core and a large hydrophobic benzamide headgroup, is shown in Fig. 1. Lead compound 1 displayed good activity in the BRAFV600E enzyme assay but showed little activity in the mechanistic (pERK) cell-based assays using the BRAFV600E mutant cell lines. Evaluation of several different headgroup linkers led to the sulfonamide-containing analog 2 which exhibited a substantial improvement in cellular potency, despite a significant drop in the enzyme assay. Replacing the imidazopyridine subunit with a thiazole core resulted in 10- and 2-fold improvement in enzyme and cellular activity, respectively. Because the thiazole analog 3 is highly lipophilic (clogP = 6.7), the N-methyl-tetrahydroisoquinoline tail was replaced by morpholinopyridine to afford 4, which showed good activity in both enzyme and cellular assays. Thiazole 4 exhibited low clearance and good overall oral systemic exposure in rats. However, 4 showed very high clearance and poor bioavailability in nonrodent species. In addition, biotransformation studies using dog and monkey liver microsomes identified several major metabolic soft spots in the tail and core regions of the molecule. Thus, significant structural modifications were implemented to reduce the molecular weight and the number of metabolic soft spots. Truncation of the tail by replacing the aminopyridine in 4 with a small alkyl group largely maintained activity but resulted in significantly higher clearance in rats. Surprisingly, the unsubstituted aminopyrimidine 5 maintained in vitro potency but suffered from rapid clearance in rats. Subsequent SAR studies revealed that C2 fluorination improved or maintained the enzyme and cellular potency and contributed to superior pharmacokinetic properties. Replacement of the isopropyl thiazole with a tert-butyl thiazole core further improved cellular potency. A combination of both structural modifications led to dabrafenib. This compound displayed potent inhibition against BRAFV600E in the enzyme and cellular assays (IC$_{50}$ = 0.7 nM; pERK EC$_{50}$ = 4 nM), BRAF (IC$_{50}$ = 5.2 nM) and CRAF (IC$_{50}$ = 6.3 nM) in the biochemical assays. Notably, dabrafenib elicited only a marginal effect in vitro on cells with wild-type BRAF (HFF IC$_{50}$ = 3.0 μM) and in tumor cells not harboring the BRAFV600E mutation. Furthermore, it showed 17-fold preference for the BRAFV600E to wild-type. Dabrafenib exhibited >500-fold selectivity over other kinases, except for ALK5 (<100-fold selectivity). Among the three FDA-approved RAF inhibitors, dabrafenib appears to be most selective (Table 1, Section 10.1).

Figure 1. Summary of the chemical optimization of dabrafenib.

Binding Mode[3–5]

Overlay of BRAFV660E protein bound with dabrafenib and vemurafenib shows comparable protein conformation and αC-helix shift (data not shown).[5] In addition, both compounds bind in a similar way at the active site of BRAFV600E.[5] Dabrafenib binds to the adenine pocket with a DFG-in, αC-out inactive conformation (type I1/2 inhibitor), and forms two hydrogen bonds between the sulfonamide moiety and Asp594 and Phe595 of the DFG sequence. The N1 of the pyrimidine forms a hydrogen bond with the amide NH of Cys532 of the hinge, and the 2-amino group attached to the pyrimidine hydrogen bonds to the amide oxygen of Cys532. The nitrogen atom of the sulfonamide moiety (pK$_a$ ca. 7, estimated) appears to be anionic at the binding site forming a hydrogen bond with the amide NH group of Asp594,[5] while one of the oxygen atoms of the sulfonamide forms a hydrogen bond with the amide NH of Phe595 (Fig. 3). The difluorophenyl group of dabrafenib and the propyl group of vemurafenib penetrate into the same interior lipophilic back pocket specific to the BRAFV600E protein, resulting in high kinase selectivity.

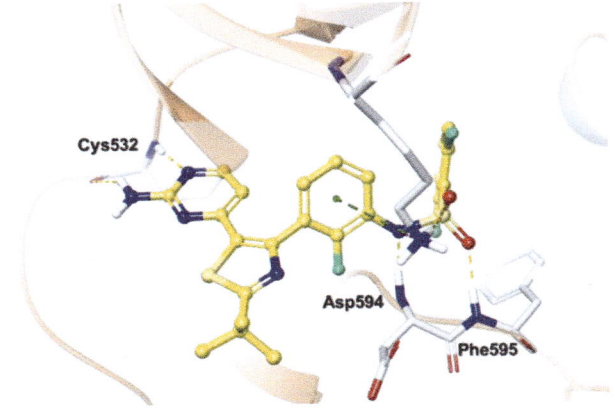

Figure 2. Co-crystal structure of dabrafenib in the active site of BRAFV600E (PDB ID: 4XV2).

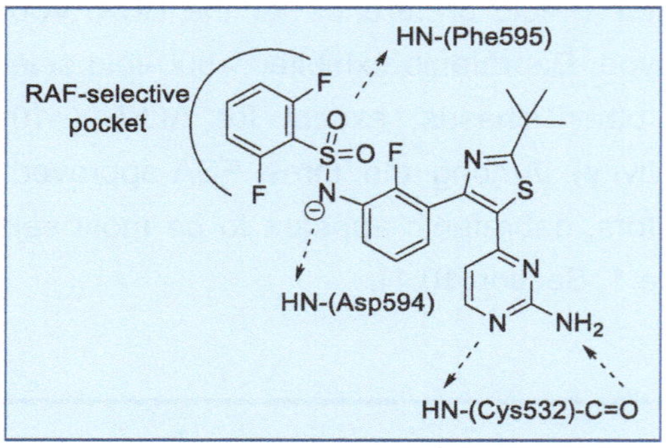

Figure 3. Dabrafenib–BRAFV600E interactions based on an X-ray co-crystal structure.

PK Properties and Metabolism[6,7]

Dabrafenib exhibits an oral bioavailability of 95%, indicative of extensive absorption and low first-pass intestinal and hepatic metabolism. The excellent oral bioavailability contributes to a much lower dosage than vemurafenib (150 mg, BID vs. 960 mg, BID). It has an elimination half-life of 8 h, resulting in twice-daily dosing regimen.

Dabrafenib undergoes metabolism primarily via oxidation of the *t*-butyl group to form hydroxy-dabrafenib **6**, which is further oxidized to carboxy-dabrafenib **7**. Subsequent decarboxylation furnishes the desmethyl-dabrafenib **8** via a pH-dependent decarboxylation (Fig. 4). The major route of elimination of dabrafenib is a combination of oxidative metabolism (48% of the dose) and biliary excretion.

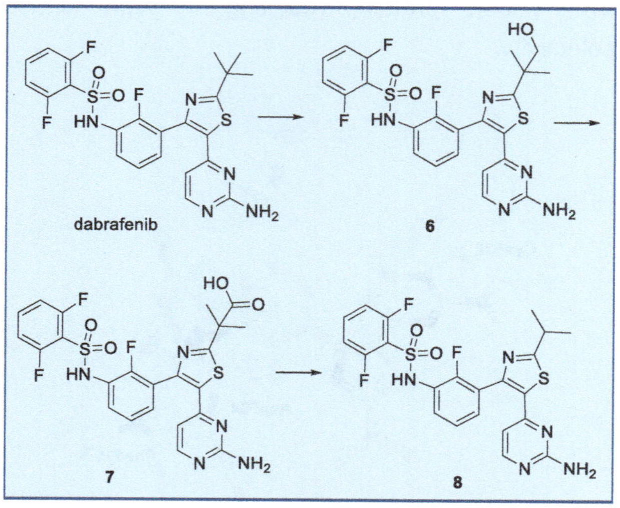

Figure 4. Major metabolic pathway of dabrafenib in humans.

1. T. R. Rheault, J. C. Stellwagen, G. W. Adjabeng, K. R. Hornberger, K. G. Petrov, A. C. Waterson, S. H. Dickerson, R. A. Mook Jr, S. G. Laquerre, A. J. King, O. W. Rossanese, M. R. Arnone, K. N. Smitheman, L. S. Kane-Carson, C. Han, G. S. Moorthy, K. G. Moss, D. E. Uehling. Discovery of dabrafenib: A selective inhibitor of Raf kinases with antitumor activity against B-Raf-driven tumors. *ACS Med. Chem. Lett.* **2013**, *4*, 358–362.

2. J. C. Stellwagen, G. M. Adjabeng, M. R. Arnone, S. H. Dickerson, C. Han, K. R. Hornberger, A. J. King, R. A. Mook Jr, K. G. Petrov, T. R. Rheault, C. M. Rominger, O. W. Rossanese, K. N. Smitheman, A. G. Waterson, D. E. Uehling. Development of potent B-RafV600E inhibitors containing an arylsulfonamide headgroup. *Bioorg. Med. Chem. Lett.* **2011**, *21*, 4436–4440.

3. C. Zhang, W. Spevak, Y. Zhang, E. A. Burton, Y. Ma, G. Habets, J. Zhang, J. Lin, T. Ewing, B. Matusow, G. Tsang, A. Marimuthu, H. Cho, G. Wu, H. Wang, D. Fong, S. Guyen, P. Shi, M. Womack, G. Nespi, G. Bollag. RAF inhibitors that evade paradoxical MAPK pathway activation. *Nature.* **2015**, *526*, 583–586.

4. Roskoski R., Jr. Targeting oncogenic Raf protein-serine/threonine kinases in human cancers. *Pharmacol. Res.* **2018**, *135*, 239–258.

5. A. Millet, A. R. Martin, C. Ronco, S. Rocchi, R. Benhida. Metastatic melanoma: Insights into the evolution of the treatments and future challenges. *Med. Res. Rev.* **2017**, *37*, 98–148.

6. H. Ellens, M. Johnson, S. K. Lawrence, C. Watson, L. Chen, L. E. Richards-Peterson. Prediction of the transporter-mediated drug-drug interaction potential of dabrafenib and its major circulating metabolites. *Drug Metab. Dispos.* **2017**, *45*, 646–656.

7. D. A. Bershas, D. Ouellet, D. B. Mamaril-Fishman, N. Nebot, S. W. Carson, S. C. Blackman, R. A. Morrison, J. L. Adams, K. E. Jurusik, D. M. Knecht, P. D. Gorycki, L. E. Richards-Peterson. Metabolism and disposition of oral dabrafenib in cancer patients: proposed participation of aryl nitrogen in carbon-carbon bond cleavage via decarboxylation following enzymatic oxidation. *Drug Metab. Dispos.* **2013**, *41*, 2215–2224.

10.3 Encorafenib (Braftovi™) (57)

Structural Formula	Space-filling Model	Brief Information
$C_{22}H_{27}N_7O_4FClS$; MW = 539 clogP = 5.5; tPSA = 120		Year of discovery: 2010 Year on Introduction: 2018 Discovered by: Novartis Developed by: Array BioPharma Primary target: BRAF Binding type: I1/2 Class: dual threonine/tyrosine kinase Treatment: melanoma with BRAFV600E/K Other name: LGX818 Oral bioavailability = 85% Elimination half-life = 6 h Protein binding = 86%

● = C   ○ = H   ● = O   ● = N   ● = S   ● = F   ● = Cl   ● = Br   ● = I

Encorafenib, a second-generation BRAF inhibitor, was approved by the FDA in 2018 as a combination therapy with binimetinib (a MEK inhibitor) for the treatment of unresectable or metastatic melanoma with a BRAFV600E or V600K mutation as detected by an FDA-approved test. This regimen is not indicated for the treatment of patients with BRAF wild-type melanoma.

Encorafenib is highly active against BRAFV600E, wild-type (wt) BRAF and CRAF. In non-cellular biochemical assays, it prevented BRAFV600E, wt BRAF and CRAF-mediated activation of MEK1 with IC_{50} values of 0.35, 0.47, and 0.30 nM, respectively, which are more potent than vemurafenib (Table 1, Section 10.1). In cellular antiproliferation assays, encorafenib inhibited growth of cell lines expressing BRAFV600E with EC_{50} value of 4 nM, and inhibited phosphorylation of downstream targets ERK1/2 and MEK1/2 with EC_{50} values of 3 and 2 nM, respectively. Encorafenib is also a highly specific BRAF inhibitor, exhibiting inhibition against only one kinase out of a panel of 99 different kinases.[1-3]

Encorafenib is characterized by a prolonged dissociation half-life from BRAFV600E mutant of more than 30h, as compared with 2 and 0.5 h for dabrafenib and vemurafenib, respectively. This very slow off-rate translates into sustained target suppression and improved potency over dabrafenib and vemurafenib.[2,3]

The binding mode of encorafenib has not been established due to the lack of a co-crystal bound structure. Nevertheless, it shares the same BRAF inhibitor pharmacophore as dabrafenib and vemurafenib, and thus may bind to BRAF in a similar way (Fig. 1).[4] More specifically, the N1 of the pyrimidine could form a hydrogen bond with the amide NH of Cys532 of the hinge, and the 2-amino group attached to the pyrimidine hydrogen could also bond to the amide oxygen of Cys532. In this scenario the deprotonated nitrogen atom of the sulfonamide moiety forms a hydrogen bond with the amide NH of Asp594, and one of the oxygen atoms of the sulfonamide forms a hydrogen bond with the amide NH of Phe595. The terminal methyl group projects into a lipophilic back pocket specific to the BRAFV600E protein, resulting in high kinase selectivity.

Encorafenib shows an oral bioavailability of 85% and a half-life of 6 h. It is administered orally once a day (450 mg), whereas the other two dabrafenib and vemurafenib are taken twice a day (Table 2, Section 10.1). The primary metabolic pathway is N-dealkylation, with CYP3A4 as the main contributor (Fig. 1).[1-3]

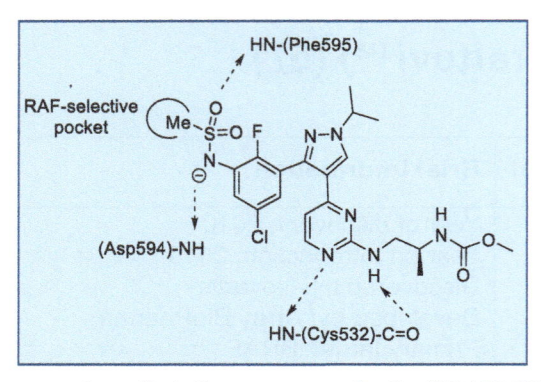

Figure 1. Putative encorafenib–BRAFV600E interactions based on BRAF inhibitor pharmacophore.

1. M. Shirley. Encorafenib and Binimetinib: First Global Approvals. *Drugs.* **2018**, *78*, 1277–1284.
2. P. Koelblinger, O. Thuerigen, R. Dummer. Development of encorafenib for BRAF-mutated advanced melanoma. *Curr. Opin. Oncol.* **2018**, *30*, 125–133.
3. J. Sun, J. S. Zager, Z. Eroglu. Encorafenib/binimetinib for the treatment of BRAF-mutant advanced, unresectable, or metastatic melanoma: design, development, and potential place in therapy. *Onco. Targets Ther.* **2018**, *11*, 9081–9089.
4. A. Millet, A. R. Martin, C. Ronco, S. Rocchi, R. Benhida. Metastatic Melanoma: Insights Into the Evolution of the Treatments and Future Challenges. *Med. Res. Rev.* **2017**, *37*, 98–148.

Chapter 11. MEK Inhibitors

Oncogenic BRAF mutations, key drivers in approximately 50% of cases of cutaneous melanoma, have been successfully targeted by three mutant specific BRAF inhibitors (vemurafenib, dabrafenib and encorafenib) for the treatment of BRAF mutant melanoma, as described in Chapter 9. However, their efficacy as monotherapy is short-lived due to the acquired resistance associated with rapid restoration of MAPK pathway signaling, suggesting that complete blockade of the pathway may be necessary to induce apoptosis in BRAF mutant melanoma. Because BRAF carries out signal transduction downstream to MEK and finally to ERK (which phosphorylates multiple targets), additional downstream MEK inhibition was thought necessary to maximize MAPK pathway inhibition and prevent resistance or delay the emergence of resistance.

Indeed, BRAF and MEK inhibitor combinations showed synergistic benefit, sufficient to justify approval for the treatment of metastatic melanoma with BRAF mutations. The three MEK inhibitors, trametinib, binimetinib, and cobimetinib (Fig. 1), are included in the combination therapies, while the last approved MEK inhibitor, selumetinib, is prescribed specifically for pediatric patients with neurofibromatosis type 1. Figure 2 shows several newer MEK inhibitors currently in clinical trials for the treatment of melanoma as well as other solid tumors.

The development of BRAF targeting drugs in combination with MEK inhibitors has radically changed clinical practice and outcomes of advanced or metastatic melanoma. This chapter describes the discovery process that led to FDA-approved MEK inhibitors **58–61** (Fig. 1).

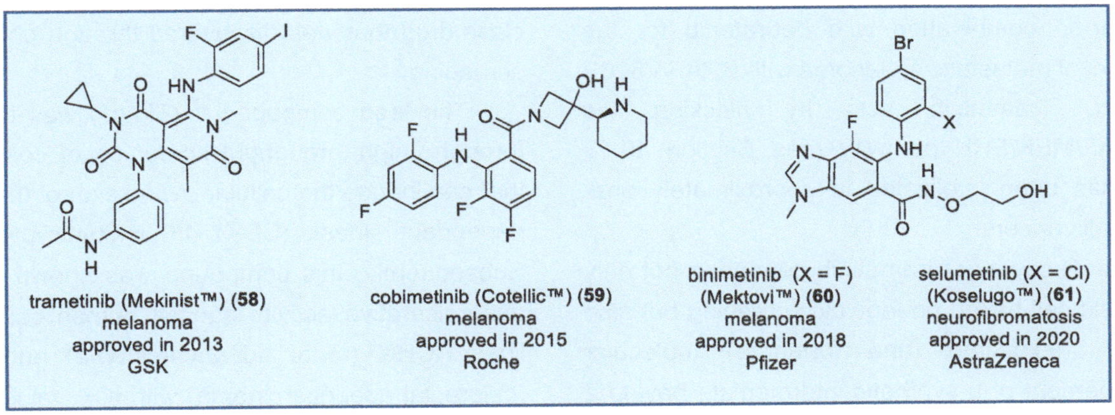

Figure 1. Approved MEK inhibitors.

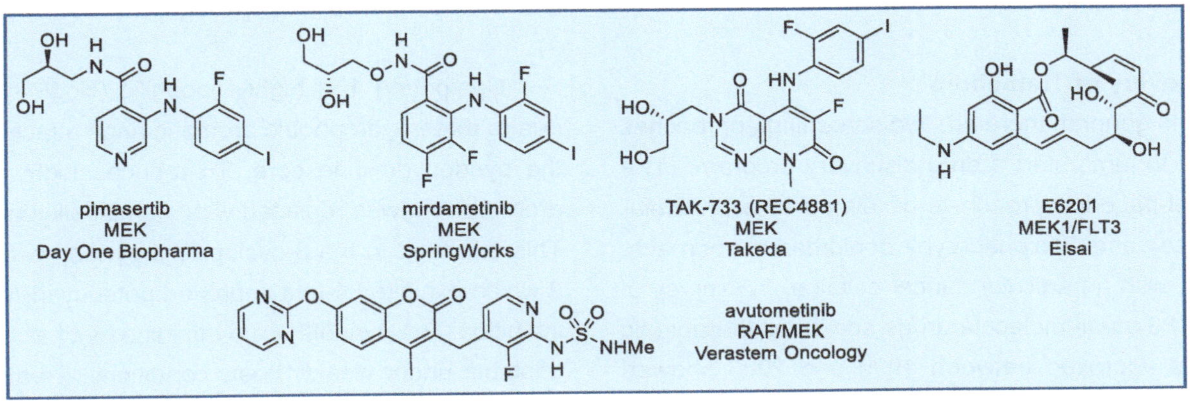

Figure 2. MEK/multikinase inhibitors in clinical trials (clinicaltrials.gov).

Molecules Engineered Against Oncogenic Proteins and Cancer, First Edition. E. J. Corey and Yong-Jin Wu.
© 2024 John Wiley & Sons Inc. Published 2024 by John Wiley & Sons Inc.

11.1 Trametinib (Mekinist™) (58)

Structural Formula	Space-filling Model	Brief Information
$C_{26}H_{23}N_5O_4FI$; MW = 615 clogP = 4.8; tPSA = 102		Year of discovery: 2004 Year of introduction: 2013 Discovered by: Japan Tobacco Developed by: GlaxoSmithKline Primary target: MEK1/2 Binding type: III Class: dual threonine/tyrosine kinase Treatment: melanoma with BRAF mutations Other name: JTP-74057, GSK1120212 Oral bioavailability = 72% Elimination half-life = 10 days Protein binding = 97.4%

● = C　○ = H　● = O　● = N　● = S　● = F　● = Cl　● = Br　● = I

Trametinib is a first-in-class, orally bioavailable, selective inhibitor of mitogen-activated protein kinases (MAPK) kinase 1 and 2 (MEK1/2). It is used alone or in combination with dabrafenib for the treatment of metastatic melanoma with BRAF V600E mutation. Trametinib acts by blocking the RAS/RAF/MEK/ERK pathway (see Section 10.1), which has been implicated in approximately one-third of all cancers.

The discovery of trametinib depended not only on classic cell-based phenotypic screening but also on an unexpected (and fortunate) molecular rearrangement of a synthetic intermediate having a pyrido[2,3-d]pyrimidine subunit to give the bioisosteric pyrido[4,3-d]pyrimidine core that would later become the heart of trametinib.[1-3]

Discovery of Trametinib[1-3]

In general, there are two screening approaches used to jump start a drug discovery program: (1) a target-based approach (e.g., aimed at a particular kinase); and (2) a phenotypic or biomarker approach, (e.g., with a particular cancer cell line).[4] A survey of the 183 small-molecule drugs across all therapeutic areas approved between 1999 and 2008 showed that 58 (32%) were discovered using phenotype-based approaches. Furthermore, the phenotypic screening approaches accounted for 28 (56%) of the 50 small-molecule first-in-class new molecular entities, whereas 17 (34%) were discovered through target-based approaches. Trametinib is a first-in-class drug that was discovered through phenotypic screening.

The lead compound **1** (Fig. 1) was identified through high-throughput screening of compounds that can induce the cellular expression of the cyclin-dependent kinase (CDK) 4/6 inhibitor p15INK4b. Subsequently, this compound was shown to have antiproliferative activity against human cancer cell lines ACHN (renal adenocarcinoma) and HT-29 (colorectal adenocarcinoma) with IC_{50} values of 4.8 μM and 0.99 μM, respectively. A medicinal chemistry program was undertaken to improve antiproliferative activity of **1** while optimizing drug-like properties (Fig. 1).

Compound **1** is highly lipophilic (clogP = 6.1) due to three hydrophobic aromatic rings attached to the pyridopyrimidine core. To reduce clogP, each aromatic ring was replaced with a small alkyl group. This effort led to the 3-cyclopropyl pyridopyrimidine **2** which exhibited 4-fold improved potency in ACHN inhibition. The pyrido[2,3-d]pyrimidine motif as in **2** is unstable under weakly basic conditions. Treatment of **2** with potassium carbonate in methanol-chloroform at room temperature afforded pyrido[4,3-d]pyrimidine **3** in 97% yield. Compound **3** was inactive, but fortunately, the new pyrido[4,3-d]pyrimidine core served as a more bioactive bioisostere of the original pyrido[2,3-d]pyrimidine system of compound **1**, thereby giving rise to a new class of useful antiproliferative agents. Repositioning of the substituents of **3** in a manner corresponding to **2** gave compound **4**, which had potency similar to that of compound **2**. Replacing the 4-bromophenylaniline subunit with 2-fluoro-4-iodophenylaniline improved ACHN and HT-29 activity by 33- and 91-fold, respectively. Despite its potent antiproliferative activity against both human cancer cell lines, this compound displayed very low aqueous solubility due to high lipophilicity (clogP = 5.8) attributed to the iodine substituent. To this end, a polar acetamide was introduced to the meta position of the N1 phenyl ring leading to trametinib, which showed further improvement in solubility, permeability as well as activity.

The SAR studies from the lead compound **1** to trametinib were guided entirely by screening against cancer cell lines for growth inhibitory activity rather than the MEK inhibitory activity, which was evaluated later. Trametinib is an allosteric, non-ATP-competitive inhibitor with sub-nanomolar activity against purified MEK1 and MEK2 kinases (IC_{50} = 0.7 nM and 0.9 nM, respectively), and it is one of the most potent MEK inhibitors in use (Table 1).[5,6]

Table 1. In vitro IC_{50}'s of MEK inhibitors[5,6]

kinase	trametinib	cobimetinib	selumetinib	binimetinib
MEK1	0.7 nM	0.95 nM	14 nM	
MEK2	0.9 nM	199 nM	>10 μM	
MEK1/2				12 nM
A375 cell	0.74 nM	5 nM		34 nM

Binding Mode[7,8]

The two crystal structures of trametinib bound to the KSR1-MEK1 and KSR2-MEK1 (data not shown)[7] show that trametinib occupies the canonical MEK inhibitor pocket, an allosteric site adjacent to the region normally occupied by ATP. KSR proteins are RAF-family pseudokinases devoid of the enzymatic activity of other RAF family members. Notably, the acetamide engages KSR at the MEK interface. The terminal methyl group in trametinib is within a bonding distance of approximately 3 Å to KSR1 or KSR2, whereas the other MEK inhibitors are up to 10 Å from direct contact with the KSR–MEK interface (data not shown).[7] Thus, the direct engagement of trametinib with KSR differentiates it from other MEK inhibitors.

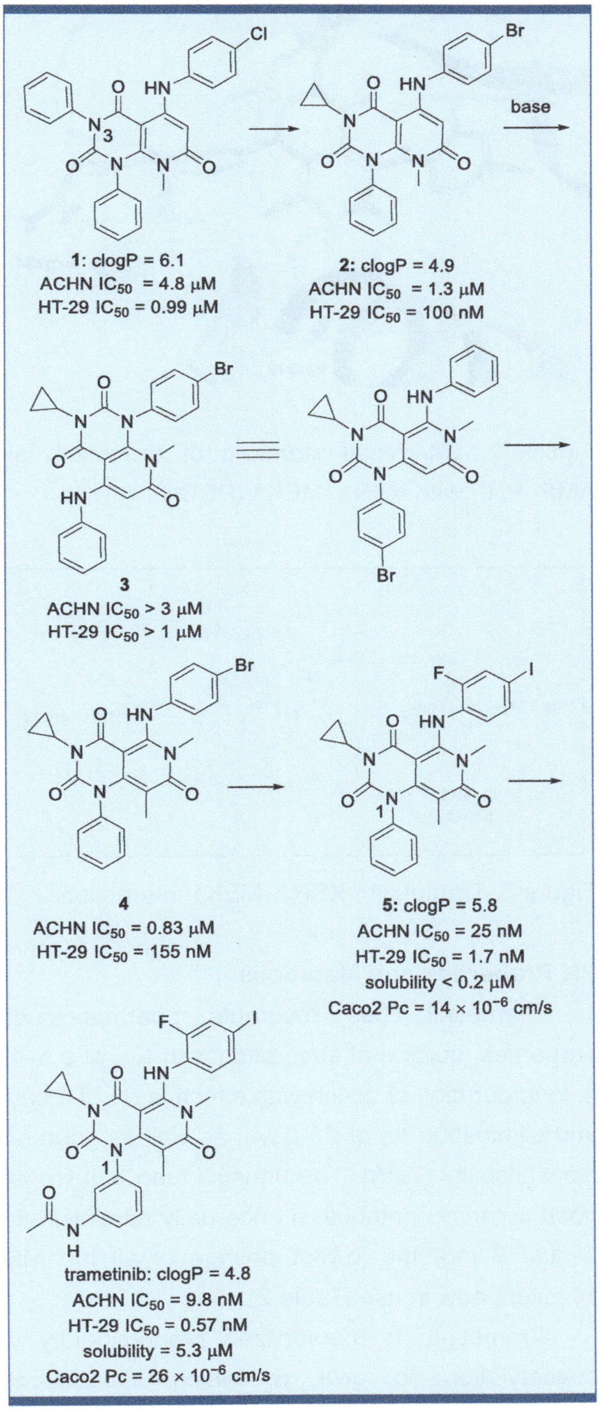

Figure 1. Structural optimization of trametinib.

As shown in the crystal structure of trametinib and AMP–PNP with KSR1-MEK1 (Fig. 2), the C7 carbonyl oxygen forms a hydrogen bond with Ser212 (MEK1), the acetamide oxygen forms a salt bridge with the guanidine residue of Arg324 (MEK1), and the terminal methyl group is in close contact with Ala825 of KSR1 (Fig. 3).

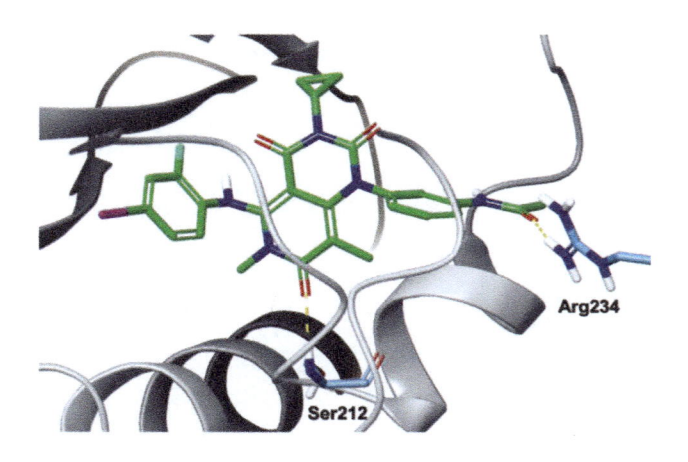

Figure 2. Co-crystal structure of trametinib and AMP–PNP with KSR1–MEK1 (PDBID: 7JUX).

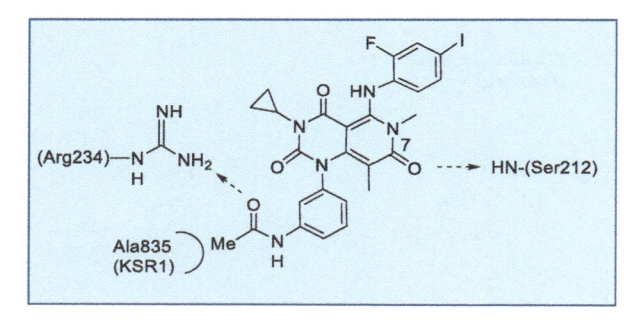

Figure 3. Trametinib–KSR1–MEK1 interactions.

PK Properties and Metabolism[5,9]

Trametinib has favorable pharmacokinetic properties: quick oral absorption with t_{max} of 0.5–1.5 h, long duration of action with effective $t_{1/2}$ of 4 days, and elimination $t_{1/2}$ of 10 days, as well as good oral bioavailability (72%). The unusual long half-life and good potency contribute a once-daily administration of just 2 mg, the lowest dosage of all the MEK inhibitors now in use (Table 2).

Trametinib is metabolized predominantly via deacetylation to give **6**, which subsequently undergoes hydroxylation to give **8** or glucuronidation to afford **7** (Fig. 4).

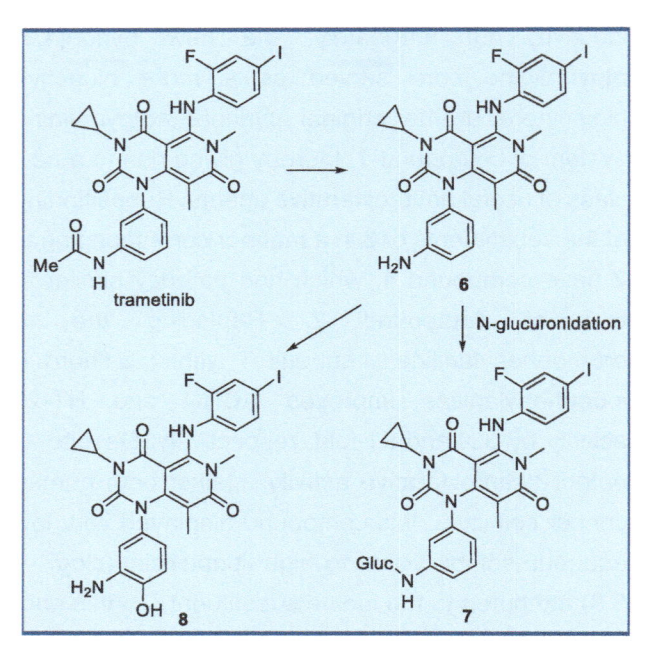

Figure 4. Major metabolic pathway of trametinib in humans.

Table 2. PK properties of MEK inhibitors.[5]

PK	trametinib	comimetinib	selumetinib	binimetinib
t_{max}	1.5 h	2.4 h	1–1.5 h	1.6 h
$t_{1/2}$	4–5 days	44 h	6.2 h	3.5 h
F	72%	46%	62%	50%
CL/F	5.4 L/h	13.8 L/h	8.8 L/h	20.2 L/h
dosage	2 mg	60 mg	25 mg	45 mg
dose	QD	QD	BID	BID

1. T. Yamaguchi, T. Yoshida, R. Kurachi, J. Kakegawa, Y. Hori, T. Nanayama, K. Hayakawa, H. Abe, K. Takagi, Y. Matsuzaki, M. Koyama, S. Yogosawa, Y. Sowa, T. Yamori, N. Tajima, T. Sakai. Identification of JTP-70902, a p15INK4b-inductive compound, as a novel MEK1/2 inhibitor. *Cancer Sci.* **2007**, *98*, 1809–1816.

2. H. Abe, S. Kikuchi, K. Hayakawa, T. Iida, N. Nagahashi, K. Maeda, J. Sakamoto, N. Matsumoto, T. Miura, K. Matsumura, N. Seki, T. Inaba, H. Kawasaki, T. Yamaguchi, R. Kakefuda, T. Nanayama, H. Kurachi, Y. Hori, T. Yoshida, J. Kakegawa, Y. Watanabe, A. G. Gilmartin, M. C. Richter, K. G. Moss, S. G. Laquerre. Discovery of a highly potent and selective MEK inhibitor: GSK1120212 (JTP-74057 DMSO solvate). *ACS Med. Chem. Lett.* **2011**, *28*, 320-4.

3. A. G. Gilmartin, M. R. Bleam, A. Groy, K. G. Moss, E. A. Minthorn, S. G. Kulkarni, C. M. Rominger, S. Erskine, K. E. Fisher, J. Yang, F. Zappacosta, R. Annan, D. Sutton, S. G. Laquerre. GSK1120212 (JTP-74057) is an inhibitor of MEK activity and activation with favorable pharmacokinetic properties for

sustained in vivo pathway inhibition. *Clin. Cancer Res.* **2011**, *17*, 989–1000.

4. J. G. Moffat, J. Rudolph, D. Bailey. Phenotypic screening in cancer drug discovery – past, present and future. *Nature reviews. Drug Discov.* **2014**, *13*, 588–602.

5. Y. Cheng, H. Tian. Current Development Status of MEK Inhibitors. *Molecules.* **2017**, *22*, 1551.

6. G. L. Gonzalez-Del Pino, K. Li, E. Park, A. M. Schmoker, B. H. Ha, M. J. Eck. Allosteric MEK inhibitors act on BRAF/MEK complexes to block MEK activation. *PNAS.* **2021**, *118*, e2107207118.

7. Z. M. Khan, A. M. Real, W. M. Marsiglia, A. Chow, M. E. Duffy, J. . Yerabolu, A. P. Scopton, A. C. Dar. Structural basis for the action of the drug trametinib at KSR-Bound MEK. *Nature.* **2020**, *588*, 509–514.

8. C. C. Ayala-Aguilera, T. Valero, A. Lorente-Macías, D. J. Baillache, S. Croke, A. Unciti-Broceta. Small molecule kinase inhibitor drugs (1995–2021): Medical indication, pharmacology, and synthesis. *J. Med. Chem.* **2022**, *65*, 1047–1131.

9. C. Leonowens, C. Pendry, J. Bauman, G. C. Young, M. Ho, F. Henriquez, L. Fang, R. A. Morrison, K. Orford, D. Ouellet. Concomitant oral and intravenous pharmacokinetics of trametinib, a MEK inhibitor, in subjects with solid tumours. *Br. J. Clin. Pharmacol.* **2014**, *78*, 524–532.

11.2 Cobimetinib (Cotellic™) (59)

Structural Formula	Space-filling Model	Brief Information
$C_{21}H_{21}N_3O_2F_3I$; MW = 531 clogP = 5.1; tPSA = 65		Year of discovery: 2005 Year of introduction: 2015 Discovered by: Exelixis Developed by: Exelixis/Roche Primary target: MEK1/2 Binding type: III Class: dual threonine/tyrosine kinase Treatment: melanoma with BRAF mutations Other name: GDC-0973, XL518 Oral bioavailability = 46% Elimination half-life = 44 h Protein binding = 95%

● = C ○ = H ● = O ● = N ● = S ● = F ● = Cl ● = Br ● = I

Cobimetinib is the second MEK inhibitor to have received FDA approval (in 2015) for use in combination with vemurafenib (BRAF inhibitor) for the treatment of patients with unresectable or metastatic melanoma with a BRAF V600E or V600K mutation.

Cobimetinib is an amide derivative, whereas the two other recent MEK inhibitors, binimetinib and selumetinib, are hydroxamate analogs.

Discovery of Cobimetinib[1,2]

The iodoanthranilic acid lead compound **1** was identified as a moderately potent MEK inhibitor (IC_{50} = 554 nM) by screening a Pfizer/Warner Lambert compound library with a cascade assay that measured phosphorylation of myelin basic protein in the presence of GST-MEK1 and GST-MAPK fusion proteins (Fig. 1). To improve the potency of the lead compound, various substituents were introduced to the N-phenylanthranilic acid template. SAR studies showed that an iodine substituent was favored at the 4′ position, and that a fluorine substituent at the 4 position was optimal. The beneficial effect of the C4 fluorine can be attributed to a dipolar interaction of the fluorine with the Ser212 amide backbone NH of the MEK1 as revealed by the co-crystal structure of a close analog in complex with MEK1. Further SAR studies around the N-phenylanthranilic acid subunit led to CI-1040, the first MEK inhibitor to demonstrate oral activity in a mouse tumor model, which subsequently became the first MEK inhibitor to enter clinical trials. However, CI-1040 did not show efficacy in phase II studies due to insufficient systemic exposure, because of poor aqueous solubility (<1 µg/mL) and rapid clearance. Adding hydroxylic polarity to the carboxylic side chain to improve solubility resulted in the discovery of PD-0325901 which showed substantial improvement in solubility (190 µg/mL vs. <1 µg/mL), cellular potency (pERK IC_{50} = 0.9 nM vs. 35 nM), and cell permeability (Caco-2 Pc = 72.7×10^{-7} cm/s vs. 16.2×10^{-7} cm/s). Despite its superior pharmacokinetic profile to CI-1040, PD-0325901 failed in the phase II clinical trials for the treatment of KRAS mutant non-small cell lung cancer (NSCLC). A phase I/II study for the treatment of melanoma, colonic neoplasms and breast neoplasms was terminated in 2007 due to musculoskeletal, neurological, and ocular toxicity.

Against the backdrop of these disappointing clinical data from Pfizer, PD-0325901 still served as a good starting point for Exelixis because of its potent and selective MEK inhibition and drug like-properties (logD = 4.6, tPSA = 91, MW = 492). Two critical issues in the clinical trials had to be addressed: (1) neurological side effects such as ataxia, confusion, and syncope (which was attributed to the brain exposure of the parent compound and/or its active metabolite(s); and (2) poor metabolic stability due to the facile N–O bond cleavage of the hydroxamate moiety. (Two other MEK inhibitors, selumetinib and binimetinib, are hydroxamate analogs, but they appear to have acceptable pharmacokinetic properties, including metabolic stability, see Table 2, Section 11.1.) Extensive SAR studies were undertaken to identify active analogs devoid of this motif. Although conversion of PD-0325901 to the corresponding carboxamide **2** led to a substantial drop in the enzyme and cellular potency, surprisingly, the azetidin-3-ol amide **3** emerged as a potent MEK inhibitor with IC_{50} values of 44 and 29 nM in the biochemical and cellular assays, respectively. The azetidine hydroxyl is critical for potency because it penetrates into the catalytic loop region, forming two hydrogen bonds with residues Asp190 and Asn195 as indicated by the co-crystal structure of azetidinol **3** in AMP-PCP bound MEK1 (data not shown). However, the oral bioavailability of **3** was still low (19%) due to metabolic inactivation and high clearance. As the azetidine secondary hydroxyl is susceptible to a CYP450 mediated oxidation, analogous tertiary alcohols were prepared. This effort led to cobimetinib with subnanomolar biochemical and cellular potency (IC_{50} = 0.9 nM; pERK IC_{50} = 0.2 nM). Introduction of a basic piperidine adjacent to the tertiary alcohol served multiple purposes: (1) improve cellular potency, solubility, and pharmacokinetic properties; (2) enhance enzyme activity via engagement with γ-phosphate of AMP-PCP and Asp190 (see binding mode); and (3) reduce brain uptake because the basic amine nitrogen leads to a protonated species which is unable to cross the lipophilic blood–brain barrier through passive diffusion. Indeed, oral administration of cobimetinib at 100 mpk resulted in only marginal brain exposure with no metabolite detected. Thus, the discovery of cobimetinib overcame two key issues with PD-0325901 that prevented its further development: the metabolic labile hydroxamate functionality and its brain uptake.

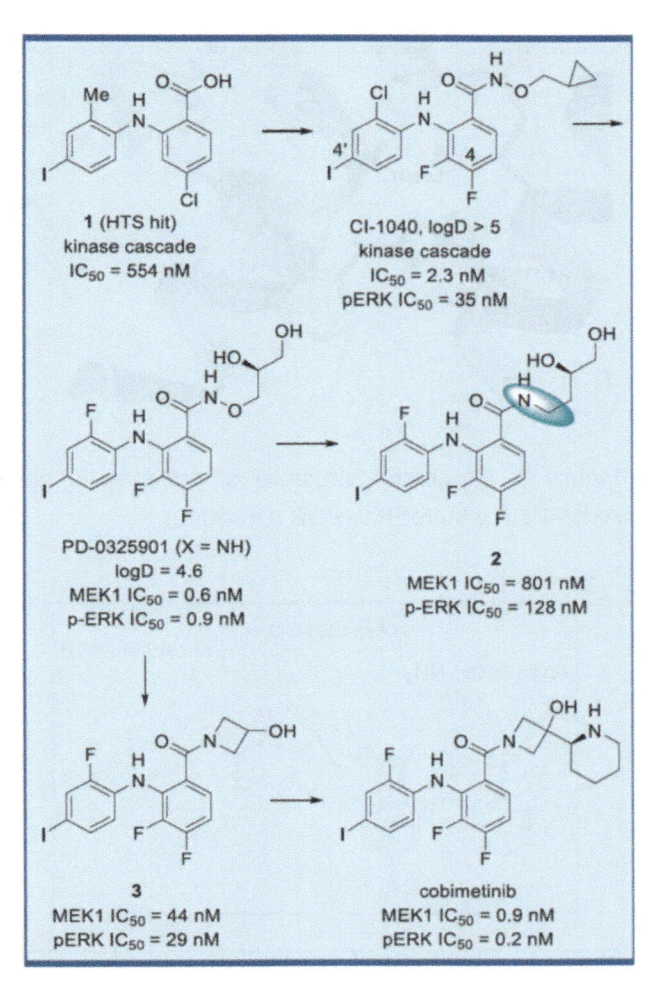

Figure 1. Summary of the chemical optimization of cobimetinib.

Binding Mode[1,3]

Cobimetinib binds to a specific allosteric pocket adjacent to the ATP site of MEK1, forming a highly complex network of interactions involving the terminal aminoethanol moiety with both the catalytic loop Asp190 and the ATP γ-phosphate (Figs. 2 and 3). The amide oxygen hydrogen bonds to the primary amine of Lys97. The piperidine nitrogen approaches the catalytic loop, thereby enabling interactions with both γ-phosphoryl oxygen of ATP and the carboxylic acid residue of Asp190. The azetidine hydroxyl also engages in interactions with Asp190 and the γ-phosphoryl oxygen of ATP.

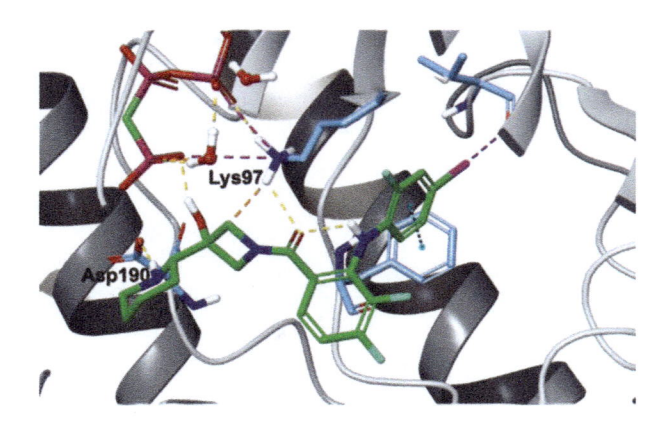

Figure 2. Co-crystal structure of cobimetinib and AMP–PCP with MEK1 (PDB ID: 4AN2).

Figure 3. Summary of cobimetinib and AMP–PCP with MEK1 interactions based on an X-ray co-crystal structure.

PK Properties and Metabolism

Cobimetinib has only moderate oral bioavailability (46%), likely due to metabolism rather than incomplete absorption.[4] However, it displays prolonged elimination half-life (44 h), which supports a once-daily dosing regimen (60 mg).

Following oral administration, the unchanged cobimetinib and metabolite **4** were the major circulating components in the plasma up to 48 hours post dose (AUC_{0-48}), accounting for 21% and 18% of all the circulating drug-related components, respectively (Fig. 4).[5]

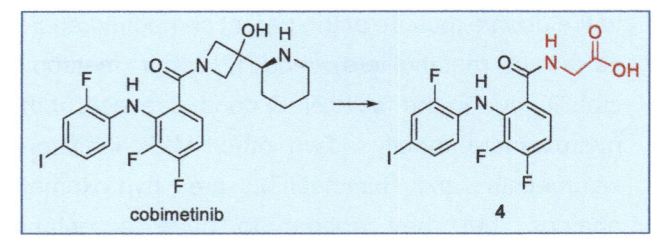

Figure 4. Major metabolic pathway of cobimetinib in humans.

1. K. D. Rice, N. Aay, N. K. Anand, C. M. Blazey, O. J. Bowles, J. Bussenius, S. Costanzo, J. K. Curtis, S. C. Defina, L. Dubenko S. Engst, A. A. Joshi, A. R. Kennedy, A. I. Kim, E. S. Koltun, J. C. Lougheed, J. C. Manalo, J. F. Martini, J. M. Nuss, C. J. Peto T. H. Tsang, P. Yu, S. Johnston. Novel carboxamide-based allosteric MEK inhibitors: Discovery and optimization efforts toward XL518 (GDC-0973). *ACS Med. Chem. Lett.* **2012**, *3*, 416-21.

2. S. D. Barrett, A. J. Bridges, D. T. Dudley, A. R. Saltiel, J. H. Fergus, C. M. Flamme, A. M. Delaney, M. Kaufman, S. LePage, W. R. Leopold, S. A. Przybranowski, J. Sebolt-Leopold, K. Van Becelaere, A. M. Doherty, R. M. Kennedy, D. Marston, W. A. Howard, Y. Smith, J. S. Warmus, H. Tecle. The discovery of the benzhydroxamate MEK inhibitors CI-1040 and PD 0325901. *Bioorg. Med. Chem. Lett.* **2008**, *18*, 6501–6504.

3. C. C. Ayala-Aguilera, T. Valero, A. Lorente-Macías, D. J. Baillache, S. Croke, A. Unciti-Broceta. Small molecule kinase inhibitor drugs (1995-2021): Medical indication, pharmacology, and synthesis. *J. Med. Chem.* **2022**, *65*, 1047–1131.

4. L. Musib, E. Choo, Y. Deng, S. Eppler, I. Rooney, I. T. Chan, M. J. Dresser. Absolute bioavailability and effect of formulation change, food, or elevated pH with rabeprazole on cobimetinib absorption in healthy subjects. *Mol. Pharm.* **2013**, *10*, 4046–4054.

5. R. H. Takahashi, E. F. Choo, S. Ma, S. Wong, J. Halladay, Y. Deng, I. Rooney, M. Gates, C. E. Hop, S. C. Khojasteh, M. J. Dresser. Absorption, metabolism, excretion, and the contribution of intestinal metabolism to the oral disposition of [^{14}C]cobimetinib, a MEK inhibitor, in humans. *Drug Metab. Dispos.* **2016**, *44*, 28–39.

11.3 Binimetinib (Mektovi™) (60)

Structural Formula	Space-filling Model	Brief Information
$C_{17}H_{15}N_4O_3BrF_2$; MW = 440 clogP = 3.2; tPSA = 86		Year of discovery: 2009 Year of introduction: 2018 Discovered by: Array BioPharma Developed by: Array BioPharma/Pfizer Primary target: MEK1/2 Binding type: III Class: dual threonine/tyrosine kinase Treatment: melanoma with BRAF mutations Other name: ARRY-162 Oral bioavailability = 50% Elimination half-life = 3.5 h Protein binding = 97%

● = C   ○ = H   ● = O   ● = N   ● = S   ● = F   ● = Cl   ● = Br   ● = I

Binimetinib, the third FDA-approved MEK inhibitor, is used in combination with encorafenib (BRAF inhibitor), for the treatment of patients with unresectable or metastatic melanoma with a BRAF V600E or V600K mutation. The encorafenib–binimetinib combination results in much less pyrexia than dabrafenib-trametinib and less photosensitivity than vemurafenib–cobimetinib.[1]

Binimetinib displays an IC_{50} of 12 nM against purified MEK1/2 enzyme, with no inhibitory activity of other protein kinases up to 20 μM.[2]

Binimetinib differs from selumetinib, another FDA-approved MEK inhibitor, only by the replacement of Cl by F (see Section 11.4).

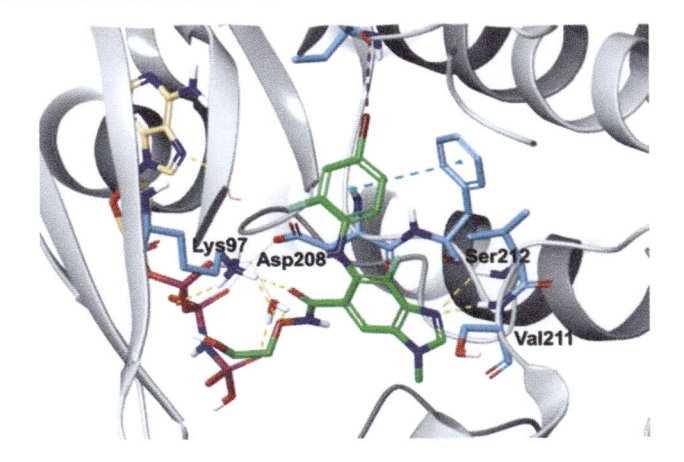

Figure 1. Co-crystal structure of binimetinib and AMP-PNP with BRAF–MEK1 (PDB ID: 7M0U).

Binding Mode[3]

As shown in the co-crystal structure of binimetinib in complex with BRAF–MEK1 kinases and AMP–PNP (Fig. 1), the imine nitrogen of the benzo[d]imidazole core hydrogen bonds to both the amide NH of Ser212 and amide NH of Val211, and the amide oxygen also forms a hydrogen bond with the primary amine of Lys97. In addition, the terminal hydroxyl group hydrogen bonds to the α-phosphate oxygen of AMP–PNP. Also, the carboxamide side chain oxygen interacts indirectly with the carboxylic acid of Asp208 and AMP–PNP via a water molecule (Fig. 2).

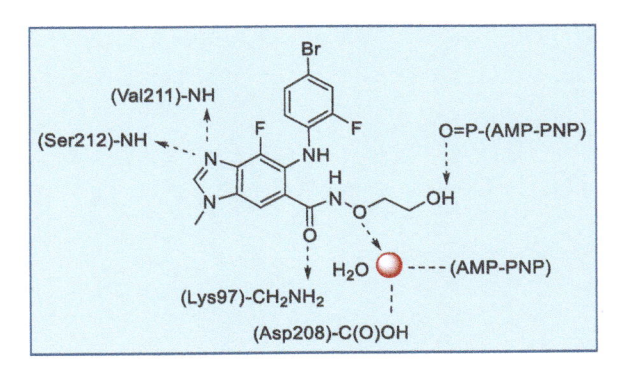

Figure 2. Summary of binimetinib and AMP–PNP with BRAF–MEK1. interactions based on an X-ray co-crystal structure.

PK Properties and Metabolism[4,5]

After oral administration, binimetinib is absorbed rapidly, with a median t_{max} of 1.48 h. Binimetinib is 50% orally bioavailable and exhibits a short elimination half-life of 3.5 h. Consequently, it requires twice-daily dosing regimen, vs. once a day for both trametinib and cobimetinib (Table 2, Section 11.1).

Binimetinib undergoes UGT1A1-mediated glucuronidation, which contributes up to 61% of the overall metabolism. Other metabolic pathways include N-dealkylation, amide hydrolysis, and loss of ethanediol from the side chain.

1. W. H. Sharfman. Encorafenib and binimetinib: A new benchmark in metastatic melanoma therapy? *The ASCO Post*. December 10, 2018.
2. B. Tran, M. S. Cohen. The discovery and development of binimetinib for the treatment of melanoma. *Expert Opin. Drug Discov.* **2020**, *15*, 745–754.
3. C. C. Ayala-Aguilera, T. Valero, A. Lorente-Macías, D. J. Baillache, S. Croke, A. Unciti-Broceta. Small molecule kinase inhibitor drugs (1995–2021): Medical indication, pharmacology, and synthesis. *J. Med. Chem.* **2022**, *65*, 1047–1131.
4. P. Koelblinger, J. Dornbierer, R. Dummer. A review of binimetinib for the treatment of mutant cutaneous melanoma. *Future Oncol.* **2017**, *13*, 1755–1766.
5. M. Shirley. Encorafenib and binimetinib: First global approvals. *Drugs.* **2018**, *78*, 1277–1284.

11.4 Selumetinib (Koselugo™) (61)

Structural Formula	Space-filling Model	Brief Information
$C_{17}H_{15}N_4O_3FBrCl$; MW = 456 clogP = 3.7; tPSA = 86		Year of discovery: 2002 Year of introduction: 2020 Discovered by: Array BioPharma Developed by: AstraZeneca Primary target: MEK1 Binding type: III Class: dual threonine/tyrosine kinase Treatment: children with NF1 Other name: AZD-6244, ARRY-142886 Oral bioavailability = 62% Elimination half-life = 6.2 h Protein binding = 97.7%

● = C   ○ = H   ● = O   ● = N   ● = S   ● = F   ● = Cl   ● = Br   ● = I

Selumetinib, is the fourth FDA-approved MEK inhibitor, specifically for the treatment of pediatric patients with neurofibromatosis type 1 (NF1), while the other three MEK inhibitors (trametinib, cobimetinib, and binimetinib) are prescribed for the treatment of metastatic melanoma with BRAFV600E mutation. Although it differs from binimetinib only by the replacement of F by Cl (Fig. 1), it has enabled a new therapy in a disease population lacking viable treatment options.

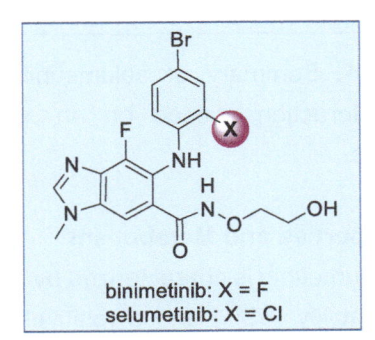

Figure 1. Binimetinib and selumetinib.

binimetinib: X = F
selumetinib: X = Cl

Selumetinib specifically inhibits the activity of purified MEK1 with an IC_{50} of 14 nM, and importantly, it shows no inhibition against a panel of >40 serine/threonine and tyrosine kinases at concentrations up to 10 μM.[1,2]

NF1 pathogenesis and MEK inhibition[3,4]

Neurofibromatosis type 1 (NF1), also known as von Recklinghausen's disease, is a genetic disorder first described by Friederich Daniel Von Recklinghausen, a German pathologist, in 1882. NF1 is one of the most frequent human genetic diseases, affecting approximately 1 out of 3,000 children at birth. It causes tumors to grow all over the body, including under the skin and along the nerves.

NF1 is caused by mutations in a tumor suppressor gene called NF1 located at the 17q11.2 chromosome. This gene provides instructions for making a protein called neurofibromin, a GTPase-activating protein (GAP) that negatively regulates RAS pathway activity by accelerating the conversion of the active, GTP-bound RAS into its inactive GDP bound form. Essentially, neurofibromin serves to counterSon of Sevenless (SOS), which catalyzes the exchange of GDP for GTP on RAS (Fig. 2). This protein prevents cell overgrowth by turning off RAS protein which stimulates cell growth and division.

Mutations in the NF1 gene lead to the production of a nonfunctional version of neurofibromin, which results in aberrant activation of the classic RAS/RAF/MEK signaling pathway, cell proliferation, and ultimately neurofibromas along nerves throughout the body. As a targeted therapy, selumetinib deactivates MEK kinase, one of the critical RAS downstream pathway components. Selumetinib has been shown to be beneficial for the treatment of NF1-associated tumors, including progressive low-grade gliomas (LGGs) and inoperable plexiform neurofibromas (PNs).

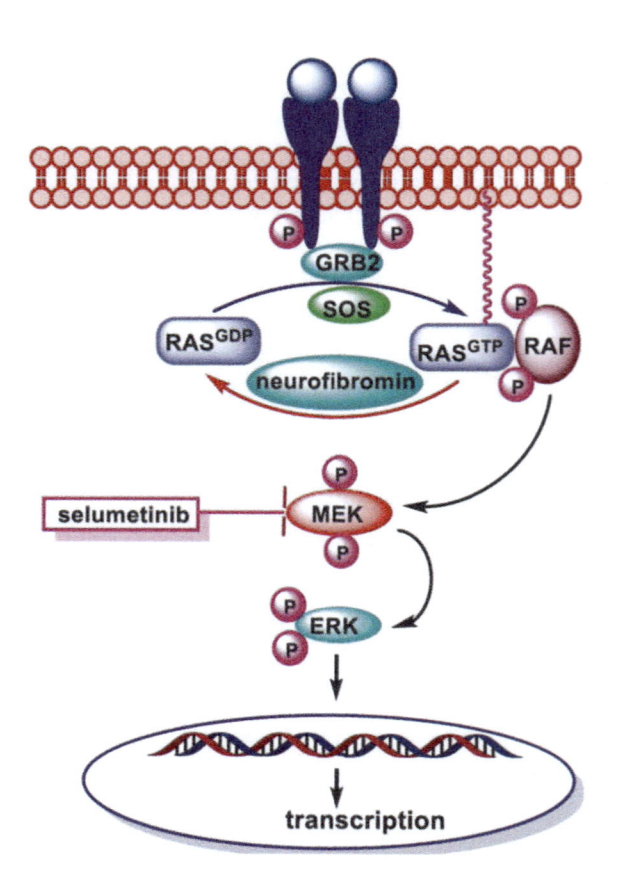

Figure 2. Neurofibromin downregulates RAS signaling pathway.

Binding mode[5]

In the co-crystal structure of selumetinib in complex with MEK1 and AMP–PNP (Fig. 3), selumetinib binds to a unique and specific allosteric pocket on the N-terminal domain of MEK1, next to a typical ATP-binding site. This binding results in a conformational change, which prevents the RAF-induced phosphorylation, and locks MEK1/2 into a catalytically inactive state, thereby blocking the RAS signaling. The imine nitrogen of the benzo[d]imidazole core hydrogen bonds to the amide NH of Ser212, and the three oxygen atoms of the amide side chain form three hydrogen bonds with the primary amine of Lys97. In addition, the terminal hydroxyl group hydrogen bonds to the α-phosphoryl oxygen of AMP–PNP (Fig. 4).

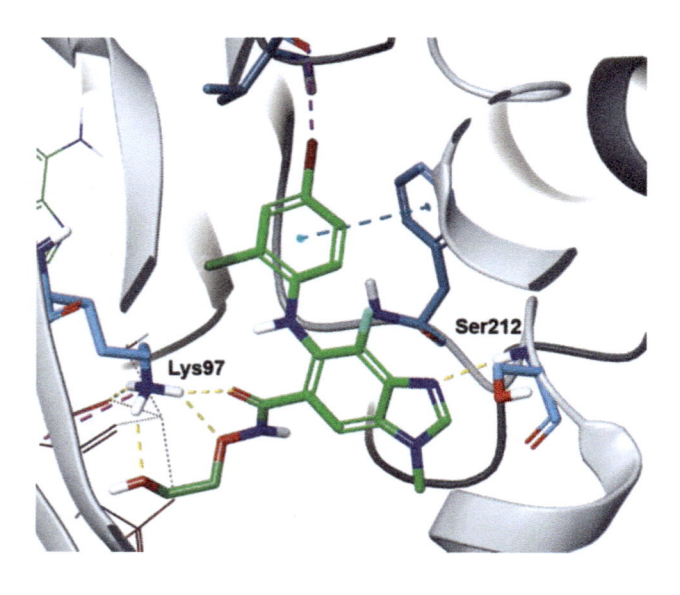

Figure 3. Co-crystal structure of selumetinib–MEK1–AMP–PNP (PDB ID: 4U7Z).

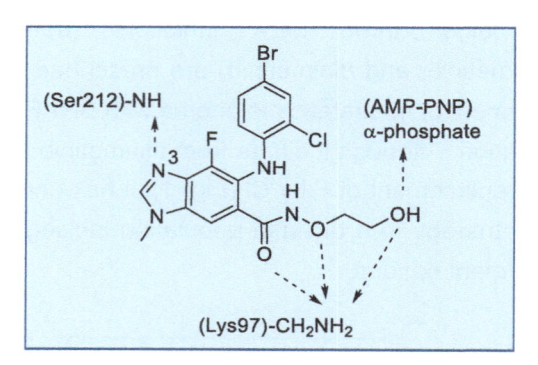

Figure 4. Summary of selumetinib–MEK1–AMP–PNP interactions based on an X-ray co-crystal structure.

PK Properties and Metabolism[6,7]

Selumetinib is characterized by a moderate oral bioavailability (62%) and a relatively short half-life (6.2 h), and these properties contribute to twice-daily dosing regimen (25 mg dosage).

Following oral administration of radiolabeled selumetinib, the most prominent drug-related component in the plasma was selumetinib, accounting for 40% of the plasma radioactivity. The major circulating metabolite was an amide glucuronide **2**, which accounted for 22% of the plasma radioactivity. This metabolite resulted from loss of the ethanediol moiety to give the primary amide **1**, which underwent glucuronidation and an additional loss of 2 mass units, most likely due to further oxidation of the *N*-methylbenzimidazole moiety (Fig. 5).

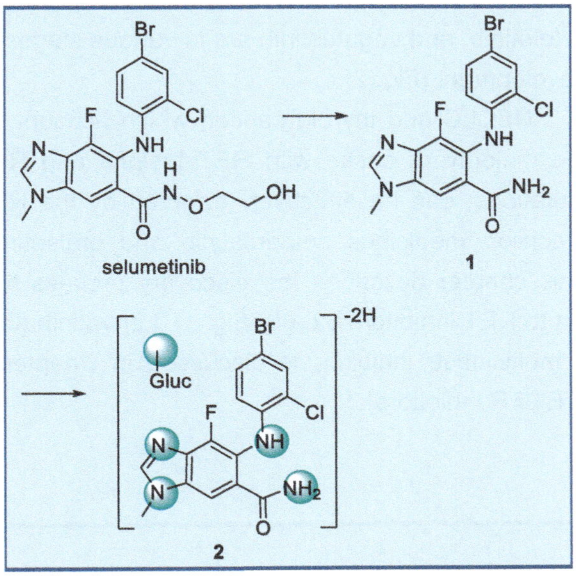

Figure 5. Major metabolic pathway of selumetinib in humans.

1. K. K. Ciombor, T. Bekaii-Saab. Selumetinib for the treatment of cancer. *Expert Opin. Investig. Drugs.* **2015**, *24*, 111–123.
2. T. C. Yeh, V. Marsh, B. A. Bernat, J. Ballard, H. Colwell, R. J. Evans, J. Parry, D. Smith, B. J. Brandhuber, S. Gross, A. Marlow, B. Hurley, J. Lyssikatos, P. A. Lee, J. D. Winkler, K. Koch, E. Wallace. Biological characterization of ARRY-142886 (AZD6244), a potent, highly selective mitogen-activated protein kinase kinase 1/2 inhibitor. *Clin.Cancer Res.* **2007**, *13*, 1576–1583.
3. L. J. Klesse, J. T. Jordan, H. B. Radtke, T. Rosser, E. Schorry N. Ullrich, D. Viskochil, P. Knight, S. R. Plotkin, K. Yohay. The use of MEK inhibitors in neurofibromatosis type 1-associated tumors and management of toxicities. *Oncologist.* **2020**, *25*, e1109–e1116.
4. L. Le, L. Parada. Tumor microenvironment and neurofibromatosis type I: Connecting the GAPs. *Oncogene.* **2007**, *26*, 4609–4616.
5. C. C. Ayala-Aguilera, T. Valero, A. Lorente-Macías, D. J. Baillache, S. Croke, A. Unciti-Broceta. Small molecule kinase inhibitor drugs (1995–2021): Medical indication, pharmacology, and synthesis. *J. Med. Chem.* **2022**, *65*, 1047–1131.
6. O. Campagne, K. K. Yeo, J. Fangusaro, C. Stewart. Clinical pharmacokinetics and pharmacodynamics of selumetinib. *Clin. Pharmacokinet.* **2021**, *60*, 283–303.
7. A. W. Dymond, C. Howes, C. Pattison, K. So, G. Mariani, M. Savage, S. Mair, G. Ford, P. Martin. Metabolism, excretion, and pharmacokinetics of selumetinib, an MEK1/2 inhibitor, in healthy adult male subjects. *Clin. Ther.*, **2016**, *38*, 2447–2458.

Chapter 12. RET/Multikinase Inhibitors

RET gene fusions or rearrangements occur in 1–2% of non-small cell lung cancer (NSCLC) and 10–20% of papillary thyroid carcinoma (PTC), while RET point mutations are frequently found in sporadic medullary thyroid cancer (MTC). Initial efforts to target RET-altered cancers with multikinase inhibitors (MKIs) (that cross-inhibit RET), cabozantinib, vandetanib, and lenvatinib, led to their approval for use in thyroid cancers. However, in NSCLC patients with RET rearrangements, these MKIs demonstrated only modest clinical benefits because the exposure required for blockage of the RET-signaling pathway was not achievable due to off-target adverse events which led to dose reduction or discontinuation of treatment. Consequently, RET-specific therapeutics have been sought. Recently, two newer RET selective inhibitors, selpercatinib and pralsetinib, have gained FDA approval for the treatment of RET-aberrant NSCLC and thyroid cancers.[1] Several other selective RET inhibitors, LOXO260 (structure undisclosed), enbenzotinib, zeteletinib, and vepafestinib are in various stages of development (Fig. 2).[2]

NSCLC and thyroid cancer, which account for the majority of cases with RET fusions and RET mutations, can be effectively targeted by the RET precision medicines selpercatinib and pralsetinib. This chapter describes the discovery process that led to RET inhibitors **62–65** (Fig. 1). Lenvatinib (**30**), a multikinase inhibitor, is discussed in Chapter 5 (VEGFR inhibitors).

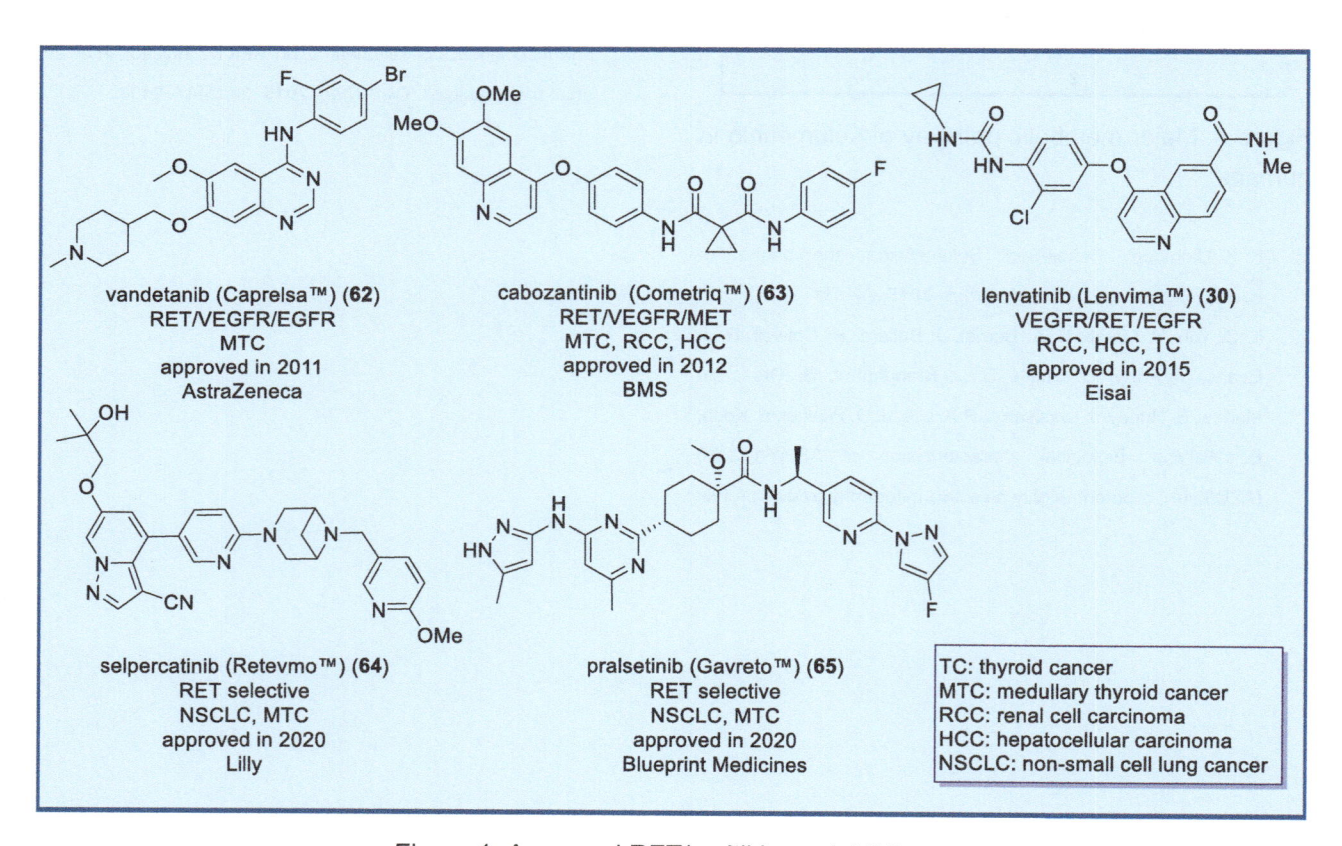

Figure 1. Approved RET/multikinase inhibitors.

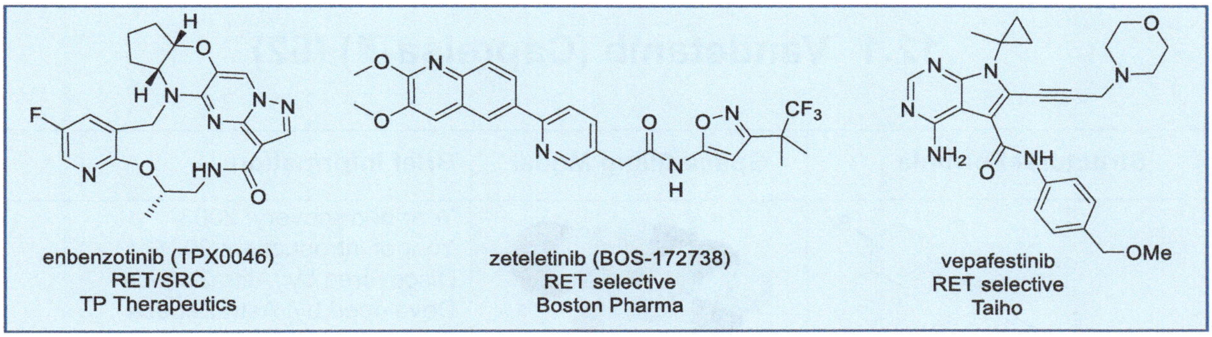

Figure 2. RET inhibitors in clinical trials (clinicaltrials.gov).

1. K. Z. Thein, V. Velcheti, B. Mooers, J. Wu, V. Subbiah. Precision therapy for RET-altered cancers with RET inhibitors. *Trends cancer.* **2021**, *7*, 1074–1088.

2. clinicaltrials.gov.

12.1 Vandetanib (Caprelsa™) (62)

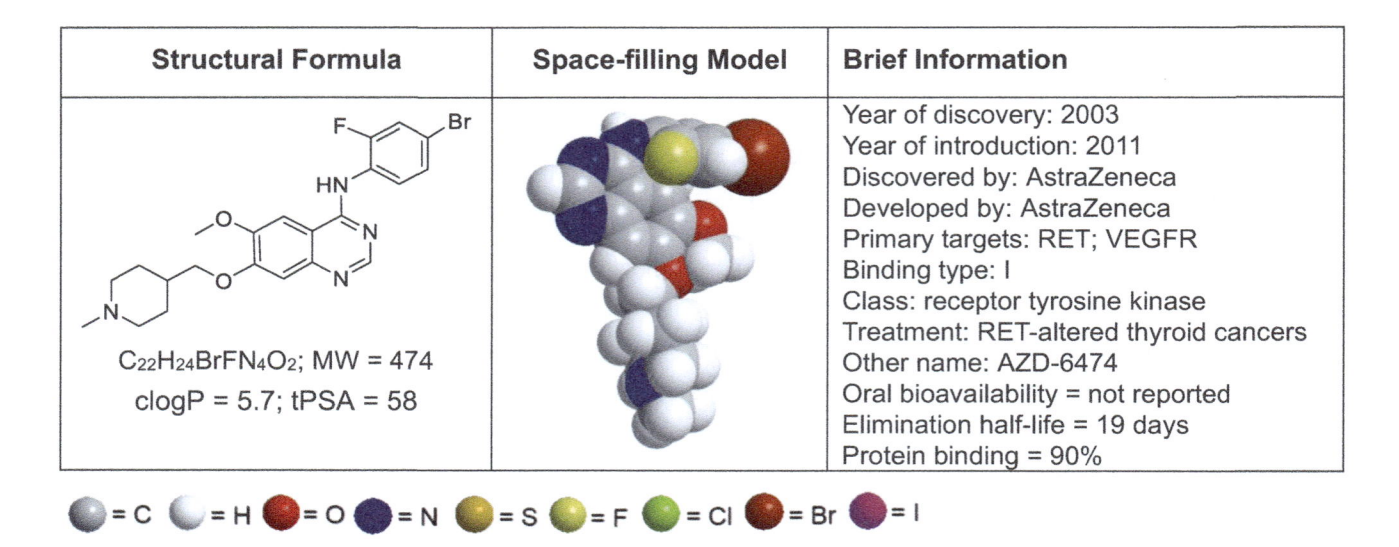

Structural Formula	Space-filling Model	Brief Information
$C_{22}H_{24}BrFN_4O_2$; MW = 474 clogP = 5.7; tPSA = 58		Year of discovery: 2003 Year of introduction: 2011 Discovered by: AstraZeneca Developed by: AstraZeneca Primary targets: RET; VEGFR Binding type: I Class: receptor tyrosine kinase Treatment: RET-altered thyroid cancers Other name: AZD-6474 Oral bioavailability = not reported Elimination half-life = 19 days Protein binding = 90%

● = C ● = H ● = O ● = N ● = S ● = F ● = Cl ● = Br ● = I

Vandetanib is the first multikinase inhibitor (MKI) approved in 2011 for the treatment of medullary thyroid cancer (MTC) that cannot be removed by surgery or that has spread to other parts of the body.[1] It does not cure advanced cancers that have spread widely throughout the body, but it can sometimes reduce or partially reverse the growth of the cancer.

Discovery of Vandetanib as a MKI for MTC

Initial efforts towards VEGF receptor tyrosine kinase inhibitors at AstraZeneca were directed as antiangiogenesis agents. Screening of Zeneca's compound collection identified a series of C4 anilinoquinazolines such as **1** and **2** as lead compounds (Fig. 1). Further optimization led to vandetanib as an orally bioavailable MKI, including rearrange during transfection (RET), vascular endothelial growth factor receptor-2 (VEGFR2) and epidermal growth factor receptor (EGFR).[2–5] As considerable cross talk occurs between various growth factor pathways, dual or multikinase inhibition appears to be an attractive strategy in the treatment of many malignancies. Initially, vandetanib was touted as a possible treatment of advanced non-small cell lung cancer (NSCLC) and several other cancers. However, it failed in phase II clinical trials for NSCLC as well as for locally advanced or metastatic pancreatic carcinoma.

Against the backdrop of these disappointing clinical data, vandetanib was developed and brought to market under orphan drug designation in the United States as the first treatment for use in patients with advanced MTC.

Vandetanib targets the activity of both wild-type and other RET alterations, including the KIF5B-RET fusion, and thereby inhibiting the growth of tumors driven by excessive RET activity. Despite its relatively high objective response rate (ORR) and survival benefits in thyroid cancers, it suffers from a variety of adverse events, which have been attributed to its activity against other receptor tyrosine kinases, such as EGFR (diarrhea and dermatologic toxicities) and VEGFR (hypertension). These off-target side effects lead to discontinuation of treatment or dose reduction, and as a result, optimal dosage to effectively suppress RET signaling pathway cannot achieved. Thus, the second-generation RET-selective inhibitors selpercatinib and pralsetinib have been developed to surpass the limitations of MKIs (see Sections 12.3 and 12.4).

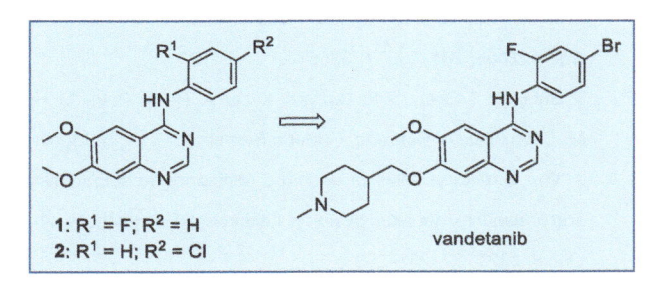

Figure 1. Structural optimization of vandetanib.

Binding Mode

Vandetanib is anchored to the hinge of RET by forming a critical hydrogen bond between the N1 nitrogen of the quinazoline group and the amide NH of Ala807 of the hinge (Figs. 2, 3). The methoxy oxygen interacts with the side chain hydroxyl of Ser811 via a bound water molecule, while the C4 NH is involved in a network of hydrogen bonds with three water molecules and the side chain carboxylate of Asp892.[6]

Vandetanib occupies both the front and back clefts of the active site of the RET kinase domain by passing through the gate that divides these two clefts, which consists of the gatekeeper Val804 and the gatewall Lys758 residue. Mutations at gatekeeper V804M/L residues result in steric hindrance that disrupts the binding of the inhibitor with RET kinase active site. As a result, vandetanib loses its potency against RETV804M/L mutants (IC$_{50}$ = 726 nM, 3.6 µM respectively, vs. 4 nM for the wt) (Table 1). In contrast, the second-generation selective RET inhibitors, selpercatinib and prasetinib, bypass the gate to occupy both the front and back clefts of the active site. Therefore, they inhibit RETV804M/L gatekeeper mutants with a similar potency to the wild-type.[7,8]

Table 1. IC$_{50}$'s of RET inhibitors.[9]

inhibitor	RET wt	RETV804L	RETV804M	RET918T	CCDC6-RET	VEGFR2
selperatinib	1 nM	2 nM	2 nM	2 nM	NA	>100 nM
praletinib	0.4 nM	0.3 nM	0.4 nM	0.4 nM	0.4	35 nM
vandetanib	4 nM	3.6 µM	726 nM	7 nM	20 nM	4 nM
cabozantinib	11 nM	45 nM	162 nM	8 nM	34 nM	2 nM

RET804L; RET804M: RET gatekeeper mutants; RET918T: other RET-activation mutant; CCDC6-RET: RET fusion

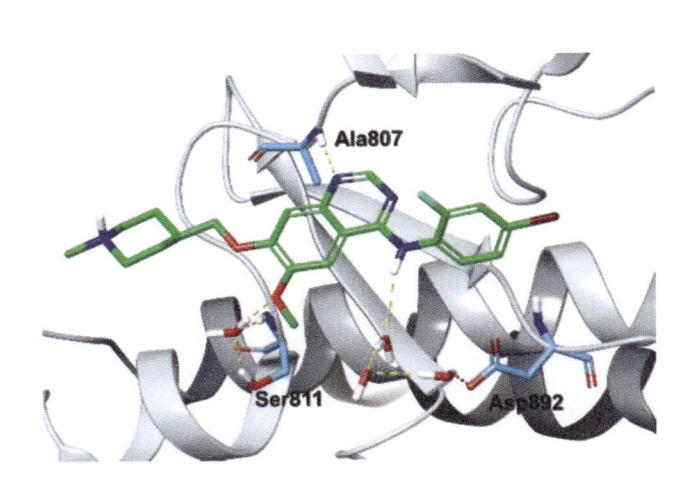

Figure 2. Co-crystal structure of vandetanib with RET (PDB ID: 2IVU).

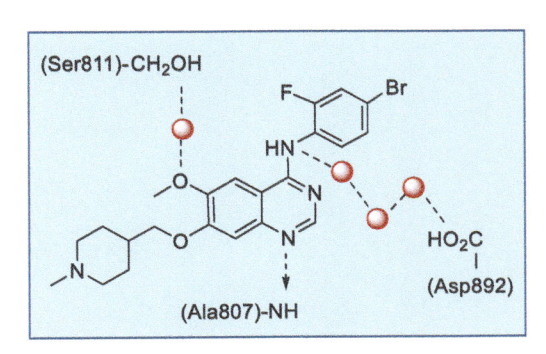

Figure 4. Summary of vandetanib–RET interactions based on their co-crystal structure.

1. H. Commander, G. Whiteside, C. Perry. Vandetanib: first global approval. *Drugs*, **2011**, *71*, 1355–1365.
2. L. F. Hennequin, A. P. Thomas, C. Johnstone, E. S. Stokes, P. A. Plé, J. J. Lohmann, D. J. Ogilvie, M. Dukes, S. R. Wedge, J. O. Curwen, J. Kendrew, C. Lambert-van der Brempt. Design and structure-activity relationship of a new class of potent VEGF

receptor tyrosine kinase inhibitors. *J. Med. Chem.* **1999**, *42*, 5369–5389.

3. L. F. Hennequin, E. S. Stokes, A. P. Thomas, C. Johnstone, P. A. Plé, D. J. Ogilvie, M. Dukes, S. R. Wedge, J. Kendrew, J. O Curwen. Novel 4-anilinoquinazolines with C-7 basic side chains: Design and structure activity relationship of a series of potent, orally active, VEGF receptor tyrosine kinase inhibitors. *J. Med. Chem.* **2002**, *45*, 1300–1312.

4. S. R. Wedge, D. J. Ogilvie, M. Dukes, J. Kendrew, R. Chester, J. A. Jackson, S. J. Boffey, P. J. Valentine, J. O. Curwen, H. L. Musgrove, G. A. Graham, G. D. Hughes, A. P. Thomas, E. S. Stokes, B. Curry, G. H. Richmond, P. F. Wadsworth, A. L. Bigley, L. F. Hennequin. ZD6474 inhibits vascular endothelial growth factor signaling, angiogenesis, and tumor growth following oral administration. *Cancer Res.* **2002**, *62*, 4645–4655.

5. A. Ryan, S. Wedge. ZD6474 – A novel inhibitor of VEGFR and EGFR tyrosine kinase activity. *Br. J. Cancer.* 2005, *92*, S6–S13.

6. P. P. Knowles, J. Murray-Rust, S. Kjær, R. P. Scott, S. Hanrahan, M. Santoro, C. F. Ibáñez, N. Q. McDonald. Structure and Chemical Inhibition of the RET Tyrosine Kinase Domain. *J. Biol. Chem.* **2006**, *281*, 33577–33587.

7. V. Subbiah, T. Shen, S. S. Terzyan, X. Liu, X. Hu, K. Patel, M. Hu, M. Cabanillas, A. Behrang, F. Meric-Bernstam, P. Vo, B. Mooers, J. Wu. Structural basis of acquired resistance to selpercatinib and pralsetinib mediated by non-gatekeeper RET mutations. *Ann. Oncol.* **2021**, *32*, 261–268.

8. K. Z. Thein, V. Velcheti, B. Mooers, J. Wu, V. Subbiah. Precision therapy for RET-altered cancers with RET inhibitors. *Trends in cancer.* **2021**, *7*, 1074–1088.

9. V. Subbiah, J. F. Gainor, R. Rahal, J. D. Brubaker, J. L. Kim, M. Maynard, W. Hu, Q. W., Cao, M. P. Sheets, D. Wilson, K. J. Wilson, L. DiPietro, P. Fleming, M. Palmer, M. I,,Hu, L. Wirth, M. S. Brose, S. I. Ou, M. Taylor, E. Garralda, E. K. Evans. Precision targeted therapy with BLU-667 for *RET*-driven cancers. *Cancer Discov.* **2018**, *8*, 836–849.

12.2 Cabozantinib (Cometriq™) (63)

Structural Formula	Space-filling Model	Brief Information
$C_{28}H_{24}FN_3O_5$; MW = 501 clogP = 4.9; tPSA = 98		Year of discovery: 2008 Year of introduction: 2012 Discovered by: Exelixis Developed by: Exelixis/BMS Primary targets: RET/VEGFR Binding type: unknown Class: receptor tyrosine kinase Treatment: MTC; RCC; HCC Other name: XL-184, BMS-907351 Oral bioavailability = not reported Elimination half-life = 110 h Protein binding > 99.7%

● = C   ○ = H   ● = O   ● = N   ● = S   ● = F   ● = Cl   ● = Br   ● = I

Cabozantinib is the second multikinase inhibitor (MKI) approved for the treatment of metastatic medullary thyroid cancer (MTC) (2012), one year after vandetanib, the other MKI for MTC. It has been approved for (1) patients with advanced renal cell carcinoma (RCC); (2) patients with hepatocellular carcinoma (HCC) who have been previously treated with sorafenib; and (3) adult and pediatric patients ≥12 years of age with locally advanced or metastatic differentiated thyroid cancer (DTC) that has progressed following prior VEGFR-targeted therapy and who are radioactive iodine-refractory or ineligible.

Cabozantinib primarily targets RET, VEGFR-2, and MET, and other receptor tyrosine kinases.[1-4] These kinases play an important role in the oncogenesis and tumor angiogenesis. By inhibiting the activity of these receptor kinases, cabozantinib shuts off oncogenesis, metastasis and supply of blood vessels and nutrients to the tumors.

Cabozantinib is active against both wild-type RET and RET fusion mutant CCDC6-RET (IC_{50} = 11 nM; 34 nM, respectively), but it is less active vs. the RETV804M/L gatekeeper mutants (IC_{50} = 162 nM and 45 nM, respectively) (for a detailed comparison with other RET inhibitors, see Section 12.1).[5] The RET inhibition is the primary driver for its efficacy in MTC.

Cabozantinib is associated with multiple adverse events, including diarrhea, fatigue, hypertension, and dermatological conditions, which arise from its inhibition of other receptor tyrosine kinases. These side effects may lead to dose reduction or interruption of treatment. To overcome the kinase selectivity issue, the second-generation RET-selective inhibitors selpercatinib and pralsetinib have been developed with a safety profile superior to MKIs (see Sections 12.3 and 12.4).

The binding mode of cabozantinib has not yet been determined due to lack of X-ray co-crystal structure with any of the receptor tyrosine kinases.

Because of its indication for multiple serious cancers, cabozantinib has become one of the most successful multikinase inhibitors, generating more than $1.5 billion sales in 2021.

1. Y. Zhang, F. Guessous, A. Kofman, D. Schiff, R. Abounader. XL-184, a MET, VEGFR-2 and RET kinase inhibitor for the treatment of thyroid cancer, glioblastoma multiforme and NSCLC. *Idrugs.* **2010**, *13*, 112–121.
2. F. M. Yakes, J. Chen, J. Tan, K. Yamaguchi, Y. Shi, P. Yu, F. Qian, F. Chu, F. Bentzien, B. Cancilla, J. Orf, A. You, A. D. Laird, S. Engst, L. Lee, J. Lesch, Y. C. Chou, A. H. Joly. Cabozantinib (XL184), a novel MET and VEGFR2 inhibitor, simultaneously suppresses metastasis, angiogenesis, and tumor growth. *Mol. Cancer Ther.* **2011**, *10*, 2298–2308.

3. R. Katayama, Y. Kobayashi, L. Friboulet, E. L. Lockerman, S. Koike, A. T. Shaw, J. A. Engelman, N. Fujita. Cabozantinib overcomes crizotinib resistance in ROS1 fusion-positive cancer. *Clin. Cancer Res.* **2015**, *21*, 166–174.

4. B. Escudier, J. C. Lougheed, L. Albiges. Cabozantinib for the treatment of renal cell carcinoma. *Expert Opin. Pharmacother.* **2016**, *17*, 2499–2504.

5. V. Subbiah, J. F. Gainor, R. Rahal, J. D. Brubaker, J. L. Kim, M. Maynard, W. Hu, Q. W., Cao, M. P. Sheets, D. Wilson, K. J. Wilson, L. DiPietro, P. Fleming, M. Palmer, M. I. Hu, L. Wirth, M. S. Brose, S. I. Ou, M. Taylor, E. Garralda, E. K. Evans. Precision targeted therapy with BLU-667 for *RET*-driven cancers. *Cancer Discov.* **2018**, *8*, 836–849.

12.3 Selpercatinib (Retevmo™) (64)

Structural Formula	Space-filling Model	Brief Information
$C_{29}H_{31}N_7O_3$; MW = 525 clogP = 2.8; tPSA = 109		Year of discovery: 2017 Year of introduction: 2020 Discovered by: Array BioPharma Developed by: LOXO/Lilly Primary target: RET Binding type: I Class: receptor tyrosine kinase Treatment: RET-altered lung/thyroid cancers Other name: ARRY-192, LOXO-292 Oral bioavailability = 73% Elimination half-life = 32 h Protein binding = 97%

● = C ○ = H ● = O ● = N ● = S ● = F ● = Cl ● = Br ● = I

Selpercatinib was granted accelerated approval in 2020 for the treatment of adult patients with (1) metastatic RET (rearranged during transfection) fusion-positive non-small cell lung cancer (NSCLC); (2) adult and pediatric patients ≥12 years of age with advanced or metastatic RET-mutant medullary thyroid cancer (MTC) who require systemic therapy; and (3) adult and pediatric patients ≥12 years of age with advanced or metastatic RET fusion-positive thyroid cancer who require systemic therapy and who are radioactive iodine-refractory.[1,2]

Selpercatinib is the first drug to selectively target the activity of both wild-type and other RET alterations, including the KIF5B-RET fusion, and thereby inhibit the growth of tumors driven by increased RET activity. It is a strong rival to Blueprint's pralsetinib, the second FDA approved selective RET inhibitor.[3]

RET Alterations in Thyroid and Lung cancer[4–8]

RET is a single-pass transmembrane protein, consisting of a typical intracellular tyrosine kinase domain and an extracellular cysteine-rich domain to interact with specific factors outside the cell and receive signals that help the cell respond to its environment (Fig. 1, top left). While the activation of a receptor tyrosine kinase (RTK) generally depends on a ligand–receptor interaction, RET is the only known RTK that does not directly bind its ligands. Instead, its activation requires a binding between its ligands (the glial cell line-derived neurotrophic factor-family ligands, GFLs) and a co-receptor (GFLs family receptor-α, GFRα). The GFL–GFRα complex interacts with the extracellular domain of RET, leading to dimerization and autophosphorylation (at the intracellular tyrosine kinase domain) and consequent activation of downstream intracellular signaling pathways (Fig. 1, top right).

One major type of genetic alterations of RET is RET rearrangement, a gene fusion, that occurs when the RET gene hybridizes with various upstream fusion partners such as KIF5B (kinesin family 5B gene) to form the KIF5B-RET fusion gene through chromosomal rearrangements. The RET gene fusion has been identified in 1–2% of NSCLCs. The KIF5B-RET is the most common fusion gene in NSCLC (approximately 70% of cases). In the process of RET rearrangement, the RET intracellular kinase domain combines with the coiled coil domain of the partner gene KIF5B to form a KIF5B-RET hybrid gene (Fig. 1, bottom left). The RET kinase domain is maintained after fusion (despite the breakpoint) with no change in downstream intracellular kinase activity. However, the newly incorporated coiled-coil domain functions as a dimerization unit, which induces constitutive homodimerization, subsequent autophosphorylation and activation of the oncogenic protein tyrosine kinase domain which triggers several signaling pathways (Fig. 1, bottom right). The increased downstream signaling of these pathways is associated with tumor invasion, migration, and proliferation.

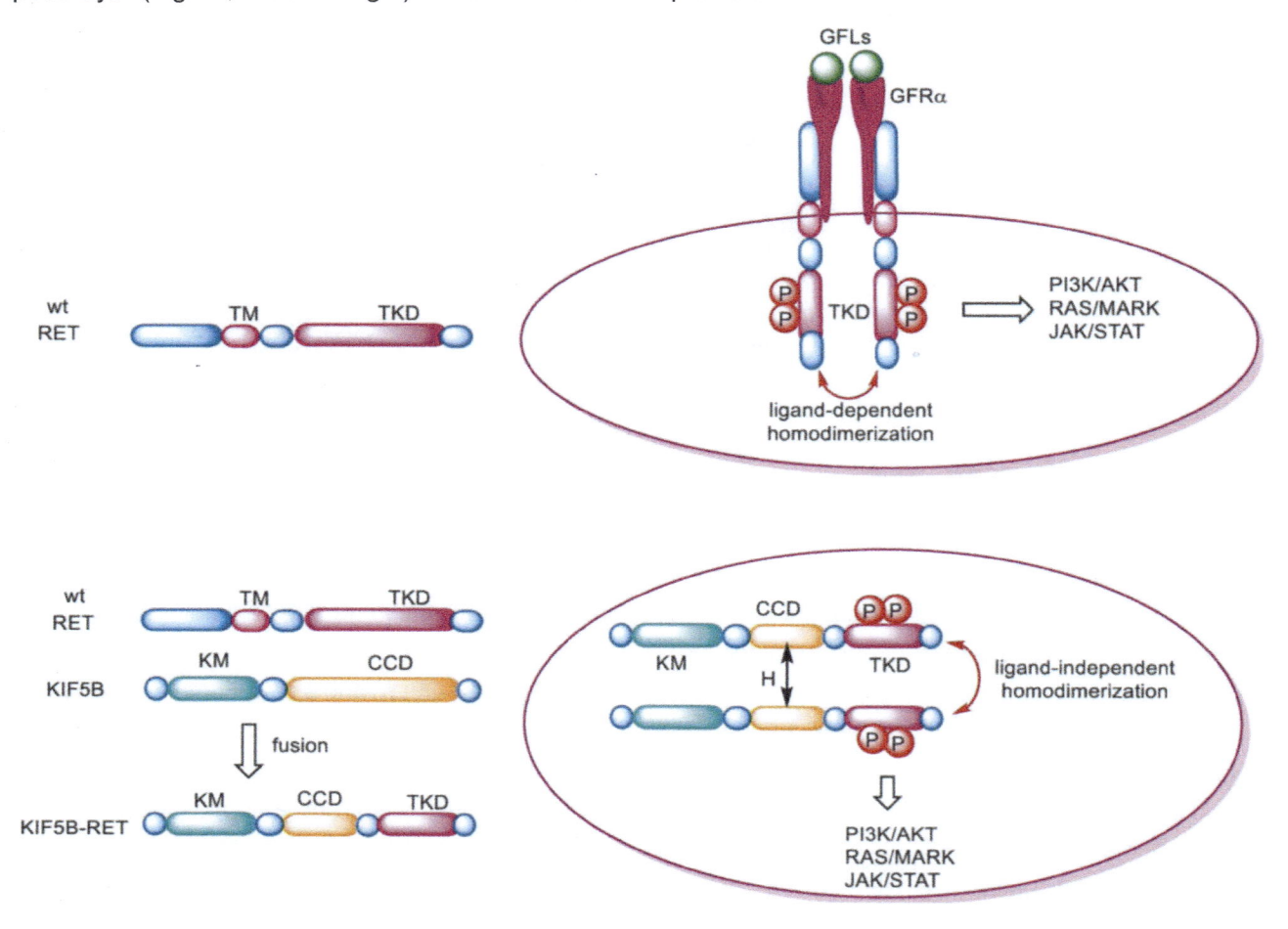

Figure 1. Top left: schematic representation of wild-type (wt) RET; Top right: ligand-dependent RET activation in wild-type; Bottom left: KIF5B-RET fusion; Bottom right: ligand-independent RET activation in KIF5B-RET. TM: transmembrane domain; TKD: tyrosine kinase domain; KM: kinesin motor; CCD: coiled-coil domain.

The other major type of RET genetic alteration is the RET point mutation which occurs in multiple endocrine neoplasias type 2 (MEN2) (Fig. 2). MEN2 is a rare, inherited polyglandular cancer syndrome, affecting approximately 1 out of 30,000 people. It has been associated with three major types of tumors: MTC, parathyroid tumors, and pheochromocytoma. There are two major MEN2 syndromes: MEN2A and MEN2B, which affect 95% and 5% of MEN2 families, respectively. MEN2A/B syndromes result from germline mutations in the RET proto-oncogene. More specifically, the MEN2A disorder originates from the cysteine mutations, particularly C634R, in the extracellular domain. These mutations result in an unpaired cysteine residue in a RET monomer, forming an aberrant intermolecular disulfide bond with another mutated molecule. Consequently, the two mutated RET monomers undergo constitutive dimerization in the absence of the GFL ligand, leading to a sustained activation of the intracellular signaling pathways (Fig. 2B). The MEN2B mutation takes place at M918T in the intracellular tyrosine-kinase domain, causing structural changes in the domain and altering RET substrate specificity. As a result, the mutated RET does not require dimerization to produce autophosphorylation, with constitutive activation of intracellular signaling pathways (Fig. 2C). The continuous activation brings about dysregulation of the parafollicular C cells in the medulla leading to the overproduction and release of hormones and the uncontrolled growth characteristic of MTC.

Selpercatinib is a potent and specific RET inhibitor with low nanomolar IC_{50} values against wt RET and a variety of RET mutations implicated in the pathogenesis of lung, thyroid, and other cancers. It is a relatively new treatment option for NSCLC and two types of thyroid cancer.

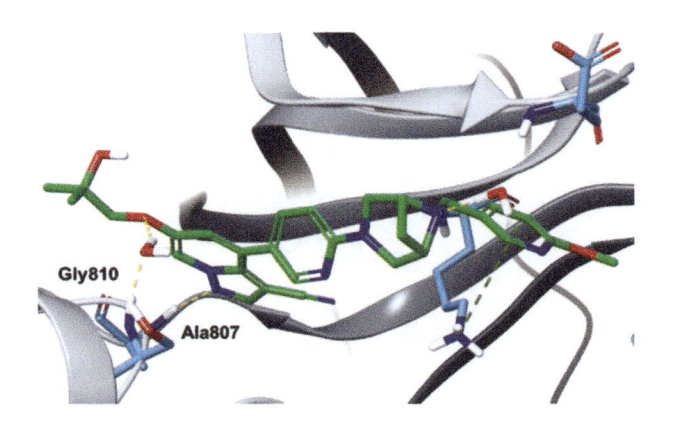

Figure 3. Co-crystal structure of selpercatinib–RET (PDB ID: 7JU6).

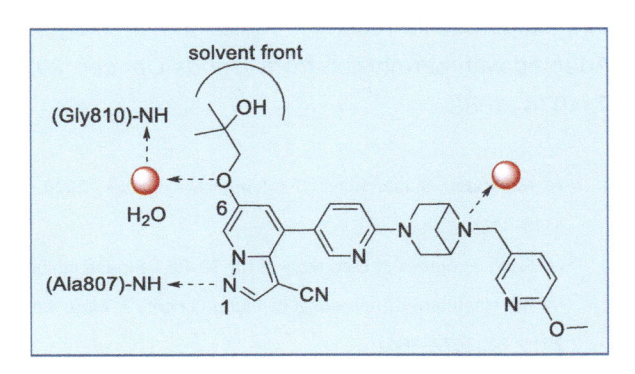

Figure 4. Summary of selpercatinib–RET interactions based on their co-crystal structure.

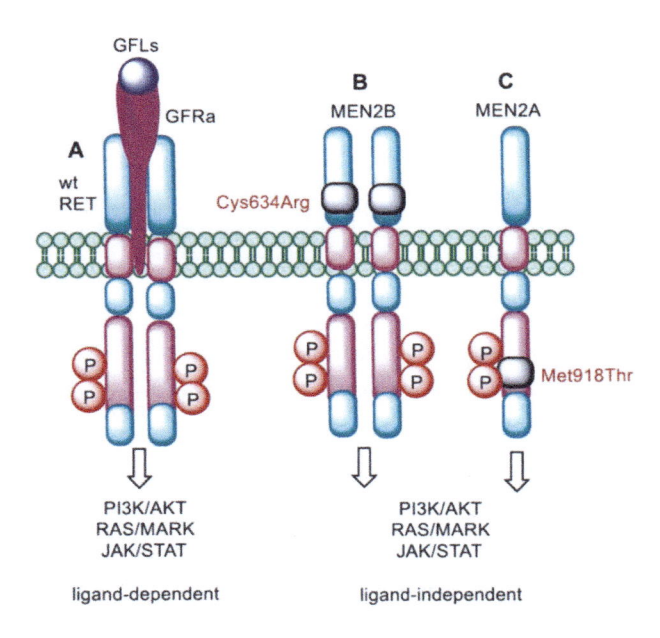

Figure 2. A: ligand-dependent RET activation in wild-type (wt); B and C: ligand-independent RET activation in MEN2A/B.

Binding Mode[9,10]

In the co-crystal structure of selpercatinib-RET kinase,[9,11] the pyrazolopyridine occupies the adenosine pocket, forming a direct hydrogen bond between N1 and amide NH of Ala807 and an indirect hydrogen bond between C6 ether oxygen and amide NH of Gly810 via a bound water molecule. In addition, one piperazine nitrogen also interact with one water molecule (Figs. 3, 4). The hydroxymethylpropoxyl group (attached to the pyrazolopyridine core) protrudes through the solvent front, and the pyridine ring engages in van der Waals interactions with the side chain of Val738 on the β2 strand.

Selpercatinib penetrates both the front and back clefts in the active site of RET kinase, but bypasses the gate that separates these two clefts, which is formed by the gatekeeper Val804 and the gatewall Lys758 residues.[10,11] It wraps around the Lys758 gatewall residue to access both the front and back clefts (Fig. 5). This unique binding avoids any potential interference that might result from gatekeeper mutations, and leads to potent inhibition of the RETV804M/L gatekeeper mutants (IC_{50} = 2 nM for both mutants, vs. 1 nM for wt in the biochemical assays).[3]

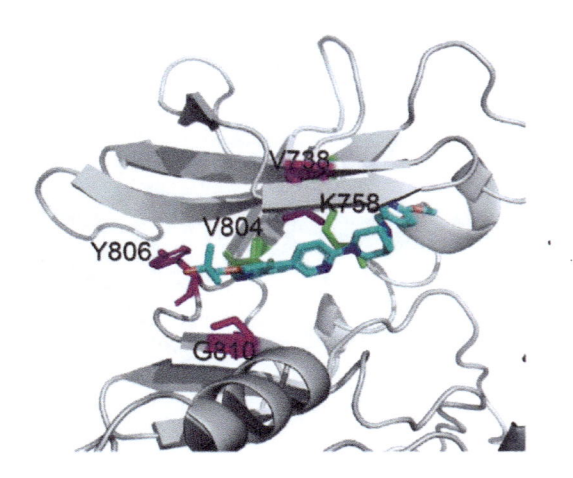

Figure 5. Co-crystal structure of selpercatinib–RET. Gatekeeper Val804 and gatewall Lys758 in green, and selpercatinib-resistant mutations in magenta. Adapted with permission from *Trends Cancer.* **2021**, *7*, 1074–1088.

1. A. Markham. Selpercatinib: First approval. *Drugs.* **2020**, *80*, 1119–1124.
2. M. Song. Progress in Discovery of KIF5B-RET Kinase inhibitors for the treatment of non-small-cell lung cancer. *J. Med. Chem.* **2015**, *58*, 3672–3681.
3. B. Brandhuber, J. Haas, B. Touch, K. Ebata, K. Bouhana, E. McFaddin, L. Williams, S. Winski, E. Brown, M. Burkhard, N. Nanda, R. Hamor, F. Sullivan, L. Hanson, T. Morales, G. Vigers, R. D. Wallace. J. Blake, S. Andrews, S. M. Rothenberg. ENA-0490. The development of a potent Of LOXO-292, a potent, KDR/EGFR2-sparing RET kinase inhibitor for treating patients with RET-dependent cancers. AACR-NCI-EORTC International Conference on Molecular Targets and Cancer Therapeutics, 2016, Poster No. 441.
4. T. E. Stinchcombe. Current management of *RET* rearranged non-small cell lung cancer. *Ther. Adv. Med. Oncol.* **2020**, *12*, 1758835920928634.
5. R. Ferrara, N. Auger, E. Auclin, B. Besse. Clinical and translational implications of RET rearrangements in non-Small cell lung cancer. *J. Thorac. Oncol.* **2018**, *13*, 27–45.
6. L. Ceolin, D. R. Siqueira, M. Romitti, C. V. Ferreira, A. L. Maia. Molecular basis of medullary thyroid carcinoma: The role of RET polymorphisms. *Int. J. Mol. Sci.* **2012**, *13*, 221–239.
7. C. Romei, C., Ciampi, R. A. Elisei. Comprehensive overview of the role of the *RET* proto-oncogene in thyroid carcinoma. *Nat Rev. Endocrinol.* **2016**, *12*, 192–202.
8. A. Drilon, Z. Hu, G. Lai, D. Tan. Targeting Ret-driven cancers: Lessons from evolving preclinical and clinical landscapes. *Nat. Rev. Clin. Oncol.* **2018**, *15*, 151−167.
9. C. Ayala-Aguilera, T. Valero, Á. Lorente-Macías, D. J. Baillache, S. Croke, A. Unciti-Broceta. Small molecule kinase inhibitor drugs (1995–2021): Medical indication, pharmacology, and synthesis. *J. Med. Chem.* **2022**, *65*, 1047-1131.
10. K. Z. Thein, V. Velcheti, B. Mooers, J. Wu, V. Subbiah. Precision therapy for RET-altered cancers with RET inhibitors. *Trends Cancer.* **2021**, *7*, 1074–1088.

12.4 Pralsetinib (Gavreto™) (65)

Structural Formula	Space-filling Model	Brief Information
$C_{27}H_{32}FN_9O_2$; FW = 533 clogP = 2.9; tPSA = 127		Year of discovery: 2016 Year of introduction: 2020 Discovered by: Blueprint Medicines Developed by: Blueprint Medicines Primary target: RET Binding type: I Class: receptor tyrosine kinase Treatment: RET-altered lung, thyroid cancers Other name: BLU-667 Oral bioavailability = not reported Elimination half-life = 22 h Protein binding = 97%

● = C ◯ = H ● = O ● = N ● = S ● = F ● = Cl ● = Br ● = I

Pralsetinib is the second selective RET inhibitor approved by the FDA (following Lilly's selpercatinib) for the treatment of (1) adult patients with metastatic RET (rearranged during transfection) fusion-positive non-small cell lung cancer (NSCLC); (2) adult and pediatric patients ≥12 years of age with advanced or metastatic RET-mutant medullary thyroid cancer (MTC) who require systemic therapy; and (3) adult and pediatric patients ≥12 years of age with advanced or metastatic RET fusion-positive thyroid cancer who require systemic therapy and who are radioactive iodine-refractory.[1]

Pralsetinib is a highly potent RET-selective kinase inhibitor with subnanomolar IC_{50} values against wild-type RET and oncogenic RET mutants including M918T, V804L, and V804M. It displayed >100-fold RET selectivity over 96% of a panel of 371 kinases.[2]

In the co-crystal structure of pralsetinib in complex with RET kinase (Fig. 1),[3,4] the aminopyrimidinyl and methylaminopyrazolyl subunits fit into the adenosine pocket, forming three critical hydrogen bonds with hinge residues Ala807 and Glu805. In addition, one pyrimidine nitrogen, methoxy oxygen, and one fluoropyrazole nitrogen also interact with the kinase residues indirectly via three bound water molecules (Fig. 2). As observed with selpercatinib-RET kinase complex, pralsetinib protrudes around the Lys758 gatewall residue to occupy both the front and back clefts in the active site, thereby bypassing the gate composed of the gatekeeper Val804 and the gatewall Lys758 residues (Fig. 3). This binding pattern obviates interference due to gatekeeper mutations, and thus provides molecular insight into its persistent potent inhibition of the RETV804M/L gatekeeper mutants (IC_{50} = 0.4 nM vs. wt and both mutants in the biochemical assays, see Table 1 in Section 12.1).[4,5]

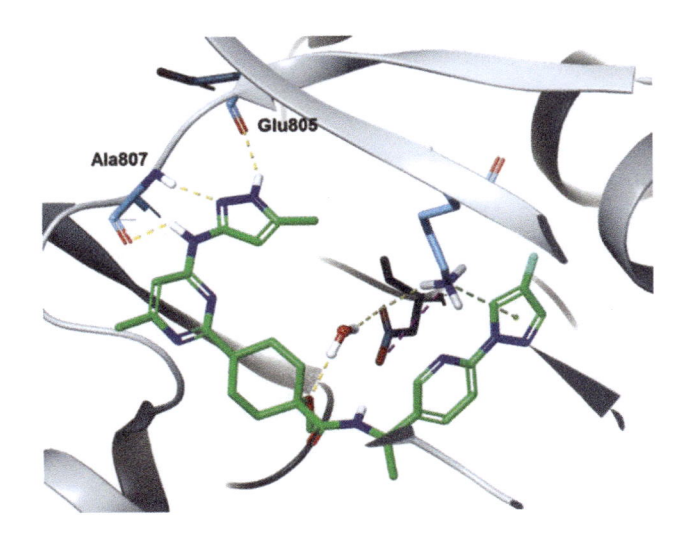

Figure 1. Co-crystal structure of pralsetinib–RET (PDB ID: 7JU5).

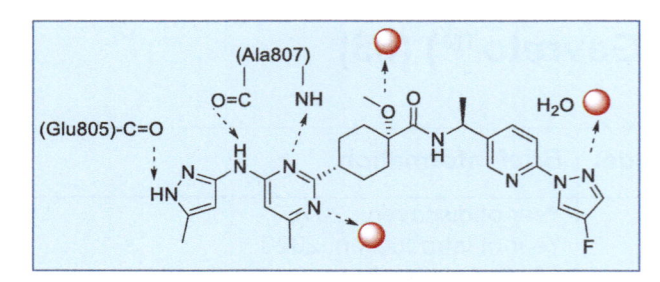

Figure 2. Summary of pralsetinib–RET interactions based on the co-crystal structure.

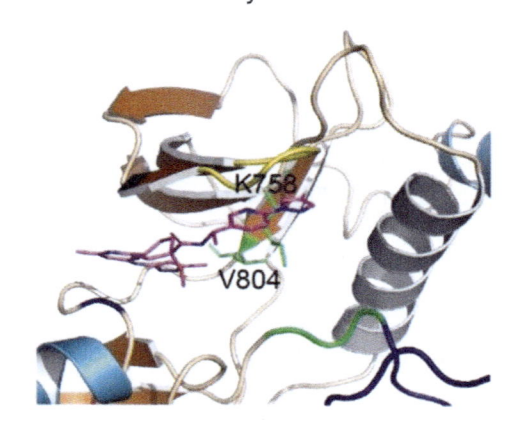

Figure 3. Co-crystal structure of pralsetinib–RET. Gatekeeper Val804 and gatewall Lys758 are in green. Adapted with permission from *Trends Cancer.* **2021**, *7*, 1074–1088.

1. A. Markham. Pralsetinib: First approval. *Drugs.* **2020**, *80*, 1865–1870.
2. V. Subbiah, J. F. Gainor, R. Rahal, J. D. Brubaker, J. L. Kim, M. Maynard, W. Hu, Q. Cao, M. P. Sheets, D. Wilson, K. J. Wilson, L. DiPietro, P. Fleming, M. Palmer, M. I. Hu, L. Wirth, M. S. Brose, S. I. Ou, M. Taylor, E. Garralda, E. K. Evans. Precision targeted therapy with BLU-667 for *RET*-driven cancers. *Cancer Discov.* **2018**, *8*, 836–849.
3. C. Ayala-Aguilera, T. Valero, Á. Lorente-Macías, D. J. Baillache, S. Croke, A. Unciti-Broceta. Small molecule kinase inhibitor drugs (1995–2021): Medical indication, pharmacology, and synthesis. *J. Med. Chem.* **2022**, *65*, 1047–1131.
4. V. Subbiah, T. Shen, S. S. Terzyan, X. Liu, X. Hu, K. Patel, M. Hu, M. Cabanillas, A. Behrang, F. Meric-Bernstam, P. Vo, B. Mooers, J. Wu. Structural basis of acquired resistance to selpercatinib and pralsetinib mediated by non-gatekeeper RET mutations. *Ann. Oncol.* **2021**, *32*, 261–268.
5. K. Z. Thein, V. Velcheti, B. Mooers, J. Wu, V. Subbiah. Precision therapy for RET-altered cancers with RET inhibitors. *Trends Cancer.* **2021**, *7*, 1074–1088.

Chapter 13. FGFR Inhibitors

FGFR2 gene fusions are found in 15% of intrahepatic cholangiocarcinomas (ICCA), while activating FGFR3 mutations account for 10–60% of urothelial carcinomas. Identification of these molecular alterations sets the stage for therapeutic intervention. Unfortunately, the first-generation multikinase inhibitors (ponatinib, lenvatinib, and nintedanib) that cross-inhibit FGFRs were unsuitable for this purpose because the exposure required to suppress the FGFR-signaling pathway was not attainable due to off-target toxicity. This prompted an intensive search for inhibitors specifically and selectively acting on the tyrosine kinase domain of FGFR family members, resulting in the discovery of the anti-FGFR therapeutics erdafitinib (**66**), pemigatinib (**67**), infigratinib (**68**), and futibatinib (**69**) (Fig. 1). Patients receiving these molecular-matched therapies had significantly longer survival than patients receiving conventional chemotherapy, justifying FDA approval for urothelial cancer and cholangiocarcinoma. The most recently approved futibatinib is the sole irreversible inhibitor that forms a covalent adduct with a cysteine side chain within the phosphate-binding loop of FGFR proteins.[1,2] This chapter describes the discovery process that led to the four FDA-approved anti-FGFR drugs **66**–**69** (Fig. 1).

Pharmaceutical research on FGFRs has been intense and has resulted in a dozen drug candidates in clinical trials, several of which are shown in Figure 2.[3] The development of so many diverse molecular matched anti-FGFR agents portends a further advance in precision molecular oncology.

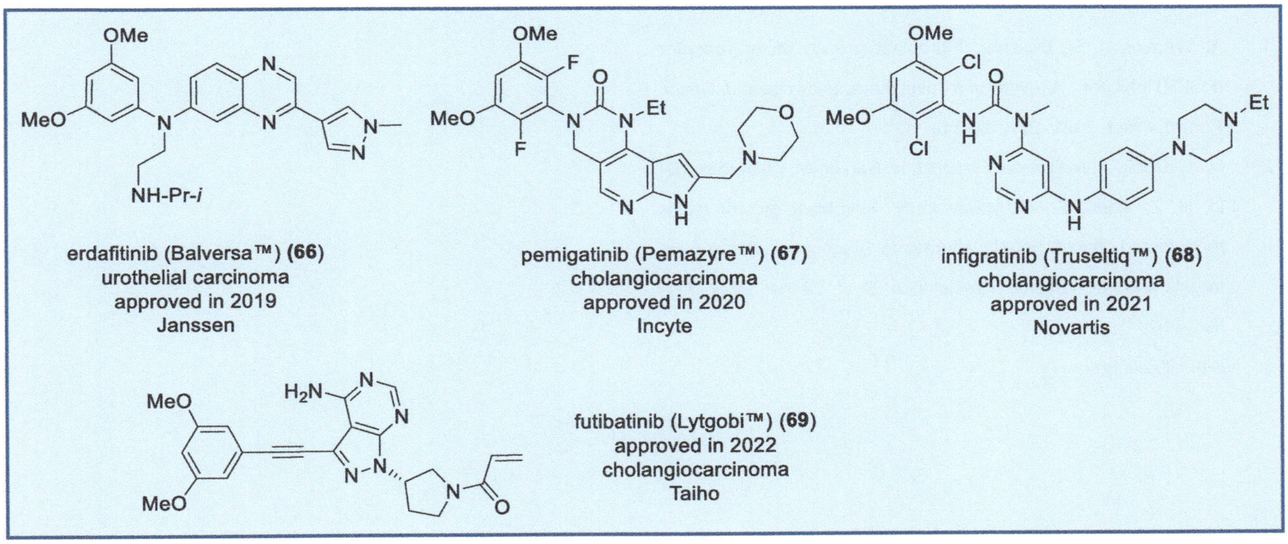

Figure 1. Approved pan-FGFR inhibitors.

Molecules Engineered Against Oncogenic Proteins and Cancer, First Edition. E. J. Corey and Yong-Jin Wu.

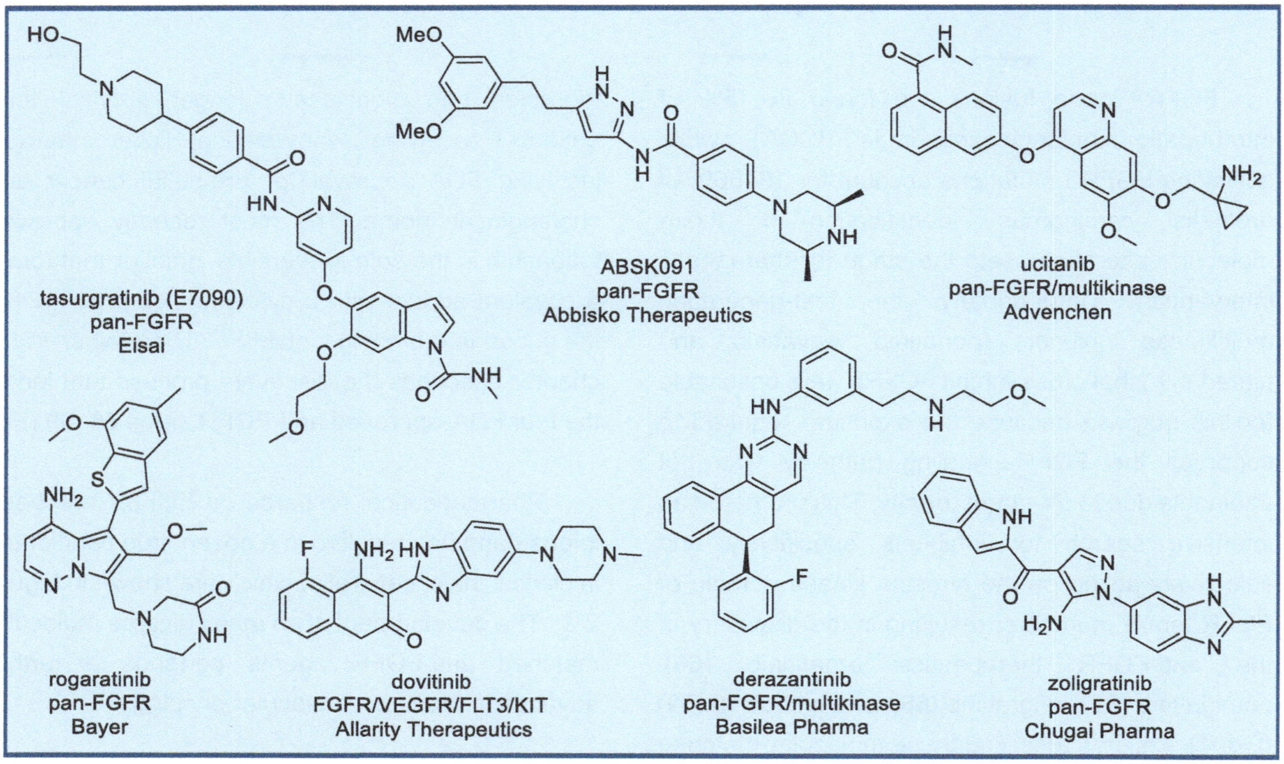

Figure 2. Pan-FGFR/multikinase inhibitors in clinical trials.

1. A. Weaver, J. B. Bossaer. Fibroblast growth factor receptor (FGFR) inhibitors: A review of a novel therapeutic class. *J. Oncol. Pharm. Pract.* **2021**, *27*, 702–710.
2. M. A. Krook, J. W. Reeser, G. Ernst, H. Barker, M. Wilberding, G. Li, H. Z. Chen, S. Roychowdhury. Fibroblast growth factor receptors in cancer: genetic alterations, diagnostics, therapeutic targets and mechanisms of resistance. *Br. J. Cancer.* **2021**, *124*, 880–892.
3. clinicaltrials.gov.

13.1 Erdafitinib (Balversa™) (66)

Structural Formula	Space-filling Model	Brief Information
$C_{25}H_{30}N_6O_2$; FW = 446 clogP = 4.2; tPSA = 74		Year of discovery: 2010 Year on Introduction: 2019 Discovered by: Janssen Developed by: Janssen Primary targets: Pan-FGFR Binding type: I1/2 Class: receptor tyrosine kinase Treatment: metastatic urothelial carcinoma Other name: JNJ-4275693 Oral bioavailability = not reported Elimination half-life = 59 h Protein binding = 99.8%

● = C ○ = H ● = O ● = N ● = S ● = F ● = Cl ● = Br ● = I

Erdafitinib is a once-daily, orally bioavailable first-in-class fibroblast growth factor receptor (FGFR) inhibitor approved in 2019 as the second-line treatment for patients with locally advanced or metastatic urothelial carcinoma (UC) with susceptible FGFR3 or FGFR2 genetic alterations, that has progressed during or following platinum-containing chemotherapy, including within 12 months of neoadjuvant or adjuvant platinum-containing chemotherapy.[1]

Bladder cancer accounts for ~4.6% of all new cancer cases annually in the United States with 80,470 new diagnoses in 2019. A total of 90% of bladder cancers are categorized as UC, and ~5% of UC patients have metastasis even at the time of the initial diagnosis. For metastatic or advanced UC, platinum-based chemotherapy is considered standard treatment, and erdafitinib offers a new therapeutic option for patients who fail to respond or have progressed after platinum-based chemotherapy.[2,3]

As many as 20% of metastatic UC patients carry FGFR alterations, and the frequency is much higher (37%) in patients with upper tract UC. The most common alterations in FGFR are FGFR2/3 activating mutations and fusions, which cause constitutive FGFR signaling that may contribute to carcinogenesis. Erdafitinib works by blocking the action of the abnormal FGFR proteins that signal cancer cells to multiply.[2-7]

Although erdafitinib contains an electron-rich 3,5-dimethoxyaniline subunit, which is prone to CYP450 oxidation to give quinone imine-type reactive metabolites, in erdafitinib this subunit appears to be metabolically stable.[8]

The development of erdafitinib as the first approved targeted therapy for metastatic UC provides an instructive example of importance of collaboration between academia (NICR, UK), biotech (Astex) and pharma (Janssen).[9]

FGFR Signaling[2-7]

The FGFR family of receptor tyrosine kinases (RTK) consists of four transmembrane receptors, FGFR1–4. Each receptor contains three extracellular immunoglobulin (Ig)-like binding domains, a transmembrane domain and two intracellular tyrosine kinase domains (TKDs). Whereas the unstimulated FGFR receptor exists as monomers in the plasma membrane, binding of the receptor to its ligand FGF induces the dimerization of the receptor, stimulation of its kinase activity, and activation of downstream signaling pathways through the intracellular domain. These pathways include the extracellular signal-regulated kinase (ERK)/mitogen-activated protein kinase (MAPK) pathway, which promotes cell survival, proliferation, development, angiogenesis and differentiation.

In normal cells, FGFR signaling is tightly controlled via ligand binding with regulation of downstream signaling. However, point mutations or gene fusions create oncogenic FGFR's which undergo aberrant activation in urothelial tumor cells resulting in tumor cell proliferation.

FGFR Point Mutations in UC[2–7]

UC is characterized by frequent gene mutations of which activating mutations in FGFR3 are the most frequent (80%), followed by FGFR2 (2.5%).

Activating FGFR point mutations can lead to constitutive activation of the kinase by: (1) receptor dimerization without activating ligands; (2) increased kinase activity; and (3) enhanced affinity for FGF ligands. Unlike other kinases, such as epidermal growth factor receptor (EGFR) and vascular endothelial growth factor receptor (VEGFR), in which activating mutations tend to occur exclusively within the kinase domain, point mutations in FGFR can occur in the extracellular domain, the transmembrane domain and the intracellular kinase domain (Fig. 1).

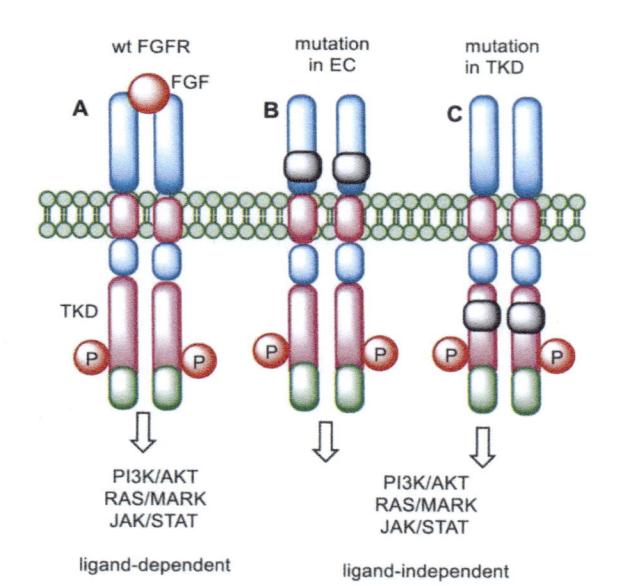

Figure 1. A: Ligand-dependent activation of FGFR in wild-type (wt); B and C: ligand-independent FGFR activation in point mutations. TKD: tyrosine kinase domain; EC: extracellular domain.

FGFR2-activating mutations predominantly take place in the transmembrane (Y375C, C382Y/R) and extracellular domains (S252W, W290C, P253R). These mutations result in increased receptor–ligand binding affinity and receptor dimerization, which contribute to their oncogenic potential.

The activating FGFR3 mutations occur most frequently at S249C in the extracellular domain, followed by Y375C in the transmembrane domain, together accounting for >80% of all FGFR3 mutations detected. These mutations, which result in introduction of a cysteine residue, are thought to enable formation of a disulfide bond, leading to covalently driven constitutive, ligand-independent receptor dimerization with consequent aberrant activation.

FGFR Fusions in UC[2–7,10,11]

FGFR can be fused with genes that encode other signaling proteins via chromosomal rearrangements or translocations, resulting in enhanced receptor dimerization and aberrant activation. Fusions of FGFR with various partner genes account for ~8% of all FGFR genetic alterations-related cancers, especially in UC, and they can be divided into type I or type II (Fig. 2), depending on whether the FGFR N terminus or C terminus is involved in the rearrangement, respectively. Type I fusions are predominant in hematological malignancies, while type II fusions are found more frequently in solid malignancies. In type I fusions, FGFR hybridizes with the N terminal regions of the fusion partner. More specifically, the extracellular and the transmembrane part of the FGFR receptors are replaced by the fusion partners, and the FGFR kinase domain is maintained after fusion (despite the breakpoint) with no change in downstream intracellular kinase activity. FGFR type I fusions are similar to those involving RET, ALK, ROS1 and NTRK. The type II fusions occur at the C-terminal regions of the fusion partner with the entire sequence of FGFR receptors remaining intact except the final exon. For both type I and type II fusions, the fusion partner domains provide self-association domains (e.g., coil-coiled domain) that induce ligand-independent dimerization or oligomerization via specific protein–protein interactions. Consequently, the fusion proteins bring about aberrant activation of multiple downstream oncogenic signaling cascades in a ligand-independent manner.

The most common FGFR fusions in metastatic UC are FGFR3 fusions, which represent ~11% of all FGFR mutations (vs. 84% for point mutations).[10] The FGFR3 fusions are found predominantly with the C-terminal regions of TACC3 (transforming acidic coiled-coil containing 3) via type II pathway, giving rise to the FGFR3-TACC3 fusion proteins. These abnormal proteins, which retain nearly all the domains of FGFR3 as well as the coiled-coil domain of TACC3, induce constitutive receptor activation with consequent downstream signaling, leading to tumourigenesis.[11]

Discovery of Erdafitinib[4,9]

Previous FGFR inhibitors in clinical development also showed potent VEGFR2 inhibition, a cause of hypertension and a liability limiting drug exposure below that required for blocking the FGFR pathway during dose escalation. Thus, a joint venture was established between the National Institute of Clinical Research, Astex and Janssen to seek inhibitors selective for FGFR over VEGFR2. A fragment-based screen of the Astex compound library identified quinoxaline 1 as a very weak dual inhibitor of FGFR3/VEGFR2. Although replacement of the amine with chlorine and incorporation of a pyrazole onto C2 improved the FGFR3 potency by 18-fold, the 7-chloro-2-pyrazole analog 2 showed only modest FGFR1 potency (IC$_{50}$ = 6.5 µM) and lacked VEGFR2 selectivity. Both issues were overcome by replacing the 7-chlorine with 3,5-dimethoxyaniline, a group which extends beyond the gatekeeper to occupy a special selectivity pocket. To optimize potency and improve physicochemical and pharmacokinetic properties, the aniline nitrogen was alkylated with a variety of substituents, leading to erdafitinib (Fig. 3).

Erdafitinib inhibited FGFR1-4 in biochemical assays with IC$_{50}$ values of 1.2, 2.5, 3.0, and 5.7 nM, respectively, vs. 36.8 nM for VEGFR2 (~30-fold selectivity relative to FGFR1). In cellular assays, erdafitinib inhibited proliferation of FGFR1, 3, and 4 expressing cells with IC$_{50}$ values of 22.1, 13.2, and 25 nM, respectively, vs 1.2 µM for VEGFR2. In addition, erdafitinib was at least 10 times more potent on BaF3 cells expressing FGFR kinases than other kinases (RET, PDGFRα/β, KIT, TIE1, LCK, LYN, ABL1, and BLK). Taken together, these data indicate that erdafitinib is a potent pan-FGFR inhibitor with limited off-target activity against VEGFR2 and other kinases.[4]

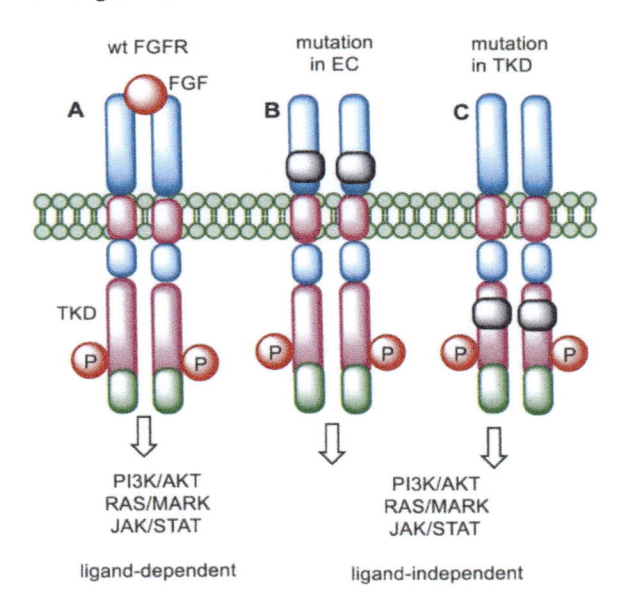

Figure 2. A. ligand-dependent activation in wt. FGFR; B: ligand-independent activation in type I and type II FGFR fusions.

Figure 3. Structural optimization of erdafitinib.

Binding Mode[3,12,13]

Erdafitinib, a type I1/2 inhibitor, is bound to the inactive conformation of FGFR1 where the activation loop exhibits a DFG-in conformation (Fig. 4). The quinoxaline N1 forms a hydrogen bond with Ala564 (the third hinge residue), and one of the methoxy oxygen atoms engages in a hydrogen bond with the backbone NH of the DFG aspartate (Asp641) (Fig. 5). The dimethoxyphenyl ring, which is perpendicular relative to the quinoxaline core, occupies a specific hydrophobic pocket (FGFR selectivity pocket) located behind the gatekeeper residue (Val561). The methyl-pyrazole subunit projects towards the solvent-front area but makes no specific interactions with the protein. In addition, the terminal isopropyl group of this side chain penetrates into a shallow hydrophobic pocket, thus contributing to the overall binding affinity and specificity.

All the three FDA-approved reversible pan-FGFR therapies (pemigatinib, erdafitinib, and infigratinib) bind similarly to a DFG-in, αC-out inactive conformation of FGFR1 protein as type I1/2 inhibitors (Figs. 6A–C).[13] The common 3,5-dimethoxyphenyl ring packs into the FGFR specificity pocket, with one of the methoxy oxygen atoms forming a hydrogen bond with Asp641. However, the hinge-binding pattern is different: erdafitinib forms one hydrogen bond with Ala564, whereas both pemigatinib and the infigratinib form two hydrogen bonds with Ala564 (Figs. 6D–F). In addition, all three inhibitors contain a water solubilizing group in the solvent-front area: a morpholine in pemigatinib, ethyl piperazine in infigratinib, and methyl pyrazole in erdafitinib (Fig. 7). Overall, the three inhibitors share similar binding mode in complex with FGFR1.[13]

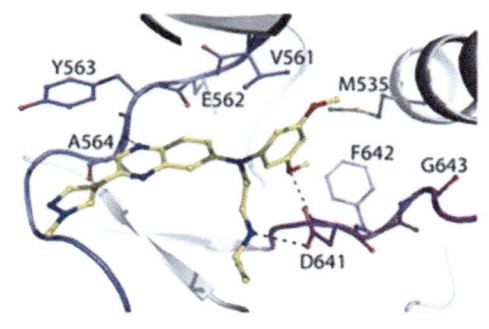

Figure 4. Co-crystal structure of erdafitinib–FGFR1. Adapted from *Oncotarget*, **2016**, *7*, 24252–24268.

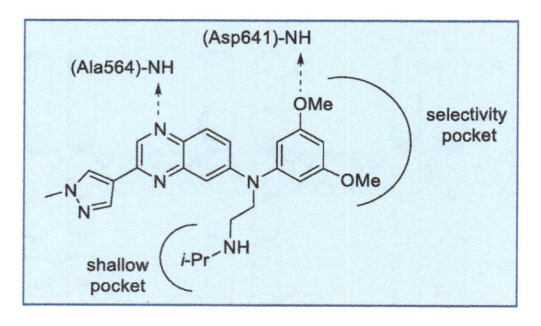

Figure 5. Erdafitinib–FGFR1 interactions.

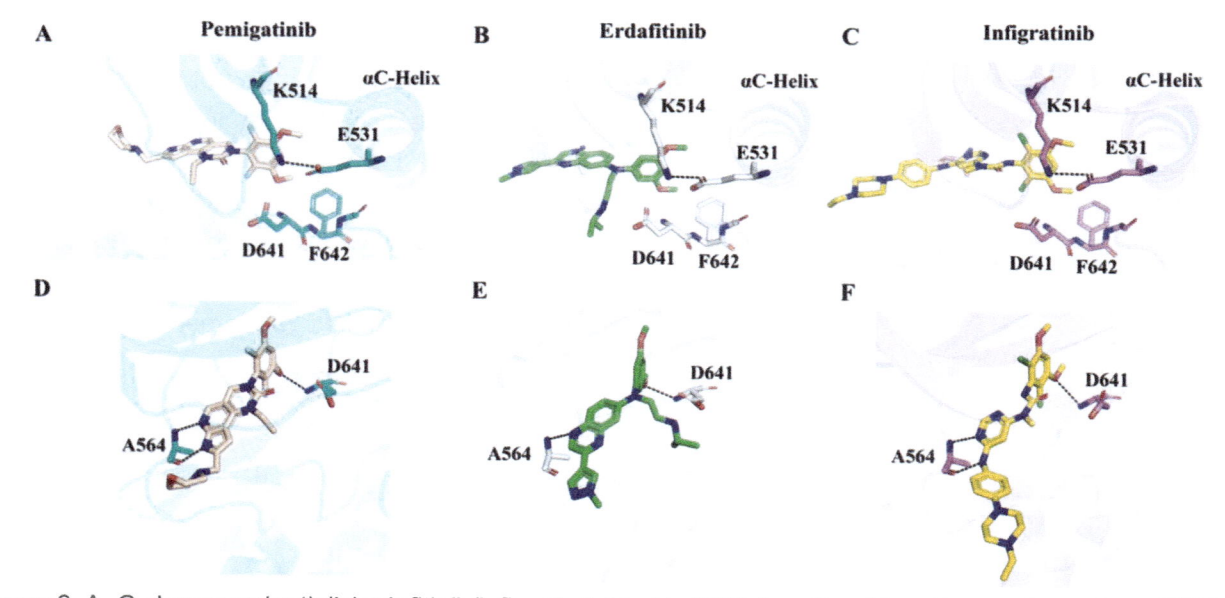

Figure 6. A–C show pemigatinib/erdafitinib/infigratinib bind to DFG-in, αC-out inactive conformation of FGFR1 (type I1/2 inhibitors). D–F: H-bond interactions. Adapted from *Commun. Chem.* **2022**, *5*, 100.

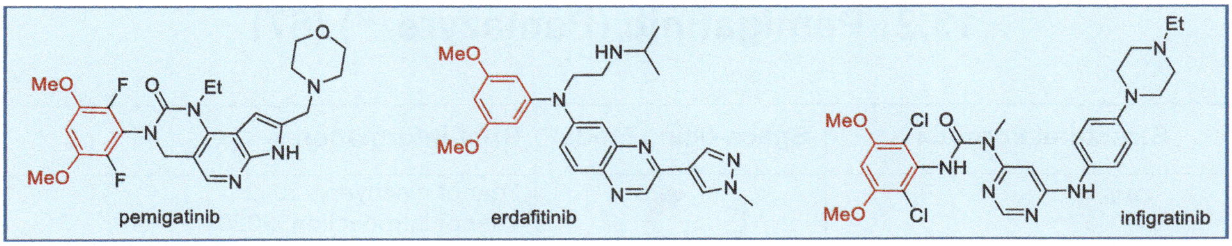

Figure 7. Structures of 3 FGFR inhibitors.

1. A. Markham. Erdafitinib: First global approval. *Drugs*. **2019**, *79*, 1017–1021.
2. M. A. Krook, J. W. Reeser, G. Ernst, H. Barker, M. Wilberding, G. Li, G., H. Z. Chen, S. Roychowdhury. Fibroblast growth factor receptors in cancer: genetic alterations, diagnostics, therapeutic targets and mechanisms of resistance. *Br. J. Cancer*. **2021**, *124*, 880–892.
3. H. Patani, T. D. Bunney, N. Thiyagarajan, R. A. Norman, D. Ogg, J. Breed, P. Ashford, A. Potterton, M. Edwards, S. V. Williams, G. S. Thomson, C. S. Pang, M. A. Knowles, A. L. Breeze, C. Orengo, C., Phillips, M. Katan. Landscape of activating cancer mutations in FGFR kinases and their differential responses to inhibitors in clinical use. *Oncotarget*. **2016**, *7*, 24252–24268.
4. T. Perera, E. Jovcheva, L. Mevellec, J. Vialard, D. De Lange, T. Verhulst, C. Paulussen, K. Van De Ven, P. King, E. Freyne, D. C. Rees, M. Squires, G. Saxty, M. Page, C. W. Murray, R. Gilissen, G. Ward, N. T. Thompson, D. R. Newell, N. Cheng, M. V. Lorenzi. Discovery and Pharmacological Characterization of JNJ-42756493 (Erdafitinib), a Functionally Selective Small-Molecule FGFR Family Inhibitor. *Mol. Cancer Ther.* **2017**, *16*, 1010–1020.
5. M. Katoh. Fibroblast growth factor receptors as treatment targets in clinical oncology. *Nat. Rev. Clin. Oncol.* **2019**, *16*, 105–122.
6. Y. Loriot, A. Necchi, S. H. Park, J. Garcia-Donas, R. Huddart, E. Burgess, M. Fleming, A. Rezazadeh, B. Mellado, S. Varlamov, M. Joshi, I. Duran, S. T. Tagawa, Y. Zakharia, B. Zhong, K. Stuyckens, A. Santiago-Walker, P. De Porre, A. O'Hagan. Erdafitinib in locally advanced or metastatic urothelial carcinoma. *N. Engl. J. Med.* **2019**, *381*, 338–348.
7. M. Katoh. Fibroblast growth factor receptors as treatment targets in clinical oncology. *Nat. Rev. Clin. Oncol.* **2019**, *16*, 105–122.
8. L. Tang, J. W. Teng, S. K. Koh, L. Zhou, M. L. Go, E. Chan. Mechanism-based inactivation of cytochrome P450 3A4 and 3A5 by the fibroblast growth factor receptor inhibitor erdafitinib. *Chem. Res. Toxicol.* **2021**, *34*, 1800–1813.
9. C. W. Murray, D. R. Newell, P. Angibaud. A successful collaboration between academia, biotech and pharma led to discovery of erdafitinib, a selective FGFR inhibitor recently approved by the FDA. *Med. Chem. Commun.* **2019**, *10*, 1509-1511.
10. M. A. Knowles. 'FGFR3 – a central player in bladder cancer pathogenesis?' *Bladder Cancer.* **2020**, 6, 403–423.
11. S. Sarkar, E. L. Ryan, S. J. Royle. FGFR3-TACC3 cancer gene fusions cause mitotic defects by removal of endogenous TACC3 from the mitotic spindle. *Open Biol.* **2017**, *7*, 170080.
12. R. Roskosk, Jr. Properties of FDA-approved small molecule protein kinase inhibitors: A 2020 update. *Pharmacol. Res.* **2020**, *152*, 104609.
13. Q. Lin, X. Chen, L. Qu, M. Guo, H. Wei, S. Dai, L. Jiang, Y. Chen Characterization of pemigatinib as a selective and potent FGFR inhibitor. *Commun. Chem.* **2022**, *5*, 100.

13.2 Pemigatinib (Pemazyre™) (67)

Structural Formula	Space-filling Model	Brief Information
$C_{24}H_{27}N_5F_2O_4$; MW = 488 clogP = 2.7; tPSA = 79		Year of discovery: 2012 Year of introduction: 2020 Discovered by: Incyte Developed by: Incyte Primary targets: Pan-FGFR Binding type: I1/2 Class: receptor tyrosine kinase Treatment: metastatic cholangiocarcinoma Other name: INCB054828 Oral bioavailability = not reported Elimination half-life = 15 h Protein binding = 90.6%

● = C ○ = H ● = O ● = N ● = S ● = F ● = Cl ● = Br ● = I

Pemigatinib was granted accelerated approval in 2020 for the treatment of adults with previously treated, unresectable locally advanced or metastatic cholangiocarcinoma (CCA) with a fibroblast growth factor receptor 2 (FGFR2) fusion or other rearrangements as detected by an FDA-approved test.[1]

CCA is a rare cancer of the bile duct, with ~8,000 people in the United States. being diagnosed every year. CCA can be subdivided into intrahepatic (ICCA) and extrahepatic (ECCA): the former occurs in the bile duct inside the liver, whereas the latter in the bile duct outside the liver. Approximately 13–17% of CCA and 10–15% of ICCA patients carry FGFR2 fusions that signal bile duct cancer cells to multiply, and pemigatinib is the first FGFR-targeted therapy for CCA patients.[2–4]

FGFR2 Fusions in CCA[2–4]

FGFR2 fusions in CCA involve a wide range of fusion partners, with BICC1 being the most frequent partner, accounting for ~28% of FGFR2 rearrangements. In the FGFR3-BICC1 fusion, the entire FGFR2 receptor is maintained except the final exon (exon 18) (corresponding to the cytoplasmic post-kinase region), which is replaced with the C-terminal portion of BICC1 via type II fusion pathway (see Section 13.1). The resulting oncogenic fusion proteins contain an intact tyrosine kinase domain of FGFR2 and a dimerization domain of the BICC1 (Fig. 1), which induces constitutive receptor activation with consequent downstream signaling, leading to tumourigenesis in the bile duct.

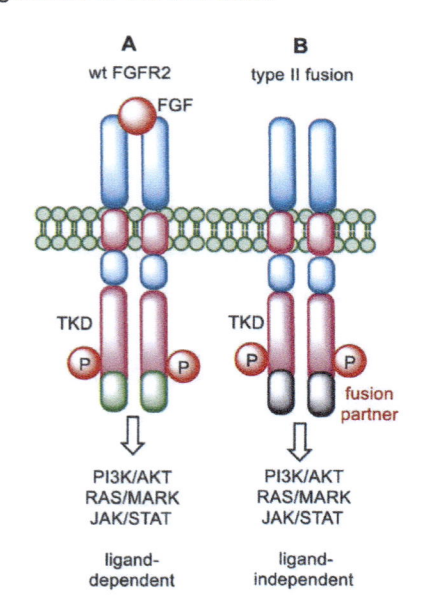

Figure 1. A. ligand-dependent activation in wt FGFR; B: ligand-independent activation in type II FGFR fusions.

Discovery of Pemigatinib [5]

High-throughput screening of the Incyte compound library identified compound **1** as a potent FGFR1 inhibitor (IC$_{50}$ = 6 nM) with good selectivity over VEGFR2 (IC_{50} = 2.1 μM) (Fig. 2). However, compound **1** also potently inhibited JAK2 (IC_{50} < 1 nM) and several other kinases (Fig. 2). At the time, the kinase selectivity was a major challenge for this series of FGFR inhibitors. Analysis of the co-crystal structure of PD173074 (a potent and specific FGFR inhibitor) with FGFR1 led to the hypothesis that the 3,5-dimethoxyphenyl substituent contributed to the high affinity and FGFR selectivity because it occupied a specific hydrophobic pocket in the vicinity of the FGFR gatekeeper region. Consequently, it was reasoned that incorporation of the 3,5-dimethoxyphenyl substituent onto the lead compound **1** could improve its kinase selectivity. Superimposition of pyrrolopyrimidine **1** with PD173074 and subsequent hybridization gave rise to a simple composite structure **2** consisting of two critical binding elements: pyrrolopyridine hinge binder from **1** and 3,5-dimethoxyphenyl FGFR specificity pocket binder from PD173074. Indeed, Hybrid **2** exhibited >100-fold selectivity over JAK2 and ~20-fold selectivity over VEGFR2, despite an 83-fold drop in FGFR1 potency relative to **1**. The FGFR potency was restored by modifying the pyrrolopyridine core of **2** to give the tetracyclic compound **3** whose rigid conformation enables the 3,5-dimethoxyphenyl group to fill the specificity pocket. Indeed, compound **3** displayed good selectivity over JAK2 and VEGFR2. Further optimization of potency, physicochemical, and pharmacokinetic properties led to pemigatinib.

In biochemical assays, pemigatinib potently inhibited FGFR1, FGFR2, and FGFR3 (IC_{50} = 0.4, 0.5, and 1 nM, respectively), and, to a much lesser extent, FGFR4 (IC_{50} = 30 nM). It also potently blocked the growth of FGFR1-, FGFR2-, and FGFR3-dependent tumor cell lines with IC_{50} values in the low nM range. Furthermore, it displayed good selectivity over other kinases with only VEGFR2 (IC_{50} = 71 nM) and KIT (IC_{50} = 266 nM) being identified with IC_{50} values less than 1 μM.

Although pemigatinib contains an electron-rich 3,5-dimethoxyaniline moiety, which is potentially vulnerable to CYP450 oxidation to give quinone or quinone imine reactive metabolites, detailed metabolite profiling of pemigatinib showed that thanks to the two fluorine substituents such reactive metabolites are not formed.[5]

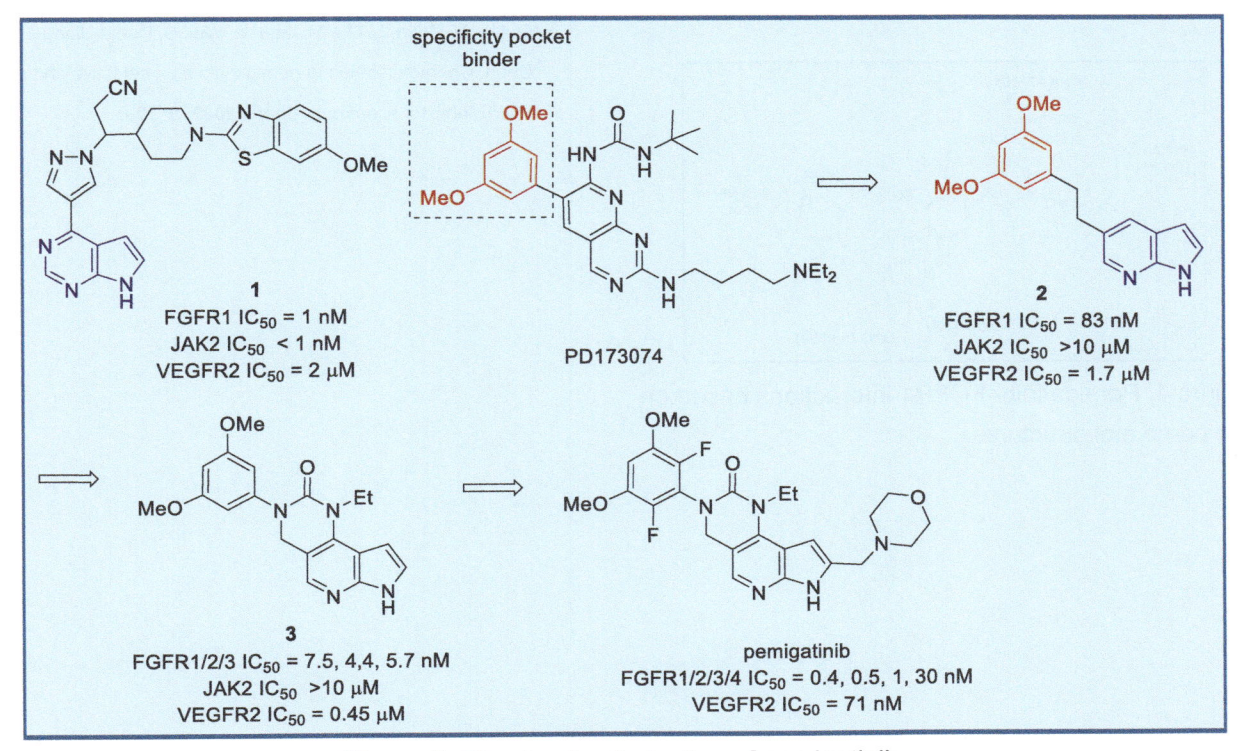

Figure 2. Structural optimization of pemigatinib.

Binding Mode[6]

Pemigatinib binds to the inactive conformation of FGFR1 with the activation loop adopting a DFG-in conformation (type I1/2 inhibitor). The pyrrolopyridine moiety forms two hydrogen bonds with the NH and the carbonyl of Ala564 in the hinge region (Figs. 3, 4). The difluoromethoxyphenyl ring, which is perpendicular relative to the tricyclic ring, fits the specificity pocket containing the gatekeeper residue Val561. One of the methoxy oxygen atoms hydrogen bonds to the backbone NH of Asp641. The terminal morpholine extends towards the solvent-front region, and makes only minor contacts with the protein.

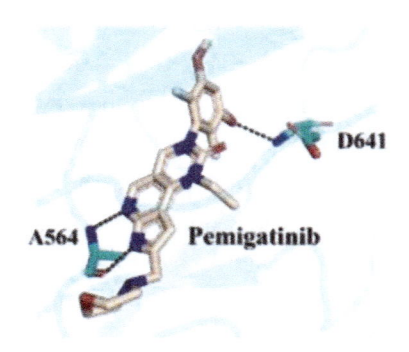

Figure 3. Co-crystal structure of pemigatinib–FGFR1. Adapted from *Commun. Chem.* **2022**, *5*, 100.

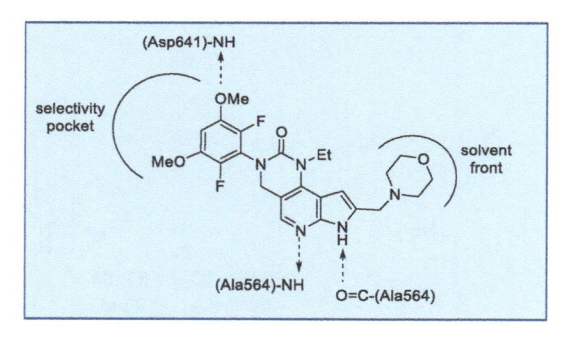

Figure 4. Pemigatinib–FGFR1 interactions based on the co-crystal structure.

1. S. M. Hoy. Pemigatinib: First Approval. *Drugs.* **2020**, *80*, 923–929.
2. F. Li, M. N. Peiris, D. J. Donoghue. Functions of FGFR2 corrupted by translocations in intrahepatic cholangiocarcinoma. *Cytokine Growth Factor Rev.* **2020**, 52, 56–67.
3. M. A. Lowery, R. Ptashkin, E. Jordan, M. F. Berger, A. Zehir, M. Capanu, N. E. Kemeny, E. M. O'Reilly, I. El-Dika, W. R. Jarnagin, J. J. Harding, M. I. D'Angelica, A. Cercek, J. F. Hechtman, D. B. Solit, N. Schultz, D. M. Hyman, D. S. Klimstra, L. B. Saltz, G. K. Abou-Alfa. Comprehensive molecular profiling of intrahepatic and extrahepatic cholangiocarcinomas: Potential targets for intervention. *Clin. Cancer Res.* **2018**, *24*, 4154–4161.
4. I. M. Silverman, A. Hollebecque, L. Friboulet, S. Owens, R. C. Newton, H. Zhen, L. Féliz, C. Zecchetto, D. Melisi, T. C. Burn. Clinicogenomic analysis of FGFR2-rearranged cholangiocarcinoma identifies correlates of response and mechanisms of resistance to pemigatinib. *Cancer Discov.* **2021**, *11*, 326–339.
5. L. Wu, C. hang, C. He, D. Qian, L. Lu, Y. Sun, M. Xu, J. Zhuo, P. Liu, R. Klabe, R. Wynn, M. Covington, K. Gallagher, L. Leffet, K. Bowman, S. Diamond, H. Koblish, Y. Zhang, M. Soloviev, G. Hollis, W. Yao. Discovery of pemigatinib: A potent and selective fibroblast growth factor receptor (FGFR) inhibitor. *J. Med. Chem.* **2021**, *64*, 10666–10679.
6. Q. Lin, X. Chen, L. Qu, M. Guo, H. Wei, S. Dai, L. Jiang, Y. Chen. Characterization of pemigatinib as a selective and potent FGFR inhibitor. *Commun. Chem.* **2022**, *5*, 100.

13.3 Infigratinib (Truseltiq™) (68)

Structural Formula	Brief Information
$C_{26}H_{31}Cl_2N_7O_3$; MW = 560 clogP = 5.6; tPSA = 94	Year of discovery: 2009 Year of introduction: 2021 Discovered by: Novartis Developed by: QED/Helsinn Primary targets: Pan-FGFR Binding type: I1/2 Class: receptor tyrosine kinase Treatment: cholangiocarcinoma Other name: NVP-BGJ398 Oral bioavailability = not reported Elimination half-life = 34 h Protein binding = 96.8%

Infigratinib was granted accelerated approval in 2021 for the treatment of adults with previously treated, unresectable locally advanced or metastatic cholangiocarcinoma (CCA) with a fibroblast growth factor receptor 2 (FGFR2) fusion or other rearrangements as detected by an FDA-approved test.[1] It is the second FDA-approved FGFR inhibitor for CCA, one year after pemigatinib.[1]

Discovery of Infigratinib[2,3]

PD166285, a broad-spectrum tyrosine kinase inhibitor (TKI), including FGFR1 (Fig. 1), contains a pyrido[2,3-*d*]pyrimidin-7-one scaffold, which is present in numerous protein kinase inhibitors. This structure was modified to generate other possible TKIs in the following way: the pyrimidine N1 of PD 166285 was moved to position 5, and the pyrimidone ring was replaced by a urea moiety, to allow an intramolecular hydrogen bond with the newly rearranged nitrogen, thus resulting in compound **1**, having a stable, planar, conjugated pseudo six-membered ring as a mimic of the pyrimidone ring (Fig. 1). Despite a substantial drop in FGFR1 potency, compound **1** served as a new FGFR inhibitor lead compound. Superimposition of pyrrolopyrimidine **1** with PD173074 (a potent, specific FGFR inhibitor), revealed that the former lacked two methoxy groups for potency and selectivity because 3,5-dimethoxyphenyl of PD173074 occupies the hydrophobic specificity pocket, and furthermore, one of the methoxy oxygen atoms hydrogen bonds to the protein. Consequently, the 3,5-dimethoxy groups were attached onto the dichlorophenyl, and the solvent-front (diethylamino)ethoxy group was substituted with a piperazine to improve pharmacokinetic properties. These two structural modifications led to infigratinib as a potent and selective pan-FGFR inhibitor.

In biochemical assays, infigratinib inhibited FGFR1–4 with IC$_{50}$ values of 0.9, 1.4, 1.0, and 60 nM, vs. 180 nM for VEGFR2. The potent inhibition of FGFR resulted in reduced cell proliferation in tumors with these FGFR mutations or rearrangements. Infigratinib also exhibited high selectivity against other kinases.

Binding Mode[3,4]

Infigratinib is a close analog of pemigatinib with the pyrimidinone cleaved (Fig. 1), and they show similar binding modes. Infigratinib is bound to the inactive conformation of FGFR1, where the activation loop adopts a DFG-in conformation. The pyrimidine nitrogen binds to the amide NH group and the anilino NH group binds to the backbone carbonyl group of the third hinge residue (Ala564) (Fig. 2). The dichloromethoxyphenyl ring, which is perpendicular relative to the aminopyrimidine ring, occupies the specificity pocket containing the gatekeeper residue

Val561. One of the methoxy oxygen atoms hydrogen bonds to the NH group of DFG-Asp641. The terminal piperazine extends toward the solvent-front region, and the ethylamine of the piperazine moiety participates in ionic interactions with the carboxylic acid side chain of Glu571. In addition, the water-mediated hydrogen bond between the urea carbonyl group of the inhibitor and the side chain of Lys514 is shown in the X-ray structure (Figs. 2, 3).

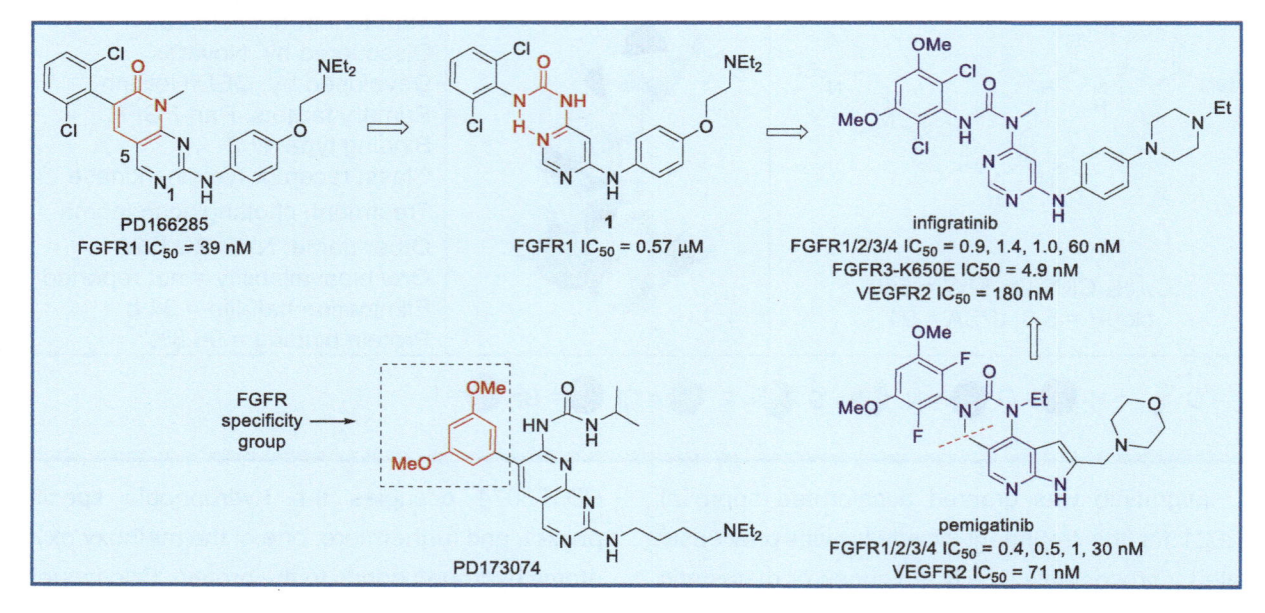

Figure 1. Structural optimization of infigratinib.

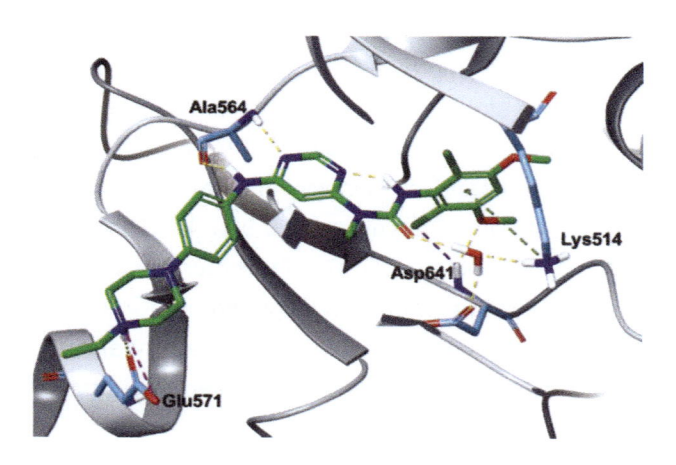

Figure 2. Co-crystal structure of infigratinib–FGFR1 (PDB ID: 3TT0).

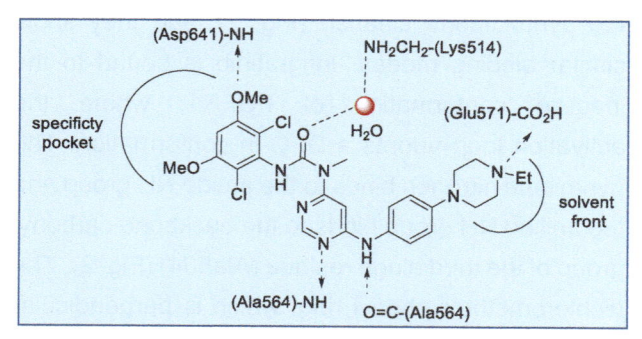

Figure 3. Infigratinib–FGFR1 interactions.

1. C. Kang. Infigratinib: First approval. *Drugs.* **2021**, *81*, 1355–1360.
2. P. Furet, G. Caravatti, V. Guagnano, M. Lang, T. Meyer, J. Schoepfer. Entry into a new class of protein kinase inhibitors by pseudo ring design. *Bioorg. Med. Chem. Lett.* **2008**, *18*, 897–900.
3. V. Guagnano, P. Furet, C. Spanka, V. Bordas, M. Le Douget, C. Stamm, J. Brueggen, M. R. Jensen, C. Schnell, H. Schmid, M. Wartmann, J. Berghausen, P. Drueckes, A. Zimmerlin, D. Bussiere, J. Murray, D. Graus Porta. Discovery of 3-(2,6-dichloro-3,5-dimethoxy-phenyl)-1-{6-[4-(4-ethyl-piperazin-1-yl)-phenylamino]-pyrimidin-4-yl}-1-methyl-urea (NVP-BGJ398), a potent and selective inhibitor of the fibroblast growth factor receptor family of receptor tyrosine kinase. *J. Med. Chem.* **2011**, *54*, 7066–7083.
4. R. Roskoski, Jr. Properties of FDA-approved small molecule protein kinase inhibitors: A 2022 update. *Pharmacol. Res.* **2022**, *175*, 106037.

13.4 Futibatinib (Lytgobi™) (69)

Structural Formula	Space-filling Model	Brief Information
$C_{22}H_{22}N_6O_3$; MW = 418 clogP = 2.0; tPSA = 105		Year of discovery: 2012 Year of introduction: 2022 Discovered by: Taiho Developed by: Taiho Primary target: Pan-EGFR Binding type: covalent Class: receptor tyrosine kinase Treatment: Cholangiocarcinoma Other name: TAS 120 Oral bioavailability = not reported Elimination half-life = 2.9 h Protein binding = 95%

● = C ○ = H ● = O ● = N ● = S ● = F ● = Cl ● = Br ● = I

Futibatinib, the first irreversible tyrosine kinase inhibitor (TKI) of pan-FGFR, was approved in 2022 for the treatment of intrahepatic cholangiocarcinoma (ICCA) harboring fibroblast growth factor receptor 2 (FGFR2) gene fusions or other rearrangements. It is the third FDA-approved TKI for ICCA, following infigratinib and pemigatinib, but its covalent, irreversibly binding mode distinguishes it from these earlier TKIs.[1]

Most irreversible inhibitors carry a reactive Michael acceptor warhead, especially an acrylamide group, to attack the thiol function of a cysteine residue at the kinase active site and form a covalent bond. In the FGFRs, a cysteine residue at the tip of the P-loop (Cys488 in FGFR1), conserved across all four FGFR isoforms, serves as the target. Futibatinib potently inhibits the P-loop active site cysteine of the FGFR family (IC_{50} = 3.9, 1.3, 1.6, and 8.3 nM for FGFR 1–4, respectively), thus blocking their signaling pathways that promote tumor survival and growth.

However, as with other kinase inhibitors, drug resistance is also a problem with FGFR inhibitors because of mutation of the "gatekeeper" residue located in the hinge region of the kinases ATP-binding pocket (a common mode of acquired resistance). For example, the FGFR2 gatekeeper mutation V564F was detected in several ICCA patients treated with pemigatinib. Despite similar IC_{50} values for wild-type FGFR2 and some mutants (wt FGFR2 = 0.9 nM; V565I = 1–3 nM; N550H = 3.6 nM; E566G = 2.4 nM), futibatinib is still ineffective against V564F, for which next-generation of FGFR inhibitors will be needed.

In the co-crystal structure with FGFR1,[2,3] the pyrazolopyrimidine core forms two hydrogen bonds to the amide NH of Ala564 and amide oxygen of Glu562 of the hinge region. The dimethoxyphenyl ring, similarly to other FGFR1-selective inhibitors, occupies a conserved hydrophobic pocket, forming a single hydrogen bond to Asp641. Most importantly, the inhibitor is irreversibly attached to the protein via a covalent bond between the acrylamide group and the reactive P-loop Cys488 (Figs. 1, 2).

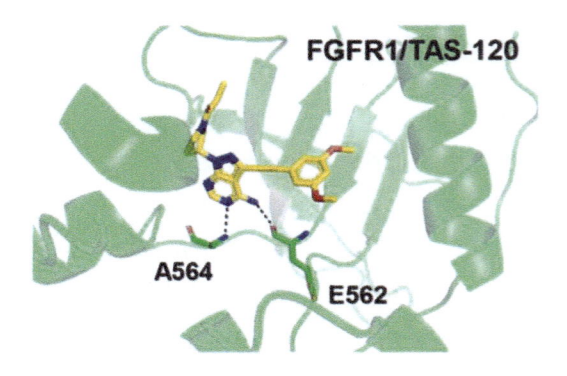

Figure 1. Co-crystal structure of futibatinib (TAS120)–FGFR1. Adapted from *Commun. Chem.* **2022**, *5*, 5. https://doi.org/10.1038/s42004-021-00623-x.

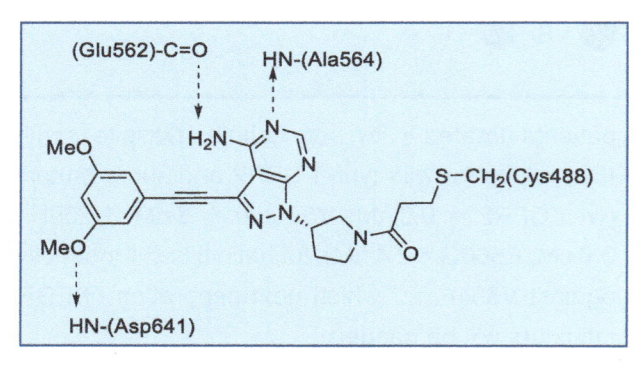

Figure 2. Summary of futibatinib–FGFR1 interactions based on the co-crystal structure.

1. T. Sagara, S. Ito, S. Otsuki, H. Sootome. 3,5-Disubstituted alkynylbenzene compound and salt thereof. WO2013108809.
2. M. Kalyukina, Y. Yosaatmadja, M. J. Middleditch, A. V. Patterson, J. B. Smaill, C. J. Squire. TAS-120 cancer target binding: Defining reactivity and revealing the first fibroblast growth factor receptor 1 (FGFR1) irreversible structure. *ChemMedChem.* **2019**, *14*, 494–500.
3. L. Qu, X. Chen, H. Wei, M. Guo, S. Dai, L. Jiang, J. Li, S. Yue, L. Chen, Y. Chen. Structural insights into the potency and selectivity of covalent pan-FGFR inhibitors. *Commun. Chem.* **2022**, *5*, 5. https://doi.org/10.1038/s42004-021-00623-x.

Chapter 14. PI3K Inhibitors

The PI3K signaling pathway is frequently aberrantly activated in a variety of cancer types, thus reducing apoptosis and promoting tumor proliferation. Blockage of this pathway with PI3K inhibitors with various isoform selectivity has been extensively explored for cancer treatment in the past two decades. Early pan-class I PI3K and dual PI3Kα/PI3Kδ inhibitors showed only marginal clinical benefits and but poor tolerability.[1] Subsequently, the discovery of isoform-selective PI3K ushered in a new era in PI3K drug development. In 2014, the PI3Kδ inhibitor idelalisib became the first PI3K inhibitor approved for use in specific B cell malignancies. Three year later, the pan-class I PI3K inhibitor copanlisib and the dual PI3Kδ/PI3Kγ inhibitor duvelisib were approved for the same indications. Umbralisib, a dual inhibitor of PI3Kδ and casein kinase-1-ε (CK1ε), was approved in 2021 to treat two types of blood cancer: relapsed or refractory marginal zone lymphoma (MZL). In 2019, the PI3Kα isoform-selective inhibitor alpelisib (Fig. 1) was approved for the treatment of advanced breast cancer, in combination with the estrogen receptor (ER) downregulator fulvestrant.

Despite these significant advances, tolerability remains a challenge with PI3K inhibitors, with off-target adverse events including hyperglycemia and diarrhea as well as a variety of immune-related toxicities and infections.[1] Hence, ongoing research continues to focus on improving PI3K isoform selectivity. Several newer PI3K inhibitors in clinical trials for various cancer indications are shown in Figure 2.[2]

This chapter describes the discovery process that led to the five FDA-approved PI3K inhibitors **70–74** (Fig. 1). Leniolisib is not included as it was approved right after completion of this book.

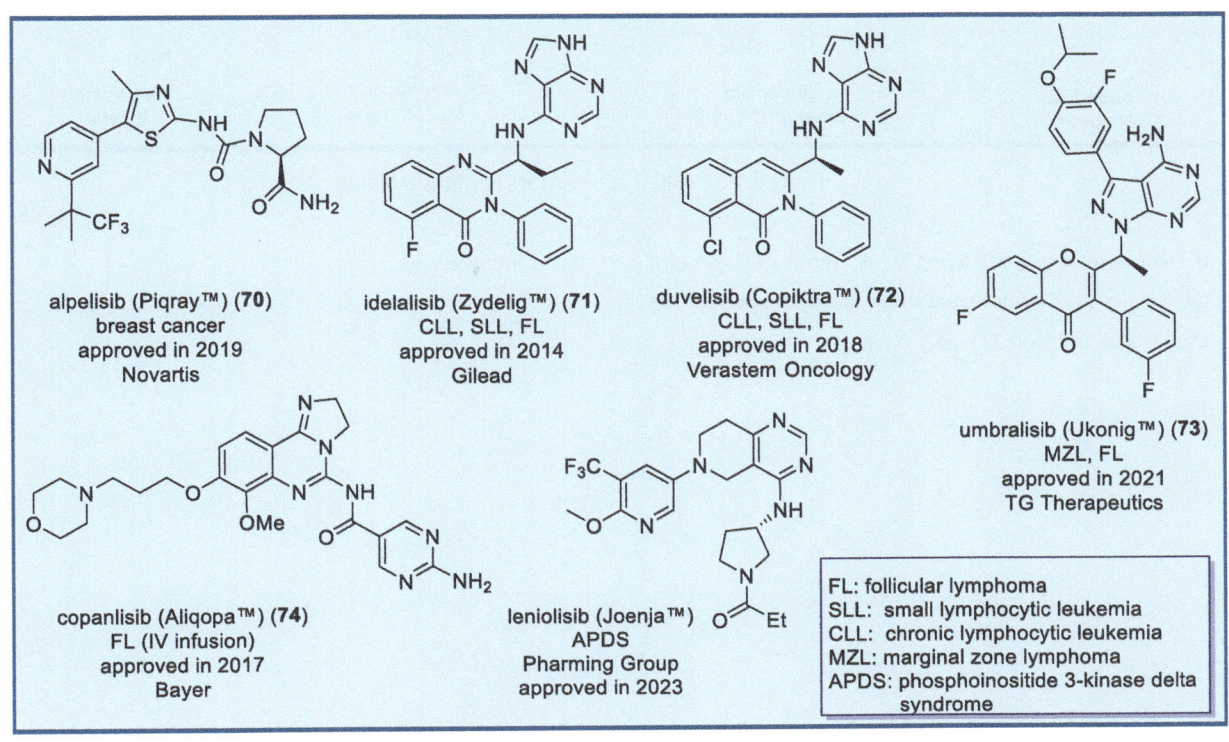

Figure 1. Approved PI3K inhibitors.

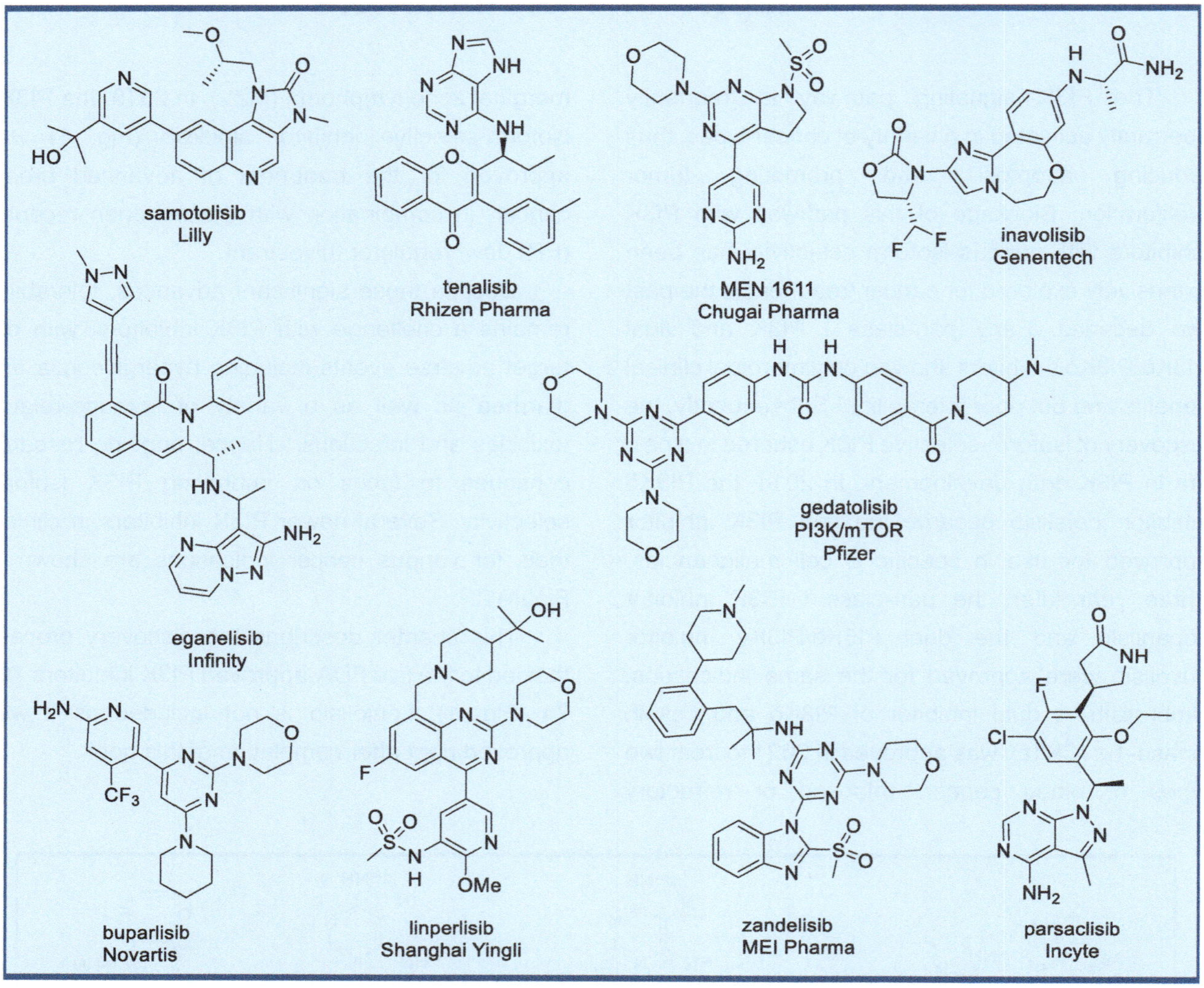

Figure 2. PI3K inhibitors in clinical trials.

1. B. Vanhaesebroeck, M. Perry, J. R. Brown, F. André, K. Okkenhaug. PI3K inhibitors are finally coming of age. *Nat. Rev. Drug Discov.* **2021**, *20*, 741–769.

2. clinicaltrials.gov.

14.1 Alpelisib (Piqray™) (70)

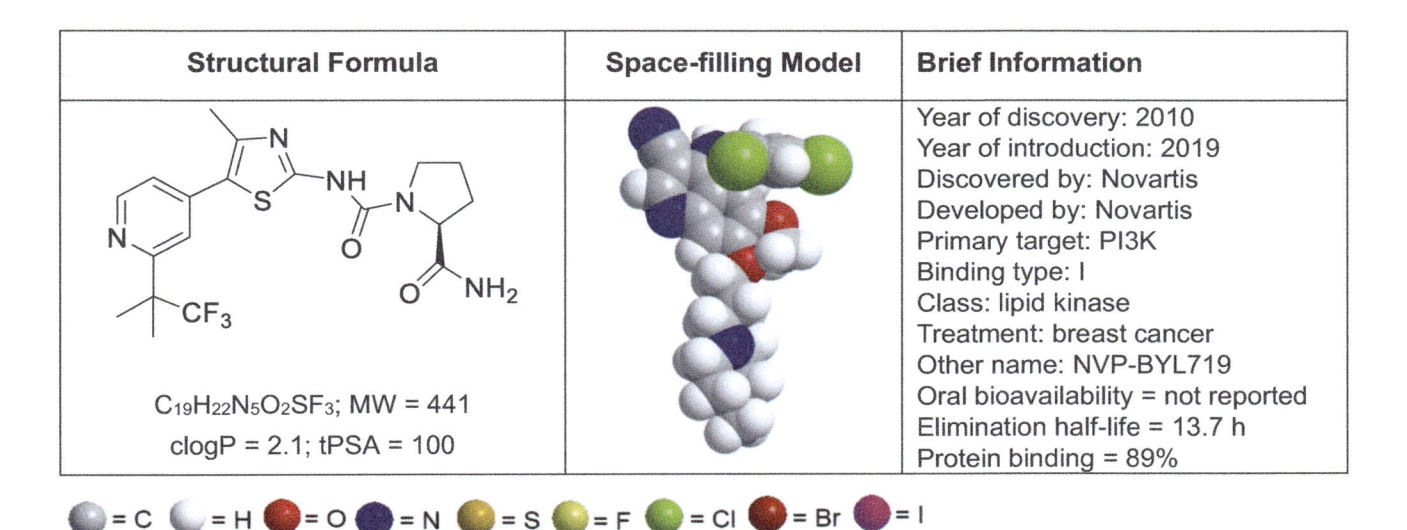

Structural Formula	Space-filling Model	Brief Information
$C_{19}H_{22}N_5O_2SF_3$; MW = 441 clogP = 2.1; tPSA = 100		Year of discovery: 2010 Year of introduction: 2019 Discovered by: Novartis Developed by: Novartis Primary target: PI3K Binding type: I Class: lipid kinase Treatment: breast cancer Other name: NVP-BYL719 Oral bioavailability = not reported Elimination half-life = 13.7 h Protein binding = 89%

● = C ○ = H ● = O ● = N ● = S ● = F ● = Cl ● = Br ● = I

Currently, there are more than 3.8 million women diagnosed with breast cancer in the United States. More than 70% of all breast cancers are both hormone receptor-positive (HR+) and human epidermal growth factor receptor 2-negative (HER2−). Approximately 40% of HR+, HER2− breast cancers carry activating mutations in the PIK3CA gene, which induce hyperactivation of phosphatidylinositol 3-kinase isoform α (PI3Kα).

Alpelisib is a specific PI3Kα inhibitor approved by the FDA in combination with fulvestrant (an estrogen receptor antagonist) for the treatment of patients with HR+, HER2−, PIK3CA-mutated, advanced or metastatic breast cancer who have received endocrine therapy previously.[1,2] Alpelisib carries no black box warning, unlike the other PI3K inhibitors, idelalisib and duvelisib, which bear black box warnings for side effects including infections, diarrhea or colitis, cutaneous reactions and pneumonitis.

Alpelisib acts by selectively targeting PI3Kα, thereby blocking PI3Kα/AKT/mTOR signaling in breast cancers.

PI3K/AKT/mTOR Signaling in Breast Cancer[1–4]

In breast cancers, the PI3K signaling pathway can be activated in two ways: (1) alterations (amplification or activating mutations) in genes that encode downstream proteins which participate in the PI3K pathway (Table 1); and (2) activation of upstream receptor tyrosine kinases (RTKs) (Table 2).[1]

The most prevalent gene alterations are the PIK3CA activating mutations, which are found in 28–47%, 23–33%, and 8–25% in HR+/HER2−, HER2+, and triple-negative breast cancer subtypes, respectively. In contrast, the PIK3CB mutations are uncommon. Inactivation and reduced expression of the PTEN tumor suppressor gene occur in 29–44%, 22%, and 67% in HR+/HER2−, HER2+, and triple-negative subtypes, respectively. PTEN dephosphorylates PIP3 to give PIP2, which lacks a binding dock for AKT and is thus unable to recruit AKT to the cell membrane for phosphorylation at Thr308 by PDK1. Consequently, AKT cannot be activated, which causes suppression of PI3K/AKT/mTOR signaling. In essence, PTEN functions as a tumor suppressor by counteracting the action of PI3K. However, the mutated PTEN gene in breast cancers causes the plasmatic levels of PIP3 to rise, leading to a constitutive PI3K/AKT signaling. PI3K signaling can also be triggered by amplification or mutations of PDK1 by hyperactivating AKT and its downstream proteins (Fig. 1).[1]

The other common stimuli to activate PI3K pathway are amplification and overexpression or activation mutations of some receptor tyrosine kinases such as HER2, epidermal growth factor receptor (EGFR), insulin-like growth factor-1 receptor (IGF1R), and fibroblast growth factor receptor (FGFR).

As PI3KCA mutations represent one of the most common genetic aberrations in breast cancer, specific inhibition of its encoded protein, PI3Kα, provides a unique opportunity to target PI3Kα mutant tumors without potential side effects associated with the inhibition of other isoforms.

Table 1. Mutation-induced PI3K pathway alterations in breast cancers[1]

Gene (protein)	Mutation type	Frequency (%)		
		Luminal	HER-2	TN
PIK3CA (PI3Kα)	activation	28–47%	23–33%	8–25%
PIK3CB (PI3Kβ)	amplification	5%		
PTEN	loss-of-function or reduced expression	29–44%	22%	67%
PDK1	Amplification or overexpression	22%	22%	38%
KRAS	activation		4–6%	

Table 2. RTK-activated PI3K pathway alterations in breast cancers[1]

RTK	Mutation type	Frequency (%)		
		Luminal	HER-2	TN
HER-2	gene amplification	10%	100%	0%
IGF1R & INSR	receptor activation & IGF1R amplification	41–48%	18–64%	42%
FGFR1	amplification	8–12%	5%	6%

Discovery of Alpelisib[5,6]

High-throughput screening of the Novartis compound library using a PI3Kγ biochemical assay identified aminothiazole **1** as a moderately potent pan-PI3K inhibitor that showed good selectivity against a panel of protein kinases but lacked cellular activity in the fMLP stimulated neutrophil assay (fMLP IC$_{50}$ = 2.6 μM) (Fig. 2). In an effort to improve cellular potency by reducing the total polar surface area of **1** (tPSA = 102), the primary sulfonamide was replaced with a pyrazole to give **2** (tPSA = 62) that showed 8-fold improvement in cellular activity (fMLP IC$_{50}$ = 0.32 μM) while maintaining PI3Kα binding affinity. As part of the structure-α isoform selectivity relationship studies, the acetamide was replaced with various urea subunits which then led to the (S)-pyrrolidine-1,2-dicarboxamide **3** as a highly potent and selective PI3Kα inhibitor (K_i = 5 nM). The high isoform specificity originates from the carboxamide moiety, which projects into a unique α isoform specificity pocket, forming two hydrogen bonds to the amide side chain of Gln859 (see binding mode). These interactions are not enabled with the other three isoforms due to the absence of this residue, thereby resulting in high α isoform specificity.

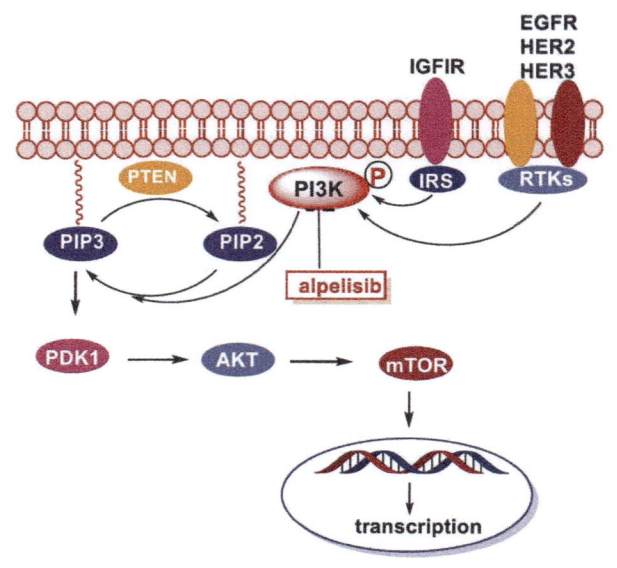

Figure 1. PI3K/AKT/mTOR signaling pathway in breast cancers.

Despite its favorable in vitro profile, compound **3** showed only modest oral bioavailability in rats (21%) in part due to relatively high clearance (21 mL/min/kg). Subsequent structural modifications were focused on the pyrazole substituted phenyl ring, which sits in the less solvent accessible affinity pocket. That approach led to the pyridine analog **4** which emerged as a potent and isoform selective PI3Kα inhibitor (IC$_{50}$ = 15 nM). However, its rat clearance remained high (21 mL/min/kg) in part due to its metabolic instability. Incubation of **4** with rat liver microsomes revealed two major metabolic pathways: hydrolysis of the primary amide to the carboxylic acid and C-hydroxylation of either the *tert*-butyl group or the methyl substituent in the 4-position of the thiazole ring. To block one of the identified metabolic pathways, one of the methyls of the *tert*-butyl group was replaced by a trifluoromethyl substituent to give alpelisib which showed a 2-fold reduction in rat clearance in (10 vs. 21 mL/min/kg for **4**). The lower clearance translated into excellent oral bioavailability in mice (106%), rats (58%), and dogs (140%). Substitution of the methyl with trifluoromethyl also resulted in a 3-fold increase in PI3Kα potency (IC_{50} = 5 vs. 14 nM for **4**), which is consistent with modelling studies. In this model, the *tert*-butyl group did not occupy all the space in the small hydrophobic cavity, and the remaining space accommodated a slightly larger trifluoromethyl group, thereby contributing to improved activity.

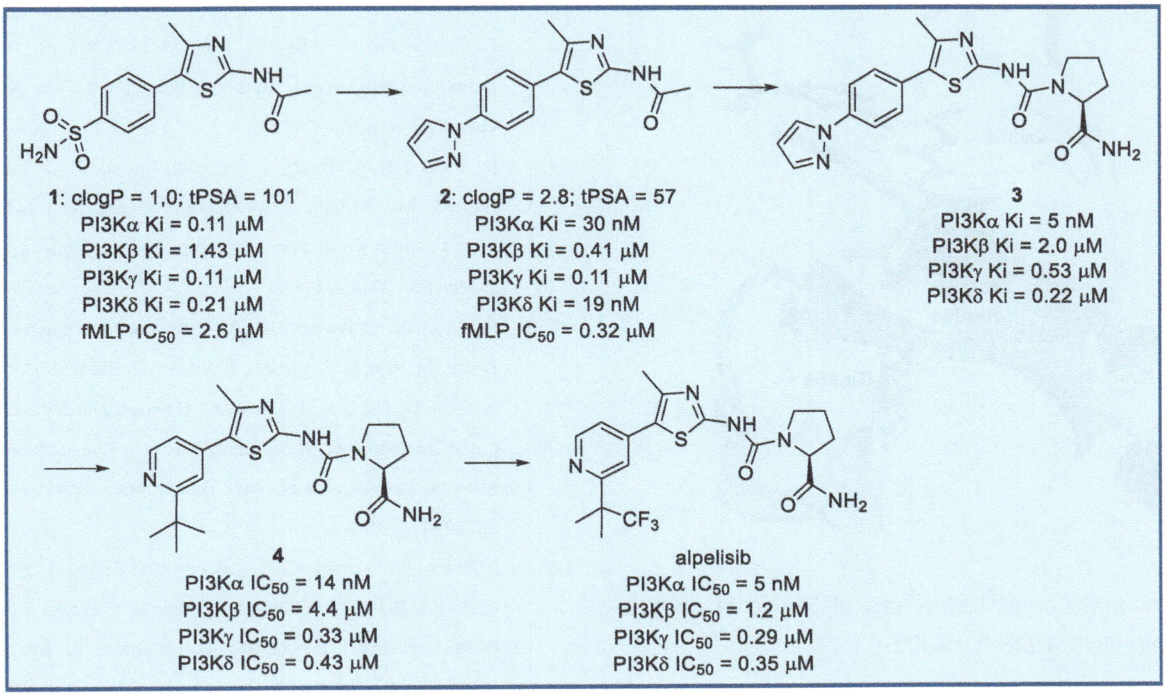

Figure 2. Structural optimization of alpelisib.

Binding Mode[5]

As shown in the co-crystal structure of alpelisib with PI3Kα (Fig. 3), the thiazole nitrogen and the 2-NH group form bidentate hydrogen bonds with the backbone NH and carbonyl oxygen of V851, while its pyridine moiety packs into the less solvent accessible section of the cavity, the so-called affinity pocket (Fig. 4). The primary amide takes advantage of its synergistic hydrogen bonding potential, forming three hydrogen bonds with PI3Kα: the amide NH_2 provides two hydrogen bonds to the backbone amide carbonyl of Ser854 and the side chain amide carbonyl of Gln859, while the amide carbonyl serves as an acceptor to the side chain amide NH of Gln859. Although the hydrogen bonds between apelisib and the backbone amide carbonyl of Ser854 can also occur in the other PI3K isoforms, the interactions involving the amide carbonyl cannot form because Gln859 is replaced by Asp, Asn, and Lys in isoforms β, δ, and γ, respectively, at this position. Obviously, both Asp and Lys residues of the β and γ isoforms do not permit the same hydrogen bond donor–acceptor interactions with the primary amide group of the inhibitor as Gln. Theoretically, the side chain of the δ isoform Asn could participate in such interactions. This possibility was eliminated by a modeling study in which Gln859 was mutated into an Asn, which demonstrated that the shorter side chain of Asn was not conducive to the formation of the donor–acceptor hydrogen bonds. Thus, the PI3Kα selectivity of alpelisib appears to originate from the two crucial hydrogen bonds between the primary amide of the inhibitor and the amide side chain of Gln859, interactions that are not attainable with the other PI3K isoforms. Of special note is that the structural basis for the PI3Kα selectivity with alpelisib is different from that of the PI3Kδ selectivity with idelalisib, which depends on the inhibitor induced opening of a new hydrophobic pocket (the specificity pocket) (see Section 14.2).

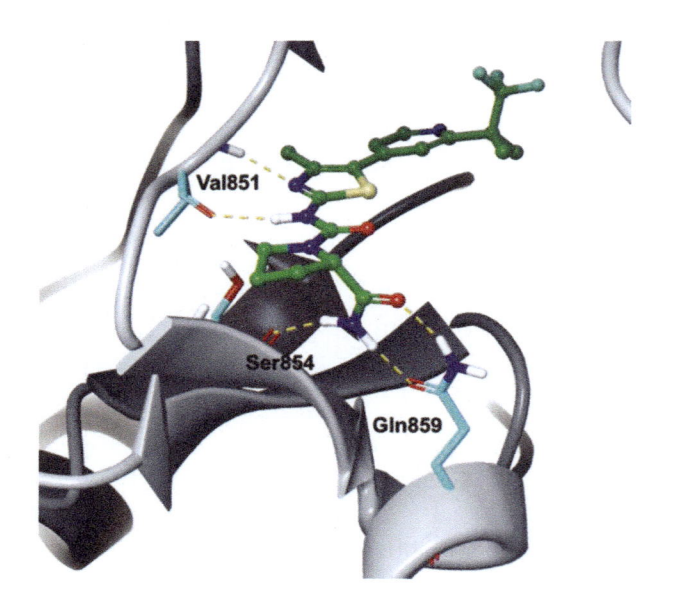

Figure 3. Co-crystal structure of PI3Kα in complex with alpelisib (PDB ID: 4JPS).

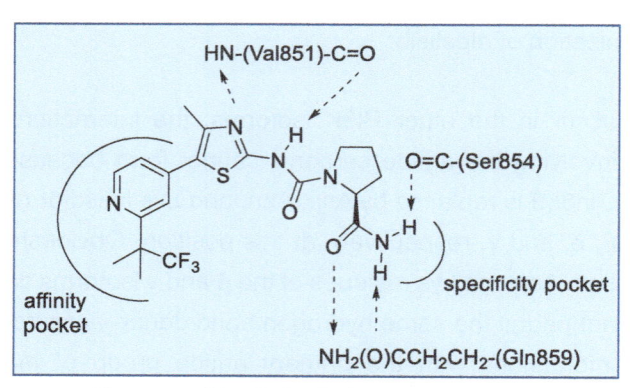

Figure 4. Alpelisib–PI3Kδ interactions.

PK Properties and Metabolism[7,8]

A single oral dose of [^{14}C]-labeled alpelisib showed rapid absorption with a t_{max} of 2 h and an elimination half-life from plasma of 13.7 h. The recommended dose is 300 mg once a day. The major metabolism of alpelisib was the hydrolysis of the carboxamide to give the carboxylic acid (structure not shown).

1. D. Y. Chang, W. L. Ma, Y. S. Lu. Role of alpelisib in the treatment of PIK3CA-mutated breast cancer: Patient selection and clinical perspectives. *Ther. Clin. Risk Manag.* **2021**, *17*, 193–207.

2. A. Markham. Alpelisib: First Global Approval. *Drugs.* **2019**, *79*, 1249–1253.

3. E. Paplomata, R. O'Regan. The PI3K/AKT/mTOR pathway in breast cancer: Targets, trials and biomarkers. *Ther. Adv. Med. Oncol.* **2014**, *6*, 154-166.

4. D. Miricescu, A. Totan, I. Stanescu-Spinu, S. C. Badoiu, C. Stefani, M. Greabu. PI3K/AKT/mTOR signaling pathway in breast cancer: From molecular landscape to clinical aspects. *Int. J. Mol. Sci.* **2010**, *22*, 173.

5. P. Furet, V. Guagnano, R. A. Fairhurst, P. Imbach-Weese, I. Bruce, M. Knapp, C. Fritsch, F. Blasco, J. Blanz, R. Aichholz, J. Hamon, D. Fabbro, G. Caravatti. Discovery of NVP-BYL719 a potent and selective phosphatidylinositol-3 kinase alpha inhibitor selected for clinical evaluation. *Bioorg. Med. Chem. Lett.* **2013**, *23*, 3741–3748.

6. I. Bruce, M. Akhlaq, G. C. Bloomfield, E. Budd, B. Cox, B. Cuenoud, P. Finan, P. Gedeck, J. Hatto, J. F. Hayler, D. Head, T. Keller, L. Kirman, C. Leblanc, D. Le Grand, C. McCarthy, D. O'Connor, C. Owen, M. S. Oza, G. Pilgrim, L. Whitehead. Development of isoform selective PI3-kinase inhibitors as pharmacological tools for elucidating the PI3K pathway. *Bioorg. Med. Chem. Lett.* **2012**, *22*, 5445–5450.

7. A. Mehta, L. Trandafir, P. Swart. Absorption, distribution, metabolism, and excretion of [(14)C]BYL719 (alpelisib) in healthy male volunteers. *Cancer Chemother. Pharmacol.* **2015**, *76*, 751–760.

8. D. Juric, J. Rodon, J. Tabernero, F. Janku, H. A. Burris, J. Schellens, M. R. Middleton, J. Berlin, M. Schuler, M. Gil-Martin, H. S. Rugo, R. Seggewiss-Bernhardt, A. Huang, D. Bootle, D. Demanse, L. Blumenstein, C. Coughlin, C. Quadt, J. Baselga. Phosphatidylinositol 3-kinase α-selective inhibition with alpelisib (BYL719) in PIK3CA-altered solid tumors: Results from the first-in-human study. *J. Clin. Oncol.* **2018**, *36*, 1291–1299.

14.2 Idelalisib (Zydelig™) (71)

Structural Formula	Space-filling Model	Brief Information
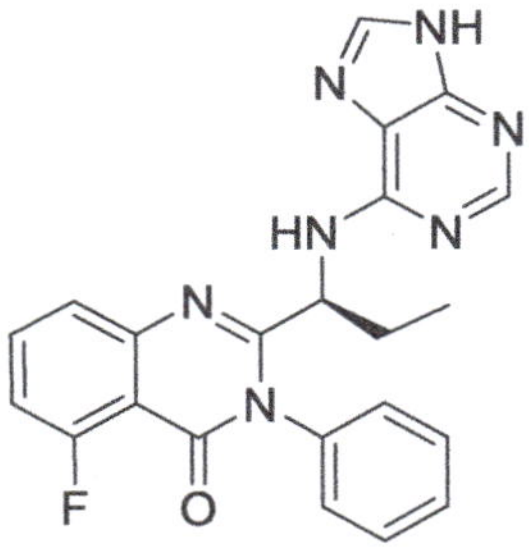 $C_{22}H_{18}N_7OF$; MW = 415 clogP = 3.6; tPSA = 94	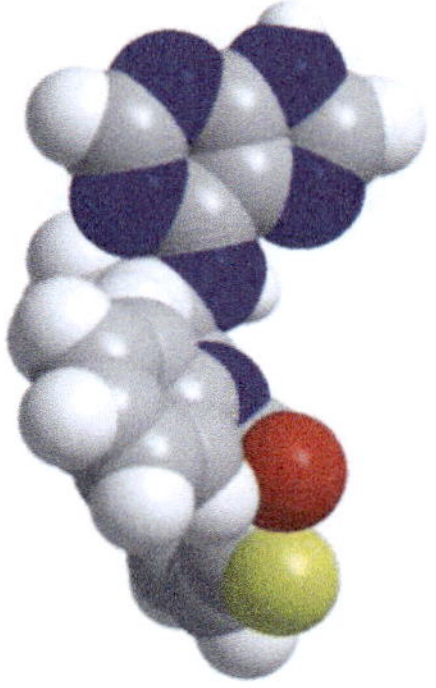	Year of discovery: 2004 Year of introduction: 2014 Discovered by: ICOS Developed by: Calistoga/Gilead Sciences Primary target: PI3Kδ Binding type: I Class: lipid kinase Treatment: CLL, SLL, FL Other name: CAL-101, GS-1101 Oral bioavailability = not reported Elimination half-life = 8.2 h Protein binding > 84%

● = C ○ = H ● = O ● = N ● = S ● = F ● = Cl ● = Br ● = I

Idelalisib is the first-in-class PI3Kδ Inhibitor approved by the FDA in 2014 to treat relapsed follicular B-cell non-Hodgkin lymphoma (FL) and relapsed small lymphocytic leukemia (SLL) along with relapsed chronic lymphocytic leukemia (CLL). The approval, however, came with a boxed warning of fatal and/or severe diarrhea or colitis, hepatotoxicity, pneumonitis, and intestinal perforation after reports of multiple deaths in its clinical trials. These side effects are associated with PI3Kδ inhibition and consequent disruption of the PI3K/AKT/mTOR pathway. In January 2022, Gilead Sciences voluntarily withdrew the use of idelalisib for two types of cancer, FL and SLL, after halting follow up clinical trials to determine efficacy and safety. However, it remains available for the treatment of CLL.

PI3K class I Isoforms[1]

PI3Ks are divided into three classes, class I, II, and III according to their structural and functional characteristics. Class I PI3Ks, which function as catalysts for the intracellular conversion of the plasma membrane lipid phosphatidylinositol (4,5)-bisphosphate (PIP$_2$) into phosphatidylinositol (3,4,5)-trisphosphate (PIP$_3$), have been associated with oncogenesis. Class I PI3K is a heterodimer composed of two subunits, one catalytic and one regulatory. The catalytic subunit, p110, has four isoforms, α, β, γ, and δ, encoded by the genes PIK3CA, PIK3CB, PIK3CG and PIK3CD, respectively (Fig. 1). The p110-α isoform (PI3Kα) is involved in insulin-dependent signaling; p110-β (PI3Kβ) in platelet aggregation, thrombosis and insulin signaling; while p110-γ (PI3Kγ) and p110-δ (PI3K δ), abundantly expressed in leukocytes, play an important role in lymphocyte activation, mast cell degranulation, and chemotaxis.

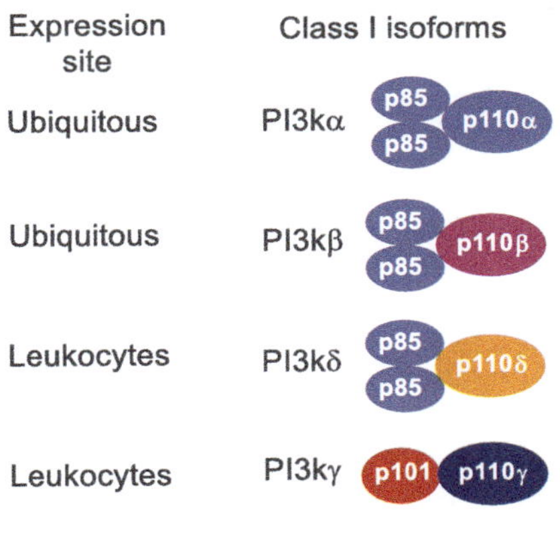

Expression site	Class I isoforms	
Ubiquitous	PI3kα	p85 / p85 / p110α
Ubiquitous	PI3kβ	p85 / p85 / p110β
Leukocytes	PI3kδ	p85 / p85 / p110δ
Leukocytes	PI3kγ	p101 / p110γ

Figure 1. Class I PI3K isoforms and the sites of their expression and corresponding inhibitors.

BCR-activated PI3K/AKT Signaling[1–5]

Hematologic B-cell malignancies display constitutively active PI3K/AKT signaling activated by the B cell receptor (BCR). When BCR binds to specific antigens, the SRC family protein tyrosine kinase LYN phosphorylates tyrosine residues in the cytoplasmic tail of the BCR co-receptor CD19, thereby activating PI3K, which is recruited to the BCR complex at the plasma membrane by the cytoplasmic B cell adapter (BCAP). The catalytic subunit p110 of PI3K catalyzes the phosphorylation of PIP2 to generate PIP3. The formation of PIP$_3$ is also negatively regulated by the tumor suppressor protein PTEN (phosphatase and tensin homolog deleted on chromosome 10), a phosphatase that dephosphorylates PIP3 to PIP2. The membrane bound PIP3 recruits AKT and PDK1 to the plasma membrane where PDK1 phosphorylates AKT at Thr308. Further phosphorylation at Ser473 by the TORC2 complex (consisting of mTOR, Rictor, Sin1, and mLST8) fully activates AKT. The activated AKT dissociates from the PIP3 and phosphorylates various cytosolic and nuclear proteins that regulate cell survival and metabolism (Fig. 2).

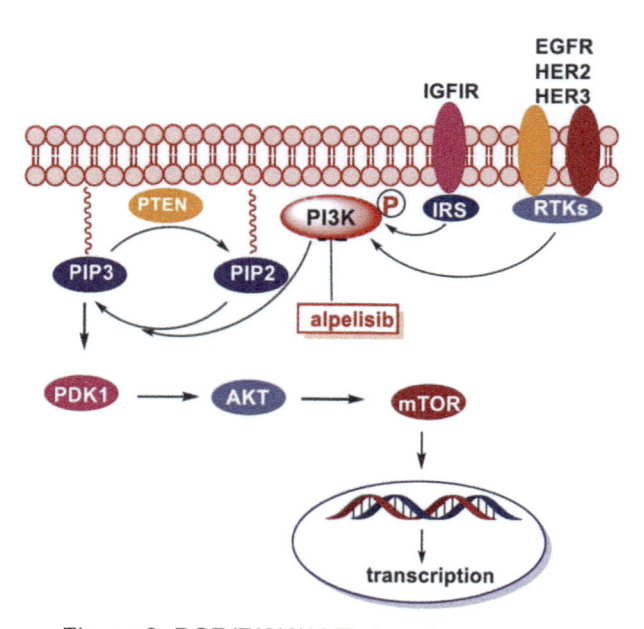

Figure 2. BCR/PI3K/AKT signaling pathway.

As BCR signaling is a crucial pathway for the pathogenesis of B cell lymphoproliferative diseases, extensive efforts have been undertaken towards the development of therapeutic agents that inhibit critical downstream pathway components, such as PI3K and mTOR or dual inhibition of both. Indeed, PI3K isoforms have been successfully targeted for the treatment of various forms of lymphoma and breast cancer.

One major challenge in the development of clinically useful PI3K inhibitors is to achieve sufficient blockage on the signaling pathway without incurring drug-related toxicities, which derive from lack of isoform specificity. For example, PI3Kα inhibitors are largely responsible for rash and hyperglycemia, whereas PI3Kδ inhibition results in mostly gastrointestinal, transaminitis, and myelosuppression.

Discovery of Idelalisib[6–8]

High-throughput screening of the Icos compound library against PI3Kδ identified a lead compound with the 4-quinazolin-4-one scaffold (Fig. 3). Subsequent analog synthesis led to IC87114 as a moderately potent PI3Kδ inhibitor with IC$_{50}$ of 0.5 µM. This compound also exhibits 58-fold selectivity against PI3Kγ and over 100-fold against PI3Kα and PI3Kβ (IC$_{50}$ = 100, 75 and 29 µM for α, β, γ isoforms, respectively). Further structural optimization resulted in the discovery of idelalisib with substantial improvement in potency and isoform selectivity. More specifically, it potently inhibits PI3Kδ with an IC$_{50}$ value of 19 nM, whereas the IC$_{50}$ values for PI3Kα, PI3Kβ, and PI3Kγ are much higher with IC$_{50}$ values of 8.6, 4.0, and 2.1 µM, respectively. Thus, idelalisib demonstrates 453-, 210- and 110-fold PI3Kδ selectivity over the α, β, and γ isoforms, respectively, the highest PI3Kδ selectivity among all the PI3K inhibitors in use (Table 1). When tested

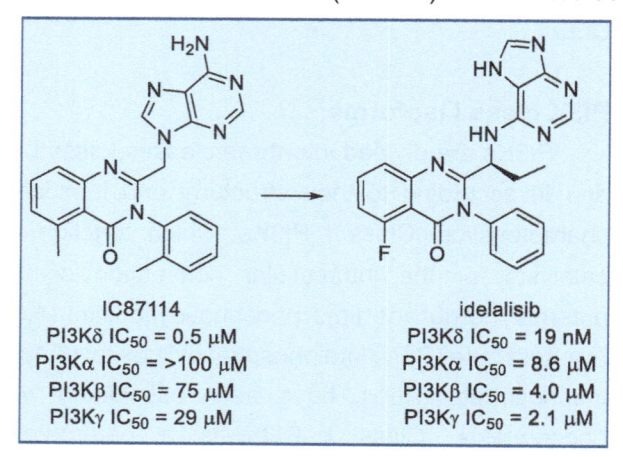

IC87114	idelalisib
PI3Kδ IC$_{50}$ = 0.5 µM	PI3Kδ IC$_{50}$ = 19 nM
PI3Kα IC$_{50}$ = >100 µM	PI3Kα IC$_{50}$ = 8.6 µM
PI3Kβ IC$_{50}$ = 75 µM	PI3Kβ IC$_{50}$ = 4.0 µM
PI3Kγ IC$_{50}$ = 29 µM	PI3Kγ IC$_{50}$ = 2.1 µM

Figure 3. Structural optimization of idelalisib.

against a broad panel of 351 kinases and 44 mutant kinases at 10 μM of idelalisib, the most potent level of inhibition seen for any non-PI3K kinase was 47%.

Table 1. In vitro IC$_{50}$'s of PI3K inhibitors[9,10]

inhibitor	PI3Kα	PI3Kβ	PI3Kγ	PI3Kδ
idelalisib	8.6 μM	4.0 μM	2.1 μM	19 nM
alpelisib	4.6 nM	1.2 μM	250 nM	290 nM
copanlisib	0.5 nM	3.7 nM	6.4 nM	0.7 nM
duvelisib	1.6 μM	85 nM	27 nM	2.5 nM
umbralisib	>10 μM	1.1 μM	1.1 μM	22 nM

Binding Mode[6–8,11]

Idelalisib binds to the PI3Kδ ATP-binding site with a propeller-shaped conformation, accompanied by the opening of a specificity hydrophobic pocket in the ATP site to accommodate the inhibitor. The purine group is anchored to the hinge, forming two hydrogen bonds between the purine N3 and the backbone amide NH of Val828 and between N9 and the backbone carbonyl oxygen of Glu826 (Figs. 4, 5). Both the purine N7 and the N1 of the quinazolinone interact indirectly with the carboxylic acid residue of Asp911 through a network of hydrogen bonds via three bound water molecules. The ethyl group of the inhibitor sits in a hydrophobic pocket formed by Ile910, Met900, and Met752. The flat quinazolinone ring system adopts a parallel configuration to the sides of the pocket, engaging in a hydrophobic interaction with the kinase.

Overlay of the idelalisib-bound and apo structures reveals that a dramatic conformational change occurs in the ATP-binding site and beyond to accommodate idelalisib (Figs. 6A, 6B). In the apo enzyme, Met752 and Trp760 are in close contact, but binding to idelalisib shifts these two residues apart by 6.5 Å, thereby opening a new hydrophobic pocket (the specificity pocket) to fit in the fluoroquinazolinone core. Because a similar pocket at this exact location is rare among other kinases, access to this specificity pocket plays an important role in the selectivity against other kinases. There are also apparent differences in accessibility of the specificity pocket between the four isoforms, which contribute to the overwhelming preference for the δ isoform. The overall kinase selectivity profile of idelalisib results not only from the opening of the specificity pocket, but also from a combination of multiple molecular contacts throughout the entire binding site.

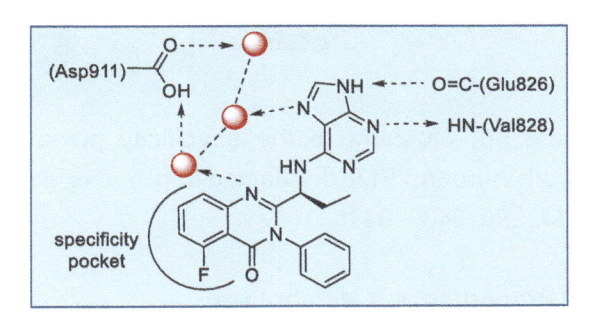

Figure 4. Summary of idelalisib–PI3Kδ interactions based on co-crystal structure.

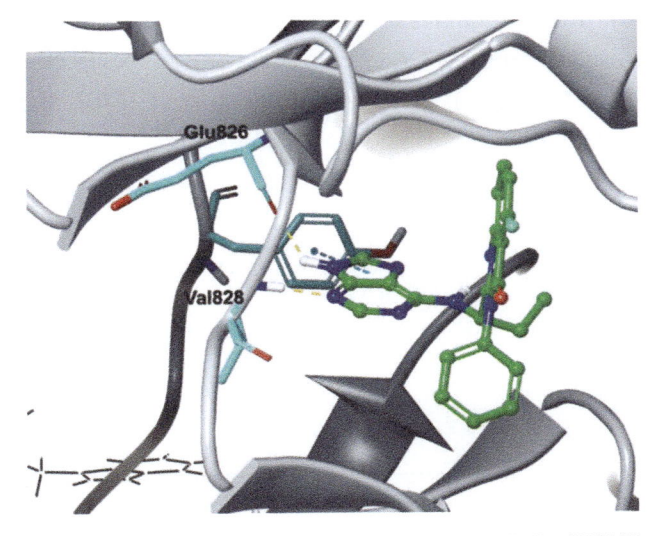

Figure 5. Co-crystal structure of idelalisib–PI3Kδ (PDB ID: 4XE0).

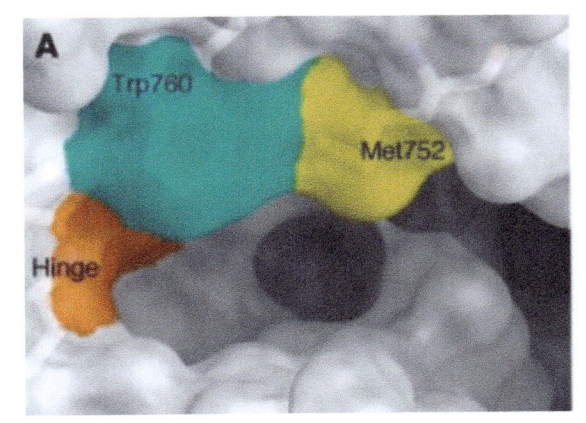

Figure 6A. Crystal structure of Apo PI3Kδ.

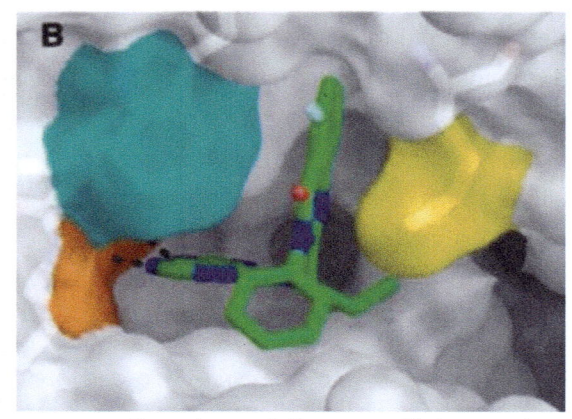

Figure 6B. Opening of the specificity pocket in idelalisib-bound PI3Kδ. Adapted from *J. Biol. Chem.* **2015**, *290*, 8439–8446.

PK Properties and Metabolism[12]

The recommended dose of idelalisib is 150 mg orally twice a day, consistent with its elimination half-life is 8.2 h (Table 2). It is absorbed rapidly with a t_{max} of 1.5 h. Idelalisib is metabolized by aldehyde oxidase and CYP3A to give a major metabolite GS-563117 (Fig. 7), which is inactive against P110δ and other isoforms.

Table 2. PK properties of oral PI3K inhibitors[9]

Inhibitor	t_{max}	$t_{1/2}$	F	Dosage	Dose
Idelalisib	1.5 h	8.2 h	NA	150 mg	BID
Alpelisib	2–4 h	7.6 h	NA	300 mg	BID
Duvelisib	1.5 h	4.7 h	42%	25 mg	BID
Umbralisib	4 h	91 h	NA	800 mg	QD

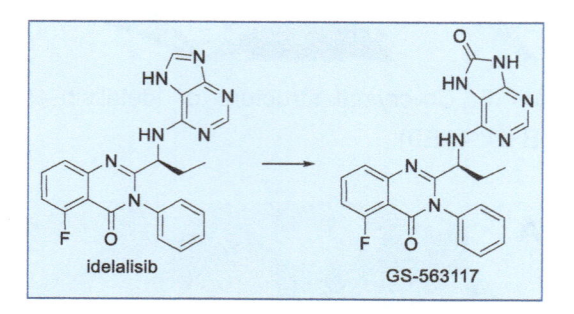

Figure 7. Metabolic pathway of idelalisib in humans.

1. A. Visentin, F. Frezzato, F. Severin, S. Imbergamo, S. Pravato, L. Romano Gargarella, S. Manni, S. Pizzo, E. Ruggieri, M. Facco, A. M. Brunati, G. Semenzato, F. Piazza, L. Trentin. Lights and shade of next-generation Pi3k inhibitors in chronic lymphocytic leukemia. *Onco.Targets Ther.* **2020**, *13*, 9679–9688.

2. J. J. Castillo, M. Iyengar, B. Kuritzky, K. D. Bishop. Isotype-specific inhibition of the phosphatidylinositol-3-kinase pathway in hematologic malignancies. *Onco. Targets Ther.* **2014**, *7*, 333–342.

3. A. Carnero, J. M. Paramio. The PTEN/PI3K/AKT pathway in vivo, cancer mouse models. *Front. Oncol.* **2014**, *4*, 252.

4. V. Ortiz-Maldonado, M. García-Morillo, J. Delgado. The biology behind PI3K inhibition in chronic lymphocytic leukaemia. *Ther. Adv. Hematol.* **2015**, *6*, 25–36.

5. Q. Yang, P. Modi, T. Newcomb, C. Quéva, V. Gandhi. Idelalisib: First-in-class PI3K delta inhibitor for the treatment of chronic lymphocytic leukemia, small lymphocytic leukemia, and follicular lymphoma. *Clin. Cancer Res.* **2015**, *21*, 1537-42.

6. A. Berndt, S. Miller, O. Williams, D. D. Le, B. T. Houseman, J. I. Pacold, F. Gorrec, W. C. Hon, Y. Liu, C. Rommel, P. Gaillard, T. Rückle, M. K. Schwarz, K. M. Shokat, J. P. Shaw, R. L. Williams. The p110 delta structure: Mechanisms for selectivity and potency of new PI(3)K inhibitors. *Nat. Chem. Biol.* **2010**, *6*, 117–124.

7. J. R. Somoza, D. Koditek, A. G. Villaseñor, N. Novikov, M. H. Wong, A. Liclican, W. Xing, L. Lagpacan, R. Wang, B. E. Schultz, G. A. Papalia, D. Samuel, L. Lad, M. E. McGrath. Structural, biochemical, and biophysical characterization of idelalisib binding to phosphoinositide 3-kinase δ. *J. Biol. Chem.* **2015**, *290*, 8439–8446.

8. J. Hoellenriegel, S. A. Meadows, M. Sivina, W. G. Wierda, H. Kantarjian, M. J. Keating, N. Giese, S. O'Brien, A. Yu, L. L. Miller, B. J. Lannutti, J. A. Burger. The phosphoinositide 3'-kinase delta inhibitor, CAL-101, inhibits B-cell receptor signaling and chemokine networks in chronic lymphocytic leukemia. *Blood.* **2011**, *118*, 3603–3612.

9. T. J. Phillips, J. M. Michot, V. Ribrag. Can next-generation PI3K inhibitors unlock the full potential of the class in patients with B-cell lymphoma? *Clin. Lymphoma Myeloma Leuk.* **2021**, *21*, 8–20.e3.

10. D. A. Rodrigues, F. S. Sagrillo, C. Fraga. Duvelisib: A 2018 novel FDA-approved small molecule inhibiting phosphoinositide 3-kinases. *Pharmaceuticals.* **2019**, *12*, 69.

11. M. S. Miller, P. E. Thompson, S. B. Gabelli. Structural determinants of isoform selectivity in PI3K inhibitors. *Biomolecules.* **2019**, *9*, 82.

12. S. Ramanathan, F. Jin, S. Sharma, B. P. Kearney. Clinical Pharmacokinetic and Pharmacodynamic Profile of Idelalisib. *Clin. Pharmacokinet.* **2016**, *55*, 33–45.

14.3 Duvelisib (Copiktra™) (72)

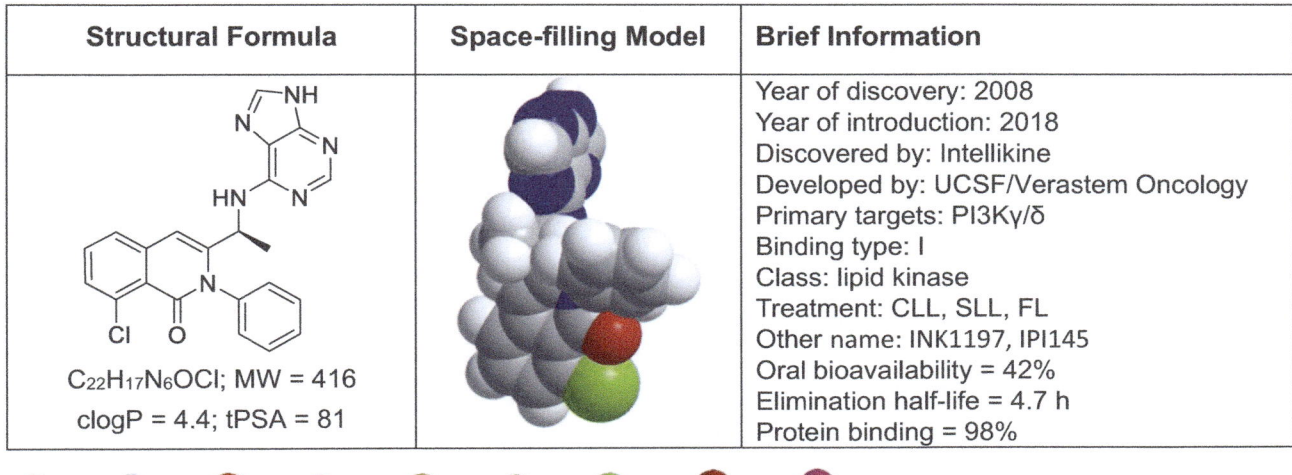

Structural Formula	Space-filling Model	Brief Information
$C_{22}H_{17}N_6OCl$; MW = 416 clogP = 4.4; tPSA = 81		Year of discovery: 2008 Year of introduction: 2018 Discovered by: Intellikine Developed by: UCSF/Verastem Oncology Primary targets: PI3Kγ/δ Binding type: I Class: lipid kinase Treatment: CLL, SLL, FL Other name: INK1197, IPI145 Oral bioavailability = 42% Elimination half-life = 4.7 h Protein binding = 98%

● = C ○ = H ● = O ● = N ● = S ● = F ● = Cl ● = Br ● = I

Duvelisib, a dual phosphoinositide kinase PI3Kγ/δ (PI3Kγ/δ) inhibitor, was approved in 2018 for patients with relapsed or refractory chronic lymphocytic leukemia (CLL), small lymphocytic lymphoma (SLL), and follicular lymphoma (FL) who have received at least two prior therapies. As with idelalisib, duvelisib carries a black box warning for infections such as diarrhea/colitis, cutaneous reactions, and pneumonitis. Its use has been limited not only because of toxicities that can lead to permanent discontinuation but also the availability of other drugs (such as BTK inhibitors) which have a better safety profile.

Duvelisib has exactly the same hinge binder as idelalisib (a highly selective PI3Kδ inhibitor) and differs only in two areas: (1) methyl vs. ethyl in the linker; and (2) a chloroisoquinolinone vs. a fluoroquinazolinone core to fit in the specificity pocket (Fig. 1). These minor structural modifications led to a potent dual PI3Kγ/δ inhibitor (Table 1). Duvelisib inhibits PI3Kβ with an IC$_{50}$ value of 85 nM but PI3Kα only at much higher concentrations. It shows no significant binding to the protein kinases outside the PI3K class I isoforms.[1,2]

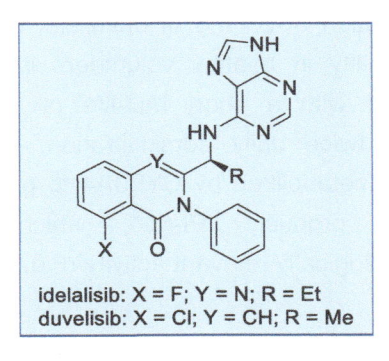

idelalisib: X = F; Y = N; R = Et
duvelisib: X = Cl; Y = CH; R = Me

Figure 1. Structural differences between idelalisib and duvelisib.

Table 1. In vitro IC$_{50}$'s of duvelisib vs. idelalisib.[3]

inhibitor	PI3Kα	PI3Kβ	PI3Kγ	PI3Kδ
idelalisib	8.6 µM	4.0 µM	2.1 µM	19 nM
duvelisib	1.6 µM	85 nM	27 nM	2.5 nM

Binding Mode[4,5]

The binding mode of duvelisib has not been established due to the lack of an X-ray co-crystal structure with PI3Kδ. Nevertheless, because it is such a close analog of idelalisib, it may bind to PI3Kδ in a similar way (Fig. 2).[4] More specifically, the purine N3 and N9 hydrogen bonds to the backbone amide NH of Val828 and the backbone carbonyl oxygen of Glu826 in the hinge region, respectively. The chloroisoquinolinone core packs into a newly opened specificity pocket in the ATP-binding pocket, thus giving rise to high kinase selectivity (Fig. 2).

Figure 2. Putative duvelisib–PI3Kδ interactions.

PK Properties and Metabolism[6,7]

The recommended dose of duvelisib is 25 mg twice daily, much lower than that of idelalisib (150 mg, bid). Duvelisib is rapidly absorbed, with a peak concentration after 1–2 h. Following the administration of 25 mg of duvelisib, the absolute bioavailability in healthy volunteers is 42%. It is eliminated with a short half-life of 4.7 h, thus requiring twice daily administration. Duvelisib is primarily metabolized by CYP3A4 to give a mono-oxidation product, IPI-656, which has no pharmacologically relevant activity (Fig. 3).

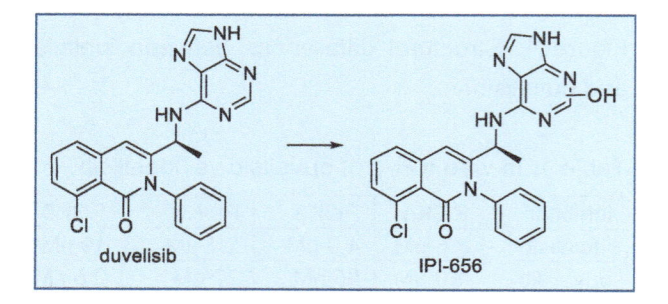

Figure 3. Metabolic pathway of duvelisib in humans.

1. D. A. Rodrigues, F. S. Sagrillo, C. Fraga. Duvelisib: A novel FDA-approved small molecule inhibiting phosphoinositide 3-kinases. *Pharmaceuticals*. **2019**, *12*, 69.
2. H. V. Vangapandu, N. Jain, V. Gandhi. Duvelisib: A phosphoinositide-3 kinase δ/γ inhibitor for chronic lymphocytic leukemia. *Expert Opin. Investig. Drugs*. **2017**, *26*, 625–632.
3. T. J. Phillips, J. M. Michot, V. Ribrag. Can next-generation PI3K inhibitors unlock the full potential of the class in patients with B-cell lymphoma? *Clin. Lymphoma Myeloma. Leuk*. **2021**, *21*, 8–20. e3.
4. M. S. Miller, P. E. Thompson, S. B. Gabelli. Structural determinants of isoform selectivity in PI3K inhibitors. *Biomolecules*. **2019**, *9*, 82.
5. J. R. Somoza, D. Koditek, A. G. Villaseñor, N. Novikov, M. H. Wong, A. Liclican, W. Xing, L. Lagpacan, R. Wang, B. E. Schultz, G. A. Papalia, D. Samuel, L. Lad, M. E. McGrath. Structural, biochemical, and biophysical characterization of idelalisib binding to phosphoinositide 3-kinase δ. *J. Biol. Chem*. **2015**, *290*, 8439–8446.
6. R. A. Dick. Refinement of in vitro methods for identification of aldehyde oxidase substrates reveals metabolites of kinase inhibitors. *Drug Metab. Dispo*. **2018**, *46*, 846–859.
7. https://www.ema.europa.eu/en/documents/assessment-report/copiktra-epar-public-assessment-report_en.pdf.

14.4 Umbralisib (Ukonig™) (73)

Structural Formula	Space-filling Model	Brief Information
$C_{31}H_{24}N_5O_3F_3$; MW = 571 logD = 4.9 (pH 7.4); tPSA = 102		Year of discovery: 2013 Year of introduction: 2021 Discovered by: Rhizen Pharma Developed by: TG Therapeutics Primary targets: PI3Kδ, CK1ε Binding type: I Class: lipid kinase Treatment: MZL, FL Other name: TGR-1202 Oral bioavailability = not reported Elimination half-life = 91 h Protein binding > 99.7%

● = C ○ = H ● = O ● = N ● = S ● = F ● = Cl ● = Br ● = I

Umbralisib, a dual inhibitor of PI3Kδ and casein kinase-1ε (CK1ε), has been approved to treat two types of blood cancer: relapsed or refractory marginal zone lymphoma (MZL), after at least one prior anti-CD20-based regimen and follicular lymphoma (FL) in patients who have failed at least three previous treatments.[1,2] Umbralisib carries no black box warning, unlike the other PI3K inhibitors, idelalisib and duvelisib, which bear black box warnings for side effects including infections, diarrhea or colitis, cutaneous reactions, and pneumonitis. Although these problems may apply to umbralisib, the FDA only describes them in the safety information section of the drug's label and so far there is no black-boxed warning. However, on February 3, 2022, the FDA launched an investigation into whether treatment with umbralisib results in an increased risk of death for its approved indications in the treatment of lymphomas. It remains to be seen if umbralisib is a safer option than other PI3Kδ inhibitors.

Umbralisib is a potent PI3Kδ inhibitor with an IC$_{50}$ value of 22 nM, and it is >1500-fold more selective for PI3Kδ than for PI3Kα and PI3Kβ and 225-fold more selective for PI3Kδ than PI3Kγ.[1,2] It is the only PI3K inhibitor in use that blocks CK1ε kinase (implicated in the pathogenesis of cancer cells, including lymphoid malignancies) with an EC$_{50}$ of 6.0 µM in the cell-free kinase assay. Despite its low potency, umbralisib effectively blocked autophosphorylation of CK1ε and phosphorylation of 4E-BP1 at 15 µM in lymphoma cells. Once-daily 1200 mg oral administration of umbralisib in the phase 1 clinical study produced a maximal plasma concentration (C_{max}) of 13.5 µM and steady-state concentration of 9.1 µM, comparable to the concentration shown to inhibit CK1ε in lymphoma cells. Thus, umbralisib as a dual PI3Kd/CK1ε inhibitor may be adequate for targeting CK1ε in patients, which would enhance its antitumor activity and widen its therapeutic window.[3,4] Although early clinical studies showed it to have improved efficacy with fewer side effects than other PI3Kδ inhibitors, long-term follow-up is needed.

Binding Mode

The binding mode of umbralisib with PI3Kδ has not been established due to the lack of a co-crystal bound structure. Nevertheless, it shares the same PI3Kδ inhibitor pharmacophore as idelalisib and duvelisib, and thus may bind to PI3Kδ in a similar way (Fig. 1).[5] More specifically, the pyrazolopyrimidine N5 and the 4-amine hydrogen bond to the backbone amide NH of Val828 and the backbone carbonyl oxygen of Glu826 in the hinge region, respectively. The 6-fluoro-chromenone core fills a newly opened specificity pocket in the ATP-binding pocket, thereby contributing to high kinase selectivity. Like idelalisib, umbralisib binds to the PI3Kδ ATP-binding site with a propeller-shaped conformation.

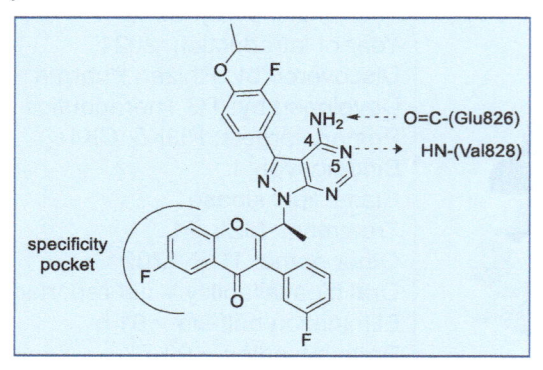

Figure 1. Putative umbralisib–PI3Kδ interactions.

The binding mode of umbralisib to CK1ε has also not been determined due to the lack of a co-crystal bound structure. However, umbralisib and PF4800567, a selective CK1ε inhibitor, share a similar overall architecture with a central pyrazolopyrimidine amine scaffold.[4] The co-crystal structure of PF4800567 in complex with CK1ε shows that pyrazolopyrimidine amine forms two hydrogen bonds with the carbonyl oxygen of Glu83 and the main chain NH of Leu85 from the hinge region.[6] Molecular docking studies showed that umbralisib may bind to CK1ε in a similar way as PF4800567 (Figs. 2, 3).

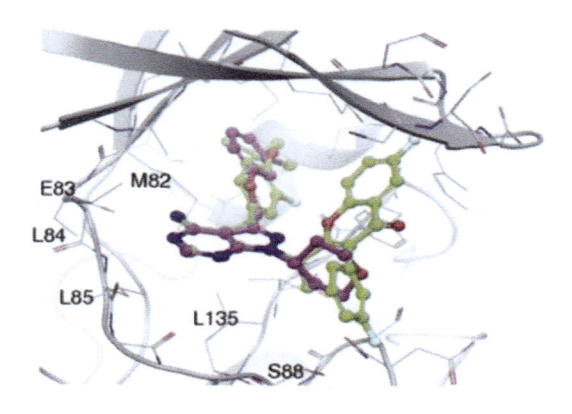

Figure 2. Overlay of the co-crystal structure of PF-4800567 (magenta) with CK1ε with the docking of umbralisib (green) in the CK1ε ATP-binding site. Adapted with permission from *Blood.* **2017**, *129*, 88–99.

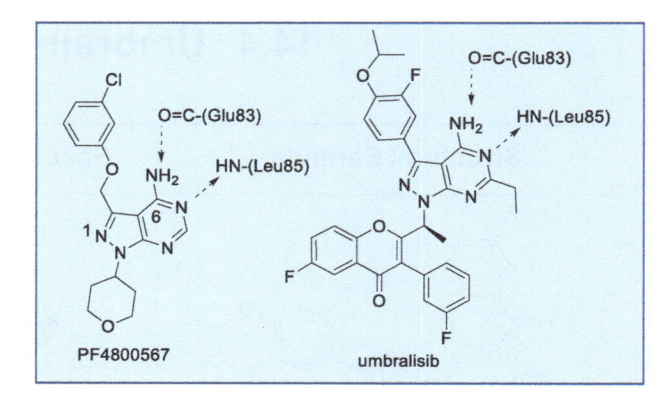

Figure 3. PF-4800567–CK1ε interactions based on co-crystal structure and putative umbralisib–CK1ε interactions.

PK Properties and Metabolism[1]

Umbralisib has a prolonged half-life of 91 h, which enables once-daily dosing. However, the recommended dosage of 800 mg is the highest among all the PI3K inhibitors in use, presumably due to its limited oral bioavailability.

1. S. Dhillon, S. J. Keam. Umbralisib: First Approval. *Drugs.* **2021**, *81*, 857–866.
2. A. Visentin, F. Frezzato, F. Severin, S. Imbergamo, S. Pravato, L. Romano Gargarella, S. Manni, S. Pizzo, E. Ruggieri, M. Facco, A. M. Brunati, G. Semenzato, F. Piazza, L. Trentin. Lights and shade of next-generation PI3k inhibitors in chronic lymphocytic leukemia. *Onco.Targets Ther.* **2020**, *13*, 9679–9688.
3. P. Janovská, E. Normant, H. Miskin, V. Bryja. Targeting casein kinase 1 (CK1) in hematological cancers. *Int. J. Mol. Sci.* **2020**, *21*, 9026.
4. C. Deng, M. R. Lipstein, L. Scotto, X. O. Jirau Serrano, M. A. Mangone, S. Li, J. Vendome, Y. Hao, X. Xu, S. X. Deng, R. B. Realubit, N. P. Tatonetti, C. Karan, S. Lentzsch, D. A. Fruman, B. Honig, D. W. Landry, O. A. O'Connor. Silencing c-Myc translation as a therapeutic strategy through targeting PI3Kδ and CK1ε in hematological malignancies. *Blood.* **2017**, *129*, 88–99.
5. M. S. Miller, P. E. Thompson, S. B. Gabelli. Structural Determinants of Isoform Selectivity in PI3K Inhibitors. *Biomolecules.* **2019**, *9*, 82.
6. A. M. Long, H. Zhao, X. Huang. Structural basis for the potent and selective inhibition of casein kinase 1 epsilon. *J. Med. Chem.* **2012**, *55*, 10307–10311.

14.5 Copanlisib (Aliqopa™) (74)

Structural Formula	Space-filling Model	Brief Information
$C_{23}H_{28}N_8O_4$; MW = 480 clogP = 1.1; tPSA = 138		Year of discovery: 2007 Year of introduction: 2017 Discovered by: Bayer Developed by: Bayer Primary targets: pan-PI3K Binding type: I Class: lipid kinase Treatment: FL (IV infusion) Other name: BAY 80-6946 Oral bioavailability = not reported Elimination half-life = 52 h (IV) Protein binding = 84.2%

● = C   ● = H   ● = O   ● = N   ● = S   ● = F   ● = Cl   ● = Br   ● = I

Copanlisib, a potent pan-PI3K inhibitor, was approved for use in patients with advanced follicular lymphoma (FL) that has returned after at least two prior systemic treatments.[1] Copanlisib is administered by a 1 h intravenous (IV) infusion with a fixed dose of 60 mg on days 1, 8, and 15 of a 28-day cycle, whereas the rest of the PI3K inhibitors are taken by mouth.

Discovery of Copanlisib[2,3]

Initial efforts toward PI3K inhibitors at Bayer were directed towards the development of PI3Kγ inhibitors as anti-inflammatory agents. High-throughput screening of the Bayer compound library against PI3Kγ led to the dihydroimidazoquinazoline **1** with modest PI3Kγ activity and 5-fold PI3Kβ selectivity (PI3Kγ IC_{50} = 0.81 µM; PI3Kβ IC_{50} = 4.0 µM) (Fig. 1). Replacement of the potentially reactive enol moiety with an amide maintained potency, and a change from phenyl to pyridyl gave **2** with >10-fold increase in potency against both β and γ isoforms. Adding 8,9-dimethoxy substitution afforded **3** with 28-fold β/γ selectivity (vs. 5-fold for the unsubstituted compound **2**), whereas the 7,8-dimethoxy counterpart **4** showed potent dual PI3Kα/β inhibition with respective IC_{50} values of 3.1 and 35 nM. Clearly, the A ring substitution exerts a profound impact on the isoform selectivity, and this finding prompted a shift in isoform selection for potency optimization from PI3Kγ inhibitors as anti-inflammatory agents to PI3Kα/β dual inhibitors for cancer therapies.

Molecular modelling of 2,3-dihydroimidazo[1,2-c]quinazoline scaffold with the published structure of PI3Kγ suggested that the dihydroimidazole N1 nitrogen binds to the hinge, and the C5 heterocyclic amide moiety occupies the so-called affinity pocket, which can be utilized to provide potential hydrogen bonding interactions to enhance potency. In addition, the C7 substituent projects toward the ATP sugar pocket and the C8 substituent toward a channel leading to the aqueous phase. Based on this docking model, structural modifications were carried out primarily in two regions: the solvent exposed C8 substitution and the heterocyclic amide moiety in the affinity pocket.

The aromatic C5 amide at the affinity pocket served as the key driver for potency. Inclusion of a hydrogen bond donor not only resulted in significantly enhanced potency but also led to reduced aqueous solubility and oral bioavailability. Ultimately, the aminopyrimidine moiety was shown to be ideal. To fine-tune aqueous solubility and PK properties, a variety of polar groups were incorporated at C8. Aminoalkoxy substituents were generally adequate, but 3-morpholinylpropoxy moiety at C8 provided the optimal balance between activity and PK properties. Taken together, these modifications led to copanlisib as a potent pan-PI3K inhibitor that showed subnanomolar to nanomolar IC_{50}'s with preferential inhibition against α and δ as compared with β and γ. Copanlisib exhibited potent cellular activity in blocking both IGF-1-stimulated AKT phosphorylation and basal AKT phosphorylation. The cellular activity was also demonstrated in the potent inhibition of proliferation in many cell lines.

Figure 1. Structural optimization of copanlisib.

Binding Mode[2]

As shown in the co-crystal structure of copanlisib bound to PI3Kγ, the inhibitor binds with only one critical hydrogen bond to the amide NH of Val882 in the adenine pocket, employing the imidazoline N1 nitrogen (Figs. 2 and 3). In addition, the C5 aminopyrimidine group fills the affinity pocket, forming hydrogen bonds with two carboxylic residues of Asp836 and Asp841 through the amino group. Finally, the solvent exposed morpholine lies over Trp812, presumably providing additional attractive molecular contacts.

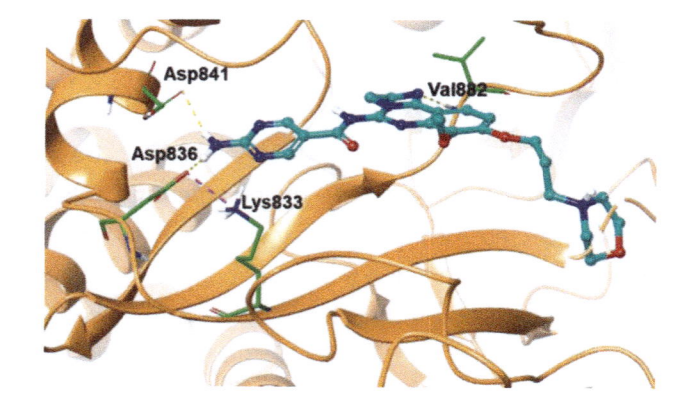

Figure 2. X-ray crystal structure of copanlisib bound to PI3Kγ (PDB ID: 5G2N).

Figure 3. Summary of copanlisib–PI3Kγ interactions based on a co-crystal structure.

PK Properties and Metabolism[4]

Copanlisib has poor oral bioavailability due to low permeability and poor solubility, and therefore, it is not suitable for an oral medication. Consequently, it was developed as an IV drug. IV infusion of copanlisib resulted in rapid distribution throughout the body and a prolonged elimination half-life (52 h). Copanlisib was the predominant component in human plasma, accounting for 84% of total radioactivity AUC, and the morpholinone metabolite **6** was the only circulating metabolite (about 5%) (Fig. 4).

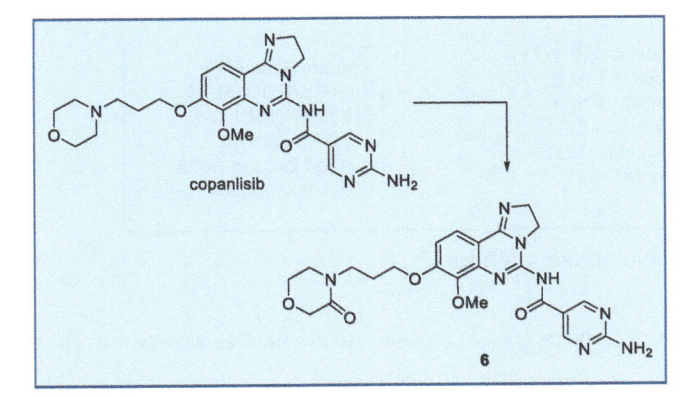

Figure 4. Metabolic pathway of copanlisib in humans.

1. A. Markham. Copanlisib: First global approval. *Drugs*. **2017**, *77*, 2057–2062.

2. W. J. Scott, M. F. Hentemann, R. B. Rowley, C. O. Bull, S. Jenkins, A. M. Bullion, J. Johnson, A. Redman, A. H. Robbins, W. Esler, R. P. Fracasso, T. Garrison, M. Hamilton, M. Michels, J. E. Wood, D. P. Wilkie, H. Xiao, J. Levy, E. Stasik, N. Liu, J. Lefranc. Discovery and SAR of novel 2,3-dihydroimidazo[1,2-c]quinazoline PI3K inhibitors: Identification of copanlisib (BAY 80-6946). *ChemMedChem*. **2016**, *11*, 1517–1530.

3. N. Liu, B. R. Rowley, C. O. Bull, C. Schneider, A. Haegebarth, C. A. Schatz, P. R. Fracasso, D. P. Wilkie, M. Hentemann, S. M., Wilhelm, W. J. Scott, D. Mumberg, K. Ziegelbauer. BAY 80-6946 is a highly selective intravenous PI3K inhibitor with potent p110α and p110δ activities in tumor cell lines and xenograft models. *Mol. Cancer Ther*. **2013**, *12*, 2319–2330.

4. M. Gerisch, T. Schwarz, D. Lang, G. Rohde, S. Reif, I. Genvresse, S. Reschke, D. van der Mey, C. Granvil. Pharmacokinetics of intravenous pan-class I phosphatidylinositol 3-kinase (PI3K) inhibitor [^{14}C]copanlisib (BAY 80-6946) in a mass balance study in healthy male volunteers. *Cancer Chemother. Pharmacol*. **2017**, *80*, 535–544.

Chapter 15. TRK/Multikinase Inhibitors

Neurotrophic tropomyosin receptor kinase (NTRK) gene fusion, is an oncogenic driver of many types of solid tumors. Tropomyosins are a very large family of structural proteins that regulate cell filaments and cytoskeleton. Oncogenic NTRK cancer is an uncommon disease family occurring, for example, with a frequency of <1% in non-small cell lung cancer (NSCLC). The tropomyosin receptor kinase (TRK) inhibitors, larotrectinib and entrectinib, are the first two FDA-approved small-molecule, tumor-agnostic therapeutics for the treatment of any solid tumors that have a NTRK gene fusion. However, these two inhibitors differ in their kinase selectivity, with larotrectinib being a selective pan-TRK inhibitor and entrectinib a multikinase inhibitor. In addition to pan-TRK, entrectinib also inhibits ALK and ROS1, which led to its approval for ROS1+ NSCLC, as an alternative to the ALK inhibitor crizotinib (see Section 9.1).[1]

Because resistance to first-generation TRKs ultimately develops, the second-generation TKI repotrectinib (a cyclized version of larotrectinib) was developed to overcome on-target resistance due to kinase domain mutations, such as solvent-front mutations.[2] This drug candidate is currently under FDA regulatory review.

Targeted therapies guided by NTRK-related predictive biomarkers provide new options for patients with this type of solid cancer. This chapter describes the discovery process that led to the three TRK inhibitors **75–77** (Fig. 1). This field is poised to grow in importance as newer and more selective inhibitors are discovered.

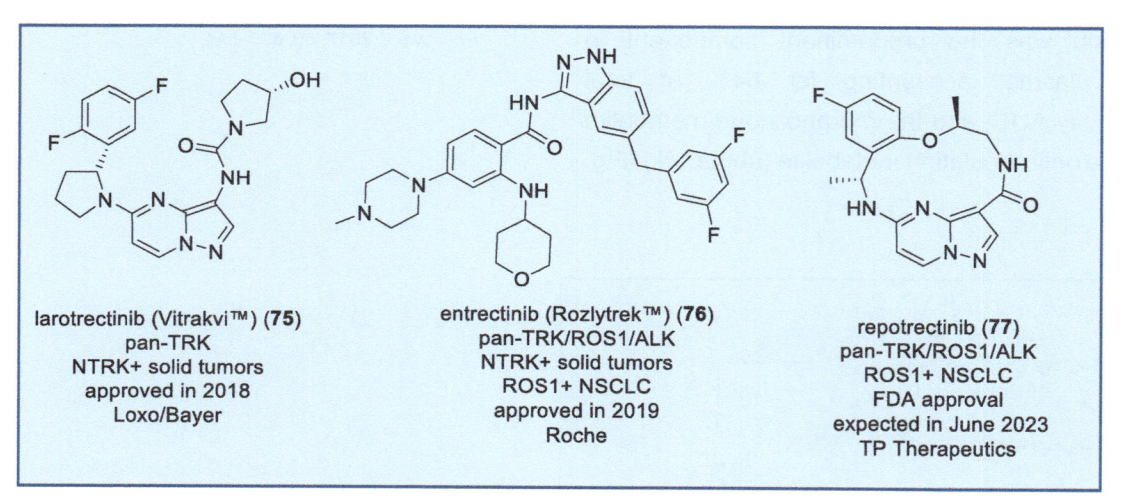

Figure 1. Pan-TRK/multikinase inhibitors.

1. A. Drilon. TRK inhibitors in TRK fusion-positive cancers. *Ann Oncol.* **2019**, 30(Suppl. 8), viii23-viii30.
2. A. Drilon, S. Ou, B. C. Cho, D. W. Kim, L. Lee, J. Li, V. W. Zhu, M. J. Ahn, D. R. Camidge, J. Nguyen, D. Zhai, W. Deng, Z. Huang, E. Rogers, J. Liu, J. Whitten, J. Lim, S. Stopatschinskaja, D. M. Hyman, R. C. Doebele, A. T. Shaw. Repotrectinib (TPX-0005) Is a next-generation ROS1/TRK/ALK inhibitor that potently inhibits ROS1/TRK/ALK solvent-front mutations. *Cancer Discov.* **2018**, *8*, 1227–1236.

15.1 Larotrectinib (Vitrakvi™) (75)

Structural Formula	Space-filling Model	Brief Information
$C_{21}H_{22}F_2N_6O_2$; MW = 428 clogP = 2.0; tPSA = 84		Year of discovery: 2009 Year of introduction: 2018 Discovered by: Array BioPharma Developed by: Loxo/Bayer Primary targets: pan-TRK Binding type: I Class: receptor tyrosine kinase Treatment: NTRK-altered solid tumors Other name: ARRY-470, LOXO-101 Oral bioavailability = 25% Elimination half-life = 2.9 h Protein binding = 70%

● = C   ○ = H   ● = O   ● = N   ● = S   ● = F   ● = Cl   ● = Br   ● = I

Larotrectinib was granted accelerated FDA approval in 2018 for adult and pediatric patients with solid tumors that have a neurotrophic receptor tyrosine kinase (NTRK) gene fusion without a known acquired resistance mutation, that is either metastatic, unsuitable for surgical resection, without any satisfactory alternative treatments, or whose cancer has progressed following treatment – regardless of the tumor's organ, tissue, or location.[1] It is the second tissue-agnostic FDA approval for the treatment of cancer, one year after Merck's pembrolizumab (marketed as Keytruda for patients with microsatellite instability or high tumor mutational burden), and the first small molecule with a tissue-agnostic breakthrough therapy designation by the FDA. Tissue-agnostic cancer drugs treat tumors defined by a specific genetic alteration, rather than the cancer's location in the body. The discovery of larotrectinib illustrates the value of biomarkers to guide drug development and precisely targeted delivery in modern medicine.

Larotrectinib is the first drug to selectively target the activity of tropomyosin receptor kinases (TRKs), and thereby inhibit the growth of tumors driven by excessive TRK signaling. It is a direct rival to Roche's entrectinib,[2] the second FDA-approved small-molecule, tissue-agnostic TRK inhibitor for use in patients harboring tumors with NTRK fusions.[3]

TRK Signaling Pathway[3]

The TRK receptor tyrosine kinases, TRKA/B/C, are encoded by the NTRK genes, NTRK1/2/3. They are single-pass transmembrane proteins, consisting of an extracellular ligand binding domain, a transmembrane domain, and an intracellular kinase domain. As with other receptor tyrosine kinases (RTKs), activation of TRK proteins requires ligand binding to the extracellular domain of the receptor (Fig. 1, top right). The ligands for TRK proteins are the neurotrophins, and each TRK receptor is activated with a distinct neurotrophin, for example, TRKA with nerve growth factor (NGF); and TRKB with brain-derived growth factor (BDGF). Binding of a specific neurotrophin to a TRK receptor leads to receptor dimerization and phosphorylation, triggering the activation of downstream signaling pathways mediated by PI3K, RAS/MAPK/ERK. Through these cascades, neurotrophins ultimately signal particular cells to survive, differentiate, or grow.

Although TRK proteins are predominantly expressed in the nervous system and play an important role in neuronal growth and development, the abnormal TRK fusion proteins have been identified in various non-neuronal solid tumors, including those of the lung, gastrointestinal tract, thyroid, and also sarcomas. These proteins are created through NTRK gene fusions, which function as a primary oncogenic driver of various cancers.

NTRK Fusions in Cancer[3,4]

The most common oncogenic NTRK molecular aberration is NTRK rearrangement, a gene fusion, that occurs when the 3' region of an NTRK gene combines with the 5' region of a dimerization domain of an unrelated upstream gene partner through chromosomal rearrangements (Fig. 1, bottom left). So far, more than 80 different fusion gene partners have been identified in various tumor types.[5] The resultant NTRK gene fusion retains the kinase domain of the TRK receptor (despite the breakpoint), leaving the downstream intracellular kinase activity intact. However, the newly incorporated dimerization domain (e.g., coiled-coil domains and zinc-finger domains) serves as a dimerization unit, which results in the spontaneous ligand-independent dimerization, subsequent autophosphorylation and activation of the oncogenic protein tyrosine kinase domain. This triggers several signaling pathways (Fig. 1, bottom right) which drive tumor cell proliferation, survival, invasion, and angiogenesis.

The NTRK gene fusions have been identified in ~0.3% of all solid tumors with an annual incidence of 1500–5000 cases in the United States.[6] However, in certain rare cancers such as secretory breast carcinoma and mammary analogue secretory carcinoma of the salivary gland (MASC), their prevalence can be >90%.

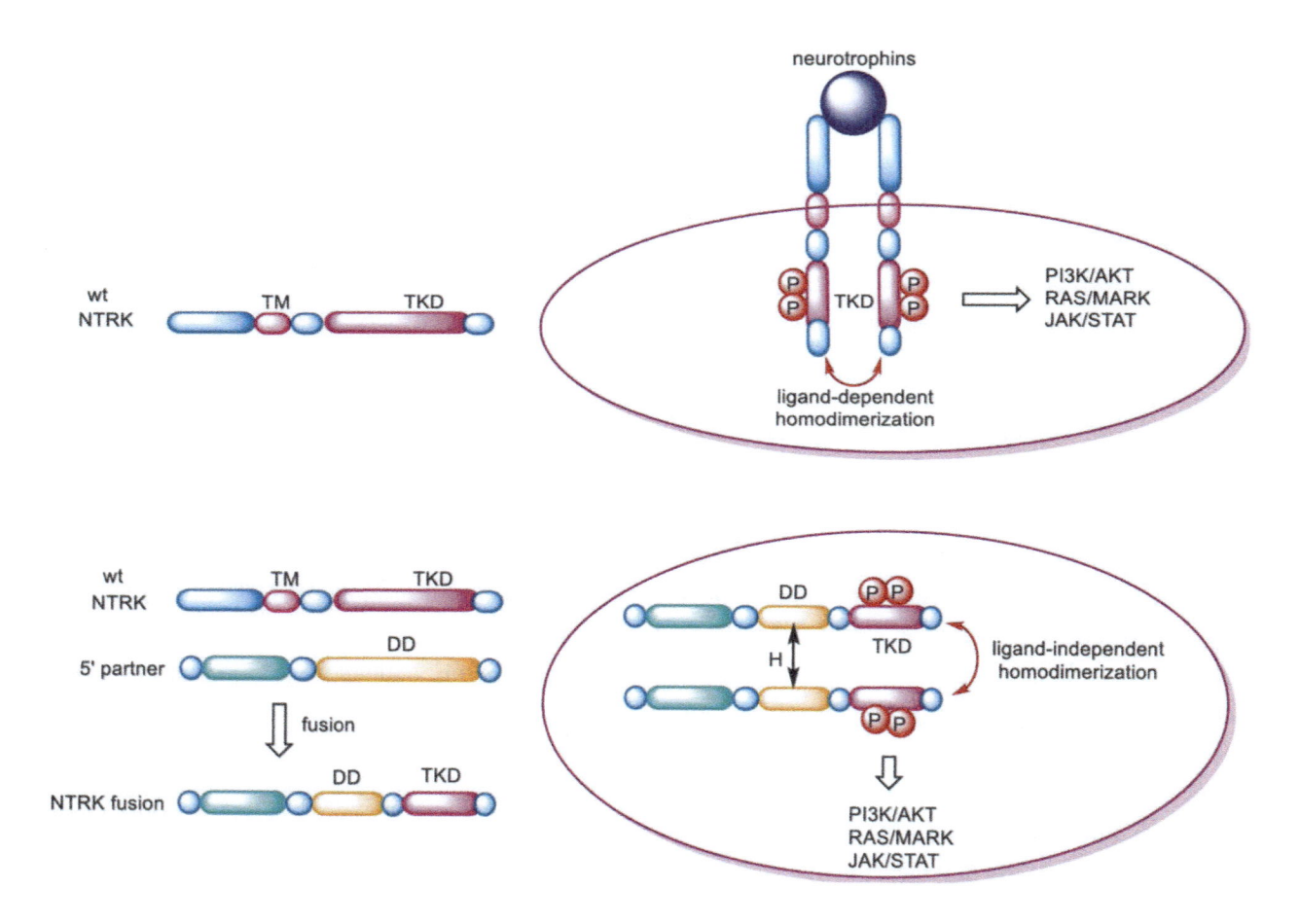

Figure 1. Top left: schematic representation of wt NTRK; Top right: ligand-dependent TRK activation in wt NTRK; Bottom left: NTRK fusion; Bottom right: ligand-independent TRK activation in altered NTRK. TM: transmembrane domain; TKD: tyrosine kinase domain; DD: dimerization domain.

Larotrectinib: a Selective Pan-TRK Inhibitor[7,8]

Larotrectinib was initially developed preclinically as a potential pain reliever to inhibit NGF/TRKA signaling which mediates the sensation of pain. Subsequently, it was developed clinically for the treatment of TRK-driven malignancies. It exhibited IC_{50} values of 5.3–11.5 nM against all TRK isoforms in biochemical assays and 9.8–25 nM in cellular assays. It also displayed ≥100-fold selectivity against 229 other kinases and ≥1000-fold selectivity against 80 nonkinases. It elicited clinical efficacy in adult and pediatric patients with NTRK fusion-positive solid tumor malignancies.

The binding mode of larotrectinib has not yet been determined due to lack of X-ray co-crystal structures with any of the TRK isoforms.

Like some other cancer medications, larotrectinib as a precision medicine is expensive. The wholesale price for one adult patient in the United States comes to $32,800 per month, or $393,600 annually. Fortunately, the high cost can be offset by a rare guarantee from the manufacturer (Bayer) to refund the cost of treatment in the event eligible patients do not experience clinical benefit within 90 days of treatment.[9]

1. L. J. Scott. Larotrectinib: First global approval. *Drugs*, **2019**, *79*, 201–206.
2. Z. T. Al-Salama, S. J. Keam. Entrectinib: First global approval. *Drugs*. **2019**, *79*, 1477–1483.
3. A. M. Lange, H. W. Lo. Inhibiting TRK proteins in clinical cancer therapy. *Cancers*, **2018**, *10*, 105.
4. A. Vaishnavi, R. C. Doebele. TRKing down an old oncogene in a new era of targeted therapy. *Cancer Discov*. **2015**, *5*, 25-34.
5. J. F. Hechtman. NTRK insights: Best practices for pathologists. *Mod. Pathol.* **2022**, *35*, 298–305.
6. E. S. Kheder, D. S. Hong. Emerging targeted therapy for tumors with *NTRK* fusion proteins. *Clin. Cancer Res.* **2018**, *24*, 5807–5814.
7. T. W. Laetsch, D. S. Hawkins. Larotrectinib for the treatment of TRK fusion solid tumors. *Expert Rev. Anticancer Ther.* **2019**, *19*, 1–10.
8. N. Federman, R. McDermott. Larotrectinib, a highly selective tropomyosin receptor kinase (TRK) inhibitor for the treatment of TRK fusion cancer. *Expert Rev. Clin. Pharmacol.* **2019**, *12*, 931–939.
9. P. Hofland. FDA Approves Larotrectinib – the First tumor-agnostic cancer treatment. Business & economics. November 28, 2018.

15.2 Entrectinib (Rozlytrek™) (76)

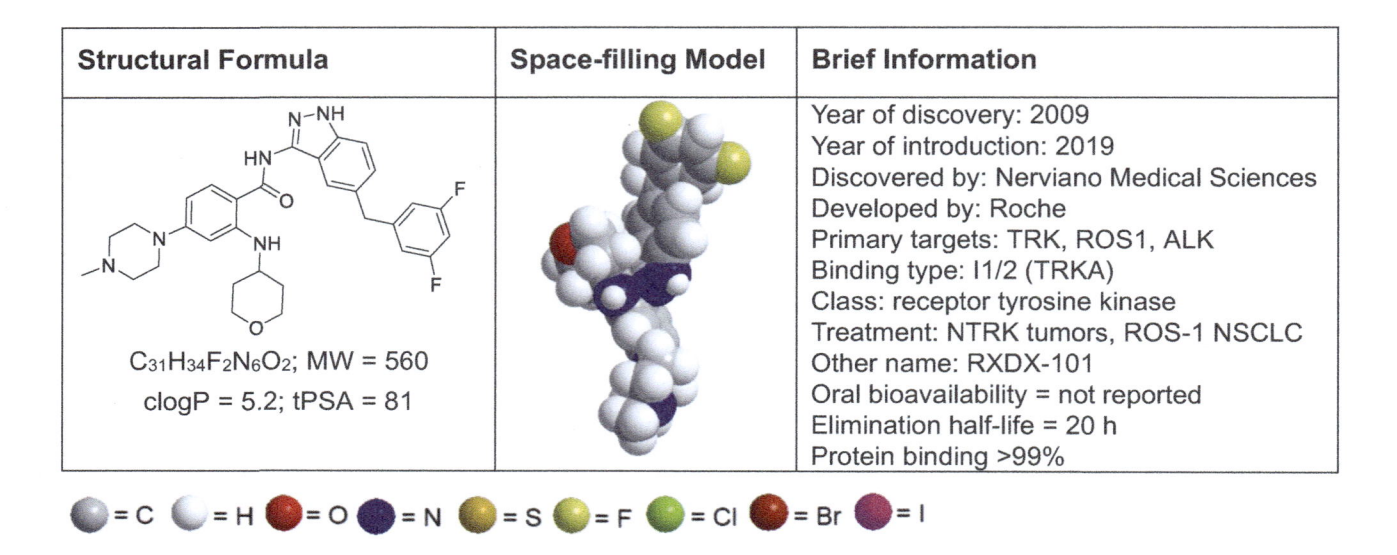

Structural Formula	Space-filling Model	Brief Information
$C_{31}H_{34}F_2N_6O_2$; MW = 560 clogP = 5.2; tPSA = 81		Year of discovery: 2009 Year of introduction: 2019 Discovered by: Nerviano Medical Sciences Developed by: Roche Primary targets: TRK, ROS1, ALK Binding type: I1/2 (TRKA) Class: receptor tyrosine kinase Treatment: NTRK tumors, ROS-1 NSCLC Other name: RXDX-101 Oral bioavailability = not reported Elimination half-life = 20 h Protein binding >99%

● = C ● = H ● = O ● = N ● = S ● = F ● = Cl ● = Br ● = I

Entrectinib was granted accelerated approval in 2019 for adults and pediatric patients ≥12 years of age with solid tumors that have a neurotrophic tyrosine receptor kinase (NTRK) gene fusion without a known acquired resistance mutation. It was also approved for adults with metastatic non-small cell lung cancer (NSCLC) whose tumors are mutated ROS1-positive.[1,2] ROS1 is a receptor tyrosine kinase which is structurally related to ALK. Abnormal ROS1 genes have been detected in multiple tumor types, e.g., lung and brain.

Following the approval of larotrectinib,[3,4] entrectinib became the second small molecule with a tissue-agnostic breakthrough therapy designation by the FDA. Interestingly, Roche's entrectinib is priced at about $17,050 per month, nearly half of the monthly price of its rival, larotrectinib, which costs $32,800 per month. However, entrectinib does not offer a 90-day efficacy guarantee as larotrectinib does.

Discovery of Entrectinib: A Potent ALK, ROS1, and Pan-TRKs Inhibitor[5]

Initial efforts at Nerviano were directed toward the second-generation ALK (anaplastic lymphoma kinase) inhibitors. High-throughput screening of Nerviano compound collection identified 3-aminoindazole **1** that showed good potency against ALK in the biochemical assay (IC_{50} = 73 nM) and moderate activity in the cellular assay (IC_{50} = 0.253 µM) (Fig. 1). Modelling studies suggested that a C2' substituent may occupy the ATP sugar pocket while displacing an unfavorable water molecule. Furthermore, a judiciously selected amino group (NHR) at C2' could form intramolecular hydrogen bond to stabilize the bioactive conformation. To test this hypothesis, a variety of amino substituents were incorporated onto C2' to give a series of analogs **2**, and this effort led to entrectinib as a potent ALK inhibitor (IC_{50} = 12 nM). Later, it was found that entrectinib also potently inhibited TRKA/B/C as well as ROS1 (IC_{50} = 1, 3, 5, 7 nM, respectively).

As compared with larotrectinib, entrectinib is a weak P-gp substrate, which contributes to its sustained CNS exposure. The brain penetration is responsible for the preclinical and clinical efficacy of entrectinib in NTRK and ROS1 fusion-positive CNS tumors and secondary CNS metastases. Thus, entrectinib may have potential for the treatment of CNS diseases such as brain metastasis.[6]

Figure 1. Structural optimization of entrectinib.

Binding Mode[7]

As shown in the co-crystal structure of entrectinib with ALK (Fig. 2), the inhibitor is anchored to the hinge region through typical three donor–acceptor–donor hydrogen bonds between the aminoindazole core and the main chain atoms of Glu1197 and Met1199 of the hinge region (Fig. 3). As suggested by the modelling studies, the C2′ NH hydrogen bonds to the amide carbonyl to lock the six-membered bioactive conformation in place, which contributes to enhanced activity. The glycine-rich loop is oriented in a unique "collapsed" conformation, with the 3,5-difluorobenzyl moiety lying between Leu1256 and Phe1127 from the glycine-rich loop. The methylpiperazine moiety penetrates into the solvent, while the partially solvent-exposed tetrahydropyranyl lies perpendicular to the aminoindazole hinge binder to optimally occupy a small hydrophobic pocket.

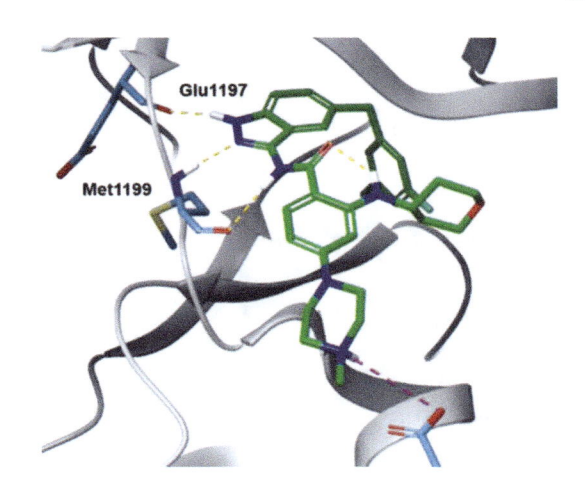

Figure 2. Co-crystal structure of entrectinib–ALK (PDB ID: 5FTO).

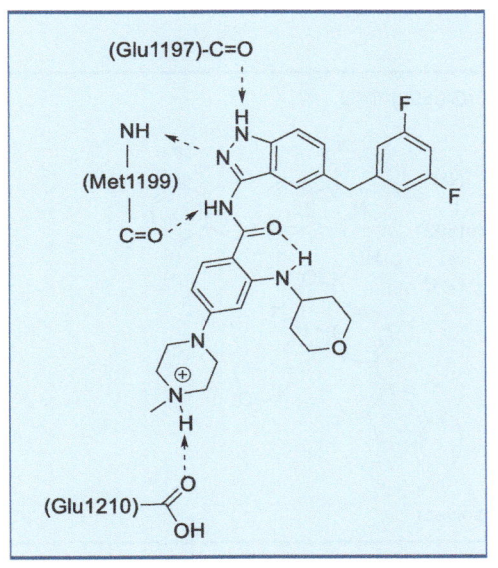

Figure 3. Summary of entrectinib–ALK interactions based on their co-crystal structure.

The co-crystal structure of entrectinib with TRKA shows that the inhibitor binds to the hinge region via the same donor–acceptor–donor hydrogen bonds as in the complex with ALK (Fig. 4). More specifically, the indazole N1 NH forms a hydrogen bond with the amide carbonyl of Glu590, N2 hydrogen bonds with the backbone amide NH of Met592, and the amino group attached to the indazole ring forms a hydrogen bond with the amide carbonyl of Met592 (Fig. 5).

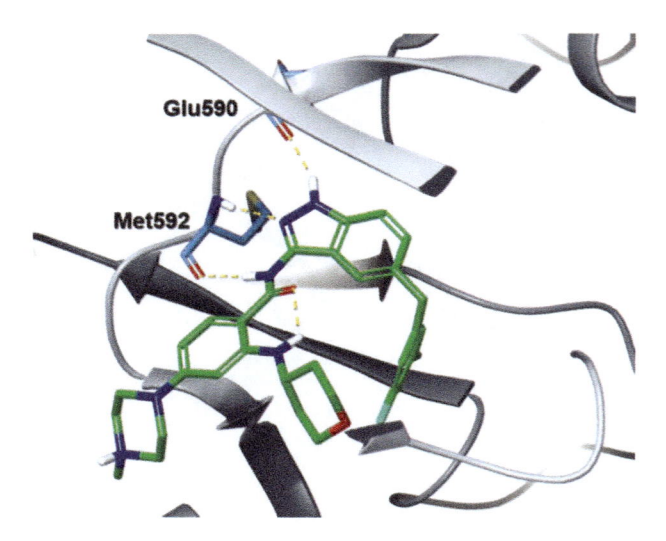

Figure 4. Co-crystal structure of entrectinib–TRKA (PDB ID: 5KVT).

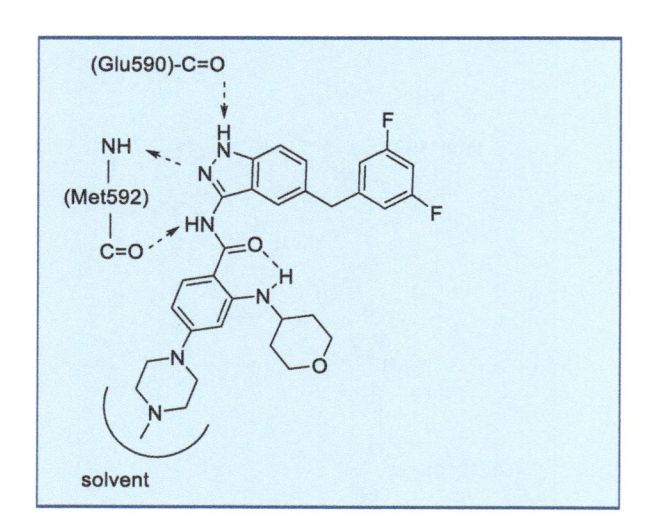

Figure 5. Summary of entrectinib–TRKA interactions based on their co-crystal structure.

1. Z. T. Al-Salama, S. J. Keam. Entrectinib: First global approval. *Drugs*. **2019**, *79*, 1477–1483.
2. L. Marcus, M. Donoghue, S. Aungst, C. E. Myers, W. S. Helms, G. Shen, H. Zhao, O. Stephens, P. Keegan, R. Pazdur. FDA approval summary: Entrectinib for the treatment of *NTRK* gene fusion solid tumors. *Clin. Cancer Res*. **2021**, *27*, 928–932.
3. L. J. Scott. Larotrectinib: First global approval. *Drugs*. **2019**, *79*, 201–206.
4. A. M. Lange, H. W. Lo. Inhibiting TRK proteins in clinical cancer therapy. *Cancers*. **2018**, *10*, 105.
5. M. Menichincheri, E. Ardini, P. Magnaghi, N. Avanzi, P. Banfi, R. Bossi, L. Buffa, G. Canevari, L. Ceriani, M. Colombo, L. Corti, D. Donati, M. Fasolini, E. Felder, C. Fiorelli, F. Fiorentini, A. Galvani, A. Isacchi, A. L. Borgia, C. Marchionni, P. Orsini. Discovery of entrectinib: A new 3-aminoindazole as a potent anaplastic lymphoma kinase (ALK), c-ros oncogene 1 kinase (ROS1), and pan-tropomyosin receptor kinases (pan-TRKs) inhibitor. *J. Med. Chem*. **2016**, *59*, 3392–3408.
6. H. Fischer, M. Ullah, C. C. de la Cruz, T. Hunsaker, C. Senn, T. Wirz, B. Wagner, D. Draganov, F. Vazvaei, M. Donzelli, A. Paehler, M. Merchant, L. Yu. Entrectinib, a TRK/ROS1 inhibitor with anti-CNS tumor activity: Differentiation from other inhibitors in its class due to weak interaction with P-glycoprotein. *Neurooncol*. **2020**, *22*, 819–829.
7. R. Roskoski, Jr. Properties of FDA-approved small molecule protein kinase inhibitors: A 2020 update. *Pharmacol. Res*. **2020**, *152*,104609.

15.3 Repotrectinib (77)

Structural Formula	Space-filling Model	Brief Information
$C_{18}H_{18}FN_5O$; FW = 355 clogP = 2.9; tPSA = 78		Year of discovery: 2014 Year on Introduction: expected in 2023 Current status: phase II Discovered by: TP Therapeutics Developed by: TP Therapeutics Primary targets: NTRK, ROS, ALT Binding type: I Class: receptor tyrosine kinase Treatment: NSCLC Other name: TPX-0005

● = C ○ = H ● = O ● = N ● = S ● = F ● = Cl ● = Br ● = I

Repotrectinib, which is currently being developed by Turning Point (TP) Therapeutics in collaboration with China's Zai Lab, was granted a breakthrough therapy designation in 2022 for the treatment of patients with ROS1-positive (ROS1+) metastatic non–small cell lung cancer (NSCLC) who have been previously treated with one ROS1 tyrosine kinase inhibitor (TKI) and have not received prior platinum-based chemotherapy. Previously, it had received two breakthrough therapy designations for patients with ROS1+ metastatic NSCLC who have not received a ROS1 TKI, and patients with advanced solid tumors that have an NTRK gene fusion who have progressed after treatment with one or two prior TRK TKIs, with or without prior chemotherapy, and without satisfactory alternative treatments.

Repotrectinib specifically targets solvent-front mutations (SFMs) of ROS1, pan-TRK, and ALK. In the biochemical and cellular assays, it inhibits the kinase activity of wt ROS1, TRKA–C, and ALK, and their SFMs with IC$_{50}$ values in the range of 0.071–4.46 nM. It exhibits high potency against ROS1 and TRKA–C with ~15-fold selectivity over ALK.[1]

In early clinical studies, repotrectinib showed a longer duration of response compared to existing ROS1 agents in first-line NSCLC. Therefore, it may offer an effective treatment option for ROS1-, NTRK1–3, or ALK-rearranged malignancies which have progressed on earlier-generation TKIs.[2]

Repotrectinib has also been shown to penetrate the blood-brain barrier. Consequently, it may be effective for the treatment of brain metastases frequently developed in patients with metastatic NSCLC.[1]

Repotrectinib (whose approval is expected in 2023) has recently been acquired by Bristol Myers Squibb as a result of the $4.1 billion purchase of the discovery oncology company Turning Point Therapeutics. It is noteworthy that selitrectinib (Fig. 1),[3,4] the other macrocyclic compound, once touted as a promising next-generation TRK kinase inhibitor, was terminated as of November 2021 for undisclosed reasons (Bayer pipeline, November 2021), another illustration that modest structural variations in a drug molecule can unpredictably affect its ultimate value and utility.

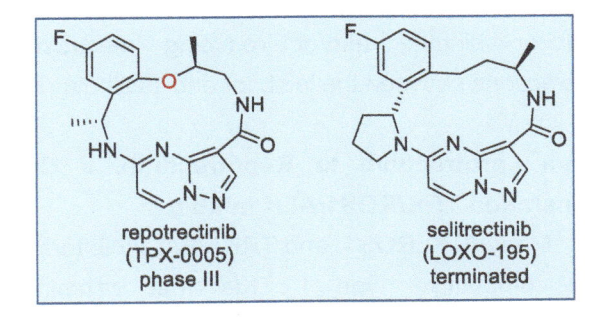

Figure 1. Structures of repotrectinib and selitrectinib.

Larotrectinib and Entrectinib: First-Generation TRK Inhibitors

The TRK receptor tyrosine kinases, TRKA/B/C, which are encoded by the NTRK genes, NTRK1/2/3, are predominantly expressed in the nervous system where they play an important role in neuronal growth and development. However, the abnormal TRK fusion proteins have been identified in ~0.3% of all solid tumors with annual incidence of 1500–5000 cases in the United States.[5] In some (rare) cancers, such as secretory breast carcinoma and mammary analogue secretory carcinoma of the salivary gland (MASC), their prevalence may be >90%.

Larotrectinib and entrectinib are the first small-molecule tissue-agnostic TRK inhibitors for use in patients harboring tumors with NTRK fusions.[5–7] They are applied as treatment for NTRK-rearranged cancers regardless of age and tumor origin.

Solvent-Front Mutations (SFMs) in TRK, ROS, and ALK[1,3]

Despite dramatic and durable disease control in patients with NTRK-rearranged malignancies, advanced-stage NTRK fusion-positive tumors ultimately become resistant to TRK kinase-directed therapy through recalcitrant pathways of on-target resistance. In particular, two solvent-front mutations, TRKAG595R and TRKCG623R, have been identified in NTRK1- and NTRK3-rearranged tumors. These mutations are homologous to the ALKG1202R in ALK-rearranged tumors, and ROS1G2032R in ROS1-rearranged tumors. In general, these particular solvent-front mutations confer resistance by sterically interfering with inhibitor binding and/or reducing electrostatic interactions between the inhibitor and its binding site.

From Larotrectinib to Repotrectinib, a Next-Generation TRK/ROS1/ALK inhibitor[1]

Many ALK, ROS1, and TRKA/B/C inhibitors are ATP-competitive type I TKIs that contain a hydrophilic segment extending to the solvent-exposed region of the kinase. That segment is purposely incorporated to modulate physical properties (e.g., lipophilicity, solubility). Unfortunately, the solvent-front motif is a frequent site of on-target resistance mediated by the acquisition of SFMs. To mitigate the SFM-mediated resistance, the solvent-front hydroxypyrrolidine moiety of larotrectinib was eliminated, and the terminal phenyl is directly connected to the urea scaffold to form the 13-membered macrocycle repotrectinib (Fig. 2), a process reminiscent of the transformation of crizotinib to lorlatinib (see Section 9.5). The macrocyclic structure of repotrectinib was inspired by the U-shaped binding conformation of larotrectinib. It avoids steric clashes with the solvent-front substitutions in ROS1G2032R TRKAG595R, TRKCG623R, and ALKG1202R (data not shown).[1] As a result, repotrectinib binding is not affected by the SFM.

The rigid macrocyclic inhibitor, locked in a bioactive conformation, fits precisely in the ATP-binding site, thus incurring little entropy penalty during binding. Its compact size, small molecular weight, modest lipophilicity (clogP = 2.90), and high CNS MPO score (4.65) result in favorable pharmacokinetic properties including moderate brain penetration (brain/plasma = 5.2% and 3.8% after single and repeated dosing in mice, respectively).[1]

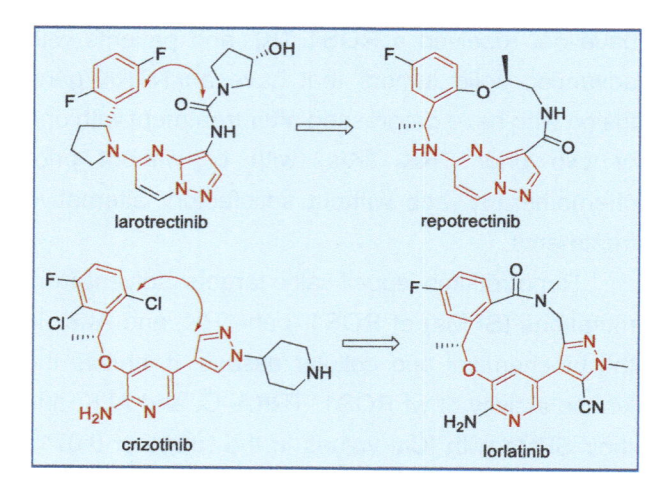

Figure 2. Macrocyclization of larotrectinib/crizotinib to repotrectinib/lorlatinib.

Binding Mode[4]

The cocrystal structures of repotrectinib with TRK proteins have not been reported. A docking model with the TRKA protein indicates that the pyrazole nitrogen hydrogen bonds with Met592 in the hinge area. The fluorophenyl, which packs into a hydrophobic pocket, adopts a perpendicular orientation with the macrocycle, and the lactam oxygen atom also forms a water mediated hydrogen bond with Met592 in the hinge region (Figs. 3, 4).

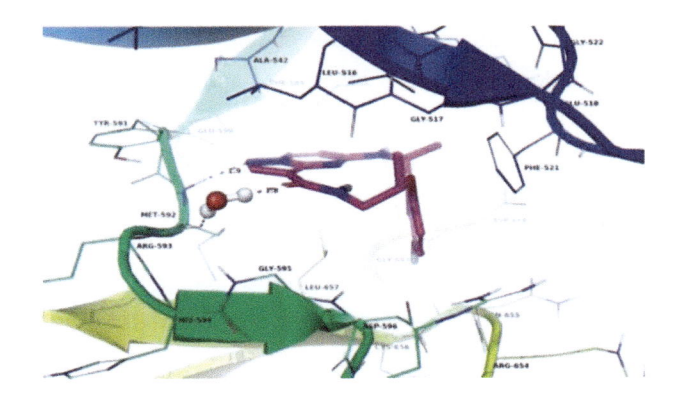

Figure 3. Docking of repotrectinib with TRKA. Adapted with permission from *J. Med. Chem.* **2021**, *64*, 14, 10286–10296.

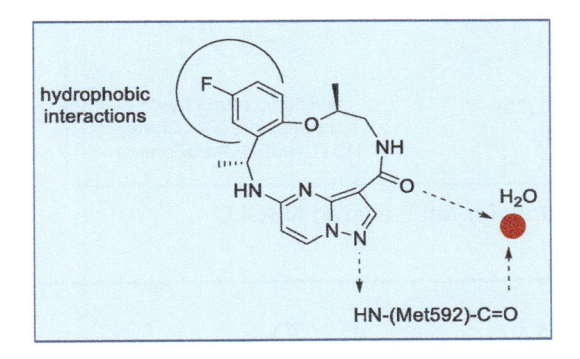

Figure 4. Putative interactions of repotrectinib–TRKA.

1. A. Drilon, S. Ou, B. C. Cho, D. W. Kim, L. Lee, J. Li, V. W. Zhu, M. J. Ahn, D. R. Camidge, J. Nguyen, D. Zhai, W. Deng, Z. Huang, E. Rogers, J. Liu, J. Whitten, J. Lim, S. Stopatschinskaja, D. M. Hyman, R. C. Doebele, A. T. Shaw. Repotrectinib (TPX-0005) Is a next-generation ROS1/TRK/ALK inhibitor that potently inhibits ROS1/TRK/ALK solvent- front mutations. *Cancer Discov.* **2018**, *8*, 1227–1236.

2. B. C. Cho, R. C. Doebele, J. J. Lin *et al.* Phase 1/2 TRIDENT-1 study of repotrectinib in patients with ROS1+ or NTRK+ advanced solid tumors. Presented at: International Association for the Study of Lung Cancer 2020 World Conference on Lung Cancer; January 28–31, 2021; Virtual. Abstract MA11.07.

3. A. Drilon, R. Nagasubramanian, J. F. Blake, N. Ku, B. Tuch, K. Ebata, S. Smith, V. Lauriault, G. R. Kolakowski, B. J. Brandhuber, P. D. Larsen, K. S. Bouhana, S. L. Winski, S. Hamor, W. Wu, A. Parker, T. H. Morales, F. X. Sullivan, A. Wollenberg, D. M. Hyman. A next-generation TRK kinase inhibitor overcomes acquired resistance to prior TRK kinase inhibition in patients with TRK fusion-positive solid tumors. *Cancer Discov.* **2017**, *7*, 963–972.

4. Z. Liu, P. Yu, L. Dong, W. Wang, S. Duan, B. Wang, X. Gong, L. .Ye, H. Wan, J. Tian. Discovery of the next-generation pan-TRK kinase inhibitors for the treatment of cancer. *J. Med. Chem.* **2021**, *64*, 10286–10296.

5. E. S. Kheder, D. S. Hong. Emerging targeted therapy for tumors with *NTRK* fusion proteins. *Clin. Cancer Res.* **2018**, *24*, 5807–5814.

6. A. M. Lange, H. W. Lo. Inhibiting TRK proteins in clinical cancer therapy. *Cancers.* **2018**, *10*, 105.

7. N. Federman, R. McDermott. Larotrectinib, a highly selective tropomyosin receptor kinase (TRK) inhibitor for the treatment of TRK fusion cancer. *Expert Rev. Clin. Pharmacol.* **2019**, *12*, 931–939.

Chapter 16. MET Inhibitors

Oncogene-directed therapies have significantly improved the treatment of advanced non-small cell lung cancer (NSCLC) over the past decade. FDA-approved molecular-matched agents are available for NSCLC with specific oncogenic drivers such as EGFR, ALK, BRAF, ROS1, NTRK, RET, and MET exon 14 skipping mutations (METex14). These targeted biomarker-guided therapeutics have substantially improved survival outcomes.

METex14 skipping mutations account for 3–4% of NSCLC cases, typically occurring without other driver mutations. Aberrant MET signaling has been targeted for cancer treatment over the past three decades, but early MET inhibitors failed to provide robust efficacy in clinical trials, primarily because no biomarker test was available at the time to accurately match therapies with the genetic alterations of the individual cancer patients. The identification of biomarkers to identify patients harboring MET mutations was crucial to the clinical success of the first selective MET inhibitors, capmatinib and tepotinib (Fig. 1), which were approved by the FDA in 2020–2021 for the treatment of patients with advanced NSCLC harboring METex14 skipping mutations.[1] Figure 2 shows several newer MET inhibitors in clinical trials for various cancers.

This chapter describes the research that led to the two FDA-approved MET inhibitors **78** and **79**.

Figure 1. Approved selective MET inhibitors for MET-altered NSCLC

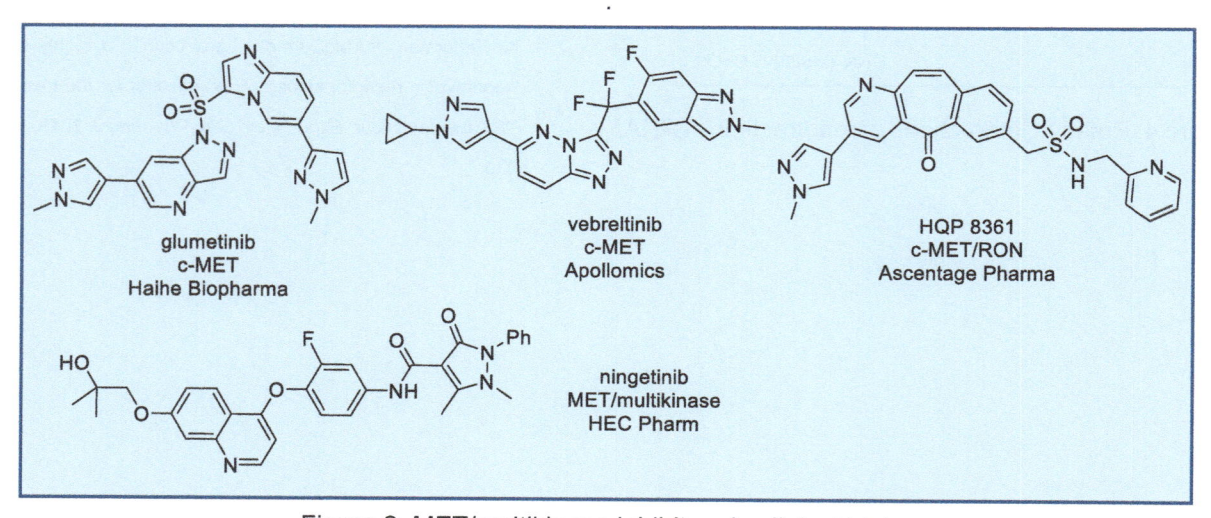

Figure 2. MET/multikinase inhibitors in clinical trials.

1. M. Santarpia, M. Massafra, V. Gebbia, A. D'Aquino, C. Garipoli, G. Altavilla, R. Rosell. A narrative review of MET inhibitors in non-small cell lung cancer with *MET* exon 14 skipping mutations. *Transl. Lung Cancer Res.* **2021**, *10*, 1536–1556A.

16.1 Capmatinib (Tabrecta™) (78)

Structural Formula	Space-filling Model	Brief Information
$C_{29}H_{17}FN_6O$; MW = 412 clogP = 1.9; tPSA = 82		Year of discovery: 2009 Year of introduction: 2020 Discovered by: Incyte Developed by: Novartis Primary target: MET Binding type: type I Class: receptor tyrosine kinase Treatment: NSCLC with METex14 Other name: INCB28060 Oral bioavailability >70% Elimination half-life = 7.8 h Protein binding = 96%

● = C   ○ = H   ● = O   ● = N   ● = S   ● = F   ● = Cl   ● = Br   ● = I

Capmatinib was granted accelerated approval in 2020 for the treatment of adult patients with metastatic non-small cell lung cancer (NSCLC) who harbor a mutation that leads to mesenchymal epithelial transition (MET) exon 14 skipping, as detected by an FDA-approved diagnostic test.

Capmatinib is the first drug to selectively target MET kinase, thereby resulting in the inhibition of cell growth of tumor cells that exhibit increased MET activity.[1] It competes with tepotinib, the second FDA-approved selective MET inhibitor.[2]

MET Signaling Pathway[3–6]

MET, also known as hepatocyte growth factor receptor (HGFR), is a transmembrane receptor tyrosine kinase, having hepatocyte growth factor (HGF) as its ligand. Binding of HGF to MET induces receptor dimerization and autophosphorylations of Tyr1234 and Tyr1235 in the intracellular tyrosine kinase domain within the cytoplasm, triggering several downstream signaling pathways including PI3K/AKT/mTOR and MAPK pathways (Fig. 1), which regulate multiple biological functions, such as cell proliferation, survival, motility, and invasion.

METex14 Alterations in NSCLC[3–6]

The MET signaling pathway can be dysregulated by (1) overexpression of the receptor itself or of its ligand, HGF; and (2) MET structural aberrations such as rare MET gene fusions with upstream partners, or MET mutations, typically exon 14 mutations, known as or METex14 or METex14-skipping mutations. These mutations represent 3%–4% of all NSCLC cases, on par with or greater than ROS1 (~1–2%), NTRK13 (<1%), RET (~1–2%), BRAF(~1–5%), and ALK (~5–7%).

METex14 mutations affect the translation of the MET gene into the MET protein, which involves a maturation step to transform the pre-mRNA to mature mRNA. During the maturation, exon 14 mutations cause errant splicing, leading to the loss of exon 14, which is located in the juxtamembrane region. Exon 14 contains a crucial Tyr1003, which functions as the binding site for the E3 ligase. The E3 ligase is one of the three key enzymes required for ubiquitination and subsequent degradation of the MET receptor via the proteasome. As a result of a lack of the E3 ligase for degradation, exon 14 deletion gives rise to a MET receptor with enhanced stability and kinase activity (Fig. 1).

Aberrant MET excessive activation has been frequently found in many hematologic malignancies and human solid tumors including NSCLC. The overactive MET kinase results in dysregulation of the NSCLC cells, which then grow and divide without control, leading to NSCLC. Thus, the MET kinase is an oncogenic driver and a therapeutically relevant target, and its inhibition represents a precisely targeted intervention for NSCLC with METex14 skipping alterations.

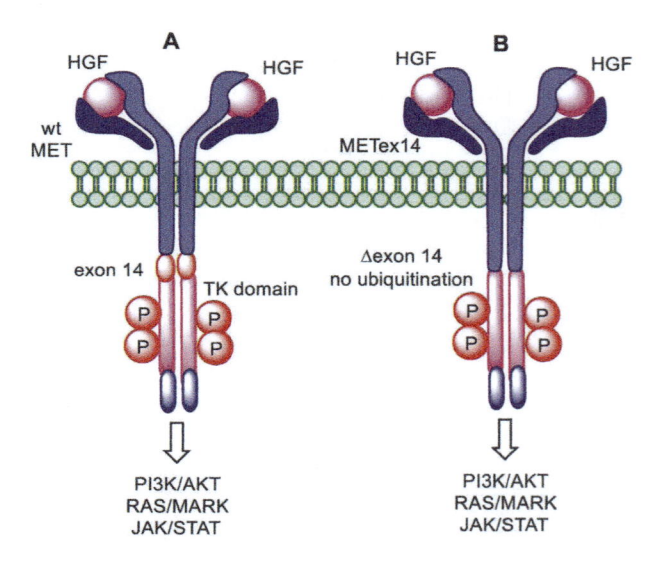

Figure 1. A. Normal MET signaling in wild-type (wt); B. Excessive MET signaling in METex14.

Capmatinib: A Selective MET Inhibitor

There are two classes of small-molecule MET kinase inhibitor molecules, both of which are ATP-competitive, as they function as analogs of ATP, the substrate for phosphorylation of MET tyrosine kinase.[6] The first class refers to selective MET inhibitors, such as capmatinib and tepotinib, which have been recently approved by the FDA. These U-shaped inhibitors specifically bind to the ATP-binding site of MET (likely via hydrogen bonding with the residue Met1160), thus displaying higher selectivity than the other class. The second class of inhibitors, which include cabozantinib and crizotinib, target multiple kinases.

Capmatinib is a highly specific MET kinase inhibitor with an IC_{50} value of 0.13 nM in a biochemical assay (vs. 4 nM for tepotinib, the other selective MET drug).[7] Capmatinib also exhibits high selectivity against other kinases (<30% inhibition against a panel of 57 structurally diverse human kinases at 2 µM). Furthermore, it showed no inhibition against RONb, another member of the c-MET RTK family, as well as EGFR and HER-3, members of the EGFR RTK family.

Although capmatinib is an ATP competitive and reversible inhibitor of MET, its binding mode remains to be determined for lack of a co-crystal X-ray structure.

1. S. Dhillon. Capmatinib: First approval. *Drugs.* **2020**, *80*, 1125–1131.
2. A. Markham. Tepotinib: First approval. *Drugs.* **2020**, *80*, 829–833.
3. M. Z. Guo, K. A. Marrone, A. Spira, D. M. Waterhouse, S. C. Scott. Targeted treatment of non-small cell lung cancer: Focus on capmatinib with companion diagnostics. *Onco.Targets Ther.* **2021**, *14*, 5321–5331.
4. H. Liang, M. Wang. MET oncogene in non-small cell lung cancer: Mechanism of MET dysregulation and agents targeting the HGF/c-Met axis. *Onco. Targets Ther.* **2020**, *13*, 2491-2510.
5. X. Huang, E. Li, H. Shen, X. Wang, T. Tang, X. Zhang, J. Xu, Z. Tang, C. Guo, X. Bai, T. Liang. Targeting the HGF/MET axis in cancer therapy: Challenges in resistance and opportunities for improvement. *Front Cell Dev. Biol.* **2020**, *8*, 152.
6. A. Puccini, N. I. Marín-Ramos, F. Bergamo, M. Schirripa, S. Lonardi, H. J. Lenz, F. Loupakis, F. Battaglin. Safety and tolerability of c-MET inhibitors in cancer. *Drug Saf.* **2019**, *42*, 211–233.
7. X. Liu, Q. Wang, G. Yang, C. Marando, H. K. Koblish, L. M. Hall, J. S. Fridman, E. Behshad, R. Wynn, Y. Li, J. Boer, S. Diamond, C. He, M. Xu, J. Zhuo, W. Yao, R. C. Newton, P. A. Scherle. A novel kinase inhibitor, INCB28060, blocks c-MET-dependent signaling, neoplastic activities, and cross-talk with EGFR and HER-3. *Clin. Cancer Res.* **2011**, *17*, 7127–7138.

16.2 Tepotinib (Tepmetko™) (79)

Structural Formula	Space-filling Model	Brief Information
$C_{29}H_{28}N_6O_2$; MW = 492 clogP = 3.4; tPSA = 94		Year of discovery: 2008 Year of introduction: 2021 Discovered by: EMD Serono Developed by: EMD Serono Primary target: MET Binding type: type I1/2 Class: receptor tyrosine kinase Treatment: NSCLC with METex14 Other name: EMD-1214063 Oral bioavailability = 72% Elimination half-life = 32 h Protein binding = 94%

● = C ○ = H ● = O ● = N ● = S ● = F ● = Cl ● = Br ● = I

Tepotinib is the second selective mesenchymal-epithelial transition (MET) inhibitor to have been granted accelerated approval by the FDA (nine months after Novartis' capmatinib) for the treatment of adult patients with metastatic non-small cell lung cancer (NSCLC) who harbor a mutation that leads to MET (mesenchymal epithelial transition) exon 14 skipping alterations, as detected by an FDA-approved diagnostic test.[1]

Tepotinib is a highly specific MET kinase inhibitor with an IC_{50} value of 3 nM in a biochemical assay (vs. 0.13 nM for capmatinib, the other MET drug).[2] It also exhibited high selectivity >1,000-fold selectivity for MET over 236 of 241 kinases tested, and >50-fold selectivity over the other five.[3,4] Such high selectivity mitigates the risk of off-target toxicities.

Both tepotinib and capmatinib have comparable oral bioavailability; however, tepotinib displays much longer elimination half-life (32 vs. 7.8 h), enabling once-daily dosing regimen (1 × 225 mg), vs. twice-daily for capmatinib (2 × 400 mg).[2]

In the co-crystal structure of tepotinib bound to MET (Fig. 1),[5] the pyrimidine nitrogen hydrogen bonds with the amide NH group of Met1160 (the third hinge residue) and the pyridazinone oxygen hydrogen bonds to the amide NH group of DFG-Asp1222 (Fig. 2). The inhibitor binds to a DFG-in and αC-out inactive conformation as a type I½ inhibitor.

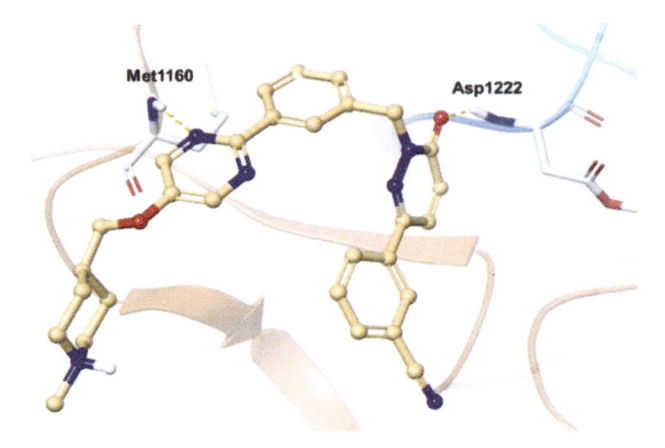

Figure 1. Co-crystal structure of tepotinib in complex with MET (PDB ID: 4R1V).

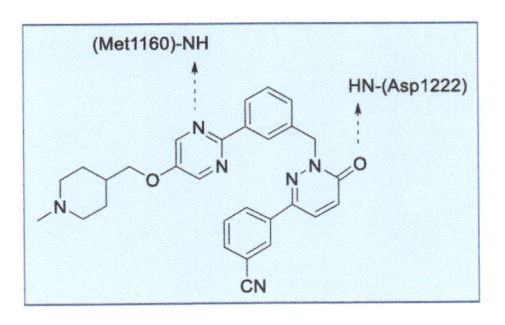

Figure 2. Key interactions of tepotinib–MET.

1. A. Markham. Tepotinib: First approval. *Drugs*. **2020**, *80*, 829–833.
2. S. Dhillon. Capmatinib: First approval. *Drugs*. **2020**, *80*, 1125–1131.
3. F. Bladt, B. Faden, M. Friese-Hamim, C. Knuehl, C. Wilm, C. Fittschen, U. Gradler, M. Meyring, D. Dorsch, F. Jaehrling, U. Pehl, F. Stieber, O. Schadt, A. Blaukat. EMD 1214063 and EMD 1204831 constitute a new class of potent and highly selective c-Met inhibitors. *Clin. Cancer Res*. **2013**, *19*, 2941–2951.
4. Friese-Hamim M, Bladt F, Locatelli G, Stammberger U, Blaukat A. The selective c-Met inhibitor tepotinib can overcome epidermal growth factor receptor inhibitor resistance mediated by aberrant c-Met activation in NSCLC models. *Am. J. Cancer. Res*. **2017**, *7*, 962–972.
5. R. Roskoski, Jr. Properties of FDA-approved small molecule protein kinase inhibitors: A 2022 update. *Pharmacol. Res*. **2022**, *175*, 106037.

Chapter 17. KIT/PDGFR/Multikinase Inhibitors

Activating mutations in platelet-derived growth factor receptor A (PDGFRA) and KIT (a tyrosine kinase encoded by the proto-oncogene c-KIT) account for 82–87% of gastrointestinal stromal tumors (GISTs), the most common type of mesenchymal tumors of the digestive tract. Overall, KIT is the most commonly mutated oncogene in GISTs, followed by PDGFR. The discovery of activating mutations in the KIT/PDGFR tyrosine kinase receptors and their role in the pathogenesis of GIST has revolutionized the treatment for this disease. Imatinib, the first multikinase inhibitor (MKI) that cross-inhibits KIT/PDGFR, was approved by the FDA in 2002 for the treatment of advanced GIST (Fig. 1). Subsequently, two additional MKIs were approved: sunitinib as a second-line therapy and regorafenib as a third-line therapy (Fig. 1). However, resistance to all these MKIs inevitably develops over time, primarily due to secondary kinase mutations, and unfortunately, sunitinib and regorafenib are ineffective against KIT/PDGFR resistance mutants.[1] To overcome these resistant mutations and mitigate off-target adverse effects associated with MKIs, more selective second-generation tyrosine kinase inhibitors (TKIs) were sought, leading to two new FDA drug approvals in 2020: ripretinib, a broad-spectrum KIT/PDGFRA inhibitor as a second-line treatment for advanced GIST; and avapritinib for GIST harboring a PDGFRA exon 18 mutation (including PDGFRA D842V mutations), as well as for advanced systemic mastocytosis (SM). Figure 2 shows several newer TKIs currently in clinical trials.[2]

This chapter describes the discovery process that led to the two FDA-approved second-generation TKIs, **80** and **81** (Fig. 1). Three MKIs, imatinib (**1**), regorafenib (**24**) and sunitinib (**25**) are discussed in Chapter 2 (BCR-ABL inhibitors) and Chapter 5 (VEGFR inhibitors), respectively.

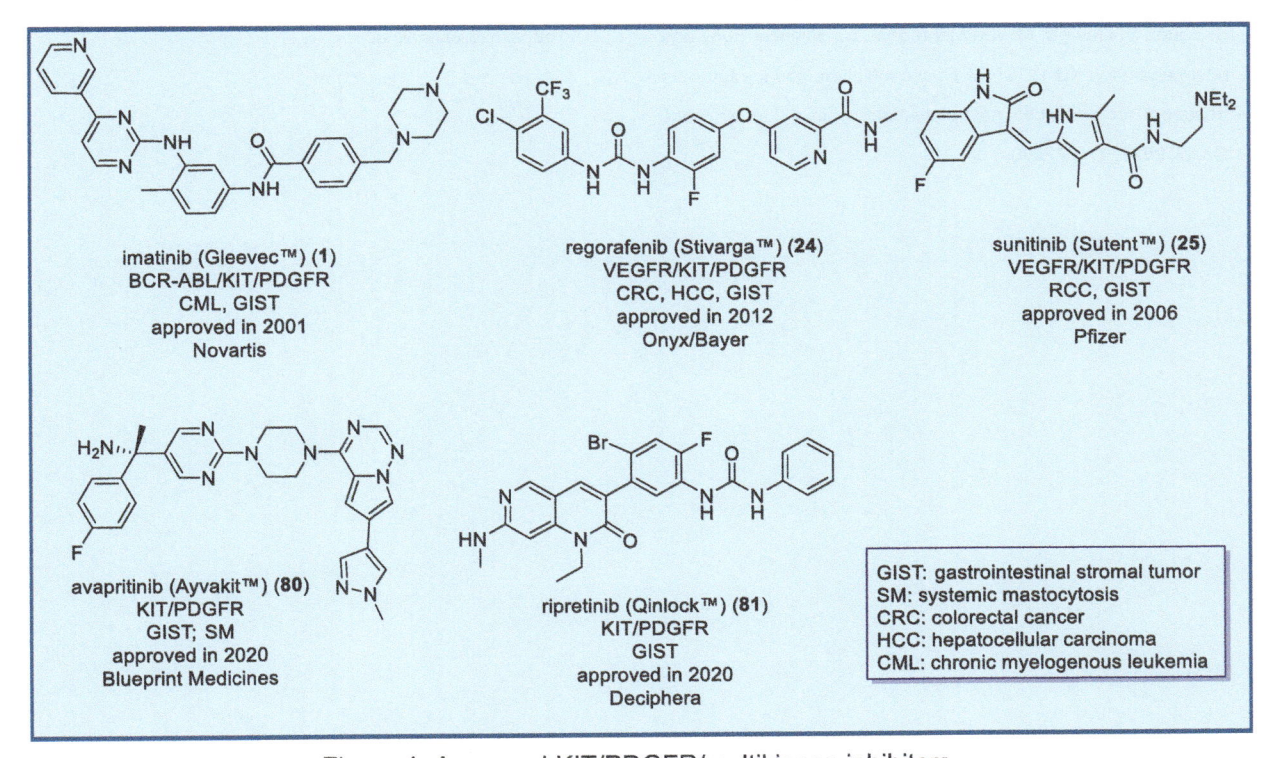

Figure 1. Approved KIT/PDGFR/multikinase inhibitors.

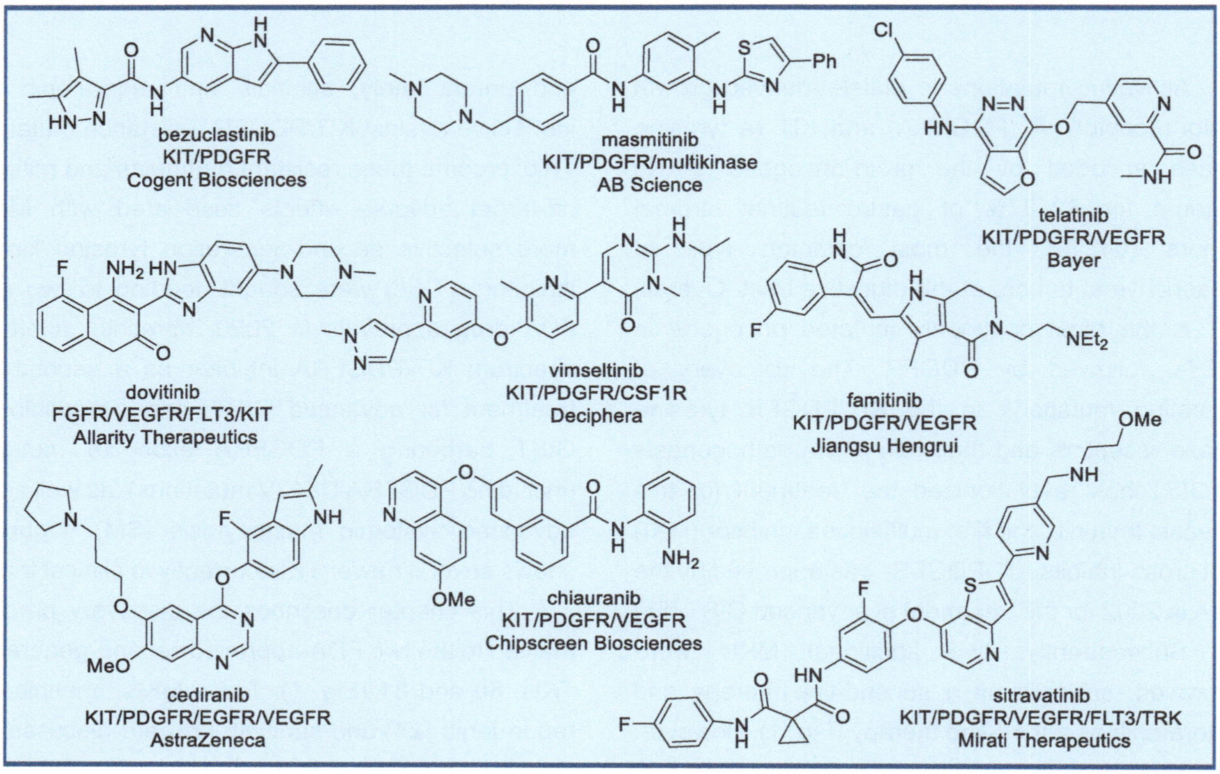

Figure 2. KIT/PDGFR/multikinase inhibitors in clinical trials.

1. S. Bauer, S. George, M. von Mehren, M. C. Heinrich. Early and next-generation KIT/PDGFRA kinase inhibitors and the future of treatment for advanced gastrointestinal stromal tumor. *Front. Oncol.* **2021**, *11*, 672500.

2. https://clinicaltrials.gov.

17.1 Avapritinib (Ayvakit™) (80)

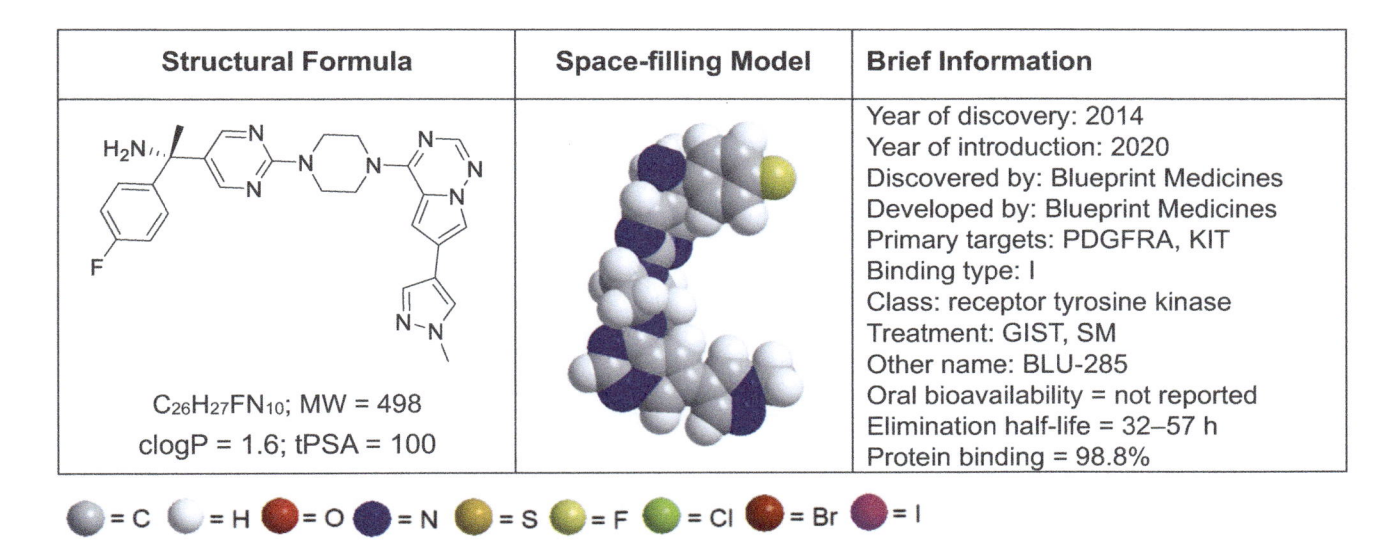

Structural Formula	Space-filling Model	Brief Information
$C_{26}H_{27}FN_{10}$; MW = 498 clogP = 1.6; tPSA = 100		Year of discovery: 2014 Year of introduction: 2020 Discovered by: Blueprint Medicines Developed by: Blueprint Medicines Primary targets: PDGFRA, KIT Binding type: I Class: receptor tyrosine kinase Treatment: GIST, SM Other name: BLU-285 Oral bioavailability = not reported Elimination half-life = 32–57 h Protein binding = 98.8%

● = C ○ = H ● = O ● = N ● = S ● = F ● = Cl ● = Br ● = I

Avapritinib is the first precision kinase inhibitor approved in 2020 for adults with unresectable or metastatic gastrointestinal stromal tumor (GIST) harboring a platelet-derived growth factor receptor A (PDGFRA) exon 18 mutation, including D842V mutations. These exon 18 mutants are resistant to the previously approved drugs such as imatinib. Avapritinib was also approved as an initial, or first-line therapy. Subsequently, it also won approval in 2021 for adult patients with advanced systemic mastocytosis (ASM), including patients with aggressive systemic mastocytosis, systemic mastocytosis with an associated hematological neoplasm, and mast cell leukemia.[1,2] For ASM indications, avapritinib is challenging Novartis' midostaurin because of its stronger efficacy and a more favorable side effect profile.[3]

GIST, although rare, is the most common sarcoma of the gastrointestinal tract, with annual occurrences from 10 and 15 cases per million. The majority of GISTs carry activating mutations in either of two growth hormone receptor tyrosine kinases, KIT (a tyrosine kinase encoded by the proto-oncogene c-KIT) (69–83%), or PDGFRA (5%–10%).[4] The identification of KIT/PDGFRA oncogenic mutations in GIST was essential to research that eventually led to dual KIT/PDGFRA kinase inhibitors to treat advanced GIST.

Systemic mastocytosis is another rare disease characterized by mast cell proliferation, and mutated KIT activation. KIT normally plays a key role in mast cell growth, differentiation and survival. Activating mutations in KIT are detected in most cases of systemic mastocytosis, with KIT D816V being the most prevalent.[5] Thus, KIT inhibition has emerged as an attractive approach for mastocytosis treatment.

Avapritinib inhibits PDGFRA and PDGFRA mutants such as D842V (IC_{50} = 0.24 nM), as well as various KIT mutants including D816V (IC_{50} = 0.27 nM). It is 10-fold more potent against PDGFRA D842V mutant than ripretinib, the other FDA-approved PDGFRA inhibitor. Avapritinib demonstrated efficacy in the treatment of GIST with PDGFRA D842V mutation and ASM, and its approval has changed the landscape of both diseases.[1]

KIT and PDGFA Receptors[4–10]

KIT and PDGFRA belong to the same subfamily of type III receptor tyrosine kinases, and as such, they are highly homogeneous cell surface proteins which are coded by similar genes found in close proximity on chromosome 4q12. These receptors consist of an extracellular domain, a juxtamembrane domain (JMD) (the portion just inside of the cell membrane), and two catalytic parts of the kinase domain. One part of the kinase domain is required for the ATP-binding and the other is involved in the phosphate group transfer to trigger kinase activation.

Both KIT and PDGFRA proteins are receptors specifically for their respective growth factors, stem cell factor (SCF), and PDGF-AA. Binding of the growth factor to the extracellular portion of its specific receptor brings two receptors of the same kind together (i.e., homodimerization) and promotes autophosphorylation, resulting in kinase activation. Subsequently, the receptor exposes its docking sites for signal-transducing effector molecules, which are involved in the activation of several downstream signaling pathways, such as the JAK/STAT3, PI3K/AKT/mTOR, and RAS/MAPK. Due to similarities between PDGFRA mutant and KIT-mutant GISTs at the molecular level, activation of either mutant KIT or mutant PDGFRA promotes analogous downstream signaling pathways. These pathways modulate cellular functions such as cell proliferation, differentiation, and apoptosis. However, mutations of KIT and PDGFRA lead to constitutive activation of the receptor in the absence of any ligand, resulting in uncontrolled tumor growth.

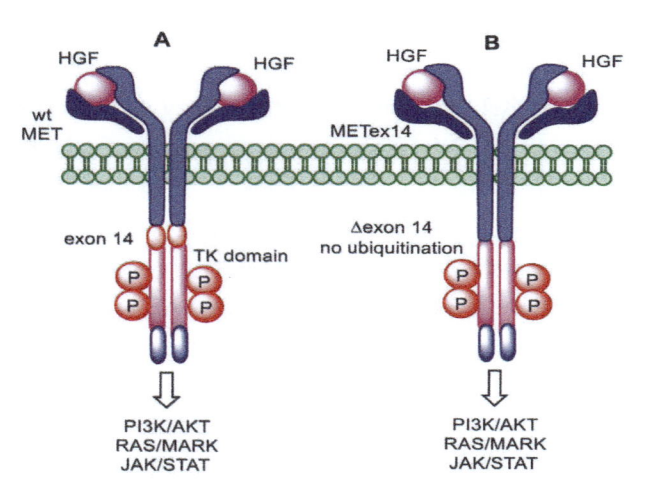

Figure 1. Mutation frequencies of KIT/PDGFRA. JMD: juxtamembrane domain; TKD: tyrosine kinase domain.[6]

Two-thirds of all KIT gene mutations in GISTs take place in the JMD (exon 11), but only 9.3% of PDGFRA mutations occur in this region (exon 12). In contrast, mutations in the activation loop of KIT (exon 17) are rare in GISTs (less than 1%), but they predominate in PDGFRA mutated tumors (89.6% exon 18). The relative frequencies of GISTs with mutations of the exons coding for various regions of the KIT and PDGFRA receptors are shown in Figure 1.[6]

KIT Mutations in GIST and SM[4–10]

The majority of KIT mutations occur in exon 11 (70%–80%), which encodes the JMD. In the normal setting, the JMD auto-inhibits receptor activation. As shown in the X-ray structure of the autoinhibited KIT kinase (Fig. 2),[11] the JMD inserts into the active site of the KIT kinase to prevent the kinase activation loop from swinging into active conformation. Thus, KIT mutations arising in exon 11 relieve the autoinhibitory constraint of the JMD, resulting in ligand-independent dimerization, autophosphorylation and constitutive activation of the receptor. The second most common KIT mutations (10%) occur in exon 9, which encodes a part of the extracellular domain of KIT that is involved in receptor dimerization. These mutations cause conformational changes similar to those when the extracellular KIT receptor is bound to its ligand, SCF, thus leading to ligand-free dimerization and constitutive activation.[9] Even though primary mutations at exon 17 (which encodes the activation loop) are rare (~1%), a single KITD816V mutation at exon 17 is prevalent in ~90% of systemic mastocytosis cases. This mutation occurs near the activation loop of the KIT kinase and also leads to constitutive ligand-independent activation of the KIT receptor.

PDGFR mutations in GIST[4–10]

Mutations in PDGFRA are detected only in 5–10% of newly diagnosed GISTs, much less frequently than KIT mutations (69–83%). The PDGFRA mutations predominantly occur in exons 18 (~5%), which encode the activation loop. The most common exon 18 mutation is a single mutation, D842V, which is found in 63% of PDGFRA-mutated tumors. It is widely believed that the kinase domain in the mutated form never adopts the inactive conformation and thus no dimerization is needed for activation.[10] Mutations in exon 12 (encoding JMD) and exon 14 (encoding the ATP-binding domain) are rare, identified in only about 1–2% and <0.1% of GISTs, respectively.

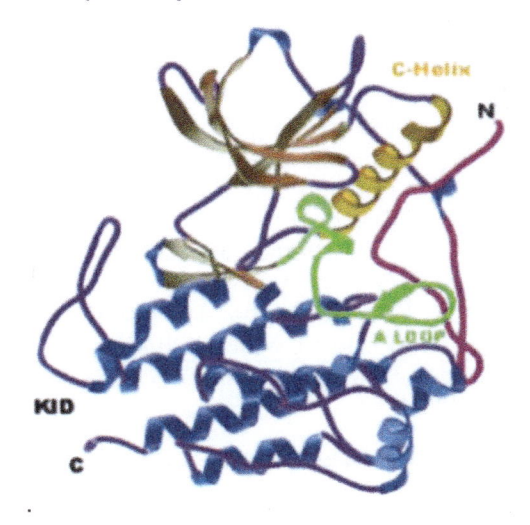

Figure 2. Structure of autoinhibited KIT kinase. KID: kinase insertion domain; α-C helices (blue), β-strands (amber), and loops (purple). JMD: juxtamembrane domain (purple). Adapted with permission from *J. Biol. Chem.* **2004**, *279*, 31655–31663.

Binding Mode

Avapritinib binds to active conformation of KIT and PDGFRA as a type I kinase inhibitor. However, its binding mode has not yet been determined due to lack of X-ray co-crystal structures with either receptor tyrosine kinase.

1. S. Dhillon. Avapritinib: First approval. *Drugs*. **2020**, *80*, 433–439.
2. S. Bauer, S. George, M. von Mehren, M. C. Heinrich. Early and Next-Generation KIT/PDGFRA Kinase Inhibitors and the Future of Treatment for Advanced Gastrointestinal Stromal Tumor. *Front. Oncol.* **2021**, *11*, 672500.
3. Sean Rai-Roche. Blueprint's Ayvakit expected to take Novartis' Rydapt crown in advanced systemic mastocytosis but off-label use in indolent unlikely. *Clinical Trials Arena*. May 18, 2021.
4. Mutation Analysis: KIT and PDGFRA. https://www.gistsupport.org.
5. M. Piris-Villaespesa, I. Alvarez-Twose. Systemic mastocytosis: Following the tyrosine kinase inhibition roadmap. *Front. Pharmacol.* **2020**, *11*, 443.
6. W. C. Foo, B. Liegl-Atzwanger, A. J. Lazar. Pathology of gastrointestinal stromal tumors. *Clin. Med. Insights Pathol.* **2012**, *5*, 23–33.
7. A. Rizzo, M. A. Pantaleo, A. Astolfi, V. Indio, M. Nannini. The identity of PDGFRA D842V-mutant gastrointestinal stromal tumors (GIST). *Cancers*. **2021**, *13*, 705.
8. A. K. Gardino, E. K. Evans, J. L. Kim, N. Brooijmans, B. L. Hodous, B. Wolf, C. Lengauer. Targeting kinases with precision. *Mol. Cell Oncol.* **2018**, *5(3)*, e1435183.
9. L. Mei, S. C. Smith, A. C. Faber, J. Trent, S. R. Grossman, C. A. Stratakis, S. A. Boikos. Gastrointestinal stromal tumors: The GIST of precision medicine. *Trends Cancer*. **2018**, *4*, 74–91.
10. F. Toffalini, J. B. Demoulin. New insights into the mechanisms of hematopoietic cell transformation by activated receptor tyrosine kinases. *Blood*. **2010**, *116*, 2429–2437.
11. C. D. Mol, D. R. Dougan, T. R. Schneider, R. J. Skene, M. L. Kraus, D. N. Scheibe, G. P. Snell, H. Zou, B. C. Sang, K. P. Wilson. Structural basis for the autoinhibition and STI-571 inhibition of c-Kit tyrosine kinase. *J. Biol. Chem.* **2004**, *279*, 31655–31663.

17.2 Ripretinib (Qinlock™) (81)

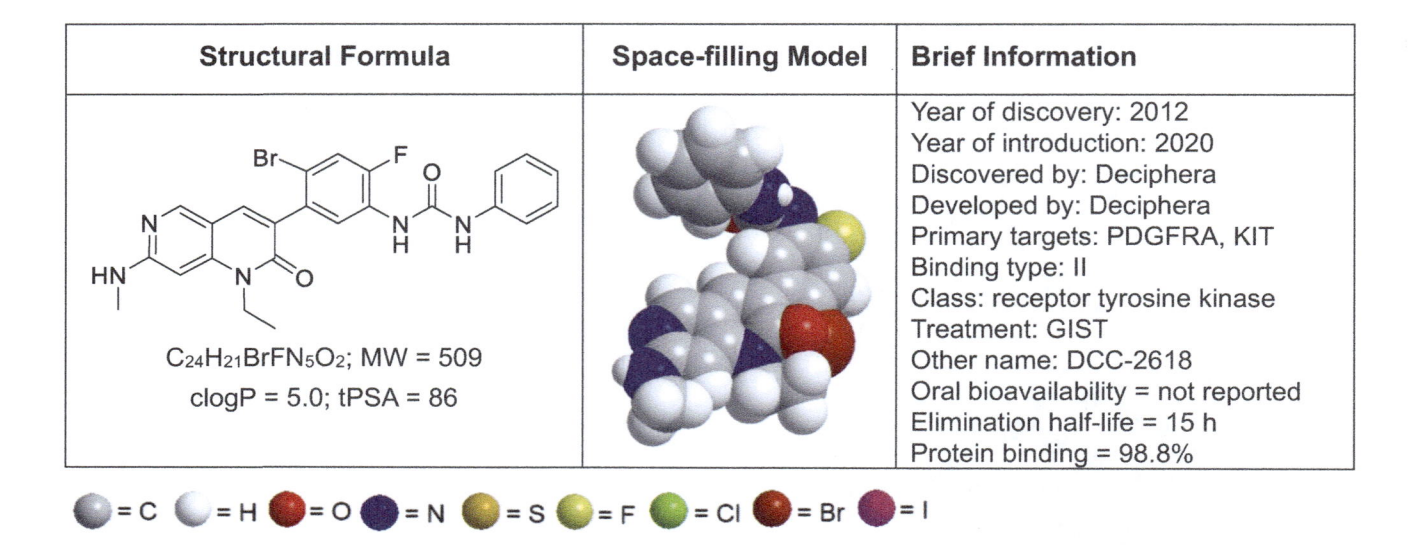

Structural Formula	Space-filling Model	Brief Information
$C_{24}H_{21}BrFN_5O_2$; MW = 509 clogP = 5.0; tPSA = 86		Year of discovery: 2012 Year of introduction: 2020 Discovered by: Deciphera Developed by: Deciphera Primary targets: PDGFRA, KIT Binding type: II Class: receptor tyrosine kinase Treatment: GIST Other name: DCC-2618 Oral bioavailability = not reported Elimination half-life = 15 h Protein binding = 98.8%

● = C ○ = H ● = O ● = N ● = S ● = F ● = Cl ● = Br ● = I

Ripretinib was approved in 2020 as a back-up treatment for advanced gastrointestinal stromal tumor (GIST) which is independent of the mutational status of KIT or PDFGRA. It is intended for adult patients who have received prior treatment with three or more kinase inhibitor therapies, including imatinib.[1]

Although there were four FDA-approved targeted GIST therapies – imatinib in 2002, sunitinib in 2006, regorafenib in 2013, and avapritinib in 2020, some patients failed to respond to any of these. Ripretinib provides a new treatment option for those individuals with GIST.

Ripretinib is a potent inhibitor of both wild-type KIT and wild-type PDGFRA with IC_{50} values of ~3 nM. It also inhibits a broad spectrum of KIT and PDGFRA mutants found in GIST. Furthermore, it targets multiple tumor-driving kinases, including PDGFRβ, TIE2, VEGFR2 and BRAF.[2]

The X-ray co-crystal structure of ripretinib bound to KIT is still lacking, but a chlorine analogue,[3] DP2976 bound to the enzyme is known. This inhibitor is anchored to the hinge through two critical hydrogen bonds: the methylamino with amide oxygen of Cys673; and the N6 of naphthyridinone with amide NH of Cys673, the third hinge residue (Fig. 1). The naphthyridinone amide oxygen hydrogen bonds to the side chain amine of Lys623. Moreover, the urea oxygen atom forms a hydrogen bond with the amide NH of DFG-Asp810, and one urea NH group hydrogen bonds with the carboxyl group of Glu640 (Fig. 2). The inhibitor binds to KIT inactive conformation with DFG-out, penetrating into the back pocket (type II inhibitor). Because the only differences in the structures of DP2976 and ripretinib is the substitution of chlorine for bromine, it is highly likely that both bind in the same way.

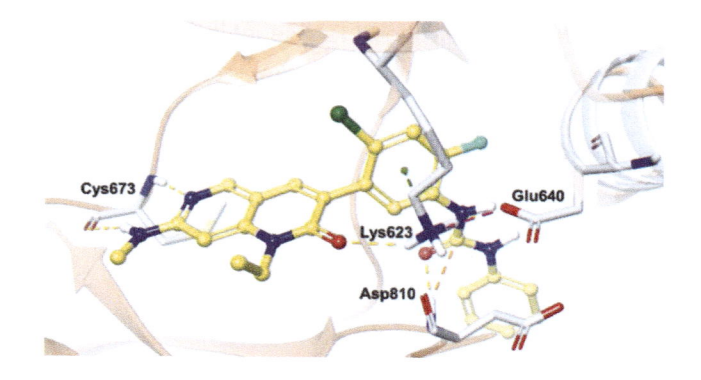

Figure 1. Co-crystal structure of DP-2976 (a chloro analog of ripretinib) in complex with KIT (PDB ID: 6M0B).

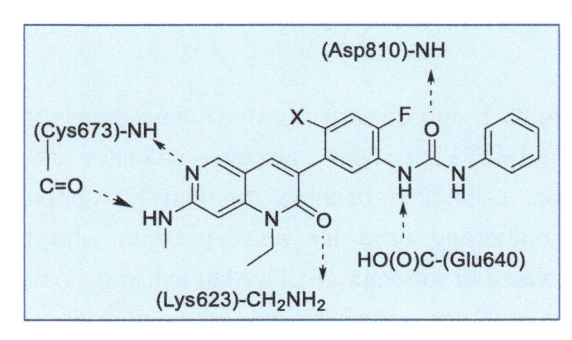

Figure 2. Summary of KIT–DP-2976 (X = Cl) interactions based on the co-crystal structure. Ripretinib: X = Br.

1. S. Dhillon. Ripretinib: First approval. *Drugs*. **2020**, *80*, 1133–1138.
2. B. D. Smith, M. D. Kaufman, W. P. Lu, A. Gupta, C. B. Leary, S. C. Wise, T. J. Rutkoski, Y. M. Ahn, G. Al-Ani, S. L. Bulfer, T. M. Caldwell, L. Chun, C. L. Ensinger, M. M. Hood, A. McKinley, W. C. Patt, R. Ruiz-Soto, Y. Su, H. Telikepalli, A. Town, D. L. Flynn. Ripretinib (DCC-2618) Is a switch control kinase inhibitor of a broad spectrum of oncogenic and drug-resistant KIT and PDGFRA variants. *Cancer Cell*. **2019**, *35*, 738–751.e9.
3. R. Roskoski, Jr. Properties of FDA-approved small molecule protein kinase inhibitors: A 2021 update. *Pharmacol. Res.* **2021**, *165*, 105463.

Chapter 18. FLT3 Inhibitors

Fms-like tyrosine kinase 3 (FLT3) mutations are found in 30% of patients with acute myeloid leukemia (AML), ~23% of whom harbor FLT3 internal tandem duplication (FLT3-ITD), while ~7% carry FLT3 tyrosine kinase domain (FLT3-TKD) point mutations or deletions. Patients with FLT3-ITD are subject to higher risk of relapse and lower cure rates. The first-generation multikinase inhibitors (that cross-inhibit FLT3) such as sorafenib, sunitinib and midostaurin as monotherapy provided only marginal clinical benefits and suffered from substantial off-target adverse effects and poor tolerability. However, midostaurin in combination with chemotherapy was shown to be efficacious and safe for the treatment of newly diagnosed FLT3+ AML, justifying its approval by the FDA in 2017.[1]

The second-generation FLT3 inhibitors, gilteritinib and quizartinib, which were developed to target FLT3 specifically, are more selective and less toxic. Gilteritinib became the first FDA-approved monotherapy drug for adult patients who have relapsed or refractory FLT+ AML (while midostaurin can only be administered in combination with chemotherapy). Subsequently, quizartinib was approved in Japan for the same indication (Fig. 1).

Because FLT3-mutated AML with FLT3-targeted TKIs remains challenging due to the development of resistance, there is an urgent need for more potent and selective FLT3 inhibitors, several of which are being evaluated in clinical trials (Fig. 2).

This chapter describes the discovery process that led to the two FDA-approved FLT3 inhibitors, **82** and **83** (Fig. 1).

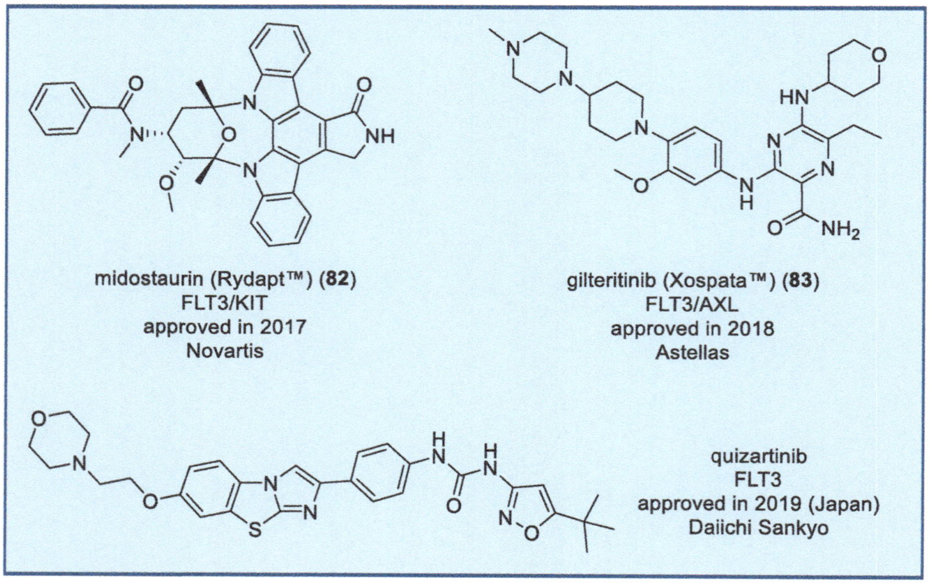

Figure 1. Approved FLT3 inhibitors for FLT3+ AML.

Molecules Engineered Against Oncogenic Proteins and Cancer, First Edition. E. J. Corey and Yong-Jin Wu.

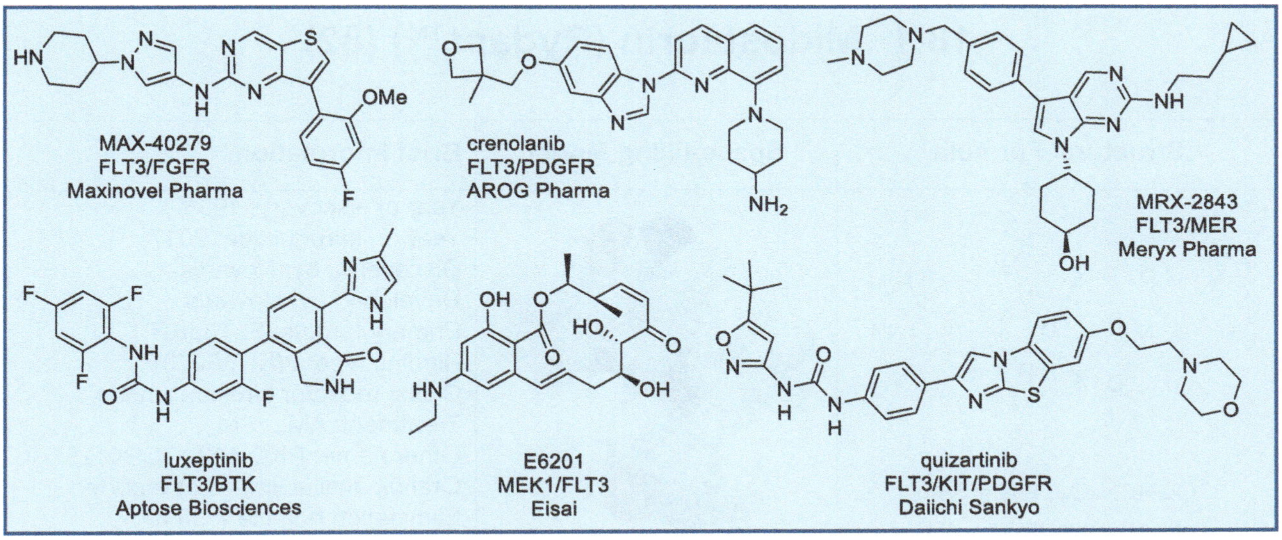

Figure 2. FLT3/multikinase inhibitors in clinical trials.

1. J. C. Zhao, S. Agarwal, H. Ahmad, K. Amin, J. P. Bewersdorf, A. M. Zeidan. A review of FLT3 inhibitors in acute myeloid leukemia. *Blood Rev.* **2022**, *52*, 100905.

18.1 Midostaurin (Rydapt™) (82)

Structural Formula	Space-filling Model	Brief Information
$C_{36}H_{32}N_4O_4$; FW = 584 clogP = 6.0; tPSA = 74		Year of discovery: 1998 Year on Introduction: 2017 Discovered by: Novartis Developed by: Novartis Primary targets: FLT3, KIT Binding type: I (KIT, FLT3) Class: receptor tyrosine kinase Treatment: AML, SM Other name: PKC-412, CGP41251 Oral bioavailability = not reported Elimination half-life = 20 h Protein binding > 99%

● = C ○ = H ● = O ● = N ● = S ● = F ● = Cl ● = Br ● = I

Midostaurin is the first-in-class targeted therapy approved in 2017 for the treatment of adults with newly diagnosed acute myeloid leukemia (AML) who harbor a specific genetic mutation in the tyrosine kinase FLT3, as detected by an FDA-approved test, in combination with chemotherapy. It was also approved to treat aggressive systemic mastocytosis (ASM), systemic mastocytosis with associated haematological neoplasm (SM-AHN) or mast cell leukaemia (MCL), which are collectively referred to as advanced SM.[1] However, the SM indication has been seriously challenged by Blueprint Medicines' recently approved avapritinib, a PDGFRA tyrosine kinase inhibitor, which has shown stronger efficacy and a more favorable side effect profile.[2]

Midostaurin is a natural product-derived pan-tyrosine kinase inhibitor including fms-like tyrosine kinase 3 (FLT3) mutants, which have been found in 35% of newly diagnosed patients with CML. It is also active against KIT (a proto-oncogene receptor tyrosine kinase) mutants, which are detected in most cases of SM.

Midostaurin represents an early breakthrough for the treatment of FLT3-mutant AML, once touted as the "boulevard of broken dreams,"[3] and it has brought new hope to the field of AML, which had remained a graveyard for drug development for 40 years.[4] The ultimately successful repositioning of a twice failed drug for AML provides an instructive example of the importance of patience, persistence, and close collaborations between pharmaceutical industry and academia in the discovery of precision medicines and breakthrough therapies.[5,6]

Discovery and Development of Midostaurin[5–9]

The long journey of midostaurin to bedside started with staurosporine, a basic natural product first isolated from the bacterium *Streptomyces staurosporeus* in 1977 at the Kitasato Institute in Japan. This novel polycyclic compound did not attract significant attention until 1986 when it was shown to exhibit potent inhibition of protein kinase C (PKC) (IC_{50} = 2.7 nM) and a strong cytotoxic effect on cancer cells. As a broad-spectrum protein kinase inhibitor, staurosporine itself is too toxic to be used as a drug. A few years later, sizable amounts of staurosporine were obtained by Ciba-Geigy via fermentation, thus allowing medicinal chemists to seek more selective inhibitors of PKC, which was considered to be an attractive therapeutic target in oncology as well as for several other indications. Comprehensive SAR studies on staurosporine to improve PKC selectivity were carried out, leading to the finding that acylation of the methylamino group increased kinase selectivity against other serine/threonine and tyrosine kinases. This effort led to the finding in 1986 that the N-benzoyl derivative, now called midostaurin, is a relatively selective PKC inhibitor (Fig. 1). In a purified enzyme assay, it exhibited IC_{50} values ranging from 0.02 to 0.03 µM against a panel of PKC subtypes, including cPKC-α, cPKC-β1, cPKC-β2, and cPKC-γ, and it was less active against other PKC subtypes, including nPKC-δ, n-PKC-η, and nPKC-ε (IC_{50} values ranging from 0.16 to 1.25 µM). Subsequent in vitro studies identified other inhibitory activities toward multiple targets, including platelet-derived growth factor receptors (PDGFRs), cyclin-dependent kinase 1 (CDK1), vascular endothelial growth factor (VEGF), and spleen tyrosine kinase (SYK). Because of its broad kinase inhibition profile, and its efficacy to shrink tumors in murine tumor xenograft models, midostaurin was initially predicted as a versatile anticancer agent. However, at that time, very few pharmaceutical companies were genuinely interested in developing kinase inhibitors. After all, there are 518 protein kinases in the human body, and yet so many different kinases are expressed at various levels in individual cells. Thus, the general perception had been: (1) it would be too challenging to move any kinase inhibitor from bench to clinic, if not impossible, due to the difficulty in achieving desirable level of selectivity for a particular kinase; and (2) multikinase inhibitors would elicit too many off-target side effects to become a viable drug. Despite the overwhelming skepticism towards developability of kinase inhibitors, the natural product rapamycin (also known as sirolimus), a mammalian target of rapamycin (mTOR) kinase inhibitor, was approved in 1999 for the prophylaxis of organ rejection in kidney transplant patients. More importantly, imatinib became the first tyrosine kinase inhibitor approved in 2001 for the treatment of cancer. It was touted as a "magical bullet" to target CML because it changed a once deadly blood cancer into a manageable chronic disease and enabled the patients to live a nearly normal lifespan. Imatinib works by inhibiting a BCR-ABL tyrosine kinase, the constitutive abnormal tyrosine kinase created by the Philadelphia chromosome abnormality in CML. The Philadelphia chromosome (BCR-ABL fusion) is prevalent in the blood cells of 90 percent of CML patients.

Although imatinib was the first tyrosine kinase inhibitor to win the approval from the FDA for cancer treatment, midostaurin was four years ahead of imatinib (in 1994) to enter into first-in-human clinical trials for the treatment of solid tumors. Unfortunately, its potent inhibition of solid tumor growth in murine xenograft models did not transfer into clinical outcomes, either as monotherapy or in combination with chemotherapy. Despite the disappointing clinical outcome, two pieces of data proved to be valuable for future clinical trials. First, the human pharmacokinetics showed that adequate drug exposures could be achieved at tolerated doses; Second, midostaurin underwent CYP3A4-mediated metabolism in the liver to give two major metabolites: CGP52421 due to mono-oxidation (as a mixture of two epimers) and CGP62221 due to de-methylation. Both metabolites were pan-kinase inhibitors, and both showed much longer plasma half-lives than midostaurin (20.3 h; CGP52421: 495 h, CGP62221: 33.4 h).

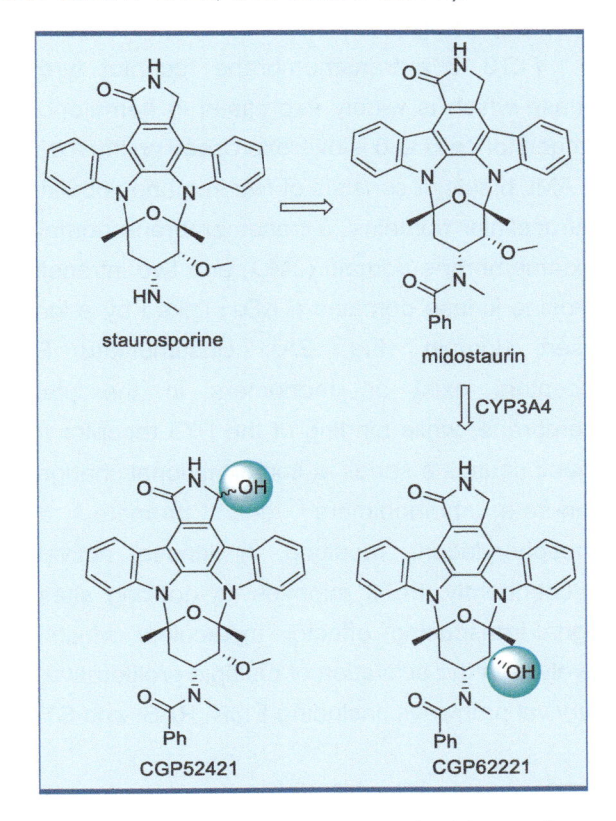

Figure 1. Structural optimization of midostaurin and its major human metabolites.

Apart from solid tumors, midostaurin was also evaluated in the clinic from 1999 to 2002 for the treatment of diabetic retinopathy, but further development in diabetes was halted due to gastrointestinal toxicity and lack of robust efficacy.

While at the final stages of the clinical trials for diabetes, midostaurin was identified as a potent FLT3 tyrosine kinase inhibitor in 2001 in a collaborative effort between the Dana-Farber Cancer Institute and Novartis to discover inhibitors of mutant FLT3-positive AML in cellular assays. It was this finding that led to the repurposing of a twice failed drug for FLT3 mutant AML. Finally, after another decade-long clinical development through continuous collaborations between Novartis and academia, midostaurin was evaluated in a pivotal trial in combination with standard cytotoxic chemotherapy, for adults with FLT3-ITD and FLT3-TKD mutant AML. The significant improvement on the 4-year overall survival compared with placebo led to the approval of midostaurin by the FDA in 2017 for newly diagnosed FLT3-mutated AML.

FLT3 Signaling[10–13]

FLT3 is a transmembrane receptor tyrosine kinase which is widely expressed in hematopoietic progenitor cells and is overexpressed on the majority of AML blasts. It consists of five immunoglobulin-like extracellular domains, a transmembrane domain, a juxtamembrane domain (JMD) and two intracellular tyrosine kinase domains (TKDs) linked by a kinase insert domain (Fig. 2A). Unstimulated FLT3 receptors exist as monomers in the plasma membrane, while binding of the FT3 receptor to its ligand causes a series of conformational changes to generate homodimers and promote auto-phosphorylation, resulting in kinase activation. Subsequently, FLT3 exposes its docking sites for signal-transducing effector molecules, which are involved in the activation of multiple proliferative and survival pathways, including PI3K, RAS, and STAT.

FLT3 mutations in CML[10–13]

One of the most frequent genetic alterations in AML is FLT3 length mutation (FLT3-LM or FLT3-ITD for "internal tandem duplication"). FLT3-ITD occurs in a frequency of 20–27% in AML in adults, and of 10–16% in children. This mutation takes place via internal tandem duplications (ITDs) and/or insertion–deletion mutations in exons 11 and 12 of the human FLT3 gene on chromosome 13q12, which codes for the JMD of the FLT3 protein. The mutated FLT3 protein contains an elongated JMD with ITDs of 6–30 amino acids. In wild-type (wt) FLT3, the FLT3 JMD auto-inhibits receptor activation. To further understand the nature of FLT3 autoinhibition, a crystal structure of the autoinhibited form of a wt FLT3 with a complete JMD has been elucidated (Fig. 3). The structure shows that JMD is bound to the active site of FLT3, stabilizing the autoinhibited form.[14] However, the insertion of ITDs in the mutated FLT3 disrupts this inhibitory effect, resulting in abnormal, ligand-independent dimerization, auto-phosphorylation, and constitutive activation of the receptor. In other words, the ITD mutation eliminates the intrinsic negative regulatory activity of the JMD that prevents ligand-independent dimerization (Fig. 2B). Interestingly, FLT3–ITD receptors can undertake not only homodimerization but also heterodimerization with the wt FLT3 receptors.

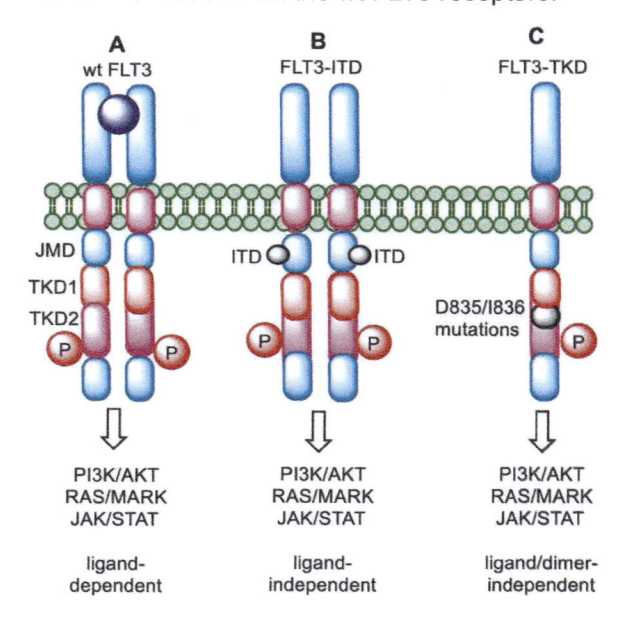

Figure 2. A: ligand-dependent FLT3 activation in wt; B: ligand-independent FLT3-ITD activation; C: ligand/dimer-independent FLT3-TKD activation. JMD: juxtamembrane domain; ITD: internal tandem duplication; TKD: tyrosine kinase domain.

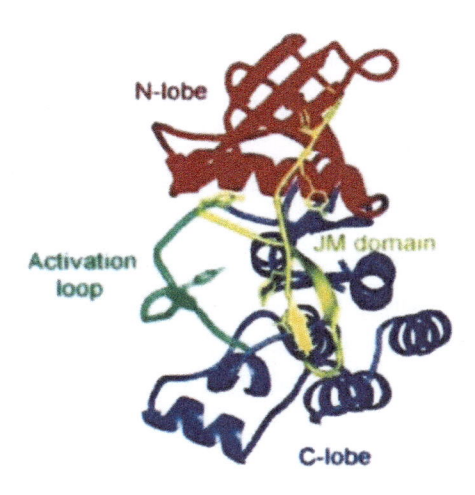

Figure 3. Structure of autoinhibited wt FLT3 in ribbon diagram. The N-terminal kinase domain in red; the C-terminal kinase domain in blue. The activation loop (green) is folded up between the two kinase domains. The JM domain (yellow) nearly spans the length of molecule. Adapted with permission from *Mol. Cell*, **2004**, *13*, 169–178.

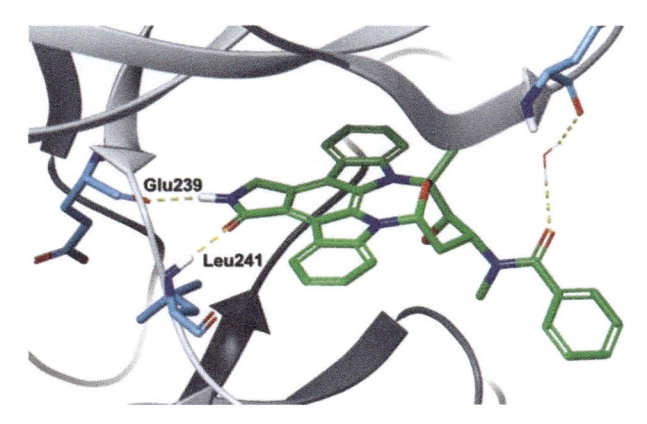

Figure 4. Co-crystal structure of midostaurin in complex with human DYRK1A (PDB ID: 4NCT).

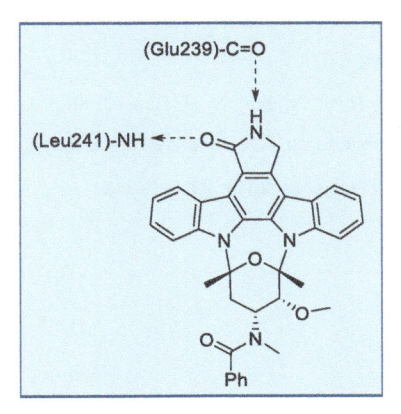

Figure 5. Summary of midostaurin–human DYRK1A interactions based on their co-crystal structure.

The second common type of FLT3 alteration is FLT3-TKD mutation, a point mutation either in codon Asp835 or as a deletion of codon Ile836 in the intracellular tyrosine kinase activation loop (Fig. 2C). Substitution or deletion causes structural changes in the loop and alters FLT3 substrate specificity. Consequently, the mutated FLT3-TKD undergoes both ligand- and dimer-independent autophosphorylation, thus constitutively activating several intracellular pathways to result in proliferation and survival of AML. Therefore, FLT3 inhibition represents an attractive approach to target CML with specific FLT3 mutations.

Binding Mode[15]

The only available X-ray co-crystal structure of midostaurin is in complex with DYRK1A (dual specificity tyrosine phosphorylation regulated kinase 1A) (Fig. 4). Midostaurin is anchored to the hinge with two hydrogen bonds: the lactam NH with amide carbonyl of Glu239; and the lactam carbonyl with amide NH of Leu241 (Fig. 5).

1. K. S. Kim. Midostaurin: First global approval. *Drugs.* **2017**, *77*, 1251–1259.

2. Sean Rai-Roche. Blueprint's Ayvakit expected to take Novartis' Rydapt crown in advanced systemic mastocytosis but off-label use in indolent unlikely. *Clinical Trials Arena.* May 18, 2021.

3. M. A. Sekeres D. P. Steensma. Boulevard of broken dreams: Drug approval for older adults with acute myeloid leukemia. *J Clin. Oncol.* **2012**, 30, 4061–4063.

4. A. H. Wei, I. S. Tiong. Midostaurin, enasidenib, CPX-351, gemtuzumab ozogamicin, and venetoclax bring new hope to AML. *Blood*, **2017**, *130*, 2469–2474.

5. R. W. Stone, P. W. Manley, R. A. Larson, R. Capdeville. Midostaurin: Its odyssey from discovery to approval for treating acute myeloid leukemia and advanced systemic mastocytosis. *Blood Adv.* **2018**, 2, 444–453.

6. P. W. Manley, E. Weisberg, M. Sattler, J. D. Griffin. Midostaurin, a natural product-derived kinase inhibitor recently approved for the treatment of hematological malignancies. *Biochem.* **2018**, *57*, 477–478.

7. E. Weisberg, M. Sattler, P. W. Manley, J. D. Griffin. Spotlight on midostaurin in the treatment of FLT3-mutated acute myeloid leukemia and systemic mastocytosis: Design, development, and potential place in therapy. *Onco Targets Ther.* **2017**, *11*, 175–182..

8. G. Caravatti, T. Meyer, A. Fredenhagen, U. Trinks, H. Mett, D. Fabbro, Inhibitory activity and selectivity of staurosporine derivatives towards protein kinase C. *Bioorg. Med. Chem. Lett.* **1994**, *4*, 399–404,

9. P. Valent , C. Akin, K. Hartmann, P. Valent, C. Akin, K. Hartmann T. I. George, K. Sotlar, B. Peter, K. V. Gleixner, K. Blatt, W. R. Sperr, P. W. Manley, O. Hermine, H. C. Kluin-Nelemans, M. Arock, H. P. Horny, A. Reiter, J. Gotlib. Midostaurin: A magic bullet that blocks mast cell expansion and activation. *Ann. Oncol.* **2017**, *28*, 2367–2376.

10. N. Daver, R. F. Schlenk, N. H. Russell, M. J. Levis. Targeting FLT3 mutations in AML: Review of current knowledge and evidence. *Leukemia.* **2019**, *33*, 299–312.

11. U. Bacher, C. Haferlach, W. Kern, T. Haferlach, S. Schnittger. Prognostic relevance of FLT3-TKD mutations in AML: The combination matters--an analysis of 3082 patients. *Blood.* **2008**, *111*, 2527–2537.

12. V. E. Kennedy, C. C. Smith. *FLT3* mutations in acute myeloid leukemia: Key concepts and emerging controversies. *Front. Oncolol.* **2020**, *10*, 612880.

13. M. M. Patnaik. The importance of FLT3 mutational analysis in acute myeloid leukemia. *Leuk. Lymphoma.* **2018**, *59*, 2273–2286.

14. J. Griffith, J.Black, C. Faerman, L. Swenson, M. Wynn, F. Lu, J. Lippke, K. Saxena. The structural basis for autoinhibition of FLT3 by the juxtamembrane domain. *Mol. Cell.* **2004**, *13*, 169–178.

15. M. Alexeeva, E. Åberg, R. A. Engh, U. Rothweiler. The Structure of a dual-specificity tyrosine phosphorylation-regulated kinase 1A-PKC412 complex reveals disulfide-bridge formation with the anomalous catalytic loop HRD(HCD) cysteine. *Acta. Crystallogr., Sect. D: Biol. Crystallogr.* **2015**, *71*, 1207−1215.

18.2 Gilteritinib (Xospata™) (83)

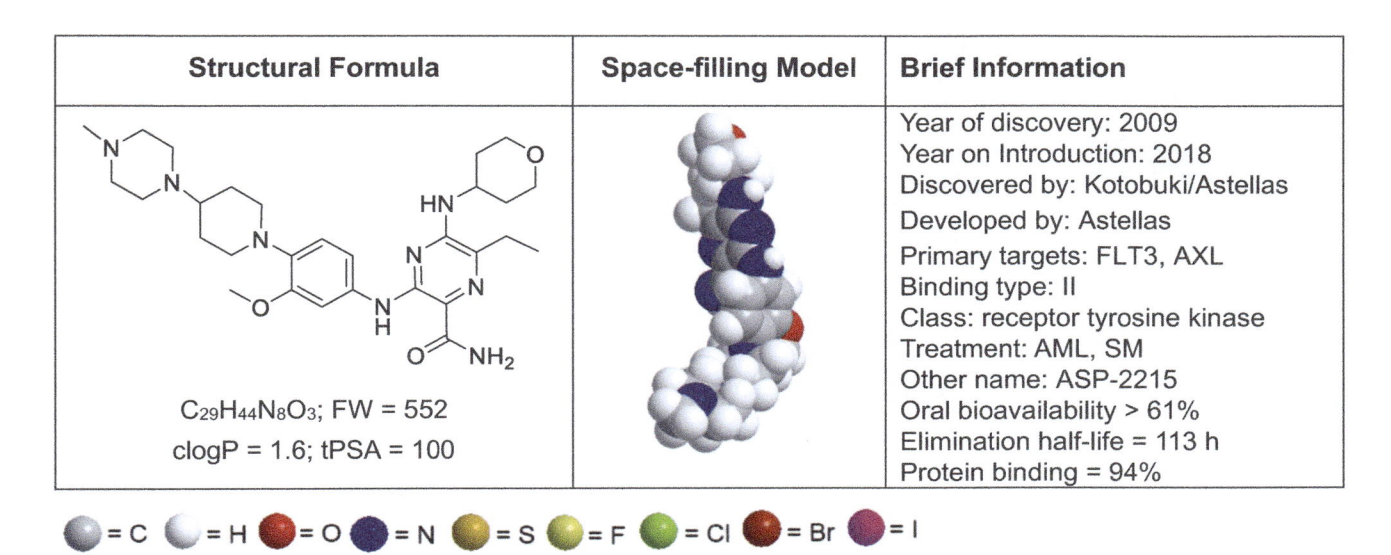

Structural Formula	Space-filling Model	Brief Information
$C_{29}H_{44}N_8O_3$; FW = 552 clogP = 1.6; tPSA = 100		Year of discovery: 2009 Year on Introduction: 2018 Discovered by: Kotobuki/Astellas Developed by: Astellas Primary targets: FLT3, AXL Binding type: II Class: receptor tyrosine kinase Treatment: AML, SM Other name: ASP-2215 Oral bioavailability > 61% Elimination half-life = 113 h Protein binding = 94%

● = C ○ = H ● = O ● = N ● = S ● = F ● = Cl ● = Br ● = I

Gilteritinib is the second-approved therapy (2018, one year after midostaurin) for treatment of adult patients who have relapsed or refractory acute myeloid leukemia (AML) with a FLT3 mutation as detected by an FDA-approved test.[1] It is the first drug that can be used as a single agent without chemotherapy for this patient population (midostaurin is used in combination with chemotherapy).[2,3]

Gilteritinib is a leading second-generation FLT3 tyrosine kinase inhibitor (TKI) and is more selective for FLT3 with potent activity against both FLT3-ITD and tyrosine kinase domain (TKD) mutations. These mutations are the most common genetic alterations in AML, identified in approximately one-third of newly diagnosed patients. Gilteritinib is also highly active against AXL (derived from the Greek word "anexelekto", which means uncontrolled), another receptor tyrosine kinase (RTK) which is activated via homodimerization upon binding to its ligand growth arrest–specific 6 (GAS6). The GAS6/AXL pathway regulates various aspects of tumor progression and metastasis. In addition, AXL is overexpressed in a broad range of cancers including AML and is associated with poor overall survival. Taken together, dual inhibition of FLT3 and AXL may provide a more effective treatment option for AML. Indeed, gilteritinib has demonstrated significant activity as monotherapy in relapsed/refractory (R/R) AML with prolongation of overall survival compared with other standards of care.[4-6] In contrast, midostaurin must be combined with standard chemotherapy to provide significant improvement on the overall survival.[2]

The crystal structure of the complex between FLT3 and gilteritinib was obtained using the kinase domain of a wild-type (wt) FLT3, including the juxtamembrane domain (JMD).[7] As shown in Figure 1, JMD is bound to the active site of FLT3, indicating that the wt FLT3 adopts the autoinhibited form. Gilteritinib binds to the ATP pocket of FLT3 with Asp829 and Phe830 embedded in the DFG-out conformation. The inhibitor is anchored to the hinge with two crucial hydrogen bonds between the carbaoxamide group and the main chain atoms of Glu692 and Cys694 (Figs. 2, 3). No distinct interaction occurs between gilteritinib and the activation loop including Asp829 and Phe830.

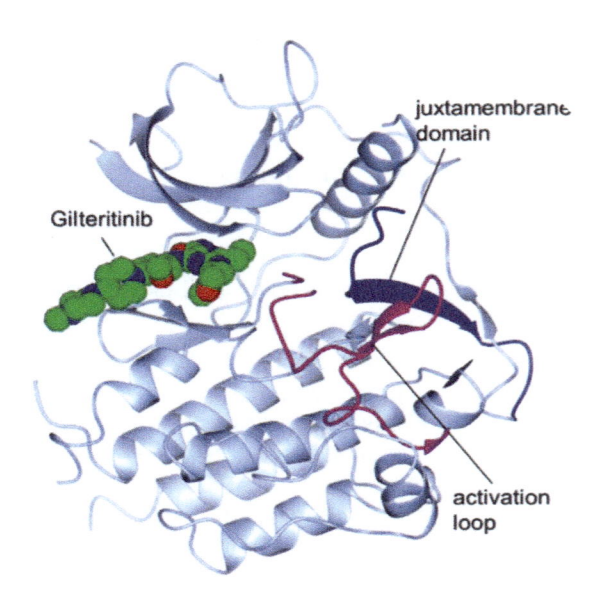

Figure 1. Overall co-crystal structure of gilteritinib–FLT3 (wt). FLT3 is shown in light blue, the activation loop is shown in magenta, and the juxtamembrane domain is shown in blue. Adapted from *Oncotarget.* **2019**, *10*, 6111−6123.

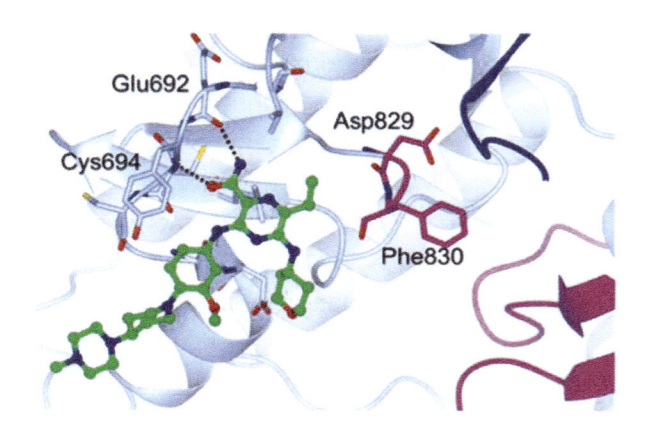

Figure 2. Co-crystal structure of gilteritinib–FLT3 (wt). FLT3 is shown in light blue, and the activation loop is shown in magenta. Adapted from *Oncotarget.* **2019**, *10*, 6111−6123.

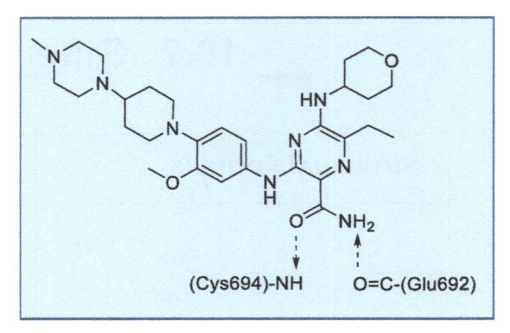

Figure 3. Summary of gilteritinib–FLT3 (wt) interactions based on their co-crystal structure.

1. S. Dhillon. Gilteritinib: First global approval. *Drugs.* **2019**, *79*, 331–339.
2. G. Keiffer, K. L. Aderhold, N. D. Palmisiano. Upfront treatment of FLT3-mutated AML: A look back at the RATIFY trial and beyond. *Front Oncol.* **2020**, *10*, 562219.
3. K. S. Kim. Midostaurin: First global approval. *Drugs.* **2017**, *77*, 1251–1259.
4. L. Y. Lee, D. Hernandez, T. Rajkhowa, S. C. Smith, J. R. Raman, B. Nguyen, D. Small, M. Levis. Preclinical studies of gilteritinib, a next-generation FLT3 inhibitor. *Blood.* **2017**, *129*, 257–260.
5. M. Mori, N. Kaneko, Y. Ueno, M. Yamada, R. Tanaka, R. Saito, I. Shimada, K. Mori, S. Kuromitsu. Gilteritinib, a FLT3/AXL inhibitor, shows antileukemic activity in mouse models of FLT3 mutated acute myeloid leukemia. *Invest. New Drugs.* **2017**, *35*, 556–565.
6. P. Y. Dumas, A. Villacreces, A. V. Guitart, A. El-Habhab, L. Massara, O. Mansier, A. Bidet, D. Martineau, S. Fernandez, T. Leguay, A. Pigneux, I. Vigon, J. M. Pasquet, V. Desplat. Dual inhibition of FLT3 and AXL by gilteritinib overcomes hematopoietic niche-driven resistance mechanisms in *FLT3*-ITD acute myeloid leukemia. *Clin. Cancer Res.* **2021**, *27*, 6012–6025.
7. T. Kawase, T. Nakazawa, T. Eguchi, H. Tsuzuki, Y. Ueno, Y. Amano, T. Suzuki, M. Mori, T. Yoshida. Effect of Fms-like tyrosine kinase 3 (FLT3) ligand (FL) on antitumor activity of gilteritinib, a FLT3 inhibitor, in mice xenografted with FL-overexpressing cells. *Oncotarget.* **2019**, *10*, 6111–6123.

Chapter 19. mTOR Inhibitors

Rapamycin (best known as sirolimus), was isolated at Wyeth in 1975 and characterized initially as an antifungal antibiotic. It was later shown to be not only a potent immunosuppressant, but also a powerful antitumor agent. It was approved by the FDA to treat post-renal transplantation (in 1999) and lymphangioleiomyomatosis (LAM) (in 2015). Subsequent SAR studies on rapamycin led to temsirolimus (an ester prodrug of rapamycin) for the treatment of advanced kidney cancer and renal cell carcinoma (RCC), and everolimus (a hydroxyethyl ether analog) for advanced kidney cancer, RCC, and breast cancer in combination with exemestane.

While medicinal chemists were modifying rapamycin's structure to make novel analogs with optimal PK properties, academic researchers carried out an intensive search for the target of rapamycin, resulting in the discovery of mammalian target of rapamycin (mTOR). This highly conserved serine/threonine protein kinase, a critical component of the PI3K/AKT/mTOR signaling pathway, is commonly overactivated in a variety of cancer types. Rapamycin and analogs work by blocking the activity of mTOR and preventing its constitutive signaling that promotes tumor growth and survival.[1]

The first-generation mTOR inhibitors such as rapamycin and analogs (Fig. 1) are canonical allosteric kinase inhibitors which suffer from strong immunosuppressive activities. This limitation has prompted the development of the second-generation of ATP-competitive mTOR kinase inhibitors, which directly bind the mTOR kinase domain and block its catalytic activity. These inhibitors were also shown to be active against rapamycin-insensitive cell lines. Figure 2 shows three second-generation mTOR kinase inhibitors currently in clinical trials.

Everolimus used to be one of Novartis's top selling oncology drugs before expiration of its patent in 2019, generating $1.6 billion in revenues in 2015.

This chapter walks through the three-decade journey of rapamycin (**84**) from Easter Island of Chile to bedside.

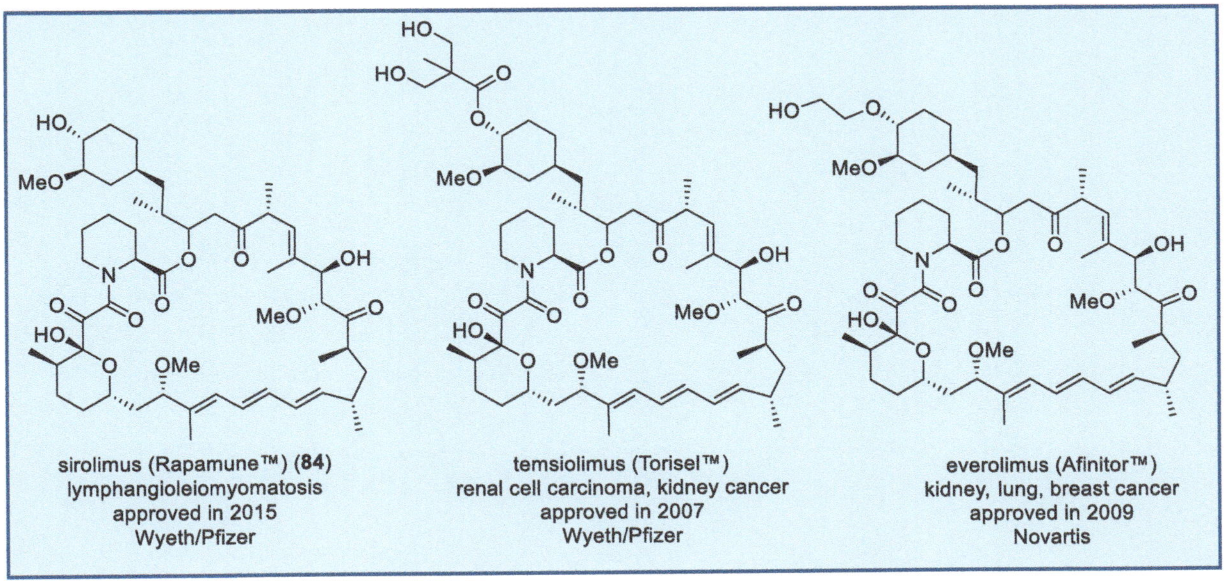

Figure 1. Approved mTOR inhibitors.

Molecules Engineered Against Oncogenic Proteins and Cancer, First Edition. E. J. Corey and Yong-Jin Wu.

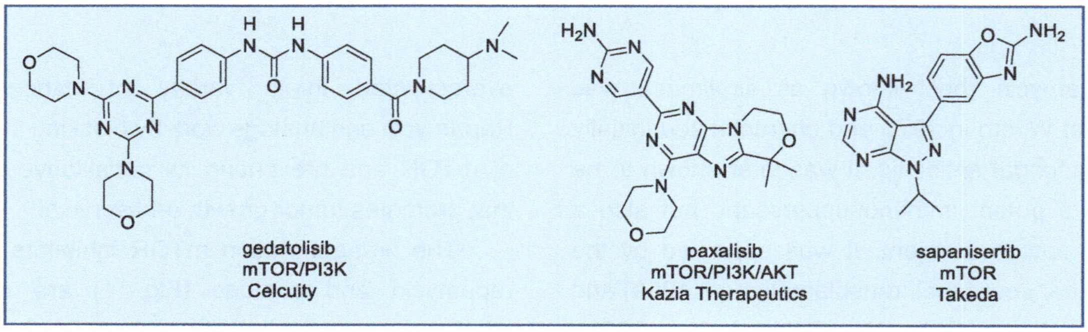

Figure 2. mTOR/multikinase inhibitors in clinical trials.

1. D. Wang, H. J. Eisen. Mechanistic target of rapamycin (mTOR) inhibitors. *Handb. Exp. Pharmacol.* **2022**, *272*, 53–72.

19.1 Sirolimus (Rapamune™) (84) and Analogs

Structural Formula	Space-filling Model	Brief Information
$C_{51}H_{79}NO_{13}$; FW = 914 clogP = 1.8; tPSA = 112		Year of discovery: 1972 Year of introduction: 1999, 2015 Discovered by: Ayerst Developed by: Ayerst/Pfizer Primary target: mTOR Binding type: IV Class: serine/threonine Treatment: organ rejection, LAM Other name: Rapamycin, AY-22989 Oral bioavailability = 14% Elimination half-life = 63 h Protein binding = 92%

● = C   ○ = H   ● = O   ● = N   ● = S   ● = F   ● = Cl   ● = Br   ● = I

Sirolimus, also known as rapamycin (which is used throughout this section), was first approved in 1999 for the prophylaxis of organ rejection in kidney transplant patients over the age of 13. As an immunosuppressant, it blocks antibody production and binds to immunophilin to elicit an immunosuppressive effect. In 2015, it also became the first approved medicine in the United States for the treatment of lymphangioleiomyomatosis (LAM), a rare, progressive cystic lung disease affecting approximately 3.4–7.8 per million women worldwide, especially women of child-bearing age. LAM is caused by unregulated activation of mammalian target of rapamycin (mTOR) signaling pathway, and sirolimus blocks its activation and proliferation of T-lymphocyte.

Development of Rapamycin and Analogs[1–5]

The three-decade journey of rapamycin from Easter Island of Chile to bedside started in 1964 when a Canadian scientific expedition team collected a soil sample from the island of Rapa Nui (a Chilean island in the southeastern Pacific Ocean). Eight years later, Suren Sehgal et al. at Ayerst in Montreal finally isolated a 31-membered polyketide macrolactone called rapamycin (reminiscent of the Rapa Nui island) from the soil sample.[6] It was initially identified as a fungicidal agent, but its potent immunosuppression prevented its further development as an antifungal drug. As the biologists were testing rapamycin's antifungal activity, they found that this compound prevented cells from growing and dividing to form a lump. Intuitively speculating about its anticancer activity, Suren Sehgal, a microbiologist, sent a sample to the U.S. National Cancer Institute (NCI) for screening for antitumor activity, and the results were astonishing. Against the 60 tumor cell line panel, rapamycin potently inhibited the growth of a number of solid tumors. This finding was significant at the time because it was the very first cytostatic agent (that keep cells from dividing), while all other existing chemotherapies were cytotoxic (deadly to cells). Because of its unique anticancer properties, NCI designated it as a priority drug. Unfortunately, in 1983, Ayerst was in the process of shutting down its Montreal research laboratories, and many programs were terminated including the rapamycin project. The scientists were asked to destroy everything without a yet defined productive use, and rapamycin fell onto the list designated for disposal because an intravenous formulation for clinical trials was not attainable, and its mechanism of action was completely unknown. Suddenly, rapamycin was abandoned by Ayerst as a drug candidate. Instead of destroying the valuable rapamycin sample, Suren

Sehgal took it home without approval from his management and stored it in his home freezer. It was six years later that he won approval from the newly merged Wyeth-Ayerst management to send rapamycin to outside investigators for evaluation in organ transplant animal models, The impressive in vivo results convinced the new management to resurrect rapamycin as an immunosuppressant drug candidate. It was ultimately approved by the FDA in 1999. In the meantime, NCI resumed anticancer work, which contributed to its approval as a LAM medication in 2015.

As a notable exception to the Lipinski's rule of five, rapamycin has 14% oral bioavailability, which is sufficient for oral administration to treat autoimmune diseases and LAM. Nevertheless, the modest oral bioavailability and poor solubility are inadequate for oral dose in cancer treatment. Subsequently, a more water-soluble prodrug temsirolimus (Fig. 1) was developed as an intravenous medication. In the meantime, extensive medicinal chemistry efforts by Novartis led to everolimus (Fig. 1) with PK properties superior to rapamycin, and the orally dosed everolimus has been widely used for kidney, lung, and breast cancers with peak sales of 1.6 billion in 2015.

More recently, rapamycin has been touted to have potential to become the first drug to slow the effects of aging – the ultimate goal for preventative medicine.[4,5]

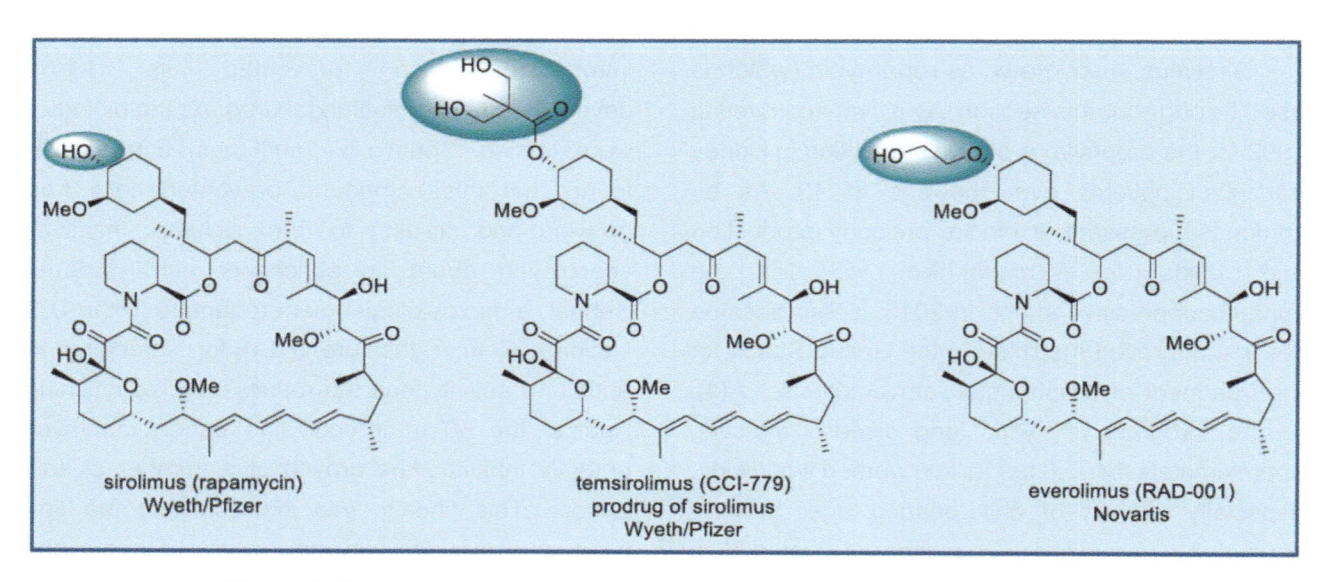

Figure 1. Structures of rapamycin (sirolimus), temsirolimus and everolimus.

Prior to its FDA approval, the mechanism of action of rapamycin remained completely unknown. Approximately 20 years after the discovery of its antitumor activity, it was determined that the target of rapamycin is the protein mTOR and that binding blocks the PI3K/AKT/mTOR signal transduction pathway, and the complete structure of mTORC1 was elucidated in 2005.[4,7]

PI3K/AKT/mTOR Pathway in LAM and Cancer[7–11]

The PI3K/AKT/mTOR signaling pathway is activated upon binding of ligands (e.g., growth factors, cytokines and hormones) to various receptors, the most important of which are receptor tyrosine kinases (RTKs) and B-cell receptors (BCRs) (Fig. 2). Ligand binding to RTKs or BCRs results in autophosphorylation of tyrosine residues on the intracellular domain of the receptors, and the activated receptors induce the dimerization and phosphorylation of PI3K. The p110 catalytic subunit of PI3K converts phosphatidylinositol 4,5-bisphosphate (PIP2) into phosphatidylinositol 3,4,5-trisphosphate (PIP3). The formation of PIP_3 is also negatively regulated by PTEN (phosphatase and tensin homolog deleted on chromosome 10) tumor suppressor protein, a phosphatase that dephosphorylates PIP3 to PIP2. The membrane bound PIP3 recruits AKT and PDK1 to the plasma membrane where PDK1 phosphorylates AKT at the Thr308. Further phosphorylation at Ser473 by the TORC2 complex fully activates AKT, and the activated AKT dissociates from the PIP3 and phosphorylates Ser939 and Thr1462 of TSC2. These phosphorylations inactivate the TSC1/TSC2 complex (tuberous sclerosis complex 1/2), a molecular switchboard that switches Rheb (RAS homologue enriched in brain) from its mTORC1 inactive GDP-bound state to its mTORC1-activating, GTP-bound state by acting as a GTPase-activating protein (GAP) towards Rheb. Upon activation, mTORC1 undertakes phosphorylation of downstream effectors, such as S6K1 and 4EBP1, both of which are essential modulators of cap-dependent and cap-independent translation. Thus, mTORC1 signaling pathway plays an important role in regulating cell survival, metabolism, growth and protein synthesis in response to upstream signals in both normal physiological and pathological conditions, especially in cancer.

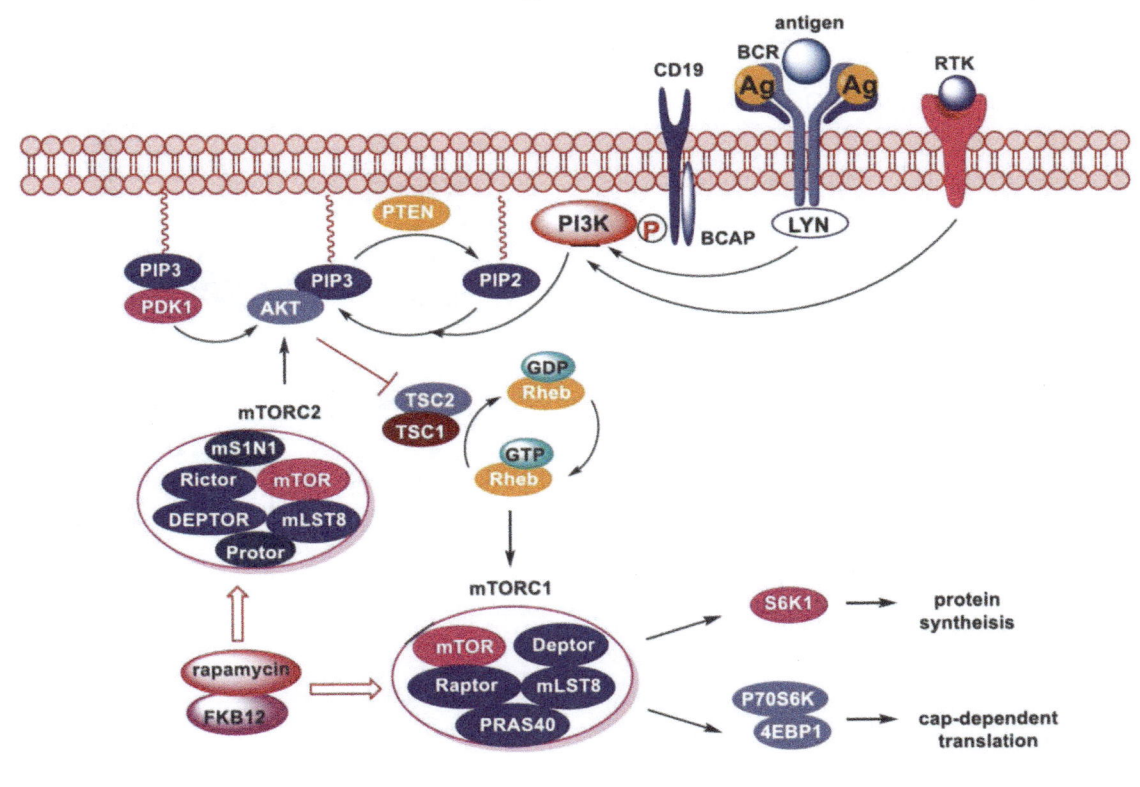

Figure 2. The PI3K/AKT/mTOR signaling pathway.

The mTOR signaling pathway has been linked to LAM, a rare lung disease caused by the mutated TSC1/2 genes.[12] The abnormal genes encode the dysfunctional TSC1/2 proteins, which form inactive TSC1-TSC2 complex, leading to abnormal activation of the RAS homologue Rheb, followed by the constitutive mTOR signaling. The sustained mTOR activation causes dysregulated cell growth and the spread of abnormal cells in the body. These abnormal cells grow uncontrollably in certain organs or tissues, such as the lungs, kidney, and lymph nodes, and give rise to the symptoms and complications of LAM.

Aberrant mTOR signaling due to genetic alterations at different levels of the signal cascade is commonly observed in a variety of cancers in addition to LAM. Therefore, mTOR inhibition represents a viable approach for the treatment of various cancers.

mTOR and mTORC1/2[7–11]

mTOR is a member of the phosphoinositide kinase-related kinase (PIKK) family. Despite its homology to PI3K (a lipid kinase), mTOR is exclusively a Ser/Thr protein kinase devoid of any lipid kinase activity. mTOR contains multiple domains including an FKBP12-rapamycin binding domain (FRB) and a C-terminal kinase domain (KIN) (Fig. 3).

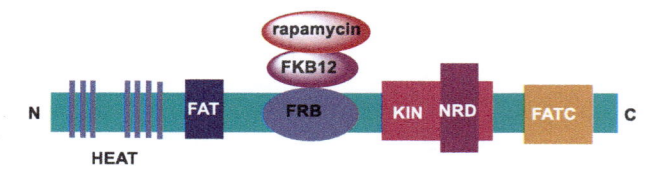

Figure 3. Structure of the mTOR backbone.

mTOR associates with various proteins to form two structurally and functionally distinct complexes called the mammalian target of rapamycin complex 1 (mTORC1) and mammalian target of rapamycin complex 2 (mTORC2) (Fig. 2). mTORC1 consists of mTOR, raptor, GβL, and deptor, while mTORC2 is composed of mTOR, Rictor, GβL, PRR5, deptor, and SIN1. mTORC1 transduces molecular signals from various growth factors, cytokines, nutrients, and energy supply to promote cell growth under high nutrient and high-energy conditions and catabolism under non-permissive circumstances, a function analogous to gas and brake pedals in car. mTORC1 is involved in the regulation of cell growth and metabolism, whereas mTORC2 mainly regulates cell proliferation and survival.

mTORC1 is highly susceptible to inhibition by rapamycin with an IC_{50} of 0.1 nM. It interacts with the immunophilin FKBP12 to form the FKBP12–sirolimus complex, which then binds to the FRB domain of mTOR. Upon docking to the FRB domain, which is in close proximity to the catalytic site of mTOR, the FKBP12–sirolimus complex allosterically inhibits the kinase activity of mTOR, thus blocking mTOR signaling pathway.

Binding Mode[13–16]

The FKBP12–rapamycin binary complex shows five hydrogen bonds within the binding site of FKBP12 domain of mTOR: (1) C40-OH to the amide carbonyl of Gln53; (2) C28-OH to the amide carbonyl of Glu54; (3) C1 carbonyl oxygen to the amide NH of Ile56; (4) C10-OH to the carboxylic acid side chain of Asp37; and (5) C8 carbonyl oxygen to the tyrosine phenol hydroxyl of Tyr82 (Figs. 4, 5). In addition, the C16 methoxy oxygen interacts indirectly with the carboxylic acid side chain of Asp37 via a bound water molecule.[13-15] These interactions contribute to the high-binding affinity of rapamycin (K_d = 0.2 nM) to FKBP12, and are the basis of potent inhibition of endogenous mTOR in HEK293 cells (IC_{50} = 0.1 nM).

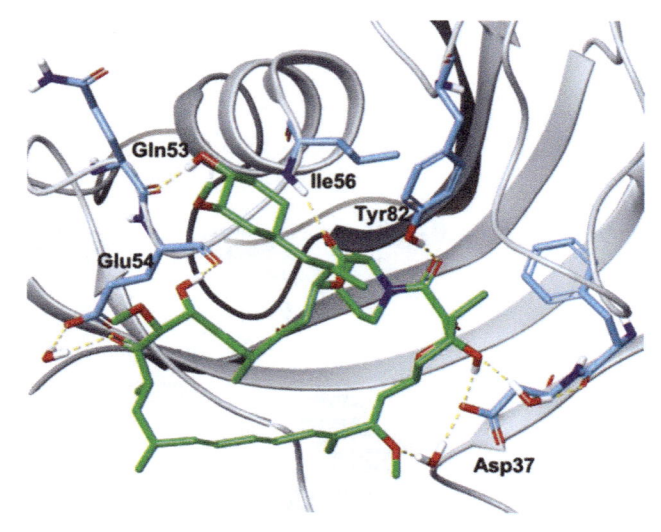

Figure 4. Co-crystal structure of rapamycin–FKBP12 (PDB ID: 1C9H).

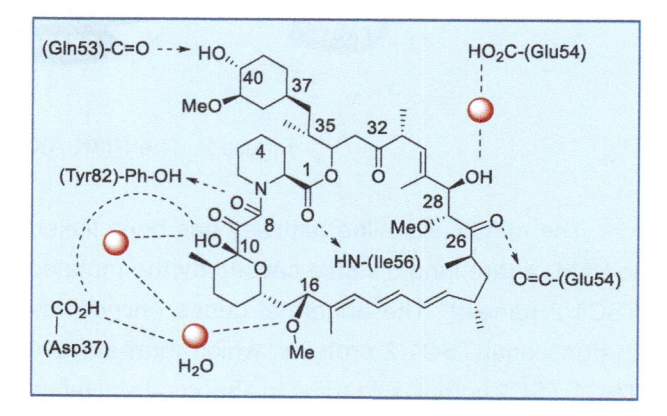

Figure 5. Summary of rapamycin–FKBP12 interactions based on their co-crystal structure.

The conformation of rapamycin in the ternary complex is very similar to that observed in the FKBP12–rapamycin binary complex (data not shown).[16,17] In the ternary complex, rapamycin binds to FK506-binding protein (FKBP12) and the FKBP12-rapamycin binding (FRB) domain of mTOR simultaneously to form the FKBP12–rapamycin–FRB ternary complex, causing subsequent perturbation of mTORC1 and producing biological activity.

1. B. Seto. Rapamycin and mTOR: A serendipitous discovery and implications for breast cancer. *Clin. Trans. Med.* **2012**, *15*, 29.
2. B. D. Kahan. Discoverer of the treasure from a barren island: Suren Sehgal. *Transplantation*. **2003**, *76*, 623–624.
3. D. Samanta. Surendra Nath Sehgal: A pioneer in rapamycin discovery. *Indian J. Cancer*. **2017**, *54*, 697–698.
4. B. Halford. Rapamycin's secrets unearthed. From its exotic origins to its revival as a potential antiaging compound, rapamycin continues to fascinate. *C & EN*. July 18, 2016.
5. K. Loria. A rogue doctor saved a potential miracle drug by storing samples in his home after being told to throw them away. *Business Insider*. February 20, 2015.
6. C. Vezina, A. Kudelski, S. N. Sehgal. Rapamycin (AY-22, 989), a new antifungal antibiotic. *J. Antibiot*. **1975**, *28*, 721-726.
7. C. H. Aylett, E. Sauer, S. Imseng, D. Boehringer, M. N. Hall, N. Ban, T. Maier. Architecture of human mTOR complex 1. *Science*. **2016**, *351*, 48–52.
8. T. Xu, D. Sun, Y. Chen, L. Ouyang. Targeting mTOR for fighting diseases: A revisited review of mTOR inhibitors. *Euro. J. Med. Chem.* **2020**, *199*, 112391.
9. D. Benjamin, M. Colombi, C. Moroni, M. N. Hall. Rapamycin passes the torch: A new generation of mTOR inhibitors. *Nat Rev Drug Discov.* **2011**, *10*, 868–880.
10. Y. Feng, X. Chen, K. Cassady, K. Zhou, S. Yang, Z. Wang, X. Zhang. The Role of mTOR inhibitors in hematologic disease: From bench to bedside. *Front. Oncol.* **2021**, *10*, 611690.
11. E. Paplomata, R. O'Regan. The PI3K/AKT/mTOR pathway in breast cancer: Targets, trials and biomarkers. *Ther. Adv. Med. Oncol.* **2014**, *6*, 154–166.
12. T. A. Smolarek, L. L. Wessner, F. X. McCormack, J. C. Mylet, A. G. Menon, E. P. Henske. Evidence that lymphangiomyomatosis is caused by TSC2 mutations: Chromosome 16p13 loss of heterozygosity in angiomyolipomas and lymph nodes from women with lymphangiomyomatosis. *Am. J. Hum. Genet.* **1998**, *62*, 810–815.
13. C. S. Deivanayagam, M. Carson, A. Thotakura, S. V. Narayana, R. S. Chodavarapu. Biological crystallography structure of FKBP12.6 in complex with rapamycin. *Acta Crystallogr., Sect. D: Biol. Crystallogr.* **2000**, *56*, 266−271.
14. J. Choi, J. Chen, S. L. Schreiber, J. Clardy. Structure of the FKBP12-rapamycin complex interacting with the binding domain of human FRAP. *Science*. **1996**, *273*, 239–242.
15. J. Liu, Y. Hu, D. Waller, J. Wang, Q. Liu, Qingsong. Natural products as kinase inhibitors. *Nat. Prod. Rep.* **2012**, *29*. 392-403.
16. J. Liang, J. Choi and J. Clardy, *Acta Crystallogr., Sect. D: Biol. Crystallogr.* **1999**, *55*, 736–744.

Chapter 20. Other Kinase Inhibitors

This chapter describes the research that led to the four FDA-approved kinase inhibitors: netarsudil (**85**), belumosudil (**86**), fostamatinib (**87**), and pexidartinib (**88**) (Fig. 1).

Netarsudil is the first trabecular outflow drug approved in 2017 for lowering elevated intraocular pressure (IOP) in patients with open-angle glaucoma or ocular hypertension. It is a disease-modifying therapy which controls IOP primarily by increasing fluid outflow through trabecular meshwork via ROCK (Rho-associated protein kinase) inhibition.

Belumosudil is an orally active ROCK2 inhibitor approved in 2021 to treat chronic graft-versus-host disease (chronic GVHD). It works by blocking the ROCK-mediated fibrotic pathways in GVHD patients. It is the third FDA-approved kinase inhibitor for GVHG, following ibrutinib (BTK inhibitor) and ruxolitinib (JAK2 inhibitor).

Fostamatinib is a first-in-class spleen tyrosine kinase (SYK) inhibitor approved by the FDA in 2018 for the treatment of adults with chronic immune thrombocytopenia (ITP), an autoimmune disorder. Several newer SYK inhibitors in clinical trials are shown in Figure 2.

Pexidartinib is the first systemic therapy approved in 2019 for adult patients with symptomatic tenosynovial giant cell tumor (TGCT) associated with severe morbidity or functional limitations and not amenable to improvement with surgery. TGCT is a complex tumor with neoplastic cells overexpressing colony-stimulating factor 1 (CSF1). Pexidartinib potently inhibits the signaling between CSF1 and CSF1 receptor (CSF1R) to target the underlying cause of the disease. Vimseltinib is a newer CSF1R inhibitor in clinical studies for TGCT (Fig. 4).

Pharmaceutical research continues to expand with other kinase targets, such as ERK and AKT, and to produce a variety of novel therapeutics in clinical trials for various cancers, autoimmune diseases and CNS disorders. Several of these are shown in Figures 3 and 4. For example, capivasertib, a promising first-in-class AKT inhibitor, in combination with fulvestrant demonstrated significant improvement in progression-free survival versus fulvestrant alone in HR+, HER2−, or low locally advanced or metastatic breast cancer patients in a recent phase III trial. AKT inhibition could become a new treatment mode in the fight against breast cancer.

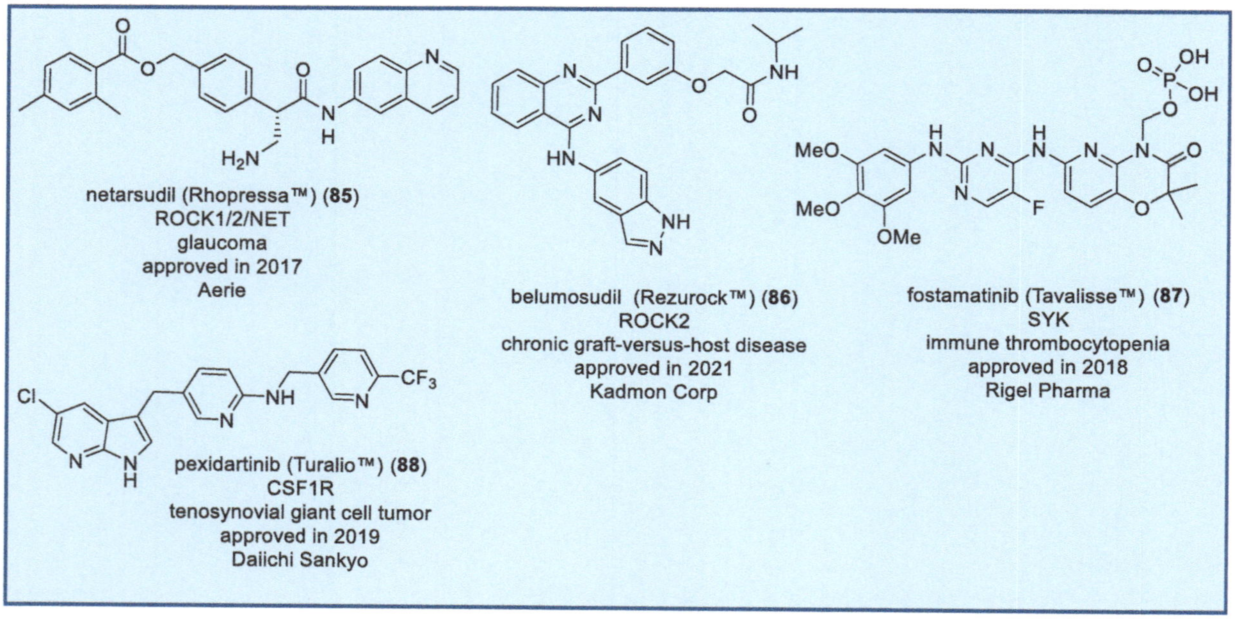

Figure 1. Approved other kinase inhibitors.

Molecules Engineered Against Oncogenic Proteins and Cancer, First Edition. E. J. Corey and Yong-Jin Wu.
© 2024 John Wiley & Sons Inc. Published 2024 by John Wiley & Sons Inc.

Figure 2. SYK inhibitors in clinical trials.

Figure 3. AKT inhibitors in clinical trials.

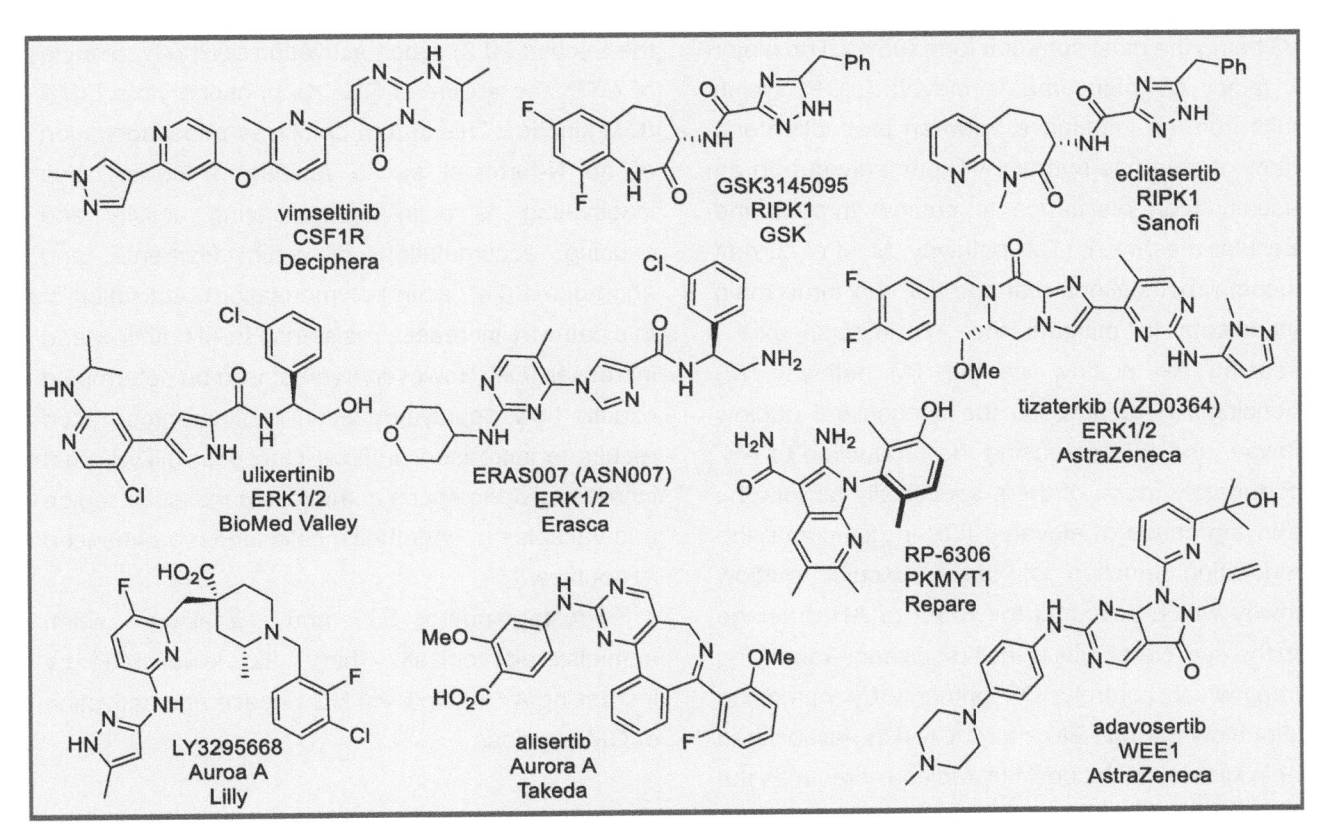

Figure 4. Other kinase inhibitors in clinical trials.

20.1 Netarsudil (Rhopressa™) (85)

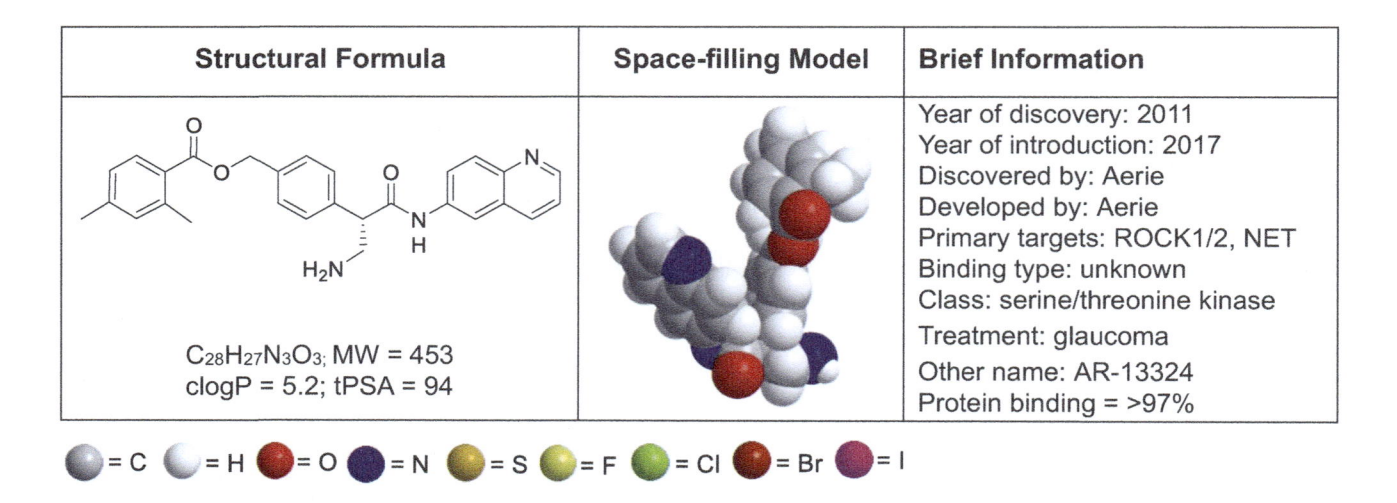

Structural Formula	Space-filling Model	Brief Information
$C_{28}H_{27}N_3O_3$; MW = 453 clogP = 5.2; tPSA = 94		Year of discovery: 2011 Year of introduction: 2017 Discovered by: Aerie Developed by: Aerie Primary targets: ROCK1/2, NET Binding type: unknown Class: serine/threonine kinase Treatment: glaucoma Other name: AR-13324 Protein binding = >97%

● = C   ○ = H   ● = O   ● = N   ● = S   ● = F   ● = Cl   ● = Br   ● = I

Netarsudil is the first trabecular outflow drug approved in 2017 for lowering elevated intraocular pressure (IOP) in patients with open-angle glaucoma (OAG) or ocular hypertension. The ophthalmic solution (0.02%) is instilled one drop daily into the affected eye(s).[1]

More than three million Americans and 80 million people worldwide suffer from glaucoma, with OAG being the most common form (95%). The major risk factor for glaucoma is elevated IOP, which results from an imbalance between production and outflow of aqueous humor (AH) primarily due to an unusually high resistance to outflow through the trabecular meshwork (TM) pathway. Most of current glaucoma medications manage IOP via three main mechanisms to mitigate the AH imbalance: (1) increasing AH outflow via the TM pathway, (2) enhancing AH outflow via the uveoscleral outflow pathway, and (3) decreasing the production of AH. Unfortunately, none of them specifically targets the underlying cause of elevated IOP in glaucoma: the deteriorating function of the trabecular outflow pathway which accounts for ~85% of AH drainage from the eye. Netarsudil is the first disease-modifying therapy which controls IOP primarily by increasing fluid outflow through TM via ROCK (Rho-associated protein kinase) inhibition.[2-4] In addition, it reduces the production of eye fluid through norepinephrine transporter (NET) inhibition.

ROCK has been detected in the optic nerve of glaucomatous eyes, and the activation of ROCK signaling in the outflow tissue has been implicated in the abnormal AH outflow (Fig. 1).[5,6] AH outflow through TM follows this route: the corneoscleral TM, the juxtacanalicular TM, Schlemm canal, and finally the episcleral venous system. An important regulator of TM outflow is ROCK signaling (which is detailed in the Section 20.2). Upon activation of Rho by binding of GTP, the activated ROCKs phosphorylate LIMK (LIM kinase). This in turn catalyzes phosphorylation of an N-terminal serine residue of cofilin, thus inactivating its actin-depolymerizing activity and causing accumulation of actin filaments and aggregates (i.e., actin polymerization). Activation of this pathway increases resistance to AH outflow and increases IOP. However, it remains to be determined exactly how depolymerization of filamentous actin results in increased outflow. One possibility is that increasing open space in the juxtacanalicular region and vacuoles in endothelial cells leads to enhanced AH outflow.[3,7]

Prostaglandin E_2 and analogs when administered topically (bid) also lower IOP by increasing AH outflow via NO release and activation of GMP cyclase.

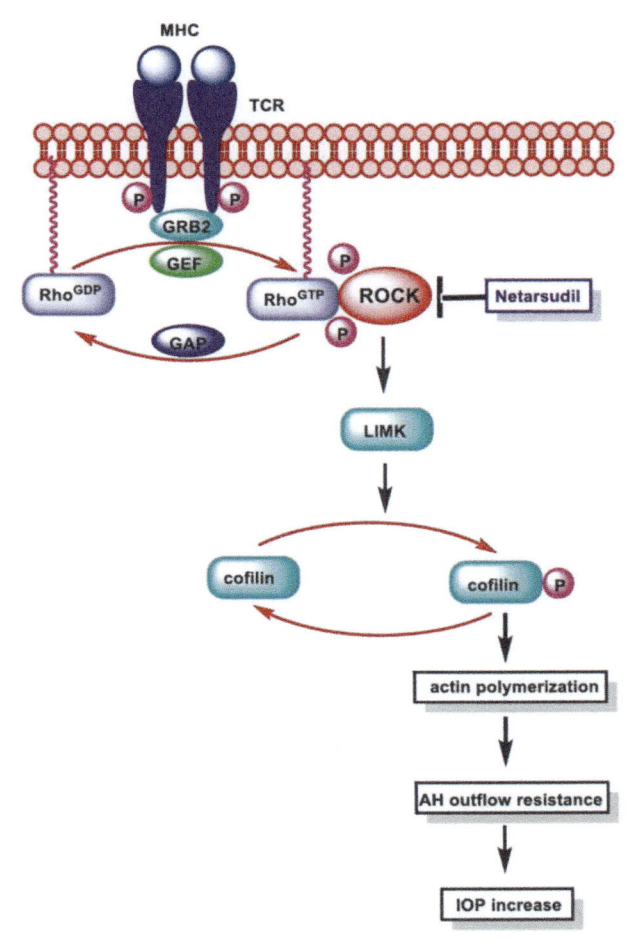

Figure 1. ROCK signaling in AH outflow pathway.

Figure 2. Metabolism of netarsudil to give active drug.

Netarsudil is a 2,4-dimethylbenzoate ester prodrug of the active molecule AR-13503 (Fig. 2) with increased permeability (vs. the active drug) into the eye. Following ocular instillation, netarsudil is absorbed through the cornea and then metabolized by corneal esterases to release its more active metabolite AR-13503 (Fig. 2), which exhibits 5-fold greater ROCK inhibitory activity than netarsudil.[1,2]

Netarsudil, which belongs to the aminoisoquinoline amide chemotype of ROCK inhibitors, reversibly binds to the ATP-binding pocket of the two isoforms, ROCK1 and ROCK2, with low nanomolar affinity. However, no co-crystal structure of netarsudil in complex with ROCK1/2 has been disclosed, so its precise binding mode remains unclear.

1. S. M. Hoy. Netarsudil ophthalmic solution 0.02%: First global approval. *Drugs*. **2018**, *78*, 389–396.
2. C. W. Lin, B. Sherman, L. A. Moore, C. L. Laethem, D. W. Lu, P. P. Pattabiraman, P. V. Rao, M. A. deLong, C. C. Kopczynski. Discovery and preclinical development of netarsudil, a novel ocular hypotensive agent for the treatment of glaucoma. *J Ocul. Pharmacol. Ther.* **2018**, *34*, 40-51.
3. S. K. Wang, R. T. Chang. An emerging treatment option for glaucoma: Rho kinase inhibitors. *Clin. Ophthalmol.* **2014**, *8*, 883–890.
4. Y. Feng, P. V. LoGrasso, O. Defert, R. Li. Rho kinase (ROCK) inhibitors and their therapeutic potential. *J. Med. Chem.* **2016**, *59*, 2269–2300.
5. P. V. Rao, P. P. Pattabiraman, C. Kopczynski. Role of the Rho GTPase/Rho kinase signaling pathway in pathogenesis and treatment of glaucoma: Bench to bedside research. *Exp. Eye. Res.* **2016**. pii: S0014-483530240-8.
6. E. Berrino, C. T. Supuran. Rho-kinase inhibitors in the management of glaucoma. *Expert opin. Ther. Pat.* **2019**, *29*, 817–827.
7. J. Buffault, F. Brignole-Baudouin, E. Reboussin, K. Kessal, A. Labbé, S. Parsadaniantz, C. Baudouin. The dual effect of Rho-kinase inhibition on trabecular meshwork cells cytoskeleton and extracellular matrix in an in vitro model of glaucoma. *J Clin. Med.* **2022**, *11*, 1001.

20.2 Belumosudil (Rezurock™) (86)

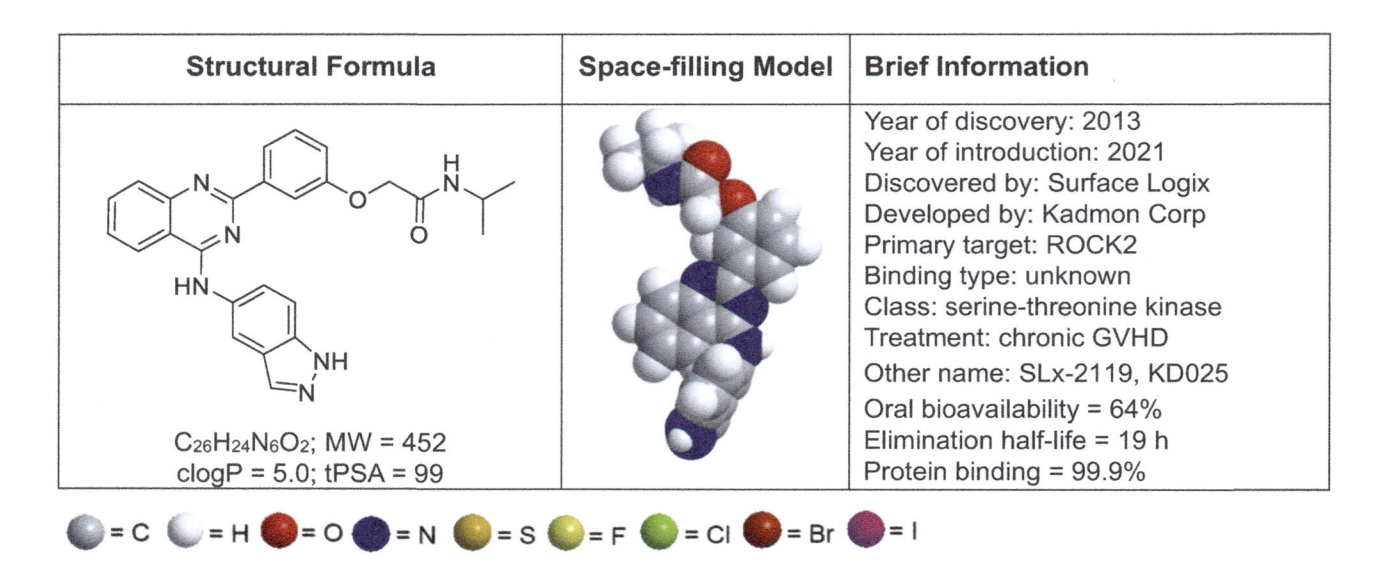

Structural Formula	Space-filling Model	Brief Information
$C_{26}H_{24}N_6O_2$; MW = 452 clogP = 5.0; tPSA = 99		Year of discovery: 2013 Year of introduction: 2021 Discovered by: Surface Logix Developed by: Kadmon Corp Primary target: ROCK2 Binding type: unknown Class: serine-threonine kinase Treatment: chronic GVHD Other name: SLx-2119, KD025 Oral bioavailability = 64% Elimination half-life = 19 h Protein binding = 99.9%

● = C   ○ = H   ● = O   ● = N   ● = S   ● = F   ● = Cl   ● = Br   ● = I

Belumosudil is an orally active Rho kinase 2 (ROCK2) inhibitor approved in 2021 for adult and pediatric patients over 12 years of age with chronic graft-versus-host disease (chronic GVHD) after failure of at least two prior lines of systemic therapy.[1]

Chronic graft-versus-host disease (cGVHD) is an immune-mediated inflammatory and fibrotic disorder, affecting ~14,000 patients in the United States (as of 2016). Glucocorticoids which have been the standard of care for chronic cGVHD provide a response rate of only 40–60%. More recently, AbbVie's ibrutinib and Incyte's ruxolitinib have also been found to be useful. Belumosudil, the third kinase inhibitor to reach the GVHD market, afforded excellent response rates among patients who had previously received the other two drugs.[2]

GVHD is associated with Th17 cells and regulatory T cells (Tregs), with the Th17/Treg ratio being higher in patients with GVHD.[3] This ratio largely depends on the signal transducer and activator of transcription 3 (STAT3) activity, which is mediated by Rho/ROCK2 signaling pathway.[4]

The Rho-ROCK pathway plays an important role in the regulation of T cell responses. Upon stimulation of the T-cell receptors (TCRs) by major histocompatibility complex (MHC), an adaptor protein GRB2 recruits GEF (guanine nucleotide exchange factor) to the cell membrane. GEF is responsible for catalyzing the exchange of GDP for GTP on Rho, leading to the activated Rho^{GTP}. The Rho is a subfamily of the RAS superfamily and has GTPase activity. It functions as a molecular switch between an activated state (bound to GTP) and an inactive state (bound to GDP), and the activated Rho protein elicits various biological effects by binding to its downstream target effector molecules. Among them, ROCK2, a serine/threonine kinase, is one of the most important effector molecules. The activated ROCK2 phosphorylates STAT3 molecules, which are translocated into the nucleus to trigger the transcription of Th17-specific transcription factors, thereby increasing the numbers of Th17 cells. Inhibition of ROCK2 blocks the differentiation of T-cells into Th17 cells and shifts the Th17/Treg balance towards regulatory T-cells (Fig. 1). Thus, ROCK2 has been shown to be a promising target in GVHD.[5]

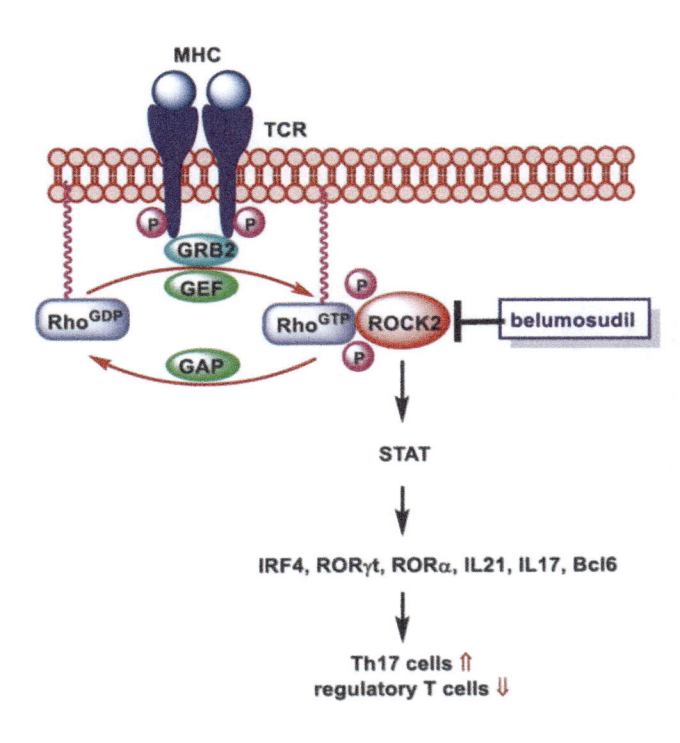

Figure 1. ROCK2 mediates Th17 differentiation in GVHD. MHC: major histocompatibility complex; TCR: T-cell receptor.

Belumosudil is a selective ROCK2 inhibitor (IC$_{50}$ ≈ 100 nM) with 240-fold selectivity against ROCK1 (IC$_{50}$ = 24 µM).[6] Its ROCK2 inhibitory activity has been demonstrated by blockage of the ROCK-mediated fibrotic pathways in GVHD patients.

There is no reported co-crystal structure of belumosudil in complex with ROCK2, and its binding mode remains unknown.

1. H. A. Blair. Belumosudil: First approval. *Drugs*. **2021**, *81*, 1677–1682.
2. C. Cutler, S. J. Lee, S. Arai, M. Rotta, B. Zoghi, A. Lazaryan, A. Ramakrishnan, Z. DeFilipp, A. Salhotra, W. Chai-Ho, R. Mehta, T. Wang, M. Arora, I. Pusic, A. Saad, N. N. Shah, S. Abhyankar, C. Bachier, J. Galvin, A. Im, S. Pavletic. Belumosudil for chronic graft-versus-host disease after 2 or more prior lines of therapy: The ROCKstar Study. *Blood*. **2021**, *138*, 2278–2289.
3. P. Ratajczak, A. Janin, R. Peffault de Latour, C. Leboeuf, A. Desveaux, K. Keyvanfar, M. Robin, E. Clave, C. Douay, A. Quinquenel, C. Pichereau, P. Bertheau, J. Y. Mary, G. Socié. Th17/Treg ratio in human graft-versus-host disease. *Blood*. **2010**, *116*, 1165–1171.
4. M. J. Park, S. H. Lee, E. K. Kim, E. J. Lee, Y. M. Moon, M. La Cho. GRIM19 ameliorates acute graft-versus-host disease (GVHD) by modulating Th17 and Treg cell balance through down-regulation of STAT3 and NF-AT activation. *J. Transl. Med.* **2016**, *14*, 206.
5. L. M. Braun, R. Zeiser. Kinase inhibition as treatment for acute and chronic graft-*versus*-host disease. *Front. Immunol.* **2021**, 12, 760199
6. N. Tran, K. H. Chun. ROCK2-specific inhibitor KD025 suppresses adipocyte differentiation by inhibiting casein kinase 2. *Molecules*. **2021**, *26*, 4747.

20.3 Fostamatinib (Tavalisse™) (87)

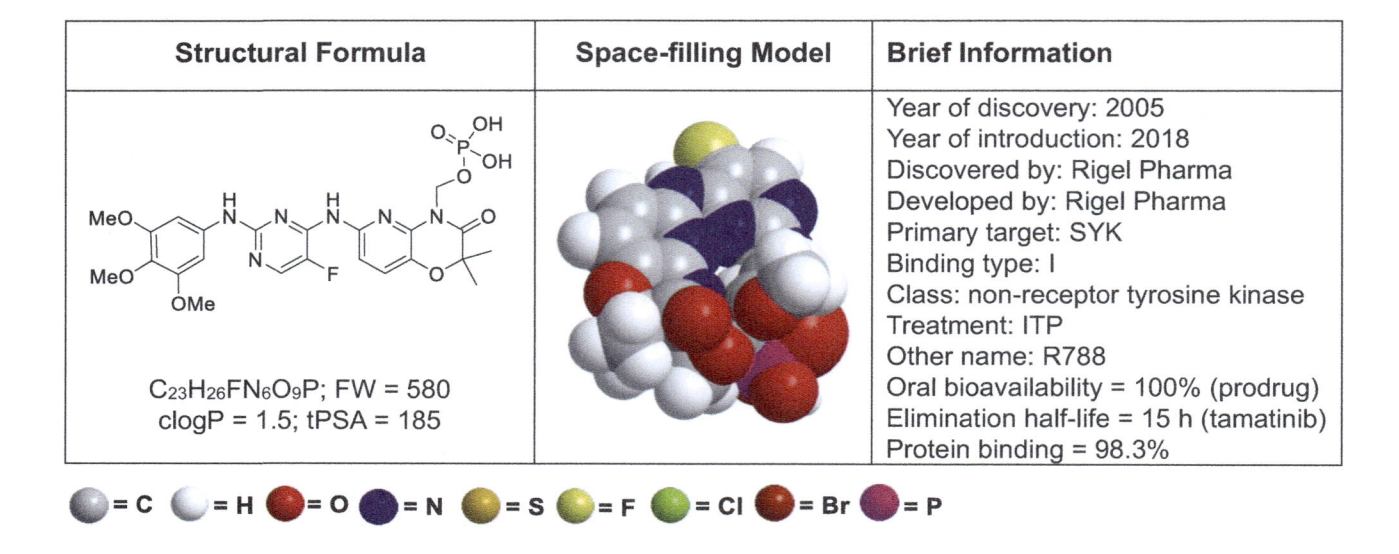

Structural Formula	Space-filling Model	Brief Information
$C_{23}H_{26}FN_6O_9P$; FW = 580 clogP = 1.5; tPSA = 185		Year of discovery: 2005 Year of introduction: 2018 Discovered by: Rigel Pharma Developed by: Rigel Pharma Primary target: SYK Binding type: I Class: non-receptor tyrosine kinase Treatment: ITP Other name: R788 Oral bioavailability = 100% (prodrug) Elimination half-life = 15 h (tamatinib) Protein binding = 98.3%

= C　= H　= O　= N　= S　= F　= Cl　= Br　= P

Fostamatinib is the first-in-class spleen tyrosine kinase (SYK) inhibitor approved by the FDA for the treatment of adults with chronic immune thrombocytopenia (ITP) who had an insufficient response to previous treatment.[1]

ITP, also known as idiopathic thrombocytopenic purpura, is an autoimmune disorder, affecting ~53,000 individuals in the United States, with 2–3 times more prevalence in women than in men. It is caused primarily by excessive platelet destruction in the spleen and liver to result in low platelet counts. More specifically, antibody-coated platelets are destroyed by macrophages in the spleen and/or liver through SYK-dependent phagocytosis of Fc receptor (FcR)-bound platelets. Fostamatinib is a prodrug, which releases its active metabolite tamatinib (also known as R406) in the intestinal mucosa (Fig. 1). As a potent SYK inhibitor, tamatinib reduces autoantibody-mediated accelerated platelet clearance, thereby resulting in durable platelet responses in the clinic; therefore, fostamatinib represents a new therapy targeting excessive platelet destruction.[2–5]

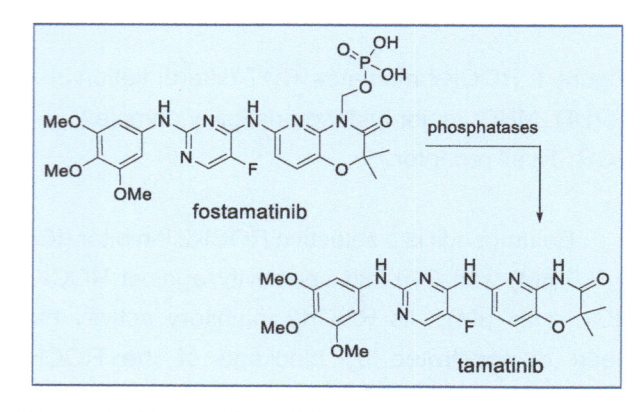

Figure 1. Conversion of fostamatinib to tamatinib in the intestine.

SYK Activation in Platelet Destruction[6,7]

SKY is a cytoplasmic protein tyrosine kinase that is abundantly expressed in hematopoietic cells, including macrophages and platelets. The SYK protein is characterized by a pair of SRC homology 2 (SH2) domains which serve as the two binding sites to immunoreceptors such as immunoglobulin Fc receptors (FcRs) for kinase activation (Fig. 2).

FcRs belong to the ITAM (immunoreceptor tyrosine-based activation motif)-associated receptor family,[8] and they play an essential role in antibody-dependent immune responses to infections and prevention of chronic inflammation or auto-immune diseases.

Cross-linking of immune complexes to

macrophage FcRs induces phosphorylations by LYN (a SRC family kinase) of the two ITAMs contained in the cytoplasmic portion of the receptors. The doubly phosphorylated ITAMs recruit the SYK through its tandem SH2 domains in a head-to-tail orientation, thus causing conformational changes and activating the kinase. The activated SYK catalyzes phosphorylations of multiple downstream effectors on their tyrosines, triggering the activation of several signaling cascades within the macrophage cytoplasm. These events bring about a cytoskeletal change in the macrophage, leading to phagocytosis of autoantibody-coated platelets and increased platelet clearance associated with ITP. Because SYK signaling is a key process in platelet destruction, SYK has become an attractive target to inhibit phagocytosis in ITP.

Tamatinib: A Potent SYK Inhibitor[2–5]

Fostamatinib is cleaved by alkaline phosphatase at the apical brush-border membranes of intestinal enterocytes to give its active metabolite tamatinib (Fig. 1). The parent drug has poor aqueous solubility, and fostamatinib was designed as a methylene-phosphate prodrug. It is rapidly absorbed into circulation, with a terminal half-life of ~15 h. Tamatinib is a competitive inhibitor for ATP-binding with a K_i of 30 nM, and it potently inhibited SYK kinase activity in vitro with an IC_{50} of 41 nM. It is active against SYK in multiple cell types involved in inflammatory and autoimmune diseases, thereby blocking signal transduction through Fc-activating receptors involved in autoimmune responses. As a result, it prevents antibody-mediated platelet phagocytosis.

Tamatinib also targets vascular endothelial growth factor receptor 2 (VEGF2), and thus fostamatinib is associated with increases in blood pressure. In addition, it also inhibits many other kinases at higher concentrations, including FLT3, JAK LCK, EGFR, FGFR3, IRAK1, PDGFRβ, RET, ROS1, and SIK2.

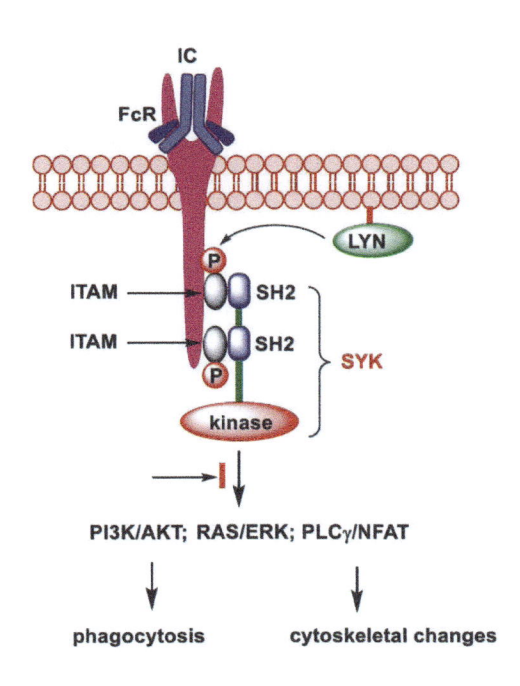

Figure 2. SYK-mediated signaling in platelets. IC: immune complex; FcR: Fc receptor.

Binding Mode[9]

As shown in the co-crystal structure of tamatinib-SYK (Fig. 3), the aminopyrimidine moiety forms two hydrogen bonds with Ala451 of the hinge region. In addition, there is an atypical hydrogen bond (CH···O=C) between CH at 6 and amide carbonyl of Glu449 (Fig. 4).

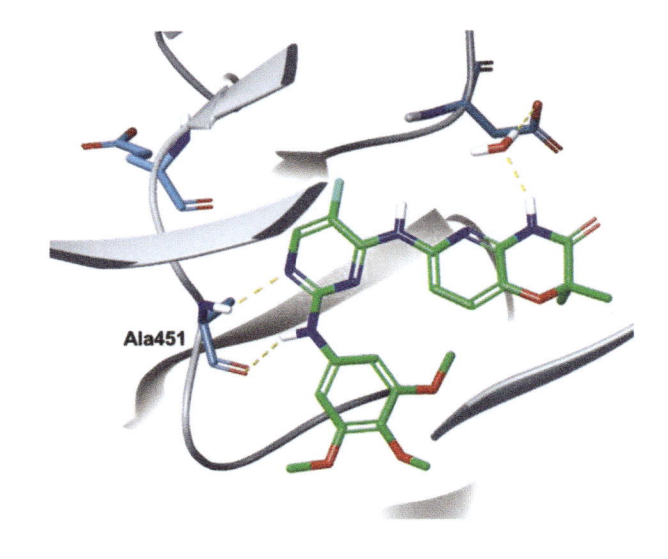

Figure 3. Co-crystal structure of tamatinib (parent drug of fostamatinib) with SYK (PDB ID: 3FQS).

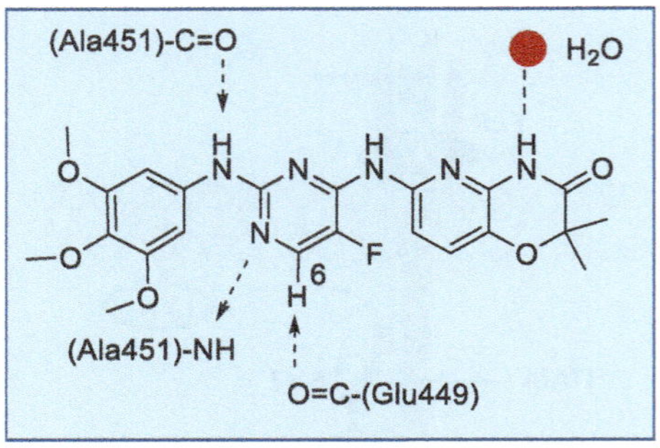

Figure 4. Summary of tamatinib–SYK interactions based on their co-crystal structure.

1. A. Markham. Fostamatinib: First global approval. *Drugs.* **2018**, *78*, 959–963
2. S. Braselmann, V. Taylor, H. Zhao, S. Wang, C. Sylvain, M. Baluom, K. Qu, E. Herlaar, A. Lau, C. Young, B. Wong, S. Lovell, T. Sun, G. Park, A. Argade, S. Jurcevic, P. Pine, R. Singh, E. B. Grossbard, D. G. Payan, E. Masuda. R406, an orally available spleen tyrosine kinase inhibitor blocks Fc receptor signaling and reduces immune complex-mediated inflammation. *J. Pharmacol. Exp. Ther.* **2006**, *319*, 998–1008.
3. A. Newland, E. J. Lee, V. McDonald, J. B. Bussel. Fostamatinib for persistent/chronic adult immune thrombocytopenia. *Immunother.* **2018**, *10*, 9–25.
4. K. McKeage, K. A. Lyseng-Williamson. Fostamatinib in chronic immune thrombocytopenia: A profile of its use in the USA. *Drugs Ther. Perspect.* **2018**, *34*, 451–456.
5. A. Newland, V. McDonald. Fostamatinib: A review of its clinical efficacy and safety in the management of chronic adult immune thrombocytopenia. *Immunother.* **2020**, *12*, 1325–1340.
6. G. M. Deng, V. C. Kyttaris, G. C. Tsokos. Targeting Syk in autoimmune rheumatic diseases. *Front. Immunol.* **2016**, *7*, 78.
7. M. Swinkels, M. Rijkers, J. Voorberg, G. Vidarsson, F. Leebeek, A. Jansen. Emerging concepts in immune thrombocytopenia. *Front. immunol.* **2018**, *9*, 880.
8. S. Mkaddem, M. Benhamou, R. C. Monteiro. Understanding Fc receptor involvement in inflammatory diseases: From mechanisms to new therapeutic tools. *Front. Immunol.* **2019**, *10*, 811.
9. Z. Xie, X. Yang, Y. Duan, J. Han, C. Liao. Small-molecule kinase inhibitors for the treatment of nononcologic diseases. *J. Med Chem.* **2021**, *64*, 1283–1345.

20.4 Pexidartinib (Turalio™) (88)

Structural Formula	Space-filling Model	Brief Information
$C_{20}H_{15}ClN_5F_3$; MW = 417 clogP = 4.0; tPSA = 61		Year of discovery: 2007 Year of introduction: 2019 Discovered by: Plexxikon Developed by: Daiichi Sankyo Primary targets: CSF1R, KIT, FLT3 Binding type: II Class: receptor tyrosine kinase Treatment: TGCT Other name: PLX3397 Oral bioavailability = not reported Elimination half-life = 26.6 h Protein binding = >99%

● = C   ○ = H   ● = O   ● = N   ● = S   ● = F   ● = Cl   ● = Br   ● = I

Pexidartinib is the first systemic therapy approved in 2019 for adult patients with symptomatic tenosynovial giant cell tumor (TGCT) associated with severe morbidity or functional limitations and not amenable to improvement with surgery.[1]

TGCT is a rare, benign tumor, occurring in or around a joint, and it can reduce movement in the affected joint and cause pain or stiffness. TGCT is subdivided into two types: localized and diffuse, affecting ~39 and ~4 per million, respectively.

TGCT is a complex tumor with neoplastic cells overexpressing colony-stimulating factor 1 (CSF1), also known as macrophage-colony stimulating factor, and pexidartinib potently inhibits the signaling between CSF1 and CSF1 receptor (CSF1R) to target the underlying cause of the disease. Treatment of TGCTs with pexidartinib has demonstrated a prolonged regression in tumor volume in most patients. However, due to the risk of serious and potentially fatal liver injury (boxed warning from FDA), it is only available in the USA to patients through a restricted Risk Evaluation and Mitigation Strategy (REMS) program.[1,2]

CSF1/CSF1R in TGCT[3–5]

TGCT originates from a small number of neoplastic cells within the tumor, which overproduce CSF1 due to a chromosomal translocation involving specific regions on chromosome 1 and chromosome 2, more specifically, a t(1;2) translocation that links the CSF1 gene on chromosome 1p13 to the COL6A3 gene on chromosome 2q37, formulated as [t (1;2) (p13;q37)] (Fig. 1). Albeit a small minority, these neoplastic cells attract other cells in the body, including the cells that have a CSF1R, particularly the white blood cells called macrophages which make up the primary cell type within these giant cell tumors. It is the high levels of CSF1 in neoplastic cells that drive the expansion of the tumor mass by recruiting and inducing local proliferation of CSF1R-dependent macrophages. Because CSF1 signaling through its receptor CSF1R controls the growth and differentiation of macrophages which constitute the bulk of TGCT, blockage of the CSF1/CSF1R signaling is an effective approach in the treatment of TGCT.

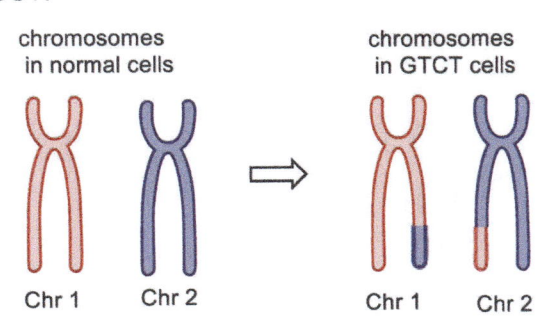

Figure 1. Chromosomal translocation from 1p13 (CSF1 gene) to 2q37 (COL6A3 gene).

CSF1R Signaling[3,5]

CSF1R (also known as c-FMS), which is a type III receptor tyrosine kinase (RTK), belongs to the platelet-derived growth factor (PDGF) receptor family consisting of PDGFα/β, the FMS-like tyrosine kinase 3 (FLT3) and the receptor for stem cell factor (KIT). Like other family members, CSF1R is composed of the extracellular domain, a single-pass transmembrane segment and the intracellular cytoplasmic domain (Fig. 2). In the extracellular domain, there are five immunoglobulin (Ig)-like domains D1 to D5: D1–D3 for ligand recognition, while D4–D5 for stabilizing the ligand–receptor complex. The cytoplasmic domain contains two kinase domains, a kinase insert (KI), a juxtamembrane domain (JMD), and a carboxyterminal tail. The JMD plays a crucial role in maintaining CSF1R in an inactive autoinhibitory state (in the absence of a ligand) by folding under the regulatory α-helix of the N-terminal lobe of the kinase domain to prevent the activation loop from adopting an active conformation.[6]

In normal cells, CSF1R activity is tightly controlled via ligand binding which releases autoinhibition and permits full activation of CSF1R with regulation of downstream signaling (Fig. 2). However, overproduction of CSF1 from neoplastic cells due to chromosome abnormality persistently activates its signaling in CSF1R-dependent macrophages, contributing tumor growth in TGCT.

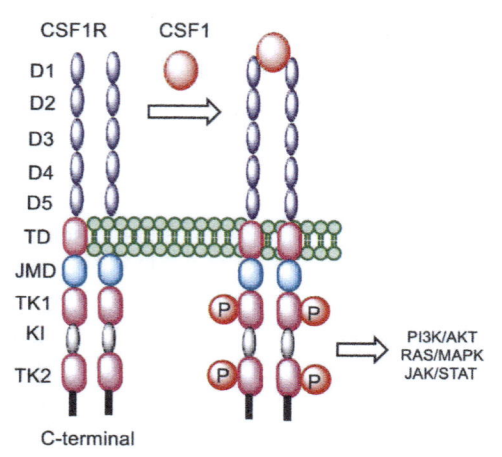

Figure 2. CSF1R signaling pathway. TD: transmembrane domain; TK: tyrosine kinase domain; KI: kinase insert; JMD: juxtamembrane domain.

Pexidartinib directly interacts with the JM residues embedded in the ATP-binding pocket of CSF1R and locks the kinase in the autoinhibited conformation, thus shutting down its downstream signaling.

From Azaindole Fragment to Pexidartinib[3,5]

As detailed in the Section 10.1, azaindole **1** was identified as a unique kinase inhibitor scaffold through high-throughput crystallography. The 3-benzyl analog **2** (an early analog) which led to vemurafenib as a highly selective BRAF inhibitor for the treatment of medullary thyroid cancer (MTC), was used once again as a lead compound toward dual CSF1R and KIT inhibitors (Fig. 3). The co-crystal structure of **2** with FGFR1 revealed a key hydrogen bond interaction between the ether oxygen and the backbone NH of the aspartate residue of the conserved DFG motif. This structure was modified to generate PLX647 in the following way: the methoxylphenyl was replaced with a pyridine whose endocyclic nitrogen served as the hydrogen bond acceptor in place of the exocyclic ether oxygen, and attachment of (trifluoromethyl)benzyl to the pyridine to occupy the kinase back pocket. PLX647 exhibited potent and selective dual inhibition of CSF1R and KIT with IC_{50} values of 28 and 16 nM, respectively, in biochemical assays.

Pexidartinib (also known as PLX3397) was derived from PLX647 by incorporating a polar nitrogen atom into the trifluoromethylphenyl tail (i.e., changing phenyl to pyridine), and a chloride at the 5-position of the azaindole hinge binder. In biochemical assays, pexidartinib was 2-fold more potent than PLX647 in inhibiting CSF1R (IC_{50} = 13 nM, vs. 28 nM), but less potent against KIT (IC_{50} = 27 nM, vs. 16 nM).

Although both PLX647 and pexidartinib bind the inactive, DFG-out conformation of CSF1R, PLX647 displaces the JM domain which sterically blocks its binding (data not shown),[3] while pexidartinib recruits the JM domain[3] to stabilize the autoinhibited conformation primarily because the polar nitrogen of the terminal pyridine participates in a water-mediated network of hydrogen bonds with Asp796 and Tyr546 (see binding mode). By directly engaging the JM residues, pexidartinib targets a more physiologically common state of the full-length CSF1R protein, resulting in significantly improved potency over PLX647 in CSF1R-dependent cellular assays.

Figure 3. Discovery of pexidartinib through fragment-based approach.

A potential drawback of using pexidartinib arises from its potent inhibition of closely related kinases such as KIT, PDGFRA, PDGFRB, and FLT3. Such off-target liabilities may limit its ability to achieve sufficient exposure for optimal CSF1R suppression. This has inspired the development of second-generation CSF1R inhibitors such as vimseltinib with improved selectivity and safety profile for the treatment of TGCT and other types of cancer.[7,8]

Novelty and Selectivity in Fragment-based Approach[9]

Pexidartinib represents one of the three launched kinase drugs (vemurafenib, erdafitinib) discovered through fragment-based approach. In this platform, one fragment can give rise to multiple different drug molecules. For example, azaindole **1** led to three drugs: vemurafenib, pexidartinib, and peficitinib (Fig. 4), a pan-JAK inhibitor for rheumatoid arthritis (see Section 7.3). Thus, novelty is desired but not required in initial fragments. The same applies to selectivity. The fact that 7-azaindole has been claimed in so many patent applications as a kinase hinge-binding element has not prevented medicinal chemists from growing it into remarkably selective inhibitors, and pexidartinib is a case in point: the added pyridyl nitrogen in pexidartinib is crucial for stabilizing the autoinhibited state of CSF1R through direct interactions with JM residues neighboring the ATP-binding pocket, thus resulting in high anti-kinase potency.

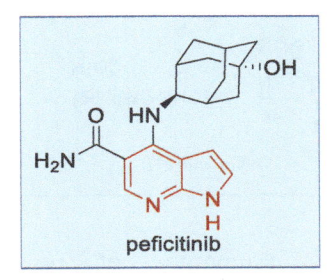

Figure 4. Structure of peficitinib.

Binding mode[3,5]

Pexidartinib is bound to the inactive, autoinhibited conformation of CSF1R where the active loop adopts a DFG-out conformation (Fig. 5). The pyrrolo NH hydrogen bonds with the amide carbonyl of Glu664 and the N7 pyridine nitrogen forms a hydrogen bond with the NH group of Cys666 (the third hinge residue). In addition, the middle pyridine nitrogen forms a hydrogen bond with the amide NH of DFG-Asp796 (Fig. 6). Most importantly, the terminal trifluoromethylpyridine directly engages the JM region primarily through two critical interactions: π–π stacking between the pyridine and Trp550; and a network of water-mediated hydrogen bonds between the polar pyridine nitrogen, Asp796 and Tyr546.[3] Taken together, these interactions clamp the kinase in the autoinhibited state.

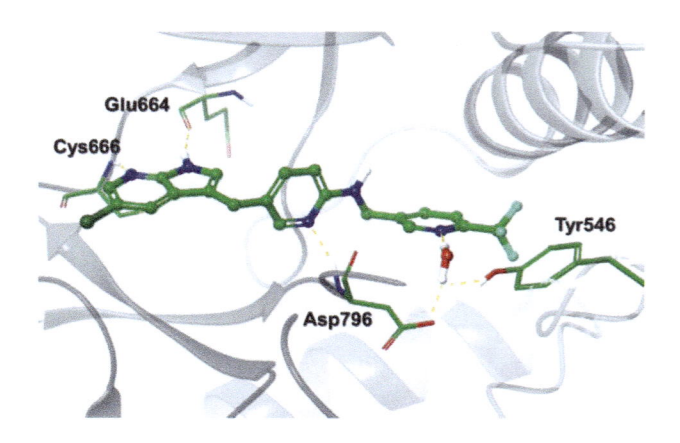

Figure 5. Crystal structure of pexidartinib in complex with CSF1R (PDB ID: 4R7H).

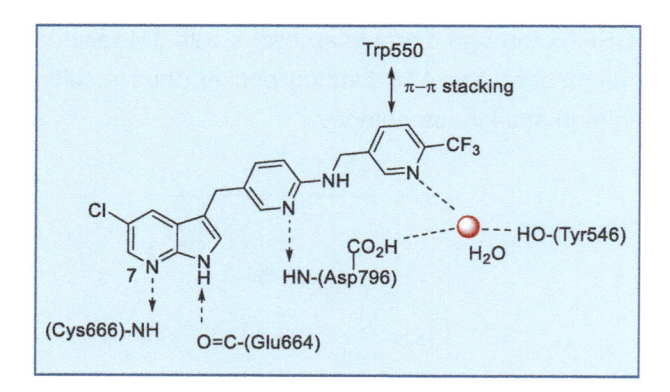

Figure 6. Summary of pexidartinib–CSF1R interactions based on the co-crystal structure.

1. Y. N. Lamb. Pexidartinib: First approval. *Drugs.* **2019**, *79*, 1805–1812.
2. B. Benner, L. Good, D. Quiroga, T. E. Schultz, M. Kassem, W. E. Carson, M. A. Cherian, S. Sardesai, R. Wesolowski. Pexidartinib, a novel small molecule CSF-1R inhibitor in use for tenosynovial giant cell tumor: A systematic review of pre-clinical and clinical development. *Drug Des. Devel. Ther.* **2020**, *14*, 1693–1704.
3. W. D. Tap, Z. A. Wainberg, S. P. Anthony, P. N. Ibrahim, C. Zhang, J. H. Healey, B. Chmielowski, A. P. Staddon, A.L. Cohn, G. I. Shapiro, V. L. Keedy, A. S. Singh, I. Puzanov, E. L. Kwak, A. J. Wagner, D. D. Von Hoff, G. J. Weiss, R. K. Ramanathan, J. Zhang, G. Habets, Y. Zhang, E. A. Burton, G. Visor, L. Sanftner, P. Severson, H. Nguyen, M. J. Kim, A. Marimuthu, G. Tsang, R. Shellooe, C. Gee, B. L. West, P. Hirth, K. Nolop, M. van de Rijn, H. H. Hsu, C. Peterfy, P. S. Lin, S. Tong-Starksen, G. Bollag. Structure-guided blockade of CSF1R kinase in tenosynovial giant-cell tumor. *N. Engl. J. Med.* **2015**, *373*, 428–437.
4. E. R. Stanley, V. Chitu. CSF-1 receptor signaling in myeloid cells. *Cold Spring Harb Perspect Biol.* **2014** June 2; 6(6), a021857.
5. C. Zhang, P. N. Ibrahim, J. Zhang, E. A. Burton, G. Habets, Y. Zhang, B. Powell, B. L. West, B. Matusow, G. Tsang, R. Shellooe, H. Carias, H. Nguyen, A. Marimuthu, K. Y. Zhang, A. Oh, R. Bremer, C. R. Hurt, D. R. Artis, G. Wu, G. Bollag. Design and pharmacology of a highly specific dual FMS and KIT kinase inhibitor. *PNAS*, **2013**, *110*, 5689–5694.
6. M. Walter, I. S. Lucet, O. Patel, S. E. Broughton, R. Bamert, N. K. Williams, E. Fantino, A. F. Wilks, J. Rossjohn. The 2.7 Å crystal structure of the autoinhibited human c-Fms kinase domain. *J. Mol. Biol.* **2007**, *367*, 839–847.
7. B. D. Smith, M. D. Kaufman, S. C. Wise, Y. M. Ahn, T. M. Caldwell, C. B. Leary, W. P. Lu, G. Tan, L. Vogeti, S. Vogeti, B. A. Wilky, L. E. Davis, M. Sharma, R. Ruiz-Soto, D. L. Flynn. Vimseltinib: A precision CSF1R therapy for tenosynovial giant cell tumors and diseases promoted by macrophages. *Mol. Cancer Ther.* **2021**, *20*, 2098–2109.
8. M. I. El-Gamal, S. K. Al-Ameen, D. M. Al-Koumi, M. G. Hamad, N. A. Jalal, C. H. Oh. Recent advances of colony-stimulating factor-1 receptor (CSF-1R) kinase and its Inhibitors. *J. Med. Chem.* **2018**, *61*, 5450–5466.
9. J. D. St Denis, R. J. Hall, C. W. Murray, T. D. Heightman, D. C. Rees. Fragment-based drug discovery: Opportunities for organic synthesis. *RSC Med. Chem.* **2021**, *12*, 321–329.

Chapter 21. KRAS Inhibitors

The RAS protein, a non-kinase, binds to and activates several families of downstream target protein kinases including RAF and PI3K, thereby regulating their signaling pathways. This protein, which serves as a key component of the RAS/RAF/MEK/ERK cascade, is the most frequent oncogenic alteration in human cancers, with KRAS mutations being particularly prevalent in some of the deadliest cancers. Despite being a logic target for cancer treatment, KRAS had been deemed "'undruggable" for nearly four decades because it has no binding sites for a small molecule

Although inhibition of oncogenic KRAS mutants remains a formidable challenge, substantial progress has been made on targeting the KRASG12C mutant (simply G12C), a common KRAS mutant found in lung tumors (glycine at position 12 replaced by cysteine). The 12 position is not located at the active site, but rather at a separate allosteric pocket ("switch II") where covalent modification can affect the protein's activity.[1,2] This prompted extensive screening to find such irreversible inhibitors and develop them into targeted therapeutics. This effort had led to (1) both sotorasib and adagrasib approved for G12C-driven NSCLC; and (2) at least 16 additional inhibitors (most of which have no disclosed structures) in 200 clinical studies. It appears that the era of targeted therapy for KRAS G12C-driven human cancers may be near.[3]

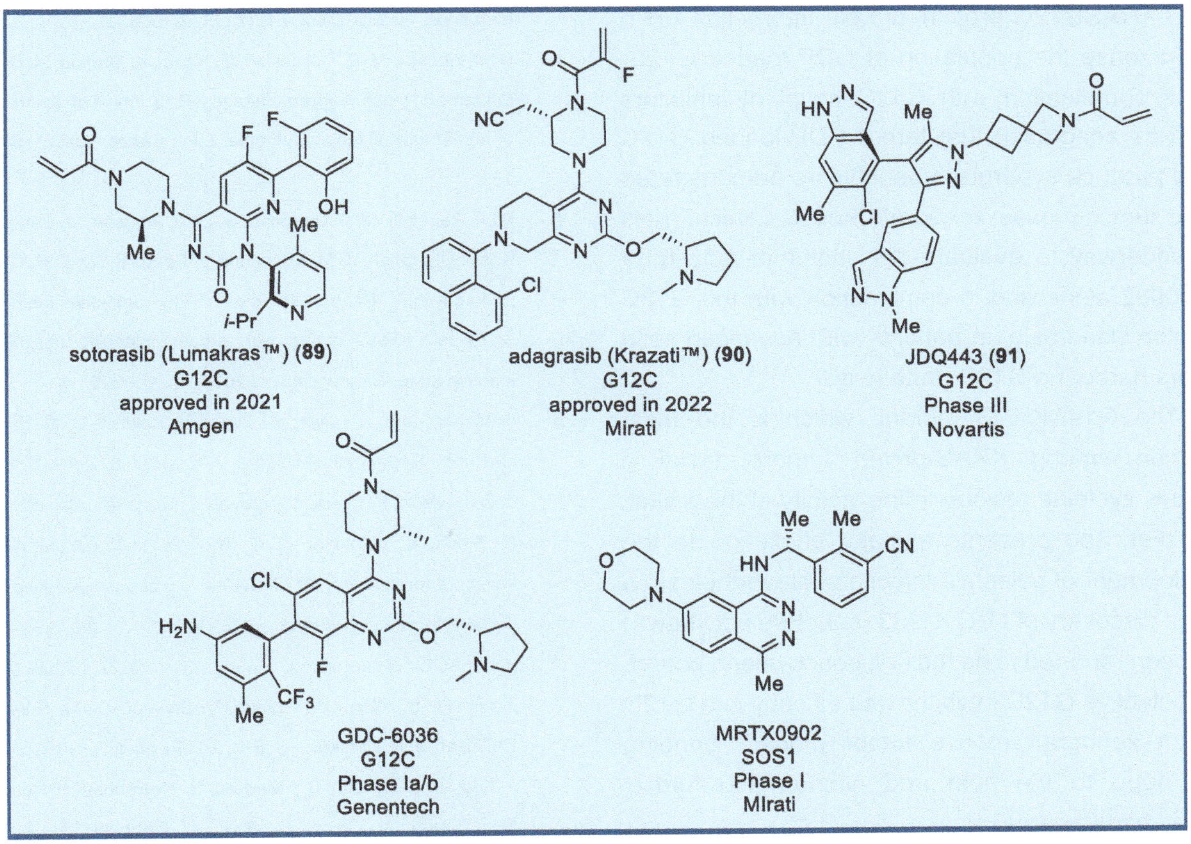

Figure 1. G12C and SOS1 inhibitors for NSCLC.

Since all current G12C drugs target the inactive GDP-bound KRAS state, whereas the active state GTP-bound KRAS binds downstream protein partners such as RAF and PI3K,[4,5] directly targeting GTP-bound KRAS may offer better clinical possibilities. However, there is no inhibitor co-crystal

Molecules Engineered Against Oncogenic Proteins and Cancer, First Edition. E. J. Corey and Yong-Jin Wu.

structure of G12C in its active GTP-bound state, and development of such inhibitors remains a formidable challenge.

The G12C inhibitors are limited by their short-lived responses. For example, in advanced KRAS-driven cancers lacking responses to other therapies, sotorasib suppressed tumor growth for just over six months. In addition, a sizable population of G12C lung cancer and colorectal cancer patients failed to respond to treatment with sotorasib. To overcome these limitations, MRTX0902 (Fig. 1), a potent, selective, and brain-penetrant SOS1 binder, was developed to disrupt the SOS1:KRAS protein–protein interaction (PPI) and block their signaling pathway.[6] SOS1 is largely responsible for catalyzing the exchange of GDP for GTP on RAS, leading to a RAS-GTP complex which then triggers RAF activation through dimerization. Disrupting the SOS1:KRASG12C protein–protein interaction (PPI) can increase the population of GDP-loaded G12C, and a combination with G12C covalent inhibitors (such as adagrasib) that target GDP-loaded G12C could produce synergistic benefits as demonstrated in the tumor mouse xenograft model. Clinical trials are underway to evaluate the antitumor activity of MTX0902 alone and in combination with the G12C inhibitor, adagrasib, in patients with advanced solid tumors harboring G12C mutations.

The G12DKRAS mutant, which is the most common among KRAS-driven tumors, lacks a reactive cysteine residue in the vicinity of the switch II pocket, and presents a major challenge to the development of selective inhibitors. Nevertheless, a recent discovery of MRTX1133 (structure not shown) has been reported to be the first noncovalent, potent, and selective G12D inhibitor with efficacy in a G12D mutant xenograft mouse tumor model,[7] bringing new hope to the field and encouraging further research in the area.

In addition to small-molecule KRAS inhibitors, proteolysis-targeted chimeras (PROTACs) have also been utilized to target tumorigenic proteins generated by G12C/D mutations. However, the poor permeability of these relatively high molecular weight degraders has complicated further in vivo studies.

In summary, the discovery of sotorasib as a molecular-matched therapy to target G12C mutants has increased the pace of KRAS research to find effective G12D/V inhibitors,

This chapter describes the decades-long research that led to the three G12C inhibitors: sotorasib (**89**), adagrasib (**90**), and JDQ443 (**91**) (Fig. 1).

1. J. M. Ostrem, U. Peters, M. L. Sos, J. A. Wells and K. M. Shokat, K-Ras(G12C) inhibitors allosterically control GTP affinity and effector interactions. *Nature.* **2013**, *503*, 548–551.
2. J. M. Ostrem, K. M. Shokat. Direct small-molecule inhibitors of KRAS: From structural insights to mechanism-based design. *Nat. Rev. Drug Discov.* **2016**, *15*, 771–85.
3. H. Ledford. Cancer drugs are closing in on some of the deadliest mutations. *Nature.* **2022**, *610*(7933), 620–622.
4. C. R. Lindsay, M. C. Garassino, E. Nadal, K. Öhrling, M. Scheffler. On target: Rational approaches to KRAS inhibition for treatment of non-small cell lung carcinoma. *Lung Cancer.* **2021**, *160*, 152–165.
5. M. P. Patricelli, M. R. Janes, L. S. Li, R. Hansen, U. Peters, L. V. Kessler, Y. Chen, J. M. Kucharski, J. Feng, T. Ely, J. H. Chen, S. J. Firdaus, A. Babbar, P. Ren, Y. Liu. Selective inhibition of oncogenic KRAS output with small molecules targeting the inactive state. *Cancer Discov.* **2016**, *6*, 316–329.
6. J. M. Ketcham, J. Haling, S. Khare, V. Bowcut, D. M. Briere, A. C. Burns, R. J. Gunn, A. Ivetac, J. Kuehler, S. Kulyk, J. Laguer, J. D. Lawson, K. Moya, N. Nguyen, L. Rahbaek, B. Saechao, C. R. Smith, N. Sudhakar, N. C. Thomas, L. Vegar, M. A. Marx. Design and discovery of MRTX0902, a potent, selective, brain-penetrant, and orally bioavailable inhibitor of the SOS1:KRAS protein-protein interaction. *J. Med. Chem.* **2022**, *65*, 9678–9690.
7. X. Wang, S. Allen, J. F. Blake, V. Bowcut, D. M. Briere, A. Calinisan, J. R. Dahlke, J. B. Fell, J. P. Fischer, R. Gunn, J, Hallin, J. Laguer, J. Lawson, J. , Medwid, B. Newhouse, P, Nguyen, J. M. O'Leary, P. Olson, S. Pajk, L. Rahbaek, M. A. Marx. Identification of MRTX1133, a noncovalent, potent, and selective KRAS^{G12D} inhibitor. *J. Med. Chem.* **2022**, *65*, 3123–3133.

21.1 Sotorasib (Lumakras™) (89)

Structural Formula	Space-filling Model	Brief Information
$C_{30}H_{30}N_6O_3F_2$; MW = 561 clogP = 5.6; tPSA = 101		Year of discovery: 2018 Year on Introduction: 2021 Discovered by: Amgen Developed by: Amgen Primary target: KRAS(G12C) Binding type: covalent Class: non-kinase Treatment: KRAS(G12C) NSCLC Other name: AMG 510 Oral bioavailability: not reported Elimination half-life: 5.5 h Protein binding = 89%

Sotorasib is the first-in-class RAS GTPase family covalent inhibitor approved in 2021 for adult patients with KRAS(G12C) (designated as G12C throughout this section) mutated locally advanced or metastatic non-small cell lung cancer (NSCLC), as determined by an FDA-approved test, who have received at least one prior systemic therapy.[1]

RAS mutation is the most frequent oncogenic alteration in human cancers, with KRAS being the most mutated isoform. More specifically, this mutation accounts for 14% of all new cancer diagnoses worldwide each year (~18 million) and results in ~1 million deaths per year.[2] The most common oncogenic KRAS mutations are colorectal adenocarcinoma (43%), followed by pancreatic adenocarcinoma (20%) and NSCLC adenocarcinoma subtype (14%).

One of the most common KRAS mutations is G12C, which accounts for 29,700 new cancer diagnoses annually in the United States (lung, colon and, pancreatic), In comparison, the other two most common KRAS mutations, G12D and G12V, are found in 53,700 and 39,100 new cancer diagnoses, respectively.[2]

The G12C mutation generates a non-native cysteine, which contains a reactive SH group that easily forms covalent bonds with a variety of electrophilic chemical warheads. Because the wild-type (wt) KRAS lacks any cysteines in the active site, covalent attachment of an inhibitor to the mutant cysteine residue spares the wt KRAS and is thus mutant-specific. Such targeted cysteine active site inhibition was applied earlier to finding the Bruton tyrosine kinase (BTK) inhibitor ibrutinib. Sotorasib inhibition occurs via a mechanism similar to that for ibrutinib: attacking the mutant cysteine of G12C with an acrylamide Michael-acceptor, forming an irreversible, covalent bond with the protein and locking it in an inactive state that prevents downstream signaling without affecting wt KRAS.[1]

RAS proteins are not kinases per se, but they bind to and activate several families of downstream target protein kinases including RAF and PI3K, thereby regulating their signaling pathways.

It is worth noting that Amgen's sotorasib as the first G12C-targeted therapy was derived from Wellspring's ARS-1620 (Fig. 1), the first orally bioavailable small-molecule G12C covalent inhibitor that induced tumor regression in patient-derived tumor models. ARS-1620 was built upon the earlier discovery of electrophilic G12C inhibitors at the switch II pocket by Shokat et al. in 2013.

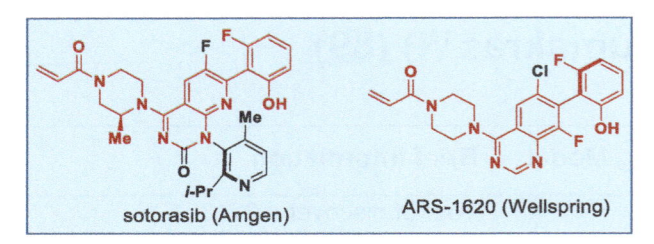

Figure 1. Structures of sotorasib and ARS-1620.

RAS function and activation[3–7]

The major RAS pathway consists of RAS, RAF(rapidly accelerated fibrosarcoma (RAF), mitogen-activated extracellular signal-regulated kinase (MEK), and extracellular signal-related kinase 1 and 2 (ERK1/2), which together comprise the RAS/RAF/MEK/ERK cascade (Fig. 2).[3] Upon activation of the cell-membrane receptors, an adaptor protein GRB2 recruits guanine nucleotide exchange factor (GEF) to the cell membrane. GEF is responsible for catalyzing the exchange of GDP for GTP on RAS, leading to a RAS–GTP complex which then triggers RAF activation through dimerization. The activated RAF proteins directly phosphorylate and activate MEK1/2, which in turn activate ERK1/2 proteins. Activated ERK1/2 then phosphorylates diverse substrates involved in cell proliferation, survival, growth, metabolism, migration, and differentiation, as summarized in Fig. 2.

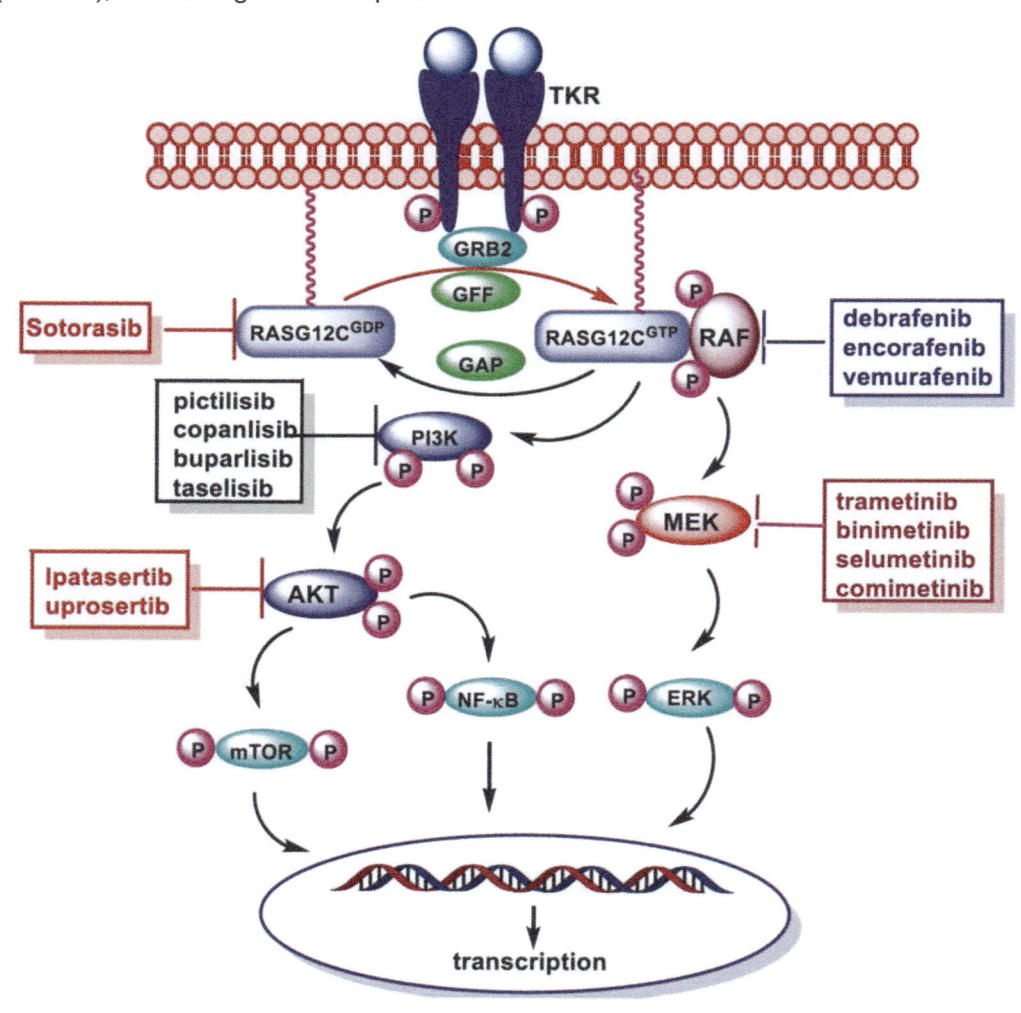

Figure 2. RAS signaling pathway.

RAS proteins serve as binary molecular switches cycling between the GTP-bound active state and the GDP-bound inactive state to transduce cell survival and growth signals. The activation of RAS proteins for further signaling occurs in the so-called G domain which consists of three regions: switch I (residues 25–40), switch II (residues 57–76), and the P loop (residues 10–17), which binds guanine nucleotides and activates signaling by interacting with effectors (Fig. 3).[6] The conformation of the two switch regions depends on whether the ligand is GDP or GTP. Thus, GDP and GTP control all nucleotide-dependent interactions with RAS-binding partners. In the GTP-bound state, Thr35 and Gly60 coordinate with the γ-phosphate/Mg^{++}, bringing the switch I and switch II regions together in the active conformation (Fig. 3A). Hydrolysis of GTP releases phosphate and returns these two regions to the inactive GDP conformation (Fig. 3B).[5,6]

In normal cells, the RAS activation via this mechanism is tightly controlled through a balance between GDP to GTP activation by GEF and GTP to GDP inactivation by GTPase-activating protein (GAP), but in RAS-mutated cancers, single nucleotide point mutations, particularly at codon 12, impair the regulated cycling between these two forms by disrupting the association of GAPs and driving the RAS pathway into a constitutively active, cancer-inducing state.

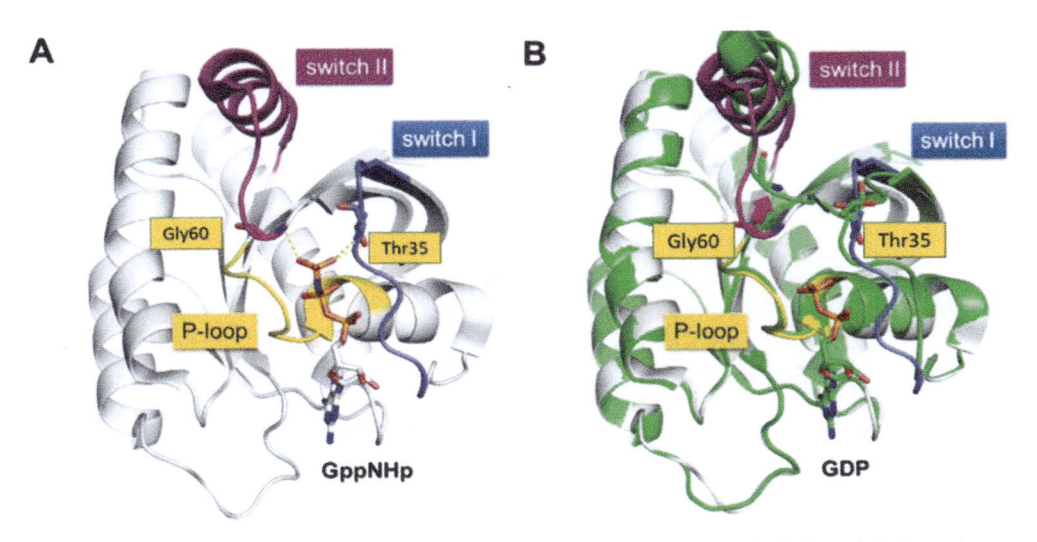

Figure 3. A: KRAS GppNHp-bound active conformation in which the switch I and II domains are bound to the γ-phosphate via the main chain NH groups of Thr35 and Gly60. B: Overlay of KRAS active (gray) and inactive (green) conformations. Dissociation of the γ-phosphate switches to the inactive state. Adapted with permission from *J. Med. Chem.* **2020**, *63*, 14404–14424.

KRAS: Undruggable for Four Decades[2]

Despite being one of the first oncogenes identified (1979), KRAS was often described as undruggable because KRAS is a small protein with a relatively smooth surface except for its GTP/GDP binding pocket and seemed to lack any other well-defined docking sites for small molecules. Furthermore, under the physiological conditions, GTP almost exclusively occupies the GTP/GDP-binding pocket due to its picomolar binding affinity along with a high cellular GTP concentration (~500 μM). Taken together, it appeared extremely challenging to identify a small-molecule inhibitor with sufficient intracellular concentration to compete with GTP at the GTP/GDP-binding pocket. For comparison, with typical tyrosine kinase inhibitors, the binding affinity with ATP is in the micromolar range (a million-fold difference in binding affinity), and yet ATP cellular concentrations are much lower, ranging from 1 to 10 μM. Consequently, a successful reversible binding competitive RAS inhibitor must possess low femtomolar binding affinity, which is even higher than that of streptavidin for biotin (K_D ~10^{-13} M): one of the strongest non-covalent interactions known in nature.

Because of these challenges with targeting RAS, therapeutic approaches had then shifted to its downstream effectors such as BRAF, PI3K, MEK and AKT, which have yielded modest efficacy due to various resistance mechanisms (Fig. 2). However, in

2013, Shokat et. al. discovered the switch II pocket of G12C and several electrophilic G12C irreversible inhibitors, which set the stage for sotorasib to become the first G12C inhibitor approved for NSCLC.

Discovery of G12C Inhibitors and Switch II Pocket[4,5]

To identify a lead compound for G12C inhibition, Shokat et. al. applied a disulfide-fragment-based screening approach, called tethering which covalently links the disulfide fragment with cysteine residues of G12C via rapid thiol exchange under partially reducing conditions (Fig. 4). This work ultimately led to a co-crystal structure of G12C disulfide conjugate **2** in the GDP state (Fig. 5B). The crystal structure shows the ligand binds to a new allosteric pocket beneath the effector binding switch II region near the mutant cysteine, termed the switch II pocket (S-IIP). This distinct allosteric pocket was not apparent in the apo-crystal structure (Fig. 5A) but was induced between the central β-sheet and switch II region upon ligand binding.

Figure 4. Tethering of G12C with disulfide ligand **1**.

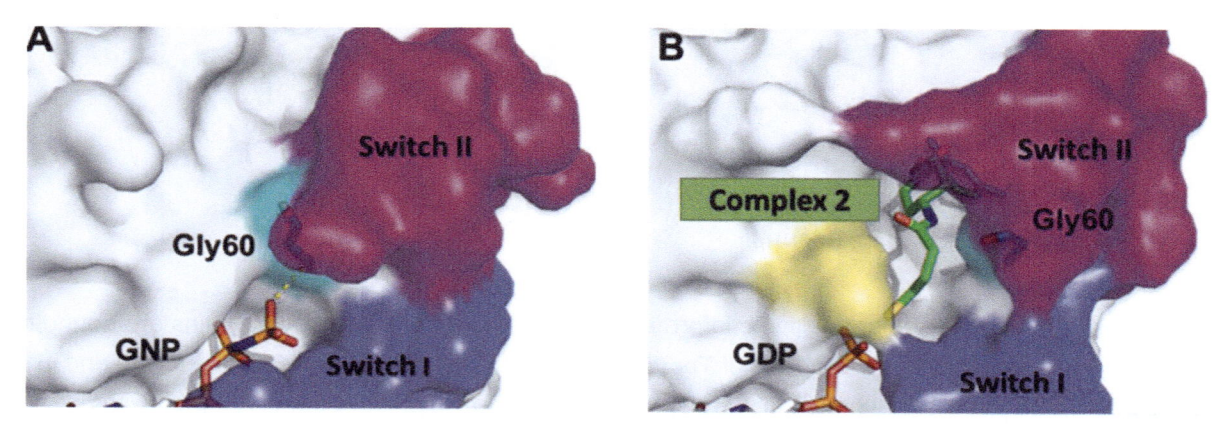

Figure 5. A: KRAS GppNHp-bound active conformation; B: KRAS G12C GDP complex **2**. The ligand penetrates into the switch II region and pushes Gly60 away from the nucleotide binding site. Adapted with permission from *J. Med. Chem.* **2020**, *63*, 14404–14424.

The covalent binding of an electrophile to the mutant cysteine sulfur in the S-IIP leads to the displacement of Gly60 in the switch II region (residues 60–76), toward the switch I region (residues 30–38), causing the disordering of the switch I region and loss of Mg^{2+}. Because the GTP-bound state of RAS is highly sensitive to conformational changes in the regions of Gly60 (switch II region) and Thr35 (switch I region), the inhibitor binding to the S-IIP is expected to exert significant impact on the orientations of these two key residues and the subsequent RAS activation. Overall, through covalent binding, the protein is locked in the inactive GDP-bound conformation to prevent GEF-catalyzed nucleotide exchange and block subsequent effector pathways.

To identify more potent inhibitors, Shokat et al. replaced the disulfide ligands with acrylamides and vinyl sulfonamides for direct covalent linkage to the mutant cysteine. Two compounds, **3** and **4** (Fig. 6), provided the highest degree of protein modification, and a co-crystal structure of compound **4** bound to G12C showed a similar binding mode to disulfide conjugate **2**.

The identification of the previously unknown targetable binding pocket S-IIP containing G12C and first two covalent switch II pocket inhibitors **3** and **4** via a 6-year fragment-based screening effort represented a major breakthrough in molecular targeting of RAS-driven cancers.

Figure 6. Structures of first two G12C specific irreversible inhibitors.

ARS-1620: the First Orally Active G12C Inhibitor[8,9]

Although the early lead compound **3** showed only weak cellular activity (IC_{50} > 100 μM), it could be improved to low micromolar potency by structural modification to ARS-853 (IC_{50} = 1.5 μM) (Fig. 7) which had poor oral bioavailability (F < 2% in mice) due to low metabolic stability in plasma. Unfortunately, tight structure–activity relationships prevented the ARS-853 chemotype from further potency optimization. Subsequent scaffold hopping and truncation of the flexible 2-amino-1-(piperazin-1-yl)ethan-1-one linker gave rise to a quinazoline-based S-atropisomer ARS-1620 that showed 10-fold improved potency over ARS-853 (in a biochemical assay) and 63% oral bioavailability in mice (Fig. 7).

The co-crystal structure of ARS-1620 bound to G12C revealed a unique binding mode and trajectory from earlier S-IIP G12C inhibitors to reach Cys12 (Fig. 8). The additional hydrogen bond between N1 and the His95 side chain of KRAS enables a more rigid conformation for the covalent bond formation, while the hydroxyl group of the fluorophenol is involved in a network of water-mediated hydrogen bonds with Arg68, Asp69, and Gln99 (Fig. 8). All of these hydrogen bonds contribute to its enhanced activity over ARS-853. The G12C protein covalently bound to ARS-1620 lost SOS- and EDTA-mediated nucleotide exchange capability, consistent with a G12C GDP-bound inactive state.

Because of its improved potency and pharmacologic properties, ARS-1620 alone elicited significant oral efficacy in multiple human cancer cell line- and patient-derived mouse xenograft tumor models, thereby providing the first in vivo evidence to support the S-IIP targeted therapy for G12C-driven cancers.

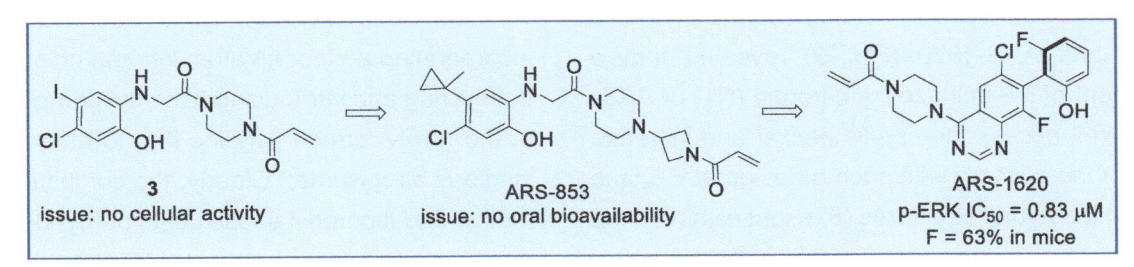

Figure 7. Structural optimization of ARS-1620.

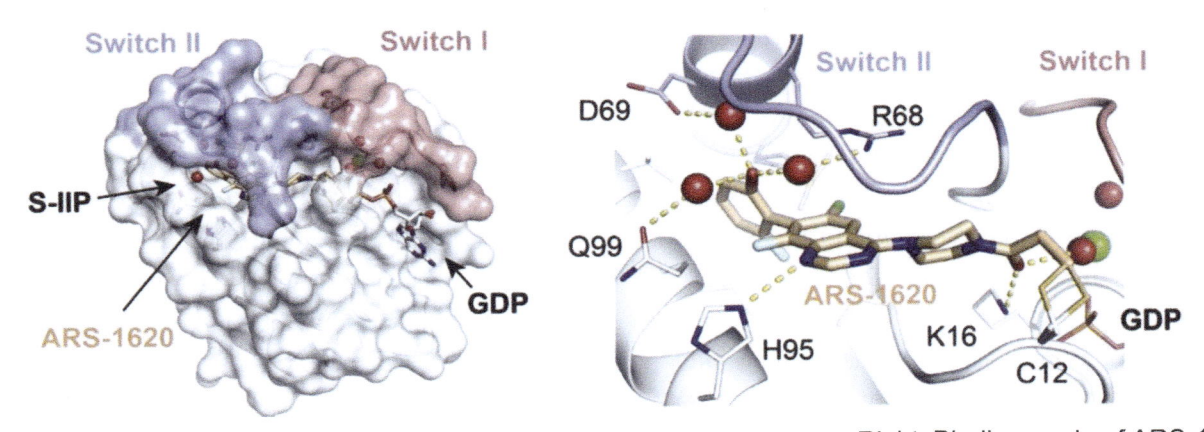

Figure 8. Left: ARS-1620 (gold) and GDP-bound G12C co-crystal structure, Right: Binding mode of ARS-1620 and G12C. Adapted with permission from *Cell*, **2018**, *172*, 578–589.e17.

From ARS-1620 to Sotorasib[10,11]

Despite its favorable pharmacokinetic profile with 63% oral bioavailability in mice, ARS-1620 showed only modest potency in a cell-based phospho-ERK1/2 immunoassay (p-ERK IC_{50} = 0.83 µM). In the meantime, Amgen in collaboration with Carmot Therapeutics independently identified an indole/acrylamide lead compound **5** (Fig. 9) which potently inactivated G12C in biochemical and cellular assays (p-ERK IC_{50} = 0.22 µM). As compared with ARS-1620, the tetrahydroisoquinoline ring of **5** occupies a previously unexploited cryptic pocket on the "lid" of the switch II pocket of KRAS formed by a cluster of residues including His95, Tyr96, and Gln99 (Fig. 9), and engagement of this cryptic pocket contributes to its potency. However, the high clearance and low oral bioavailability in rodent model systems blocked further in vivo studies with **5**. Because ARS-1620 and **5** are complimentary in terms of potency and PK properties, it was thought that hybrid molecules that exploit the His95/Tyr96/Gln99 cryptic pocket might provide potent inhibitors with favorable oral bioavailability.

Superposition of the binding modes of indole lead **5** and ARS-1620 (Fig. 9) revealed that a substituent of the quinazoline nitrogen (N1) of ARS-1620 might access the cryptic pocket and increase activity. One concern with such a design at the time was that the hybrid structures (**6**) might eliminate the hydrogen bond between N1 and the His95 side chain of KRAS present in ARS-1620 and compromise potency. Nonetheless, various hybrid analogs were synthesized leading to quinazolinone **7** as a new lead with a ~6-fold improvement over ARS-1620 in cellular potency (p-ERK IC_{50} = 0.13 µM vs. 0.83 µM), apparently due to the isopropylphenyl substituent which is in close contact with the Tyr96, His95, and Gln99 residues of the cryptic pocket as shown in the co-crystal structure of **7** with GDP-G12C (Fig. 9).

Although **7** was much more potent than ARS-1620 in inhibiting G12C signaling, it showed no measurable oral bioavailability in mice due to (1) low membrane permeability; (2) relatively poor aqueous solubility; and (3) metabolism of the piperazine moiety. To improve membrane permeability, a nitrogen was incorporated at the C8 position of the quinazolinone ring (to form azaquinazolinone), which serves as an internal hydrogen-bond acceptor to mask the adjacent phenol hydrogen bond donor (i.e., to eliminate one hydrogen bond donor which can have a deleterious effect on the drug's membrane partition and permeability). The newly formed quinazolinone ring also improved aqueous solubility. The metabolic instability of the piperazine ring was mitigated by introducing a methyl group at the piperazine C2 position, which also improved the inhibitory activity. Taken together, these modifications led to azaquinazolinone *R*-atropisomer **8** which demonstrated significantly enhanced oral bioavailability (33%) and excellent cellular activity (p-ERK IC_{50} = 44 nM). The C2 methyl group of the piperazine is critical for oral bioavailability as the unsubstituted analog showed only 1.3% oral bioavailability. However, the two separable atropisomers **8** and **9** underwent slow interconversion at 25 °C with a half-life of 8 days, compromising the further development of **8** as a single drug molecule.

The atropisomerism issue was overcome by incorporating a nitrogen atom into the cryptic pocket phenyl ring and introducing a methyl group to the C4 of the newly formed pyridine ring to afford **10** as a single *R* atropisomer. Clearly, the combined bulk of methyl and isopropyl substituents on pyridine raises the energetic barrier to atropisomer interconversion to allow a stable *R* atropisomer. Compound **10** showed enhanced aqueous solubility but substantially reduced MDCK permeability, which led to no measurable oral bioavailability in mice. Fortunately, a single replacement of the C6 chlorine with a fluorine (to give sotorasib) increased oral bioavailability from 0% to 22–40% (as a crystalline form) due to improvement in solubility and permeability, which in turn arises from reduction in lipophilicity (clogP from 6.0 to 5.6). Sotorasib, the *R*-atropisomer, was found to be configurationally stable with a calculated racemization half-life of >180 years at 25 °C. Although sotorasib was not one of the most potent analogs in the series (p-ERK IC_{50} = 68 nM), the overall balance of in vitro, in vivo, and PK properties, combined with a favorable safety profile, made it the top drug candidate.

Sotorasib exhibits a relatively short half-life of 5.5 hours, but it is still approved as a once-daily dosing administration because of its relatively high dosage (960 mg).

Although sotorasib contains an electron-rich phenol which is potentially vulnerable to CYP450 oxidation to give reactive quinonoid metabolites, metabolite profiling of sotorasib in rats showed that such reactive metabolites were not formed from the phenol moiety.[12]

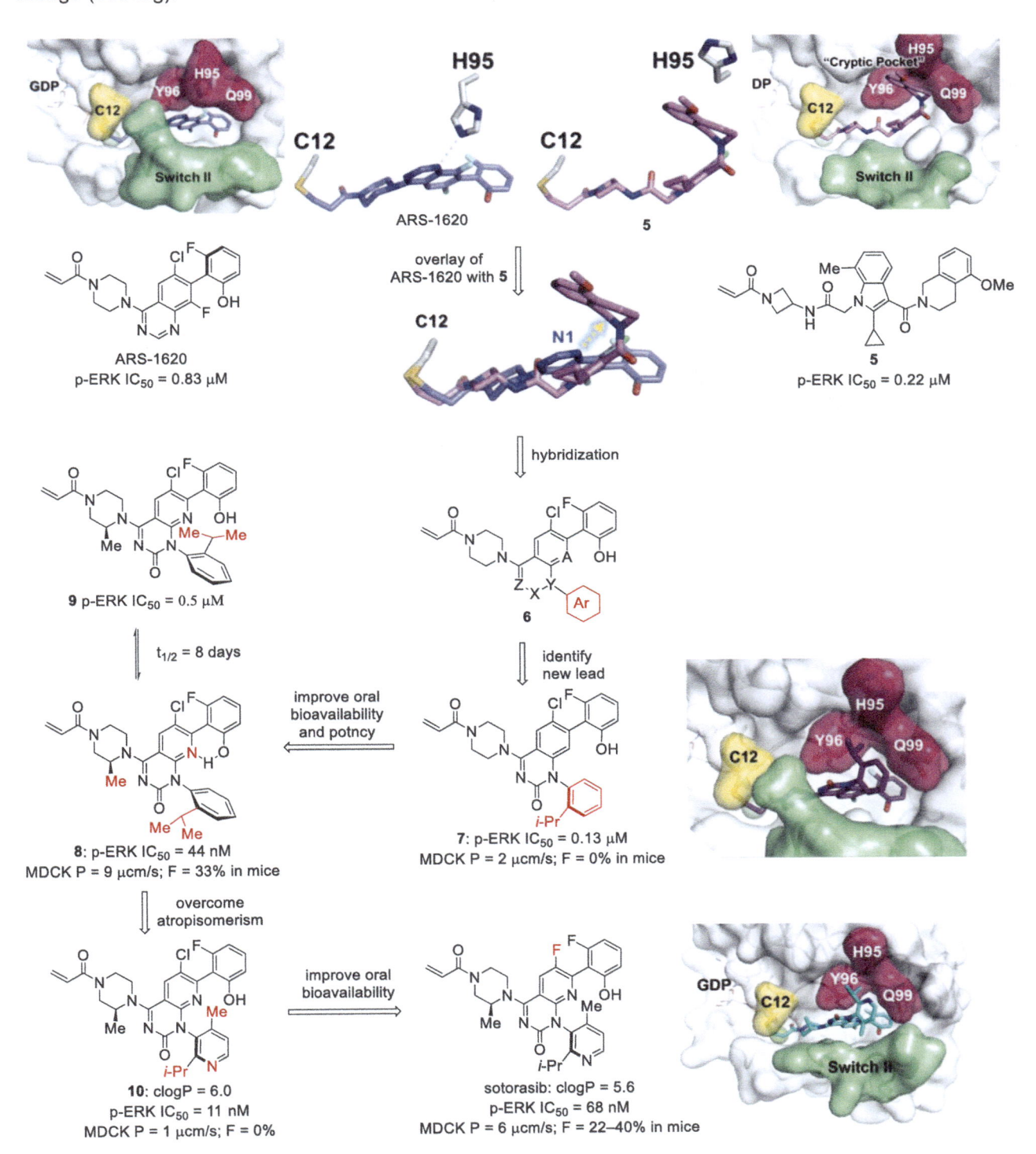

Figure 9. Design and structural optimization of sotorasib. X-ray structures adapted with permission from *J. Med. Chem.* **2020**, *63*, 52–65.

Binding Mode of Sotorasib[10,13]

In the co-crystal structure of sotorasib in complex with G12C (Fig. 11), the quinazolinone core occupies the KRAS switch II pocket, and the acrylamide moiety forms a covalent bond with the active site Cys12 (Figs. 10, 11). The (S)-methylpiperazine ring is oriented in a twist-boat conformation, with the C2 methyl substituent making hydrophobic interactions with Cys12 and Tyr96. In addition, the isopropyl substituent of the pyridyl ring fits the cryptic pocket consisting of a cluster of residues from Tyr96, His95, and Gln99, contributing to the high potency. Importantly, unlike adagrasib (see Section 21.2), sotorasib avoids direct interaction with His95, a recognized site mutation for resistance.[14]

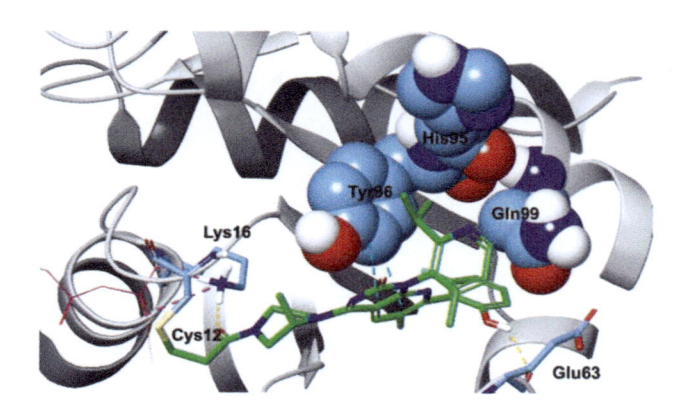

Figure 10. Co-crystal structure of sotorasib-G12C (PDB ID: 6OIM).

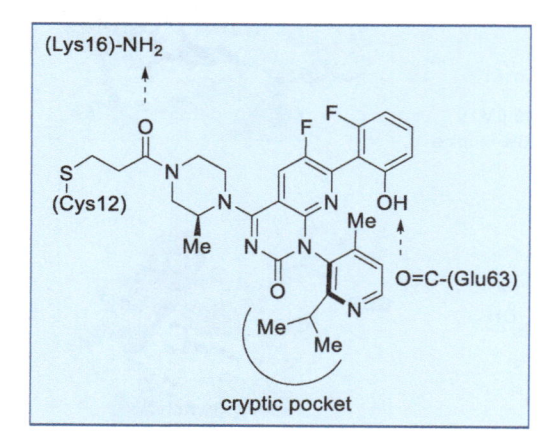

Figure 11. Sotorasib–G12C interactions based on the co-crystal structure.

1. H. A. Blair. Sotorasib: First Approval. Drugs. **2021**, *81*, 1573–1579.
2. A. K. Kwan, G. A. Piazza, A. B. Keeton, C. A. Leite. The path to the clinic: A comprehensive review on direct KRAS[G12c] inhibitors. *J. Exp. Clin. Cancer Res.* **2022**, *41*, 27.
3. M. Trott. The ras pathway and cancer: Regulation, challenges and the therapeutic progress. Technology Networks. May 04, 2021.
4. J. M. Ostrem, U. Peters, M. L. Sos, J. A. Wells and K. M. Shokat. K-Ras(G12C) inhibitors allosterically control GTP affinity and effector interactions. *Nature.* **2013**, *503*, 548–551.
5. J. M. Ostrem, K. M. Shokat. Direct small-molecule inhibitors of KRAS: from structural insights to mechanism-based design. *Nat. Rev. Drug Discov.* **2016**, *15*, 771–785.
6. H. Chen, J. B. Smaill, T. Liu, K. Ding, X. Lu. Small-molecule inhibitors directly targeting KRAS as anticancer therapeutics. *J. Med. Chem.* **2020**, *63*, 14404–14424.
7. C. R. Lindsay, M. C. Garassino, E. Nadal, K. Öhrling, M. Scheffler. On target: Rational approaches to KRAS inhibition for treatment of non-small cell lung carcinoma. *Lung Cancer.* **2021**, *160*, 152–165.
8. M. P. Patricelli, M. R. Janes, L. S. Li, R. Hansen, U. Peters, L. V. Kessler, Y. Chen, J. M. Kucharski, J. Feng, T. Ely, J. H. Chen, S. J. Firdaus, A. Babbar, P. Ren, Y. Liu. Selective inhibition of oncogenic KRAS output with small molecules targeting the inactive state. *Cancer Discov.* **2016**, *6*, 316–329.
9. M. R. Janes, J. Zhang, L. Li, R. Hansen, U. Peters, X. Guo, Y. Chen, A. Babbar, S. J. Firdaus, L. Darjania, J. Feng, J. H. Chen, S. Li, Y. O. Long, C. Thach, Y. Liu, A. Zarieh, T. Ely, J. M. Kucharski, L. V. Kessler, T. Wu, K. Yu, Y. Wang, Y. Yao, X. Deng, P. P. Zarrinkar, D. Brehmer, D. Dhanak, M. V. Lorenzi, D. Hu-Lowe, M. P. Patricelli, P. Ren, Y. Liu. Targeting KRAS mutant cancers with a covalent G12C-specific inhibitor. *Cell.* **2018**, *172*, 578–589.
10. B. A. Lanman, J. R. Allen, J. G. Allen, A. K. Amegadzie, K. S. Ashton, S. K. Booker, J. J. Chen, N. Chen, M. J. Frohn, G. Goodman, D. J. Kopecky, L. Liu, P. Lopez, J. D. Low, V. Ma, A. E. Minatti, T. T. Nguyen, N. Nishimura, A. J. Pickrell, A. B. Reed, V. J. Cee. Discovery of a covalent inhibitor of KRAS[G12C] (AMG 510) for the treatment of solid Tumors. *J. Med. Chem.* **2020**, *63*, 52–65.
11. Y. Shin, J. W. Jeong, R. P. Wurz, P. Achanta, T. Arvedson, M. D. Bartberger, I. D. Campuzano, R. Fucini, S. K. Hansen, J. Ingersoll, J. Iwig, J. R. Lipford, V. Ma, D. J. Kopecky, T. San Miguel, C. Mohr, S. Sabet, A. Y. Saiki, A. Sawayama, S. Sethofer,

C. M. Tegley, L. P. Volak, K. Yang, B. A. Lanman, D. A. Erlanson, V. J. Cee. Discovery of N-(1-aryloylazetidin-3-yl)-2-(1H-indol-1-yl)acetamides as a covalent inhibitor of KRASG12C. *ACS Med. Chem. Lett.* **2019**, *10*, 1302−1308.

12. J. A. Werner, R. Davies, J. Wahlstrom, U. P. Dahal, M. Jiang, J. Stauber, B. David, W. Siska, B. Thomas, K. Ishida, W. G. Humphreys, J. R. Lipford, T. M. Monticello. Mercapturate pathway metabolites of sotorasib, a covalent inhibitor of KRAS^{G12C}, are associated with renal toxicity in the Sprague Dawley rat. *Toxicol. Appl. Pharmacol.* **2021**, *423*, 115578.

13. Y. Li, L. Han, Z. Zhang. Understanding the influence of AMG 510 on the structure of KRAS^{G12C} empowered by molecular dynamics simulation. *Comput. Struct. Biotechnol. J.* **2022**, *20*, 1056–1067.

14. N. S. Persky, D. E. Root, K. E. Lowder, H. Feng, S. S. Zhang, K. M. Haigis, Y. P. Hung, L. M. Sholl, A. J. Aguirre. Acquired resistance to KRAS^{G12C} inhibition in cancer. *New Engl. J. Med.* **2021**, *384*, 2382–2393.

21.2 Adagrasib (Krazati™) (90)

Structural Formula	Space-filling Model	Brief Information
$C_{32}H_{35}ClFN_7O_2$; FW = 604 clogP = 5.8; tPSA = 87		Year of discovery: 2017 Year of Introduction: 2022 Discovered by: Array/Mirati Developed by: Mirati Primary target: KRAS(G12C) Binding type: covalent Class: non-kinase Treatment: KRAS(G12C) NSCLC Other name: MRTX849 Oral bioavailability: not reported Elimination half-life: 24 h Protein binding = 98.3%

● = C   ○ = H   ● = O   ● = N   ● = S   ● = F   ● = Cl   ● = Br   ● = I

Adagrasib, an irreversible inhibitor of the RAS GTPase family, was granted accelerated approval on December 12, 2022, for adult patients with KRAS(G12C)-mutated locally advanced or metastatic non-small cell lung cancer (NSCLC), as determined by an FDA-approved test. It became the second G12C-targeted therapy, following 19 months after Amgen's sotorasib.

The two inhibitors appear to be comparable for NSCLC, but adagrasib may have an advantage for other cancers, including colorectal cancer, pancreatic and gastrointestinal tumors.[1]

Both inhibitors have also demonstrated efficacy for the treatment of central nervous system metastases, which occur in 27–42% of patients with G12C-mutated NSCLC at diagnosis. Whereas adagrasib was tested in NSCLC patients who had never received treatment for brain lesions, sotorasib was tested in individuals who had received prior treatment, so direct comparison is mute.[1]

Both inhibitors are associated with liver toxicity, and sotorasib carries a label warning of interstitial lung disease. Adagrasib exhibited greater risk of sudden cardiac death due to QT prolongation, which occurred in 14% of patients treated with adagrasib, but that was not observed with sotorasib.[1]

Although adagrasib has a much longer half life than sotorasib (24 vs. 5.5 h), it is dosed twice a day (2 × 600 mg) vs. once a day for sotorasib (900 mg).

Discovery of Adagrasib[2,3]

Following the discovery of the previously unknown targetable switch II binding pocket containing G12C and the first two covalent switch II pocket inhibitors by Shokat et. al. in 2013,[4,5] Array BioPharma carried out a covalent fragment screening in a G12C protein modification assay. Subsequent hit elaboration and a structure-based drug design effort identified **1** as a very weak G12C inhibitor (Fig. 1). Two structural changes to **1** resulted in compound **2** with cellular IC_{50} value of 142 nM: (1) incorporation of a hydroxyl onto C3 of the naphthyl; and (2) C2 substitution of the tetrahydropyridopyrimidine with (1-methylpyrrolidin-2-yl)methoxy. The naphthyl hydroxyl hydrogen bonds with the carboxylate side chain of Asp69, while the pyrrolidine amine forms a salt-bridge with the carboxylate of Glu62 as well as a cation–π interaction with His95. These interactions resulted in the high potency of **2**.

Although **2** demonstrated rapid tumor regressions in rodent models, its high clearance and poor oral bioavailability (2.4%) in mice hampered its further development. Incubation of **2** with mouse hepatocytes showed the following major metabolites: (1) O-glucuronides or O-sulfate, and (2) glutathione adduct of conjugation of the acrylamide. Given the metabolite profile, removal of the hydroxyl moiety was expected to reduce its phase II metabolism and overall clearance. Indeed, the des-hydroxyl analog **3** showed a much lower clearance and 14% bioavailability in mice. Not surprisingly, it lost activity because it lacks the hydroxyl to form a hydrogen bond with the carboxylate side chain of Asp69.

Figure 1. Structural optimization of adagrasib.

The co-crystal structure of **3** bound to G12C (Fig. 2) revealed a bound water molecule complexed to Gly10 and Thr58 which are in the neighborhood of the piperazine ring. This water molecule forms hydrogen bonds with the side chain hydroxyl of Thr58 and the carbonyl of Gly10 as indicated in Fig. 1. Because solvent organization can play a decisive role in the binding affinity, exploiting water may provide an opportunity to optimize ligand binding. Displacement of an energetically favorable water molecule from the binding site by the ligand can reduce the binding affinity, while it is desirable to remove energetically unfavorable water molecules. Although no WaterMap studies were carried out at the time to assess this particular water molecule, it was thought to be a high-energy trapped water and its displacement could lead to a significant potency boost.[6,7] Modelling studies suggested that an appropriately installed substituent on piperazine might cause exclusion of this bound water molecule.

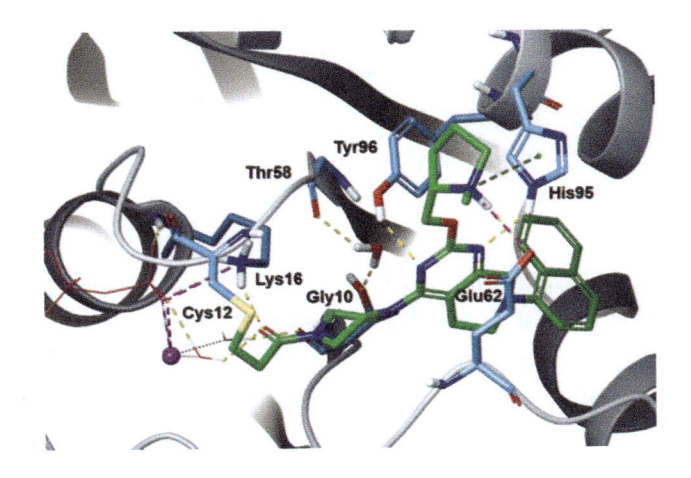

Figure 2. Compound **3** bound to G12C (6USX).

Based on the bound water hypothesis, various substituents were introduced at the piperazine subunit, leading to **4**, substituted at the C1 position with CH_2CN, which showed a 440-fold increase in cellular potency (IC_{50} = 10 nM), which can be attributed to water displacement as well as the enhanced electrophile reactivity. Incubation of **3** with GSH at 37 °C showed the disappearance $t_{1/2}$ (a measure of electrophile reactivity) of 17 h (based on extrapolation), whereas the $t_{1/2}$ for **4** was only 4 h, indicating that the electron-withdrawing cyanomethyl substituent increased the reactivity of the acrylamide toward GSH. However, too much reactivity can cause excessive clearance due to glutathione S-transferase (GST)-mediated glutathione (GSH) conjugation, which was addressed later (*vide infra*).

The co-crystal structure of **4** bound to G12C (Fig. 3) revealed a narrow hydrophobic pocket near the C8 of the naphthyl (highlighted with a red arrow in Fig. 3), which was exploited by a chlorine substituent to give **5** as one of the most potent inhibitors in the series with cellular IC_{50} of 1 nM.

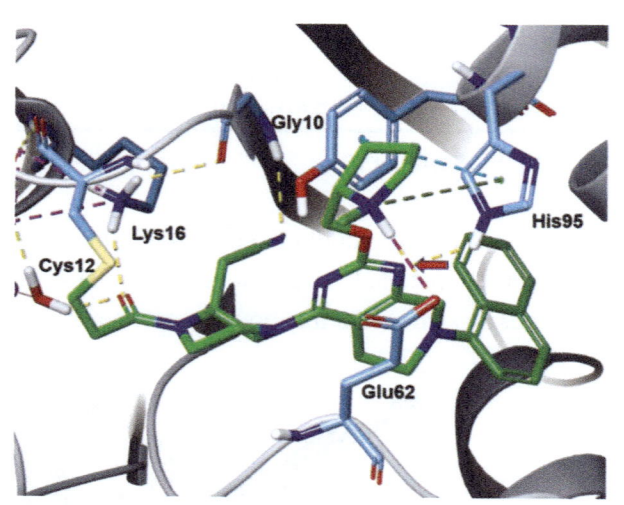

Figure 3. X-ray crystal structure of compound **4** to G12C (6USZ).

Despite its potent G12C inactivation in biochemical and cellular assays as well as rapid tumor regression in vivo, compound **5** exhibited only modest oral bioavailability (17% in mice and 4% in dogs) due to extra-hepatic clearance from the GST-mediated GSH conjugation. To reduce reactivity toward GSH, substituted acrylamide analogs were prepared, leading to the α-fluoro substituted acrylamide derivative, adagrasib, with significantly improved oral bioavailability across species (63% in mice; 30% in rats; 26% in dogs).

Adagrasib selectively modified mutant cysteine 12 in GDP-bound G12C and inhibited KRAS-dependent signaling. It demonstrated significant tumor regression in 17 of 26 (65%) G12C-positive cell line- and patient-derived xenograft models from multiple tumor types.[8]

Binding Mode[3]

In the co-crystal structure of adagrasib in complex with G12C (Fig. 4), the tetrahydropyridopyrimidine core occupies the KRAS switch II pocket, forming a covalent bond between the active site Cys12 sulfur and the acrylamide β-carbon. The acrylamide carbonyl hydrogen bonds to the side chain amine of Lys16, while the N1 of the pyrimidine ring interacts directly with the imidazole NH of His95, which is a recognized site mutation for resistance.[9] The cyanomethyl substituent displaces the Gly10 bound water and forms a hydrogen bond to the backbone NH of Gly10. In addition, the pyrrolidine amine forms a salt-bridge with the carboxylate side chain of Glu62, and the 8-chloronaphthyl fills the lipophilic pocket (Figs. 4, 5).

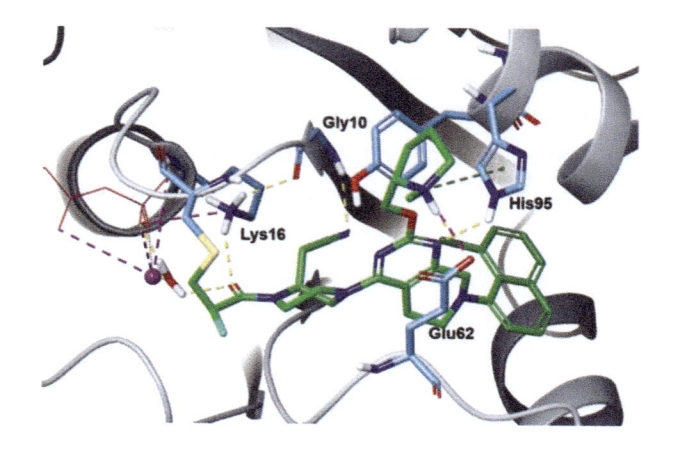

Figure 4. Co-crystal structure of adagrasib-G12C (PDB ID: 6UT0).

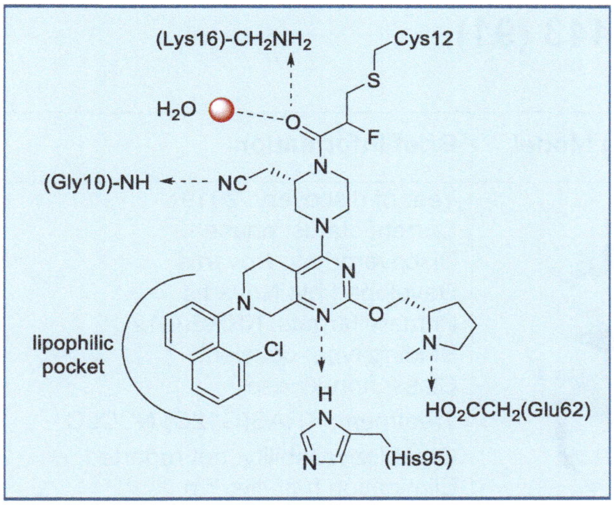

Figure 5. Summary of adagrasib–G12C interactions based on the co-crystal structure.

1. M. Terry. Mirati's KRAS inhibitor looks good, but is it good enough to beat Amgen? Biospace. June 06, 2022.
2. J. B. Fell, J. P. Fischer, B. R. Baer, J. Ballard, J. F. Blake, K. Bouhana, B. J. Brandhuber, D. M. Briere, L. E.Burgess, M. R. Burkard, H. Chiang, M. J. Chicarelli, K. Davidson, J. Gaudino, J. Hallin, L. Hanson, K. Hee, E. J. Hicken, R. J. Hinklin, M. A. Marx, J. G. Christensen. Discovery of tetrahydropyrido-pyrimidines as irreversible covalent inhibitors of KRAS-G12C with in vivo activity. *ACS Med. Chem. Lett.* **2018**, *9*, 1230–1234.
3. J. B. Fell, J. P. Fischer, B. R. Baer, J. F. Blake, K. Bouhana, D. M. Briere, K. D. Brown, L. E. Burgess, A. C. Burns, M. R. Burkard, H. Chiang, M. J. Chicarelli, A. W. Cook, J. J. Gaudino, J. Hallin, L. Hanson, D. P. Hartley, E. J. Hicken, G. P. Hingorani, R. J. Hinklin, M. A. Marx. Identification of the clinical development candidate MRTX849, a Covalent KRAS[G12C] inhibitor for the treatment of cancer. *J. Med. Chem.* **2020**, *63*, 6679–6693.
4. J. M. Ostrem, U. Peters, M. L. Sos, J. A. Wells and K. M. Shokat. K-Ras(G12C) inhibitors allosterically control GTP affinity and effector interactions. *Nature.* **2013**, *503*, 548–551.
5. J. M. Ostrem, K. M. Shokat. Direct small-molecule inhibitors of KRAS: from structural insights to mechanism-based design. *Nat. Rev. Drug Discov.* **2016**, *15*, 771–785.
6. J. Michel, J.Tirado-Rives, W. L. Jorgensen. Energetics of displacing water molecules from protein binding sites: consequences for ligand optimization. *J. Am. Chem. Soc.* **2009**, *131*, 15403–15411.
7. S. Andreev, T. Pantsar, R. Tesch, N. Kahlke, A. El-Gokha, F. Ansideri, L. Grätz, J. Romasco, G. Sita, C. Geibel, M. Lämmerhofer, A. Tarozzi, S. Knapp, S. A. Laufer, P. Koch. Addressing a trapped high-energy water: Design and synthesis of highly potent pyrimidoindole-based glycogen synthase kinase-3β inhibitors. *J. Med. Chem.* **2022**, *65*, 1283–1301.
8. J. Hallin, L. D. Engstrom, L. Hargis, A. Calinisan, R. Aranda, D. M. Briere, N. Sudhakar, V. Bowcut, B. R. Baer, J. A. Ballard, M. R. Burkard, J. B. Fell, J. P. Fischer, G. P. Vigers, Y. Xue, S. Gatto, J. Fernandez-Banet, A. Pavlicek, K. Velastagui, R. C. Chao, J. G. Christensen. The KRAS[G12C] inhibitor MRTX849 provides insight toward therapeutic susceptibility of KRAS-mutant cancers in mouse models and patients. *Cancer Discov.* **2020**, *10*, 54–71.
9. N. S. Persky, D. E. Root, K. E. Lowder, H. Feng, S. S. Zhang, K. M. Haigis, Y. P. Hung, L. M. Sholl, A. J. Aguirre. Acquired resistance to KRAS[G12C] inhibition in cancer. *New Engl. J. Med.* **2021**, *384*, 2382–2393.

21.3 JDQ443 (91)

Structural Formula	Space-filling Model	Brief Information
$C_{29}H_{28}ClN_7O$; MW = 526 clogP = 3.2; tPSA = 76		Year of discovery: 2019 Current status: phase II Discovered by: Novartis Developed by: Novartis Primary targets: KRAS(G12C) Binding type: covalent Class: non-kinase Treatment: KRAS(G12C) NSCLC Oral bioavailability: not reported Elimination half-life: 7 h Protein binding = not reported

● = C ● = H ● = O ● = N ● = S ● = F ● = Cl ● = Br ● = I

JDQ443 is a RAS GTPase family covalent inhibitor currently in phase II trial in patients with non-small cell lung cancer (NSCLC) harboring the KRASG12C (designated as G12C throughout this section) mutation who have received at least one prior systemic therapy.

The data from phase 1/2 trials suggest that JDQ443 is at least as potent as Amgen's sotorasib for NSCLC patients. The challenge for Novartis is that both sotorasib and adagrasib were approved in 2021–2022, while JDQ443's phase III studies will not even start until later in 2022. Furthermore, the phase III trial will compare JDQ443 against docetaxel, not sotorasib; therefore, it remains unclear if a pivotal trial in NSCLC without a head-to-head against sotorasib will be sufficient.[1]

One advantage of JDQ443 is its much lower dosage: 200 mg twice daily vs. 600 mg twice daily for adagrasib and 900 mg once daily for sotorasib.

Discovery of JDQ443[2,3]

Compound **1**, a quinazoline-based G12C inhibitor from an Araxes Pharma patent application, was used in the design of Novartis' hybrid inhibitors (Fig. 1). The co-crystal structure of **1** revealed the critical role of the indazole fragment, which forms a hydrogen bond with the Asp69 side chain and the Ser65 backbone via a bound water molecule. Acrylamide **2** is a weak HTS hit from Novartis identified by a mass spectrometry–based screening of G12C protein modification. The hybrid inhibitor **3** retained the indazole fragment of **1** and phenyl acrylamide of **2** with the middle linker to be determined by in silico modelling. This process led to a dimethylpyrazole analog **4** with modest G12C activity. This compound exists as a stable, single atropisomer because the two methyl groups block the free rotation around the indazole–pyrazole bond.

In addition to its weak activity, **4** also suffered from high chemical reactivity with endogenous glutathione as indicated by a disappearance $t_{1/2}$ of 84 min in the glutathione (GSH) incubation assay. The acrylamide reactivity was reduced by replacing the aniline linker with a spirocyclic azetidine moiety to give **5** which showed significantly improved stability toward GST conjugation ($t_{1/2}$ = 241 min). Extensive structural optimizations on potency and PK properties led to JDQ443 with a balanced profile of the G12C activity and GST stability (Fig. 1).

JDQ443 demonstrated covalent, irreversible G12C binding and low-reversible binding affinity to the KRAS switch II pocket in the GDP-bound state. The mutant selectivity of JDQ443 extended into cellular systems as shown by its potent and selective inhibition of effector recruitment to G12C. In addition, JDQ443 resulted in dose-dependent reductions of phosphorylated ERK (pERK) levels with IC_{50} of 18 nM, vs. 14 and 11 nM for sotorasib and adagrasib, respectively.

Table 1 shows the overall biochemical and cellular profiles of three G12C inhibitors. In biochemical assays, adagrasib showed 2.4 μM reversible binding affinity to wt KRAS in the GDP bound state, much lower than JDQ443 and sotorasib. In cellular assays, unlike adagrasib, both JDQ443 and sotorasib inhibited the proliferation of G12C/His95 double mutants, consistent with the lack of direct interactions between the inhibitor and His95 in their co-crystal structures. Overall, all three inhibitors exhibit similar in vitro profiles.

In phases I and II trials, oral administration of JDQ443 at 200 mg, twice a day, demonstrated tumor responses across all cancer types including NSCLC, colorectal cancer, pancreatic cancer and ovarian cancer. Since it is likely to enter phase III by the end of 2022 it could become the third G12C-targeted therapy, following after adagrasib.

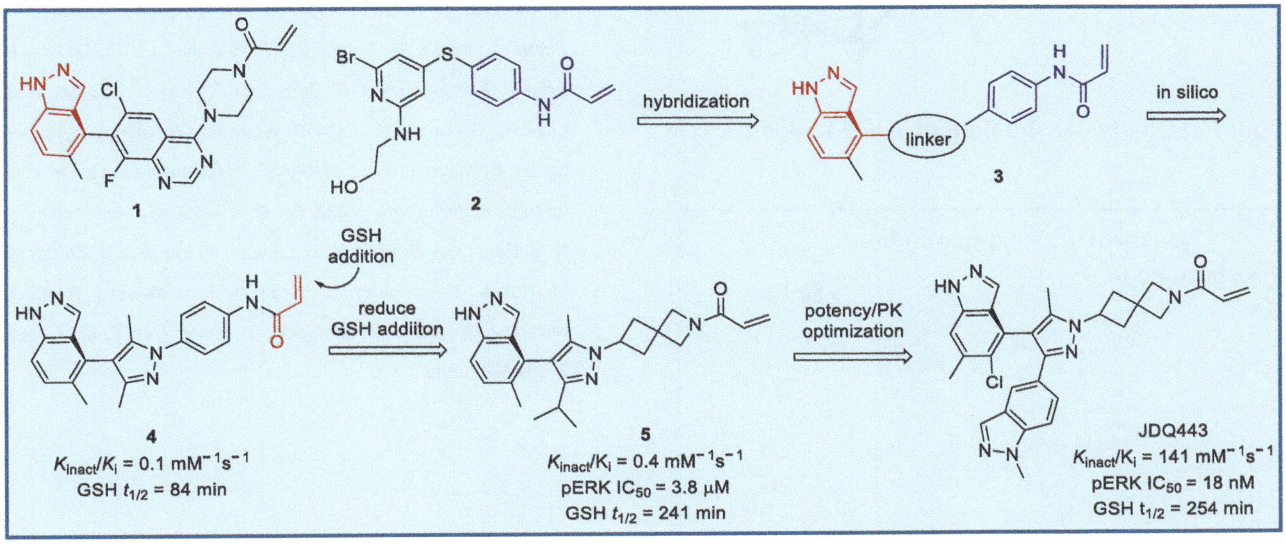

Figure 1. Structural optimization of JDQ443.–

Table 1. Biochemical and cellular profiles of JDQ443, sotorasib, and adagrasib[2]

G12C inhibitor	JDQ443	sotorasib	adagrasib
K_{inact}/K_i (mM^{-1}s^{-1}) (biochemical)	141	54	446
KRAS wt K_D (μM) (biochemical)	30.6	76.2	2.4
CRAF recruitment IC$_{50}$ for G12C (cellular)	12 nM	17 nM	10 nM
CRAF recruitment IC$_{50}$ for wt KRAS (cellular)	>1 μM	>1μM	>1 μM
pERK IC$_{50}$ (NCI-H3558, cellular)	18 nM	14 nM	11 nM
proliferation GI$_{50}$ (NCI-H3558, cellular)	19 nM	36 nM	39 nM
proliferation GI$_{50}$ (G12C, cellular)	115 nM	389 nM	136 nM
proliferation GI$_{50}$ (G12C/H95R, cellular)	24 nM	33 nM	>1 μM
proliferation GI$_{50}$ (G12C/H95Q, cellular)	284 nM	233 nM	>1 μM
proliferation GI$_{50}$ (G12C/H95D, cellular)	612 nM	262 nM	>1 μM

Binding Mode[2,3]

In the cocrystal structure of JDQ443 in complex with G12C (Fig. 2). The rigid spirocyclic linker enables the acrylamide to project toward Cys12 to form a covalent bond, while the acrylamide oxygen hydrogen bonds to the side chain amine of Lys16 and also engages in a network of water-mediated interaction with the Mg^{2+} and the GDP β-phosphoryl group (Fig. 3). The N-methyl indazole is almost planar with the pyrazole and is sandwiched between the switch II and the Gln99 side chain. As compared with other known G12C inhibitors, these interactions are unique to JDQ443 binding due to its distinct pyrazole linker scaffold. It should be noted that, like sotorasib and in contrast to adagrasib, JDQ443 does not interact directly with the imidazole NH of His95, which is a recognized site mutation for resistance.[4]

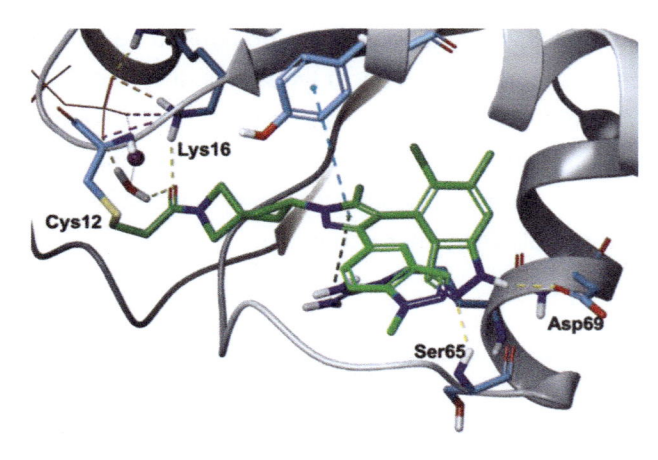

Figure 2. Co-crystal structure of JDQ443–G12C.

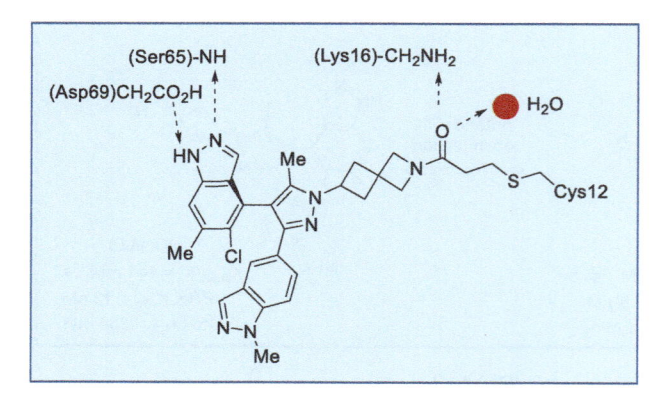

Figure 3. Summary of JDQ443–G12C interactions based on the co-crystal structure.

1. J. Pileth. AACR 2022 – Novartis's late Kras challenge to Amgen. Evaluate Vantage. April 12, 2022.

2. A. Weiss, E. Lorthiois, L. Barys, K. S. Beyer, C. Bomio-Confaglia, H. Burks, X. Chen, X. Cui, R. de Kanter, L. Dharmarajan, C. Fedele, M. Gerspacher, D. A. Guthy, V. Head, A. Jaeger, E. J. Núñez, J. D. Kearns, C. Leblanc, S. M. Maira, J. Murphy, S. M. Brachmann. Discovery, preclinical characterization, and early clinical activity of JDQ443, a structurally novel, potent, and selective covalent oral inhibitor of KRASG12C. *Cancer Discov.* **2022**, *12*, 1500–1517.

3. E. Lorthiois, M. Gerspacher, K. S. Beyer, A. Vaupel, C. Leblanc, R. Stringer, A. Weiss, R. Wilcken, D. A. Guthy, A. Lingel, C. Bomio-Confaglia, R. Machauer, P. Rigollier, J. Ottl, D. Arz, P. Bernet, G. esjonqueres, S. Dussauge, M. Kazic-Legueux, M. A. Lozac'h, S. Cotesta. JDQ443, a structurally novel, pyrazole-based, covalent inhibitor of KRAS^{G12C} for the treatment of solid tumors. *J. Med. Chem.* **2022**, *65*, 16173–16203.

4. N. S. Persky, D. E. Root, K. E. Lowder, H. Feng, S. S. Zhang, K. M. Haigis, Y. P. Hung, L. M. Sholl, A. J. Aguirre. Acquired resistance to KRAS^{G12C} inhibition in cancer. *New Engl. J. Med.* **2021**, *384*, 2382–2393.

Chapter 22. An Overview of the Discovery Process for Medically Useful Inhibitors of Oncogenic Protein Kinases

The many new anticancer therapeutics that are described in the foregoing sections of this book were enabled by decades of progress in many areas of science. Here are just a few of the advances that led to this new approach to the treatment of malignant disease, once a death sentence, but increasingly a manageable or even curable condition: (1) recognition that many clinically distinct forms of cancer result from mutated genes (oncogenes) that code for cell-growth/regulating proteins (oncogenic regulatory proteins); (2) genome, gene, and protein sequencing technology; (3) the development of high-throughput micro bioassays for screening large compound library collections to identify molecules – 'hits' – that potently inhibit oncoprotein-induced cell signaling; (4) the identification of the most promising inhibitors for further development based on multiple criteria of modern drug discovery, termed high-quality leads; (5) analysis by X-ray crystallography or cryo-electron microscopy to determine at atomic resolution of the 3D structure of the oncoprotein alone or as a co-crystal complex with a lead compound candidate; and (6) structural modification of the lead compound by chemical synthesis and bio-evaluation to find molecules with superior properties and the highest chance of clinical utility.

As the field has progressed, many new insights have been developed with regard to the features of kinase structure and signaling, modes of kinase inhibition, and also the ability of malignant cells to evade inhibition by further oncogene mutation. In addition, the daunting challenge of attaining selectivity for the target onco-kinase over the other 517 or so human kinases has been shown not to be insurmountable despite the fact that all these enzymes utilize ATP as the phosphorylating cofactor and often have various structural similarities. These include: (1) N-terminal and C-terminal lobes connected by a flexible amide- containing backbone (or hinge region), and (2) an ATP-binding pocket in between the two lobes with a Mg(II) binding loop having a common Asp-Phe-Gly triad (DFG) either at the ATP-binding pocket (DFG-in) or away from it (DFG-out), the latter being an inactive conformation of the kinase.

The design of kinase inhibitors often takes advantage of binding to a specific CONH (amide) subunit in the protein chain within the hinge region – either with the DFG in or out arrangement. Most utilize the ATP-adenine binding pocket and contain an adenine-like 2-amino pyridine type structural element together with one or two nearby binding sites, often hydrophobic.

X-ray cocrystal structures of the target kinase and synthetic compounds with promising activity starting with a high-quality lead provide enormously helpful Information and guidance in the search for a useful drug candidate.

A number of specific tactics that facilitate the discovery of useful anti-kinase drugs are described in the sections that follow.

1. **High-quality Leads**

The wide availability of large chemical compound libraries and automated high-throughput screening technologies (HTS) have made it relatively easy to identify bioactive hits, the most promising of which can later be exploited as lead compounds by medicinal chemists in search of a high-quality lead. Such a lead must satisfy criteria for a future drug candidate, using e.g., receptor profiling, kinase or isoform selectivity and lack of CYP inhibition. It is desirable that the molecular weight be under 400. In fact, the overwhelming majority of useful kinase therapeutics started as lead compounds with molecular weight (MW) less than 400, which subsequently increased because several additional subunits were needed to improve various properties such as PK and solubility (Fig. 1). A good lead compound should exhibit at least modest oral

Molecules Engineered Against Oncogenic Proteins and Cancer, First Edition. E. J. Corey and Yong-Jin Wu.
© 2024 John Wiley & Sons Inc. Published 2024 by John Wiley & Sons Inc.

bioavailability and solubility, which can be improved through SAR studies. High-quality leads should also allow further structural diversification. A high-quality lead is key to a successful discovery program.

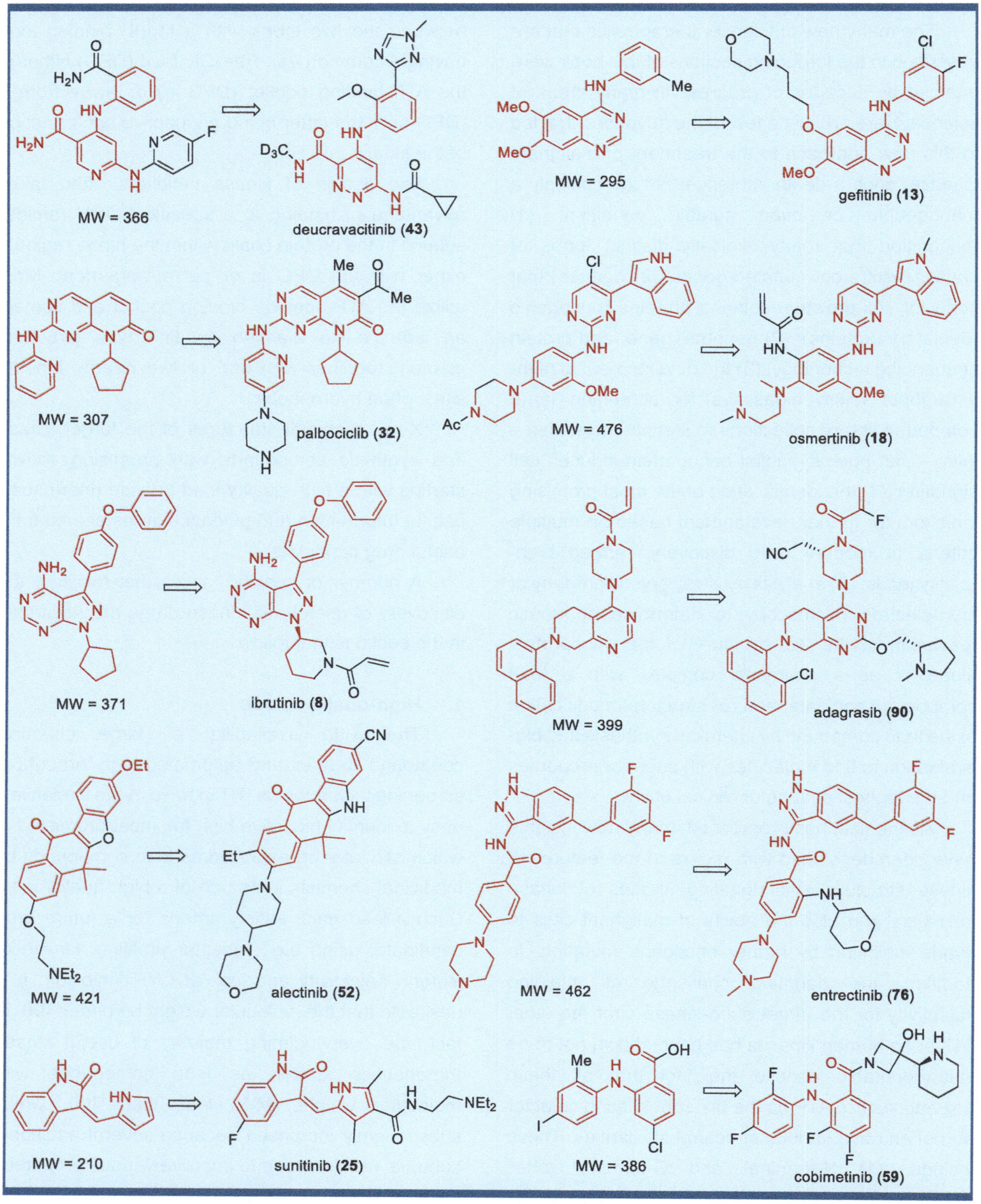

Figure 1. Examples of high-quality HTS hits to drugs.

2. Integrating Substructures from Different High-quality Leads or Established Inhibitors

The most effective way to move a lead compound forward is to determine its binding mode to the target protein either through an X-ray structure co-crystal or by computational modeling. Comparison of co-crystal structures of the lead compound(s) with those reported for known inhibitors of the target kinase could reveal other binding sites and suggest molecular modifications of the lead to generate new, patentable chemical entities that incorporate beneficial substructures from different chemotypes. Essentially, it is a strategy to improve the properties of one molecule by incorporating information on another molecule's complementary advantages.

The recent discovery of sotorasib is a good illustration (Fig. 2). ARS-1620, a preclinical candidate from Wellspring, exhibits good PK properties but poor cellular potency, while Amgen's lead compound **A** is exactly the opposite. In contrast to ARS-1620, the tetrahydroisoquinoline ring of **A** occupies a previously unexploited cryptic pocket on the "lid" of the switch II pocket of KRAS, and that engagement of this cryptic pocket contributes substantially to its potency. Because both inhibitors are complimentary in potency and PK properties, adding a cryptic pocket binder to ARS-1620 might increase the potency while retaining its favorable PK properties.

Modelling of the binding modes of indole lead **A** and ARS-1620 revealed that a substituent of the quinazoline nitrogen (N1) of ARS-1620 could fit the cryptic pocket. This hybrid design concept led to various hybrid analogs of formula **B**, which were further modified to give sotorasib.

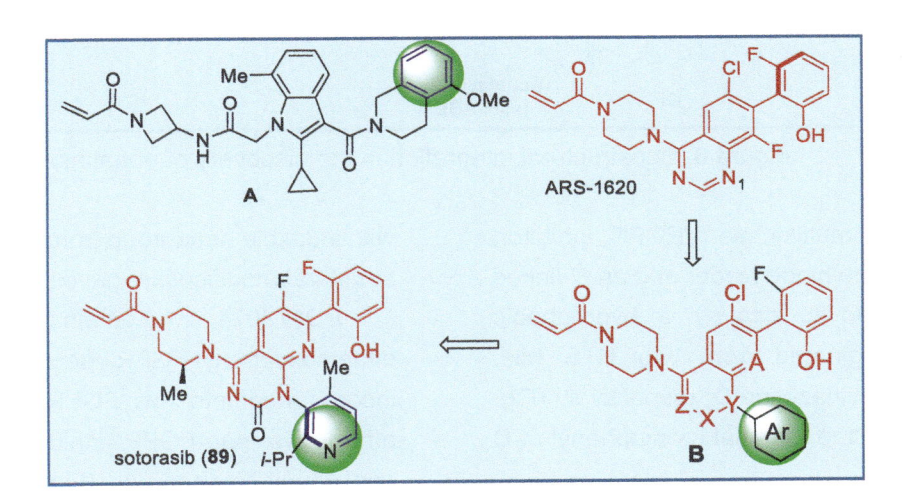

Figure 2. Substructural integration in the discovery of sotorasib.

Ponatinib, a BCR-ABL inhibitor, originated from AP23464, a purine-based dual SRC/ABL inhibitor that strongly inhibits SRC, wild-type BCR-ABL as well as BCR-ABL mutants (except T315I) with subnanomolar potencies (Fig. 3). To target the DFG-out conformation of the T315I mutant kinase, the N-hydroxyphenethyl part of AP23464 was replaced by a DFG-out targeting diarylcarboxamide fragment taken from nilotinib, and the connecter was changed from CH_2CH_2 to CH=CH to avoid contact with the isoleucine side chain of the mutant T315I. In addition, the C2 cyclopentyl group of the purine core was removed because it is not accommodated in a DFG-out targeted conformation due to clashes with residues located in the glycine-rich P-loop. Furthermore, the non-pharmacophore phenyl-dimethylphosphine oxide group was replaced by a much smaller cyclopropyl to reduce lipophilicity and improve PK properties. Taken together, these structural modifications led to AP24163.

To further reduce steric repulsion with the bulky isoleucine side chain of the T315I mutant, the vinyl linker was replaced by $C \equiv C$, and bioisosteric replacement of the terminal headgroup resulted in ponatinib.

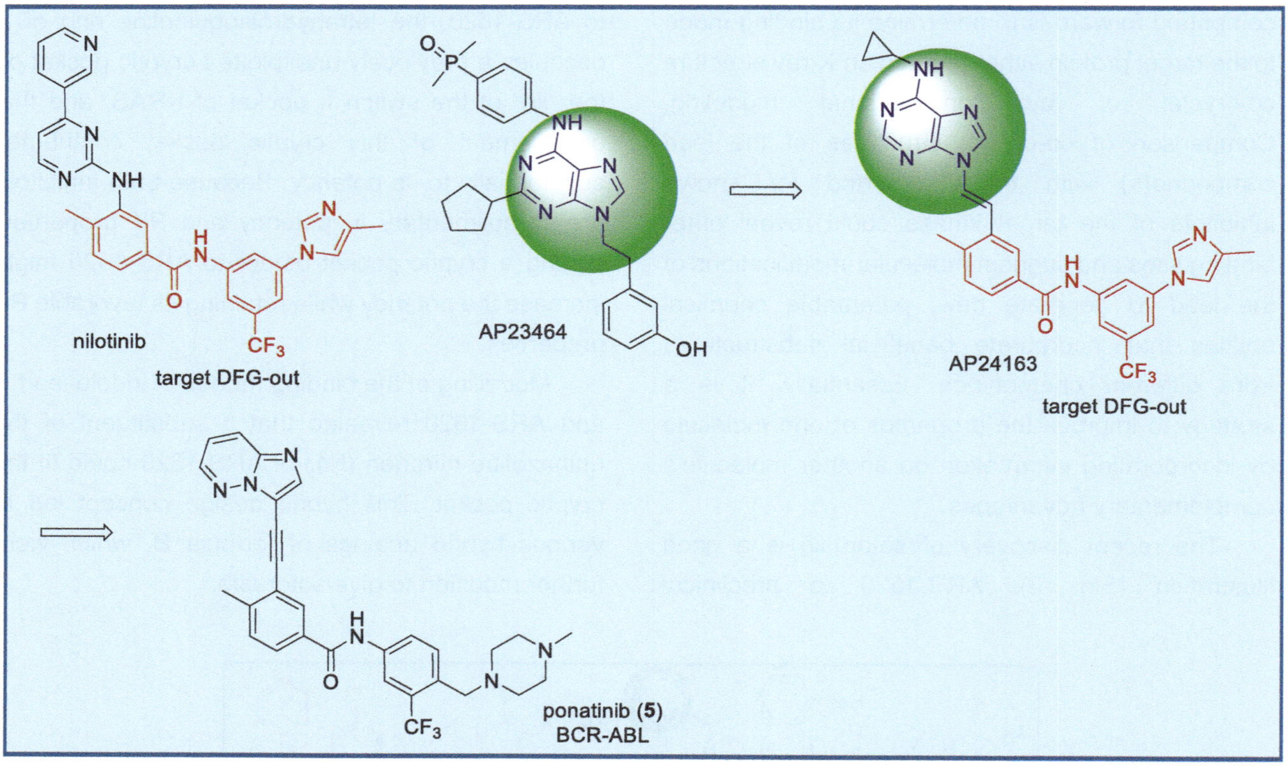

Figure 3. Substructural integration in the discovery of ponatinib.

Pazopanib, a multikinase VEGFR inhibitor, contains a pyrimidine hinge binder and an *N*-linked indazole headgroup to occupy a hydrophobic pocket, which originated from two HTS hits, pyrimidine **C** and quinazoline **D**, respectively (Fig. 4). Replacement of the terminal bromophenyl of **C** with indazole headgroup from **D**, followed by minor structural modification, gave pazopanib.

PD173074 contains an optimal dimethoxyphenyl FGFR specific binder, which was incorporated into two FGFR leads **E** and **F**, to afford two pan-FGFR inhibitors erdafitinib and pemigatinib, respectively (Fig. 5).

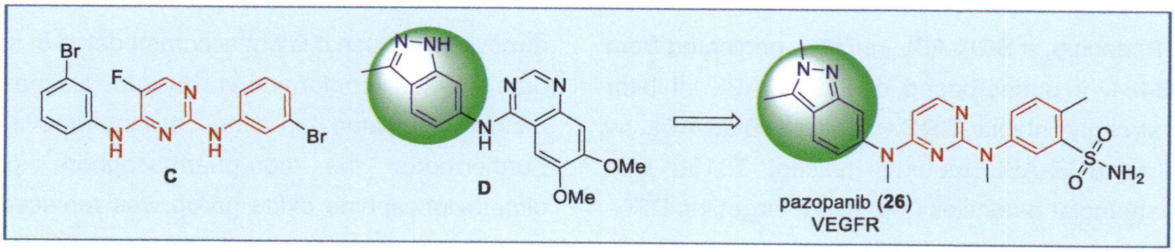

Figure 4. Substructural integration in the discovery of pazopanib.

Figure 5. Substructural integration in the discovery of two pan-FGFR inhibitors.

3. Variation of Hinge-binding Nucleus

Kinase structures consist of an N-terminal and C-terminal lobe composed primarily of β sheets and α helices, respectively. The lobes are joined by a linear hinge region containing three conserved residues. These provide a binding pocket for ATP involving several key H-bonds with the bicyclic heteroaromatic adenine. The hinge region is targeted by most ATP-competitive kinase inhibitors via a hinge-binding motif involving the same key H-bond acceptor and donor interactions as those of the adenine of ATP and thus displacing ATP. Because such hinge-binding motifs are key to the potency of a kinase inhibitor, their modification can provide avenues for innovation. More importantly, variation of hinge-binding elements alone or in conjunction with structural changes in other parts of the inhibitor can profoundly influence selectivity and potency. For example, nilotinib, a second-generation BCR-ABL kinase inhibitor, forms a single H-bond with the hinge through the pyridine nitrogen and another H-bond with the side chain hydroxyl of the gatekeeper Thr315 (located just before the hinge) (Fig. 6). It lacks activity against T315I mutants because binding of the pyrimidinyl NH to the 315 hydroxyl becomes sterically unfavorable due to repulsion with the bulkier isoleucine sidechain. Replacing the pyridyl hinge-binding motif with a more compact bicyclic imidazopyridine and simplifying the aminopyridine linker to a linear acetylene led to the discovery of ponatinib, a third-generation of BCR-ABL inhibitor that circumvents the T315I gatekeeper resistance by bypassing the mutated larger isoleucine residue.[1]

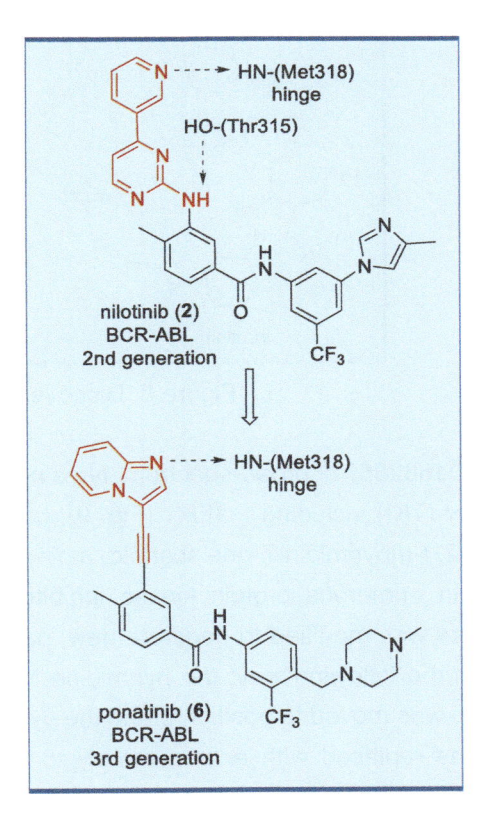

Figure 6. Discovery of ponatinib via hinge-binding variation.

Variation of the hinge-binding core was a key tactic in the discovery of crizotinib (Fig. 7). A tricyclic coplanar H-bonded pyrrole-oxindole hinge-binding scaffold was judiciously replaced with a substantially simpler aminopyridine **G** (which maintained the same key hinge interactions). Crizotinib was derived from **G** in a few steps.

Zanubrutinib was derived from ibrutinib, which has the 4-aminopyrazolopyrimidine as the donor/acceptor hinge-binding motif (Fig. 8). Mimicking the pyrimidinone ring through intramolecular hydrogen bond led to 5-aminopyrazole-4-carboxamide scaffold as in **H** with similar shape and hinge-binding elements. Further modification of the linker between the Michael acceptor warhead and the bicyclic core provided zanubrutinib.

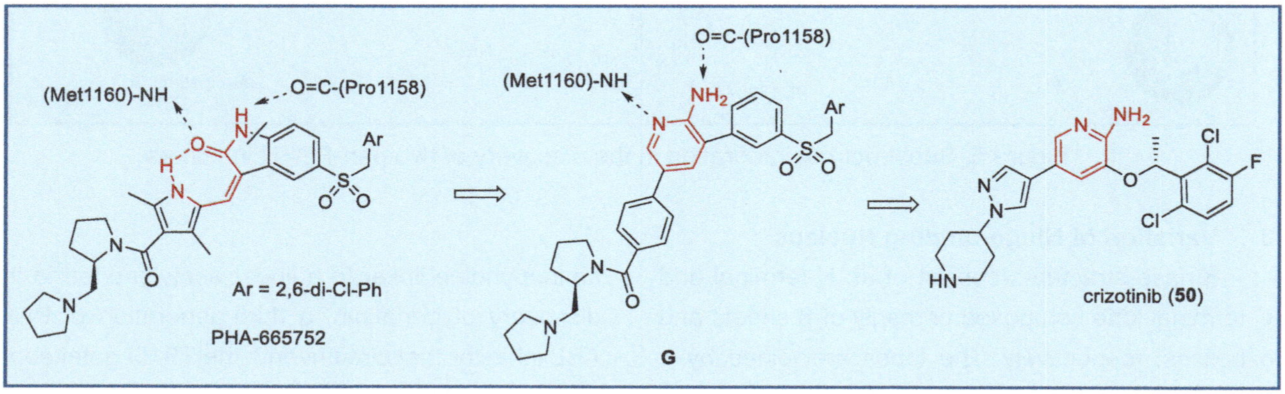

Figure 7. Discovery of crizotinib via hinge-binding variation.

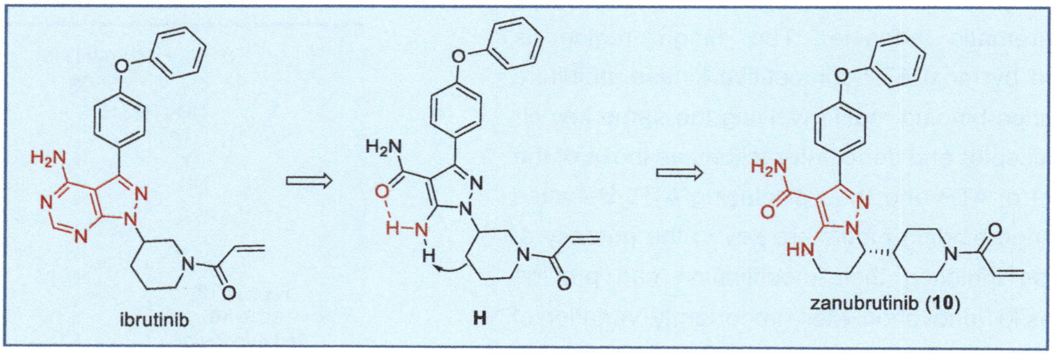

Figure 8. Discovery of zanubrutinib via hinge-binding variation.

PD166285, a broad-spectrum tyrosine kinase inhibitor (TKI) including FGFR1 (Fig. 9), contains a pyrido[2,3-d]pyrimidin-7-one scaffold, a hinge binder found in numerous protein kinase inhibitors. This structure was modified to generate new, patentable TKIs in the following way: the pyrimidine N1 of PD 166285 was moved to position 5 and the pyrimidone ring was replaced with a urea moiety to allow an intramolecular hydrogen bond with the newly rearranged nitrogen. This structural modification resulted in compound **I**, which has a stable, planar, conjugated pseudo six-membered ring as a mimic of the pyrimidone ring (Fig. 9). Compound **I** served as a new lead FGFR inhibitor that was key to the discovery of infigratinib (**68**), a potent and selective pan-FGFR inhibitor.

Figure 10 describes several other applications of hinge-binding variations.

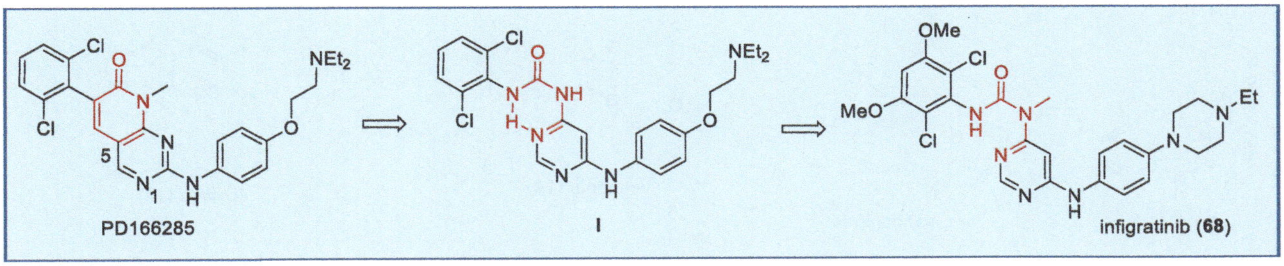

Figure 9. Discovery of infigratinb via hinge-binding variation.

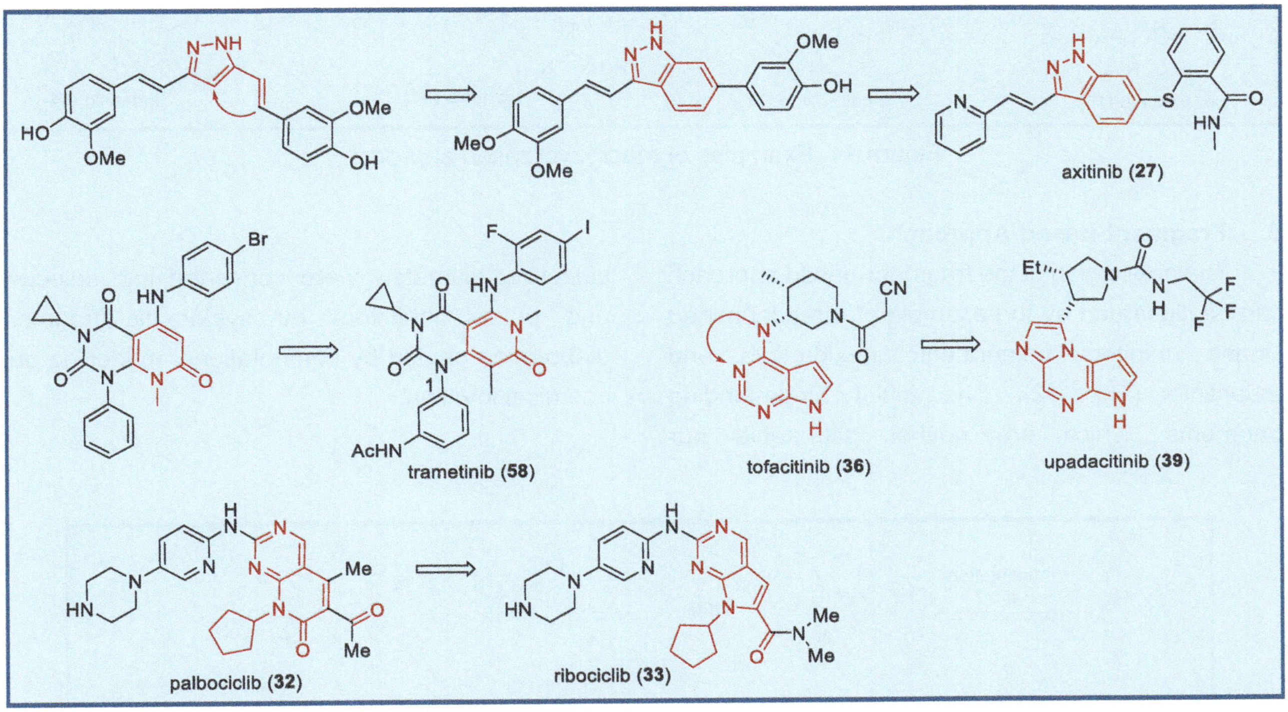

Figure 10. Other examples of hinge-binding variation.

4. Macrocyclization

Macrocyclization of U-shaped molecules could provide new patentable inhibitors with improved properties (Fig. 11). For example, the JAK2 HTS hit **J** that led to pacritinib contains a ubiquitous pyrimidine-based motif claimed in numerous issued patents and patent applications. To develop novel and proprietary candidates, macrocyclic analogs of varying ring size were synthesized and evaluated, leading eventually to pacritinib (**46**), a potent JAK2/FLT3 inhibitor. Similarly, icotinib was obtained as a new, patentable macrocyclic derivative of erlotinib.

Repotrectinib, a multikinase NTRK/ROS/ALK inhibitor, was inspired by the U-shaped binding conformation of larotrectinib. The compact structure of repotrectinib avoids steric clashes with the solvent-front substitutions in ROS1/TRK/ALK mutants, and as a result, its binding is not affected by these mutations.

The rigid macrocyclic inhibitor, locked in a bioactive conformation, fits precisely in the ATP-binding site, thus incurring little entropy penalty during binding. Its compact size, minimal molecular weight, and modest lipophilicity result in favorable PK properties including moderate brain penetration.

Macrocyclization of crizotinib served as a critical step in the discovery of lorlatinib.

The macrocyclization strategy is appropriate with inhibitors that bind to the target proteins in a U-shaped molecular conformation.

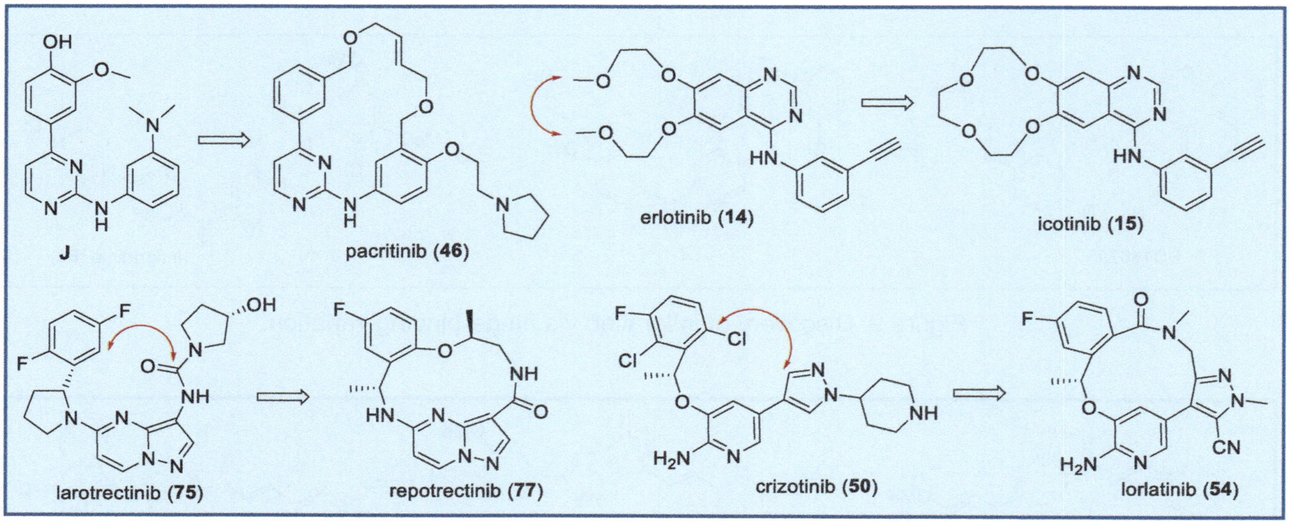

Figure 11. Examples of macrocyclization approach.

5. Fragment-based Approach

The application of the fragment-based approach can be illustrated by the example of three launched kinase inhibitors: vemurafenib, pexidartinib, and erdafitinib (Fig. 12). The initial hinge-binding fragments, which are neither patentable nor selective, ultimately were converted into selective and potent inhibitors by systematic fragment elaboration guided by computational modelling and chemical intuition.

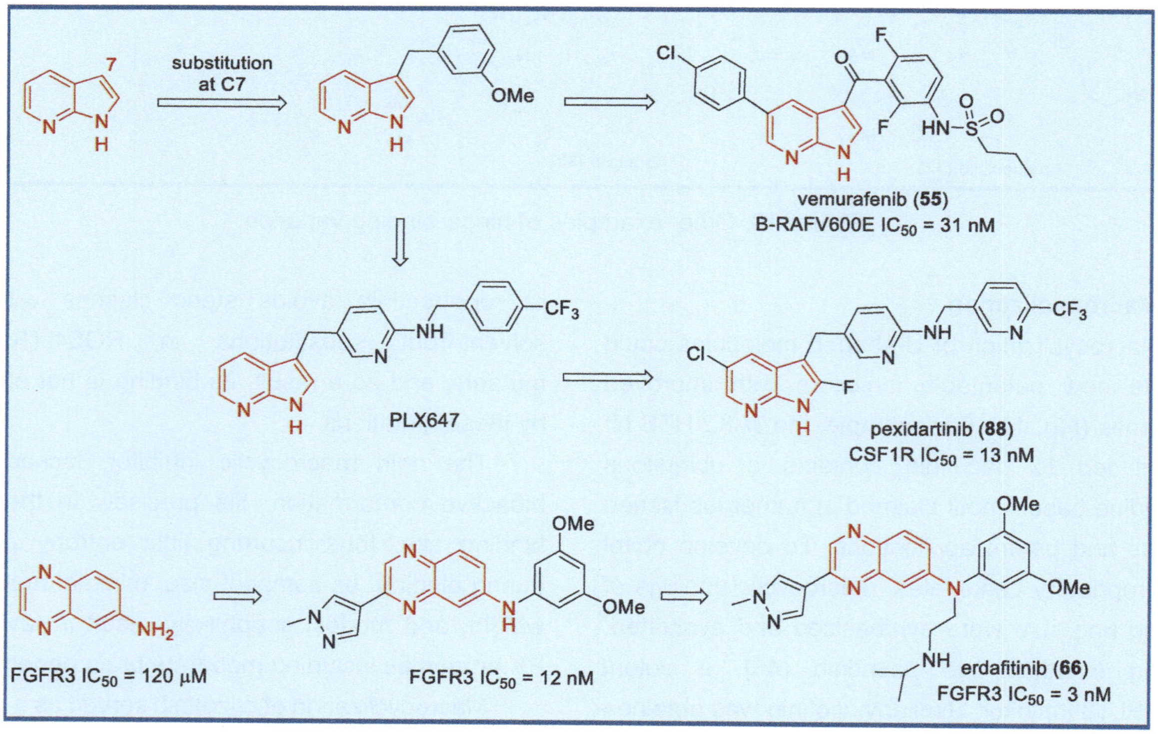

Figure 12. Discovery of three inhibitors via fragment-based approach.

The lead compound **K** that led to asciminib, a first-in-class allosteric inhibitor of BCR-ABL1 kinase activity, was obtained via fragment-based approach (Fig. 13). Upon blocking the ATP-site with imatinib, the ABL1-imatinib complex was subjected to fragment screening using NMR spectroscopy. This effort identified ~30 hits from ~500 fragments, and then representative fragments were crystallized as ternary fragment–imatinb–ABL1 complexes. Subsequently, similarity and pharmacophore searches were conducted on multiple ternary crystal structures, and the resulting hits were docked onto the bent α-helix I conformation of the myristate pocket, resulting in compound **K** with modest binding affinity. Subsequently, **K** was elaborated into asciminib through multi-step incremental optimization.

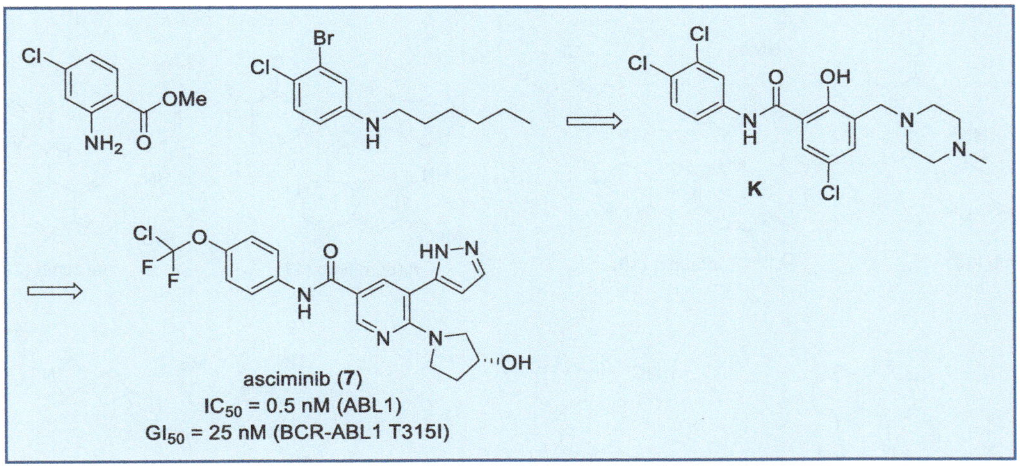

Figure 13. Discovery of asciminib via fragment-based approach.

6. Covalent Inhibitors

Nearly half of eukaryotic kinases contain at least one cysteine residue in the neighborhood of the ATP pocket (e.g., EGFR and BTK), which can potentially be used as reactive sites for covalent attachment and irreversible inhibition. Covalent kinase inhibitors (CKIs) contain an electrophilic warhead (typically a Michael acceptor) suitably positioned to react with a cysteine in the kinase and form a covalent bond (Fig. 14). The non-ATP competitive irreversible interactions usually translate to sustained target inhibition because of very long off rates which generally reduce the chance for acquired resistance relative to reversible inhibitors. CKIs may also provide improved selectivity for the primary target over other members of the kinome that do not contain reactive cysteine near the ATP-binding site. However, there are 11 kinases including EGFR, ITK, BTK, JAK3, and others that share an equivalently positioned cysteine residue which may also be targeted through the covalent binding approach. Therefore, it remains challenging to achieve high selectivity against them. In addition, CKIs may suffer from off-target toxicity associated with immunogenicity of the many protein adducts that can form.[2]

Nonetheless, CKI's have had remarkable success. Two covalent inhibitors, ibrutinib and osimertinib, have become top oncology drugs, with 2021 sales coming in at $9.8 and $5 billion, respectively.

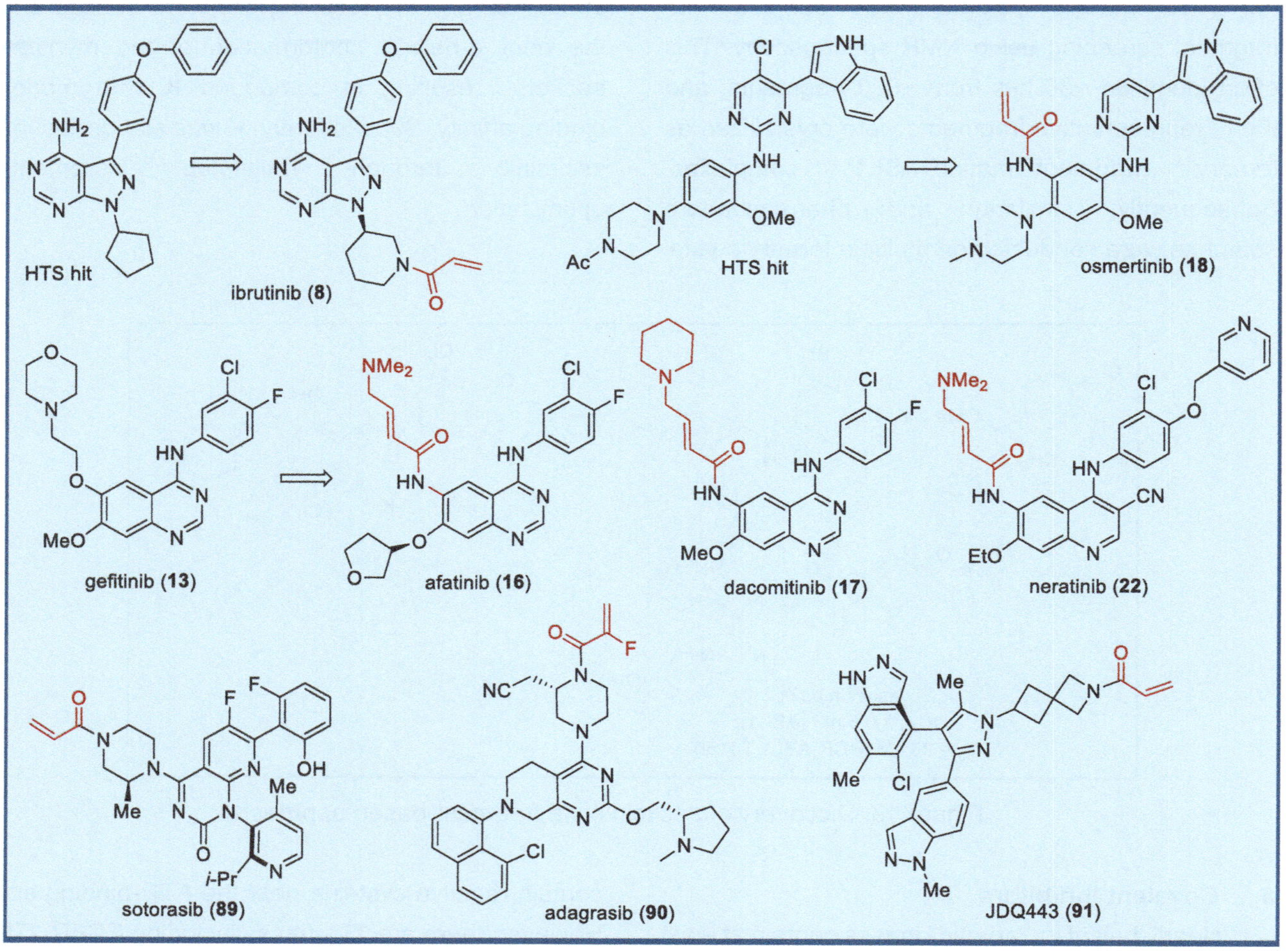

Figure 14. Irreversible covalent inhibitors.

7. Strategic Structural Modification of Prior Drugs

"The most fruitful basis for the discovery of a new drug is to start with an old drug," Sir James Whyte Black (1998 Nobel Laureate in Physiology & medicine) once said. Indeed, subtle changes on earlier problematic drugs can eliminate their liabilities and, with luck, quickly produce successful next-generation therapies. For example, TAE684, a highly potent ALK inhibitor which had been terminated due to toxicological liabilities, could be modified to brigatinib which emerged to become the best-in-class ALK inhibitor for the treatment of crizotinib-resistant NSCLC cancers. TAE684 and brigatinib are structurally identical except that the isopropylsulfonyl group of the former is replaced by dimethylphosphoryl (DMPO) (Fig. 15).

The DMPO subunit of brigatinib plays a critical role in its superior kinase selectivity profile because the basic oxygen of the P=O bond forms a 6-membered intramolecular hydrogen bond with the adjacent NH, which stabilizes and rigidifies the bioactive conformation. The DMPO analog, which is more potent and substantially less lipophilic than the sulfone, exhibits lower plasma protein binding and better aqueous solubility. Overall, the DMPO group proves to be key to mitigating the reactive metabolite issue caused by the electron-rich benzene-1,4-diamine subunit.

Figure 15. From TAE-684 to brigatinib.

Gefitinib and erlotinib are the first two EGFR-targeted TKIs which share a 4-anilinoquinazoline scaffold. The ortho-fluoro activated electrophilic chlorine atom in gefitinib attracts the nucleophilic amide carbonyl oxygen atom of Leu788 within the active site of EGFR. Both the aromatic fluorine activated C–Cl of gefitinib and aryl C≡CH of erlotinib show electron affinity at the tip of these groups that allows van der Waal's bonding (Fig. 16). The weakly acidic acetylenic CH moiety in erlotinib can act as a hydrogen bond donor to the amide carbonyl oxygen of Leu788 of the EGFR protein as shown from the co-crystal structures with EGFR.

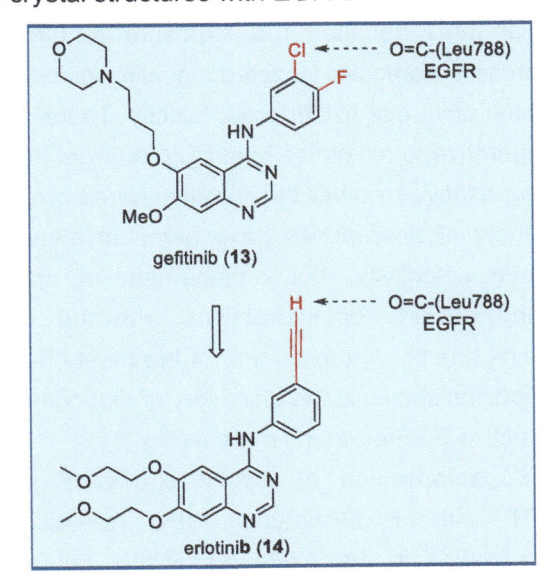

Figure 16. From gefitinib to erlotinib.

Regorafenib, a follow-up to sorafenib, is identical to sorafenib with the exception of a fluorine atom added ortho to the urea at the central phenyl group (Fig. 17). This electron-attracting fluorine results in a ~20-fold increase in potency against VEFGR2 relative to sorafenib, probably because that fluorine increases the acidity of the urea hydrogens, thus resulting in stronger H-bonds to the side chain carboxylate oxygen of Glu885 (Fig. 17). The increased VEGFR2 potency translates to much smaller single oral dose: 160 vs. 400 mg for the des-fluorine analog, sorafenib.

Figure 17. From sorafenib to regorafenib.

Mobocertinib is derived from osimertinib, a second-generation TKI which exhibits only limited activity against several resistant EGFRex20ins mutants. Both inhibitors are structurally identical except that the pyrimidine ring of moboceritinib incorporates a snugly fitting isopropyl ester group (Fig. 18) that targets a previously unoccupied pocket, resulting in expanded coverage of EGFRex20ins mutations as well as improved selectivity over wild-type EGFR vs. osimertinib.

Figure 18. From osimertinib to mobocertinib.

Rapamycin has an oral bioavailability of 14%, which is sufficient for oral administration to treat autoimmune diseases and LAM. However, the modest oral bioavailability and poor solubility are insufficient for oral dosing in cancer treatment. Fortunately, attachment of a hydroxyethyl ether to the cyclohexyl side chain led to everolimus with PK properties superior to rapamycin, and the orally dosed everolimus has been widely used for kidney, lung, and breast cancer, with peak sales of $1.6 billion in 2015 (Fig. 19).

Figure 19. From sirolimus to everolimus.

Nilotinib, a second-generation BCR-ABL inhibitor, was developed starting with the imatinib structure by replacing the N-methylpiperazine group, retaining the rest of the molecule (i.e., A, B, C, D rings, Fig. 20) and reversing the amide functionality, a classic strategy in drug design. Subtle conformational changes in the protein induced by the reverse amide inhibitor enable the D ring substituents to bind at sites different from that occupied by the N-methylpiperazine ring of imatinib, thereby resulting in enhanced binding. This design approach gave rise to nilotinib in which the 4-methylimidazole is attached to the D ring, to reduce lipophilicity and enhance aqueous solubility. Novartis' nilotinib had sales of $2.1 billion in 2021.

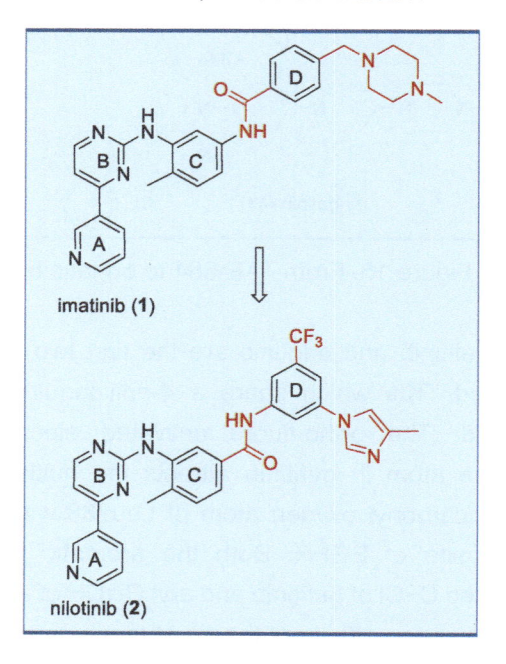

Figure 20. From imatinib to nilotinib.

8. Exploiting a Specific Kinase Pocket to Optimize Selectivity

Multikinase inhibitors can be ineffective in clinical trials because the exposure required to suppress a particular targeted signaling pathway is not attainable due to off-target toxicity. Therefore, it is imperative to maximize binding to a target protein kinase relative to other structurally related proteins. A variety of approaches have been developed to achieve selectivity, but engagement of specific binding pockets or interactions with the target proteins has proved to be one of the most effective methods as shown in the discovery of moboceritinib, a selective EGFRex20ins mutant (Fig. 21).

Superimposition of crystal structures of wt EGFR with EGFRex20ins NPG reveals no differentiation in the ATP-binding site, but subtle structural variations between the two proteins in the αC-helix region, which could be exploited for selectivity. A docking model of osimertinib in complex with the EGFRex20ins NPG mutant showed an unoccupied pocket (close to the αC-helix). This pocket can be accessed by a substituent attached to the pyrimidine hinge binder. Consequently, an isopropyl ester group was incorporated onto osimeritinib (to give mobocertinib) to interact with the gatekeeper residue within this selectivity pocket. It is this isopropyl ester group that gives rise to the improved activity for the EGFRex20ins mutants over osimertinib.

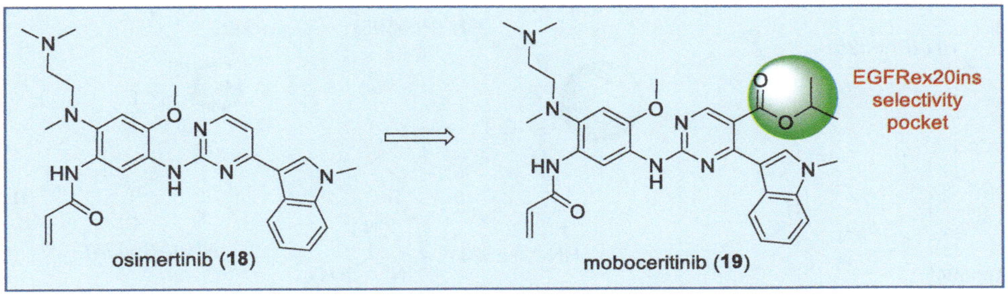

Figure 21. EGFRex20ins selectivity pocket in moboceritinib.

The serendipitous finding of the "alanine pocket" of the TYK2 JH2 domain during SAR studies was a major breakthrough in the discovery of deucravacitinib, a highly selective TYK2 inhibitor (Fig. 22). As part of the SAR efforts, the primary amide **L**, a HTS hit, was converted to the N-methyl derivative **M**, which surprisingly showed substantial improvement in JAK isoform as well as overall kinome selectivity.

The profound effect of the N-methyl group on selectivity has been attributed to the N-methyl, which approaches the proximal Ala671 of the target protein. The alanine in this particular position, which is found in only nine other kinases, is substituted with a much larger residue such as valine, leucine, or isoleucine in a majority of other kinases. Conceivably, the C3 primary amide of **L** can be tolerated by larger residues of other kinases, thus leading to very modest selectivity profile. In contrast, the limited space provided by an adjacent alanine ("alanine pocket") of the TYK2 JH2 domain can only tightly fit the N-methyl of **M**, consistent with the high selectivity. Compound **M** was further modified to give deucravacitinib upon several structural optimizations.

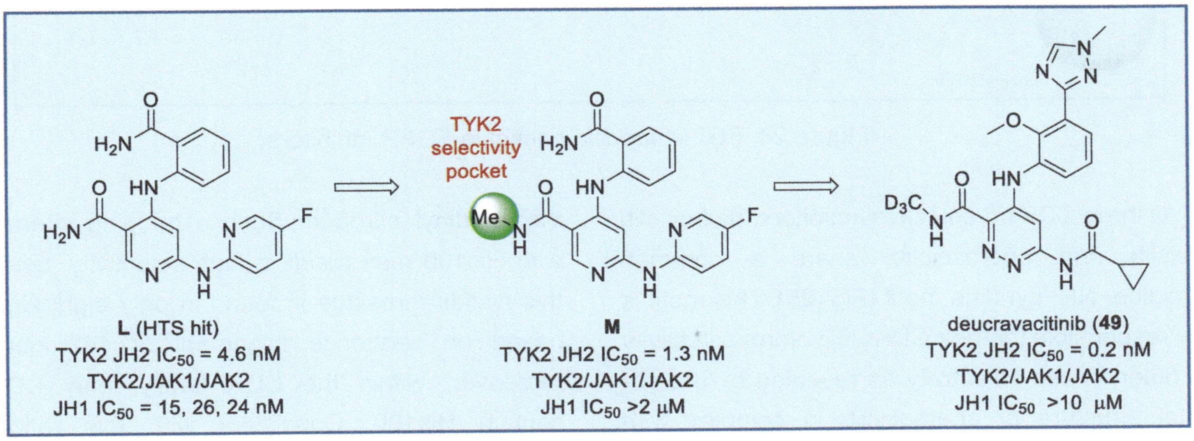

Figure 22. Special "alanine pocket" of TYK2 in deucravacitinib.

Access to a RAF-selective pocket induced by inhibitors plays an important role in the selectivity of vemurafenib against non-RAF kinases (Fig. 23). The nitrogen atom of the sulfonamide moiety hydrogen bonds to the main chain NH group of Asp594 in the BRAF protein, while one of the oxygen atoms of the sulfonamide forms H-bond with the backbone NH of Phe595. Both hydrogen bonds enable the tail n-propyl group to fit an interior pocket specific to the mutant BRAF protein. A similar pocket at this exact location is rare among other non-RAF kinases, and therefore, access to the RAF-selective pocket is essential to its selectivity. A similar pocket is found in the other two BRAF-selective inhibitors: dabrafenib and encorafenib.

Figure 23. Specific RAF-selective pocket in BRAF inhibitors.

The selective EGFR inhibition of erdafitinib results from the engagement of a specific hydrophobic FGFR pocket. The dimethoxyphenyl ring, which is perpendicular relative to the quinoxaline core, fits tightly within this pocket. This pocket is exploited in three other FGFR inhibitors (Fig. 24).

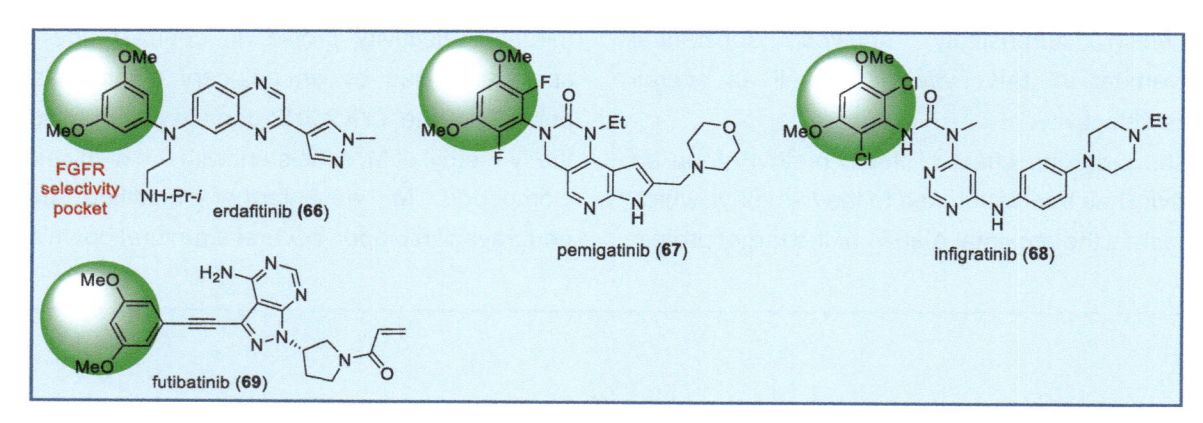

Figure 24. FGFR specific pocket in FGFR inhibitors.

All three CDK4/6 selective inhibitors, palbociclib, ribociclib and abemaciclib, share a common pyrimidine–NH–pyridine motif (Fig. 25). This motif is likely responsible for their CDK4/6 isoform selectivity and other kinase selectivity as revealed by the co-crystal structure of abemaciclib in complex with CDK6 which shows an ordered water molecule bridging the imidazole of hinge residue His100 and the pyridinyl nitrogen. Such a bridging interaction with His100 may result in high selectivity, because this histidine residue is found in only eight kinases based on sequence alignment of 442 kinases. Moreover, within the CDK family, only CDK4/6 contain His100. Consistent with this rationale, removal of the pyridyl nitrogen gave nonselective analogs.

Figure 25. Bridging interaction with His100 in CDK4/6 inhibitors.

9. Solvent-exposed Appendages to Enhance Solubility and PK Properties

Optimization of PK properties and physical properties of lead compounds is an integral part of any medicinal chemistry program, and the solvent-exposed regions adjacent to the ATP-binding sites of targeted proteins provide significant opportunities for extensive structural modifications without affecting binding activity. For example, a variety of water-soluble polar groups were incorporated at the solvent exposed region to improve solubility and reduce lipophilicity as demonstrated by three ALK inhibitors (Fig. 26).

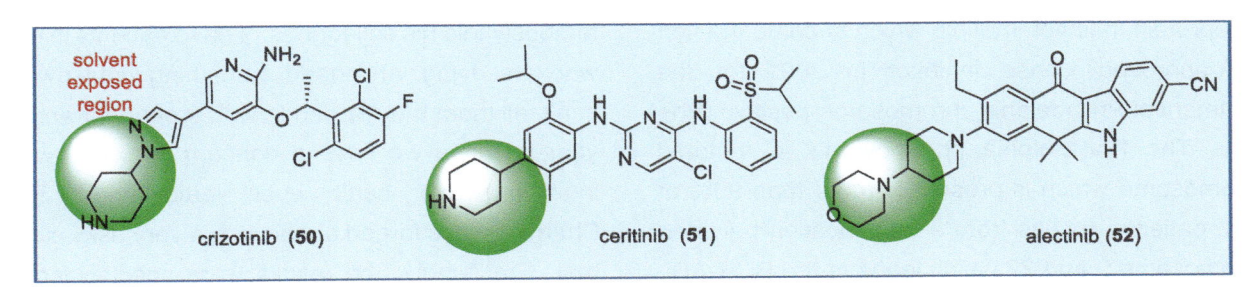

Figure 26. Examples of solvent-front substituents to improve PK and solubility.

1. P. Mukherjee, J. Bentzien, T. Bosanac, W. Mao, M. Burke, I. Muegge. Kinase crystal miner: A powerful approach to repurposing 3D hinge binding fragments and its application to finding novel Bruton tyrosine kinase inhibitors. *J. Chem. Inf. Model.* **2017**, *57*(9), 2152–2160.

2. T. O'Hare, R. Pollock, E. P. Stoffregen, J. A. Keats Wang B, H. Wu, C. Hu, H. Wang, J. Liu, W. Wang, Q. Liu Q. An overview of kinase downregulators and recent advances in discovery approaches. *Signal Transduct. Target Ther.* **2021**, *6*, 423.

Chapter 23. Targeted Molecular Anticancer Therapies – Successes and Challenges

1. The Beginning

The modern era of protein kinase-centered drug discovery started in 1988 with a collaboration between scientists at Ciba-Geigy and Brian Druker of Oregon Health and Science University to develop targeted chemotherapies for chronic myelogenous leukemia (CML). Back then, targeting a protein kinase was considered foolhardy because it was considered impossible to achieve the necessary high level of selectivity for one particular kinase out of so many, and a non-selective kinase inhibitor would generate too many off-target side effects to become a viable drug. Against all odds, the joint effort ultimately led to the antileukemic BCR-ABL oncoprotein inhibitor imatinib which became the first FDA-approved kinase inhibitor in 2001 for the treatment of Philadelphia chromosome-positive (Ph+) CML. The Philadelphia chromosome, a mutated chromosome which is present in more than 90% of CML patients, results from a translocation between chromosome 9 and 22 which leads to the BCR-ABL fusion gene. This gene produces an abnormal BCR-ABL tyrosine kinase protein, which enables CML cells to proliferate uncontrollably. Imatinib was the first drug to specifically target this protein. It was hailed as a "magic bullet" against CML because it changed a once deadly blood cancer into a manageable chronic disease. Even though it does not cure cancer, simply taking one pill a day by mouth for life allows patients to live a nearly normal lifespan. Imatinib generated peak sales of $4.7 billion in 2012 before it went off patent in 2015.

Targeted anti-kinase therapy of CML with imatinib introduced a paradigm shift in cancer treatment towards precision medicine.[1] In 2013 – 12 years after the introduction of imatinib – ibrutinib, a BTK inhibitor, often touted as "the imatinib of CLL," revolutionized the treatment of chronic lymphocytic leukemia (CLL). BTK is a pivotal protein for B cell receptor signaling and tissue homing of CLL cells. Interestingly, ibrutinib was originally designed and used as a tool compound to target the SH group of Cys481 in the ATP-binding domain of BTK through irreversible covalent attachment of that SH to the Michael acceptor warhead. This tool compound was used to compare ibrutinib with compounds that could bind BTK tightly, but not covalently. At the time it was believed that covalent kinase inhibitors were likely to be dangerous because of severe off-target toxicities. Given its covalent, irreversible inhibition, this compound would have been shelved by most pharmaceutical companies. However, Pharmacyclics' CEO and co-founder, Richard Miller, an entrepreneur and clinical oncologist, once famously told his colleagues, "I have patients in clinic who are dying, and need something right away. I can't tell them they'll need to wait around for another year because we have a concern (about covalent inhibition) we can't even articulate."[2] Thus, Pharmacyclics forged ahead with a very risky clinical trial. Gratifyingly, the results were spectacular: the drug shrank certain tumors, especially some B-cell cancers, and yet caused no serious decrease in the production of normal healthy cells. This bold move enabled ibrutinib to be one of the first two FDA-approved covalent inhibitors in 2013. Like imatinib, ibrutinib does not cure cancer, but one pill a day for life can manage the disease for two to three years, or occasionally longer, thus extending the lives of terminally ill patients. Thanks to its efficacy and favorable safety profile, ibrutinib has become a very successful kinase inhibitor with $9.8 billion sales in 2021, ranking as the seventh top-selling drug worldwide and fourth top-selling cancer drug in the same year. The discovery of ibrutinib not only paved the way for a new wave of BTK inhibitors but also stimulated the development of covalent inhibitors.

2. Further Developments

A dramatic expansion of oncogene/oncoprotein

Molecules Engineered Against Oncogenic Proteins and Cancer, First Edition. E. J. Corey and Yong-Jin Wu.
© 2024 John Wiley & Sons Inc. Published 2024 by John Wiley & Sons Inc.

targets and antikinase-directed therapies has greatly changed the landscape of cancer management, for instance, treatment of non-small cell lung cancer (NSCLC). FDA-approved molecular-matched agents are available for NSCLC with specific oncogenic drivers such as EGFR, ALK, BRAF, ROS1, NTRK, RET, and MET exon 14 skipping mutations (METex14). These targeted biomarker-guided therapeutics have substantially improved survival outcomes. For example, osimeritinib, an irreversible third-generation TKI, is efficacious in patients with EGFR-mutant NSCLC regardless of the difficult-to-treat gatekeeper T790M mutation. As a result, it has quickly become a drug of choice for oncologists, generating $5 billion in 2021 and $6.4 billion in 2022.

The treatment of patients with melanoma has also been significantly advanced by the development of molecular-targeted therapies. For example, the BRAFV660E mutation is found in more than half of the patients diagnosed with cutaneous melanoma, making BRAF a logical target. In 2011–2018, three BRAF kinase inhibitors (vemurafenib, dabrafenib, and encorafenib) were approved for the treatment of BRAF mutant melanoma. Subsequently, dual inhibitors of BRAF and MEK (a downstream signaling target of BRAF in the MAPK pathway) showed synergistic benefit and therefore have been approved for the treatment of melanoma. The combination therapy of dabrafenib (BRAF) with trametinib (MEK) brought in revenues of $1.69 billion in 2021.

The discovery of activating mutations in the KIT/PDGFR tyrosine kinase receptors and their role in the pathogenesis of gastrointestinal stromal tumors (GIST) has profoundly altered the treatment of this disease. Three multikinase inhibitors (MKIs) that cross-inhibit KIT/PDGFR were approved by the FDA, and two selective second-generation tyrosine kinase inhibitors (TKIs), ripretinib and avapritinib, were recently launched to overcome various resistance mechanisms observed in earlier MKI's.

The advent of JAK inhibitors has transformed the landscape of treatment options for patients with myelofibrosis (MF), an uncommon type of bone marrow cancer caused by a JAK2 V617F mutation. Ruxolitinib was the first to win FDA approval in 2011 for the treatment of MF patients, and more recently, two more selective JAK2 inhibitors (fedratinib and pacritinib) were approved for the same indication.

Since the discovery of imatinib in 1995, the FDA has approved 78 small-molecule kinase inhibitors and 2 KRAS inhibitors mainly for cancer treatment. In the meantime, there are more than 253 new kinase inhibitors[3] (most of these are listed in Appendix 2) in various stages of clinical trials, most of which are for genomically defined subsets of cancer patients. Because of further mutation of oncogenes, new biomarker-guided therapies have become even more important for cancer treatment.

3. Biomarker-driven Drug Development

The development of targeted therapeutics requires robust and validated biomarkers to guide the recruitment of the right patients for clinic trials and to monitor drug activity and therapeutic response. For example, midostaurin failed twice in the clinic for the treatment of solid tumors due to lack of the biomarker. In 2001, midostaurin was identified as a potent FLT3 tyrosine kinase inhibitor by a collaborative effort between the Dana-Farber Cancer Institute and Novartis to discover inhibitors of mutant FLT3-positive AML in cellular assays. It was this finding that led to the repurposing of midostaurin for FLT3-mutant AML. It was approved by the FDA in 2017 for newly diagnosed FLT3-mutated AML.

Another example is palbociclib, a selective CDK4/6 inhibitor, which was ineffective in patients with Rb-intact advanced solid tumors and non-Hodgkin lymphoma. However, three years later, it was shown to be highly active toward the ER+ cell lines, a significant finding as more than 60% of all breast cancers are ER+. This ultimately led to the establishment of CDK4/6 inhibitors (palbociclib, abemaciclib, and ribociclib) as the standard of care in the treatment of HR+, HER2− metastatic breast cancer. These inhibitors generated $8.8 billion in total sales in 2022: $5.1 billion for palbociclib, $2.5 billion for abemaciclib, and $1.2 billion for ribociclib.

Identification of predictive biomarkers is key to the treatment of targeted anti-cancer drugs. The TRK inhibitors, larotrectinib and entrectinib, are the first two FDA-approved small-molecule, tumor-agnostic therapeutics for the treatment of any solid tumors with NTRK gene rearrangements, regardless of cancer type and patient age. These rearrangements can be found at high frequencies (up to 90%) in certain types of cancer, such as infantile fibrosarcoma (a rare disease); however, oncogenic NTRK cancer is an uncommon disease family occurring, for example, with a frequency of <1% in NSCLC. Thus, the validated biomarkers ensure the treatment of patients with certain types of cancer with genomically matched agents across tumor types.

4. Mitigation of Drug Resistance

Despite high initial response, targeted therapies do not cure cancer, and they only delay tumor progression for a few months to several years. Ultimately, they lose efficacy in advanced cancers because tumors develop resistance and relapse in part due to mutations in the kinase domain that affect the binding of inhibitors to the target protein. Such resistance necessitates development of the next-generation of inhibitors. For example, resistance in most patients treated with imatinib predominantly occurs through point mutations of the BCR-ABL gene, and subsequently, nilotinib, a second-generation inhibitor, overcomes most of these resistance mechanisms by targeting different back pockets (Fig. 1). It is active against all imatinib resistant mutants except for the gatekeeper T315I mutant, offering an option for the treatment of CML or Ph+ ALL patients resistant to prior therapy.

For a large number of tyrosine kinases and serine/threonine kinases, the gatekeeper is typically threonine with a hydroxyl substituted small side chain, which serves as a handle for H-bond with the target proteins and also forms a small hydrophobic pocket near the ATP-binding site. This tight pocket has been frequently exploited for potency and selectivity of ATP-competitive inhibitors. However, it is this threonine gatekeeper that readily undergoes mutation to an amino acid with a larger side chain such as isoleucine, thereby disrupting drug binding. More specifically, a single gatekeeper amino acid alteration, in which Thr315 of BCR-ABL is changed to isoleucine (T315I), greatly decreases imatinib sensitivity and is the cause of resistance in many patients. To evade such a steric clash with the bulky isoleucine side chain of T315I mutant, a "skinny" ethynyl linkage was built into earlier inhibitors to give ponatinib, a third-generation inhibitor to restore T315I potency (Fig. 1). Alternatively, asciminib (Fig. 1), a type IV inhibitor, binds to a myristoyl site of the BCR-ABL1 protein and locks the protein into an inactive conformation. Because it binds remotely from ATP-binding site, it avoids drug resistance arising from ATP-binding site mutations including the gatekeeper mutant T315I.

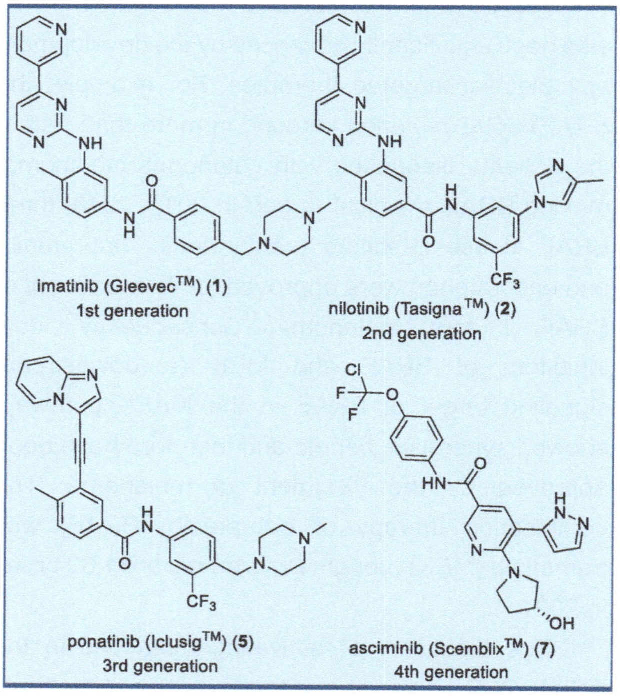

Figure 1. Four generations of BCR-ABL inhibitors.

A similar single gatekeeper mutation T790M (substitution of threonine 790 with a bulky methionine) is highly prevalent in NSCLC treated with the EGFR inhibitors such as gefitinib (Fig. 2). Interestingly, unlike BCR-ABL T315I, the EGFRT790M gatekeeper mutation also enhances the ATP-binding affinity, making it more difficult for gefitinib to out-compete ATP and thus resulting in diminished activity. However, the increased ATP-binding has less pronounced impact on the covalent, irreversible inhibitors because they permanently block the active binding site and no longer compete with ATP once covalently bound to the ATP-binding site. Based on this concept, osimertinib (Fig. 2), a covalent and irreversible third-generation TKI, was developed to combat T790M resistance.

Figure 2. First- and third-generations of EGFR inhibitors.

Other mutations can also occur in the kinase domain to cause drug resistance, e.g., solvent-front mutations (SFMs) and DFG-loop mutations. The in-depth molecular mechanisms of these mutations enable more rapid and effective drug discovery efforts as illustrated by repotrectinib. To address the SFM-mediated resistance, the solvent-front hydroxypyrrolidine moiety of larotrectinib was eliminated, and the terminal phenyl is directly connected to the urea scaffold to form a 13-membered macrocycle known as repotrectinib (Fig. 3). The macrocyclic structure of repotrectinib avoids steric clashes with the solvent-front substitutions in ROS1G2032R TRKAG595R, TRKCG623R, and ALKG1202R. As a result, repotrectinib binding is not affected by the SFM.

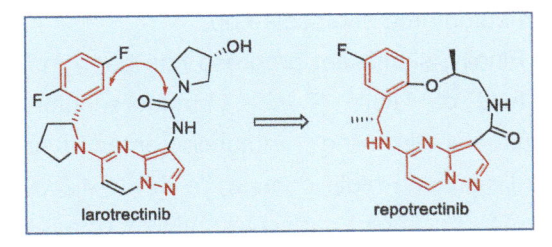

Figure 3. From larotrectinib to repotrectinib.

Combination therapy such as BRAF plus MEK inhibitors for melanoma represents another approach to prevent acquired resistance. Oncogenic BRAF mutations, key drivers in approximately 50% of cases of cutaneous melanoma, have been successfully targeted by three mutant-specific BRAF inhibitors (vemurafenib, dabrafenib, and encorafenib) for the treatment of BRAF mutant melanoma. However, their efficacy as monotherapy is insufficient due to the acquired resistance, and additional downstream MEK inhibition ensures complete MAPK pathway inhibition. Indeed, BRAF and MEK inhibitor combinations showed synergistic benefit, the three MEK inhibitors, trametinib, binimetinib, and cobimetinib are added to the combination therapies for the treatment of metastatic melanoma with BRAF mutations.

The combination therapy has also been expanded to immunotherapy. For example, lenvatanib (an antiangiogenic multikinase inhibitor) in combination with pembrolizumab (a PD-1 antibody) has been approved for the treatment of patients with advanced endometrial carcinoma. In addition, the combination therapy of pembrolizumab with axitinib (a relatively selective VEGFR inhibitor) has also won approval for the first-line treatment for patients with advanced RCC.

In addition to drug-induced target modifications, there are a variety of other resistance mechanisms, including non-drug-induced gene mutations, gene amplifications, cancer stem cells, efflux transporters, apoptosis dysregulation, and autophagy. Clearly, significant challenges lie ahead in tackling such diversified pathways of resistance.[1,4]

5. Miscellaneous Approaches
5.1 Previously "Undruggable" Targets

There has been considerable interest in exploring previously "undruggable" targets such as KRAS, the most frequent oncogenic alteration in human cancers. Despite being a logical target for cancer treatment, KRAS had been deemed undruggable for nearly four decades due to the high binding affinity of the intrinsic ligand GTP and lack of binding sites for a small molecule. Nevertheless, substantial progress has been made on targeting the KRASG12C mutant (simply G12C), a common KRAS mutant found in lung tumors. The 12 position is not located at the active site, but rather at a separate allosteric pocket ("switch II") where covalent modification can affect the protein's activity. This finding led to sotorasib and adagrasib as the first two FDA-approved drugs for G12C-driven NSCLC.

In addition to KRAS, other types of such previously undruggable cancer targets include c-MYC, protein tyrosine phosphatases (PTPs), and aberrant protein–protein interactions.

5.2 PROTAC

PROTAC (proteolysis-targeted chimera) employs small molecules to recruit target proteins for ubiquitination and removal by the proteasome. It reduces the activity of target proteins by catalyzing their degradation.[4] Indeed, PROTACs have been shown to successfully degrade three RTKs – EGFR, HER2, and MET, including multiple mutants of EGFR and Met.[5] In addition, several kinase degraders (e.g., NX-2127, NX-5948, HSK29116) are being tested in clinical trials, and NX-2127, an orally bioavailable BTK degrader, has shown encouraging results. It demonstrated clinically meaningful responses regardless of prior treatments or BTK mutational status, and therefore, it may serve as a back-up therapy for patients who have exhausted all other treatment options. PROTACs have also been utilized to target tumorigenic proteins generated by G12C/D mutations.

PROTAC may have the potential to address various challenges confronting ATP-competitive inhibitors, such as selectivity and drug-induced mutations (e.g., ATP-binding site modifications). However, the poor permeability and low oral bioavailability of these relatively high molecular weight degraders present a major challenge to their clinical use.

5.3 Immunotherapies

Another important field in cancer treatment that lies outside the province of this book is the application of biomolecules/immunology to the treatment of human malignant disease – for example, humanized monoclonal antibodies (mabs), antibody-drug conjugates, and vaccines. There are now more than 20 FDA-approved mabs that are available to treat cancer. These include mabs targeting VEGF (bevacizumab), EGFR (cetuximab, panitumab) VEGFR2 (ramucirumab), HER2 (trastuzumab), PD-1 (nivolumab and pembrolizumab), and CD20 (rituxumab). Unfortunately, these agents are not effective in all patients and also suffer from the development of tumor resistance. In addition, they are much less effective in the treatment of solid tumors.[6,7]

The accelerating advances in immunology and cell biology bode well for the development of other types of anticancer treatment. For example, there is recent evidence that a mRNA vaccine against the neoantigens that accumulate on the surface of cancer cells as a result of mutation (i.e., tumor surface oncoproteins) can activate T cells to attack and clear pancreatic tumors that were hitherto untreatable.[8]

Another form of immunotherapy that shows promise is based on T cells that are engineered to recognize an antigen specific to a cell surface protein expressed by an individual patient's tumor cells.[8] These engineered "Car T cells" may be derived from the patient's own T cells or from those of a healthy donor. After administration Car T cells bind specifically to the targeted tumor cells and become cytotoxic and proliferate – eventually if all goes well leading to cure of disease. There are currently six FDA-approved Car T cell lines which target blood cancers, e.g., lymphomas and multiple myeloma.[8] Although this kind of personalized, curative therapy is currently expensive, it is likely to become less so when automated and scaled up.

Finally, significant progress has been made in the field of TIGIT (T cell immunoreceptor with immunoglobulin and immunoreceptor tyrosine-based inhibition motif domain), a newly discovered immune checkpoint molecule mainly expressed on the surface of T cells and natural killer (NK) cells. Binding of TIGIT to cluster of differentiation 155 (CD155) and other ligands inhibits T cell and NK cell-mediated immune responses, and the TIGIT/CD155 pathway has been shown to play an important role in a variety of solid and hematological tumors.[9,10] So far, more than 15 anti-TIGIT checkpoint inhibitors are being evaluated in clinical trials either alone or in combination with programmed cell death 1 (PD-1)/programmed cell death ligand 1 (PD-L1) inhibitors for various cancers. A recent NSCLC clinical study on a combination therapy of two checkpoint inhibitors, domvanalimab (an anti-TIGIT) and zimberelimab (an anti-PD-1), has demonstrated additional immunotherapy benefit beyond anti-PD-1 monotherapy.[11] The TIGIT immunotherapy combination could lead to a new standard of care for patients with advanced lung cancer and other types of cancer.

6. Discovery Chemistry

The creation of new molecules that selectively inhibit a specific targeted oncogenic protein kinase depends critically on the capabilities and power of modern synthetic chemistry. This is a vast topic, well beyond the scope of this book. Fortunately, this area of medicinal chemistry has been reviewed recently.[12] It is extraordinary that this review has been downloaded more than 45,000 times within just a few months of publication.

1. P. Cohen, D. Cross, P. A. Jänne. Kinase drug discovery 20 years after imatinib: Progress and future directions. *Nat. Rev. Drug Discov.* **2021**, *20*, 551–569.
2. D. Shaywitz. The wild story behind a promising experimental cancer Drug. *Forbes.* Apr 5, 2013.
3. M. M. Attwood, D. Fabbro, A. V. Sokolov, S. Knapp, H. B. Schiöth. Trends in kinase drug discovery: Targets, indications and inhibitor design. *Nat. Rev. Drug Discov.* **2021,** *20*, 839–861.
4. L. Zhong, Y. Li, L. Xiong, W. Wang, M. Wu, T. Yuan, W. Yang, C. Tian, Z. Miao, T. Wang, S. Yang. Small molecules in targeted cancer therapy: Advances, challenges, and future perspectives. *Signal Transduct. Target. Ther.* **2021**, *6*, 201.
5. G. M. Burslem, B. E. Smith, A. C. Lai, S. Jaime-Figueroa, D. C. McQuaid, D. P. Bondeson, M. Toure, H. Dong, Y. Qian, J. Wang, A. P. Crew, J. Hines, C. M. Crews. The Advantages of targeted protein degradation over inhibition: An RTK case study. *Cell Chem Biol.* **2018**, *25*, 67–77.
6. D. Zahevi and L. Weiner. Monoclonal Antibodies in Cancer Therapy. Antibodies, **2020**, 9, 34–53.
7. A. Heine, S. Juranek, P. Brossart. Clinical and immunological effects of mRNA vaccines in malignant diseases. *Mol. Cancer.* **2021**, *20*, 52.
8. R. C. Larson, M. V. Maus. Recent advances and discoveries in the mechanisms and functions of CAR T cells. *Nat. Rev. Cancer.* **2021**, *21*, 145–161.
9. Z. Ge, M. P. Peppelenbosch, D. Sprengers, J. Kwekkeboom. TIGIT, the next step towards successful combination immune checkpoint therapy in cancer. *Front. Immunol.* **2021**, *12*, 699895.
10. H. Harjunpää, C. Guillerey. TIGIT as an emerging immune checkpoint. *Clin. Exp. Immunol.* **2020**, *200*, 108–119.
11. C. Helwick. In Stage IV NSCLC, Anti-TIGIT Antibody Boosts Immunotherapy Benefit. The ASCO Post. December 21, 2022.
12. C. C. Ayala-Aguilera, T. Valero, A. Lorente-Marcias, S.Croke, A. Unciti-Broceta. Small molecule kinase inhibitor drugs synthesis (1995-2021). *J. Med. Chem.* **2022**, *65*, 1047–1131.

Appendix 1. First FDA Approvals by Year

1999 1. Sirolimus (mTOR)
2001 2. Imatinib (BCR-ABL)
2003 3. Gefitinib (EGFR)
2004 4. Erlotinib (EGFR)
2005 5. Sorafenib (VEGFR)
2006 6. Dasatinib (BCR-ABL)
7. Sunitinib (VEGFR)
2007 8. Nilotinib (BCR-ABL)
9. Lapatinib (EGFR)
10. Temsirolimus (mTOR)
2009 11. Pazopanib (VEGFR)
12. Everolimus (mTOR)
2011 13. Ruxolitinib (JAK1/2)
14. Crizotinib (ALK)
15. Vemurafenib (BRAF)
16. Vandetanib (RET/VEGFR)
2012 17. Bosutinib (BCR-ABL)
18. Ponatinib (BCR-ABL)
19. Regorafenib (VEGFR)
20. Axitinib (VEGFR)
21. Tofacitinib (pan-JAK)
22. Cabozantinib (RET/VEGFR)
2013 23. Ibrutinib (BTK)
24. Afatinib (EGFR)
25. Mobocertinib (EGFR)
26. Dabrafenib (BRAF)
27. Trametinib (MEK)
2014 28. Nintedanib (PDGFR/VEGFR/FGFR)
29. Ceritinib (ALK)
30. Idelalisib (PI3K)
2015 31. Osimertinib (EGFR)
32. Lenvatinib (VEGFR2)
33. Palbociclib (CDK4/6)
34. Alectinib (ALK/RET)
35. Cobimetinib (MEK)
2017 36. Acalabrutinib (BTK)
37. Neratinib (HER2)
38. Ribociclib (CDK4/6)
39. Abemaciclib (CDK4/6)
40. Brigatinib (ALK/EGFR)
41. Copanlisib (PI3K)
42. Midostaurin (FLT3/KIT)
43. Netarsudil (ROCK1/2/NET)
2018 44. Dacomitinib (EGFR)
45. Baricitinib (JAK1/2)
46. Lorlatinib (ALK/ROS1)
47. Encorafenib (BRAF)
48. Binimetinib (MEK)
49. Duvelisib (PI3K)
50. Larotrectinib (TRK)
51. Gilteritinib (FLT3/AXL)
52. Fostamatinib (SYK)
2019 53. Upadacitinib (JAK1)
54. Fedratinib (JAK2)
55. Erdafitinib (FGFR)
56. Alpelisib (PI3K)
57. Entrecitinib (TRK/ROS1/ALK)
58. Pexidartinib (CSF1R)
2020 59. Tucatinib (HER2)
60. Selumetinib (MEK1)
61. Selpercatinib (RET)
62. Pralsetinib (RET)
63. Pemigatinib (FGFR)
64. Capmatinib (MET)
65. Avapritinib (PDGFR/KIT)
66. Ripretinib (PDGFR/KIT)
2021 67. Asciminib (BCR-ABL)
68. Zanubrutinib (BTK)
69. Tovozanib (VEGFR)
70. Trilaciclib (CDK4/6)
71. Infigratinib (FGFR)
72. Umbralisib (PI3K)
73. Tepotinib (MET)
74. Belumosudil (ROCK2)
75. Sotorasib (KRAS)
2022 76. Abrocitinib (JAK1)
77. Deucravacitinib (TYK2)
78. Pacritinib (JAK2/FLT3)
79. Futibatinib (FGFR)
80. Adagrasib (KRAS)
2023 81. Pirtobrutinib (BTK)
82. Leniolisib (PI3K)

Appendix 2. Kinase/KRAS Inhibitors in Development[1,2]

#	Drug	Target
1.	Remibrutinib (Novartis)	BTK
2.	Tolebrutinib (Sanofi)	BTK
3.	Rilzabrutinib (Sanofi)	BTK
4.	Evobrutinib (Merck Serono)	BTK
5.	Branebrutinib (BMS)	BTK
6.	Elsubrutinib (AbbVie)	BTK
7.	Memtabrutinib (Merck)	BTK
8.	Fenebrutinib (Genentech)	BTK
9.	TT01488 (TransThera)	BTK
10.	Abivertinib (ACEA)	EGFR
11.	Olafertinib (Checkpoint)	EGFR
12.	Zipalertinib (Cullinan)	EGFR
13.	BLU-525 (Blueprint)	EGFR
14.	BLU-451 (Blueprint)	EGFR
15.	BLU-945 (Blueprint)	EGFR
16.	Poziotinib (Hanmi)	EGFR
17.	Sunvozertinib (Dizai)	EGFR
18.	Avitinib (ACEA)	EGFR
19.	NX 019 (Nalo Thera.)	EGFR
20.	Nazartinib (Novartis)	EGFR
21.	THE-349 (Theseus)	EGFR
22.	ERAS801 (Erasca)	EGFR
23.	SI B001 (Sichuan Baili)	HER3
24.	Lucitanib (Clovis)	VEGFR
25.	Tesevatinib (Exelixis)	VEGFR
26.	Dovitinib (Allarity)	VEGFR/multikinase
27.	Sitravatinib (Mirati)	VEGFR
28.	OQL011 (OnQuality)	VEGFR
29.	Lerociclib (G1 Therapeutics)	CDK4/6
30.	BPI-16350 (Betta Pharma)	CDK4/6
31.	Dalpiciclib (Jiangsu Hengrui)	CDK4/6
32.	Dinaociclib (Merck)	CDK4/6
33.	Ebvaciclib (pfizer)	CDK2/4/6
34.	PF-07220060 (Pfizer)	CDK4
35.	PF-07104091 (Pfizer)	CDK2
36.	Fadraciclib (Cyclacel)	CDK4/6
37.	SY5609 (Syros)	CDK7
38.	Mevociclib (Syros)	CDK7
39.	Samuraciclib (Carrick)	CDK7
40.	LY3405105 (Lilly)	CDK7
41.	XL102 (Exelixis)	CDK7
42.	Ritlecitinib (Pfizer)	JAK1/TYK2
43.	Brepocitinib (Pfizer)	JAK1/TYK2
44.	CPL409116 (Celon)	JAK/ROCK1
45.	Momelotinib (Sierra)	JAK1/2
46.	Itacitinib (Incyte)	JAK1
47.	Ropsacitinib (Pfizer)	TYK2
48.	NDI-034858 (Takeda)	TYK2
49.	VTX958 (Ventyx)	TYK2
50.	ESK001 (Esker)	TYK2
51.	TPX-0131 (BMS)	ALK
52.	Foritinib (Fochon Pharma)	ALK/ROS
53.	Repotrectinib (BMS)	NTRK/ROS/ALK
54.	Unecritinib (Chia Tai)	ALK/ROS/MET
55.	Naporafenib (Novartis)	RAF
56.	Lifirafenib (BeiGene)	RAF
57.	Belvarafenib (Genentech)	RAF
58.	Exarafenib (Kinnate)	RAF
59.	Tinlorafenib (Pfizer)	RAF
60.	Tovorafenib (Biogen)	RAF
61.	DCC-3084 (Deciphera)	RAF
62.	PF-07799933 (Pfizer)	RAF
63.	Pimasertinib (Day One)	MEK
64.	Mirdametinib (SpringWorks)	MEK
65.	TAK-733 (Takeda)	MEK
66.	E6201 (Eisai)	MEK/FLT3
67.	Avutometinib (Verastem)	MEK/RAF
68.	IK-595 (Ikena)	MEK/RAF
69.	Enbezotinib (BMS)	RET/SRC
70.	Zeteletinib (Boston Pharma)	RET
71.	Vepafestinib (Taiho)	RET
72.	LOXO-260 (Lilly)	RET
73.	Tasugratinib (Eisai)	FGFR
74.	ABSK091 (Abbisko)	FGFR
75.	Ucitinib (Advenchen)	FGFR
76.	Rogaratinib (Bayer)	FGFR
77.	Zoligratinib (Chugai)	FGFR
78.	HMPL453 (Hutchison)	FGFR
79.	Tasurgratinib (Eisai)	FGFR
80.	Gunagratinib (InnoCare)	FGFR
81.	RLY4008 (Relay Thera.)	FGFR2
82.	TYRA300 (TYRA Bio.)	FGFR3
83.	KIN3248 (Kinnate)	FGFR2/3
84.	Dovitinib (Allarity)	FGFR/multikinase
85.	Derazanitinib (Basilea)	FGFR/multikinase
86.	Samotolisib (Lilly)	PI3K
87.	Tenalisib (Rhizen)	PI3K
88.	MEN 1611 (Chugai)	PI3K

#	Compound (Company)	Target	#	Compound (Company)	Target
89.	Inavolisib (Genentech)	PI3K	134.	XRD 0394 (XRad)	ATM/DNA-PK
90.	Eganelisib (Infinity)	PI3K	135.	Peposertib (Merck KGaA)	DNA-PK
91.	Buparlisib (Novartis)	PI3K	136.	ART 0380 (ShangPharma)	ART
92.	Linperlisib (Shanghai Yingli)	PI3K	137.	AZD1390 (AstraZeneca)	ATM
93.	Zandelisib (MEI Pharma)	PI3K	138.	Ceralasertib (AstraZeneca)	ATR
94.	Parsaclisib (Incyte)	PI3K	139.	SLC391 (SignalChem)	AXL
95.	Linperlisib (Shanghai Yingli)	PI3K	140.	INCB81776 (Incyte)	AXL/MER
96.	LOXO-783 (Lilly)	PI3K	141.	TTI 622 (Pfizer)	AXL/MER
97.	Gedatolisib (Pfizer)	PI3K/mTOR	142.	GSK3145095 (GSK)	RIPK1
98.	Glumetinib (Haihe)	MET	143.	Eclitasertib (Sanofi)	RIPK1
99.	Vebreltinib (Apollomics)	MET	144.	Ulixertinib (BioMed)	ERK1/2
100.	BPI 9016 (Betta Pharma)	MET	145.	ERAS007 (Erasca)	ERK1/2
101.	HQP 8361 (Ascentage)	MET/RON	146.	Tizaterkib (AstraZeneca)	ERK1/2
102.	Ringetinib (HEC)	MET/multikinase	147.	LY3295668 (Lilly)	Aurora A
103.	Elzovantinib (BMS)	MET/CSF1R/multikinase	148.	Alisertib (Takeda)	Aurora A
104.	Bezuclastinib (Cogent)	KIT/PDGFR	149.	VIC1911 (Taiho)	Aurora A
105.	BLU-263 (Blueprint)	KIT/PDGFR	150.	JAB-2485 (Jacobio)	Aurora A
106.	THE-630 (Theseus)	KIT/PDGFR	151.	Zimlovisertib (Pfizer)	IRAK4
107.	Telatinib (Bayer)	KIT/PDGFR/VEGFR	152.	Adavosertib (AstraZeneca)	WEE1
108.	Vimseltinib (Deciphera)	KIT/PDGFR/CSF1R	153.	IMP7068 (Impact Thera.)	WEE1
109.	Famitinib (Jiangsu)	KIT/PDGFR/VEGFR	154.	Azenosertib (Zentalis)	WEE1
110.	Cediranib (AstraZeneca)	KIT/PDGFR/VEGFR	155.	RP 6306 (Repare)	PKMYT1
111.	Chiauranib (Chipsreen)	KIT/PDGFR/VEGFR	156.	Camonsertib (Repare)	ATR
112.	Sitravatinib (Mirati)	KIT/PDGFR/FLT3/TRK	157.	PF-07265028 (Pfizer)	HPK1
113.	Masmitinib (AB Sci.)	KIT/PDGFR/multikinase	158.	DCC3116 (Deciphera)	ULK1/2
114.	MAX-40279 (Maxinovel)	FLT3/FGFR	159.	Opaganib (RedHill)	SK-2
115.	Crenolanib (AROG Pharma)	FLT3/PDGFR	160.	Saracatinib (AstraZeneca)	SRC/ABL
116.	Luxeptinib (Aptose)	FLT3/BTK	161.	NXP900 (Nuvectis)	SRC/YES1
117.	MRX-2843 (Meryx)	FLT3/MER	162.	XL092 (Exelixix)	MET/VEGFR2/AXL/MER
118.	Quizartinib (Daichi Sankyo)	FLT3/multikinase	163.	JDQ443 (Novartis)	KRAS(G12C)
119.	Sapanisertib (Takeda)	mTOR	164.	GDC-6036 (Genentech)	KRAS(G12C)
120.	Onatasertib (BMS)	mTOR	165.	JNJ-74699157 (J & J)	KRAS(G12C)
121.	Gedatolisib (Celtuicy)	mTOR/PI3K	166.	HBI-2438 (Jemincare)	KRAS(G12C)
122.	Paxalisib (Kazia Thera.)	mTOR/PI3K/AKT	167.	JAB21822 (Jacobio)	KRAS(G12C)
123.	RMC552 (Revolution Med.)	mTOR/4EBP1	168.	MK-1084 (Merck)	KRAS(G12C)
124.	Entospletinib (Kronos)	SYK	169.	BI-1823911 (BI)	KRAS(G12C)
125.	Lanraplenib (Kronos)	SYK	170.	GFH925 (GenFleet)	KRAS(G12C)
126.	Sovleplenib (HutchMed)	SYK	171.	D3S001 (D3 Bio)	KRAS(G12C)
127.	Cevidoplenib (Genosco)	SYK	172.	YL15293 (Shanghai Yingli)	KRAS(G12C)
128.	Mivavotinib (Takeda)	SYK/FLT3	173.	LY3637982 (Lilly)	KRAS(G12C)
129.	Capivasertib (AstraZeneca)	AKT	174.	D1553 (Inventis Bio)	KRAS(G12C)
130.	Ipatasertib (Genetech)	AKT	175.	HS 10370 (Jiangsu Hansoh)	KRAS(G12C)
131.	Afuresertib (Laekna)	AKT	176.	RMC6291 (Revolution Med.)	KRAS(G12C)
132.	Miransertib (Merck)	AKT	177.	ERAS-3490 (Erasca)	KRAS(G12C)
133.	TAS0612 (Taiho)	AKT/RSK/S6K	178.	LY3537982 (Lilly)	KRAS(G12C)

179. HRS-4642 (Jiangsu Hengrui)   KRAS(G12D)
180. MRTX1133 (Mirati)   KRAS(G12D)

[1]https://clinicaltrials.gov
[2]list incomprehensive

Appendix 3. Visualization of Differentially Expressed Kinases in Cancer[1,2]

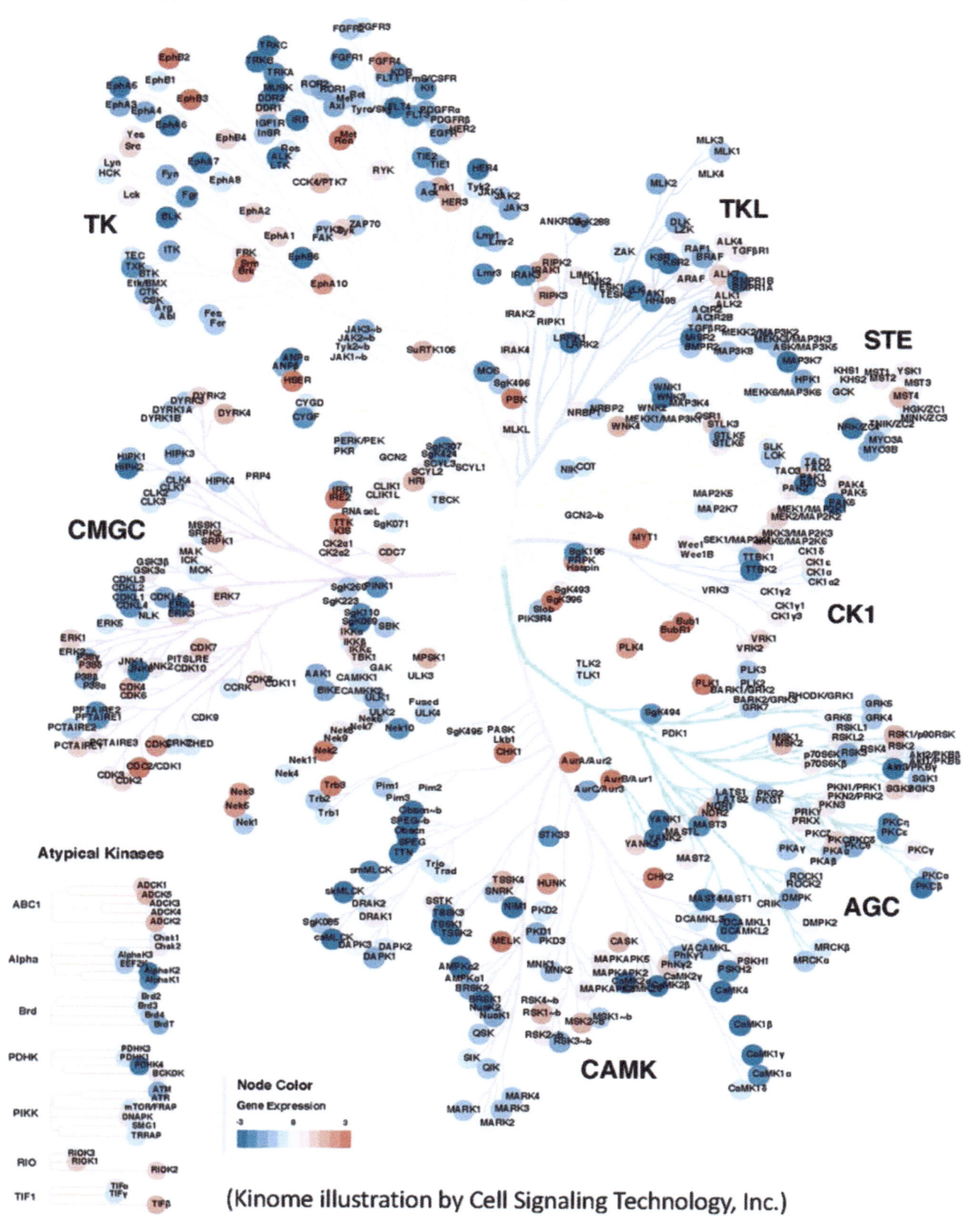

(Kinome illustration by Cell Signaling Technology, Inc.)

[1] Differential expression of kinases in Colon Adenocarcinoma was estimated using 100 TCGA-COAD RNA-Seq tumor samples (cases) and 165 normal samples (controls) derived from various human tissues.

[2] Adpated from *Front. Cell Dev. Biol.* **2022**, *10*, 942500.

Appendix 4. M&A Transactions Driven by Oncology-focused Kinase and KRAS inhibitors[1,2]

Date	Acquirer	Target	Deal value	Lead assets	Kinase
06/99	Pharmacia	Sugen	645 million	SU5416 (D) SU6668 (D) sunitinib (A)	VEGFR
03/10	Astellas	OSI Pharma	~$4 billion	erlotinib (A)	EGFR
06/10	Sanofi	TargaGen	Up to ~$560 million	fedratinib (A)	JAK2
02/11	Gilead	Calistoga Pharma	Up to ~$600 million	idelalisib (A)	PI3K
02/11	Daiichi	Plexxikon	Up to ~$935 million	vemurafenib (A)	BRAF
12/11	Takeda	Intellikine	Up to ~$310 million	sapanisertinib (R) serabelisib (D)	mTOR, PI3K
01/12	Celgene	Avila Thera.	Up to ~925 million	spebrutinib (D)	BTK
12/12	Gilead	YM Biosciences	~$510 million	momelotinib (R)	JAK1/2
08/13	Amgen	Onyx Pharma	~$10.4 billion	sorafenib (A) regorafenib (A) carfilzomib (PI)	VEGFR, FLT3/RAF, PI (non-kinase)
11/13	Clovis	EOS	Up to ~$420 million	lucitanib (D)	VEGFR, PDGFR, FGFR
04/14	Novartis	GSK Oncology	Up to ~$16 billion	dabrafenib (A) trametinib (A) pazopanib (A) lapatinib (A)	BRAF, MEK, VEGFR, HER2
09/14	Daiichi	Ambit Biosciences	Up to ~$410 million	quizartinib (R)	FLT3
03/15	AbbVie	Pharmacyclics	~$21 billion	brutinib (A)	BTK
12/15	AstraZeneca	Acerta Pharma	Up to ~$7 billion	acalabrutinib (A)	BTK
01/17	Takeda	Ariad	~$5.2 billion	ponatinib (A) brigatinib (A)	BCR-ABL, ALK
11/17	Bayer	Loxo Oncology	400 million	selitrectinib (D)	TRK
12/17	Roche	Ignyta	~$1.7 billion	entrectinib (A)	ROS1, NTRK
01/18	Celgene	Impact Biosciences	Up to ~$7 billion	fedratinib (A)	JAK2
01/18	SeaGen	Cascadian	~$614 million	tucatinib (A)	HER2
05/18	Lilly	AurKa Pharma	Up to ~$575 million	AK-01 (D)	Aurora kinase A
01/19	Lilly	Loxo Oncology	~$8 billion	Larotrectinib (A) selpercatinib (A) pirtobrutinib (T)	NTRK, RET, BTK
06/19	Pfizer	Array BioPharma	~$11.4 billion	encorafenib (A) binimetinib (A)	BRAF, MEK
12/19	Merck	ArQule	$2.7 billion	nemtabrutinib (T)	BTK
01/20	Merck	Taiho/Astex	$2.6 billion	preclinical	KRAS
07/20	Roche	Blueprint	$1.7 billion	pralsetinib (A)	RET, multikinase
08/20	Sanofi	Princilpia Biopharma	$3.7 billion	rilzabrutinib (T)	BTK
07/22	GSK	Sierra Oncology	$1.9 billion	momelotinib (R)	JAK1/2
06/22	BMS	TP Therapeutics	$4.1 billion	repotrectinib (R)	NTRK, ROS, ALK
12/22	Takeda	Nimbus	$4 billion	NDI-034858 (T)	TYK2

[1] 2010–2019 data from J. Bell. BiopharmaDive. August 16, 2019. Meet the protein responsible for nearly $100B in cancer drug deals.

[2] Current status: A (FDA-approved); D (discontinued); R: (FDA review); T: (in clinical trials).

Appendix 5. Alphabetic List of Oncogenic Protein Inhibitors

Abemaciclib	139	Erlotinib	67	Pexidartinib	331
Abrocitinib	166	Everolimus	317	Pirtobrutinib	44
Acalabrutinib	51	Fedratinib	173	Ponatinib	33
Adagrasib	346	Filgotinib	163	Pralsetinib	251
Afatinib	74	Fostamatinib	328	Pyrotinib	60
Alectinib	205	Fruquintinib	97	Quizartinib	306
Alpelisib	269	Furmonertinib	84	Regorafenib	104
Anlotinib	97	Futibatinib	265	Repotrectinib	291
Apatinib	121	Gefitinib	61	Ribociclib	136
Asciminib	38	Gilteritinib	313	Ripretinib	304
Aumolertinib	83	Ibrutinib	45	Ritlecitinib	177
Avapritinib	301	Icotinib	72	Ropsacitinib	184
Axitinib	114	Idelalisib	273	Ruxolitinib	170
Baricitinib	151	Imatinib	19	Savolitinib	294
Belumosudil	326	Infigratinib	263	Selpercatinib	247
Binimetinib	235	JDQ443	350	Selumetinib	237
Bosutinib	30	Lapatinib	90	Sirolimus	317
Brepocitinib	181	Larotrectinib	285	Sorafenib	99
Brigatinib	207	Lazertinib	59	Sotorasib	337
Cabozantinib	245	Leniolisib	267	Sunitinib	106
Capmatinib	295	Lenvatinib	122	Surufatinib	97
Ceritinib	202	Lorlatinib	210	Tamatinib	328
Cobimetinib	232	Midostaurin	308	Temsirolimus	317
Copanlisib	281	Mobocertinib	86	Tepotinib	297
Crizotinib	197	Neratinib	95	Tirabrutinib	57
Dabrafenib	222	Netarsudil	324	Tovozanib	125
Dacomitinib	77	Nilotinib	24	Trametinib	228
Dasatinib	27	Nintedanib	117	Trilaciclib	142
Delgocitinib	161	Olvermbatinib	37	Tofacitinib	147
Deucravacitinib	189	Orelabrutinib	58	Tucatinib	93
Deuruxolitinib	172	Osimertinib	80	Umbralisib	279
Duvelisib	277	Pacritinib	175	Upadacitinib	158
Encorafenib	225	Palbociclib	129	Vandetanib	242
Ensartinib	195	Pazopanib	112	Vemurafenib	216
Entrectinib	288	Peficitinib	153	Zanubrutinib	54
Erdafitinib	255	Pemigatinib	260		